DYNAMICS OF THE ATMOSPHERE: A COURSE IN THEORETICAL METEOROLOGY

Dynamics of the Atmosphere is a textbook with numerous exercises and solutions, written for senior undergraduate and graduate students of meteorology and related sciences. It may also be used as a reference source by professional meteorologists and researchers in atmospheric science. In order to encourage the reader to follow the mathematical developments in detail, the derivations are complete and leave out only the most elementary steps.

The book consists of two parts, the first presenting the mathematical tools needed for a thorough understanding of the second part. Mathematical topics include a summary of the methods of vector and tensor analysis in generalized coordinates; an accessible presentation of the method of covariant differentiation; and a brief introduction to nonlinear dynamics. These mathematical tools are used later in the book to tackle such problems as the fields of motion over different types of terrain, and problems of predictability.

The second part of the book begins with the derivation of the equation describing the atmospheric motion on the rotating earth, followed by several chapters that consider the kinematics of the atmosphere and introduce vorticity and circulation theorems. Weather patterns can be considered as superpositions of waves of many wavelengths, and the authors therefore present a discussion of wave motion in the atmosphere, including the barotropic model and some Rossby physics. A chapter on inertial and dynamic stability is presented and the component form of the equation of motion is derived in the general covariant, contravariant, and physical coordinate forms. The subsequent three chapters are devoted to turbulent systems in the atmosphere and their implications for weather-prediction equations. At the end of the book newer methods of weather prediction, such as the spectral technique and the stochastic dynamic method, are introduced in order to demonstrate their potential for extending the forecasting range as computers become increasingly powerful.

WILFORD ZDUNKOWSKI received B.S. and M.S. degrees from the University of Utah and was awarded a Ph.D. in meteorology from the University of Munich in 1962. He then returned to the Department of Meteorology at the University of Utah, where he was later made Professor of Meteorology. In 1977, he took up a professorship, at the Universität Mainz, where for twenty years he taught courses related to the topics presented in this book. Professor Zdunkowski has been the recipient of numerous awards from various research agencies in the USA and in Germany, and has travelled extensively to report his findings to colleagues around the world.

ANDREAS BOTT received a Diploma in Meteorology from the Universität Mainz in 1982, and subsequently worked as a research associate under Professor Paul Crutzen at the Max-Planck-Institut für Chemie in Mainz, where he was awarded a Ph.D. in Meteorology in 1986. He held a variety of positions at the Institute for Atmospheric Physics in the Universität Mainz between 1986 and 1999, and during this time he also spent periods as a guest scientist at institutions in the USA, Norway, and Japan. Since 2000, Dr Bott has been a University Professor for Theoretical Meteorology at the Rheinische Friedrich-Wilhelms-Universität, in Bonn. Professor Bott teaches courses in theoretical meteorology, atmospheric thermodynamics, atmospheric dynamics, cloud microphysics, atmospheric chemistry, and numerical modeling.

DYNAMICS OF THE ATMOSPHERE: A COURSE IN THEORETICAL METEOROLOGY

WILFORD ZDUNKOWSKI
and
ANDREAS BOTT

CAMBRIDGE UNIVERSITY PRESS
Cambridge, New York, Melbourne, Madrid, Cape Town, Singapore,
São Paulo, Delhi, Dubai, Tokyo, Mexico City

Cambridge University Press
The Edinburgh Building, Cambridge CB2 8RU, UK

Published in the United States of America by Cambridge University Press, New York

www.cambridge.org
Information on this title: www.cambridge.org/9780521006668

First published 2003

A catalogue record for this publication is available from the British Library

ISBN 978-0-521-80949-8 Hardback
ISBN 978-0-521-00666-8 Paperback

This book is dedicated to the memory of
Professor K. H. Hinkelmann (1915–1989)
and Dr J. G. Korb (1928–1991)
who excelled as theoretical meteorologists and as
teachers of meteorology at the University of Mainz, Germany.

Contents

Preface

This book has been written for students of meteorology and of related sciences at the senior and graduate level. The goal of the book is to provide the background for graduate studies and individual research. The second part, *Thermodynamics of the Atmosphere*, will appear shortly. To a considerable degree we have based our book on the excellent lecture notes of Professor Karl Hinkelmann on various topics in dynamic meteorology, including Prandtl-layer theory and turbulence. Moreover, we were fortunate to have Dr Korb's outstanding lecture notes on kinematics of the atmosphere and on mathematical tools for the meteorologist at our disposal.

Quite early on during the writing of this book, it became apparent that we had to replace various topics treated in their notes by more modern material in order to give a reasonably up-to-date account of theoretical meteorology. We were guided by the idea that any topic we have selected for presentation should be treated in some depth in order for it to be of real value to the reader. Insofar as space would permit, all but the most trivial steps have been included in every development. This is the reason why our book is somewhat more bulky than some other books on theoretical meteorology. The student may judge for himself whether our approach is profitable.

The reader will soon recognize that various interesting and important topics have been omitted from this textbook. Including these and still keeping the book of the same length would result in the loss of numerous mathematical details. This, however, might discourage some students from following the discussion in depth. We believe that the approach we have chosen is correct and smoothes the path to additional and more advanced studies.

This book consists of two separate parts. In the first part we present the mathematical techniques needed to handle the various topics of dynamic meteorology which are presented in the second part of the book. The modern student of meteorology and of related sciences at the senior and the graduate level has accumulated a sufficient working knowledge of vector calculus applied to the Cartesian coordinate system. We are safe to assume that the student has also encountered the important integral theorems which play a dominant role in many branches of physics and engineering. The required extension to more general coordinate systems is not difficult. Nevertheless, the reader may have to deal with some unfamiliar topics. He should not be discouraged since often unfamiliarity is mistaken for inherent difficulty. The unavoidable formality presented in the introductory chapters on first reading looks worse than it really is. After overcoming some initial difficulties, the student will soon gain confidence in his ability to handle the new techniques. The authors came to the conclusion, as the result of many years of learning and teaching, that a mastery of the mathematical introduction is surely worth what it costs in effort.

All mathematical operations have been restricted to three dimensions in space. However, many important formulas can be easily extended to higher-order spaces. Some knowledge of tensor analysis is required for our studies. Since three-dimensional tensor analysis in generalized coordinates can be handled very effectively with the help of dyadics, we have introduced the necessary operations. Only as the last step do we write down the tensor components. By proceeding in this manner, we are likely to avoid errors that may occur quite easily with use of the index notation throughout. We admit that dyadics are quite dispensable when one is working with Cartesian tensors, but they are of great help when one is working with generalized coordinate systems.

The second part of the book treats some of the major topics of dynamic meteorology. As is customary in many textbooks, the introductory chapters discuss some basic topics of thermodynamics. We will depart from this much-trodden path. The reason for this departure is that modern thermodynamics cannot be adequately dealt with in this manner. If formulas from thermodynamics are required, they will be carefully stated. Detailed derivations, however, will be omitted since these will be presented in part II of A Course in Theoretical Meteorology. When reference to this book on thermodynamics is made we will use the abbreviation TH.

We will now give a brief description of the various chapters of the dynamics part of the book. Chapter 1 presents the laws of atmospheric motion. The method of scale analysis is introduced in Chapter 2 in order to show which terms in the component form of the equation of motion may be safely neglected in large-scale flow fields. Chapters 3–10 discuss some topics that traditionally belong to the kinematics part of theoretical meteorology. Included are discussions on the material and the local description of flow, the Navier–Stokes stress tensor, the Helmholtz theorem, boundary surfaces, circulation, and vorticity theorems. Since atmospheric flow, particularly in the air layers near the ground, is always turbulent, in Chapters 11 and 12 we present a short introduction to turbulence theory. Some important aspects of boundary-layer theory will be given in Chapter 13. Wave motion in the atmosphere, some stability theory, and early weather-prediction models are introduced in Chapters 14–17. Lagrange's and Hamilton's treatments of the equation of motion are discussed in Chapter 18.

The following chapters consider flow fields in various coordinate systems. In Chapters 19 and 20 we give a fairly detailed account of the air motion described with the help of the geographic and the stereographic coordinate systems. This description and the following topics are of great importance for numerical weather prediction. In order to study the airflow over irregular terrain, the orography-following coordinate system is introduced in Chapter 21. The air motion in stereographic coordinate systems with a generalized vertical coordinate is discussed in Chapter 22.

Some earlier baroclinic weather-prediction models employed the so-called quasi-geostrophic theory which is discussed in some detail in Chapters 23 and 24. Modern numerical weather prediction, however, is based on the numerical solutions of the primitive equations, i.e. the scale-analyzed original equations describing the flow field. Nevertheless, the quasi-geostrophic theory is still of great value in discussing some major features of atmospheric motion. We will employ this theory to construct weather-prediction models and we show the operational principle.

A brief and very incomplete introduction of numerical methods is given in Chapter 25 to motivate the modeling of atmospheric flow by spectral techniques. Some basic theory of the spectral method is given in Chapter 26. The final chapter of this book, Chapter 27, introduces the problems associated with atmospheric predictability. The famous Lorenz equations and the strange attractor are discussed. The method of stochastic dynamic prediction is introduced briefly.

Problems of various degrees of difficulty are given at the end of most chapters. The almost trivial problems were included to provide the opportunity for the student to become familiar with the new material before he is confronted with more demanding problems. Some answers to these problems are provided at the end of the book. To a large extent these problems were given to the meteorology students of the University of Mainz in their excercise classes. We were very fortunate to be assisted by very able instructors, who conducted these classes independently. We wish to express our sincere gratitude to them. These include Drs G. Korb, R. Schrodin, J. Siebert, and T. Trautmann. It would be impossible to name all contributors to the excercise classes. Our special gratitude goes to Dr W.-G. Panhans for his splendid cooperation with the authors in organizing and conducting these classes. Whenever asked, he also taught some courses to lighten the burden.

It seems to be one of the unfortunate facts of life that no book as technical as this one can be published free of error. However, we take some comfort in the thought that any errors appearing in this book were made by the co-author. To remove these, we would be grateful to anyone pointing out to us misprints and other mistakes they have discovered.

In writing this book we have greatly profited from Professor H. Fortak, whose lecture notes were used by K. Hinkelmann and G. Korb as a guide to organize their manuscripts. We are also indebted to the late Professor G. Hollmann and to Professor F. Wippermann. Parts of their lecture notes were at our disposal.

We also wish to thank our families for their constant support and encouragement.

Finally, we express our gratitude to Cambridge University Press for their effective cooperation in preparing the publication of this book.

W. Zdunkowski
A. Bott

Part 1

Mathematical tools

M1

Algebra of vectors

M1.1 Basic concepts and definitions

A *scalar* is a quantity that is specified by its sign and by its magnitude. Examples are temperature, the specific volume, and the humidity of the air. Scalars will be written using Latin or Greek letters such as $a, b, \ldots, A, B, \ldots, \alpha, \beta, \ldots$. A *vector* requires for its complete characterization the specification of magnitude and direction. Examples are the velocity vector and the force vector. A vector will be represented by a boldfaced letter such as $\mathbf{a}, \mathbf{b}, \ldots, \mathbf{A}, \mathbf{B}, \ldots$. A *unit vector* is a vector of prescribed direction and of magnitude 1. Employing the unit vector $\mathbf{e}_A$, the arbitrary vector $\mathbf{A}$ can be written as

$$\mathbf{A} = |\mathbf{A}|\,\mathbf{e}_A = A\mathbf{e}_A \implies \mathbf{e}_A = \frac{\mathbf{A}}{|\mathbf{A}|} \tag{M1.1}$$

Two vectors $\mathbf{A}$ and $\mathbf{B}$ are equal if they have the same magnitude and direction regardless of the position of their initial points,

that is $|\mathbf{A}| = |\mathbf{B}|$ and $\mathbf{e}_A = \mathbf{e}_B$. Two vectors are *collinear* if they are parallel or antiparallel. Three vectors that lie in the same plane are called *coplanar*. Two vectors always lie in the same plane since they define the plane. The following rules are valid:

the commutative law: $\mathbf{A} \pm \mathbf{B} = \mathbf{B} \pm \mathbf{A}, \quad \mathbf{A}\alpha = \alpha\mathbf{A}$

the associative law: $\mathbf{A} + (\mathbf{B} + \mathbf{C}) = (\mathbf{A} + \mathbf{B}) + \mathbf{C}, \quad \alpha(\beta\mathbf{A}) = (\alpha\beta)\mathbf{A}$

the distributive law: $(\alpha + \beta)\mathbf{A} = \alpha\mathbf{A} + \beta\mathbf{A}$

(M1.2)

The concept of linear dependence of a set of vectors $\mathbf{a}_1, \mathbf{a}_2, \ldots, \mathbf{a}_N$ is closely connected with the dimensionality of space. The following definition applies: A set of N vectors $\mathbf{a}_1, \mathbf{a}_2, \ldots, \mathbf{a}_N$ of the same dimension is linearly dependent if there exists a set of numbers $\alpha_1, \alpha_2, \ldots, \alpha_N$, not all of which are zero, such that

$$\alpha_1\mathbf{a}_1 + \alpha_2\mathbf{a}_2 + \cdots + \alpha_N\mathbf{a}_N = 0 \tag{M1.3}$$

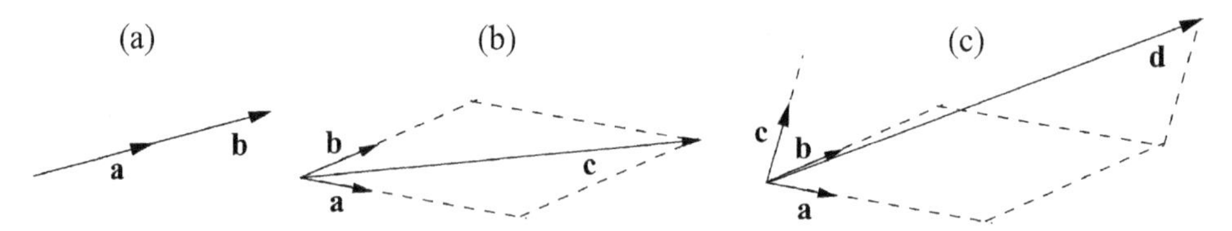

Fig. M1.1 Linear vector spaces: (a) one-dimensional, (b) two-dimensional, and (c) three-dimensional.

If no such numbers exist, the vectors $\mathbf{a}_1, \mathbf{a}_2, \ldots, \mathbf{a}_N$ are said to be linearly independent. To get the geometric meaning of this definition, we consider the vectors **a** and **b** as shown in Figure M1.1(a). We can find a number $k \neq 0$ such that

$$\mathbf{b} = k\mathbf{a} \tag{M1.4a}$$

By setting $k = -\alpha/\beta$ we obtain the symmetrized form

$$\alpha\mathbf{a} + \beta\mathbf{b} = 0 \tag{M1.4b}$$

Assuming that neither α nor β is equal to zero then it follows from the above definition that two collinear vectors are linearly dependent. They define the one-dimensional *linear vector space*. Consider two noncollinear vectors **a** and **b** as shown in Figure M1.1(b). Every vector **c** in their plane can be represented by

$$\mathbf{c} = k_1\mathbf{a} + k_2\mathbf{b} \quad \text{or} \quad \alpha\mathbf{a} + \beta\mathbf{b} + \gamma\mathbf{c} = 0 \tag{M1.5}$$

with a suitable choice of the constants k_1 and k_2. Equation (M1.5) defines a two-dimensional linear vector space. Since not all constants α, β, γ are zero, this formula insures that the three vectors in the two-dimensional space are linearly dependent. Taking three noncoplanar vectors **a**, **b**, and **c**, we can represent every vector **d** in the form

$$\mathbf{d} = k_1\mathbf{a} + k_2\mathbf{b} + k_3\mathbf{c} \tag{M1.6}$$

in a three-dimensional linear vector space, see Figure M1.1(c). This can be generalized by stating that, in an N-dimensional linear vector space, every vector can be represented in the form

$$\mathbf{x} = k_1\mathbf{a}_1 + k_2\mathbf{a}_2 + \cdots + k_N\mathbf{a}_N \tag{M1.7}$$

where the $\mathbf{a}_1, \mathbf{a}_2, \ldots, \mathbf{a}_N$ are linearly independent vectors. Any set of vectors containing more than N vectors in this space is linearly dependent.

Table M1.1. *Extensive quantities of different degrees for the N-dimensional linear vector space*

Extensive quantity	Degree υ	Symbol	Number of vectors	Number of components
Scalar	0	B	0	$N^0 = 1$
Vector	1	$\mathbf{B}$	1	N^1
Dyadic	2	$\mathbb{B}$	2	N^2

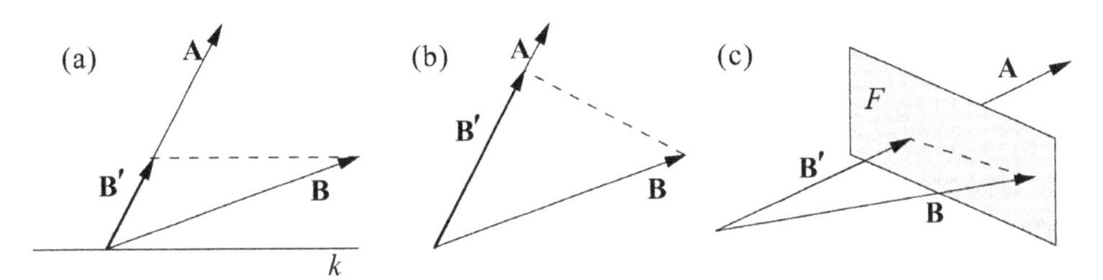

Fig. M1.2 Projection of a vector **B** onto a vector **A**.

We call the set of N linearly independent vectors $\mathbf{a}_1, \mathbf{a}_2, \ldots, \mathbf{a}_N$ the *basis vectors* of the N-dimensional linear vector space. The numbers $k_1, k_2, \ldots, k_N$ appearing in (M1.7) are the *measure numbers* associated with the basis vectors. The term $k_i\mathbf{a}_i$ of the vector $\mathbf{x}$ in (M1.7) is the *component of this vector* in the direction $\mathbf{a}_i$.

A vector **B** may be projected onto the vector **A** parallel to the direction of a straight line k as shown in Figure M1.2(a). If the direction of the straight line k is not given, we perform an orthogonal projection as shown in part (b) of this figure. A projection in three-dimensional space requires a plane F parallel to which the projection of the vector **B** onto the vector **A** can be carried out; see Figure M1.2(c).

In vector analysis an *extensive quantity* of degree υ is defined as a homogeneous sum of general products of vectors (with no dot or cross between the vectors). The number of vectors in a product determines the degree of the extensive quantity. This definition may seem strange to begin with, but it will be familiar soon. Thus, a scalar is an extensive quantity of degree zero, and a vector is an extensive quantity of degree one. An extensive quantity of degree two is called a *dyadic*. Every dyadic $\mathbb{B}$ may be represented as the sum of three or more *dyads*. $\mathbb{B} = \mathbf{p}_1\mathbf{P}_1 + \mathbf{p}_2\mathbf{P}_2 + \mathbf{p}_3\mathbf{P}_3 + \cdots$. Either the *antecedents* $\mathbf{p}_i$ or the *consequents* $\mathbf{P}_i$ may be arbitrarily assigned as long as they are linearly independent. Our practical work will be restricted to extensive quantities of degree two or less. Extensive quantities of degree three and four also appear in the highly specialized literature. Table M1.1 gives a list of extensive quantities used in our work. Thus, in the three-dimensional linear vector space with $N = 3$, a vector consists of three and a dyadic of nine components.

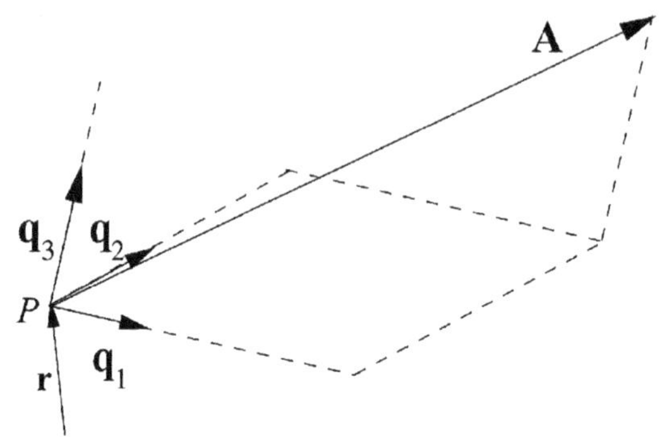

Fig. M1.3 The general vector basis $\mathbf{q}_1, \mathbf{q}_2, \mathbf{q}_3$ of the three-dimensional space.

M1.2 Reference frames

The representation of a vector in component form depends on the choice of a particular coordinate system. A *general vector basis* at a given point in three-dimensional space is defined by three arbitrary linearly independent basis vectors $\mathbf{q}_1, \mathbf{q}_2, \mathbf{q}_3$ spanning the space. In general, the basis vectors are neither orthogonal nor unit vectors; they may also vary in space and in time.

Consider a position vector $\mathbf{r}$ extending from an arbitrary origin to a point P in space. An arbitrary vector $\mathbf{A}$ extending from P is defined by the three basis vectors $\mathbf{q}_i, i = 1, 2, 3$, existing at P at time t, as shown in Figure M1.3 for an oblique coordinate system. Hence, the vector $\mathbf{A}$ may be written as

$$\mathbf{A} = A^1\mathbf{q}_1 + A^2\mathbf{q}_2 + A^3\mathbf{q}_3 = \sum_{k=1}^{3} A^k\mathbf{q}_k \tag{M1.8}$$

where it should be observed that the so-called *affine measure numbers* A^1, A^2, A^3 carry superscripts, and the basis vectors $\mathbf{q}_1, \mathbf{q}_2, \mathbf{q}_3$ carry subscripts. This type of notation is used in the *Ricci calculus*, which is the tensor calculus for nonorthonormal coordinate systems. Furthermore, it should be noted that there must be an equal number of upper and lower indices.

Formula (M1.8) can be written more briefly with the help of the familiar *Einstein summation convention* which omits the summation sign:

$$\mathbf{A} = A^1\mathbf{q}_1 + A^2\mathbf{q}_2 + A^3\mathbf{q}_3 = A^n\mathbf{q}_n \tag{M1.9}$$

We will agree on the following notation: Whenever an index (subscript or superscript) m, n, p, q, r, s, t, is repeated in a term, we are to sum over that index from 1 to 3, or more generally to N. In contrast to the summation indices m, n, p, q, r, s, t, the letters i, j, k, l are considered to be "free" indices that are used to enumerate equations. Note that summation is not implied even if the free indices occur twice in a term or even more often.

A special case of the general vector basis is the *Cartesian vector basis* represented by the three orthogonal unit vectors **i**, **j**, **k**, or, more conveniently, $\mathbf{i}_1$, $\mathbf{i}_2$, $\mathbf{i}_3$. Each of these three unit vectors has the same direction at all points of space. However, in rotating coordinate systems these unit vectors also depend on time. The arbitrary vector **A** may be represented by

$$\begin{aligned} \mathbf{A} &= A_x\mathbf{i} + A_y\mathbf{j} + A_z\mathbf{k} = A^n\mathbf{i}_n = A_n\mathbf{i}_n \\ \text{with} \quad A_x &= A^1 = A_1, \quad A_y = A^2 = A_2, \quad A_z = A^3 = A_3 \end{aligned} \tag{M1.10}$$

In the Cartesian coordinate space there is no need to distinguish between upper and lower indices so that (M1.10) may be written in different ways. We will return to this point later.

Finally, we wish to define the *position vector* **r**. In a Cartesian coordinate system we may simply write

$$\mathbf{r} = x\mathbf{i} + y\mathbf{j} + z\mathbf{k} = x^n\mathbf{i}_n = x_n\mathbf{i}_n \tag{M1.11}$$

In an oblique coordinate system, provided that the same basis exists everywhere in space, we may write the general form

$$\mathbf{r} = q^1\mathbf{q}_1 + q^2\mathbf{q}_2 + q^3\mathbf{q}_3 = q^n\mathbf{q}_n \tag{M1.12}$$

where the q^i are the measure numbers corresponding to the basis vectors $\mathbf{q}_i$. The form (M1.12) is also valid along the radius in a spherical coordinate system since the basis vectors do not change along this direction.

A different situation arises in case of curvilinear coordinate lines since the orientations of the basis vectors change with position. This is evident, for example, on considering the coordinate lines (lines of equal latitude and longitude) on the surface of a nonrotating sphere. In case of curvilinear coordinate lines the position vector **r** has to be replaced by the differential expression $d\mathbf{r} = dq^n\,\mathbf{q}_n$. Later we will discuss this topic in the required detail.

M1.3 Vector multiplication

M1.3.1 The scalar product of two vectors

By definition, the coordinate-free form of the scalar product is given by

$$\mathbf{A} \cdot \mathbf{B} = |\mathbf{A}|\,|\mathbf{B}|\cos(\mathbf{A}, \mathbf{B}) \tag{M1.13}$$

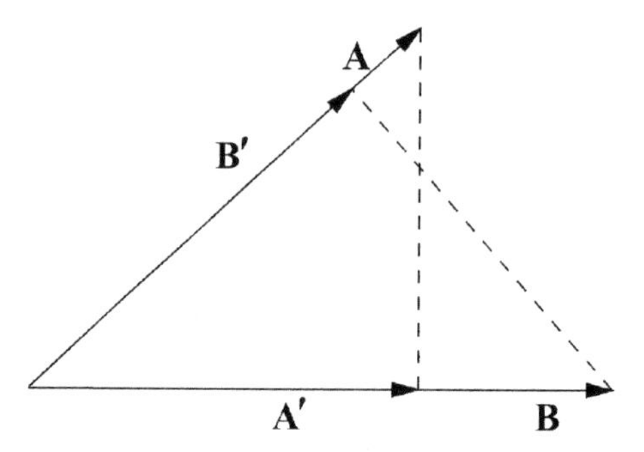

Fig. M1.4 Geometric interpretation of the scalar product.

If the vectors $\mathbf{A}$ and $\mathbf{B}$ are orthogonal the expression $\cos(\mathbf{A}, \mathbf{B}) = 0$ so that the scalar product vanishes. The following rules involving the scalar product are valid:

$$
\begin{aligned}
&\textit{the commutative law:} && \mathbf{A}\cdot\mathbf{B} = \mathbf{B}\cdot\mathbf{A} \\
&\textit{the associative law:} && (k\mathbf{A})\cdot\mathbf{B} = k(\mathbf{A}\cdot\mathbf{B}) = k\mathbf{A}\cdot\mathbf{B} \\
&\textit{the distributive law:} && \mathbf{A}\cdot(\mathbf{B}+\mathbf{C}) = \mathbf{A}\cdot\mathbf{B} + \mathbf{A}\cdot\mathbf{C}
\end{aligned}
\tag{M1.14}
$$

Moreover, we recognize that the scalar product, also known as the dot product or inner product, may be represented by the orthogonal projections

$$
\mathbf{A}\cdot\mathbf{B} = |\mathbf{A}'||\mathbf{B}|, \qquad \mathbf{A}\cdot\mathbf{B} = |\mathbf{A}||\mathbf{B}'| \tag{M1.15}
$$

whereby the vector $\mathbf{A}'$ is the projection of $\mathbf{A}$ on $\mathbf{B}$, and $\mathbf{B}'$ is the projection of $\mathbf{B}$ on $\mathbf{A}$; see Figure M1.4.

The component notation of the scalar product yields

$$
\begin{aligned}
\mathbf{A}\cdot\mathbf{B} = {} & A^1B^1\mathbf{q}_1\cdot\mathbf{q}_1 + A^1B^2\mathbf{q}_1\cdot\mathbf{q}_2 + A^1B^3\mathbf{q}_1\cdot\mathbf{q}_3 \\
& + A^2B^1\mathbf{q}_2\cdot\mathbf{q}_1 + A^2B^2\mathbf{q}_2\cdot\mathbf{q}_2 + A^2B^3\mathbf{q}_2\cdot\mathbf{q}_3 \\
& + A^3B^1\mathbf{q}_3\cdot\mathbf{q}_1 + A^3B^2\mathbf{q}_3\cdot\mathbf{q}_2 + A^3B^3\mathbf{q}_3\cdot\mathbf{q}_3
\end{aligned}
\tag{M1.16}
$$

Thus, in general the scalar product results in nine terms. Utilizing the Einstein summation convention we obtain the compact notation

$$
\boxed{\mathbf{A}\cdot\mathbf{B} = A^m\mathbf{q}_m\cdot B^n\mathbf{q}_n = A^mB^n\mathbf{q}_m\cdot\mathbf{q}_n = A^mB^ng_{mn}} \tag{M1.17}
$$

The quantity g_{ij} is known as the covariant *metric fundamental quantity* representing an element of a covariant tensor of rank two or order two. This tensor is called the *metric tensor* or the *fundamental tensor*. The expression “covariant” will be described later. Since $\mathbf{q}_i\cdot\mathbf{q}_j = \mathbf{q}_j\cdot\mathbf{q}_i$ we have the identity

$$
\boxed{g_{ij} = g_{ji}} \tag{M1.18}
$$

On substituting for **A**, **B** the unit vectors of the Cartesian coordinate system, we find the well-known orthogonality conditions for the Cartesian unit vectors

$$\mathbf{i} \cdot \mathbf{j} = 0, \qquad \mathbf{i} \cdot \mathbf{k} = 0, \qquad \mathbf{j} \cdot \mathbf{k} = 0 \tag{M1.19}$$

or the normalization conditions

$$\mathbf{i} \cdot \mathbf{i} = 1, \qquad \mathbf{j} \cdot \mathbf{j} = 1, \qquad \mathbf{k} \cdot \mathbf{k} = 1 \tag{M1.20}$$

For the special case of Cartesian coordinates, from (M1.16) we, therefore, obtain for the scalar product

$$\mathbf{A} \cdot \mathbf{B} = A_x B_x + A_y B_y + A_z B_z \tag{M1.21}$$

When the basis vectors **i**, **j**, **k** are oriented along the (x, y, z)-axes, the coordinates of their terminal points are given by

$$\mathbf{i}: \quad (1, 0, 0), \qquad \mathbf{j}: \quad (0, 1, 0), \qquad \mathbf{k}: \quad (0, 0, 1) \tag{M1.22}$$

This expression is the Euclidian three-dimensional space or the space of ordinary human life. On generalizing to the N-dimensional space we obtain

$$\mathbf{e}_1: \quad (1, 0, \ldots, 0), \qquad \mathbf{e}_2: \quad (0, 1, \ldots, 0), \qquad \ldots \qquad \mathbf{e}_N: \quad (0, 0, \ldots, 1) \tag{M1.23}$$

This equation is known as the Cartesian reference frame of the N-dimensional Euclidian space. In this space the generalized form of the position vector **r** is given by

$$\mathbf{r} = x^1 \mathbf{e}_1 + x^2 \mathbf{e}_2 + \cdots + x^N \mathbf{e}_N \tag{M1.24}$$

The length or the magnitude of the vector **r** is also known as the *Euclidian norm*

$$|\mathbf{r}| = \sqrt{\mathbf{r} \cdot \mathbf{r}} = \sqrt{(x^1)^2 + (x^2)^2 + \cdots + (x^N)^2} \tag{M1.25}$$

M1.3.2 The vector product of two vectors

In coordinate-free or invariant notation the vector product of two vectors is defined by

$$\mathbf{A} \times \mathbf{B} = \mathbf{C} = |\mathbf{A}|\,|\mathbf{B}| \sin(\mathbf{A}, \mathbf{B})\, \mathbf{e}_C \tag{M1.26}$$

The unit vector $\mathbf{e}_C$ is perpendicular to the plane defined by the vectors **A** and **B**. The direction of the vector **C** is defined in such a way that the vectors **A**, **B**, and **C**

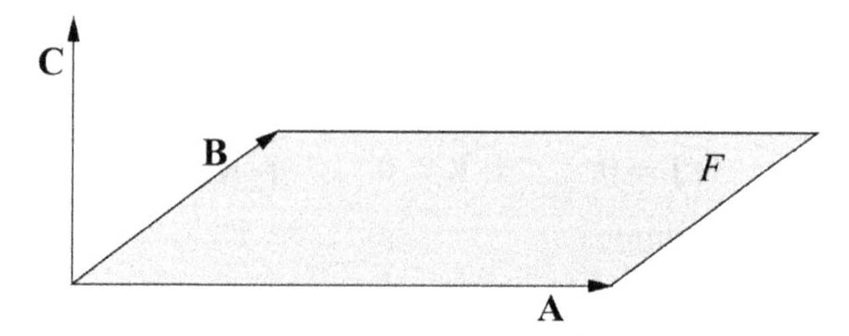

Fig. M1.5 Geometric interpretation of the vector or cross product.

form a right-handed system. The magnitude of **C** is equal to the area F of a parallelogram defined by the vectors **A** and **B** as shown in Figure M1.5. Interchanging the vectors **A** and **B** gives $\mathbf{A}\times\mathbf{B} = -\mathbf{B}\times\mathbf{A}$. This follows immediately from (M1.26) since the unit vector $\mathbf{e}_C$ now points in the opposite direction.

The following vector statements are valid:

$$\begin{aligned} \mathbf{A}\times(\mathbf{B}+\mathbf{C}) &= \mathbf{A}\times\mathbf{B}+\mathbf{A}\times\mathbf{C} \\ (k\mathbf{A})\times\mathbf{B} &= \mathbf{A}\times(k\mathbf{B}) = k\mathbf{A}\times\mathbf{B} \\ \mathbf{A}\times\mathbf{B} &= -\mathbf{B}\times\mathbf{A} \end{aligned} \tag{M1.27}$$

The component representation of the vector product yields

$$\boxed{\mathbf{A}\times\mathbf{B} = A^m\mathbf{q}_m \times B^n\mathbf{q}_n = \begin{vmatrix} \mathbf{q}_2\times\mathbf{q}_3 & \mathbf{q}_3\times\mathbf{q}_1 & \mathbf{q}_1\times\mathbf{q}_2 \\ A^1 & A^2 & A^3 \\ B^1 & B^2 & B^3 \end{vmatrix}} \tag{M1.28}$$

By utilizing Cartesian coordinates we obtain the well-known relation

$$\mathbf{A}\times\mathbf{B} = \begin{vmatrix} \mathbf{i} & \mathbf{j} & \mathbf{k} \\ A_x & A_y & A_z \\ B_x & B_y & B_z \end{vmatrix} \tag{M1.29}$$

M1.3.3 The dyadic representation, the general product of two vectors

The general or *dyadic product* of two vectors **A** and **B** is given by

$$\mathbf{\Phi} = \mathbf{AB} = (A^1\mathbf{q}_1 + A^2\mathbf{q}_2 + A^3\mathbf{q}_3)(B^1\mathbf{q}_1 + B^2\mathbf{q}_2 + B^3\mathbf{q}_3) \tag{M1.30}$$

It is seen that the vectors are not separated by a dot or a cross. At first glance this type of vector product seems strange. However, the advantage of this notation will

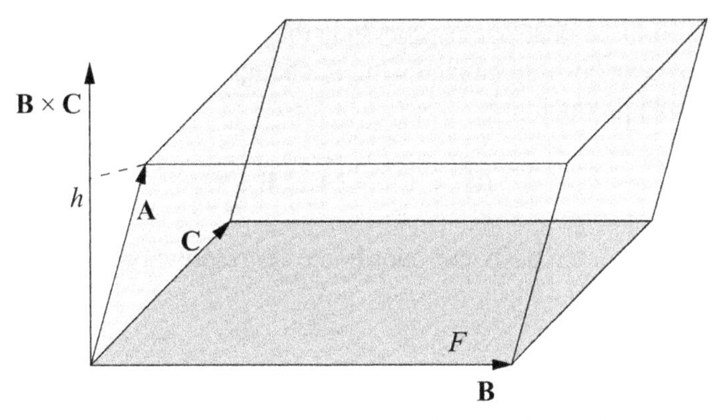

Fig. M1.6 Geometric representation of the scalar triple product.

become apparent later. On performing the dyadic multiplication we obtain

$$\boxed{\begin{aligned}\mathbf{\Phi} = \mathbf{AB} = & A^1B^1\mathbf{q}_1\mathbf{q}_1 + A^1B^2\mathbf{q}_1\mathbf{q}_2 + A^1B^3\mathbf{q}_1\mathbf{q}_3 \\ & + A^2B^1\mathbf{q}_2\mathbf{q}_1 + A^2B^2\mathbf{q}_2\mathbf{q}_2 + A^2B^3\mathbf{q}_2\mathbf{q}_3 \\ & + A^3B^1\mathbf{q}_3\mathbf{q}_1 + A^3B^2\mathbf{q}_3\mathbf{q}_2 + A^3B^3\mathbf{q}_3\mathbf{q}_3\end{aligned}} \tag{M1.31}$$

In carrying out the general multiplication, we must be careful not to change the position of the basis vectors. The following statements are valid:

$$(\mathbf{A} + \mathbf{B})\mathbf{C} = \mathbf{AC} + \mathbf{BC}, \qquad \mathbf{AB} \neq \mathbf{BA} \tag{M1.32}$$

M1.3.4 The scalar triple product

The scalar triple product, sometimes also called the box product, is defined by

$$\mathbf{A} \cdot (\mathbf{B} \times \mathbf{C}) = [\mathbf{A}, \mathbf{B}, \mathbf{C}] \tag{M1.33}$$

The absolute value of the scalar triple product measures the volume of the parallelepiped having the three vectors **A**, **B**, **C** as adjacent edges, see Figure M1.6. The height h of the parallelepiped is found by projecting the vector **A** onto the cross product $\mathbf{B} \times \mathbf{C}$. If the volume vanishes then the three vectors are coplanar. This situation will occur whenever a vector appears twice in the scalar triple product. It is apparent that, in the scalar triple product, any cyclic permutation of the factors leaves the value of the scalar triple product unchanged. A permutation that reverses the original cyclic order changes the sign of the product:

$$\begin{aligned}[\mathbf{A}, \mathbf{B}, \mathbf{C}] &= [\mathbf{B}, \mathbf{C}, \mathbf{A}] = [\mathbf{C}, \mathbf{A}, \mathbf{B}] \\ [\mathbf{A}, \mathbf{B}, \mathbf{C}] &= -[\mathbf{B}, \mathbf{A}, \mathbf{C}] = -[\mathbf{A}, \mathbf{C}, \mathbf{B}]\end{aligned} \tag{M1.34}$$

From these observations we may conclude that, in any scalar triple product, the dot and the cross can be interchanged without changing the magnitude and the sign of the scalar triple product

$$\mathbf{A}\cdot(\mathbf{B}\times\mathbf{C})=(\mathbf{A}\times\mathbf{B})\cdot\mathbf{C} \tag{M1.35}$$

For the general vector basis the coordinate representation of the scalar triple product yields

$$\mathbf{A}\cdot(\mathbf{B}\times\mathbf{C})=(A^1\mathbf{q}_1+A^2\mathbf{q}_2+A^3\mathbf{q}_3)\cdot\begin{vmatrix}\mathbf{q}_2\times\mathbf{q}_3 & \mathbf{q}_3\times\mathbf{q}_1 & \mathbf{q}_1\times\mathbf{q}_2\\ B^1 & B^2 & B^3\\ C^1 & C^2 & C^3\end{vmatrix} \tag{M1.36}$$

It is customary to assign the symbol $\sqrt{g}$ to the scalar triple product of the basis vectors:

$$\boxed{\sqrt{g}=\mathbf{q}_1\cdot\mathbf{q}_2\times\mathbf{q}_3} \tag{M1.37}$$

It is regrettable that the symbol g is also assigned to the acceleration due to gravity, but confusion is unlikely to occur. By combining equations (M1.36) and (M1.37) we obtain the following important form of the scalar triple product:

$$\boxed{[\mathbf{A},\mathbf{B},\mathbf{C}]=\sqrt{g}\begin{vmatrix}A^1 & A^2 & A^3\\ B^1 & B^2 & B^3\\ C^1 & C^2 & C^3\end{vmatrix}} \tag{M1.38}$$

For the basis vectors of the Cartesian system we obtain from (M1.37)

$$\sqrt{g}=\mathbf{i}\cdot(\mathbf{j}\times\mathbf{k})=1 \tag{M1.39}$$

so that in the Cartesian coordinate system (M1.38) reduces to

$$[\mathbf{A},\mathbf{B},\mathbf{C}]=\begin{vmatrix}A_x & A_y & A_z\\ B_x & B_y & B_z\\ C_x & C_y & C_z\end{vmatrix} \tag{M1.40}$$

In this expression, according to equation (M1.10), the components A^1, A^2, A^3, etc. have been written as A_x, A_y, A_z.

Without proof we accept the formula

$$[\mathbf{A}, \mathbf{B}, \mathbf{C}]^2 = \begin{vmatrix} \mathbf{A}\cdot\mathbf{A} & \mathbf{A}\cdot\mathbf{B} & \mathbf{A}\cdot\mathbf{C} \\ \mathbf{B}\cdot\mathbf{A} & \mathbf{B}\cdot\mathbf{B} & \mathbf{B}\cdot\mathbf{C} \\ \mathbf{C}\cdot\mathbf{A} & \mathbf{C}\cdot\mathbf{B} & \mathbf{C}\cdot\mathbf{C} \end{vmatrix} \tag{M1.41}$$

which is known as the *Gram determinant*. The proof, however, will be given later. Application of this important formula gives

$$[\mathbf{q}_1, \mathbf{q}_2, \mathbf{q}_3]^2 = \left(\sqrt{g}\right)^2 = \begin{vmatrix} \mathbf{q}_1\cdot\mathbf{q}_1 & \mathbf{q}_1\cdot\mathbf{q}_2 & \mathbf{q}_1\cdot\mathbf{q}_3 \\ \mathbf{q}_2\cdot\mathbf{q}_1 & \mathbf{q}_2\cdot\mathbf{q}_2 & \mathbf{q}_2\cdot\mathbf{q}_3 \\ \mathbf{q}_3\cdot\mathbf{q}_1 & \mathbf{q}_3\cdot\mathbf{q}_2 & \mathbf{q}_3\cdot\mathbf{q}_3 \end{vmatrix} = \begin{vmatrix} g_{11} & g_{12} & g_{13} \\ g_{21} & g_{22} & g_{23} \\ g_{31} & g_{32} & g_{33} \end{vmatrix} = |g_{ij}| \tag{M1.42}$$

which involves all elements g_{ij} of the metric tensor. Comparison of (M1.37) and (M1.42) yields the important statement

$$\boxed{\mathbf{q}_1 \cdot (\mathbf{q}_2 \times \mathbf{q}_3) = \sqrt{g} = \sqrt{|g_{ij}|}} \tag{M1.43}$$

so that the scalar triple product involving the general basis vectors $\mathbf{q}_1, \mathbf{q}_2, \mathbf{q}_3$ can easily be evaluated. This will be done in some detail when we consider various coordinate systems. Owing to (M1.43), $\sqrt{g}$ is called the *functional determinant* of the system.

M1.3.5 The vectorial triple product

At this point it will be sufficient to state the extremely important formula

$$\boxed{\mathbf{A} \times (\mathbf{B} \times \mathbf{C}) = (\mathbf{A} \cdot \mathbf{C})\mathbf{B} - (\mathbf{A} \cdot \mathbf{B})\mathbf{C}} \tag{M1.44}$$

which is also known as the *Grassmann rule*. It should be noted that, without the parentheses, the meaning of (M1.44) is not unique. The proof of this equation will be given later with the help of the so-called reciprocal coordinate system.

M1.3.6 The scalar product of a vector with a dyadic

On performing the scalar product of a vector with a dyadic we see that the commutative law is not valid:

$$\mathbf{D} = \mathbf{A} \cdot (\mathbf{BC}) = (\mathbf{A} \cdot \mathbf{B})\mathbf{C}, \qquad \mathbf{E} = (\mathbf{BC}) \cdot \mathbf{A} = \mathbf{B}(\mathbf{C} \cdot \mathbf{A}) \tag{M1.45}$$

Whereas in the first expression the vectors **D** and **C** are collinear, in the second expression the direction of **E** is along the vector **B** so that $\mathbf{D} \neq \mathbf{E}$.

M1.3.7 Products involving four vectors

Let us consider the expression $(\mathbf{A} \times \mathbf{B}) \cdot (\mathbf{C} \times \mathbf{D})$. Defining the vector $\mathbf{F} = \mathbf{C} \times \mathbf{D}$ we obtain the scalar triple product

$$(\mathbf{A} \times \mathbf{B}) \cdot (\mathbf{C} \times \mathbf{D}) = (\mathbf{A} \times \mathbf{B}) \cdot \mathbf{F} = \mathbf{A} \cdot (\mathbf{B} \times \mathbf{F}) = \mathbf{A} \cdot [\mathbf{B} \times (\mathbf{C} \times \mathbf{D})] \tag{M1.46}$$

This equation results from interchanging the dot and the cross and by replacing the vector **F** by its definition. Application of the Grassmann rule (M1.44) yields

$$(\mathbf{A} \times \mathbf{B}) \cdot (\mathbf{C} \times \mathbf{D}) = \mathbf{A} \cdot [(\mathbf{B} \cdot \mathbf{D})\mathbf{C} - (\mathbf{B} \cdot \mathbf{C})\mathbf{D}] = (\mathbf{A} \cdot \mathbf{C})(\mathbf{B} \cdot \mathbf{D}) - (\mathbf{A} \cdot \mathbf{D})(\mathbf{B} \cdot \mathbf{C}) \tag{M1.47}$$

so that equation (M1.46) can be written as

$$\boxed{(\mathbf{A} \times \mathbf{B}) \cdot (\mathbf{C} \times \mathbf{D}) = \begin{vmatrix} \mathbf{A} \cdot \mathbf{C} & \mathbf{A} \cdot \mathbf{D} \\ \mathbf{B} \cdot \mathbf{C} & \mathbf{B} \cdot \mathbf{D} \end{vmatrix}} \tag{M1.48}$$

The vector product of four vectors may be evaluated with the help of the Grassmann rule:

$$(\mathbf{A} \times \mathbf{B}) \times (\mathbf{C} \times \mathbf{D}) = (\mathbf{F} \cdot \mathbf{D})\mathbf{C} - (\mathbf{F} \cdot \mathbf{C})\mathbf{D} \quad \text{with} \quad \mathbf{F} = \mathbf{A} \times \mathbf{B} \tag{M1.49}$$

On replacing **F** by its definition and using the rules of the scalar triple product, we find the following useful expression:

$$\boxed{(\mathbf{A} \times \mathbf{B}) \times (\mathbf{C} \times \mathbf{D}) = [\mathbf{A}, \mathbf{B}, \mathbf{D}]\mathbf{C} - [\mathbf{A}, \mathbf{B}, \mathbf{C}]\mathbf{D}} \tag{M1.50}$$

M1.4 Reciprocal coordinate systems

As will be seen shortly, operations with the so-called reciprocal basis systems result in particularly convenient mathematical expressions. Let us consider two basis systems. One of these is defined by the three linearly independent basis vectors $\mathbf{q}_i$, $i = 1, 2, 3$, and the other one by the linearly independent basis vectors $\mathbf{q}^i$, $i = 1, 2, 3$. To have reciprocality for the basis vectors the following relation must be valid:

$$\boxed{\mathbf{q}_i \cdot \mathbf{q}^k = \mathbf{q}^k \cdot \mathbf{q}_i = \delta_i^k \quad \text{with} \quad \delta_i^k = \begin{cases} 0 & i \neq k \\ 1 & i = k \end{cases}} \tag{M1.51}$$

where δ_i^k is the Kronecker-delta symbol. Reciprocal systems are also called *contragredient systems*. As is customary, the system represented by basis vectors with the lower index is called *covariant* while the system employing basis vectors with an upper index is called *contravariant*. Therefore, $\mathbf{q}_i$ and $\mathbf{q}^i$ are called *covariant* and *contravariant basis vectors*, respectively.

Consider for example in (M1.51) the case $i = k = 1$. While the scalar product $\mathbf{q}_1 \cdot \mathbf{q}^1 = 1$ may be viewed as a normalization condition for the two systems, the scalar products $\mathbf{q}_1 \cdot \mathbf{q}^2 = 0$ and $\mathbf{q}_1 \cdot \mathbf{q}^3 = 0$ are conditions of orthogonality. Thus, $\mathbf{q}_1$ is perpendicular to $\mathbf{q}^2$ and to $\mathbf{q}^3$ so that we may write

$$\mathbf{q}_1 = C(\mathbf{q}^2 \times \mathbf{q}^3) \tag{M1.52a}$$

where C is a factor of proportionality. On substituting this expression into the normalization condition we obtain for C

$$\mathbf{q}^1 \cdot \mathbf{q}_1 = C\mathbf{q}^1 \cdot (\mathbf{q}^2 \times \mathbf{q}^3) = 1 \implies C = \frac{1}{\mathbf{q}^1 \cdot (\mathbf{q}^2 \times \mathbf{q}^3)} \tag{M1.52b}$$

so that (M1.52a) yields

$$\mathbf{q}_1 = \frac{\mathbf{q}^2 \times \mathbf{q}^3}{[\mathbf{q}^1, \mathbf{q}^2, \mathbf{q}^3]} \tag{M1.52c}$$

We may repeat this exercise with $\mathbf{q}_2$ and $\mathbf{q}_3$ and find the general expression

$$\boxed{\mathbf{q}_i = \frac{\mathbf{q}^j \times \mathbf{q}^k}{[\mathbf{q}^1, \mathbf{q}^2, \mathbf{q}^3]}} \tag{M1.53}$$

with i, j, k in cyclic order. Similarly we may write for $\mathbf{q}^1$, with D as the proportionality constant,

$$\mathbf{q}^1 = D(\mathbf{q}_2 \times \mathbf{q}_3), \quad \mathbf{q}_1 \cdot \mathbf{q}^1 = D\mathbf{q}_1 \cdot (\mathbf{q}_2 \times \mathbf{q}_3) = 1 \implies \mathbf{q}^1 = \frac{\mathbf{q}_2 \times \mathbf{q}_3}{[\mathbf{q}_1, \mathbf{q}_2, \mathbf{q}_3]} \tag{M1.54}$$

Thus, the general expression is

$$\boxed{\mathbf{q}^i = \frac{\mathbf{q}_j \times \mathbf{q}_k}{[\mathbf{q}_1, \mathbf{q}_2, \mathbf{q}_3]}} \tag{M1.55}$$

with i, j, k in cyclic order. Equations (M1.53) and (M1.55) give the explicit expressions relating the basis vectors of the two reciprocal systems.

Let us consider the special case of the Cartesian coordinate system with basis vectors $\mathbf{i}_1, \mathbf{i}_2, \mathbf{i}_3$. Application of (M1.55) shows that $\mathbf{i}^j = \mathbf{i}_j$ since $[\mathbf{i}_1, \mathbf{i}_2, \mathbf{i}_3] = 1$, so that in the Cartesian coordinate system there is no difference between covariant and contravariant basis vectors. This is the reason why we have written $A^i = A_i$, $i = 1, 2, 3$ in (M1.10).

Now we return to equation (M1.43). By replacing the covariant basis vector $\mathbf{q}_1$ with the help of (M1.52c) and utilizing (M1.48) we find

$$\begin{aligned} \mathbf{q}_1 \cdot (\mathbf{q}_2 \times \mathbf{q}_3) &= \frac{(\mathbf{q}^2 \times \mathbf{q}^3) \cdot (\mathbf{q}_2 \times \mathbf{q}_3)}{[\mathbf{q}^1, \mathbf{q}^2, \mathbf{q}^3]} \\ &= \frac{1}{[\mathbf{q}^1, \mathbf{q}^2, \mathbf{q}^3]} \begin{vmatrix} \mathbf{q}^2 \cdot \mathbf{q}_2 & \mathbf{q}^2 \cdot \mathbf{q}_3 \\ \mathbf{q}^3 \cdot \mathbf{q}_2 & \mathbf{q}^3 \cdot \mathbf{q}_3 \end{vmatrix} = \frac{1}{[\mathbf{q}^1, \mathbf{q}^2, \mathbf{q}^3]} \end{aligned} \tag{M1.56}$$

From (M1.51) it follows that the value of the determinant in (M1.56) is equal to 1. Since $\mathbf{q}_1 \cdot (\mathbf{q}_2 \times \mathbf{q}_3) = \sqrt{g}$ we immediately find

$$\boxed{[\mathbf{q}^1, \mathbf{q}^2, \mathbf{q}^3] = \frac{1}{\sqrt{g}}} \tag{M1.57}$$

Thus, the introduction of the contravariant basis vectors shows that (M1.43) and (M1.57) are inverse relations.

Often it is desirable to work with unit vectors having the same directions as the selected three linearly independent basis vectors. The desired relationships are

$$\mathbf{e}_i = \frac{\mathbf{q}_i}{|\mathbf{q}_i|} = \frac{\mathbf{q}_i}{\sqrt{\mathbf{q}_i \cdot \mathbf{q}_i}} = \frac{\mathbf{q}_i}{\sqrt{g_{ii}}}, \qquad \mathbf{e}^i = \frac{\mathbf{q}^i}{|\mathbf{q}^i|} = \frac{\mathbf{q}^i}{\sqrt{\mathbf{q}^i \cdot \mathbf{q}^i}} = \frac{\mathbf{q}^i}{\sqrt{g^{ii}}} \tag{M1.58}$$

While the scalar product of the covariant basis vectors $\mathbf{q}_i \cdot \mathbf{q}_j = g_{ij}$ defines the elements of the *covariant metric tensor*, the *contravariant metric tensor* is defined by the elements $\mathbf{q}^i \cdot \mathbf{q}^j = g^{ij}$, and we have

$$\boxed{\mathbf{q}^i \cdot \mathbf{q}^j = \mathbf{q}^j \cdot \mathbf{q}^i = g^{ij} = g^{ji}} \tag{M1.59}$$

Owing to the symmetry relations $g_{ij} = g_{ji}$ and $g^{ij} = g^{ji}$ each metric tensor is completely specified by six elements.

Some special cases follow directly from the definition (M1.13) of the scalar product. In case of an orthonormal system, such as the Cartesian coordinate system, we have

$$g_{ij} = g_{ji} = g^{ij} = g^{ji} = \delta_i^j \tag{M1.60}$$

As will be shown later, for any orthogonal system the following equation applies:

$$g_{ii} g^{ii} = 1 \tag{M1.61}$$

While in the Cartesian coordinate system the metric fundamental quantities are either 0 or 1, we cannot give any information about the g_{ij} or g^{ij} unless the coordinate system is specified. This will be done later when we consider various physical situations.

In the following we will give examples of the efficient use of reciprocal systems. Work is defined by the scalar product $dA = \mathbf{K} \cdot d\mathbf{r}$, where $\mathbf{K}$ is the force and $d\mathbf{r}$ is the path increment. In the Cartesian system we obtain a particularly simple result:

$$\mathbf{K} \cdot d\mathbf{r} = (K_x\mathbf{i} + K_y\mathbf{j} + K_z\mathbf{k}) \cdot (dx\,\mathbf{i} + dy\,\mathbf{j} + dz\,\mathbf{k}) = K_x\,dx + K_y\,dy + K_z\,dz \tag{M1.62}$$

consisting of three work contributions in the directions of the three coordinate axes. For specific applications it may be necessary, however, to employ more general coordinate systems. Let us consider, for example, an oblique coordinate system with contravariant components and covariant basis vectors of $\mathbf{K}$ and $d\mathbf{r}$. In this case work will be expressed by

$$\begin{aligned} \mathbf{K} \cdot d\mathbf{r} &= (K^1\mathbf{q}_1 + K^2\mathbf{q}_2 + K^3\mathbf{q}_3) \cdot (dq^1\mathbf{q}_1 + dq^2\mathbf{q}_2 + dq^3\mathbf{q}_3) \\ &= K^m\,dq^n\,\mathbf{q}_m \cdot \mathbf{q}_n = K^m\,dq^n\,g_{mn} \end{aligned} \tag{M1.63}$$

Expansion of this expression results in nine components in contrast to only three components of the Cartesian coordinate system. A great deal of simplification is achieved by employing reciprocal systems for the force and the path increment. As in the case of the Cartesian system, work can then be expressed by using only three terms:

$$\begin{aligned} \mathbf{K} \cdot d\mathbf{r} &= (K_1\mathbf{q}^1 + K_2\mathbf{q}^2 + K_3\mathbf{q}^3) \cdot (dq^1\,\mathbf{q}_1 + dq^2\,\mathbf{q}_2 + dq^3\,\mathbf{q}_3) \\ &= K_m\,dq^n\,\mathbf{q}^m \cdot \mathbf{q}_n = K_m\,dq^n\,\delta_n^m = K_1\,dq^1 + K_2\,dq^2 + K_3\,dq^3 \end{aligned} \tag{M1.64a}$$

or

$$\begin{aligned}\mathbf{K}\cdot d\mathbf{r} &= (K^1\mathbf{q}_1 + K^2\mathbf{q}_2 + K^3\mathbf{q}_3)\cdot(dq_1\,\mathbf{q}^1 + dq_2\,\mathbf{q}^2 + dq_3\,\mathbf{q}^3)\\ &= K^m\,dq_n\,\mathbf{q}_m\cdot\mathbf{q}^n = K^m\,dq_n\,\delta^n_m = K^1\,dq_1 + K^2\,dq_2 + K^3\,dq_3\end{aligned}\tag{M1.64b}$$

Finally, utilizing reciprocal coordinate systems, it is easy to give the proof of the Grassmann rule (M1.44). Let us consider the expression $\mathbf{D} = \mathbf{A}\times(\mathbf{B}\times\mathbf{C})$. According to the definition (M1.26) of the vector product, $\mathbf{D}$ is perpendicular to $\mathbf{A}$ and to $(\mathbf{B}\times\mathbf{C})$. Therefore, $\mathbf{D}$ must lie in the plane defined by the vectors $\mathbf{B}$ and $\mathbf{C}$ so that we may write

$$\mathbf{A}\times(\mathbf{B}\times\mathbf{C}) = \lambda\mathbf{B} + \mu\mathbf{C}\tag{M1.65}$$

where λ and μ are unknown scalars to be determined. To make use of the properties of the reciprocal system, we first set $\mathbf{B} = \mathbf{q}_1$ and $\mathbf{C} = \mathbf{q}_2$. These two vectors define a plane oblique coordinate system. To complete the system we assume that the vector $\mathbf{q}_3$ is a unit vector orthogonal to the plane spanned by $\mathbf{q}_1$ and $\mathbf{q}_2$. Thus, we have

$$\mathbf{B} = \mathbf{q}_1,\qquad \mathbf{C} = \mathbf{q}_2,\qquad \mathbf{e}_3 = \frac{\mathbf{q}_1\times\mathbf{q}_2}{|\mathbf{q}_1\times\mathbf{q}_2|}\tag{M1.66}$$

and

$$\mathbf{q}_1\cdot(\mathbf{q}_2\times\mathbf{e}_3) = \mathbf{e}_3\cdot(\mathbf{q}_1\times\mathbf{q}_2) = \mathbf{e}_3\cdot\mathbf{e}_3\,|\mathbf{q}_1\times\mathbf{q}_2| = |\mathbf{q}_1\times\mathbf{q}_2|\tag{M1.67}$$

According to (M1.55), the coordinate system which is reciprocal to the $(\mathbf{q}_1, \mathbf{q}_2, \mathbf{q}_3)$ system is given by

$$\mathbf{q}^1 = \frac{\mathbf{q}_2\times\mathbf{e}_3}{|\mathbf{q}_1\times\mathbf{q}_2|},\qquad \mathbf{q}^2 = \frac{\mathbf{e}_3\times\mathbf{q}_1}{|\mathbf{q}_1\times\mathbf{q}_2|},\qquad \mathbf{e}^3 = \frac{\mathbf{q}_1\times\mathbf{q}_2}{|\mathbf{q}_1\times\mathbf{q}_2|} = \mathbf{e}_3\tag{M1.68}$$

The determination of λ and μ follows from scalar multiplication of $\mathbf{A}\times(\mathbf{B}\times\mathbf{C}) = \mathbf{A}\times(\mathbf{q}_1\times\mathbf{q}_2) = \lambda\mathbf{q}_1 + \mu\mathbf{q}_2$ by the reciprocal basis vectors $\mathbf{q}^1$ and $\mathbf{q}^2$:

$$\begin{aligned}\lambda &= \left[\mathbf{A}\times(\mathbf{q}_1\times\mathbf{q}_2)\right]\cdot\mathbf{q}^1 = \mathbf{A}\times(\mathbf{q}_1\times\mathbf{q}_2)\cdot\frac{(\mathbf{q}_2\times\mathbf{e}_3)}{|\mathbf{q}_1\times\mathbf{q}_2|}\\ &= (\mathbf{A}\times\mathbf{e}_3)\cdot(\mathbf{q}_2\times\mathbf{e}_3) = \mathbf{A}\cdot\mathbf{q}_2 = \mathbf{A}\cdot\mathbf{C}\end{aligned}\tag{M1.69a}$$

Analogously we obtain

$$\mu = \left[\mathbf{A}\times(\mathbf{q}_1\times\mathbf{q}_2)\right]\cdot\mathbf{q}^2 = (\mathbf{A}\times\mathbf{e}_3)\cdot(\mathbf{e}_3\times\mathbf{q}_1) = -\mathbf{A}\cdot\mathbf{q}_1 = -\mathbf{A}\cdot\mathbf{B}\tag{M1.69b}$$

Substitution of λ and μ into (M1.65) gives the final result

$$\mathbf{A}\times(\mathbf{B}\times\mathbf{C}) = (\mathbf{A}\cdot\mathbf{C})\mathbf{B} - (\mathbf{A}\cdot\mathbf{B})\mathbf{C}\tag{M1.70}$$

M1.5 Vector representations

The vector **A** may be represented with the help of the covariant basis vectors $\mathbf{q}_i$ or $\mathbf{e}_i$ and the contravariant basis vectors $\mathbf{q}^i$ or $\mathbf{e}^i$ as

$$\mathbf{A} = A^m\mathbf{q}_m = A_m\mathbf{q}^m = \overset{*}{A}{}^m\mathbf{e}_m = \overset{*}{A}_m\mathbf{e}^m \tag{M1.71}$$

The invariant character of **A** is recognized by virtue of the fact that we have the same number of upper and lower indices. In addition to the *contravariant* and *covariant* measure numbers A^i and A_i of the basis vectors $\mathbf{q}_i$ and $\mathbf{q}^i$ we have also introduced the *physical measure numbers* $\overset{*}{A}{}^i$ and $\overset{*}{A}_i$ of the unit vectors $\mathbf{e}_i$ and $\mathbf{e}^i$. In general the contravariant and covariant measure numbers do not have uniform dimensions. This becomes obvious on considering, for example, the spherical coordinate system which is defined by two angles, which are measured in degrees, and the radius of the sphere, which is measured in units of length. Physical measure numbers, however, are uniformly dimensioned. They represent the lengths of the components of a vector in the directions of the basis vectors. The formal definitions of the physical measure numbers are

$$\overset{*}{A}{}^i = A^i\,|\mathbf{q}_i| = A^i\sqrt{g_{ii}}, \qquad \overset{*}{A}_i = A_i|\mathbf{q}^i| = A_i\sqrt{g^{ii}} \tag{M1.72}$$

Now we will show what consequences arise by interpreting the measure numbers vectorially. Scalar multiplication of $\mathbf{A} = A^n\mathbf{q}_n$ by the reciprocal basis vector $\mathbf{q}^i$ yields for A^i

$$\boxed{\mathbf{A}\cdot\mathbf{q}^i = A^m\mathbf{q}_m\cdot\mathbf{q}^i = A^m\delta^i_m = A^i} \tag{M1.73}$$

so that

$$\mathbf{A} = A^m\mathbf{q}_m = \mathbf{A}\cdot\mathbf{q}^m\mathbf{q}_m \tag{M1.74}$$

This expression leads to the introduction to the *unit dyadic* $\mathbb{E}$,

$$\mathbb{E} = \mathbf{q}^m\mathbf{q}_m \tag{M1.75a}$$

This very special dyadic or unit tensor of rank two has the same degree of importance in tensor analysis as the unit vector in vector analysis. The unit dyadic $\mathbb{E}$ is indispensable and will accompany our work from now on. In the Cartesian coordinate system the unit dyadic is given by

$$\mathbb{E} = \mathbf{ii} + \mathbf{jj} + \mathbf{kk} = \mathbf{i}_1\mathbf{i}_1 + \mathbf{i}_2\mathbf{i}_2 + \mathbf{i}_3\mathbf{i}_3 \tag{M1.75b}$$

We repeat the above procedure by representing the vector **A** as $\mathbf{A} = A_m\mathbf{q}^m$. Scalar multiplication by $\mathbf{q}_i$ results in

$$\boxed{\mathbf{A}\cdot\mathbf{q}_i = A_m\mathbf{q}^m\cdot\mathbf{q}_i = A_i} \tag{M1.76}$$

and the equivalent definition of the unit dyadic $\mathbb{E}$

$$\mathbf{A} = A_m \mathbf{q}^m = \mathbf{A} \cdot \mathbf{q}_m \mathbf{q}^m \implies \mathbb{E} = \mathbf{q}_m \mathbf{q}^m \tag{M1.77}$$

Of particular interest is the scalar product of two unit dyadics:

$$\begin{aligned} \mathbb{E} \cdot \mathbb{E} &= \mathbf{q}^m \mathbf{q}_m \cdot \mathbf{q}^n \mathbf{q}_n = \mathbf{q}^m \delta_m^n \mathbf{q}_n = \mathbf{q}^m \mathbf{q}_m = \mathbb{E} \\ \mathbb{E} \cdot \mathbb{E} &= \mathbf{q}_m \mathbf{q}^m \cdot \mathbf{q}_n \mathbf{q}^n = \mathbf{q}_m \delta_n^m \mathbf{q}^n = \mathbf{q}_m \mathbf{q}^m = \mathbb{E} \end{aligned} \tag{M1.78}$$

From these expressions we obtain additional representations of the unit dyadic that involve the metric fundamental quantities g_{ij} and g^{ij}:

$$\mathbb{E} \cdot \mathbb{E} = \mathbf{q}^m \mathbf{q}_m \cdot \mathbf{q}_n \mathbf{q}^n = g_{mn} \mathbf{q}^m \mathbf{q}^n = \mathbf{q}_m \mathbf{q}^m \cdot \mathbf{q}^n \mathbf{q}_n = g^{mn} \mathbf{q}_m \mathbf{q}_n \tag{M1.79}$$

Again it should be carefully observed that each expression contains an equal number of subscripts and superscripts to stress the invariant character of the unit dyadic. We collect the important results involving the unit dyadic as

$$\boxed{\mathbb{E} = \mathbf{q}^m \mathbf{q}_m = \delta_m^n \mathbf{q}^m \mathbf{q}_n = \mathbf{q}_m \mathbf{q}^m = \delta_n^m \mathbf{q}_m \mathbf{q}^n = g_{mn} \mathbf{q}^m \mathbf{q}^n = g^{mn} \mathbf{q}_m \mathbf{q}_n} \tag{M1.80}$$

Scalar multiplication of $\mathbb{E}$ in two of the forms of (M1.80) with $\mathbf{q}_i$ results in

$$\begin{aligned} \mathbb{E} \cdot \mathbf{q}_i &= (\mathbf{q}_m \mathbf{q}^m) \cdot \mathbf{q}_i = \mathbf{q}_m \delta_i^m = \mathbf{q}_i \\ &= (g_{mn} \mathbf{q}^m \mathbf{q}^n) \cdot \mathbf{q}_i = g_{mn} \mathbf{q}^m \delta_i^n = g_{im} \mathbf{q}^m \end{aligned} \tag{M1.81}$$

Hence, we see immediately that

$$\boxed{\mathbf{q}_i = g_{im} \mathbf{q}^m} \tag{M1.82}$$

This very useful expression is known as the *raising rule* for the index of the basis vector $\mathbf{q}_i$. Analogously we multiply the unit dyadic by $\mathbf{q}^i$ to obtain

$$\mathbb{E} \cdot \mathbf{q}^i = (\mathbf{q}^m \mathbf{q}_m) \cdot \mathbf{q}^i = \mathbf{q}^i = (g^{mn} \mathbf{q}_m \mathbf{q}_n) \cdot \mathbf{q}^i = g^{im} \mathbf{q}_m \tag{M1.83}$$

and thus

$$\boxed{\mathbf{q}^i = g^{im} \mathbf{q}_m} \tag{M1.84}$$

which is known as the *lowering rule* for the index of the contravariant basis vector $\mathbf{q}^i$.

With the help of the unit dyadic we are in a position to find additional important rules of tensor analysis. In order to avoid confusion, it is often necessary to replace a letter representing a summation index by another letter so that the letter representing a summation does not occur more often than twice. If the replacement is done properly, the meaning of any mathematical expression will not change. Let us consider the expression

$$\mathbb{E} = \mathbf{q}_r\mathbf{q}^r = g_{rm}g^{rn}\mathbf{q}^m\mathbf{q}_n = \delta^n_m\mathbf{q}^m\mathbf{q}_n \tag{M1.85}$$

Application of (M1.82) and (M1.84) gives the expression to the right of the second equality sign. For comparison purposes we have also added one of the forms of (M1.80) as the final expression in (M1.85). It should be carefully observed that the summation indices m, n, r occur twice only.

To take full advantage of the reciprocal systems we perform a scalar multiplication first by the contravariant basis vector $\mathbf{q}^i$ and then by the covariant basis vector $\mathbf{q}_j$, yielding

$$(\mathbb{E}\cdot\mathbf{q}^i)\cdot\mathbf{q}_j = g_{rm}g^{rn}\delta^i_n\delta^m_j = \delta^n_m\delta^i_n\delta^m_j \tag{M1.86}$$

from which it follows immediately that

$$g_{rj}g^{ri} = \delta^i_j \tag{M1.87a}$$

By interchanging i and j, observing the symmetry of the fundamental quantities, we find

$$g_{ir}g^{rj} = \delta^j_i \quad \text{or} \quad (g_{ij})(g^{ij}) = \begin{pmatrix} 1 & 0 & 0 \\ 0 & 1 & 0 \\ 0 & 0 & 1 \end{pmatrix} \Longrightarrow (g_{ij}) = (g^{ij})^{-1} \tag{M1.87b}$$

Hence, the matrices (g_{ij}) and (g^{ij}) are inverse to each other. Owing to the symmetry properties of the metric fundamental quantities, i.e. $g_{ij} = g_{ji}$ and $g^{ij} = g^{ji}$, we need six elements only to specify either metric tensor. In case of an orthogonal system $g_{ij} = 0$, $g^{ij} = 0$ for $i \neq j$ so that (M1.87a) reduces to

$$g_{ii}g^{ii} = 1 \tag{M1.88}$$

thus verifying equation (M1.61). At this point we must recall the rule that we do not sum over repeated free indices i, j, k, l.

Next we wish to show that, in an orthonormal system, there is no difference between contravariant and covariant basis vectors. The proof is very simple:

$$\mathbf{e}_i = \frac{\mathbf{q}_i}{\sqrt{g_{ii}}} = \frac{g_{in}}{\sqrt{g_{ii}}}\mathbf{q}^n = \sqrt{g_{ii}}\,\mathbf{q}^i = \frac{\mathbf{q}^i}{\sqrt{g^{ii}}} = \mathbf{e}^i \tag{M1.89}$$

Here use of the raising rule has been made. With the help of (M1.89) it is easy to show that there is no difference between contravariant and covariant physical measure numbers. Utilizing (M1.71) we find

$$\mathbf{A}\cdot\mathbf{e}^i = \mathbf{A}\cdot\mathbf{e}_i \Longrightarrow \overset{*}{A}{}^n\mathbf{e}_n\cdot\mathbf{e}^i = \overset{*}{A}_n\mathbf{e}^n\cdot\mathbf{e}_i \Longrightarrow \overset{*}{A}{}^i = \overset{*}{A}_i \tag{M1.90}$$

M1.6 Products of vectors in general coordinate systems

There are various ways to express the dyadic product of vector **A** with vector **B** by employing covariant and contravariant basis vectors:

$$\mathbf{AB} = A^m B^n \mathbf{q}_m \mathbf{q}_n = A_m B_n \mathbf{q}^m \mathbf{q}^n = A_m B^n \mathbf{q}^m \mathbf{q}_n = A^m B_n \mathbf{q}_m \mathbf{q}^n \tag{M1.91}$$

This yields four possibilities for formulating the scalar product $\mathbf{A} \cdot \mathbf{B}$:

$$\begin{aligned} \mathbf{A} \cdot \mathbf{B} &= A^m B^n \mathbf{q}_m \cdot \mathbf{q}_n = A^m B^n g_{mn} = A_m B_n \mathbf{q}^m \cdot \mathbf{q}^n = A_m B_n g^{mn} \\ &= A_m B^n \mathbf{q}^m \cdot \mathbf{q}_n = A_m B^m = A^m B_n \mathbf{q}_m \cdot \mathbf{q}^n = A^m B_m \end{aligned} \tag{M1.92}$$

[1] The last two forms with mixed basis vectors (covariant and contravariant) are more convenient since the sums involve the evaluation of only three terms. In contrast, nine terms are required for the first two forms since they involve the metric fundamental quantities.

There are two useful forms in which to express the vector product $\mathbf{A} \times \mathbf{B}$. From the basic definition (M1.28) and the properties of the reciprocal systems (M1.55) we obtain

$$\mathbf{A} \times \mathbf{B} = A^m \mathbf{q}_m \times B^n \mathbf{q}_n = \sqrt{g} \begin{vmatrix} \mathbf{q}^1 & \mathbf{q}^2 & \mathbf{q}^3 \\ A^1 & A^2 & A^3 \\ B^1 & B^2 & B^3 \end{vmatrix} \tag{M1.93}$$

where all measure numbers are of the contravariant type. If it is desirable to express the vector product in terms of covariant measure numbers we use (M1.53) and (M1.57). Thus, we find

$$\mathbf{A} \times \mathbf{B} = A_m \mathbf{q}^m \times B_n \mathbf{q}^n = \frac{1}{\sqrt{g}} \begin{vmatrix} \mathbf{q}_1 & \mathbf{q}_2 & \mathbf{q}_3 \\ A_1 & A_2 & A_3 \\ B_1 & B_2 & B_3 \end{vmatrix} \tag{M1.94}$$

The two forms involving mixed basis vectors are not used, in general.

On performing the scalar triple product operation (M1.33) we find

$$\begin{aligned} [\mathbf{A}, \mathbf{B}, \mathbf{C}] &= A^m \mathbf{q}_m \cdot (B^n \mathbf{q}_n \times C^r \mathbf{q}_r) \\ &= A^m \sqrt{g} \mathbf{q}_m \cdot \begin{vmatrix} \mathbf{q}^1 & \mathbf{q}^2 & \mathbf{q}^3 \\ B^1 & B^2 & B^3 \\ C^1 & C^2 & C^3 \end{vmatrix} = \sqrt{g} \begin{vmatrix} A^1 & A^2 & A^3 \\ B^1 & B^2 & B^3 \\ C^1 & C^2 & C^3 \end{vmatrix} \end{aligned} \tag{M1.95}$$

[1] For the scalar product $\mathbf{A} \cdot \mathbf{A}$ we usually write $\mathbf{A}^2$.

which is in agreement with (M1.38). All measure numbers are of the contravariant type. If covariant measure numbers of the three vectors are desired, utilizing (M1.94) we obtain

$$[\mathbf{A}, \mathbf{B}, \mathbf{C}] = A_m \mathbf{q}^m \cdot (B_n \mathbf{q}^n \times C_r \mathbf{q}^r) = \frac{1}{\sqrt{g}} \begin{vmatrix} A_1 & A_2 & A_3 \\ B_1 & B_2 & B_3 \\ C_1 & C_2 & C_3 \end{vmatrix} \tag{M1.96}$$

In this expression the covariant measure numbers may be replaced with the help of (M1.76) by $A_i = \mathbf{A} \cdot \mathbf{q}_i$, $B_i = \mathbf{B} \cdot \mathbf{q}_i$, $C_i = \mathbf{C} \cdot \mathbf{q}_i$, yielding

$$[\mathbf{A}, \mathbf{B}, \mathbf{C}][\mathbf{q}_1, \mathbf{q}_2, \mathbf{q}_3] = \begin{vmatrix} \mathbf{A} \cdot \mathbf{q}_1 & \mathbf{A} \cdot \mathbf{q}_2 & \mathbf{A} \cdot \mathbf{q}_3 \\ \mathbf{B} \cdot \mathbf{q}_1 & \mathbf{B} \cdot \mathbf{q}_2 & \mathbf{B} \cdot \mathbf{q}_3 \\ \mathbf{C} \cdot \mathbf{q}_1 & \mathbf{C} \cdot \mathbf{q}_2 & \mathbf{C} \cdot \mathbf{q}_3 \end{vmatrix} \tag{M1.97}$$

Finally, if in this equation we replace the basis vectors $\mathbf{q}_1, \mathbf{q}_2, \mathbf{q}_3$ by the arbitrary vectors $\mathbf{D}, \mathbf{F}, \mathbf{G}$, we obtain

$$[\mathbf{A}, \mathbf{B}, \mathbf{C}][\mathbf{D}, \mathbf{F}, \mathbf{G}] = \begin{vmatrix} \mathbf{A} \cdot \mathbf{D} & \mathbf{A} \cdot \mathbf{F} & \mathbf{A} \cdot \mathbf{G} \\ \mathbf{B} \cdot \mathbf{D} & \mathbf{B} \cdot \mathbf{F} & \mathbf{B} \cdot \mathbf{G} \\ \mathbf{C} \cdot \mathbf{D} & \mathbf{C} \cdot \mathbf{F} & \mathbf{C} \cdot \mathbf{G} \end{vmatrix} \tag{M1.98}$$

On setting in this expression $\mathbf{D} = \mathbf{A}$, $\mathbf{F} = \mathbf{B}$, and $\mathbf{G} = \mathbf{C}$ we obtain the Gram determinant, which was already stated without proof as equation (M1.41).

M1.7 Problems

M1.1: Are the three vectors $\mathbf{A} = 2\mathbf{q}_1 + 6\mathbf{q}_2 - 2\mathbf{q}_3$, $\mathbf{B} = 3\mathbf{q}_1 + \mathbf{q}_2 + 2\mathbf{q}_3$, and $\mathbf{C} = 8\mathbf{q}_1 + 16\mathbf{q}_2 - 3\mathbf{q}_3$ linearly dependent?

M1.2:
(a) Are the three vectors $\mathbf{A} = 4\mathbf{q}_1 - \mathbf{q}_2 + 5\mathbf{q}_3$, $\mathbf{B} = -2\mathbf{q}_1 + 3\mathbf{q}_2 + \mathbf{q}_3$, and $\mathbf{C} = -2\mathbf{q}_1 - 2\mathbf{q}_2 - 6\mathbf{q}_3$ linearly dependent?
(b) Use Cartesian basis vectors to show that the vectors $\mathbf{F} = \mathbf{A} \times \mathbf{B}$, $\mathbf{G} = \mathbf{B} \times \mathbf{C}$, and $\mathbf{H} = \mathbf{C} \times \mathbf{A}$ are collinear.
(c) The vectors $\mathbf{A}$ and $\mathbf{B}$ are the same as before but now $\mathbf{C} = 8\mathbf{q}_1 + 7\mathbf{q}_2 - 5\mathbf{q}_3$. Is the new set of vectors linearly dependent? Show that now $\mathbf{F}$ is orthogonal to $\mathbf{G}$ and $\mathbf{H}$.

M1.3: Decompose the vector $\mathbf{A} = A^1\mathbf{q}_1 + A^2\mathbf{q}_2$ into the two components $\mathbf{A}_1 = A^1\mathbf{q}_1$ and $\mathbf{A}_2 = A^2\mathbf{q}_2$ by assuming that (i) $\mathbf{A}_1$ deviates from $\mathbf{A}$ by 15° and (ii) the length $|\mathbf{A}_1| = \frac{1}{3}|\mathbf{A}|$. Determine the length $|\mathbf{A}_2|$ and the angle between $\mathbf{A}_2$ and $\mathbf{A}$.

M1.4: The vector $\mathbf{A} = 3\mathbf{i}_1 + 2\mathbf{i}_2 + 4\mathbf{i}_3$ is given.
(a) Find the measure numbers A^i of $\mathbf{A} = A^1\mathbf{q}_1 + A^2\mathbf{q}_2 + A^3\mathbf{q}_3$ if the basis vectors are given by $\mathbf{q}_1 = \mathbf{i}_1 + 2\mathbf{i}_2,\ \mathbf{q}_2 = 2\mathbf{i}_2 + \mathbf{i}_3$, and $\mathbf{q}_3 = 2\mathbf{i}_3$.
(b) By employing the reciprocal basis vectors $\mathbf{q}^i$ find the measure numbers A_i of $\mathbf{A}$.

M1.5: By direct transformation of the contravariant measure numbers in (M1.38), show that

$$[\mathbf{A}, \mathbf{B}, \mathbf{C}] = \frac{1}{\sqrt{g}} \begin{vmatrix} A_1 & A_2 & A_3 \\ B_1 & B_2 & B_3 \\ C_1 & C_2 & C_3 \end{vmatrix}$$

M1.6: Show that the lowering and raising rules do not apply to physical measure numbers.

M1.7: Show that

$$(\mathbf{A} \times \mathbf{B}) \cdot (\mathbf{B} \times \mathbf{C}) \times (\mathbf{C} \times \mathbf{A}) = (\mathbf{A} \cdot \mathbf{B} \times \mathbf{C})^2$$

M1.8: The vectors

$$\widetilde{\mathbf{A}} = \frac{\mathbf{B} \times \mathbf{C}}{[\mathbf{A}, \mathbf{B}, \mathbf{C}]}, \qquad \widetilde{\mathbf{B}} = \frac{\mathbf{C} \times \mathbf{A}}{[\mathbf{A}, \mathbf{B}, \mathbf{C}]}, \qquad \widetilde{\mathbf{C}} = \frac{\mathbf{A} \times \mathbf{B}}{[\mathbf{A}, \mathbf{B}, \mathbf{C}]}$$

with $[\mathbf{A}, \mathbf{B}, \mathbf{C}] \neq 0$ are given. Find $[\widetilde{\mathbf{A}}, \widetilde{\mathbf{B}}, \widetilde{\mathbf{C}}]$.

M1.9: The unit vector $\mathbf{e}$ is perpendicular to the plane defined by the vectors $\mathbf{B}$ and $\mathbf{C}$. Show that $[\mathbf{e}, \mathbf{B}, \mathbf{C}] = |\mathbf{B} \times \mathbf{C}|$.

M2

Vector functions

M2.1 Basic definitions and operations

In general scalars and vectors depend on the position coordinates q^i or q_i and on the time t. Therefore, we have to deal with expressions of the type

$$\mathbf{B} = B^n(q^1, q^2, q^3, t)\mathbf{q}_n(q^1, q^2, q^3, t) = B_n(q_1, q_2, q_3, t)\mathbf{q}^n(q_1, q_2, q_3, t) \quad \text{(M2.1)}$$

While in stationary Cartesian coordinates the basis vectors are independent of position, in the general case $\mathbf{q}^i$ and $\mathbf{q}_i$ are functions of the corresponding position coordinates. In rotating coordinate systems they are functions of time also.

Of particular interest to our studies are *linear vector functions* defined by

$$\mathbf{B}(\mathbf{r}^{(1)} + \mathbf{r}^{(2)}) = \mathbf{B}(\mathbf{r}^{(1)}) + \mathbf{B}(\mathbf{r}^{(2)}), \qquad \mathbf{B}(\lambda\mathbf{r}) = \lambda\mathbf{B}(\mathbf{r}) \quad \text{(M2.2)}$$

where λ is a scalar and $\mathbf{r}^{(1)}$ and $\mathbf{r}^{(2)}$ are position vectors. An example of a linear vector function is given by

$$\mathbf{B}(\mathbf{r}) = \mathbf{B}(q^n\mathbf{q}_n) = \mathbf{B}(\mathbf{q}_n)q^n = \mathbf{B}_n q^n \quad \text{with} \quad \mathbf{B}_i = \mathbf{B}(\mathbf{q}_i) \quad \text{(M2.3)}$$

Since $\mathbf{r} = q^n\mathbf{q}_n$ we have $q^i = \mathbf{q}^i \cdot \mathbf{r}$ and therefore

$$\mathbf{B}(\mathbf{r}) = \mathbf{B}_n q^n = \mathbf{B}_n \mathbf{q}^n \cdot \mathbf{r} = \mathbb{B} \cdot \mathbf{r} \quad \text{(M2.4)}$$

This is the defining equation for the *complete dyadic* $\mathbb{B} = \mathbf{B}_n\mathbf{q}^n$. In our studies the sum will always be restricted to three terms. By expressing the position vector in $\mathbf{B}(\mathbf{r})$ as $\mathbf{r} = q_n\mathbf{q}^n$, we obtain analogously

$$\mathbf{B}(\mathbf{r}) = \mathbf{B}(q_n\mathbf{q}^n) = \mathbf{B}(\mathbf{q}^n)q_n = \mathbf{B}^n q_n = \mathbf{B}^n\mathbf{q}_n \cdot \mathbf{r} = \mathbb{B} \cdot \mathbf{r} \quad \text{(M2.5)}$$

with $q_i = \mathbf{q}_i \cdot \mathbf{r}$. Hence, the complete dyadic $\mathbb{B}$ may be written as

$$\boxed{\mathbb{B} = \mathbf{B}_n\mathbf{q}^n = \mathbf{B}^n\mathbf{q}_n} \quad \text{(M2.6)}$$

Scalar multiplication of a vector by a dyadic yields a vector. In special but very important cases the summation is terminated after two terms. The resulting *planar dyadic* will be of great interest in the following work. Dyadics consisting of only one term are called dyads; those consisting of two terms only are known as *singular dyadics*. Whenever two vectors of a dyadic, such as $\mathbf{B}_i$ or $\mathbf{q}^i$ $(i = 1, 2, 3)$, are linearly dependent then the complete dyadic transforms to a singular dyadic. As we know already, the nine components of a three-term dyadic can be arranged as a square matrix. Therefore, the rules of matrix algebra can be applied to perform operations with second-order dyadics. Unless specifically stated otherwise, we will be dealing with complete dyadics.

Let us now think of the dyadic $\mathbb{B}$ as representing an operator. Scalar multiplication of the dyadic by the original vector $\mathbf{r}$ results in a new vector $\mathbf{r}'$, which is called the *image vector*:

$$\begin{aligned} \mathbf{r}' = \mathbb{B}\cdot\mathbf{r} &= (\mathbf{B}_m\mathbf{q}^m)\cdot(q^n\mathbf{q}_n) = \mathbf{B}_m q^n \delta^m_n = \mathbf{B}_m q^m \\ &= (\mathbf{B}^m\mathbf{q}_m)\cdot(q_n\mathbf{q}^n) = \mathbf{B}^m q_n \delta^n_m = \mathbf{B}^m q_m \end{aligned} \tag{M2.7}$$

There are several ways to represent a complete dyadic. Some important results are given below. As will be seen, various dyadic measure numbers occur, which will now be discussed. Let us first consider the form in which the dyadic $\mathbb{B}$ is expressed with the help of the covariant vectorial measure numbers $\mathbf{B}_n$ and contravariant basis vectors $\mathbf{q}^n$, i.e. $\mathbb{B} = \mathbf{B}_n\mathbf{q}^n$. Scalar multiplication of $\mathbb{B}$ by $\mathbf{q}_i$ gives

$$\mathbb{B}\cdot\mathbf{q}_i = \mathbf{B}_n\mathbf{q}^n\cdot\mathbf{q}_i = \mathbf{B}_n\delta^n_i = \mathbf{B}_i \tag{M2.8}$$

The vector $\mathbf{B}_i$ may be represented in the two equivalent forms

$$\mathbf{B}_i = B^n_{\ i}\,\mathbf{q}_n = B_{ni}\mathbf{q}^n \tag{M2.9}$$

Repeating this procedure by expressing $\mathbb{B}$ in the form $\mathbb{B} = \mathbf{B}^n\mathbf{q}_n$ yields analogously

$$\mathbb{B}\cdot\mathbf{q}^i = \mathbf{B}^n\delta^i_n = \mathbf{B}^i, \qquad \mathbf{B}^i = B^{ni}\mathbf{q}_n = B_n^{\ i}\mathbf{q}^n \tag{M2.10}$$

From (M2.9) and (M2.10) we obtain four possibilities for representing the dyadic $\mathbb{B}$:

$$\boxed{\mathbb{B} = B^m_{\ n}\,\mathbf{q}_m\mathbf{q}^n = B_{mn}\mathbf{q}^m\mathbf{q}^n = B^{mn}\mathbf{q}_m\mathbf{q}_n = B_m^{\ n}\,\mathbf{q}^m\mathbf{q}_n} \tag{M2.11}$$

While $B^i_{\ j}$ and $B_i^{\ j}$ are called the *mixed measure numbers* of $\mathbb{B}$, the terms B_{ij} and B^{ij} are the covariant and contravariant measure numbers of $\mathbb{B}$, respectively. At each measure number the positions of the subscripts and superscripts indicate not only the kind of the corresponding basis vectors (covariant or contravariant) of the dyadic but also the order in which they appear.

By utilizing the property (M1.51) of reciprocal systems as well as the lowering and raising rules for the basis vectors, it is a simple task to obtain relations between the different types of measure numbers of a dyadic. This will be clarified by the following two examples. Application of the raising rule (M1.82) to the first expression of (M2.11) yields, together with the second expression of (M2.11),

$$\mathbb{B} = B^r{}_n \mathbf{q}_r \mathbf{q}^n = B^r{}_n g_{rm} \mathbf{q}^m \mathbf{q}^n = B_{mn} \mathbf{q}^m \mathbf{q}^n \tag{M2.12}$$

Scalar multiplication of (M2.12) first by the covariant basis vector $\mathbf{q}_j$ and then by $\mathbf{q}_i$ results in

$$(\mathbb{B} \cdot \mathbf{q}_j) \cdot \mathbf{q}_i = B^r{}_n g_{rm} \delta^n_j \delta^m_i = B_{mn} \delta^n_j \delta^m_i \tag{M2.13a}$$

Applying the lowering rule (M1.84) to the basis vector $\mathbf{q}^r$ of the dyadic $\mathbb{B} = B_r{}^n \mathbf{q}^r \mathbf{q}_n$ and multiplying the result from the right-hand side first by $\mathbf{q}^j$ and then by $\mathbf{q}^i$ yields

$$(\mathbb{B} \cdot \mathbf{q}^j) \cdot \mathbf{q}^i = B_r{}^n g^{rm} \delta^j_n \delta^i_m = B^{mn} \delta^j_n \delta^i_m \tag{M2.13b}$$

The final evaluation of the expressions of (M2.13a) and (M2.13b) gives the raising rule and the lowering rule for the measure numbers of a dyadic:

$$\boxed{B_{ij} = g_{ri} B^r{}_j, \qquad B^{ij} = g^{ri} B_r{}^j} \tag{M2.14}$$

Since in Cartesian coordinate systems there is no difference between covariant and contravariant basis vectors, from (M2.11) it may easily be seen that, in these systems, the different measure numbers of a dyadic are identical:

$$\begin{aligned} \mathbb{B} &= B^m{}_n \, \mathbf{i}_m \mathbf{i}^n = B_{mn} \mathbf{i}^m \mathbf{i}^n = B^{mn} \mathbf{i}_m \mathbf{i}_n = B_m{}^n \mathbf{i}^m \mathbf{i}_n \\ &\text{with} \quad B^i{}_j = B_{ij} = B^{ij} = B_i{}^j \end{aligned} \tag{M2.15}$$

In the following sections, for the sake of brevity, we will not list all possible representations of the dyadic, but will usually limit ourselves to the form

$$\Phi = \mathbf{B}_n \mathbf{q}^n = B^m_n \, \mathbf{q}_m \mathbf{q}^n \tag{M2.16}$$

This does not imply that a preference should be given to this particular form.

M2.2 Special dyadics

Whenever the antecedents and the consequents of the dyadic $\mathbb{\Phi} = \mathbf{B}_m\mathbf{q}^m$ form linearly independent systems, then $\mathbb{\Phi}$ is called a complete dyadic. If the antecedents or the consequents are coplanar then the complete dyadic reduces to a planar dyadic of only two terms.

M2.2.1 The conjugate dyadic

A special but very important dyadic known as the conjugate dyadic is obtained by interchanging the positions of the first vector (antecedent) and the second vector (consequent) of the dyadic $\mathbb{B} = \mathbf{B}_m\mathbf{q}^m$:

$$\boxed{\widetilde{\mathbb{B}} = \mathbf{q}^m\mathbf{B}_m} \tag{M2.17}$$

Thus, the original antecedent becomes the consequent. The conjugate dyadic will be identified by means of the tilde. An important relationship involving the original and the conjugate dyadic is

$$\boxed{\mathbf{D}\cdot\mathbb{B} = \mathbf{D}\cdot\mathbf{B}_m\mathbf{q}^m = \mathbf{q}^m\mathbf{D}\cdot\mathbf{B}_m = \mathbf{q}^m\mathbf{B}_m\cdot\mathbf{D} = \widetilde{\mathbb{B}}\cdot\mathbf{D}} \tag{M2.18}$$

M2.2.2 The symmetric dyadic

A dyadic $\mathbb{\Phi}'$ is called symmetric if interchanging the positions of the two vectors does not change the dyadic:

$$\boxed{\widetilde{\mathbb{\Phi}}' = \mathbb{\Phi}'} \tag{M2.19}$$

An important consequence is

$$\mathbb{\Phi}'\cdot\mathbf{D} = \mathbf{D}\cdot\widetilde{\mathbb{\Phi}}' = \mathbf{D}\cdot\mathbb{\Phi}' \tag{M2.20}$$

It is easy to show that the sum of an arbitrary dyadic $\mathbb{B}$ and the corresponding conjugate dyadic $\widetilde{\mathbb{B}}$ is symmetric:

$$\mathbb{\Phi} = \mathbb{B} + \widetilde{\mathbb{B}} = \mathbf{B}_m\mathbf{q}^m + \mathbf{q}^m\mathbf{B}_m, \qquad \widetilde{\mathbb{\Phi}} = \mathbf{q}^m\mathbf{B}_m + \mathbf{B}_m\mathbf{q}^m = \mathbb{\Phi} = \mathbb{\Phi}' \tag{M2.21}$$

For the Cartesian coordinate system the explicit form of a symmetric dyadic is given by

$$\begin{aligned}\mathbb{\Phi}' = {} & B_{11}\mathbf{i}_1\mathbf{i}_1 + B_{12}\mathbf{i}_1\mathbf{i}_2 + B_{13}\mathbf{i}_1\mathbf{i}_3 \\ & + B_{12}\mathbf{i}_2\mathbf{i}_1 + B_{22}\mathbf{i}_2\mathbf{i}_2 + B_{23}\mathbf{i}_2\mathbf{i}_3 \\ & + B_{13}\mathbf{i}_3\mathbf{i}_1 + B_{23}\mathbf{i}_3\mathbf{i}_2 + B_{33}\mathbf{i}_3\mathbf{i}_3\end{aligned} \tag{M2.22}$$

since $B_{ij} = B_{ji}$, which is the condition characterizing symmetric matrices. Since $g^{ij} = g^{ji}$ the unit dyadic is also symmetric:

$$\mathbb{E} = g^{mn}\mathbf{q}_m\mathbf{q}_n, \qquad \widetilde{\mathbb{E}} = g^{mn}\mathbf{q}_n\mathbf{q}_m = g^{nm}\mathbf{q}_n\mathbf{q}_m = \mathbb{E} \tag{M2.23}$$

M2.2.3 The antisymmetric or skew-symmetric dyadic

A dyadic $\mathbb{\Phi}''$ is called antisymmetric if interchanging the positions of the two vectors results in a change of sign of the dyadic:

$$\boxed{\widetilde{\mathbb{\Phi}''} = -\mathbb{\Phi}''} \tag{M2.24}$$

From this definition we immediately see that

$$\mathbb{\Phi}''\cdot\mathbf{D} = \mathbf{D}\cdot\widetilde{\mathbb{\Phi}''} = -\mathbf{D}\cdot\mathbb{\Phi}'' \tag{M2.25}$$

The difference between a dyadic $\mathbb{B}$ and the conjugate dyadic $\widetilde{\mathbb{B}}$ is always antisymmetric:

$$\mathbb{\Phi} = \mathbb{B} - \widetilde{\mathbb{B}} = \mathbf{B}_m\mathbf{q}^m - \mathbf{q}^m\mathbf{B}_m, \qquad \widetilde{\mathbb{\Phi}} = \mathbf{q}^m\mathbf{B}_m - \mathbf{B}_m\mathbf{q}^m = -\mathbb{\Phi} \tag{M2.26}$$

From (M2.21) and (M2.26) we conclude that any dyadic may be expressed as the sum of a symmetric and an antisymmetric dyadic:

$$\boxed{\mathbb{\Phi} = \tfrac{1}{2}(\mathbb{B} + \widetilde{\mathbb{B}}) + \tfrac{1}{2}(\mathbb{B} - \widetilde{\mathbb{B}}) = \mathbb{\Phi}' + \mathbb{\Phi}''} \tag{M2.27}$$

The representation of an antisymmetric dyadic in the Cartesian coordinate system is given by

$$\begin{aligned} \mathbb{\Phi}'' = 0 &+ B_{12}\mathbf{i}_1\mathbf{i}_2 + B_{13}\mathbf{i}_1\mathbf{i}_3 \\ &- B_{12}\mathbf{i}_2\mathbf{i}_1 + 0 + B_{23}\mathbf{i}_2\mathbf{i}_3 \\ &- B_{13}\mathbf{i}_3\mathbf{i}_1 - B_{23}\mathbf{i}_3\mathbf{i}_2 + 0 \end{aligned} \tag{M2.28}$$

The elements on the main diagonal must be zero to satisfy the condition of antisymmetry $B_{ij} = -B_{ji}$ which characterizes every antisymmetric matrix.

In order to discover another property of an antisymmetric dyadic, we wish to reconsider the vectorial triple product. Using the Grassmann rule (M1.44) we find

$$\mathbf{D} \times (\mathbf{A} \times \mathbf{B}) = \mathbf{D}\cdot(\mathbf{BA} - \mathbf{AB}) \tag{M2.29}$$

which is the scalar product of a vector $\mathbf{D}$ with an antisymmetric dyadic.

M2.2.4 The adjoint dyadic

The adjoint dyadic of $\mathbf{\Phi} = \mathbf{B}_s\mathbf{q}^s$ is defined by

$$\boxed{\mathbf{\Phi}_{\mathrm{a}} = (\mathbf{B}_s\mathbf{q}^s)_{\mathrm{a}} = \tfrac{1}{2}(\mathbf{q}^m \times \mathbf{q}^n)(\mathbf{B}_m \times \mathbf{B}_n)} \tag{M2.30}$$

On expanding and rearranging the sums we obtain for the adjoint dyadic the useful form

$$\mathbf{\Phi}_{\mathrm{a}} = (\mathbf{q}^1 \times \mathbf{q}^2)(\mathbf{B}_1 \times \mathbf{B}_2) + (\mathbf{q}^2 \times \mathbf{q}^3)(\mathbf{B}_2 \times \mathbf{B}_3) + (\mathbf{q}^3 \times \mathbf{q}^1)(\mathbf{B}_3 \times \mathbf{B}_1) \tag{M2.31}$$

According to (M2.17) the conjugate of the adjoint dyadic may be written as

$$\widetilde{\mathbf{\Phi}_{\mathrm{a}}} = \tfrac{1}{2}(\mathbf{B}_m \times \mathbf{B}_n)(\mathbf{q}^m \times \mathbf{q}^n) \tag{M2.32}$$

A little reflection shows that

$$\widetilde{\mathbf{\Phi}_{\mathrm{a}}} = \left(\widetilde{\mathbf{\Phi}}\right)_{\mathrm{a}} \tag{M2.33}$$

We will now give the component form of the adjoint dyadic whose original definition (M2.30) involves the vector products $\mathbf{q}^j \times \mathbf{q}^k$ and $\mathbf{B}_j \times \mathbf{B}_k$. With the help of (M1.53) and (M1.93) we find

$$\mathbf{B}_i \times \mathbf{B}_j = B^m_{\ \ i}\mathbf{q}_m \times B^n_{\ \ j}\mathbf{q}_n = [\mathbf{q}_1, \mathbf{q}_2, \mathbf{q}_3]\begin{vmatrix} \mathbf{q}^1 & \mathbf{q}^2 & \mathbf{q}^3 \\ B^1_{\ \ i} & B^2_{\ \ i} & B^3_{\ \ i} \\ B^1_{\ \ j} & B^2_{\ \ j} & B^3_{\ \ j} \end{vmatrix} = [\mathbf{q}_1, \mathbf{q}_2, \mathbf{q}_3]\mathbf{D}_{ij} \tag{M2.34}$$

and

$$\begin{aligned}
(\mathbf{q}^1 \times \mathbf{q}^2)(\mathbf{B}_1 \times \mathbf{B}_2) &= \mathbf{q}_3[\mathbf{q}^1, \mathbf{q}^2, \mathbf{q}^3][\mathbf{q}_1, \mathbf{q}_2, \mathbf{q}_3]\mathbf{D}_{12} = \mathbf{q}_3\mathbf{D}_{12} \\
(\mathbf{q}^3 \times \mathbf{q}^1)(\mathbf{B}_3 \times \mathbf{B}_1) &= \mathbf{q}_2\mathbf{D}_{31} \\
(\mathbf{q}^2 \times \mathbf{q}^3)(\mathbf{B}_2 \times \mathbf{B}_3) &= \mathbf{q}_1\mathbf{D}_{23}
\end{aligned} \tag{M2.35}$$

By adding the three terms, as required by (M2.31), we obtain the desired result

$$\mathbf{\Phi}_{\mathrm{a}} = \mathbf{q}_1\mathbf{D}_{23} + \mathbf{q}_2\mathbf{D}_{31} + \mathbf{q}_3\mathbf{D}_{12} = M^m_{\ \ n}\mathbf{q}_m\mathbf{q}^n \tag{M2.36}$$

The $M^i_{\ \ j}$ are elements of the adjoint matrix $(M^i_{\ \ j})$ corresponding to the coefficient matrix $(B^i_{\ \ j})$ of the dyadic $\mathbf{\Phi} = B^m_{\ \ n}\mathbf{q}_m\mathbf{q}^n$. Thus, we obtain the adjoint dyadic by replacing the elements $B^i_{\ \ j}$ by the corresponding elements of the adjoint matrix which is obtained by transposing the cofactor matrix of $(B^i_{\ \ j})$.

M2.2.5 The reciprocal dyadic

The unit dyadic $\mathbb{E}$ may be expressed with the help of any three noncoplanar reciprocal basis vectors. Thus, we may choose the form

$$\mathbb{E} = \mathbf{B}_m \mathbf{B}^m \qquad \text{(M2.37)}$$

The reciprocal dyadic Φ^{-1} of Φ is defined by

$$\Phi \cdot \Phi^{-1} = \mathbb{E} \qquad \text{(M2.38)}$$

If Φ is given in the form $\Phi = \mathbf{B}_m \mathbf{q}^m$ then the reciprocal dyadic Φ^{-1} must be

$$\boxed{\Phi^{-1} = \mathbf{q}_m \mathbf{B}^m} \qquad \text{(M2.39)}$$

since

$$\Phi \cdot \Phi^{-1} = \mathbf{B}_m \mathbf{q}^m \cdot \mathbf{q}_n \mathbf{B}^n = \mathbf{B}_m \delta^n_m \mathbf{B}^n = \mathbf{B}_m \mathbf{B}^m = \mathbb{E} \qquad \text{(M2.40)}$$

In analogy to (M1.55) we may express $\mathbf{B}^i$ as

$$\mathbf{B}^i = \frac{\mathbf{B}_j \times \mathbf{B}_k}{[\mathbf{B}_1, \mathbf{B}_2, \mathbf{B}_3]} \qquad \text{(M2.41)}$$

On substituting (M1.53) and (M2.41) into (M2.39) we immediately recognize that the reciprocal dyadic Φ^{-1} may be formulated in terms of the adjoint dyadic:

$$\boxed{\Phi^{-1} = \frac{1}{2} \frac{(\mathbf{q}^m \times \mathbf{q}^n)(\mathbf{B}_m \times \mathbf{B}_n)}{[\mathbf{q}^1, \mathbf{q}^2, \mathbf{q}^3][\mathbf{B}_1, \mathbf{B}_2, \mathbf{B}_3]} = \frac{\Phi_{\mathrm{a}}}{[\mathbf{q}^1, \mathbf{q}^2, \mathbf{q}^3][\mathbf{B}_1, \mathbf{B}_2, \mathbf{B}_3]}} \qquad \text{(M2.42)}$$

Without proof we give the following expressions:

$$\boxed{\begin{aligned} (\mathbb{B} \cdot \mathbb{D} \cdot \mathbb{F})^{-1} &= \mathbb{F}^{-1} \cdot \mathbb{D}^{-1} \cdot \mathbb{B}^{-1} \\ \left(\mathbb{B}^{-1}\right)^{\alpha} &= \mathbb{B}^{-\alpha} \\ \mathbb{B}^{\alpha} \cdot \mathbb{B}^{\beta} &= \mathbb{B}^{\alpha+\beta} \\ \mathbb{B}^{0} &= \mathbb{E} \end{aligned}} \qquad \text{(M2.43)}$$

which may be generalized to include additional factors. Since dyadics are *tensors* of rank two, they can be treated as square matrices, for which the inversion formulas are derived in any textbook on the subject.

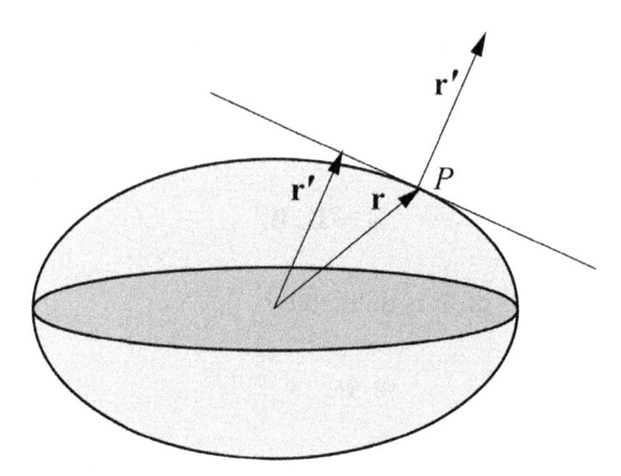

Fig. M2.1 The tensor ellipsoid $\mathbf{r}\cdot\boldsymbol{\Phi}'\cdot\mathbf{r} = 1$. $\mathbf{r}' = \boldsymbol{\Phi}'\cdot\mathbf{r}$ is perpendicular to the tangential plane at P.

M2.3 Principal-axis transformation of symmetric tensors

Premultiplication and postmultiplication of the general dyadic by the position vector $\mathbf{r} = x_m\mathbf{i}_m$ yields

$$\mathbf{r}\cdot\boldsymbol{\Phi}\cdot\mathbf{r} = \mathbf{r}\cdot\boldsymbol{\Phi}'\cdot\mathbf{r} + \mathbf{r}\cdot\boldsymbol{\Phi}''\cdot\mathbf{r} = \mathbf{r}\cdot\boldsymbol{\Phi}'\cdot\mathbf{r} \tag{M2.44}$$

The term $\mathbf{r}\cdot\boldsymbol{\Phi}''\cdot\mathbf{r}$ of this expression vanishes since the scalar premultiplication and postmultiplication of an antisymmetric dyadic by an arbitrary vector must vanish in order to avoid the following contradiction:

$$\mathbf{A}\cdot\boldsymbol{\Phi}''\cdot\mathbf{A} = (\mathbf{A}\cdot\boldsymbol{\Phi}'')\cdot\mathbf{A} = \mathbf{A}\cdot(\mathbf{A}\cdot\boldsymbol{\Phi}'') = \mathbf{A}\cdot(\widetilde{\boldsymbol{\Phi}''}\cdot\mathbf{A}) = -\mathbf{A}\cdot\boldsymbol{\Phi}''\cdot\mathbf{A} = 0 \tag{M2.45}$$

On substituting (M2.22) into (M2.44) we obtain the quadratic form

$$\begin{aligned} F = \mathbf{r}\cdot\boldsymbol{\Phi}'\cdot\mathbf{r} &= x_r\mathbf{i}_r\cdot B_{mn}\mathbf{i}_m\mathbf{i}_n\cdot x_s\mathbf{i}_s = x_r x_s B_{mn}\delta_r^m\delta_s^n = x_m x_n B_{mn} \\ &= B_{11}x_1^2 + B_{22}x_2^2 + B_{33}x_3^2 + 2B_{12}x_1x_2 + 2B_{23}x_2x_3 + 2B_{13}x_1x_3 \end{aligned} \tag{M2.46}$$

It can be seen that the constant coefficients B_{ij} of the symmetric dyadic determine the type of the quadratic surface. The *tensor surface* F may be an ellipsoid, a hyperboloid, or a paraboloid. Most applications from physics deal with ellipsoids so that collectively we speak of *tensor ellipsoids* of the symmetric tensor. As a representative of the tensor ellipsoid we select $F = 1$. The position vector $\mathbf{r}$ extends from the origin of the ellipsoid to some point P on the ellipsoidal surface; see Figure M2.1. Since the point $-\mathbf{r}$ is also located on the tensor surface, we speak of a *midpoint surface*.

Let us now investigate the transformation $\mathbf{r}' = \boldsymbol{\Phi}'\cdot\mathbf{r}$. There exist special directions of $\mathbf{r}$ for which $\mathbf{r}$ and $\mathbf{r}'$ are parallel vectors. These directions are known as the *principal-axis directions*. The corresponding mathematical statement is given by

$$\mathbf{r}' = \Phi' \cdot \mathbf{r} = \lambda \mathbf{r} \quad \text{or} \quad (\Phi' - \lambda \mathbb{E}) \cdot \mathbf{r} = 0 \tag{M2.47}$$

where the scalar λ is the eigenvalue of the operator Φ'. An equation of this form is known as an *eigenvalue equation* or *characteristic equation*. In coordinate notation this eigenvalue equation can be written as

$$(B_{mn} - \lambda \delta_{mn}) x_n = 0 \tag{M2.48}$$

For the variables x_j this is a linear homogeneous system that has nontrivial solutions only if the determinant of the system vanishes, namely

$$\begin{vmatrix} B_{11} - \lambda & B_{12} & B_{13} \\ B_{21} & B_{22} - \lambda & B_{23} \\ B_{13} & B_{23} & B_{33} - \lambda \end{vmatrix} = 0 \tag{M2.49}$$

This results in an eigenvalue equation of third order, where the λ_i are the eigenvalues of the operator Φ'. Every solution vector $\mathbf{r}_i$ of

$$(\Phi' - \lambda_i \mathbb{E}) \cdot \mathbf{r}_i = 0 \tag{M2.50}$$

that differs from the zero vector is an *eigenvector* corresponding to the eigenvalue λ_i. From linear algebra we know that the eigenvalues of a symmetric matrix are real and that the eigenvectors of such a matrix can always be chosen to be real. Furthermore, a real symmetric matrix is diagonalizable. Moreover, eigenvectors of such a matrix corresponding to distinct eigenvalues are orthogonal.

Let us now briefly consider the principal-axis directions. The symmetric dyadic $\Phi' = B_{mn}\mathbf{i}_m\mathbf{i}_n$ can be reduced to the simple form

$$\Phi' = \lambda_1 \mathbf{e}_1 \mathbf{e}_1 + \lambda_2 \mathbf{e}_2 \mathbf{e}_2 + \lambda_3 \mathbf{e}_3 \mathbf{e}_3 \tag{M2.51}$$

since symmetric matrices can be diagonalized. The basis vectors $\mathbf{e}_i$ appearing in (M2.51) are directed along the principal axes of the tensor ellipsoid as shown in Figure M2.2. By expressing the position vector $\mathbf{r}$ as

$$\mathbf{r} = \mathbf{e}_1 \xi_1 + \mathbf{e}_2 \xi_2 + \mathbf{e}_3 \xi_3 \tag{M2.52}$$

we obtain the principal-axes form of the tensor ellipsoid

$$F = \mathbf{r} \cdot \Phi' \cdot \mathbf{r} = \lambda_1 \xi_1^2 + \lambda_2 \xi_2^2 + \lambda_3 \xi_3^2 = 1 \tag{M2.53}$$

The particular tensor ellipsoid of interest to our future studies is the stress tensor or the stress dyadic. We will return to this section later when we discuss the viscous forces acting on a surface element.

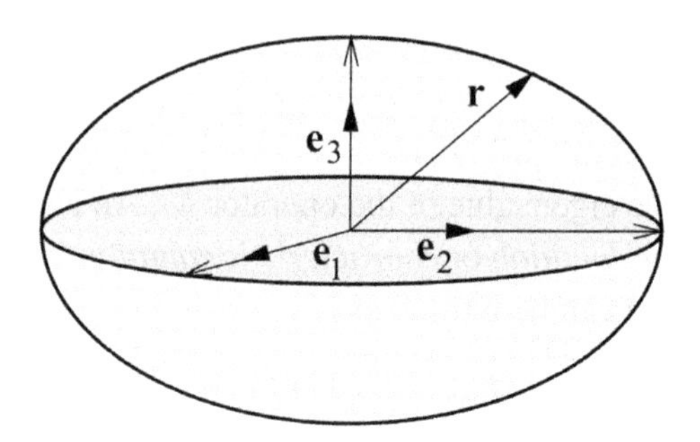

Fig. M2.2 Eigenvectors $\mathbf{e}_i$ of the tensor ellipsoid $\mathbf{r}\cdot\Phi'\cdot\mathbf{r} = 1$ corresponding to the eigenvalues λ_i.

M2.4 Invariants of a dyadic

Dyadics are extensive quantities that are independent of a coordinate system. If the general multiplication is changed to scalar or vectorial multiplication, we find expressions that again are independent of the coordinate system. Such coordinate-independent expressions are called invariants of a dyadic. The starting point of our discussion is the complete dyadic

$$\Phi = \mathbf{B}_n\mathbf{q}^n = B^m_{\ n}\mathbf{q}_m\mathbf{q}^n \quad \text{with} \quad \mathbf{B}_i = \Phi\cdot\mathbf{q}_i = B^m_{\ i}\mathbf{q}_m \tag{M2.54}$$

M2.4.1 The first scalar of a dyadic

On taking the scalar product of the vectors appearing in the dyadic we obtain the first scalar Φ_{I} of the dyadic,

$$\boxed{\Phi_{\mathrm{I}} = \mathbf{B}_n\cdot\mathbf{q}^n = B^m_{\ n}\mathbf{q}_m\cdot\mathbf{q}^n = B^m_{\ n}\delta^n_m = B^1_{\ 1} + B^2_{\ 2} + B^3_{\ 3}} \tag{M2.55}$$

This is the trace of the coefficient matrix $(B^i_{\ j})$. Owing to the validity of the commutative law for the scalar product (M1.14), we obtain

$$\Phi_{\mathrm{I}} = \mathbf{B}_n\cdot\mathbf{q}^n = \mathbf{q}^n\cdot\mathbf{B}_n = \widetilde{\Phi}_{\mathrm{I}} \tag{M2.56}$$

Application of (M2.55) to the unit dyadic yields the interesting result

$$\mathbb{E}_{\mathrm{I}} = \mathbf{q}_n\cdot\mathbf{q}^n = \mathbf{q}_1\cdot\mathbf{q}^1 + \mathbf{q}_2\cdot\mathbf{q}^2 + \mathbf{q}_3\cdot\mathbf{q}^3 = 3 \tag{M2.57}$$

According to (M2.28) the diagonal elements of an antisymmetric dyadic vanish so that

$$\Phi''_{\mathrm{I}} = 0 \implies \Phi_{\mathrm{I}} = \Phi'_{\mathrm{I}} \tag{M2.58}$$

In thermodynamics and elsewhere it is customary to decompose the symmetric dyadic Φ' into an *isotropic symmetric part* Φ'_{iso} and an *anisotropic symmetric part* Φ'_{aniso}:

$$\Phi' = \Phi'_{\text{iso}} + \Phi'_{\text{aniso}} \quad \text{requiring that} \quad (\Phi'_{\text{iso}})_{\text{I}} = \Phi'_{\text{I}} \implies (\Phi'_{\text{aniso}})_{\text{I}} = 0 \tag{M2.59}$$

We now rewrite Φ' in the form

$$\Phi' = \alpha\mathbb{E} + (\Phi' - \alpha\mathbb{E}) \tag{M2.60}$$

where α is a scalar that needs to be determined. The first term on the right-hand side of (M2.60) represents the isotropic part while the second term is the anisotropic part. Hence, α is found from

$$(\Phi' - \alpha\mathbb{E})_{\text{I}} = \Phi'_{\text{I}} - 3\alpha = 0 \implies \alpha = \Phi'_{\text{I}}/3 \tag{M2.61}$$

and the isotropic and anisotropic parts of Φ' are given by

$$\Phi'_{\text{iso}} = \frac{\Phi'_{\text{I}}}{3}\mathbb{E}, \qquad \Phi'_{\text{aniso}} = \Phi' - \frac{\Phi'_{\text{I}}}{3}\mathbb{E} \tag{M2.62}$$

Utilizing (M2.27), (M2.59), and (M2.62) we may express any complete dyadic in the form

$$\boxed{\Phi = \Phi' + \Phi'' = \Phi'_{\text{iso}} + \Phi'_{\text{aniso}} + \Phi'' = \frac{\Phi'_{\text{I}}}{3}\mathbb{E} + \left(\Phi' - \frac{\Phi'_{\text{I}}}{3}\mathbb{E}\right) + \Phi''} \tag{M2.63}$$

M2.4.2 The vector of a dyadic

The vector of a dyadic is defined by placing the cross, denoting vectorial multiplication, between the members of each pair of vectors. Application of (M1.53) and observing (M1.57) results in

$$\boxed{\begin{aligned}\Phi_\times &= \mathbf{B}_n \times \mathbf{q}^n = B_{mn}\mathbf{q}^m \times \mathbf{q}^n \\ &= \frac{1}{\sqrt{g}}\left[(B_{23} - B_{32})\mathbf{q}_1 + (B_{31} - B_{13})\mathbf{q}_2 + (B_{12} - B_{21})\mathbf{q}_3\right]\end{aligned}} \tag{M2.64}$$

It is immediately obvious that

$$\Phi_\times = \mathbf{B}_n \times \mathbf{q}^n = -\mathbf{q}^n \times \mathbf{B}_n = -\widetilde{\Phi}_\times \tag{M2.65}$$

From (M2.64) it follows immediately that the vector of a symmetric dyadic must vanish:

$$\Phi'_{\times} = 0 \implies \Phi_{\times} = \Phi''_{\times} \tag{M2.66}$$

Of particular interest is the following equation:

$$\boxed{\mathbb{E} \times \mathbf{B} = \mathbf{B} \times \mathbb{E}} \tag{M2.67}$$

indicating that vectorial multiplication of the unit dyadic $\mathbb{E}$ by an arbitrary vector $\mathbf{B}$ is commutative. In order to give a very brief proof, we assume that $\mathbf{B} = \mathbf{q}_1$, yielding

$$\begin{aligned}
&\mathbb{E} \times \mathbf{q}_1 = \mathbf{q}^n \mathbf{q}_n \times \mathbf{q}_1 = \mathbf{q}^2(\mathbf{q}_2 \times \mathbf{q}_1) + \mathbf{q}^3(\mathbf{q}_3 \times \mathbf{q}_1) = -\mathbf{q}^2\mathbf{q}^3\sqrt{g} + \mathbf{q}^3\mathbf{q}^2\sqrt{g} \\
&\mathbf{q}_1 \times \mathbb{E} = \mathbf{q}_1 \times \mathbf{q}_n \mathbf{q}^n = (\mathbf{q}_1 \times \mathbf{q}_2)\mathbf{q}^2 + (\mathbf{q}_1 \times \mathbf{q}_3)\mathbf{q}^3 = \mathbf{q}^3\mathbf{q}^2\sqrt{g} - \mathbf{q}^2\mathbf{q}^3\sqrt{g} \\
&\qquad \implies \mathbb{E} \times \mathbf{q}_1 = \mathbf{q}_1 \times \mathbb{E}
\end{aligned} \tag{M2.68}$$

In the general case the vector $\mathbf{B}$ consists of three components so that the proof is more lengthy but not more difficult.

Of some interest to our future studies is the cross product of the unit dyadic with the vector of the dyadic Φ. By employing the Grassmann rule (M1.44) we can easily carry out the operations

$$\begin{aligned}
\mathbb{E} \times \Phi_{\times} &= \mathbb{E} \times (\mathbf{B}_n \times \mathbf{q}^n) = \mathbf{q}^m\left[\mathbf{q}_m \times (\mathbf{B}_n \times \mathbf{q}^n)\right] \\
&= \mathbf{q}^m(\mathbf{q}_m \cdot \mathbf{q}^n)\mathbf{B}_n - \mathbf{q}^m(\mathbf{q}_m \cdot \mathbf{B}_n)\mathbf{q}^n \\
&= \mathbf{q}^n\mathbf{B}_n - \mathbb{E} \cdot \mathbf{B}_n\mathbf{q}^n = \mathbf{q}^n\mathbf{B}_n - \mathbf{B}_n\mathbf{q}^n = -2\Phi''
\end{aligned} \tag{M2.69}$$

This provides a new way to express an antisymmetric dyadic in the form

$$\Phi'' = -\tfrac{1}{2}\mathbb{E} \times \Phi_{\times} = -\tfrac{1}{2}\Phi_{\times} \times \mathbb{E} \tag{M2.70}$$

Taking the scalar product of an antisymmetric dyadic with a vector $\mathbf{B}$ gives an interesting and useful relation that may be helpful if complicated expressions have to be manipulated:

$$\Phi'' \cdot \mathbf{B} = -\tfrac{1}{2}(\Phi_{\times} \times \mathbb{E}) \cdot \mathbf{B} = -\tfrac{1}{2}\Phi_{\times} \times \mathbf{B} = \tfrac{1}{2}\mathbf{B} \times \Phi_{\times} \tag{M2.71}$$

M2.4.3 The second scalar of a dyadic

Another invariant is the second scalar of a dyadic, which by definition is the first scalar of the adjoint dyadic. By employing equation (M2.31) we find

$$\begin{aligned}
\Phi_{\mathrm{II}} = (\Phi_{\mathrm{a}})_{\mathrm{I}} &= (\mathbf{q}^1 \times \mathbf{q}^2) \cdot (\mathbf{B}_1 \times \mathbf{B}_2) + (\mathbf{q}^2 \times \mathbf{q}^3) \cdot (\mathbf{B}_2 \times \mathbf{B}_3) \\
&\quad + (\mathbf{q}^3 \times \mathbf{q}^1) \cdot (\mathbf{B}_3 \times \mathbf{B}_1)
\end{aligned} \tag{M2.72}$$

The cross products of the contravariant basis vectors appearing in this equation can be removed with the help of equation (M1.53). By obvious steps we obtain

$$\begin{aligned}\Phi_{\rm II} &= \frac{1}{[\mathbf{q}_1,\mathbf{q}_2,\mathbf{q}_3]}\left[\mathbf{q}_3\cdot(\mathbf{B}_1\times\mathbf{B}_2)+\mathbf{q}_1\cdot(\mathbf{B}_2\times\mathbf{B}_3)+\mathbf{q}_2\cdot(\mathbf{B}_3\times\mathbf{B}_1)\right]\\ &= \frac{1}{[\mathbf{q}_1,\mathbf{q}_2,\mathbf{q}_3]}\left([\mathbf{q}_1,\mathbf{B}_2,\mathbf{B}_3]+[\mathbf{B}_1,\mathbf{q}_2,\mathbf{B}_3]+[\mathbf{B}_1,\mathbf{B}_2,\mathbf{q}_3]\right)\end{aligned} \tag{M2.73}$$

where the rules associated with the scalar triple product have been applied to obtain the second equation. Since according to (M2.9) $\mathbf{B}_i = B^n{}_i\mathbf{q}_n$, with the help of (M1.38) we find for the first scalar triple product of (M2.73)

$$[\mathbf{q}_1,\mathbf{B}_2,\mathbf{B}_3] = [\mathbf{q}_1,\mathbf{q}_2,\mathbf{q}_3]\begin{vmatrix} 1 & 0 & 0\\ B^1{}_2 & B^2{}_2 & B^3{}_2\\ B^1{}_3 & B^2{}_3 & B^3{}_3\end{vmatrix} = [\mathbf{q}_1,\mathbf{q}_2,\mathbf{q}_3]\begin{vmatrix} B^2{}_2 & B^3{}_2\\ B^2{}_3 & B^3{}_3\end{vmatrix} \tag{M2.74}$$

Expressing the remaining terms in (M2.73) in the same way yields

$$\boxed{\Phi_{\rm II} = \begin{vmatrix} B^1{}_1 & B^2{}_1\\ B^1{}_2 & B^2{}_2\end{vmatrix} + \begin{vmatrix} B^1{}_1 & B^3{}_1\\ B^1{}_3 & B^3{}_3\end{vmatrix} + \begin{vmatrix} B^2{}_2 & B^3{}_2\\ B^2{}_3 & B^3{}_3\end{vmatrix}} \tag{M2.75}$$

showing that the second scalar of the dyadic Φ is the sum of three second-order determinants.

Let us now consider the determinant of the matrix $(B^i{}_j)$. From (M2.75) we easily recognize that the second scalar of the dyadic Φ is the sum of the cofactors of the elements forming the main diagonal of B:

$$\Phi_{\rm II} = M^3{}_3 + M^2{}_2 + M^1{}_1 \tag{M2.76}$$

Since the reversal of factors in (M2.72) cannot yield a different result, we recognize that the second scalar of the conjugate dyadic is identical with the second scalar of the dyadic Φ itself:

$$\widetilde{\Phi}_{\rm II} = \Phi_{\rm II} \tag{M2.77}$$

Without verification we accept

$$\Phi_{\rm II} = \Phi'_{\rm II} + \Phi''_{\rm II} \tag{M2.78}$$

The proof is somewhat lengthy but not particularly instructive.

Before continuing with additional operations involving the second scalar, it will be profitable to introduce the so-called *double scalar product*, which is also known as the *double-dot product*. To begin with, consider the scalar multiplications $\mathbf{A}\cdot\Phi\cdot\mathbf{B}$. Since the scalar product of a vector with a dyadic results in a new vector, the product can be formed in two ways:

$$\begin{aligned}\mathbf{A}\cdot\Phi\cdot\mathbf{B} &= \mathbf{A}\cdot(\Phi\cdot\mathbf{B}) = (\Phi\cdot\mathbf{B})\cdot\mathbf{A}\\ \mathbf{A}\cdot\Phi\cdot\mathbf{B} &= (\mathbf{A}\cdot\Phi)\cdot\mathbf{B} = \mathbf{B}\cdot(\mathbf{A}\cdot\Phi)\end{aligned} \tag{M2.79}$$

Obviously, the scalar product of vectors $\mathbf{A}$ and $\mathbf{B}$ is meaningless in these two expressions. Thus, it is customary to introduce the double-dot or double scalar product

$$(\Phi\cdot\mathbf{B})\cdot\mathbf{A} = \Phi\cdot\cdot\mathbf{BA}, \qquad \mathbf{B}\cdot(\mathbf{A}\cdot\Phi) = \mathbf{BA}\cdot\cdot\Phi \tag{M2.80}$$

On combining (M2.79) and (M2.80) we obtain the defining equation of the double scalar product:

$$\boxed{\mathbf{A}\cdot\Phi\cdot\mathbf{B} = \mathbf{BA}\cdot\cdot\Phi = \Phi\cdot\cdot\mathbf{BA}} \tag{M2.81}$$

Some authors place the two dots vertically so that the double-dot product looks like a ratio symbol.

Of considerable interest to our studies is the second scalar of an antisymmetric dyadic. From (M2.8) we have $\mathbf{B}_i = \Phi\cdot\mathbf{q}_i$. If $\Phi = \Phi''$ we find from (M2.66) and (M2.71) the relations

$$\Phi''\cdot\mathbf{q}_i = \mathbf{B}_i = \tfrac{1}{2}\mathbf{q}_i \times \Phi''_\times = \tfrac{1}{2}\mathbf{q}_i \times \Phi_\times \tag{M2.82}$$

Substituting this expression into (M2.73) yields

$$\begin{aligned}\Phi''_{\mathrm{II}} = \frac{1}{4[\mathbf{q}_1, \mathbf{q}_2, \mathbf{q}_3]}\{&\mathbf{q}_3\cdot[(\mathbf{q}_1 \times \Phi_\times) \times (\mathbf{q}_2 \times \Phi_\times)] + \mathbf{q}_2\cdot[(\mathbf{q}_3 \times \Phi_\times) \times (\mathbf{q}_1 \times \Phi_\times)]\\ &+ \mathbf{q}_1\cdot[(\mathbf{q}_2 \times \Phi_\times) \times (\mathbf{q}_3 \times \Phi_\times)]\}\end{aligned} \tag{M2.83}$$

Application of (M1.50) to each term within the braces of (M2.83) results in a total of six terms. Three terms vanish since in these terms the same vector appears twice in the scalar triple product. For the first term, this is explicitly shown in

$$(\mathbf{q}_1 \times \Phi_\times) \times (\mathbf{q}_2 \times \Phi_\times) = [\mathbf{q}_1, \Phi_\times, \Phi_\times]\mathbf{q}_2 - [\mathbf{q}_1, \Phi_\times, \mathbf{q}_2]\Phi_\times = -[\mathbf{q}_1, \Phi_\times, \mathbf{q}_2]\Phi_\times \tag{M2.84}$$

Performing the same operation with the other two terms in (M2.83) and using the rules of the scalar triple product, we find

$$\Phi''_{\mathrm{II}} = \frac{1}{4[\mathbf{q}_1, \mathbf{q}_2, \mathbf{q}_3]}(\mathbf{q}_3\cdot[\mathbf{q}_1, \mathbf{q}_2, \Phi_\times]\Phi_\times + \mathbf{q}_2\cdot[\mathbf{q}_3, \mathbf{q}_1, \Phi_\times]\Phi_\times + \mathbf{q}_1\cdot[\mathbf{q}_2, \mathbf{q}_3, \Phi_\times]\Phi_\times) \tag{M2.85}$$

Utilizing the defining relation (M1.55) for the contravariant basis vectors results in

$$\Phi''_{\rm II} = \tfrac{1}{4}\left[\mathbf{q}_3\cdot(\mathbf{q}^3\cdot\Phi_\times)\Phi_\times + \mathbf{q}_2\cdot(\mathbf{q}^2\cdot\Phi_\times)\Phi_\times + \mathbf{q}_1\cdot(\mathbf{q}^1\cdot\Phi_\times)\Phi_\times\right] \tag{M2.86}$$

In this equation we introduce the double scalar product according to (M2.81) as well as the definition of the unit dyadic (M1.80), thus obtaining the final result

$$\boxed{\begin{aligned}\Phi''_{\rm II} &= \tfrac{1}{4}\left[(\mathbf{q}_1\mathbf{q}^1 + \mathbf{q}_2\mathbf{q}^2 + \mathbf{q}_3\mathbf{q}^3)\cdot\cdot\Phi_\times\Phi_\times\right] = \tfrac{1}{4}\mathbb{E}\cdot\cdot\Phi_\times\Phi_\times\\ &= \tfrac{1}{4}\Phi_\times\cdot\mathbb{E}\cdot\Phi_\times = \frac{1}{4}\Phi_\times\cdot\Phi_\times\end{aligned}} \tag{M2.87}$$

M2.4.4 The third scalar of a dyadic

By introducing the reciprocal systems (M1.53) and (M2.41) into the definition of the adjoint dyadic (M2.31), we obtain

$$\begin{aligned}\Phi_{\rm a} &= \mathbf{q}_3\mathbf{B}^3[\mathbf{q}^1,\mathbf{q}^2,\mathbf{q}^3][\mathbf{B}_1,\mathbf{B}_2,\mathbf{B}_3] + \mathbf{q}_1\mathbf{B}^1[\mathbf{q}^1,\mathbf{q}^2,\mathbf{q}^3][\mathbf{B}_1,\mathbf{B}_2,\mathbf{B}_3]\\ &\quad + \mathbf{q}_2\mathbf{B}^2[\mathbf{q}^1,\mathbf{q}^2,\mathbf{q}^3][\mathbf{B}_1,\mathbf{B}_2,\mathbf{B}_3]\\ &= [\mathbf{q}^1,\mathbf{q}^2,\mathbf{q}^3][\mathbf{B}_1,\mathbf{B}_2,\mathbf{B}_3]\mathbf{q}_n\mathbf{B}^n = [\mathbf{q}^1,\mathbf{q}^2,\mathbf{q}^3][\mathbf{B}_1,\mathbf{B}_2,\mathbf{B}_3]\Phi^{-1}\end{aligned} \tag{M2.88}$$

where the reciprocal dyadic has been used according to (M2.39). Scalar multiplication of this expression by the dyadic Φ yields

$$\Phi\cdot\Phi_{\rm a} = [\mathbf{q}^1,\mathbf{q}^2,\mathbf{q}^3][\mathbf{B}_1,\mathbf{B}_2,\mathbf{B}_3]\mathbb{E} \tag{M2.89}$$

This is the defining equation for the third scalar of the dyadic Φ, which may be written as

$$\boxed{\Phi_{\rm III} = [\mathbf{q}^1,\mathbf{q}^2,\mathbf{q}^3][\mathbf{B}_1,\mathbf{B}_2,\mathbf{B}_3] = \frac{[\mathbf{B}_1,\mathbf{B}_2,\mathbf{B}_3]}{[\mathbf{q}_1,\mathbf{q}_2,\mathbf{q}_3]} = \begin{Vmatrix} B^1_{\ 1} & B^1_{\ 2} & B^1_{\ 3}\\ B^2_{\ 1} & B^2_{\ 2} & B^2_{\ 3}\\ B^3_{\ 1} & B^3_{\ 2} & B^3_{\ 3}\end{Vmatrix}} \tag{M2.90}$$

Hence, the third scalar of the dyadic Φ is the determinant of the coefficient matrix $(B^i_{\ j})$.

We conclude this section by stating two important special cases. From equation (M2.26) for the antisymmetric dyadic we immediately see that

$$\Phi''_{\rm III} = 0 \tag{M2.91}$$

since the determinants of $\mathbf{q}^n\mathbf{B}_n$ and $\mathbf{B}_n\mathbf{q}^n$ are equal. Finally, owing to the fact that $\mathrm{Det}(B^i_{\ j}) = \mathrm{Det}(B^{\ j}_{i})$, we obtain

$$(\widetilde{\Phi})_{\rm III} = \Phi_{\rm III} \tag{M2.92}$$

M2.5 Tensor algebra

We conclude this chapter by presenting some interesting and useful tensor operations. Let us consider a dyadic in the form $\boldsymbol{\Phi} = \mathbf{B}_n\mathbf{q}^n$. The double scalar product involving the same dyadic twice is of some interest to our studies. By connecting the exterior and interior vectors by the dots representing the scalar multiplication we find

$$\boldsymbol{\Phi}\cdot\cdot\boldsymbol{\Phi} = \mathbf{B}_r\mathbf{q}^r\cdot\cdot\mathbf{B}_n\mathbf{q}^n = (\mathbf{q}^n\cdot\mathbf{B}_r)(\mathbf{q}^r\cdot\mathbf{B}_n) \tag{M2.93}$$

Using this result together with the first scalar $\Phi_{\mathrm{I}} = \mathbf{B}_n\cdot\mathbf{q}^n = \mathbf{q}^n\cdot\mathbf{B}_n$ we may easily obtain the following determinant:

$$\boldsymbol{\Phi}\cdot\cdot\boldsymbol{\Phi} - \Phi_{\mathrm{I}}\Phi_{\mathrm{I}} = (\mathbf{q}^n\cdot\mathbf{B}_r)(\mathbf{q}^r\cdot\mathbf{B}_n) - (\mathbf{q}^n\cdot\mathbf{B}_n)(\mathbf{q}^r\cdot\mathbf{B}_r) = \begin{vmatrix} \mathbf{q}^n\cdot\mathbf{B}_r & \mathbf{q}^n\cdot\mathbf{B}_n \\ \mathbf{q}^r\cdot\mathbf{B}_r & \mathbf{q}^r\cdot\mathbf{B}_n \end{vmatrix} \tag{M2.94}$$

Applying equation (M1.48) and utilizing the definitions of the adjoint dyadic (M2.30) and of the second scalar (M2.72) yields

$$\begin{aligned} \boldsymbol{\Phi}\cdot\cdot\boldsymbol{\Phi} - \Phi_{\mathrm{I}}\Phi_{\mathrm{I}} &= (\mathbf{q}^n\times\mathbf{q}^r)\cdot(\mathbf{B}_r\times\mathbf{B}_n) \\ &= -(\mathbf{q}^n\times\mathbf{q}^r)\cdot(\mathbf{B}_n\times\mathbf{B}_r) = -2(\boldsymbol{\Phi}_{\mathrm{a}})_{\mathrm{I}} = -2\Phi_{\mathrm{II}} \end{aligned} \tag{M2.95}$$

so that

$$\boxed{\boldsymbol{\Phi}\cdot\cdot\boldsymbol{\Phi} = \Phi_{\mathrm{I}}\Phi_{\mathrm{I}} - 2\Phi_{\mathrm{II}}} \tag{M2.96}$$

Recall that the first scalar of an antisymmetric dyadic is zero; see (M2.58). Hence, we find

$$\boldsymbol{\Phi}''\cdot\cdot\boldsymbol{\Phi}'' = -2\Phi''_{\mathrm{II}}, \qquad \boldsymbol{\Phi}'\cdot\cdot\boldsymbol{\Phi}' = \Phi'_{\mathrm{I}}\Phi'_{\mathrm{I}} - 2\Phi'_{\mathrm{II}} \tag{M2.97}$$

Next let us consider the vector product of the dyadic $\boldsymbol{\Phi}$ with the vector $(\mathbf{C}\times\mathbf{D})$. By applying the Grassmann rule we find

$$\boxed{\begin{aligned} \boldsymbol{\Phi}\times(\mathbf{C}\times\mathbf{D}) &= \mathbf{B}_n[\mathbf{q}^n\times(\mathbf{C}\times\mathbf{D})] = \mathbf{B}_n[(\mathbf{q}^n\cdot\mathbf{D})\mathbf{C} - (\mathbf{q}^n\cdot\mathbf{C})\mathbf{D}] \\ &= (\boldsymbol{\Phi}\cdot\mathbf{D})\mathbf{C} - (\boldsymbol{\Phi}\cdot\mathbf{C})\mathbf{D} \end{aligned}} \tag{M2.98}$$

Changing the position of the parentheses will also change the result.

Operations involving dyadics require some care, as the following examples show. While the scalar product of two vectors is commutative, the scalar product of a vector and a dyadic is not commutative:

$$\mathbf{C}\cdot\boldsymbol{\Phi} = \mathbf{C}\cdot(\mathbf{B}_n\mathbf{q}^n) = (\mathbf{C}\cdot\mathbf{B}_n)\mathbf{q}^n, \qquad \boldsymbol{\Phi}\cdot\mathbf{C} = (\mathbf{B}_n\mathbf{q}^n)\cdot\mathbf{C} = (\mathbf{q}^n\cdot\mathbf{C})\mathbf{B}_n \tag{M2.99}$$

Moreover, the scalar product of two dyadics is not commutative either:

$$\boldsymbol{\Phi}\cdot\boldsymbol{\Psi} = (\mathbf{AB})\cdot(\mathbf{CD}) = (\mathbf{B}\cdot\mathbf{C})\mathbf{AD}, \qquad \boldsymbol{\Psi}\cdot\boldsymbol{\Phi} = (\mathbf{CD})\cdot(\mathbf{AB}) = (\mathbf{D}\cdot\mathbf{A})\mathbf{CB} \tag{M2.100}$$

Often it may be desirable to involve the scalar measure numbers in the scalar product of two dyadics. Usually it is of advantage to use the reciprocal system of basis vectors:

$$\mathbf{\Phi}\cdot\mathbf{\Psi} = A^m_{\ n}\mathbf{q}_m\mathbf{q}^n\cdot\mathbf{q}_p\mathbf{q}^r B^p_{\ r} = A^m_{\ n}B^n_{\ r}\mathbf{q}_m\mathbf{q}^r = C^m_{\ r}\mathbf{q}_m\mathbf{q}^r \tag{M2.101}$$

The two scalar quantities are summed over n to give $C^i_{\ j} = A^i_{\ n}B^n_{\ j}$. This result corresponds to the rules of matrix multiplication. The elements of the ith row are multiplied by the elements of the jth column.

Next we wish to point out that the conjugate of the scalar product of two dyadics is equal to the scalar product of the individual conjugate dyadics:

$$\widetilde{\mathbf{\Phi}\cdot\mathbf{\Psi}} = \widetilde{\mathbf{\Phi}}\cdot\widetilde{\mathbf{\Psi}} \tag{M2.102}$$

Of great importance is the following equation:

$$\mathbb{E}\cdot\mathbf{\Phi} = \mathbf{\Phi}\cdot\mathbb{E} = \mathbf{\Phi} \tag{M2.103}$$

showing that the scalar product of a dyadic with the unit dyadic results in the dyadic itself. Therefore, taking the scalar product of an abitrary dyadic with the unit dyadic is commutative. The proofs of (M2.102) and (M2.103) are straightforward and are, therefore, omitted.

Consider the double-dot product of $\mathbf{\Phi} = \mathbf{B}_n\mathbf{q}^n$ and $\mathbf{\Psi} = \mathbf{D}_r\mathbf{q}^r$. The double-dot product of the two dyadics involves the scalar product of reciprocal basis vectors:

$$\begin{aligned}\mathbf{\Phi}\cdot\cdot\mathbf{\Psi} &= \mathbf{B}_n\mathbf{q}^n\cdot\cdot\mathbf{D}_r\mathbf{q}^r = B^m_{\ n}\mathbf{q}_m\mathbf{q}^n\cdot\cdot D^p_{\ r}\mathbf{q}_p\mathbf{q}^r \\ &= B^m_{\ n}D^p_{\ r}(\mathbf{q}^n\cdot\mathbf{q}_p)(\mathbf{q}_m\cdot\mathbf{q}^r) = B^m_{\ n}D^n_{\ m} = C^m_{\ m}\end{aligned} \tag{M2.104}$$

Hence, it can be seen that the double-dot product simply gives the trace of the matrix $(C^i_{\ j})$. From (M2.81) and (M2.104) we see that

$$\mathbf{\Phi}\cdot\cdot\mathbf{\Psi} = \mathbf{\Psi}\cdot\cdot\mathbf{\Phi} = (\mathbf{\Phi}\cdot\mathbf{\Psi})_{\mathrm{I}} = C^m_{\ m} \tag{M2.105}$$

On substituting into this expression $\mathbf{\Psi} = \mathbb{E}$ we obtain the important result

$$\boxed{\mathbf{\Phi}\cdot\cdot\mathbb{E} = \mathbb{E}\cdot\cdot\mathbf{\Phi} = (\mathbf{\Phi}\cdot\mathbb{E})_{\mathrm{I}} = \mathbf{\Phi}_{\mathrm{I}}} \tag{M2.106}$$

The validity of

$$\widetilde{\mathbf{\Phi}}\cdot\cdot\widetilde{\mathbf{\Psi}} = \mathbf{\Phi}\cdot\cdot\mathbf{\Psi} \tag{M2.107}$$

may easily be verified and is left as a simple excercise.

Of particular interest, as we shall see in our study of thermodynamics, is the double scalar product of a symmetric and an antisymmetric dyadic. Using equation (M2.107) and applying the definitions of the symmetric and antisymmetric dyadics leads us to conclude that the double scalar product must vanish for consistency, as shown in

$$\boxed{\mathbf{\Phi}'\cdot\cdot\mathbf{\Psi}'' = \widetilde{\mathbf{\Phi}'}\cdot\cdot\widetilde{\mathbf{\Psi}''} = -\mathbf{\Phi}'\cdot\cdot\mathbf{\Psi}'' = 0} \tag{M2.108}$$

M2.6 Problems

M2.1: Show that

$$\mathbf{A} \times \mathbf{r} = \mathbf{r} \cdot \boldsymbol{\Phi}''$$

where $\boldsymbol{\Phi}''$ is an antisymmetric dyadic. Find $\boldsymbol{\Phi}''$.

M2.2: By utilizing (M2.97) show that $\boldsymbol{\Phi}' \cdot\cdot \boldsymbol{\Phi}'$ is positive definite.

M2.3: Show the validity of the following expressions:

$$\text{(a)} \quad \boldsymbol{\Phi}_\times \cdot \boldsymbol{\Phi} = \boldsymbol{\Phi} \cdot \boldsymbol{\Phi}_\times, \qquad \text{(b)} \quad (\widetilde{\boldsymbol{\Phi}_\mathrm{a}})_\times = \boldsymbol{\Phi}_\times \cdot \boldsymbol{\Phi}$$

M2.4: Find the eigenvalues and eigenvectors of the eigenvalue problem

$$(\boldsymbol{\Phi} - \lambda \mathbb{E}) \cdot \mathbf{r} = 0, \qquad (\Phi_{ij}) = \begin{pmatrix} 2 & -1 & 0 \\ -1 & 2 & -1 \\ 0 & -1 & 2 \end{pmatrix}$$

where (Φ_{ij}) is the matrix corresponding to $\boldsymbol{\Phi}$. Hint: One eigenvalue is $\lambda_1 = 2$.

M3

Differential relations

Many mathematical statements of the laws of physics involve differential or integral expressions. Therefore, we must be able to differentiate and integrate expressions involving vectors and dyadics. While in the Cartesian nonrotating coordinate system the basis vectors **i**, **j**, and **k** may be treated as constants, basis vectors in generalized coordinate systems depend on spatial coordinates and may depend on time.

M3.1 Differentiation of extensive functions

Let us consider a curve k in space as shown in Figure M3.1. From some origin we draw the position vector $\mathbf{r}(s)$ to the point s on the curve. Upon advancing the arclength Δs along the curve, the new position vector will be $\mathbf{r}(s + \Delta s)$. Applying the definition of the derivative

$$\frac{d\mathbf{r}}{ds} = \lim_{\Delta s \to 0} \frac{|\Delta \mathbf{r}|\, \mathbf{e}_{\Delta \mathbf{r}}}{\Delta s} = \lim_{\Delta s \to 0} \frac{\Delta s\, \mathbf{e}_{\Delta \mathbf{r}}}{\Delta s} = \mathbf{e}_{\Delta \mathbf{r}} = \mathbf{e}_{\mathrm{T}} \tag{M3.1}$$

we obtain the unit tangent vector $\mathbf{e}_{\mathrm{T}}$ on the curve at point s.

When we are differentiating products involving vectors and dyadics we must observe the sequence of the factors. An example is given by

$$\frac{d}{ds}(\mathbf{A} \cdot \mathbf{\Phi} \cdot \mathbf{B}) = \frac{d\mathbf{A}}{ds} \cdot \mathbf{\Phi} \cdot \mathbf{B} + \mathbf{A} \cdot \frac{d\mathbf{\Phi}}{ds} \cdot \mathbf{B} + \mathbf{A} \cdot \mathbf{\Phi} \cdot \frac{d\mathbf{B}}{ds} \tag{M3.2}$$

Suppose that we wish to differentiate the vector **A** with respect to the parameter s. In this case we must differentiate not only the scalar measure number A^i but also the basis vector $\mathbf{q}_i$, since this vector varies in space:

$$\frac{d\mathbf{A}}{ds} = \frac{d}{ds}(A^n \mathbf{q}_n) = \frac{dA^n}{ds}\, \mathbf{q}_n + A^n \frac{d\mathbf{q}_n}{ds} \tag{M3.3}$$

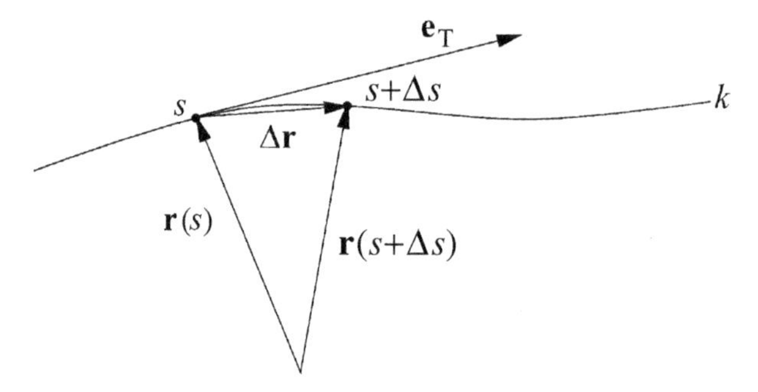

Fig. M3.1 The unit tangent vector $\mathbf{e}_T$ along a curve k in space.

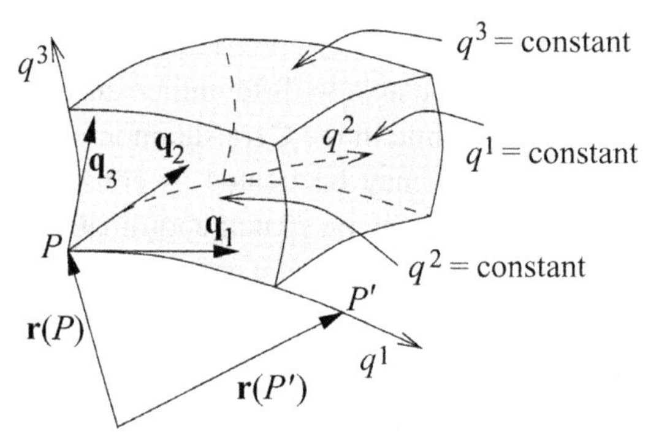

Fig. M3.2 A volume element of the generalized q^i system.

An important example is the scalar product of the unit vectors, $\mathbf{e}_A \cdot \mathbf{e}_A$. Differentiation of this expression with respect to s shows that the derivative $d\mathbf{e}_A/ds$ is perpendicular to $\mathbf{e}_A$

$$\mathbf{e}_A \cdot \mathbf{e}_A = 1 \implies \frac{d(\mathbf{e}_A \cdot \mathbf{e}_A)}{ds} = \mathbf{e}_A \cdot \frac{d\mathbf{e}_A}{ds} + \frac{d\mathbf{e}_A}{ds} \cdot \mathbf{e}_A = 2\mathbf{e}_A \cdot \frac{d\mathbf{e}_A}{ds} = 0 \qquad \text{(M3.4)}$$

Let us now consider a volume element defined by sections of the contravariant *curvilinear coordinate lines* q^i as shown in Figure M3.2. Note that the coordinate line q^1 is formed by the intersection of the coordinate surfaces $q^2 =$ constant and $q^3 =$ constant in the same way as that in which the x-axis of the orthogonal Cartesian system is formed by the intersection of the surfaces $y =$ constant and $z =$ constant. Analogously, the coordinate lines q^2 and q^3 are formed by the intersection of the two other coordinate surfaces. The covariant basis vectors $\mathbf{q}_i$ are tangent vectors along the respective coordinate lines. The orientation of the basis vectors $\mathbf{q}_i$ at a given point P differs from the orientation of the basis vectors at another point P'. All vectors existing at point P have to be defined with respect to the basis existing at that point. In general, every point P' that is separated by a finite distance from point P has a different basis so that all vectors defined at point P' must be expressed in terms of the basis existing at P'.

In the general q^i-coordinate system all quantities depend on the spatial coordinates q^i $(i = 1, 2, 3)$ and on time t. Hence, we write

$$\begin{aligned}\psi &= \psi(q^1, q^2, q^3, t) = \psi(q^i, t)\\ \mathbf{A} &= A^n \mathbf{q}_n = A^n(q^i, t)\mathbf{q}_n(q^i, t)\\ \boldsymbol{\Phi} &= \phi^m_{\ n}(q^i, t)\mathbf{q}_m(q^i, t)\mathbf{q}^n(q^i, t)\end{aligned} \tag{M3.5}$$

showing the explicit dependency of ψ, A^i, $\phi^i_{\ j}$ and of the basis vectors $\mathbf{q}_i$, $\mathbf{q}^j$ on q^i $(i = 1, 2, 3)$ and t.

According to the rules of differential calculus, we may write the individual total differentials of ψ, $\mathbf{A}$, $\boldsymbol{\Phi}$ in the forms

$$\boxed{\begin{aligned}d\psi &= \frac{\partial \psi}{\partial t}\,dt + \frac{\partial \psi}{\partial q^n}\,dq^n\\ d\mathbf{A} &= \frac{\partial \mathbf{A}}{\partial t}\,dt + \frac{\partial \mathbf{A}}{\partial q^n}\,dq^n\\ d\boldsymbol{\Phi} &= \frac{\partial \boldsymbol{\Phi}}{\partial t}\,dt + \frac{\partial \boldsymbol{\Phi}}{\partial q^n}\,dq^n\end{aligned}} \tag{M3.6}$$

We often call these expressions the *total differentials* or sometimes the *individual differentials* of the corresponding variables. It can be seen that the total differential consists of two parts. The first term represents the change with time of the variable at a fixed point in space. Usually the partial derivative with respect to time $\partial/\partial t$ is called the *local time derivative*. The second terms on the right-hand sides of (M3.6) are the so-called *geometric differentials* of the variables, which are usually abbreviated as follows:

$$\boxed{\begin{aligned}d_{\rm g}\psi &= \frac{\partial \psi}{\partial q^n}\,dq^n\\ d_{\rm g}\mathbf{A} &= \frac{\partial \mathbf{A}}{\partial q^n}\,dq^n\\ d_{\rm g}\boldsymbol{\Phi} &= \frac{\partial \boldsymbol{\Phi}}{\partial q^n}\,dq^n\end{aligned}} \tag{M3.7}$$

Now we wish to investigate the relationship between the position vector $\mathbf{r}$ and the basis vectors. Let us begin with the simple orthogonal Cartesian coordinate system. For a fixed time the differential $d\mathbf{r}$ of the position vector $\mathbf{r} = \mathbf{r}(x^1, x^2, x^3)$

can be written as

$$\begin{aligned} d\mathbf{r} &= \mathbf{i}_n\, dx^n = \mathbf{i}_1\, dx^1 + \mathbf{i}_2\, dx^2 + \mathbf{i}_3\, dx^3 \\ &= \frac{\partial \mathbf{r}}{\partial x^n}\, dx^n = \frac{\partial \mathbf{r}}{\partial x^1}\, dx^1 + \frac{\partial \mathbf{r}}{\partial x^2}\, dx^2 + \frac{\partial \mathbf{r}}{\partial x^3}\, dx^3 \end{aligned} \tag{M3.8}$$

By comparing the differentials dx^i of this equation we obtain the basic definition of the unit vectors in the Cartesian system:

$$\mathbf{i}_1 = \frac{\partial \mathbf{r}}{\partial x^1}, \qquad \mathbf{i}_2 = \frac{\partial \mathbf{r}}{\partial x^2}, \qquad \mathbf{i}_3 = \frac{\partial \mathbf{r}}{\partial x^3} \tag{M3.9}$$

Similarly, by expressing the position vector in the general curvilinear coordinate system, we find for a fixed time t

$$d\mathbf{r} = \mathbf{q}_n\, dq^n = \frac{\partial \mathbf{r}}{\partial q^n}\, dq^n \tag{M3.10}$$

and the basic definition of the covariant basis vector $\mathbf{q}_i$

$$\boxed{\mathbf{q}_i = \frac{\partial \mathbf{r}}{\partial q^i}} \tag{M3.11}$$

The unit vector $\mathbf{e}_i$ is given as

$$\mathbf{e}_i = \frac{\mathbf{q}_i}{|\mathbf{q}_i|} = \frac{\mathbf{q}_i}{\sqrt{g_{ii}}} = \frac{1}{\sqrt{g_{ii}}} \frac{\partial \mathbf{r}}{\partial q^i} \tag{M3.12}$$

Using the definition (M3.11), the scalar triple product formed by the covariant basis vectors may be written as[1]

$$\sqrt{\underset{q}{g}} = [\mathbf{q}_1, \mathbf{q}_2, \mathbf{q}_3] = \left[\frac{\partial \mathbf{r}}{\partial q^1}, \frac{\partial \mathbf{r}}{\partial q^2}, \frac{\partial \mathbf{r}}{\partial q^3} \right] \tag{M3.13}$$

and, by means of equation (M1.40), in the explicit form

$$\sqrt{\underset{q}{g}} = \sqrt{\underset{x}{g}} \begin{vmatrix} \dfrac{\partial x^1}{\partial q^1} & \dfrac{\partial x^2}{\partial q^1} & \dfrac{\partial x^3}{\partial q^1} \\ \dfrac{\partial x^1}{\partial q^2} & \dfrac{\partial x^2}{\partial q^2} & \dfrac{\partial x^3}{\partial q^2} \\ \dfrac{\partial x^1}{\partial q^3} & \dfrac{\partial x^2}{\partial q^3} & \dfrac{\partial x^3}{\partial q^3} \end{vmatrix} = \left| \frac{\partial(x^1, x^2, x^3)}{\partial(q^1, q^2, q^3)} \right| \tag{M3.14}$$

[1] Here and in the following, subscripts q or x (or others) denote the particular coordinate system in which the corresponding variable has to be evaluated. They will, however, be used only if confusion is likely.

The determinant of this expression is known as the functional determinant of the transformation from the initial Cartesian x^i-coordinate system to the general q^i system. The term $\sqrt{\underset{x}{g}}$ is the functional determinant of the Cartesian system with $\sqrt{\underset{x}{g}} = [\mathbf{i}_1, \mathbf{i}_2, \mathbf{i}_3] = 1$. The $\underset{q}{g_{ij}}$ of the square of the functional determinant are

$$\underset{q}{g_{ij}} = \mathbf{q}_i \cdot \mathbf{q}_j = \frac{\partial \mathbf{r}}{\partial q^i} \cdot \frac{\partial \mathbf{r}}{\partial q^j} = \frac{\partial \mathbf{i}_n\, x^n}{\partial q^i} \cdot \frac{\partial \mathbf{i}_m\, x^m}{\partial q^j} = \frac{\partial x^n}{\partial q^i} \frac{\partial x^n}{\partial q^j} \tag{M3.15}$$

Whenever the relationship between the Cartesian x^i-coordinate system and the q^i system is given, it is a simple matter to evaluate this equation.

In analogy to the definition of the covariant basis vector $\mathbf{q}_i$ we define the contravariant basis vector $\mathbf{q}^i$. From the basic definition

$$d\mathbf{r} = \frac{\partial \mathbf{r}}{\partial q_n} dq_n = \mathbf{q}^n\, dq_n \tag{M3.16}$$

and comparing the coefficients of the differential dq_i, we easily find

$$\mathbf{q}^i = \frac{\partial \mathbf{r}}{\partial q_i} = \underset{q}{g^{in}} \mathbf{q}_n \tag{M3.17}$$

where the second equality is the lowering rule according to (M1.84). If the three covariant basis vectors $\mathbf{q}_i$ are known as well as the contravariant metric fundamental quantities $\underset{q}{g^{ij}}$, then the basis vectors $\mathbf{q}^i$ may be computed.

There is a very important point that needs to be mentioned. The contravariant coordinate lines of the curvilinear system are continuous coordinate lines so that

$$\frac{\partial^2}{\partial q^i\, \partial q^j} = \frac{\partial^2}{\partial q^j\, \partial q^i} \tag{M3.18}$$

applies. In this case the order of the partial derivatives is immaterial. Owing to the properties of the reciprocal coordinate systems, the covariant coordinate lines can be defined only piecewise. Therefore, as is shown in textbooks on analysis, the order of the differentiation cannot be interchanged:

$$\frac{\partial^2}{\partial q_i\, \partial q_j} \neq \frac{\partial^2}{\partial q_j\, \partial q_i} \tag{M3.19}$$

For this reason we prefer partial derivatives with respect to the contravariant coordinates q^i. In a later chapter we will encounter this situation involving (M3.19) when we are dealing with natural coordinates.

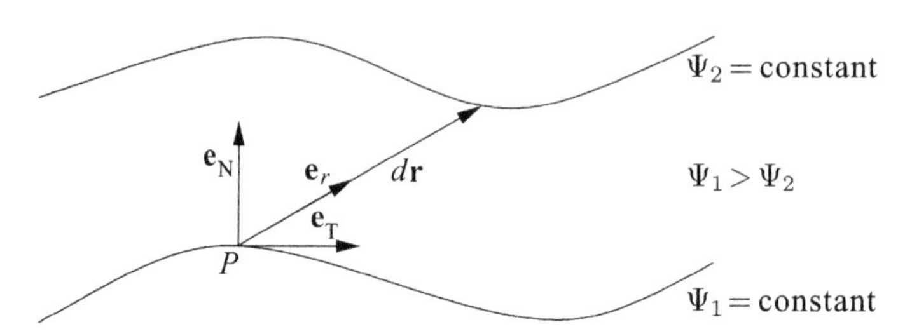

Fig. M3.3 The geometric meaning of the gradient operator.

M3.2 The Hamilton operator in generalized coordinate systems

Let us reconsider the geometric differential of a scalar field function as given in (M3.7). By inserting the scalar product $\mathbf{q}^i \cdot \mathbf{q}_i$, which does not change the meaning of $d_g\psi$, we obtain

$$d_g\psi = \frac{\partial \psi}{\partial q^n}\, dq^n = \sum_{i=1}^{3} \frac{\partial \psi}{\partial q^i}\, dq^i = \sum_{i=1}^{3} \left(\frac{\partial \psi}{\partial q^i}\, \mathbf{q}^i \right) \cdot \left(\mathbf{q}_i\, dq^i \right) = \nabla\psi \cdot d\mathbf{r} \qquad \text{(M3.20)}$$

Here we have introduced the *Hamilton operator* ∇, which is also known as the *gradient operator*, the *del operator*, and the *nabla operator*. Thus, in general coordinates we define ∇ by

$$\boxed{\nabla = \mathbf{q}^n \frac{\partial}{\partial q^n} = \mathbf{q}^n \nabla_n, \qquad \nabla_i = \frac{\partial}{\partial q^i}} \qquad \text{(M3.21)}$$

where we have also introduced the symbol ∇_i to remind the reader that the partial derivative $\partial/\partial q^i$ is a covariant expression while q^i itself is a contravariant coordinate. The meaning of the nabla operator is twofold: First, it is a vector; and second, it is a differential operator.

Now we consider the two scalar surfaces shown in Figure M3.3. At the point P we construct the displacement vector $d\mathbf{r} = \mathbf{e}_r\, dr$ as well as the unit normal vector $\mathbf{e}_\mathrm{N}$ and a unit tangent vector $\mathbf{e}_\mathrm{T}$ along the surface $\psi_1 = \text{constant}$. These two unit vectors are perpendicular to each other.

We slightly rewrite equation (M3.20) in the form

$$d_g\psi = d\mathbf{r} \cdot \nabla\psi = \mathbf{e}_r\, dr \cdot \nabla\psi \qquad \text{(M3.22)}$$

so that the geometric change $d_g\psi/dr$ is expressed by

$$\frac{d_g\psi}{dr} = \mathbf{e}_r \cdot \nabla\psi \qquad \text{(M3.23)}$$

If we select the displacement vector $d\mathbf{r}$ at P along the surface $\psi_1 =$ constant, i.e. $\mathbf{e}_r = \mathbf{e}_{\mathrm{T}}$, then the geometric change vanishes:

$$\mathbf{e}_{\mathrm{T}} \cdot \nabla\psi = 0 \tag{M3.24}$$

Therefore, the vectors $\nabla\psi$ and $\mathbf{e}_{\mathrm{T}}$ are perpendicular to each other and $\nabla\psi$ is parallel to the unit normal vector $\mathbf{e}_{\mathrm{N}}$:

$$\nabla\psi = \mathbf{e}_{\mathrm{N}} |\nabla\psi| \tag{M3.25}$$

that is $\nabla\psi$ is perpendicular to the surface $\psi =$ constant. Moreover, choosing $d\mathbf{r} = \mathbf{e}_{\mathrm{N}}\, dr_{\mathrm{N}}$ means that the magnitude of $\nabla\psi$ corresponds to the geometric change $d_{\mathrm{g}}\psi$ along $\mathbf{e}_{\mathrm{N}}$:

$$\frac{d_{\mathrm{g}}\psi}{dr_{\mathrm{N}}} = \mathbf{e}_{\mathrm{N}} \cdot \nabla\psi = |\nabla\psi| \tag{M3.26}$$

If the ψ-surfaces are closely spaced, then the ψ-gradient is strong and vice versa.

Let us return to the contravariant basis vectors $\mathbf{q}^i$, whose directions are known if the directions of the covariant basis vectors $\mathbf{q}_i$ are known, as follows from (M1.55) for the reciprocal systems. By utilizing the definition of the nabla operator we have another way to obtain the direction of the contravariant basis vectors. From (M3.21) it follows immediately that

$$\boxed{\nabla q^i = \mathbf{q}^n \frac{\partial q^i}{\partial q^n} = \mathbf{q}^i} \tag{M3.27}$$

Often it is desirable to express the gradient operator in terms of the unit vectors $\mathbf{e}^i$ instead of employing the general basis vectors $\mathbf{q}^i$. By obvious steps we introduce the unit vector $\mathbf{e}^i$, which automatically leads to the introduction of the covariant physical measure numbers $\overset{*}{\nabla}_i$ of the Hamilton operator:

$$\mathbf{q}^i \nabla_i = \frac{\mathbf{q}^i}{\sqrt{g^{ii}}} \sqrt{g^{ii}}\, \nabla_i = \mathbf{e}^i \overset{*}{\nabla}_i = \mathbf{e}^i \frac{\partial}{\partial \overset{*}{q}{}^i} \quad \text{with} \quad \overset{*}{\nabla}_i = \sqrt{g^{ii}}\, \nabla_i \tag{M3.28}$$

Whenever we are dealing with orthogonal systems, it is advantageous to employ the gradient operator in the form

$$\nabla = \mathbf{e}^n \frac{\partial}{\partial \overset{*}{q}{}^n} = \mathbf{e}^n \overset{*}{\nabla}_n \tag{M3.29}$$

Application of the gradient operator to the position vector $\mathbf{r}$ yields the unit dyadic $\mathbb{E}$:

$$\boxed{\nabla\mathbf{r} = \mathbf{q}^n \frac{\partial \mathbf{r}}{\partial q^n} = \mathbf{q}^n \mathbf{q}_n = \mathbb{E}} \tag{M3.30}$$

where use of (M3.11) has been made. In the special case of the orthogonal Cartesian system we obtain

$$\nabla \mathbf{r} = \left(\mathbf{i}\frac{\partial}{\partial x} + \mathbf{j}\frac{\partial}{\partial y} + \mathbf{k}\frac{\partial}{\partial z}\right)(\mathbf{i}x + \mathbf{j}y + \mathbf{k}z) = \mathbf{ii} + \mathbf{jj} + \mathbf{kk} = \mathbb{E} \tag{M3.31}$$

Let us now return to the total derivatives (M3.6). Introducing the velocity vector $\mathbf{v}$,

$$\mathbf{v} = \frac{d\mathbf{r}}{dt} = \frac{dq^n}{dt}\mathbf{q}_n = \dot{q}^n \mathbf{q}_n \tag{M3.32}$$

we may write

$$\frac{dq^n}{dt}\frac{\partial \psi}{\partial q^n} = \left(\frac{dq^n}{dt}\mathbf{q}_n\right)\cdot\left(\mathbf{q}^n \frac{\partial \psi}{\partial q^n}\right) = \mathbf{v}\cdot\nabla\psi \tag{M3.33}$$

The scalar product $\mathbf{v}\cdot\nabla\psi$ is known as the *advection term*. Substituting (M3.33) and corresponding expressions for $\mathbf{A}$ and $\mathbf{\Phi}$ into (M3.6) yields

$$\boxed{\begin{aligned} \frac{d\psi}{dt} &= \frac{\partial \psi}{\partial t} + \mathbf{v}\cdot\nabla\psi \\ \frac{d\mathbf{A}}{dt} &= \frac{\partial \mathbf{A}}{\partial t} + \mathbf{v}\cdot\nabla\mathbf{A} \\ \frac{d\mathbf{\Phi}}{dt} &= \frac{\partial \mathbf{\Phi}}{\partial t} + \mathbf{v}\cdot\nabla\mathbf{\Phi} \end{aligned}} \tag{M3.34}$$

These equations are known as the *Euler development* of the corresponding variables. The Euler development applies not only to scalar field functions, vectors, and dyadics, but also to higher-order tensors.

Before concluding this section let us briefly apply Euler's development to the vector $\mathbf{A}$:

$$\frac{d\mathbf{A}}{dt} = \left(\frac{\partial \mathbf{A}}{\partial t}\right)_{q^i} + \dot{q}^n \frac{\partial \mathbf{A}}{\partial q^n} = \left(\frac{\partial A^m \mathbf{q}_m}{\partial t}\right)_{q^i} + \dot{q}^n \frac{\partial A^m \mathbf{q}_m}{\partial q^n} \tag{M3.35}$$

As will be realized by now, in contrast to the Cartesian coordinate system, the basis vectors $\mathbf{q}_i$ of the general q^i system are functions of the spatial coordinates q^i and of time. Thus, in the general case the gradient of the vector $\mathbf{A}$, which is also known as the *local dyadic* of $\mathbf{A}$, must be written as

$$\begin{aligned} \nabla\mathbf{A} &= \nabla(A^n \mathbf{q}_n) = (\nabla A^n)\mathbf{q}_n + A^n(\nabla\mathbf{q}_n) \\ &= \mathbf{q}^m \frac{\partial A^n \mathbf{q}_n}{\partial q^m} = \mathbf{q}^m \mathbf{q}_n \frac{\partial A^n}{\partial q^m} + A^n \mathbf{q}^m \frac{\partial \mathbf{q}_n}{\partial q^m} \end{aligned} \tag{M3.36}$$

M3.3 The spatial derivative of the basis vectors

To begin with, we perform the partial differentiation of the covariant basis vector $\mathbf{q}_i$ with respect to the contravariant measure number q^j. Since the order of the differentiation may be interchanged, see (M3.18), we obtain the important identity

$$\frac{\partial \mathbf{q}_i}{\partial q^j}=\frac{\partial}{\partial q^j}\left(\frac{\partial \mathbf{r}}{\partial q^i}\right)=\frac{\partial^2 \mathbf{r}}{\partial q^j\,\partial q^i}=\frac{\partial^2 \mathbf{r}}{\partial q^i\,\partial q^j}=\frac{\partial}{\partial q^i}\left(\frac{\partial \mathbf{r}}{\partial q^j}\right)=\frac{\partial \mathbf{q}_j}{\partial q^i} \tag{M3.37}$$

Differentiation of the covariant metric fundamental quantity g_{ij}, using (M3.37), yields

$$\frac{\partial g_{ij}}{\partial q^k}=\frac{\partial}{\partial q^k}(\mathbf{q}_i\cdot\mathbf{q}_j)=\frac{\partial \mathbf{q}_i}{\partial q^k}\cdot\mathbf{q}_j+\mathbf{q}_i\cdot\frac{\partial \mathbf{q}_j}{\partial q^k}=\frac{\partial \mathbf{q}_k}{\partial q^i}\cdot\mathbf{q}_j+\mathbf{q}_i\cdot\frac{\partial \mathbf{q}_k}{\partial q^j} \tag{M3.38}$$

By obvious steps this equation can be rewritten in the form

$$\begin{aligned}\frac{\partial g_{ij}}{\partial q^k}&=\frac{\partial}{\partial q^i}(\mathbf{q}_j\cdot\mathbf{q}_k)-\mathbf{q}_k\cdot\frac{\partial \mathbf{q}_j}{\partial q^i}+\frac{\partial}{\partial q^j}(\mathbf{q}_i\cdot\mathbf{q}_k)-\mathbf{q}_k\cdot\frac{\partial \mathbf{q}_i}{\partial q^j}\\&=\frac{\partial g_{jk}}{\partial q^i}+\frac{\partial g_{ik}}{\partial q^j}-2\mathbf{q}_k\cdot\frac{\partial \mathbf{q}_i}{\partial q^j}\end{aligned} \tag{M3.39}$$

In order to obtain compact and concise differentiation formulas we will introduce the so-called *Christoffel symbols*. The Christoffel symbol of the first kind is denoted by Γ_{ijk} and is defined by

$$\boxed{\Gamma_{ijk}=\mathbf{q}_k\cdot\frac{\partial \mathbf{q}_i}{\partial q^j}=\frac{1}{2}\left(\frac{\partial g_{jk}}{\partial q^i}+\frac{\partial g_{ik}}{\partial q^j}-\frac{\partial g_{ij}}{\partial q^k}\right)} \tag{M3.40}$$

From this expression it is easily seen that the indices i and j of the Christoffel symbol may be interchanged, that is $\Gamma_{ijk}=\Gamma_{jik}$.

Next we perform a dyadic multiplication of (M3.40) with the contravariant basis vector $\mathbf{q}^l$, yielding

$$\mathbf{q}^l\mathbf{q}_k\cdot\frac{\partial \mathbf{q}_i}{\partial q^j}=\Gamma_{ijk}\mathbf{q}^l \tag{M3.41}$$

It is now possible to introduce the unit dyadic by a very useful mathematical manipulation called *contraction*. First we set the covariant index k equal to the contravariant index l and then we sum over k. However, according to the Einstein summation convention, we introduce the summation index n by setting $k=l=n$ and obtain

$$\mathbf{q}^n\mathbf{q}_n\cdot\frac{\partial \mathbf{q}_i}{\partial q^j}=\mathbb{E}\cdot\frac{\partial \mathbf{q}_i}{\partial q^j}=\Gamma_{ijn}\mathbf{q}^n \tag{M3.42}$$

Often it is convenient to use the Christoffel symbol of the second kind, Γ_{ij}^{k}. This symbol is obtained from the Christoffel symbol of the first kind by lowering the index of the contravariant basis vectors. Observing (M2.103), we find

$$\frac{\partial \mathbf{q}_i}{\partial q^j} = \Gamma_{ijn}\mathbf{q}^n = \Gamma_{ijn}g^{nm}\mathbf{q}_m = \Gamma_{ij}^{m}\mathbf{q}_m \implies \Gamma_{ij}^{k} = \Gamma_{ijn}g^{nk} \tag{M3.43}$$

From (M3.42) and (M3.43) we obtain the desired expression for the spatial derivative of the covariant basis vector $\mathbf{q}_i$:

$$\boxed{\frac{\partial \mathbf{q}_i}{\partial q^j} = \frac{\partial \mathbf{q}_j}{\partial q^i} = \Gamma_{ijn}\mathbf{q}^n = \Gamma_{ij}^{n}\mathbf{q}_n} \tag{M3.44}$$

With the help of this equation, the gradient of the covariant basis vector $\mathbf{q}_i$ can be written in the following concise form:

$$\boxed{\nabla \mathbf{q}_i = \mathbf{q}^s \frac{\partial \mathbf{q}_i}{\partial q^s} = \mathbf{q}^s \mathbf{q}_m \Gamma_{is}^{m}} \tag{M3.45}$$

At first glance, the Christoffel symbols have the appearance of tensors. However, they are not tensors since they do not transform according to the tensor transformation laws which will be discussed later.

Using equations (M3.36) and (M3.45), the gradient of the vector $\mathbf{A}$, expressed with the help of the contravariant measure numbers A^i, can now be written in the form

$$\boxed{\nabla \mathbf{A} = \mathbf{q}^m \mathbf{q}_n \left(\frac{\partial A^n}{\partial q^m} + A^s \Gamma_{ms}^{n} \right)} \tag{M3.46}$$

which required the renaming of the indices in order to extract the dyadic factor $\mathbf{q}^m \mathbf{q}_n$.

If we wish to express the gradient of the vector $\mathbf{A}$ with the help of covariant measure numbers A_i, we must proceed similarly. Differentiation of the scalar product of the reciprocal basis vectors with respect to q^j yields, together with (M3.44),

$$\frac{\partial}{\partial q^j}(\mathbf{q}^i \cdot \mathbf{q}_k) = 0 \implies \mathbf{q}_k \cdot \frac{\partial \mathbf{q}^i}{\partial q^j} = -\mathbf{q}^i \cdot \frac{\partial \mathbf{q}_k}{\partial q^j} = -\Gamma_{jk}^{i} \tag{M3.47}$$

Dyadic multiplication of the latter equation by the contravariant basis vector $\mathbf{q}^l$ and contraction of the mixed tensor (setting $k = l$ and summing, i.e. setting $k = l = n$) gives

$$\boxed{\mathbf{q}^n \mathbf{q}_n \cdot \frac{\partial \mathbf{q}^i}{\partial q^j} = \frac{\partial \mathbf{q}^i}{\partial q^j} = \frac{\partial \mathbf{q}^j}{\partial q^i} = -\Gamma_{jn}^{i}\mathbf{q}^n} \tag{M3.48}$$

Utilizing this expression, it is a simple matter to find the gradient of $\mathbf{q}^i$:

$$\boxed{\nabla\mathbf{q}^i = \mathbf{q}^n \frac{\partial \mathbf{q}^i}{\partial q^n} = -\mathbf{q}^n\mathbf{q}^s\Gamma^i_{ns}} \tag{M3.49}$$

Finally, we wish to find the gradient of the vector $\mathbf{A}$ if this vector is expressed in terms of covariant measure numbers A_i. This leads to

$$\nabla\mathbf{A} = \nabla(A_n\mathbf{q}^n) = \mathbf{q}^s\mathbf{q}^n \frac{\partial A_n}{\partial q^s} - A_n\mathbf{q}^s\mathbf{q}^m\Gamma^n_{sm} \tag{M3.50}$$

Again, by interchanging the summation indices we can extract the factor $\mathbf{q}^m\mathbf{q}^n$ to obtain the more compact form

$$\boxed{\nabla\mathbf{A} = \mathbf{q}^m\mathbf{q}^n\left(\frac{\partial A_n}{\partial q^m} - A_s\Gamma^s_{mn}\right)} \tag{M3.51}$$

which should be compared with (M3.46).

M3.4 Differential invariants in generalized coordinate systems

In order to obtain the required invariants, we expand the general form of the local dyadic $\nabla\mathbf{A}$ and replace the contravariant basis vectors $\mathbf{q}^i$ of the gradient operator by means of formula (M1.55). The result is

$$\begin{aligned}
\nabla\mathbf{A} &= \mathbf{q}^s \frac{\partial \mathbf{A}}{\partial q^s} = \mathbf{q}^1 \frac{\partial \mathbf{A}}{\partial q^1} + \mathbf{q}^2 \frac{\partial \mathbf{A}}{\partial q^2} + \mathbf{q}^3 \frac{\partial \mathbf{A}}{\partial q^3} \\
&= \frac{1}{\sqrt{g}}\left((\mathbf{q}_2 \times \mathbf{q}_3)\frac{\partial \mathbf{A}}{\partial q^1} + (\mathbf{q}_3 \times \mathbf{q}_1)\frac{\partial \mathbf{A}}{\partial q^2} + (\mathbf{q}_1 \times \mathbf{q}_2)\frac{\partial \mathbf{A}}{\partial q^3}\right) \\
&= \frac{1}{\sqrt{g}}\left(\frac{\partial}{\partial q^1}(\mathbf{q}_2 \times \mathbf{q}_3\mathbf{A}) + \frac{\partial}{\partial q^2}(\mathbf{q}_3 \times \mathbf{q}_1\mathbf{A}) + \frac{\partial}{\partial q^3}(\mathbf{q}_1 \times \mathbf{q}_2\mathbf{A})\right) \\
&\quad - \frac{1}{\sqrt{g}}\left(\frac{\partial}{\partial q^1}(\mathbf{q}_2 \times \mathbf{q}_3) + \frac{\partial}{\partial q^2}(\mathbf{q}_3 \times \mathbf{q}_1) + \frac{\partial}{\partial q^3}(\mathbf{q}_1 \times \mathbf{q}_2)\right)\mathbf{A}
\end{aligned} \tag{M3.52}$$

It is a little tedious to show that the sum of the three terms within the last set of large parentheses vanishes. Thus, equation (M3.52) reduces to

$$\boxed{\nabla\mathbf{A} = \frac{1}{\sqrt{g}}\left(\frac{\partial}{\partial q^1}(\mathbf{q}_2 \times \mathbf{q}_3\mathbf{A}) + \frac{\partial}{\partial q^2}(\mathbf{q}_3 \times \mathbf{q}_1\mathbf{A}) + \frac{\partial}{\partial q^3}(\mathbf{q}_1 \times \mathbf{q}_2\mathbf{A})\right)} \tag{M3.53}$$

This is the form which will be used to obtain the invariants.

M3.4.1 The first scalar or the divergence of the local dyadic $\nabla\mathbf{A}$

By expressing the vector $\mathbf{A}$ with the help of the covariant basis vectors $\mathbf{q}_i$ and then taking the scalar product, we find

$$\nabla\cdot\mathbf{A} = \frac{1}{\sqrt{g}}\left(\frac{\partial}{\partial q^1}[(\mathbf{q}_2\times\mathbf{q}_3)\cdot A^n\mathbf{q}_n] + \frac{\partial}{\partial q^2}[(\mathbf{q}_3\times\mathbf{q}_1)\cdot A^n\mathbf{q}_n] + \frac{\partial}{\partial q^3}[(\mathbf{q}_1\times\mathbf{q}_2)\cdot A^n\mathbf{q}_n]\right) \tag{M3.54}$$

Let us consider the first term in expanded form. Only the term involving $\mathbf{q}_1$ makes a contribution; the remaining terms involving $\mathbf{q}_2$ and $\mathbf{q}_3$ must vanish since the scalar triple products are zero. The same arguments hold for the remaining two terms in (M3.54), so that

$$\nabla\cdot\mathbf{A} = \frac{1}{\sqrt{g}}\left(\frac{\partial}{\partial q^1}[(\mathbf{q}_2\times\mathbf{q}_3)\cdot A^1\mathbf{q}_1] + \frac{\partial}{\partial q^2}[(\mathbf{q}_3\times\mathbf{q}_1)\cdot A^2\mathbf{q}_2] + \frac{\partial}{\partial q^3}[(\mathbf{q}_1\times\mathbf{q}_2)\cdot A^3\mathbf{q}_3]\right) \tag{M3.55}$$

By observing the definition of $\sqrt{g}$ and the rules of the scalar triple product we find the desired relation:

$$\boxed{\nabla\cdot\mathbf{A} = \frac{1}{\sqrt{g}}\left(\frac{\partial}{\partial q^n}(A^n\sqrt{g})\right) = \frac{1}{\sqrt{g}}\frac{\partial}{\partial q^n}(\mathbf{q}^n\cdot\mathbf{A}\sqrt{g})} \tag{M3.56}$$

with $A^i = \mathbf{q}^i\cdot\mathbf{A}$. In the orthogonal Cartesian system, the divergence expression is particularly simple since $\sqrt{g} = 1$ so that (M3.56) reduces to

$$\boxed{\nabla\cdot\mathbf{A} = \frac{\partial A_x}{\partial x} + \frac{\partial A_y}{\partial y} + \frac{\partial A_z}{\partial z}} \tag{M3.57}$$

The first scalar of the dyadic $\nabla\mathbb{B}$ has the same form as (M3.56). In this case, however, we have to watch the position of the dyadic as stated in

$$\boxed{\nabla\cdot\mathbb{B} = \frac{1}{\sqrt{g}}\frac{\partial}{\partial q^n}(\mathbf{q}^n\cdot\mathbb{B}\sqrt{g})} \tag{M3.58}$$

The reader is invited to verify this formula. This is an easy but somewhat tedious task since it involves the spatial differentiation of the basis vectors.

M3.4.2 The Laplacian of a scalar field function

Defining the vector $\mathbf{A} = \nabla\psi$, we find with the help of (M3.56) the desired expression

$$\boxed{\begin{aligned}\nabla\cdot\nabla\psi = \nabla^2\psi &= \frac{1}{\sqrt{g}}\frac{\partial}{\partial q^n}\left(\sqrt{g}\mathbf{q}^n\cdot\mathbf{q}^m\frac{\partial\psi}{\partial q^m}\right)\\ &= \frac{1}{\sqrt{g}}\frac{\partial}{\partial q^n}\left(\sqrt{g}\,g^{nm}\frac{\partial\psi}{\partial q^m}\right)\end{aligned}} \tag{M3.59}$$

where ∇^2 is the so-called *Laplace operator*. Some authors denote the Laplacian operator by the symbol Δ. We shall reserve the symbol Δ for representing the difference operator. Only in the Cartesian system does (M3.59) reduce to the following very simple form:

$$\boxed{\nabla\cdot\nabla\psi = \frac{\partial^2\psi}{\partial x^2} + \frac{\partial^2\psi}{\partial y^2} + \frac{\partial^2\psi}{\partial z^2}} \tag{M3.60}$$

M3.4.3 The vector of the local dyadic $\nabla\mathbf{A}$

Taking the cross product of the gradient operator with the vector $\mathbf{A}$ in equation (M3.53), we first obtain three vectorial triple cross products:

$$\begin{aligned}\nabla\times\mathbf{A} = \frac{1}{\sqrt{g}}\bigg(&\frac{\partial}{\partial q^1}[(\mathbf{q}_2\times\mathbf{q}_3)\times\mathbf{A}] + \frac{\partial}{\partial q^2}[(\mathbf{q}_3\times\mathbf{q}_1)\times\mathbf{A}]\\ &+ \frac{\partial}{\partial q^3}[(\mathbf{q}_1\times\mathbf{q}_2)\times\mathbf{A}]\bigg)\end{aligned} \tag{M3.61}$$

Application of the Grassmann rule (M1.44) immediately results in

$$\begin{aligned}\nabla\times\mathbf{A} &= \frac{1}{\sqrt{g}}\bigg(\frac{\partial}{\partial q^1}[(\mathbf{A}\cdot\mathbf{q}_2)\mathbf{q}_3 - (\mathbf{A}\cdot\mathbf{q}_3)\mathbf{q}_2] + \frac{\partial}{\partial q^2}[(\mathbf{A}\cdot\mathbf{q}_3)\mathbf{q}_1 - (\mathbf{A}\cdot\mathbf{q}_1)\mathbf{q}_3]\\ &\qquad + \frac{\partial}{\partial q^3}[(\mathbf{A}\cdot\mathbf{q}_1)\mathbf{q}_2 - (\mathbf{A}\cdot\mathbf{q}_2)\mathbf{q}_1]\bigg)\\ &= \frac{1}{\sqrt{g}}\left(\frac{\partial}{\partial q^1}(A_2\mathbf{q}_3 - A_3\mathbf{q}_2) + \frac{\partial}{\partial q^2}(A_3\mathbf{q}_1 - A_1\mathbf{q}_3) + \frac{\partial}{\partial q^3}(A_1\mathbf{q}_2 - A_2\mathbf{q}_1)\right)\end{aligned} \tag{M3.62}$$

which can be written in the convenient determinant form

$$\boxed{\nabla \times \mathbf{A} = \frac{1}{\sqrt{g}} \begin{vmatrix} \mathbf{q}_1 & \mathbf{q}_2 & \mathbf{q}_3 \\ \dfrac{\partial}{\partial q^1} & \dfrac{\partial}{\partial q^2} & \dfrac{\partial}{\partial q^3} \\ A_1 & A_2 & A_3 \end{vmatrix}} \tag{M3.63}$$

Since in the Cartesian system $\sqrt{g} = 1$, in this special case we obtain the well-known formula

$$\boxed{\nabla \times \mathbf{A} = \begin{vmatrix} \mathbf{i} & \mathbf{j} & \mathbf{k} \\ \dfrac{\partial}{\partial x} & \dfrac{\partial}{\partial y} & \dfrac{\partial}{\partial z} \\ A_x & A_y & A_z \end{vmatrix}} \tag{M3.64}$$

The vector $\nabla \times \mathbf{A}$ is commonly known as the curl of $\mathbf{A}$ or rot $\mathbf{A}$.

M3.5 Additional applications

Of particular interest are the invariants of the unit dyadic and of the gradient of the position vector. Utilizing the expressions (M3.44) and (M3.48) for the spatial derivatives of the covariant and the contravariant basis vectors, respectively, we may write

$$\begin{aligned} \nabla \mathbb{E} &= \mathbf{q}^n \frac{\partial}{\partial q^n}(\mathbf{q}^m \mathbf{q}_m) = \mathbf{q}^n \left(\frac{\partial\, \mathbf{q}^m}{\partial q^n} \mathbf{q}_m + \mathbf{q}^m \frac{\partial \mathbf{q}_m}{\partial q^n} \right) \\ &= \mathbf{q}^n \left(-\mathbf{q}^r \Gamma_{rn}^m \mathbf{q}_m + \mathbf{q}^m \Gamma_{mn}^r \mathbf{q}_r \right) = 0 \end{aligned} \tag{M3.65}$$

The two terms involving the Christoffel symbols are identical since in the last term the summation indices m and r may be interchanged without changing the value of this term. Thus, the gradient of the unit dyadic vanishes and we have

$$\boxed{\nabla \cdot \mathbb{E} = 0, \qquad \nabla \times \mathbb{E} = 0} \tag{M3.66}$$

Recalling that according to (M3.30) the gradient of the position vector $\mathbf{r}$ is the unit dyadic, that is $\nabla \mathbf{r} = \mathbb{E} = \mathbf{q}^n \mathbf{q}_n$, we immediately obtain

$$\boxed{\nabla \cdot \mathbf{r} = \mathbf{q}^n \cdot \mathbf{q}_n = 3, \qquad \nabla \times \mathbf{r} = \mathbf{q}^n \times \mathbf{q}_n = 0} \tag{M3.67}$$

It will very often be necessary to apply the Hamilton operator to products of various types of extensive functions. Sometimes this can be very complex. To avoid errors we must apply the following rules:

(I) As the first step we must apply the well-known product rules of differential calculus.
(II) As the second step we must apply the rules of vector algebra with the goal in mind to obtain a unique and an unmistakeable arrangement of operations.

The following rule helps us to accomplish this task: An arrow placed above any vector in a product implies that only this vector need be differentiated while the remaining vectors in the product are treated as constants. As an example consider the operation

$$\nabla(\mathbf{A}\cdot\overset{\downarrow}{\mathbf{B}}) = \nabla(\overset{\downarrow}{\mathbf{B}}\cdot\mathbf{A}) = (\nabla\mathbf{B})\cdot\mathbf{A} \tag{M3.68}$$

The arrow placed above the vector $\mathbf{B}$ indicates that the gradient operator is applied to this vector only. Since we are dealing with a scalar product, the operation is very simple. When the proper order of the operations is clear we omit the arrow. The following example implies that only vector $\mathbf{A}$ ahead of the gradient symbol is to be differentiated:

$$\overset{\downarrow}{\mathbf{A}}(\nabla\cdot\mathbf{B}) = (\mathbf{B}\cdot\nabla)\overset{\downarrow}{\mathbf{A}} = \mathbf{B}\cdot\nabla\mathbf{A} \tag{M3.69}$$

We must be careful that the dot between the gradient operator and the vector $\mathbf{B}$ does not change its position, otherwise the value of the entire expression will be changed. When the operation is finished, the arrow is no longer needed and may be omitted.

A somewhat more complex example is given by

$$\begin{aligned}\nabla\cdot(\mathbf{A}\cdot\mathbf{\Phi}) &= \nabla\cdot(\overset{\downarrow}{\mathbf{A}}\cdot\mathbf{\Phi}) + \nabla\cdot(\mathbf{A}\cdot\overset{\downarrow}{\mathbf{\Phi}}) \\ &= \nabla\mathbf{A}\cdot\cdot\mathbf{\Phi} + (\mathbf{A}\cdot\overset{\curvearrowleft}{\mathbf{\Phi})\cdot\nabla} = \nabla\mathbf{A}\cdot\cdot\mathbf{\Phi} + \mathbf{A}\cdot(\nabla\cdot\widetilde{\mathbf{\Phi}})\end{aligned} \tag{M3.70}$$

Note that the term $\mathbf{A}\cdot\overset{\downarrow}{\mathbf{\Phi}}$ is a vector so that $\nabla\cdot(\mathbf{A}\cdot\overset{\downarrow}{\mathbf{\Phi}})$ is a scalar product. Thus we may rewrite this expression by interchanging the positions of the two vectors ∇ and $(\mathbf{A}\cdot\overset{\downarrow}{\mathbf{\Phi}})$. The curved backward arrow occurring in (M3.70) indicates that the gradient operates on $\mathbf{\Phi}$ only. Finally, in order to obtain the usual notation without the backward arrow, in the last expression of (M3.70) we have used the rule (M2.18) for the multiplication of a vector by a dyadic. Later it will become apparent that this equation is particularly interesting if $\mathbf{A}$ is identified as the velocity vector $\mathbf{v}$ and the dyadic $\mathbf{\Phi}$ as $\nabla\mathbf{v}$. This results in

$$\nabla\cdot(\mathbf{v}\cdot\nabla\mathbf{v}) = \nabla\mathbf{v}\cdot\cdot\nabla\mathbf{v} + \mathbf{v}\cdot[\nabla\cdot(\overset{\curvearrowleft}{\mathbf{v}\nabla})] = \nabla\mathbf{v}\cdot\cdot\nabla\mathbf{v} + \mathbf{v}\cdot\nabla(\nabla\cdot\mathbf{v}) \tag{M3.71}$$

Now consider the expression $\mathbf{A} \times (\nabla \times \mathbf{B})$. First of all it should be observed that the differentiation refers to the vector $\mathbf{B}$. By using Grassmann's rule (M1.44) we find

$$\mathbf{A} \times (\nabla \times \mathbf{B}) = (\mathbf{A}\cdot\overset{\curvearrowleft}{\mathbf{B}})\nabla - \mathbf{A}\cdot\nabla\mathbf{B} = (\nabla\mathbf{B})\cdot\mathbf{A} - \mathbf{A}\cdot\nabla\mathbf{B} \tag{M3.72}$$

Replacing $\mathbf{A}$ in this equation by the nabla operator yields

$$\boxed{\nabla \times (\nabla \times \mathbf{B}) = \nabla(\overset{\curvearrowleft}{\mathbf{B}}\cdot\nabla) - \nabla\cdot\nabla\mathbf{B} = \nabla(\nabla\cdot\mathbf{B}) - \nabla^2\mathbf{B}} \tag{M3.73}$$

This particular form is often needed.

Equation (M3.72) may be rearranged as

$$\mathbf{A}\cdot\nabla\mathbf{B} = (\nabla\mathbf{B})\cdot\mathbf{A} + (\nabla \times \mathbf{B}) \times \mathbf{A} \tag{M3.74}$$

Replacing now $\mathbf{A}$ and $\mathbf{B}$ by $\mathbf{v}$, we find the so-called *Lamb transformation*

$$\boxed{\mathbf{v}\cdot\nabla\mathbf{v} = (\nabla\mathbf{v})\cdot\mathbf{v} + (\nabla \times \mathbf{v}) \times \mathbf{v} = \nabla\left(\frac{\mathbf{v}^2}{2}\right) + (\nabla \times \mathbf{v}) \times \mathbf{v}} \tag{M3.75}$$

which is of great importance in dynamic meteorology.

Now we will derive another important formula that will be used in dynamic meteorology. We start by rearranging (M3.71) into the form

$$\nabla\mathbf{v}\cdot\cdot\nabla\mathbf{v} = \nabla\cdot(\mathbf{v}\cdot\nabla\mathbf{v}) - \mathbf{v}\cdot\nabla(\nabla\cdot\mathbf{v}) \tag{M3.76}$$

Utilizing (M3.75), the first term on the right-hand side may be rewritten as

$$\nabla\cdot(\mathbf{v}\cdot\nabla\mathbf{v}) = \nabla^2\left(\frac{\mathbf{v}^2}{2}\right) - \nabla\cdot[\mathbf{v} \times (\nabla \times \mathbf{v})] \tag{M3.77}$$

In this equation the last term on the right-hand side will be evaluated:

$$\begin{aligned}\nabla\cdot[\mathbf{v} \times (\nabla \times \mathbf{v})] &= \nabla\cdot[\overset{\downarrow}{\mathbf{v}} \times (\nabla \times \mathbf{v})] - \nabla\cdot[(\nabla \overset{\downarrow}{\times} \mathbf{v}) \times \mathbf{v}] \\ &= (\nabla \times \mathbf{v})\cdot(\nabla \times \mathbf{v}) - [\nabla \times (\nabla \times \mathbf{v})]\cdot\mathbf{v} \\ &= (\nabla \times \mathbf{v})^2 - \mathbf{v}\cdot[\nabla \times (\nabla \times \mathbf{v})] \\ &= (\nabla \times \mathbf{v})^2 - \mathbf{v}\cdot[\nabla(\nabla\cdot\mathbf{v}) - \nabla^2\mathbf{v}]\end{aligned} \tag{M3.78}$$

The last equation has been obtained by making use of (M3.73) with $\mathbf{B} = \mathbf{v}$. Hence, (M3.77) may be written as

$$\nabla\cdot(\mathbf{v}\cdot\nabla\mathbf{v}) = \nabla^2\left(\frac{\mathbf{v}^2}{2}\right) - (\nabla \times \mathbf{v})^2 + \mathbf{v}\cdot[\nabla(\nabla\cdot\mathbf{v})] - \mathbf{v}\cdot\nabla^2\mathbf{v} \tag{M3.79}$$

Substituting this expression into (M3.76) yields the final result

$$\boxed{\nabla\mathbf{v}\cdot\cdot\nabla\mathbf{v} = \nabla^2\left(\frac{\mathbf{v}^2}{2}\right) - (\nabla\times\mathbf{v})^2 - \mathbf{v}\cdot\nabla^2\mathbf{v}} \tag{M3.80}$$

Finally, we consider an arbitrary field vector $\mathbf{A}$, which is assumed to depend on some scalar functions ψ and χ, whereby these functions depend on the position and on time, as shown in

$$\mathbf{A} = \mathbf{A}[\psi(q^1, q^2, q^3, t), \chi(q^1, q^2, q^3, t)] \tag{M3.81}$$

Now the gradient of $\mathbf{A}$ is given by

$$\nabla\mathbf{A} = \mathbf{q}^n \frac{\partial\mathbf{A}}{\partial q^n} = \mathbf{q}^n\left(\frac{\partial\mathbf{A}}{\partial\psi}\frac{\partial\psi}{\partial q^n} + \frac{\partial\mathbf{A}}{\partial\chi}\frac{\partial\chi}{\partial q^n}\right) = \nabla\psi\,\frac{\partial\mathbf{A}}{\partial\psi} + \nabla\chi\,\frac{\partial\mathbf{A}}{\partial\chi} \tag{M3.82}$$

It is important to observe the positions of the vectors forming the dyadics.

In the simple case that $\mathbf{A} = \mathbf{A}[\psi(q^1, q^2, q^3, t)]$, the gradient of $\mathbf{A}$, the first scalar, and the vector of $\nabla\mathbf{A}$ are given by

$$\nabla\mathbf{A} = \nabla\psi\,\frac{d\mathbf{A}}{d\psi}, \qquad \nabla\cdot\mathbf{A} = \nabla\psi\cdot\frac{d\mathbf{A}}{d\psi}, \qquad \nabla\times\mathbf{A} = \nabla\psi\times\frac{d\mathbf{A}}{d\psi} \tag{M3.83}$$

The Laplacian of $\mathbf{A}$ is obtained from

$$\boxed{\begin{aligned}\nabla^2\mathbf{A} &= \nabla\cdot\left(\nabla\psi\,\frac{d\mathbf{A}}{d\psi}\right) = \nabla^2\psi\,\frac{d\mathbf{A}}{d\psi} + \nabla\psi\cdot\nabla\left(\frac{d\mathbf{A}}{d\psi}\right)\\ &= \nabla^2\psi\,\frac{d\mathbf{A}}{d\psi} + \nabla\psi\cdot\left(\nabla\psi\,\frac{d^2\mathbf{A}}{d\psi^2}\right) = \nabla^2\psi\,\frac{d\mathbf{A}}{d\psi} + (\nabla\psi)^2\,\frac{d^2\mathbf{A}}{d\psi^2}\end{aligned}} \tag{M3.84}$$

In the final step use of the first equation of (M3.83) has been made by replacing $\mathbf{A}$ there by the derivative $d\mathbf{A}/d\psi$.

M3.6 Problems

M3.1: Use generalized coordinates to show that

$$\nabla \mathbf{r} = \mathbb{E}, \quad \nabla(\mathbf{r}\cdot\mathbf{r}) = 2\mathbf{r}, \quad \nabla r = \mathbf{e}_r, \quad \nabla\left(\frac{1}{r}\right) = -\frac{\mathbf{r}}{r^3}$$

$$\nabla\cdot\mathbf{r} = 3, \quad \nabla\times\mathbf{r} = 0, \quad \nabla\cdot\mathbf{e}_r = \frac{2}{r}, \quad \nabla\cdot\left(\frac{\mathbf{e}_r}{r^2}\right) = 0, \quad \nabla^2\left(\frac{1}{|\mathbf{r}|}\right) = 0$$

M3.2: The potential temperature of dry air is given by $\theta = T/\Pi$ where $\Pi = (p/p_0)^{R_0/c_{p,0}}$ is the so-called Exner function. p is the pressure, $p_0 = 1000$ hPa, T the temperature, R_0 the individual gas constant, and $c_{p,0}$ the specific heat at constant pressure. Find the gradient of the potential temperature.

M3.3: Assume that a vector $\mathbf{W}$ is given by $\mathbf{W} = \mathbf{r}^2\boldsymbol{\Omega}$, where $\boldsymbol{\Omega}$ is a constant rotational vector. Use generalized coordinates to find expressions for

$$\text{(a)}\quad \nabla\cdot\mathbf{W}, \qquad \text{(b)}\quad \nabla\times\mathbf{W}, \qquad \text{(c)}\quad \nabla^2\mathbf{W}, \qquad \text{(d)}\quad \mathbf{v}\cdot\nabla\mathbf{W}$$

M3.4: According to a theorem of potential theory, the velocity vector can be split according to $\mathbf{v} = -\nabla\times\boldsymbol{\Psi} + \nabla\chi$ with the additional condition $\nabla\cdot\boldsymbol{\Psi} = 0$. Here $\boldsymbol{\Psi}$ is a vector (the stream-function vector) and χ is a scalar function (the velocity potential). You do not need to know the meaning of $\boldsymbol{\Psi}$ and χ to perform the following differential operations:

$$\text{(a)}\quad \nabla\cdot\mathbf{v}, \qquad \text{(b)}\quad \nabla\times\mathbf{v}, \qquad \text{(c)}\quad \nabla^2\mathbf{v}$$

$$\text{(d)}\quad \nabla(\nabla\cdot\mathbf{v}), \qquad \text{(e)}\quad \nabla\cdot[\tfrac{1}{2}(\nabla\mathbf{v} + \overset{\curvearrowleft}{\mathbf{v}\nabla} - \tfrac{1}{3}\nabla\cdot\mathbf{v}\mathbb{E}]$$

Hint: The Laplace operator and the gradient operator can be interchanged. In generalized coordinates the validity of this operation is difficult to show, so the proof will be omitted.

M3.5: Use generalized coordinates to show that $\nabla\cdot\mathbb{E} = 0$.

M3.6: The frictional stress tensor can be written in the form

$$\mathbb{J} = \frac{\eta}{2}\left(\nabla\mathbf{v} + \overset{\curvearrowleft}{\mathbf{v}\nabla} - \frac{2}{3}\nabla\cdot\mathbf{v}\mathbb{E}\right)$$

where η is a scalar coefficient. Use Cartesian coordinates to solve the following problems.

(a) Show that the trace of $\mathbb{J}$ is zero.

(b) By assuming that $w = 0$, $\partial u/\partial x = 0$, $\partial u/\partial y = 0$, $\partial v/\partial x = 0$, $\partial v/\partial y = 0$, write $\mathbb{J}$ in matrix form.

(c) By using the simplifications stated in (b), find the eigenvalues of $\mathbb{J}$.
(d) Show that the eigenvectors corresponding to (c) form an orthogonal system.
(e) In this orthogonal system $\mathbb{J}$ can be written in the form $\mathbb{J} = \sum_{i=1}^{3} \lambda_i \boldsymbol{\Psi}^i \boldsymbol{\Psi}^i$, where $\boldsymbol{\Psi}^i$ are the eigenvectors corresponding to the eigenvalues λ_i. By assuming additionally that $\partial v/\partial z = 0$, show that the tensor ellipsoid is a hyperbola.

M3.7: By evaluating the expression $\partial g_{ij}/\partial q^k$ show the validity of the relation

$$\frac{\partial \mathbf{q}_i}{\partial q^j} = \Gamma_{ij}^m \mathbf{q}_m$$

M3.8: Show that the following relations are valid:

$$\text{(a)} \quad \Gamma_{in}^n = \frac{1}{\sqrt{g}} \frac{\partial \sqrt{g}}{\partial q^i}, \qquad \text{(b)} \quad g_{mn} \frac{\partial g^{mn}}{\partial q^i} = -\frac{2}{\sqrt{g}} \frac{\partial \sqrt{g}}{\partial q^i}$$

M3.9: Show that, in generalized coordinates, the following expression is valid:

$$\nabla \cdot \mathbb{J} = \frac{1}{\sqrt{g}} \frac{\partial}{\partial q^n} (\sqrt{g} J^{mn} \mathbf{q}_m) \quad \text{with} \quad \mathbb{J} = J^{mn} \mathbf{q}_m \mathbf{q}_n$$

Hint: Start with equation (M3.53). The properties of the vector **A** have not been used in the derivation. Can **A** be replaced by a dyadic?

M3.10: Suppose that the vectors **M** and **N** satisfy the expressions

$$\nabla^2 \mathbf{M} + \mathbf{M} = 0, \quad \nabla^2 \mathbf{N} + \mathbf{N} = 0 \quad \text{with} \quad \nabla \cdot \mathbf{M} = 0, \quad \nabla \cdot \mathbf{N} = 0$$

Show that a consequence of these relations is that

$$\mathbf{M} = \nabla \times \mathbf{N}, \qquad \mathbf{N} = \nabla \times \mathbf{M}$$

M4

Coordinate transformations

M4.1 Transformation relations of time-independent coordinate systems

M4.1.1 Introduction

As mentioned earlier, all physically relevant quantities such as mass, time, force, and work are defined without reference to any coordinate system. This implies that they are invariant with regard to coordinate transformations. On the other hand, there are other quantities such as the components of vectors and of tensors and particularly the basis vectors $\mathbf{q}_i$, $\mathbf{q}^i$ that strongly depend on the choice of a coordinate system. The question of the way in which these quantities are changed by changing the coordinate system arises quite naturally. In order to derive the desired transformation relations we will consider two time-independent but otherwise arbitrary coordinate systems q^i and a^i as stated in

$$q^i = q^i(a^1, a^2, a^3), \qquad a^i = a^i(q^1, q^2, q^3) \tag{M4.1}$$

In general, these relations are nonlinear. Extensive quantities that are invariant with regard to coordinate transformations may be expressed in either coordinate system. For instance we may write

	q^i system	a^i system
(a)	$\psi = \psi(q^i, t)$	$\psi = \psi(a^i, t)$
(b)	$\mathbf{A} = \underset{q}{A}^n(q^i, t)\mathbf{q}_n = \underset{q}{A}_n(q^i, t)\mathbf{q}^n$	$\mathbf{A} = \underset{a}{A}^n(a^i, t)\mathbf{a}_n = \underset{a}{A}_n(a^i, t)\mathbf{a}^n$
(c)	$d\mathbf{r} = \mathbf{q}_n\, dq^n = \mathbf{q}^n\, dq_n$	$d\mathbf{r} = \mathbf{a}_n\, da^n = \mathbf{a}^n\, da_n$
(d)	$\sqrt{\underset{q}{g}} = [\mathbf{q}_1, \mathbf{q}_2, \mathbf{q}_3]$	$\sqrt{\underset{a}{g}} = [\mathbf{a}_1, \mathbf{a}_2, \mathbf{a}_3]$ (M4.2)
(e)	$\dfrac{1}{\sqrt{\underset{q}{g}}} = [\mathbf{q}^1, \mathbf{q}^2, \mathbf{q}^3]$	$\dfrac{1}{\sqrt{\underset{a}{g}}} = [\mathbf{a}^1, \mathbf{a}^2, \mathbf{a}^3]$
(f)	$\nabla = \mathbf{q}^n \underset{q}{\nabla}_n = \mathbf{q}^n \dfrac{\partial}{\partial q^n}$	$\nabla = \mathbf{a}^n \underset{a}{\nabla}_n = \mathbf{a}^n \dfrac{\partial}{\partial a^n}$

While the coordinate systems are assumed to be independent of time, the scalar ψ and the vector components $\underset{q}{A}^i$, $\underset{a}{A}^i$ in general are time-dependent. The time dependency is evident on considering such scalar quantities as temperature and the components of the velocity vector.

We will always assume that the relations (M4.1) are uniquely invertible, which is guaranteed if the functional determinant is nonzero, that is

$$\left|\frac{\partial(a^1, a^2, a^3)}{\partial(q^1, q^2, q^3)}\right| = \begin{vmatrix} \dfrac{\partial a^1}{\partial q^1} & \dfrac{\partial a^1}{\partial q^2} & \dfrac{\partial a^1}{\partial q^3} \\ \dfrac{\partial a^2}{\partial q^1} & \dfrac{\partial a^2}{\partial q^2} & \dfrac{\partial a^2}{\partial q^3} \\ \dfrac{\partial a^3}{\partial q^1} & \dfrac{\partial a^3}{\partial q^2} & \dfrac{\partial a^3}{\partial q^3} \end{vmatrix} = \sqrt{\underset{q}{g}}\,[\mathbf{q}^n \underset{q}{\nabla}_n a^1, \mathbf{q}^m \underset{q}{\nabla}_m a^2, \mathbf{q}^r \underset{q}{\nabla}_r a^3]$$
$$= \sqrt{\underset{q}{g}}\,J_q(a^1, a^2, a^3) \neq 0 \tag{M4.3}$$

The form stated to the right of the second equality sign is easily established by substituting $\mathbf{A} = \mathbf{q}^n \underset{q}{\nabla}_n a^1$, $\mathbf{B} = \mathbf{q}^m \underset{q}{\nabla}_m a^2$, $\mathbf{C} = \mathbf{q}^r \underset{q}{\nabla}_r a^3$ into equation (M1.96). The symbol $J_q(a^1, a^2, a^3)$ is known as the *Jacoby operator* of the q^i system. By interchanging the symbols q^i and a^i we immediately find the inverse relation

$$\left|\frac{\partial(q^1, q^2, q^3)}{\partial(a^1, a^2, a^3)}\right| = \begin{vmatrix} \dfrac{\partial q^1}{\partial a^1} & \dfrac{\partial q^1}{\partial a^2} & \dfrac{\partial q^1}{\partial a^3} \\ \dfrac{\partial q^2}{\partial a^1} & \dfrac{\partial q^2}{\partial a^2} & \dfrac{\partial q^2}{\partial a^3} \\ \dfrac{\partial q^3}{\partial a^1} & \dfrac{\partial q^3}{\partial a^2} & \dfrac{\partial q^3}{\partial a^3} \end{vmatrix} = \sqrt{\underset{a}{g}}\,[\mathbf{a}^n \underset{a}{\nabla}_n q^1, \mathbf{a}^m \underset{a}{\nabla}_m q^2, \mathbf{a}^r \underset{a}{\nabla}_r q^3]$$
$$= \sqrt{\underset{a}{g}}\,J_a(q^1, q^2, q^3) \neq 0 \tag{M4.4}$$

The relation between the two functional determinants is given by

$$\left|\frac{\partial(a^1, a^2, a^3)}{\partial(q^1, q^2, q^3)}\right| \left|\frac{\partial(q^1, q^2, q^3)}{\partial(a^1, a^2, a^3)}\right| = 1 \tag{M4.5}$$

This formula, which is derived in many textbooks on analysis, will be verified later.

M4.1.2 Transformation of basis vectors and coordinate differentials

Let us now consider the following equations:

$$\begin{aligned} d\mathbf{r}(q^1, q^2, q^3) &= d\mathbf{r}[a^1(q^1, q^2, q^3), a^2(q^1, q^2, q^3), a^3(q^1, q^2, q^3)] \\ d\mathbf{r}(a^1, a^2, a^3) &= d\mathbf{r}[q^1(a^1, a^2, a^3), q^2(a^1, a^2, a^3), q^3(a^1, a^2, a^3)] \end{aligned} \tag{M4.6}$$

where we have used the functional relations (M4.1). After taking the total differentials of both sides of these expressions, we compare the coefficients of dq^i and da^i. This leads to

$$\frac{\partial \mathbf{r}}{\partial q^i} = \frac{\partial \mathbf{r}}{\partial a^n}\frac{\partial a^n}{\partial q^i}, \qquad \frac{\partial \mathbf{r}}{\partial a^i} = \frac{\partial \mathbf{r}}{\partial q^n}\frac{\partial q^n}{\partial a^i} \tag{M4.7}$$

By employing the definition (M3.11), we immediately obtain the important transformation rules

$$\boxed{\mathbf{q}_i = \frac{\partial a^n}{\partial q^i}\,\mathbf{a}_n, \qquad \mathbf{a}_i = \frac{\partial q^n}{\partial a^i}\,\mathbf{q}_n} \tag{M4.8}$$

for the covariant basis vectors. Using the definition (M3.27), we find from (M4.2f) the transformation rules for the contravariant basis vectors

$$\boxed{\mathbf{q}^i = \nabla q^i = \frac{\partial q^i}{\partial a^n}\,\mathbf{a}^n, \qquad \mathbf{a}^i = \nabla a^i = \frac{\partial a^i}{\partial q^n}\,\mathbf{q}^n} \tag{M4.9}$$

Now we will discuss the relations among the differentials da_i, da^i, dq_i, and dq^i. Owing to (M4.2c) and (M4.9) we may first write

$$\begin{aligned} &\text{(a)} \quad d\mathbf{r} = \mathbf{q}^n\,dq_n = \mathbf{a}^n\,da_n = \mathbf{q}^m\,\frac{\partial a^n}{\partial q^m}\,da_n \\ &\text{(b)} \quad d\mathbf{r} = \mathbf{a}^n\,da_n = \mathbf{q}^n\,dq_n = \mathbf{a}^m\,\frac{\partial q^n}{\partial a^m}\,dq_n \end{aligned} \tag{M4.10}$$

Scalar multiplication of (M4.10a) by the basis vector $\mathbf{q}_i$ and scalar multiplication of (M4.10b) by the basis vector $\mathbf{a}_i$ immediately gives the transformations

$$\boxed{dq_i = \frac{\partial a^n}{\partial q^i}\,da_n, \qquad da_i = \frac{\partial q^n}{\partial a^i}\,dq_n} \tag{M4.11}$$

Comparison of (M4.11) with (M4.8) shows that the differentials da_i and dq_i transform in exactly the same way as the covariant basis vectors $\mathbf{a}_i$ and $\mathbf{q}_i$.

We proceed analogously to find the relations for the differentials of the contravariant quantities da^i and dq^i. Starting with

$$\begin{aligned} &\text{(a)} \quad d\mathbf{r} = \mathbf{q}_n\,dq^n = \mathbf{a}_n\,da^n = \mathbf{q}_m\,\frac{\partial q^m}{\partial a^n}\,da^n \\ &\text{(b)} \quad d\mathbf{r} = \mathbf{a}_n\,da^n = \mathbf{q}_n\,dq^n = \mathbf{a}_m\,\frac{\partial a^m}{\partial q^n}\,dq^n \end{aligned} \tag{M4.12}$$

and multiplying (M4.12a) by $\mathbf{q}^i$ and (M4.12b) by $\mathbf{a}^i$ yields the transformations

$$\boxed{dq^i = \frac{\partial q^i}{\partial a^n}\,da^n, \qquad da^i = \frac{\partial a^i}{\partial q^n}\,dq^n} \tag{M4.13}$$

Comparison of (M4.13) with (M4.9) shows that the contravariant differentials are transformed in the same way as the corresponding contravariant basis vectors.

M4.1.3 Transformation of vectors and dyadics

The transformation relations for the covariant and the contravariant measure numbers of the vector **A** are found in the same way as for the various differentials of the previous section. First we write **A** in the four different forms

$$\begin{aligned}
&\text{(a)} \quad \mathbf{A} = \mathbf{q}^n \underset{q}{A}_n = \mathbf{a}^n \underset{a}{A}_n = \mathbf{q}^m \frac{\partial a^n}{\partial q^m} \underset{a}{A}_n \\
&\text{(b)} \quad \mathbf{A} = \mathbf{a}^n \underset{a}{A}_n = \mathbf{q}^n \underset{q}{A}_n = \mathbf{a}^m \frac{\partial q^n}{\partial a^m} \underset{q}{A}_n \\
&\text{(c)} \quad \mathbf{A} = \mathbf{q}_n \underset{q}{A}^n = \mathbf{a}_n \underset{a}{A}^n = \mathbf{q}_m \frac{\partial q^m}{\partial a^n} \underset{a}{A}^n \\
&\text{(d)} \quad \mathbf{A} = \mathbf{a}_n \underset{a}{A}^n = \mathbf{q}_n \underset{q}{A}^n = \mathbf{a}_m \frac{\partial a^m}{\partial q^n} \underset{q}{A}^n
\end{aligned} \tag{M4.14}$$

Scalar multiplication of (M4.14a) by $\mathbf{q}_i$, (M4.14b) by $\mathbf{a}_i$, (M4.14c) by $\mathbf{q}^i$, and (M4.14d) by $\mathbf{a}^i$ yields the desired results

$$\boxed{\begin{aligned}
\underset{q}{A}_i &= \frac{\partial a^n}{\partial q^i} \underset{a}{A}_n, \qquad & \underset{a}{A}_i &= \frac{\partial q^n}{\partial a^i} \underset{q}{A}_n \\
\underset{q}{A}^i &= \frac{\partial q^i}{\partial a^n} \underset{a}{A}^n, \qquad & \underset{a}{A}^i &= \frac{\partial a^i}{\partial q^n} \underset{q}{A}^n
\end{aligned}} \tag{M4.15}$$

Again, as expected, the covariant and the contravariant measure numbers of **A** transform in the same way as the corresponding basis vectors in (M4.8) and (M4.9).

The transformation of the measure numbers of $\mathbb{B}$ is a little more complex. We will derive in detail the transformation rules for the contravariant measure numbers. In the two coordinate systems $\mathbb{B}$ may be written as

$$\mathbb{B} = \underset{a}{B}^{mn} \mathbf{a}_m \mathbf{a}_n = \underset{q}{B}^{rs} \mathbf{q}_r \mathbf{q}_s = \underset{q}{B}^{rs} \frac{\partial a^u}{\partial q^r} \frac{\partial a^v}{\partial q^s} \mathbf{a}_u \mathbf{a}_v \tag{M4.16}$$

Scalar multiplication of this expression from the right first by $\mathbf{a}^j$ and then by $\mathbf{a}^i$ gives

$$\underset{a}{B}^{mn} \delta^i_m \delta^j_n = \underset{q}{B}^{rs} \frac{\partial a^u}{\partial q^r} \frac{\partial a^v}{\partial q^s} \delta^i_u \delta^j_v \tag{M4.17}$$

which may be further evaluated, yielding

$$\underset{a}{B}^{ij} = \underset{q}{B}^{rs} \frac{\partial a^i}{\partial q^r} \frac{\partial a^j}{\partial q^s} \tag{M4.18}$$

The transformation of the covariant and the mixed measure numbers proceeds analogously. All results are summarized in

$$
\begin{aligned}
&\underset{q}{B}^{ij} = \underset{a}{B}^{mn} \frac{\partial q^i}{\partial a^m} \frac{\partial q^j}{\partial a^n}, && \underset{a}{B}^{ij} = \underset{q}{B}^{mn} \frac{\partial a^i}{\partial q^m} \frac{\partial a^j}{\partial q^n} \\
&\underset{q}{B}_{ij} = \underset{a}{B}_{mn} \frac{\partial a^m}{\partial q^i} \frac{\partial a^n}{\partial q^j}, && \underset{a}{B}_{ij} = \underset{q}{B}_{mn} \frac{\partial q^m}{\partial a^i} \frac{\partial q^n}{\partial a^j} \\
&\underset{q}{B}^i{}_j = \underset{a}{B}^m{}_n \frac{\partial q^i}{\partial a^m} \frac{\partial a^n}{\partial q^j}, && \underset{a}{B}^i{}_j = \underset{q}{B}^m{}_n \frac{\partial a^i}{\partial q^m} \frac{\partial q^n}{\partial a^j} \\
&\underset{q}{B}_i{}^j = \underset{a}{B}_m{}^n \frac{\partial a^m}{\partial q^i} \frac{\partial q^j}{\partial a^n}, && \underset{a}{B}_i{}^j = \underset{q}{B}_m{}^n \frac{\partial q^m}{\partial a^i} \frac{\partial a^j}{\partial q^n}
\end{aligned}
\qquad \text{(M4.19)}
$$

Utilizing the transformation rules for the covariant basis vectors (M4.8), it is easy to find the transformation relations for the functional determinants of the q^i system and the a^i system. According to (M1.37) and (M4.8) we may write

$$
\begin{aligned}
\sqrt{\underset{q}{g}} &= [\mathbf{q}_1, \mathbf{q}_2, \mathbf{q}_3] = \left[\frac{\partial a^n}{\partial q^1} \mathbf{a}_n, \frac{\partial a^m}{\partial q^2} \mathbf{a}_m, \frac{\partial a^r}{\partial q^3} \mathbf{a}_r \right] \\
&= [\mathbf{a}_1, \mathbf{a}_2, \mathbf{a}_3] \left| \frac{\partial(a^1, a^2, a^3)}{\partial(q^1, q^2, q^3)} \right| = \sqrt{\underset{a}{g}} \left| \frac{\partial a^i}{\partial q^j} \right|
\end{aligned}
\qquad \text{(M4.20)}
$$

The inverse relation can be found in the same way or by simply interchanging the symbols q and a so that

$$
\sqrt{\underset{q}{g}} = \sqrt{\underset{a}{g}} \left| \frac{\partial a^i}{\partial q^j} \right|, \qquad \sqrt{\underset{a}{g}} = \sqrt{\underset{q}{g}} \left| \frac{\partial q^i}{\partial a^j} \right| \qquad \text{(M4.21)}
$$

Occasionally the *chain rule for the functional determinant* is of great usefulness. With the help of (M4.21) this rule can be derived very easily by recalling that, for the Cartesian system, the functional determinant $\sqrt{\underset{x}{g}} = 1$. Setting in the first and the second equation of (M4.21) $a^i = x^i$ and $q^i = x^i$, respectively, we obtain

$$
\sqrt{\underset{q}{g}} = \left| \frac{\partial x^i}{\partial q^j} \right|, \qquad \sqrt{\underset{a}{g}} = \left| \frac{\partial x^i}{\partial a^j} \right| \qquad \text{(M4.22)}
$$

Substitution of these expressions into (M4.21) gives

$$
\sqrt{\underset{q}{g}} = \left| \frac{\partial x^i}{\partial a^j} \right| \left| \frac{\partial a^k}{\partial q^l} \right|, \qquad \sqrt{\underset{a}{g}} = \left| \frac{\partial x^i}{\partial q^j} \right| \left| \frac{\partial q^k}{\partial a^l} \right| \qquad \text{(M4.23)}
$$

The interpretation of the chain rule is quite simple. The functional determinant $\sqrt{\underset{q}{g}}$ corresponds to an initial transformation from the Cartesian x^i system into the a^i system followed by the transformation into the q^i system. For $\sqrt{\underset{a}{g}}$ the interpretation is analogous.

We conclude this section by presenting the transformation rule for the measure numbers of the nabla operator which, according to (M4.2f), may be stated in various ways. Application of (M4.15) immediately yields the transformation rules

$$\boxed{\frac{\partial}{\partial q^i} = \underset{q}{\nabla}_i = \frac{\partial a^n}{\partial q^i}\underset{a}{\nabla}_n = \frac{\partial a^n}{\partial q^i}\frac{\partial}{\partial a^n}, \qquad \frac{\partial}{\partial a^i} = \underset{a}{\nabla}_i = \frac{\partial q^n}{\partial a^i}\underset{q}{\nabla}_n = \frac{\partial q^n}{\partial a^i}\frac{\partial}{\partial q^n}} \tag{M4.24}$$

which are quite useful for handling various problems.

M4.2 Transformation relations of time-dependent coordinate systems

M4.2.1 The addition theorem of the velocities

In order to derive the equations of air motion relative to the rotating earth, we must consider two coordinate systems. The first system is a time-independent *absolute coordinate system* or *inertial system*, which is assumed to be at rest with respect to the fixed stars.[1] This coordinate system will be described in terms of the Cartesian coordinates x^i. The second coordinate system, also known as the *relative coordinate system*, is time-dependent and is moving with respect to the absolute system. Motion in the relative system will be described with the help of the q^i-coordinates, which may be curvilinear and oblique. The transformation relation for the two coordinate systems is given by

$$x^i = x^i(q^1, q^2, q^3, t), \qquad i = 1, 2, 3 \tag{M4.25}$$

In contrast to the transformation relation (M4.1), we have now admitted an explicit time dependency.

Physical quantities (scalars, vectors, etc.) are invariant with respect to coordinate transformations and may be expressed either in the x^i system or in the q^i system. Two examples are

$$\begin{aligned} \psi &= \psi(x^1, x^2, x^3, t) = \psi(q^1, q^2, q^3, t) \\ \mathbf{A} &= \underset{x}{A}^n(x^1, x^2, x^3, t)\mathbf{i}_n = \underset{q}{A}^n(q^1, q^2, q^3, t)\mathbf{q}_n(q^1, q^2, q^3, t) \end{aligned} \tag{M4.26}$$

The reader should note that, in the q^i system, the basis vectors $\mathbf{q}_i$ depend not only on the coordinates q^i but also explicitly on time t.

[1] The system may also move with a constant velocity with respect to the fixed stars.

Let us consider the scalar quantity ψ which may be expressed by the functional relation

$$\psi = \psi(q^1, q^2, q^3, t) = \psi[x^1(q^1, q^2, q^3, t), x^2(q^1, q^2, q^3, t), x^3(q^1, q^2, q^3, t), t] \tag{M4.27}$$

Forming the partial derivative of this expression with respect to time, we obtain the local change with time of ψ which differs by one term in the two coordinate systems,

$$\left(\frac{\partial \psi}{\partial t}\right)_{q^i} = \left(\frac{\partial \psi}{\partial t}\right)_{x^i} + \frac{\partial \psi}{\partial x^n}\left(\frac{\partial x^n}{\partial t}\right)_{q^i} \tag{M4.28}$$

The total derivative of ψ with respect to time yields the two equations

$$\begin{aligned} \frac{d\psi}{dt} &= \left(\frac{\partial \psi}{\partial t}\right)_{x^i} + \frac{\partial \psi}{\partial x^n}\dot{x}^n \quad \text{with} \quad \dot{x}^n = \frac{dx^n}{dt} \\ \frac{d\psi}{dt} &= \left(\frac{\partial \psi}{\partial t}\right)_{q^i} + \frac{\partial \psi}{\partial q^n}\dot{q}^n \quad \text{with} \quad \dot{q}^n = \frac{dq^n}{dt} \end{aligned} \tag{M4.29}$$

Analogous expressions are also obtained for vectors and higher-degree extensive functions. The total time derivative d/dt is an invariant operator, i.e. it is independent of any particular coordinate system. Thus, d/dt expresses the *individual changes with time* of physical quantities such as temperature, density and velocity.

We now form the individual change with time of the position vector **r** expressed in the absolute system

$$\frac{d\mathbf{r}}{dt} = \mathbf{i}_n\frac{dx^n}{dt} = \mathbf{i}_n\left(\frac{\partial x^n}{\partial t}\right)_{x^i} + \mathbf{i}_n\frac{\partial x^n}{\partial x^m}\dot{x}^m = 0 + \mathbf{i}_n\dot{x}^n = \mathbf{v}_{\mathrm{A}} \tag{M4.30}$$

First we observe that the partial derivative with respect to time vanishes since the x^i-coordinates are held constant. The second part of (M4.30) leads to the introduction of the velocity in the absolute coordinate system $\mathbf{v}_{\mathrm{A}}$ which is also called the *absolute velocity*.

In the q^i-coordinate system the individual change with time of the position vector **r** is given by

$$\frac{d\mathbf{r}}{dt} = \left(\frac{\partial \mathbf{r}}{\partial t}\right)_{q^i} + \dot{q}^n\frac{\partial \mathbf{r}}{\partial q^n} = \left(\frac{\partial \mathbf{r}}{\partial t}\right)_{q^i} + \dot{q}^n\mathbf{q}_n = \mathbf{v}_P + \mathbf{v} \tag{M4.31}$$

where use of (M3.11) has been made. Now the local change with time of **r** does not vanish since the q^i system is time-dependent. Instead, the first term of (M4.31) expresses the motion of a fixed point in the q^i system moving with velocity $\mathbf{v}_P$ relative to the Cartesian system. The second part of (M4.31) represents the velocity

$\mathbf{v}$ of a fluid particle relative to the q^i system, which is known as the *relative velocity*. Combining (M4.30) and (M4.31) yields

$$\boxed{\mathbf{v}_{\mathrm{A}} = \mathbf{v}_P + \mathbf{v}} \tag{M4.32}$$

which is known as the addition theorem of the velocities.

The relative coordinate system in our case is attached to the solid earth surrounded by the atmosphere. A point fixed in this system will experience the velocity $\mathbf{v}_P$ which consists of three parts

$$\boxed{\mathbf{v}_P = \mathbf{v}_{\mathrm{T}} + \mathbf{v}_{\Omega} + \mathbf{v}_{\mathrm{D}}} \tag{M4.33}$$

The first part is the *translatory velocity* $\mathbf{v}_{\mathrm{T}}$ representing the motion of the earth around the sun. In general, for atmospheric systems the acceleration of this part of the total motion is ignored. Therefore, $\mathbf{v}_{\mathrm{T}}$ will be omitted from now on. The second part of $\mathbf{v}_P$ is the *rotational velocity* $\mathbf{v}_{\Omega}$ due to the rotation of the earth about its axis. Finally, in arbitrary time-dependent coordinate systems a point fixed on a coordinate surface q^i = constant may perform motions with respect to the corresponding x^i-coordinate of the inertial system. This may be easily recognized if the vertical coordinate q^3 is given by the pressure p. A point fixed on a pressure surface $q^3 = p$ = constant will perform vertical motions since usually pressure surfaces are not stationary. This velocity is known as the *deformation velocity* $\mathbf{v}_{\mathrm{D}}$ due to the deformation of the coordinate surfaces q^i = constant. A more detailed discussion of the velocity $\mathbf{v}_P$ will be given later.

We will conclude this section by considering the divergence of $\mathbf{v}_P$. Moreover, we will also show that $\mathbf{v}_P$ can be found with the help of the *absolute kinetic energy* of an atmospheric mass particle. According to equation (M3.11) we may first write

$$\left(\frac{\partial \mathbf{q}_i}{\partial t}\right)_{q^i} = \left[\frac{\partial}{\partial t}\left(\frac{\partial \mathbf{r}}{\partial q^i}\right)\right]_{q^i} = \left[\frac{\partial}{\partial q^i}\left(\frac{\partial \mathbf{r}}{\partial t}\right)_{q^i}\right] = \frac{\partial \mathbf{v}_P}{\partial q^i} \tag{M4.34}$$

With the help of (M4.2f) and (M4.34) we find for the divergence of $\mathbf{v}_P$

$$\begin{aligned}
\nabla\cdot\mathbf{v}_P &= \mathbf{q}^n\cdot\frac{\partial \mathbf{v}_P}{\partial q^n} = \mathbf{q}^n\cdot\left(\frac{\partial \mathbf{q}_n}{\partial t}\right)_{q^i} \\
&= \frac{\mathbf{q}_2\times\mathbf{q}_3}{\sqrt{g}_q}\cdot\left(\frac{\partial \mathbf{q}_1}{\partial t}\right)_{q^i} + \frac{\mathbf{q}_3\times\mathbf{q}_1}{\sqrt{g}_q}\cdot\left(\frac{\partial \mathbf{q}_2}{\partial t}\right)_{q^i} + \frac{\mathbf{q}_1\times\mathbf{q}_2}{\sqrt{g}_q}\cdot\left(\frac{\partial \mathbf{q}_3}{\partial t}\right)_{q^i} \\
&= \frac{1}{\sqrt{g}_q}\left(\frac{\partial}{\partial t}[\mathbf{q}_1\cdot(\mathbf{q}_2\times\mathbf{q}_3)]\right)_{q^i} = \frac{1}{\sqrt{g}_q}\left(\frac{\partial \sqrt{g}_q}{\partial t}\right)_{q^i}
\end{aligned} \tag{M4.35}$$

where use of the basic definition of the reciprocal systems (M1.55) has been made. As will be shown later, the divergence of $\mathbf{v}_\Omega$ is zero so that the divergence of $\mathbf{v}_P$ is given by

$$\boxed{\nabla\cdot\mathbf{v}_P = \frac{1}{\sqrt{\underset{q}{g}}}\left(\frac{\partial\sqrt{\underset{q}{g}}}{\partial t}\right)_{q^i} = \nabla\cdot\mathbf{v}_\mathrm{D}} \tag{M4.36}$$

From physical reasoning this result is also obvious since the angular velocity of the earth is a constant vector.

We will now show how $\mathbf{v}_P$ can be found from the absolute kinetic energy per unit mass

$$\begin{aligned} K_\mathrm{A} &= \frac{\mathbf{v}_\mathrm{A}^2}{2} = \frac{(\mathbf{v}+\mathbf{v}_P)\cdot(\mathbf{v}+\mathbf{v}_P)}{2} = \frac{\mathbf{v}^2}{2} + \mathbf{v}\cdot\mathbf{v}_P + \frac{\mathbf{v}_P^2}{2} \\ &= \frac{\dot{q}^m\dot{q}^n}{2}\mathbf{q}_m\cdot\mathbf{q}_n + \dot{q}^m\mathbf{q}_m\cdot\underset{q}{W}_n\mathbf{q}^n + \frac{\mathbf{v}_P^2}{2} \end{aligned} \tag{M4.37}$$

where $\underset{q}{W}_i$ are the covariant measure numbers of $\mathbf{v}_P$, i.e. $\mathbf{v}_P = \underset{q}{W}_m\mathbf{q}^m$. By raising the index, $\underset{q}{W}_i$ can also be found very easily, as will be demonstrated later. More briefly, we may write for K_A

$$\boxed{K_\mathrm{A} = K + K_P \quad \text{with} \quad K = \frac{\dot{q}^m\dot{q}^n}{2}\underset{q}{g}_{mn}, \qquad K_P = \dot{q}^n\underset{q}{W}_n + \frac{\mathbf{v}_P^2}{2}} \tag{M4.38}$$

which is composed of the two parts K and K_P. Thus, K is the *relative kinetic energy*, that is the part of the absolute kinetic energy which is quadratically homogeneous in the relative velocity $\dot{q}^i$ and does not contain any part of $\mathbf{v}_P$. The expression for K_P is linear in $\dot{q}^i$.

Inspection of (M4.38) shows that the metric fundamental quantities and, hence, the functional determinant $\sqrt{\underset{q}{g}}$ can be easily found from the kinetic energy K of the relative motion

$$\boxed{\underset{q}{g}_{ij} = \frac{\partial^2 K}{\partial\dot{q}^i\,\partial\dot{q}^j} \implies \sqrt{\underset{q}{g}} = \sqrt{\left|\underset{q}{g}_{ij}\right|} = \sqrt{\left|\frac{\partial^2 K}{\partial\dot{q}^i\,\partial\dot{q}^j}\right|}} \tag{M4.39}$$

where the notation $|\cdots|$ denotes the determinant. Similarly, it is possible to find the covariant measure numbers of $\mathbf{v}_P$ from K_P:

$$\boxed{\underset{q}{W}_i = \frac{\partial K_P}{\partial\dot{q}^i}} \tag{M4.40}$$

From (M4.39) and (M4.40) we conclude that the metric fundamental quantities and the velocity $\mathbf{v}_P$ of an arbitrary q^i-coordinate system are known if knowledge of the absolute kinetic energy in this system is available. In later chapters these expressions will be very useful for us.

M4.2.2 Orthogonal q^i systems

We recall from Section M1.5 that, in the case of orthogonal q^i systems, there exist some important simplifications, that is $g_{ij} = 0$ for $i \neq j$, $g_{ii}g^{ii} = 1$, and $\mathbf{e}^i = \mathbf{e}_i$. Furthermore, if orthogonal systems are employed, it is advantageous to use unit vectors instead of basis vectors. Whenever the time derivative of a unit or basis vector is taken, we should expect some relationship involving the velocity $\mathbf{v}_P$.

The local time derivative of the unit vector $\mathbf{e}_i$ follows from the local time derivative of the basis vector,

$$\left(\frac{\partial \mathbf{q}_i}{\partial t}\right)_{q^i} = \left(\frac{\partial \sqrt{g_{ii}}\mathbf{e}_i}{\partial t}\right)_{q^i} = \left(\frac{\partial \sqrt{g_{ii}}}{\partial t}\right)_{q^i} \mathbf{e}_i + \sqrt{g_{ii}}\left(\frac{\partial \mathbf{e}_i}{\partial t}\right)_{q^i} = \frac{\partial \mathbf{v}_P}{\partial q^i} \tag{M4.41}$$

Thus, we have

$$\boxed{\left(\frac{\partial \mathbf{e}_i}{\partial t}\right)_{q^i} = \frac{1}{\sqrt{g_{ii}}}\frac{\partial \mathbf{v}_P}{\partial q^i} - \frac{1}{\sqrt{g_{ii}}}\left(\frac{\partial \sqrt{g_{ii}}}{\partial t}\right)_{q^i} \mathbf{e}_i} \tag{M4.42}$$

In order to find the spatial derivatives of $\mathbf{e}_i$, we return to equation (M3.44) and introduce the condition of orthogonality. This results in

$$\begin{aligned}
\frac{\partial \mathbf{q}_i}{\partial q^j} &= \Gamma_{ij}^m \mathbf{q}_m = \frac{g^{mn}}{2}\left(\frac{\partial g_{jn}}{\partial q^i} + \frac{\partial g_{in}}{\partial q^j} - \frac{\partial g_{ij}}{\partial q^n}\right)\mathbf{q}_m \\
&= \frac{g^{nn}}{2}\left(\frac{\partial g_{jn}}{\partial q^i} + \frac{\partial g_{in}}{\partial q^j} - \frac{\partial g_{ij}}{\partial q^n}\right)\mathbf{q}_n \\
&= \frac{g^{jj}}{2}\frac{\partial g_{jj}}{\partial q^i}\mathbf{q}_j + \frac{g^{ii}}{2}\frac{\partial g_{ii}}{\partial q^j}\mathbf{q}_i - \delta_i^j \frac{g^{nn}}{2}\frac{\partial g_{ii}}{\partial q^n}\mathbf{q}_n \\
&= \frac{\mathbf{q}_j}{\sqrt{g_{jj}}}\frac{\partial \sqrt{g_{jj}}}{\partial q^i} + \frac{\mathbf{q}_i}{\sqrt{g_{ii}}}\frac{\partial \sqrt{g_{ii}}}{\partial q^j} - \delta_i^j \frac{\mathbf{q}_n}{\sqrt{g_{nn}}}\frac{\sqrt{g_{ii}}}{\sqrt{g_{nn}}}\frac{\partial \sqrt{g_{ii}}}{\partial q^n} \\
&= \mathbf{e}_j \frac{\partial \sqrt{g_{jj}}}{\partial q^i} + \mathbf{e}_i \frac{\partial \sqrt{g_{ii}}}{\partial q^j} - \delta_i^j \mathbf{e}_n \frac{\sqrt{g_{ii}}}{\sqrt{g_{nn}}}\frac{\partial \sqrt{g_{ii}}}{\partial q^n}
\end{aligned} \tag{M4.43}$$

Observing the identity

$$\frac{\partial \mathbf{q}_i}{\partial q^j} = \frac{\partial \sqrt{g_{ii}}\mathbf{e}_i}{\partial q^j} = \mathbf{e}_i \frac{\partial \sqrt{g_{ii}}}{\partial q^j} + \sqrt{g_{ii}}\frac{\partial \mathbf{e}_i}{\partial q^j} \tag{M4.44}$$

after some slight rearrangements we obtain from (M4.43)

$$\boxed{\frac{\partial \mathbf{e}_i}{\partial q^j} = \frac{\mathbf{e}_j}{\sqrt{g_{ii}}}\frac{\partial \sqrt{g_{jj}}}{\partial q^i} - \delta_i^j \frac{\mathbf{e}_n}{\sqrt{g_{nn}}}\frac{\partial \sqrt{g_{ii}}}{\partial q^n} = \mathbf{e}_j \overset{*}{\nabla}_i \sqrt{g_{jj}} - \delta_i^j \mathbf{e}_n \overset{*}{\nabla}_n \sqrt{g_{ii}}} \tag{M4.45}$$

In the last step we have also introduced the physical measure numbers of the nabla operator $\overset{*}{\nabla}_i = (1/\sqrt{g_{ii}})(\partial/\partial q^i)$.

Applications often require knowledge of the gradient of the unit vectors. With the help of (M4.45) we most easily find the required result as

$$\boxed{\nabla \mathbf{e}_i = \mathbf{q}^m \nabla_m \mathbf{e}_i = \frac{\mathbf{e}_m}{\sqrt{g_{mm}}}\frac{\partial \mathbf{e}_i}{\partial q^m} = \frac{\mathbf{e}_m}{\sqrt{g_{mm}}}\left(\mathbf{e}_m \overset{*}{\nabla}_i \sqrt{g_{mm}} - \delta_i^m \mathbf{e}_n \overset{*}{\nabla}_n \sqrt{g_{ii}}\right)} \tag{M4.46}$$

M4.2.3 The generalized vertical coordinate

Often it is of advantage to meteorological analysis to replace the vertical coordinate q^3 in the dynamic equations by the generalized coordinate ξ. According to the specific problem which might be considered, this coordinate will be specified as pressure, potential temperature, density, or some other suitable scalar function. A necessary condition for the use of the generalized vertical coordinate is the existence of a strictly monotonic relationship between q^3 and ξ. Usually the generalized coordinate depends on the horizontal and the vertical coordinates as well as on time. This becomes obvious on considering, for example, $\xi = p$. In general the pressure varies in the horizontal direction, with height, and with time.

For a scalar quantity ψ the transformation relation for going from the (q^1, q^2, q^3) system to the (q^1, q^2, ξ) system is given by

$$\psi(q^1, q^2, q^3, t) = \psi[q^1, q^2, \xi(q^1, q^2, q^3, t), t] \tag{M4.47}$$

From this equation we directly obtain the transformation equations for the partial derivatives

$$\begin{aligned}
\left(\frac{\partial \psi}{\partial t}\right)_{q^3} &= \left(\frac{\partial \psi}{\partial t}\right)_\xi + \frac{\partial \psi}{\partial \xi}\left(\frac{\partial \xi}{\partial t}\right)_{q^3} \\
\left(\frac{\partial \psi}{\partial q^1}\right)_{q^3} &= \left(\frac{\partial \psi}{\partial q^1}\right)_\xi + \frac{\partial \psi}{\partial \xi}\left(\frac{\partial \xi}{\partial q^1}\right)_{q^3} \\
\left(\frac{\partial \psi}{\partial q^2}\right)_{q^3} &= \left(\frac{\partial \psi}{\partial q^2}\right)_\xi + \frac{\partial \psi}{\partial \xi}\left(\frac{\partial \xi}{\partial q^2}\right)_{q^3} \\
\frac{\partial \psi}{\partial q^3} &= \frac{\partial \psi}{\partial \xi}\frac{\partial \xi}{\partial q^3}
\end{aligned} \tag{M4.48}$$

It is customary to state the transformation equations in the form

$$\boxed{\begin{aligned}
\left(\frac{\partial\psi}{\partial t}\right)_\xi &= \left(\frac{\partial\psi}{\partial t}\right)_{q^3} - \frac{\partial\psi}{\partial\xi}\left(\frac{\partial\xi}{\partial t}\right)_{q^3} \\
\left(\frac{\partial\psi}{\partial q^1}\right)_\xi &= \left(\frac{\partial\psi}{\partial q^1}\right)_{q^3} - \frac{\partial\psi}{\partial\xi}\left(\frac{\partial\xi}{\partial q^1}\right)_{q^3} \\
\left(\frac{\partial\psi}{\partial q^2}\right)_\xi &= \left(\frac{\partial\psi}{\partial q^2}\right)_{q^3} - \frac{\partial\psi}{\partial\xi}\left(\frac{\partial\xi}{\partial q^2}\right)_{q^3} \\
\frac{\partial\psi}{\partial\xi} &= \frac{\partial\psi}{\partial q^3}\frac{\partial q^3}{\partial\xi}
\end{aligned}} \tag{M4.49}$$

We may also write the transformation relation for ψ as

$$\psi(q^1, q^2, \xi, t) = \psi[q^1, q^2, q^3(q^1, q^2, \xi, t), t] \tag{M4.50}$$

Now the independent variables are q^1, q^2, ξ, and t. We obtain analogously

$$\boxed{\begin{aligned}
\left(\frac{\partial\psi}{\partial t}\right)_{q^3} &= \left(\frac{\partial\psi}{\partial t}\right)_\xi - \frac{\partial\psi}{\partial q^3}\frac{\partial q^3}{\partial t} \\
\left(\frac{\partial\psi}{\partial q^1}\right)_{q^3} &= \left(\frac{\partial\psi}{\partial q^1}\right)_\xi - \frac{\partial\psi}{\partial q^3}\left(\frac{\partial q^3}{\partial q^1}\right)_\xi \\
\left(\frac{\partial\psi}{\partial q^2}\right)_{q^3} &= \left(\frac{\partial\psi}{\partial q^2}\right)_\xi - \frac{\partial\psi}{\partial q^3}\left(\frac{\partial q^3}{\partial q^2}\right)_\xi \\
\frac{\partial\psi}{\partial q^3} &= \frac{\partial\psi}{\partial\xi}\frac{\partial\xi}{\partial q^3}
\end{aligned}} \tag{M4.51}$$

The terms $(\partial q^3/\partial q^i)_\xi$ are the inclinations of the ξ-surface along the q^i-directions.

M4.3 Problems

M4.1: Identify the a system in (M4.19) as the Cartesian system. The transformation relations between the Cartesian system and the rotating geographical system of the earth ($q^1 = \lambda$ – longitude, $q^2 = \varphi$ – latitude, $q^3 = r$ – radial distance) are given by

$$x^1 = r\cos\varphi\cos(\lambda + \Omega t), \quad x^2 = r\cos\varphi\sin(\lambda + \Omega t), \quad x^3 = r\sin\varphi$$

where Ω is the rotational speed of the earth.
(a) Find the covariant metric fundamental quantities g_{ij} of the geographical system and $\sqrt{\underset{q}{g}}$.
(b) Find the contravariant quantities g^{ij}.

M4.2: In the geographical system the absolute kinetic energy per unit mass is given by

$$K_{\mathrm{A}} = \frac{1}{2}\left\{r^2 \cos^2 \varphi \, (\dot{\lambda} + \Omega)^2 + r^2 \dot{\varphi}^2 + \dot{r}^2\right\}$$

(a) By utilizing this equation, verify the results of problem M4.1.
(b) Find the contravariant measure numbers of $\mathbf{v}_{\mathrm{p}} = \mathbf{v}_{\Omega}$.

M4.3: Find an expression for the functional determinant $\sqrt{\underset{\xi}{g}}$ in the (q^1, q^2, ξ) system, where ξ is the generalized vertical coordinate.

M5

The method of covariant differentiation

The numerical investigation of specific meteorological problems requires the selection of a suitable coordinate system. In many cases the best choice is quite obvious. Attempts to use the same coordinate system for entirely different geometries usually introduce additional mathematical complexities, which should be avoided. For example, it is immediately apparent that the rectangular Cartesian system is not well suited for the treatment of problems with spherical symmetry. The inspection of the metric fundamental quantities g_{ij} or g^{ij} and their derivatives helps to decide which coordinate system is best suited for the solution of a particular problem. The study of the motion in irregular terrain may require a terrain-following coordinate system. However, it is not clear from the beginning whether the motion is best described in terms of covariant or contravariant measure numbers.

From the thermo-hydrodynamic system of equations, consisting of the dynamic equations, the continuity equation, the heat equation, and the equation of state, we will direct our attention mostly to the equation of motion using covariant and contravariant measure numbers. We will also briefly derive the continuity equation in general coordinates. In addition we will derive the equation of motion using physical measure numbers of the velocity components if the curvilinear coordinate lines are orthogonal.

In order to proceed efficiently, it is best to extend the tensor-analytical treatment presented in the previous chapters by introducing the method of covariant differentiation. What may seem strange and difficult to begin with is, in fact, a very easy and efficient mathematical treatment that requires no additional theory. Our discussion in this section will necessarily be quite formal.

M5.1 Spatial differentiation of vectors and dyadics

The situation is particularly simple if we consider the differentiation of a vector **A** in a rectangular Cartesian coordinate system:

$$\underset{x}{\nabla}_i \mathbf{A} = \underset{x}{\nabla}_i (A^m \mathbf{i}_m) = \mathbf{i}_m \underset{x}{\nabla}_i A^m, \qquad \underset{x}{\nabla}_i = \frac{\partial}{\partial x^i} \tag{M5.1}$$

In this case the unit vector $\mathbf{i}_j$ is independent of position on a particular coordinate line or axis and thus may be extracted and placed in front of the differentiation operator. Suppose that we express the same vector $\mathbf{A}$ in terms of contravariant measure numbers and covariant basis vectors whose directions change with position on a coordinate line. In this case we cannot simply extract the basis vector $\mathbf{q}_i$ without making a serious mistake. Nevertheless, guided by (M5.1), we still go ahead and extract the basis vector and correct the error we have made by formally changing the ordinary differential operator ∇_i to the so-called *covariant differential operator* $\nabla\!\!\!\!\nabla_i$:

$$\nabla_i \mathbf{A} = \nabla_i (A^m \mathbf{q}_m) = \mathbf{q}_m \nabla\!\!\!\!\nabla_i A^m, \qquad \nabla_i = \frac{\partial}{\partial q^i} \tag{M5.2}$$

This operation is of course formal and of no help unless we find a relation between the two types of differential operators ∇_i and $\nabla\!\!\!\!\nabla_i$. To establish the relation between the two types of partial derivatives, we first carry out the ordinary differentiation, using (M3.43), and then introduce the covariant derivative as

$$\begin{aligned}\nabla_i (A^m \mathbf{q}_m) &= \mathbf{q}_m \nabla_i A^m + A^m \nabla_i \mathbf{q}_m = \mathbf{q}_m \nabla_i A^m + A^m \Gamma^n_{mi} \mathbf{q}_n \\ &= \mathbf{q}_n \left(\nabla_i A^n + A^m \Gamma^n_{mi} \right) = \mathbf{q}_n \nabla\!\!\!\!\nabla_i A^n\end{aligned} \tag{M5.3}$$

Scalar multiplication by the vector $\mathbf{q}^k$ gives the required relationship between the two differential operators operating on the contravariant measure number A^k:

$$\boxed{\nabla\!\!\!\!\nabla_i A^k = \nabla_i A^k + A^m \Gamma^k_{mi}} \tag{M5.4}$$

The result is that the covariant differential operator $\nabla\!\!\!\!\nabla_i$ applied to A^k is equal to the ordinary differential operator applied to A^k plus an additional term involving the Christoffel symbol. The covariant derivative is of tensorial character whereas the ordinary spatial derivatives are not tensorial expressions.

We may proceed in the same way if $\mathbf{A}$ is expressed in terms of covariant measure numbers. Employing (M3.48) we find

$$\begin{aligned}\nabla_i (A_m \mathbf{q}^m) &= \mathbf{q}^m \nabla_i A_m + A_m \nabla_i \mathbf{q}^m = \mathbf{q}^m \nabla_i A_m - A_m \Gamma^m_{in} \mathbf{q}^n \\ &= \mathbf{q}^n \left(\nabla_i A_n - A_m \Gamma^m_{in} \right) = \mathbf{q}^m \nabla\!\!\!\!\nabla_i A_m\end{aligned} \tag{M5.5}$$

Scalar multiplication by the covariant basis vector $\mathbf{q}_k$ leads to the desired relation

$$\boxed{\nabla\!\!\!\!\nabla_i A_k = \nabla_i A_k - A_m \Gamma^m_{ik}} \tag{M5.6}$$

Thus, the covariant operator is defined in different ways depending on the coordinate definition of the vector **A**. If we choose contravariant measure numbers to represent **A**, then we must apply (M5.4). Expressing **A** in terms of covariant measure numbers requires the use of (M5.6).

So far we have considered only individual components or measure numbers of **A**. Next we consider the local dyadic $\nabla\mathbf{A}$:

$$\nabla\mathbf{A} = \mathbf{q}^n\,\nabla_n(A^m\mathbf{q}_m) = \mathbf{q}^n\,\nabla_n(A_m\mathbf{q}^m) \tag{M5.7}$$

where we may select either covariant or contravariant measure numbers to represent **A**. Application of the covariant differential operator leads to

$$\nabla\mathbf{A} = \mathbf{q}^n\mathbf{q}_m\,\nabla\!\!\!\!\nabla_n A^m = \mathbf{q}^n\mathbf{q}^m\,\nabla\!\!\!\!\nabla_n A_m \tag{M5.8}$$

where $\nabla\!\!\!\!\nabla_i A^k$ or $\nabla\!\!\!\!\nabla_i A_k$ are measure numbers of the local dyadic $\nabla\mathbf{A}$.

Finally we wish to remark that, for the covariant and contravariant (not yet defined) derivatives, the sum and the product rules of ordinary differential calculus are valid for tensors of any rank that are free from basis vectors

$$\nabla\!\!\!\!\nabla_i(XY) = X\,\nabla\!\!\!\!\nabla_i Y + (\nabla\!\!\!\!\nabla_i X)Y \tag{M5.9}$$

where X, Y are tensors of arbitrary rank. Recall that the rank of a tensor is fixed by the numbers of indices attached to the measure number.

Now we are going to discuss the covariant differentiation of the local triadic which in our case is the gradient of a dyadic. What at first glance looks involved and difficult is, in fact, a very easy operation. The formal covariant differentiation is introduced in

$$\nabla\mathbb{B} = \mathbf{q}^s\,\nabla_s(B_{mn}\mathbf{q}^m\mathbf{q}^n) = \mathbf{q}^s\mathbf{q}^m\mathbf{q}^n\,\nabla\!\!\!\!\nabla_s B_{mn} \tag{M5.10}$$

No further comment is needed. The ordinary differentiation is carried out in

$$\mathbf{q}^s\,\nabla_s\mathbb{B} = \mathbf{q}^s\mathbf{q}^m\mathbf{q}^n\,\nabla_s B_{mn} + B_{mn}\mathbf{q}^s\,\nabla_s(\mathbf{q}^m)\mathbf{q}^n + B_{mn}\mathbf{q}^s\mathbf{q}^m\,\nabla_s(\mathbf{q}^n) \tag{M5.11}$$

making sure that the sequence of the basis vectors $\mathbf{q}^i$ is not changed. The second and the third term are rewritten yielding

$$\begin{aligned}\mathbf{q}^s\,\nabla_s\mathbb{B} &= \mathbf{q}^s\mathbf{q}^m\mathbf{q}^n\,\nabla_s B_{mn} + B_{mn}\mathbf{q}^s\left(-\mathbf{q}^r\Gamma_{sr}^{m}\right)\mathbf{q}^n + B_{mn}\mathbf{q}^s\mathbf{q}^m\left(-\mathbf{q}^r\Gamma_{sr}^{n}\right)\\ &= \mathbf{q}^s\mathbf{q}^m\mathbf{q}^n\left(\nabla_s B_{mn} - B_{rn}\Gamma_{sm}^{r} - B_{mr}\Gamma_{sn}^{r}\right)\end{aligned} \tag{M5.12}$$

In the last expression of (M5.12) some indices have been rewritten with the goal of extracting the common factor $\mathbf{q}^s\mathbf{q}^m\mathbf{q}^n$. This operation is quite valid since the summation indices may be given any name without changing the meaning of the expression. According to (M5.10) the left-hand side of (M5.12) can be written as $\mathbf{q}^s\mathbf{q}^m\mathbf{q}^n\, \overline{\nabla}_s B_{mn}$ so that the sequence of contravariant basis vectors appears on both sides of this equation. Scalar multiplication from the right with $\mathbf{q}_j$, $\mathbf{q}_i$, $\mathbf{q}_k$ then leads to

$$\overline{\nabla}_k B_{ij} = \nabla_k B_{ij} - B_{nj}\Gamma^n_{ki} - B_{in}\Gamma^n_{kj} \tag{M5.13}$$

In the same way we find the covariant derivatives of the contravariant and mixed measure numbers so that we have

$$\boxed{\begin{aligned} &\overline{\nabla}_k B_{ij} = \nabla_k B_{ij} - B_{nj}\Gamma^n_{ki} - B_{in}\Gamma^n_{kj}, && \overline{\nabla}_k B^{ij} = \nabla_k B^{ij} + B^{nj}\Gamma^i_{kn} + B^{in}\Gamma^j_{kn} \\ &\overline{\nabla}_k B_i^{\ j} = \nabla_k B_i^{\ j} - B_n^{\ j}\Gamma^n_{ki} + B_i^{\ n}\Gamma^j_{kn}, && \overline{\nabla}_k B^i_{\ j} = \nabla_k B^i_{\ j} + B^n_{\ j}\Gamma^i_{kn} - B^i_{\ n}\Gamma^n_{kj} \end{aligned}} \tag{M5.14}$$

Let us now direct our attention to some operations involving the unit dyadic $\mathbb{E}$, which is an invariant operator in space and time. Therefore we may write for the gradient of $\mathbb{E}$

$$\nabla\mathbb{E} = \mathbf{q}^s\, \nabla_s\mathbb{E} = 0, \qquad \mathbb{E} = g_{mn}\mathbf{q}^m\mathbf{q}^n = g^{mn}\mathbf{q}_m\mathbf{q}_n = \delta_m^{\ n}\mathbf{q}^m\mathbf{q}_n = \delta^m_{\ n}\mathbf{q}_m\mathbf{q}^n \tag{M5.15}$$

The various representations of $\mathbb{E}$ are reviewed in this equation. We now wish to find out how the metric fundamental quantities g_{ij} and g^{ij} are affected by covariant differentiation. By simply introducing the covariant derivative analogously to (M5.10),

$$\begin{aligned} &\mathbf{q}^s\, \nabla_s(g_{mn}\mathbf{q}^m\mathbf{q}^n) = \mathbf{q}^s\mathbf{q}^m\mathbf{q}^n\, \overline{\nabla}_s g_{mn} = 0, && \mathbf{q}^s\, \nabla_s(g^{mn}\mathbf{q}_m\mathbf{q}_n) = \mathbf{q}^s\mathbf{q}_m\mathbf{q}_n\, \overline{\nabla}_s g^{mn} = 0 \\ &\mathbf{q}^s\, \nabla_s\left(\delta_m^{\ n}\mathbf{q}^m\mathbf{q}_n\right) = \mathbf{q}^s\mathbf{q}^m\mathbf{q}_n\, \overline{\nabla}_s\delta_m^{\ n} = 0, && \mathbf{q}^s\, \nabla_s\left(\delta^m_{\ n}\mathbf{q}_m\mathbf{q}^n\right) = \mathbf{q}^s\mathbf{q}_m\mathbf{q}^n\, \overline{\nabla}_s\delta^m_{\ n} = 0 \end{aligned} \tag{M5.16}$$

we find that the metric fundamental quantities g_{ij} and g^{ij} may be treated as constants in covariant differentiation. The results, summarized in

$$\boxed{\overline{\nabla}_k g_{ij} = 0, \qquad \overline{\nabla}_k g^{ij} = 0, \qquad \overline{\nabla}_k \delta_i^{\ j} = 0, \qquad \overline{\nabla}_k \delta^i_{\ j} = 0} \tag{M5.17}$$

will be very helpful and important in our studies. For instance, often it is necessary to raise or lower the indices of measure numbers. Utilizing (M5.17) we find

$$\boxed{\begin{aligned} \overline{\nabla}_i A_k &= \overline{\nabla}_i(g_{kn}A^n) = g_{kn}\, \overline{\nabla}_i A^n \\ \overline{\nabla}_i A^k &= \overline{\nabla}_i(g^{kn}A_n) = g^{kn}\, \overline{\nabla}_i A_n \end{aligned}} \tag{M5.18}$$

We are now ready to consider the gradient of the dyadic $\mathbb{B}$ and the divergence of $\mathbb{B}$ in terms of the covariant derivatives. It is sometimes advantageous, though not necessary, to involve the mixed measure numbers in order to apply the properties of the reciprocal basis vectors. The covariant derivative is introduced in

$$\nabla \mathbb{B} = \mathbf{q}^s\, \nabla_s \left(B^m_{\ n} \mathbf{q}_m \mathbf{q}^n\right) = \mathbf{q}^s \mathbf{q}_m \mathbf{q}^n\, \forall_s B^m_{\ n} \tag{M5.19}$$

so that the divergence of the dyadic $\mathbb{B}$ is given by

$$\nabla \cdot \mathbb{B} = \mathbf{q}^s \cdot \mathbf{q}_m \mathbf{q}^n\, \forall_s B^m_{\ n} \ = \mathbf{q}^n\, \forall_m B^m_{\ n} \tag{M5.20}$$

which is a rather simple expression.

We conclude this section by applying (M5.13) to a situation of considerable importance. For the special case $B_{ij} = g_{ij}$, recalling that the metric fundamental quantity is a constant in covariant differentiation, we obtain

$$\begin{aligned} &\forall_k g_{ij} = \nabla_k g_{ij} - g_{nj} \Gamma^n_{ki} - g_{in} \Gamma^n_{kj} = 0 \\ &\Longrightarrow \boxed{\nabla_k g_{ij} = \Gamma_{kij} + \Gamma_{kji}} \end{aligned} \tag{M5.21}$$

Finally we make use of the tensor properties of the covariant derivative to introduce *contravariant differentiation*. By raising or by lowering the index we find

$$\boxed{\forall_k = g_{kn}\, \forall^n, \qquad \forall^k = g^{kn}\, \forall_n} \tag{M5.22}$$

It may be appropriate to give an example employing the contravariant derivative for two representations of the vector $\mathbf{A}$. With the help of (M5.22) we may replace the contravariant derivative by the more common covariant derivative:

$$\begin{aligned} \nabla \mathbf{A} &= \mathbf{q}_n\, \nabla^n (A_m \mathbf{q}^m) = \mathbf{q}_n \mathbf{q}^m\, \forall^n A_m = \mathbf{q}_n \mathbf{q}^m g^{ns}\, \forall_s A_m \\ \nabla \mathbf{A} &= \mathbf{q}_n\, \nabla^n (A^m \mathbf{q}_m) = \mathbf{q}_n \mathbf{q}_m\, \forall^n A^m = \mathbf{q}_n \mathbf{q}_m g^{ns}\, \forall_s A^m \end{aligned} \tag{M5.23}$$

M5.2 Time differentiation of vectors and dyadics

Let us consider the velocity of a point relative to the absolute system as defined by (M4.33) with $\mathbf{v}_{\mathrm{T}} = 0$. If the point is fixed somewhere in the atmosphere it participates in the rotation of the earth so that $\mathbf{v}_P = \mathbf{v}_\Omega$. If, for example, the vertical coordinate of this point is located on a pressure surface, then this point also participates in the vertical motion of the material surface as described by the deformation velocity $\mathbf{v}_{\mathrm{D}}$. If the measure numbers of the rotational and deformational velocities are given by $\underset{\Omega}{W_i}$ and $\underset{\mathrm{D}}{W_i}$ then the change with time of the basis vector is given by

$$\left(\frac{\partial \mathbf{q}_i}{\partial t}\right)_{q^k} = \frac{\partial}{\partial t}\left(\frac{\partial \mathbf{r}}{\partial q^i}\right)_{q^k} = \nabla_i \mathbf{v}_P = \nabla_i (\underset{\Omega}{W_n} \mathbf{q}^n + \underset{\mathrm{D}}{W_n} \mathbf{q}^n) = \mathbf{q}^n\, \forall_i (\underset{\Omega}{W_n} + \underset{\mathrm{D}}{W_n}) \tag{M5.24}$$

where we have also applied the definition (M5.2) of the covariant differentiation.

For simplicity the notation $(\ )_{q^i}$ denoting that the local time derivative is taken at constant coordinates q^i will henceforth be omitted, except where confusion may occur. Next we consider some relationships involving the time differentiation of the reciprocal basis vectors. From the definition

$$\mathbf{q}^i \cdot \mathbf{q}_j = 0 \quad \text{for} \quad i \neq j \tag{M5.25}$$

we obtain

$$\mathbf{q}^i \cdot \frac{\partial \mathbf{q}_j}{\partial t} + \mathbf{q}_j \cdot \frac{\partial \mathbf{q}^i}{\partial t} = 0 \tag{M5.26}$$

Dyadic multiplication with $\mathbf{q}^j$ and then summing over j, i.e. replacing j by n, results in the introduction of the unit dyadic $\mathbb{E}$

$$\mathbf{q}^n \mathbf{q}^i \cdot \frac{\partial \mathbf{q}_n}{\partial t} + \mathbb{E} \cdot \frac{\partial \mathbf{q}^i}{\partial t} = 0 \tag{M5.27}$$

Since $\mathbb{E}$ is invariant in space and time, we may place $\mathbb{E}$ inside of the differential operator and perform the scalar multiplication, yielding

$$\begin{aligned}
\frac{\partial \mathbf{q}^i}{\partial t} &= -\mathbf{q}^n \mathbf{q}^i \cdot \frac{\partial \mathbf{q}_n}{\partial t} = -\mathbf{q}^n \mathbf{q}^i \cdot \nabla_n \mathbf{v}_{\mathrm{P}} = -\mathbf{q}^n \mathbf{q}^i \cdot \mathbf{q}_m \nabla_n W^m \\
&= -\mathbf{q}^n \nabla_n W^i = -\mathbf{q}^n \nabla_n(\underset{\Omega}{W^i} + \underset{\mathrm{D}}{W^i}) = -\mathbf{q}_n \nabla^n(\underset{\Omega}{W^i} + \underset{\mathrm{D}}{W^i}) \\
&\text{with} \quad \mathbf{q}^n \nabla_n = g^{nm} \mathbf{q}_m \nabla_n = \mathbf{q}_m \nabla^m
\end{aligned} \tag{M5.28}$$

We will now summarize some important results. From (M5.24) we find

$$\boxed{\mathbf{q}_j \cdot \frac{\partial \mathbf{q}_i}{\partial t} = \nabla_i(\underset{\Omega}{W_j} + \underset{\mathrm{D}}{W_j}), \qquad \mathbf{q}^j \cdot \frac{\partial \mathbf{q}_i}{\partial t} = \nabla_i(\underset{\Omega}{W^j} + \underset{\mathrm{D}}{W^j})} \tag{M5.29}$$

and from (M5.28) we obtain

$$\boxed{\mathbf{q}_j \cdot \frac{\partial \mathbf{q}^i}{\partial t} = -\nabla_j(\underset{\Omega}{W^i} + \underset{\mathrm{D}}{W^i}), \qquad \mathbf{q}^j \cdot \frac{\partial \mathbf{q}^i}{\partial t} = -\nabla^j(\underset{\Omega}{W^i} + \underset{\mathrm{D}}{W^i})} \tag{M5.30}$$

Let us now perform the local time differentiation of an arbitrary vector $\mathbf{A}$. Extracting the basis vector in analogy to the covariant nabla operator

$$\frac{\partial \mathbf{A}}{\partial t} = \frac{\partial}{\partial t}(A_n \mathbf{q}^n) = \mathbf{q}^n \frac{\eth A_n}{\eth t} \tag{M5.31}$$

leads to the introduction of the *covariant time operator* $\eth/\eth t$ which must now be related to the ordinary local time derivative. This is most easily done by carrying out the differentiation in (M5.31) and replacing $\partial \mathbf{q}^n/\partial t$ by (M5.28), yielding

$$\mathbf{q}^n \frac{\eth A_n}{\eth t} = \mathbf{q}^n \frac{\partial A_n}{\partial t} + A_n \Big[-\mathbf{q}^m \nabla_m(\underset{\Omega}{W^n} + \underset{\mathrm{D}}{W^n}) \Big] \tag{M5.32}$$

The covariant spatial operators appearing in this equation have previously been defined by using the ordinary spatial derivatives so that (M5.32) can be evaluated entirely in terms of ordinary derivatives. Scalar multiplication of both sides by $\mathbf{q}_i$ results in the definition of the covariant time differentiation of the covariant measure number A_i:

$$\boxed{\frac{\partial A_i}{\partial t} = \frac{\partial A_i}{\partial t} - A_n \nabla_i(\underset{\Omega}{W^n} + \underset{\mathrm{D}}{W^n})} \tag{M5.33}$$

By precisely the same procedure we find

$$\boxed{\frac{\partial A^i}{\partial t} = \frac{\partial A^i}{\partial t} + A^n \nabla_n(\underset{\Omega}{W^i} + \underset{\mathrm{D}}{W^i})} \tag{M5.34}$$

giving the equation for the covariant time differentiation of the contravariant measure number A^i.

Of particular importance is the covariant time differentiation of the metric fundamental quantities g_{ij} and g^{ij}. The starting point of the analysis, as might be expected by now, is the time-invariant unit dyadic $\mathbb{E}$. The mathematical operation is given by

$$\frac{\partial \mathbb{E}}{\partial t} = 0 \Longrightarrow \frac{\partial}{\partial t}(\mathbf{q}^n \mathbf{q}_n) = \frac{\partial}{\partial t}(g^{nr} \mathbf{q}_r \mathbf{q}_n) = \mathbf{q}_r \mathbf{q}_n \frac{\partial g^{nr}}{\partial t} = 0 \tag{M5.35}$$

showing that the metric fundamental quantity g^{ij} is a constant in covariant time differentiation. Proceeding analogously we find the same result for g_{ij} so that

$$\boxed{\frac{\partial g^{ij}}{\partial t} = 0, \qquad \frac{\partial g_{ij}}{\partial t} = 0} \tag{M5.36}$$

From these expressions follow the important relations

$$\boxed{\begin{aligned} \frac{\partial A_i}{\partial t} &= \frac{\partial}{\partial t}(g_{in} A^n) = g_{in} \frac{\partial A^n}{\partial t} \\ \frac{\partial A^i}{\partial t} &= \frac{\partial}{\partial t}(g^{in} A_n) = g^{in} \frac{\partial A_n}{\partial t} \end{aligned}} \tag{M5.37}$$

By now it should be apparent that operations with covariant derivatives are easy and very convenient. Otherwise, the many differential relations involving basis vectors would be very tedious to apply.

Now we need to introduce an expression for the individual covariant differential time operator. This expression is very simply obtained since we have already provided the necessary ingredients. In analogy to the definition of the individual time derivative

$$\frac{d}{dt} = \frac{\partial}{\partial t} + \dot{q}^n \frac{\partial}{\partial q^n} = \frac{\partial}{\partial t} + v^n \frac{\partial}{\partial q^n} = \frac{\partial}{\partial t} + v^n \nabla_n \tag{M5.38}$$

we may define the individual covariant time derivative by

$$\frac{\mathcal{d}}{dt} = \frac{\boldsymbol{\partial}}{\partial t} + v^n \boldsymbol{\nabla}_n \tag{M5.39}$$

Using (M5.6) and (M5.33) we immediately obtain

$$\frac{\mathcal{d} A_k}{dt} = \frac{\boldsymbol{\partial} A_k}{\partial t} + v^n \boldsymbol{\nabla}_n A_k = \frac{\partial A_k}{\partial t} - A_n \boldsymbol{\nabla}_k (\underset{\Omega}{W^n} + \underset{\mathrm{D}}{W^n}) + v^n \big(\nabla_n A_k - A_m \Gamma_{nk}^m\big) \tag{M5.40}$$

which is the desired expression. Finally, replacing the covariant spatial derivative, using (M5.4), gives

$$\boxed{\frac{\mathcal{d} A_k}{dt} = \frac{\partial A_k}{\partial t} - A_n \Big[\nabla_k (\underset{\Omega}{W^n} + \underset{\mathrm{D}}{W^n}) + (\underset{\Omega}{W^m} + \underset{\mathrm{D}}{W^m}) \Gamma_{mk}^n\Big] + v^n \big(\nabla_n A_k - A_m \Gamma_{nk}^m\big)} \tag{M5.41}$$

The involvement of the measure numbers of $\mathbf{v}_P$ should not be surprising in view of (M5.33). Similarly, the application of (M5.39) to the contravariant measure number A^k leads to

$$\boxed{\frac{\mathcal{d} A^k}{dt} = \frac{\partial A^k}{\partial t} + A^n \Big[\nabla_n (\underset{\Omega}{W^k} + \underset{\mathrm{D}}{W^k}) + (\underset{\Omega}{W^m} + \underset{\mathrm{D}}{W^m}) \Gamma_{nm}^k\Big] + v^n \big(\nabla_n A^k + A^m \Gamma_{nm}^k\big)} \tag{M5.42}$$

M5.3 The local dyadic of $\mathbf{v}_P$

The representation of the motion in general coordinate systems requires a suitable description of the local dyadic of $\mathbf{v}_P = \mathbf{v}_\Omega + \mathbf{v}_\mathrm{D}$. First, we split the local dyadic of $\mathbf{v}_P$ into its symmetric and antisymmetric parts. Utilizing the conjugated dyadic $\widetilde{\nabla \mathbf{v}_P}$ of $\nabla \mathbf{v}_P$, we may write according to (M2.27)

$$\nabla \mathbf{v}_P = \tfrac{1}{2}(\nabla \mathbf{v}_P + \widetilde{\nabla \mathbf{v}_P}) + \tfrac{1}{2}(\nabla \mathbf{v}_P - \widetilde{\nabla \mathbf{v}_P}) = \mathbb{D}_P + \mathfrak{R}_P \tag{M5.43}$$

The symmetric part $\mathbb{D}_P$ of (M5.43) is given by

$$\begin{aligned}\mathbb{D}_P &= \tfrac{1}{2}(\nabla \mathbf{v}_\Omega + \widetilde{\nabla \mathbf{v}_\Omega}) + \tfrac{1}{2}(\nabla \mathbf{v}_\mathrm{D} + \widetilde{\nabla \mathbf{v}_\mathrm{D}}) = \tfrac{1}{2}(\nabla \mathbf{v}_\mathrm{D} + \widetilde{\nabla \mathbf{v}_\mathrm{D}}) \\ &= \tfrac{1}{2}\mathbf{q}^m\,\nabla_m(\underset{\mathrm{D}}{W_n}\mathbf{q}^n) + \tfrac{1}{2}\widetilde{\mathbf{q}^n\,\nabla_n(\underset{\mathrm{D}}{W_m}\mathbf{q}^m)} = \tfrac{1}{2}(\mathbf{q}^m\mathbf{q}^n\,\forall_m \underset{\mathrm{D}}{W_n} + \widetilde{\mathbf{q}^n\mathbf{q}^m}\,\forall_n \underset{\mathrm{D}}{W_m}) \\ &= \tfrac{1}{2}(\mathbf{q}^m\mathbf{q}^n\,\forall_m \underset{\mathrm{D}}{W_n} + \mathbf{q}^m\mathbf{q}^n\,\forall_n \underset{\mathrm{D}}{W_m}) = \tfrac{1}{2}\mathbf{q}^m\mathbf{q}^n(\forall_m \underset{\mathrm{D}}{W_n} + \forall_n \underset{\mathrm{D}}{W_m}) = d_{mn}\mathbf{q}^m\mathbf{q}^n\end{aligned} \tag{M5.44}$$

The term $\nabla \mathbf{v}_\Omega + \widetilde{\nabla \mathbf{v}_\Omega}$ vanishes because $\nabla \mathbf{v}_\Omega$ is an antisymmetric tensor. The measure number d_{ij} is best obtained from (M5.44) by scalar multiplication first by $\mathbf{q}_i$ and then by $\mathbf{q}_j$. The result is

$$d_{ij} = \tfrac{1}{2}(\forall_i \underset{\mathrm{D}}{W_j} + \forall_j \underset{\mathrm{D}}{W_i}) \tag{M5.45}$$

Using scalar multiplication to obtain d_{ij} insures against errors, while by comparison the suffices i and j might easily be misplaced.

The antisymmetric dyadic $\mathfrak{R}_P$ is given by

$$\mathfrak{R}_P = \tfrac{1}{2}(\nabla \mathbf{v}_\Omega - \widetilde{\nabla \mathbf{v}_\Omega}) + \tfrac{1}{2}(\nabla \mathbf{v}_\mathrm{D} - \widetilde{\nabla \mathbf{v}_\mathrm{D}}) = \omega_{mn}\mathbf{q}^m\mathbf{q}^n \tag{M5.46}$$

Analogously to d_{ij} we find the measure numbers ω_{ij}:

$$\begin{aligned}\omega_{ij} &= \tfrac{1}{2}(\forall_i \underset{\Omega}{W_j} - \forall_j \underset{\Omega}{W_i}) + \tfrac{1}{2}(\forall_i \underset{\mathrm{D}}{W_j} - \forall_j \underset{\mathrm{D}}{W_i}) \\ &= \tfrac{1}{2}(\nabla_i \underset{\Omega}{W_j} - \nabla_j \underset{\Omega}{W_i}) + \tfrac{1}{2}(\nabla_i \underset{\mathrm{D}}{W_j} - \nabla_j \underset{\mathrm{D}}{W_i}) = \underset{\Omega}{\omega_{ij}} + \underset{\mathrm{D}}{\omega_{ij}}\end{aligned} \tag{M5.47}$$

The covariant derivatives in (M5.47) have been replaced with the help of (M5.6). The abbreviations $\underset{\Omega}{\omega_{ij}}$ and $\underset{\mathrm{D}}{\omega_{ij}}$, which have been introduced in (M5.47), are given by

$$\boxed{\begin{aligned}\underset{\Omega}{\omega_{ij}} &= \tfrac{1}{2}(\forall_i \underset{\Omega}{W_j} - \forall_j \underset{\Omega}{W_i}) = \tfrac{1}{2}(\nabla_i \underset{\Omega}{W_j} - \nabla_j \underset{\Omega}{W_i}) \\ \underset{\mathrm{D}}{\omega_{ij}} &= \tfrac{1}{2}(\forall_i \underset{\mathrm{D}}{W_j} - \forall_j \underset{\mathrm{D}}{W_i}) = \tfrac{1}{2}(\nabla_i \underset{\mathrm{D}}{W_j} - \nabla_j \underset{\mathrm{D}}{W_i})\end{aligned}} \tag{M5.48}$$

Furthermore, from (M5.47) we obtain the useful relations

$$\begin{aligned}\forall_i \underset{\Omega}{W_j} &= 2\underset{\Omega}{\omega_{ij}} + \forall_j \underset{\Omega}{W_i} \quad \text{or} \quad \nabla_i \underset{\Omega}{W_j} = 2\underset{\Omega}{\omega_{ij}} + \nabla_j \underset{\Omega}{W_i} \\ \forall_i \underset{\mathrm{D}}{W_j} &= 2\underset{\mathrm{D}}{\omega_{ij}} + \forall_j \underset{\mathrm{D}}{W_i} \quad \text{or} \quad \nabla_i \underset{\mathrm{D}}{W_j} = 2\underset{\mathrm{D}}{\omega_{ij}} + \nabla_j \underset{\mathrm{D}}{W_i}\end{aligned} \tag{M5.49}$$

M5.4 Problems

M5.1: In (M5.14) only the first expression has been derived. Verify the remaining expressions.

M6

Integral operations

M6.1 Curves, surfaces, and volumes in the general q^i system

An arbitrary curve in space may be viewed as a coordinate line q^1 of a curvilinear coordinate system. Therefore, the differential increment $d\mathbf{r}$ along this curve can be written in the form

$$d\mathbf{r} = dq^1\,\mathbf{q}_1 = dq^1\,\sqrt{g_{11}}\mathbf{e}_1 = d\overset{*}{q}{}^1\,\mathbf{e}_1 = dr\,\mathbf{e}_1 \tag{M6.1}$$

where the basis vector $\mathbf{q}_1$ is tangential to the coordinate line q^1 at a given point P as shown in Figure M6.1. Instead of using the arbitrary basis vector $\mathbf{q}_1$, we may also employ the unit vector $\mathbf{e}_1$ which is identical to the unit tangent vector at P. This leads to the introduction of the physical measure number $d\overset{*}{q}{}^1$ of the vector $d\mathbf{r}$ which is equivalent to the differential arclength dr.

Now we wish to discuss very briefly the geometrical meaning of a coordinate line. For simplicity let us first consider the Cartesian coordinate system. It is immediately apparent that the intersection of the surfaces y = constant and z = constant produces a straight line that may be chosen as the x-axis. The y- and z-axes are found by intersecting the surfaces x = constant, z = constant and x = constant, y = constant, respectively. Analogously, in the curvilinear coordinate system the intersection of two surfaces q^i = constant, q^j = constant yields the coordinate line of the corresponding third coordinate q^k.

An arbitrary surface in three-dimensional space may be defined by two coordinate lines q^1 and q^2 with q^3 = constant; see Figure M6.2. Since $\mathbf{q}_1$ and $\mathbf{q}_2$ are tangent vectors to the coordinate lines q^1 and q^2, they span a tangential plane to the surface q^3 = constant. The vector $\mathbf{q}_3$ is a tangent vector to the q^3-coordinate line and $\mathbf{e}_3 = \mathbf{q}_3/\sqrt{g_{33}}$ is the corresponding unit vector. In our investigation we set the unit vector $\mathbf{e}_3$ equal to the unit normal $\mathbf{e}_N$ of the surface q^3 = constant. Hence, we are dealing with a *semi-orthogonal coordinate system*. In this system $\mathbf{e}_3$ is perpendicular to the basis vectors $\mathbf{q}_1$ and $\mathbf{q}_2$ so that $\mathbf{e}_3$ is also parallel to the

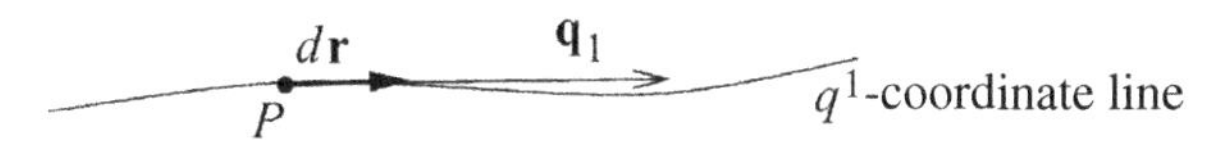

Fig. M6.1 The q^1-coordinate line in space, corresponding basis vector $\mathbf{q}_1$, and position vector $d\mathbf{r}$.

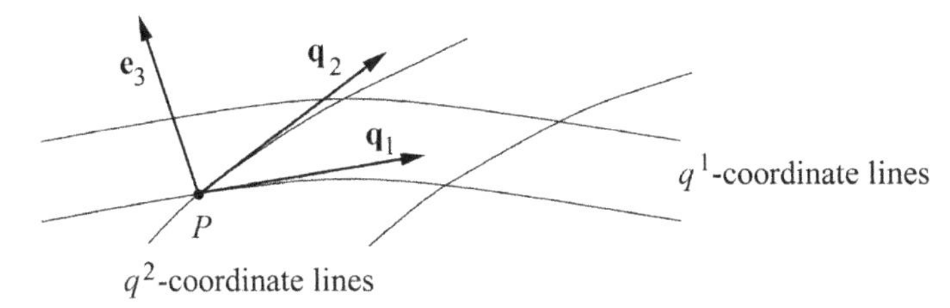

Fig. M6.2 The coordinate surface q^3 = constant in three-dimensional space. The coordinate lines drawn on the surface are q^1 and q^2. The corresponding basis vectors are $\mathbf{q}_1$ and $\mathbf{q}_2$, respectively. The unit vector of the q^3-coordinate line $\mathbf{e}_3$ is the normal vector to the surface.

direction defined by the cross product $\mathbf{q}_1 \times \mathbf{q}_2$. In summary, we may write the unit normal vector $\mathbf{e}_3$ in the following forms:

$$\mathbf{e}_3 = \frac{\mathbf{q}_3}{\sqrt{g_{33}}} = \frac{\mathbf{q}_3}{|\mathbf{q}_3|} = \frac{\mathbf{q}_1 \times \mathbf{q}_2}{|\mathbf{q}_1 \times \mathbf{q}_2|} = \frac{\mathbf{q}^3\sqrt{g}}{|\mathbf{q}^3\sqrt{g}|} = \frac{\mathbf{q}^3}{|\mathbf{q}^3|} = \frac{\mathbf{q}^3}{\sqrt{g^{33}}} = \mathbf{e}^3 \tag{M6.2}$$

where we have used the properties of the reciprocal systems (M1.55). Hence, in the semi-orthogonal system we have $\mathbf{e}_3 = \mathbf{e}^3$ and $g_{33}g^{33} = 1$. It should not be concluded, however, that $g_{ii}g^{ii} = 1$, $i = 1, 2$, since we are dealing with a semi-orthogonal coordinate system only, so that the general formula (M1.88) for orthogonal systems does not apply. Moreover, from the general expressions (M1.37) and (M1.57) of the scalar triple product we obtain for the semi-orthogonal system

$$\begin{aligned}[\mathbf{q}_1, \mathbf{q}_2, \mathbf{q}_3][\mathbf{q}^1, \mathbf{q}^2, \mathbf{q}^3] &= \left[\mathbf{q}_1, \mathbf{q}_2, \frac{\mathbf{q}_3}{\sqrt{g_{33}}}\right]\left[\mathbf{q}^1, \mathbf{q}^2, \frac{\mathbf{q}^3}{\sqrt{g^{33}}}\right] \\ &= [\mathbf{q}_1, \mathbf{q}_2, \mathbf{e}_3][\mathbf{q}^1, \mathbf{q}^2, \mathbf{e}^3] = 1\end{aligned} \tag{M6.3}$$

We will now consider a directed surface area element $d\mathbf{S}$ that is spanned by the two vectors $\mathbf{q}_1\, dq^1$ and $\mathbf{q}_2\, dq^2$ having the direction $\mathbf{q}_1 \times \mathbf{q}_2$. Utilizing (M6.2) we may write

$$d\mathbf{S} = \mathbf{q}_1 \times \mathbf{q}_2 \, dq^1 \, dq^2 = \mathbf{q}^3 \sqrt{g} \, dq^1 \, dq^2 = \mathbf{e}_3 \sqrt{g^{33}} \sqrt{g} \, dq^1 \, dq^2 = |d\mathbf{S}| \, \mathbf{e}_3$$

$$\text{with} \quad |d\mathbf{S}| = \sqrt{g^{33}} \sqrt{g} \, dq^1 \, dq^2 = \left[\mathbf{q}_1, \mathbf{q}_2, \frac{\mathbf{q}_3}{\sqrt{g_{33}}} \right] dq^1 \, dq^2 = [\mathbf{q}_1, \mathbf{q}_2, \mathbf{e}_3] \, dq^1 \, dq^2 \tag{M6.4}$$

Of some importance is the metric fundamental form for the given surface $q^3 =$ constant. Since an arbitrary line element on this surface is defined by $d\mathbf{r} = \mathbf{q}_1 \, dq^1 + \mathbf{q}_2 \, dq^2$, the metric fundamental form in the two-dimensional case is given by

$$\begin{aligned} d\mathbf{r} \cdot d\mathbf{r} &= (\mathbf{q}_1 \, dq^1 + \mathbf{q}_2 \, dq^2) \cdot (\mathbf{q}_1 \, dq^1 + \mathbf{q}_2 \, dq^2) \\ &= g_{11}(dq^1)^2 + 2g_{12} \, dq^1 \, dq^2 + g_{22}(dq^2)^2 \end{aligned} \tag{M6.5}$$

If the coordinate lines on the surface are orthogonal, the quantity g_{12} vanishes.

In general, the metric fundamental form is given by

$$\boxed{d\mathbf{r} \cdot d\mathbf{r} = g_{mn} \, dq^m \, dq^n \quad \text{with} \quad d\mathbf{r} = \mathbf{q}_n \, dq^n} \tag{M6.6}$$

In the Cartesian system this form reduces to

$$d\mathbf{r} \cdot d\mathbf{r} = dx^m \, dx^m \tag{M6.7}$$

At this point it is opportune to present a few remarks on the Riemannian space and the Euclidian space. In order not to disrupt our train of thought, we present some basic ideas to the interested reader in the appendix to this chapter. On the surface $q^3 =$ constant the two-dimensional Hamilton operator may be written as

$$\nabla_2 = \mathbf{q}^1 \frac{\partial}{\partial q^1} + \mathbf{q}^2 \frac{\partial}{\partial q^2} \tag{M6.8}$$

The contravariant and the covariant basis vectors are related in the following forms:

$$\begin{aligned} \mathbf{q}^1 &= \frac{\mathbf{q}_2 \times \mathbf{e}_3}{[\mathbf{q}_1, \mathbf{q}_2, \mathbf{e}_3]}, \qquad & \mathbf{q}^2 &= \frac{\mathbf{e}_3 \times \mathbf{q}_1}{[\mathbf{q}_1, \mathbf{q}_2, \mathbf{e}_3]} \\ \mathbf{q}_1 &= \frac{\mathbf{q}^2 \times \mathbf{e}^3}{[\mathbf{q}^1, \mathbf{q}^2, \mathbf{e}^3]}, \qquad & \mathbf{q}_2 &= \frac{\mathbf{e}^3 \times \mathbf{q}^1}{[\mathbf{q}^1, \mathbf{q}^2, \mathbf{e}^3]} \end{aligned} \tag{M6.9}$$

with $\mathbf{e}_3 = \mathbf{e}^3$. Finally, the relationship between the three- and the two-dimensional Hamilton operators is given by

$$\nabla = \nabla_2 + \mathbf{q}^3 \frac{\partial}{\partial q^3} \tag{M6.10}$$

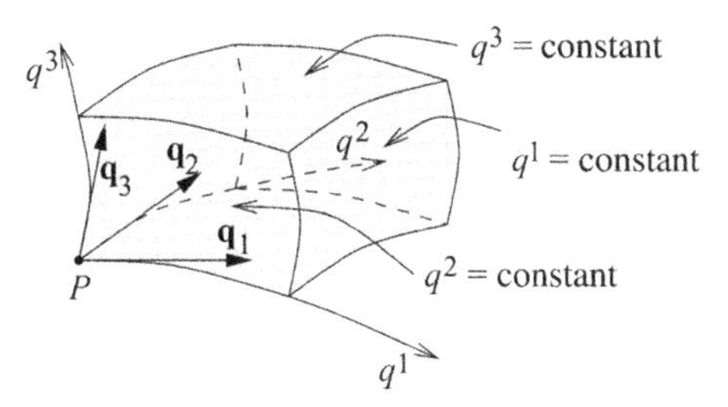

Fig. M6.3 A volume element $d\tau$ in the general curvilinear nonorthogonal coordinate system defined by the three coordinate surfaces q^i = constant.

Now we consider an arbitrary surface area element $d\mathbf{S}$ in the q^i-space, which may be decomposed as

$$\begin{aligned} d\mathbf{S} &= \mathbf{q}_2 \times \mathbf{q}_3 \, dq^2 \, dq^3 + \mathbf{q}_3 \times \mathbf{q}_1 \, dq^3 \, dq^1 + \mathbf{q}_1 \times \mathbf{q}_2 \, dq^1 \, dq^2 \\ &= \mathbf{q}^1 \sqrt{g} \, dq^2 \, dq^3 + \mathbf{q}^2 \sqrt{g} \, dq^3 \, dq^1 + \mathbf{q}^3 \sqrt{g} \, dq^1 \, dq^2 \\ &= \mathbf{q}^1 \, dS_1 + \mathbf{q}^2 \, dS_2 + \mathbf{q}^3 \, dS_3 = (d\mathbf{S})_1 + (d\mathbf{S})_2 + (d\mathbf{S})_3 \end{aligned} \tag{M6.11}$$

Reference to (M6.4) shows that this expression is a special case of (M6.11). Clearly, the surface area elements $\mathbf{q}^i \, dS_i$ are projections of $d\mathbf{S}$ in the directions q^i parallel to the planes spanned by $(\mathbf{q}_j \times \mathbf{q}_k)$.

In order to define a volume element in the generalized q^i-coordinate system we draw three coordinate surfaces as shown in Figure M6.3. The intersection of these surfaces results in the three coordinate lines. The differential volume element $d\tau$ is now given by

$$d\tau = [\mathbf{q}_1 \, dq^1, \mathbf{q}_2 \, dq^2, \mathbf{q}_3 \, dq^3] = \mathbf{q}_1 \, dq^1 \cdot (\mathbf{q}_2 \, dq^2 \times \mathbf{q}_3 \, dq^3) = \sqrt{g} \, dq^1 \, dq^2 \, dq^3 \tag{M6.12}$$

M6.2 Line integrals, surface integrals, and volume integrals

As before we consider a curve k in space, which is formed by the coordinate line q^1. Along this curve a field vector $\mathbf{A}$ may be written as $\mathbf{A}(q^1)$ while a vectorial line element $d\mathbf{r}$ is given by (M6.1). The line integral of $\mathbf{A}$ along the curve k between the points $(q^1)_1$ and $(q^1)_2$ is defined by

$$\boxed{\mathbb{L} = \int_{(q^1)_1}^{(q^1)_2} d\mathbf{r} \, \mathbf{A} = \int_{(q^1)_1}^{(q^1)_2} dq^1 \, (\mathbf{q}_1 \mathbf{A})} \tag{M6.13}$$

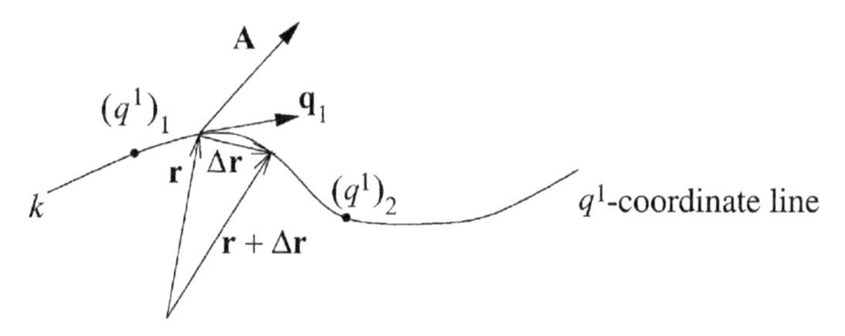

Fig. M6.4 An illustration of the curve k in space, showing the parameters used in the definition of the line integral (M6.13).

This situation is illustrated in Figure M6.4. In general $\mathbf{A}$ is a three-dimensional field vector and the line integral a dyadic.

The line integral over the tangential component $A_{\mathrm{T}} = A_1$ of $\mathbf{A}$ along k is given by the first scalar of the dyadic $\mathbb{L}$:

$$\mathbb{L}_{\mathrm{I}} = \int_{(q^1)_1}^{(q^1)_2} d\mathbf{r}\cdot\mathbf{A} = \int_{(q^1)_1}^{(q^1)_2} dq^1\,(\mathbf{q}_1\cdot\mathbf{A}) = \int_{(q^1)_1}^{(q^1)_2} dq^1\,(\sqrt{g_{11}}\mathbf{e}_1\cdot\mathbf{A}) = \int_{(q^1)_1}^{(q^1)_2} dr\,A_1 \tag{M6.14}$$

This integral is known as the *flow integral* whenever the field vector $\mathbf{A}$ represents the velocity vector $\mathbf{v}$ of the flow field along the curve k. If the curve k is closed then we speak of the *circulation integral*. This important concept will be discussed in some detail later.

If it is required to compute the normal component of the vector $\mathbf{A}$ along k then we have to take the vector of $\mathbb{L}$, which is easily obtained from (M6.13) as

$$\mathbb{L}_{\times} = \int_{(q^1)_1}^{(q^1)_2} d\mathbf{r}\times\mathbf{A} = \int_{(q^1)_1}^{(q^1)_2} dq^1\,(\sqrt{g_{11}}\mathbf{e}_1\times\mathbf{A}) = \int_{(q^1)_1}^{(q^1)_2} dr\,(\mathbf{e}_1\times\mathbf{A}) \tag{M6.15}$$

Let us reconsider a surface in space such as the surface $q^3 =$ constant. The vectorial surface element $(d\mathbf{S})_3$ was defined previously by equation (M6.4). According to (M6.11) we have added the number 3 to indicate the direction of the surface normal vector. On this surface we draw the coordinate lines q^1 and q^2 as depicted in Figure M6.5. We assume that a field vector $\mathbf{A}(q^1, q^2)$ is defined at every point on the surface. The surface integral which is a dyadic is defined by

$$\boxed{\mathbb{S} = \int_S (d\mathbf{S})_3\mathbf{A} = \int_{(q^1)_1}^{(q^1)_2} dq^1 \int_{(q^2)_1}^{(q^2)_2} dq^2\,\sqrt{g}\sqrt{g^{33}}\mathbf{e}_3\mathbf{A}(q^1, q^2)} \tag{M6.16}$$

While $(q^1)_1$ and $(q^1)_2$ are fixed boundary coordinates, the boundary coordinates $(q^2)_1$ and $(q^2)_2$ of the second integral are functions of q^1; see Figure M6.5.

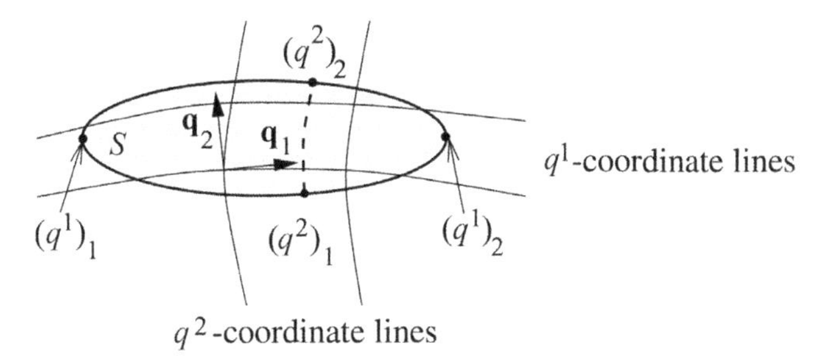

Fig. M6.5 An illustration of a surface S in space, showing the basic parameters needed to define the surface integral (M6.16).

The surface integral over the normal component of **A** is found by taking the first scalar of the dyadic $\mathbb{S}$:

$$\mathbb{S}_{\mathrm{I}} = \int_{(q^1)_1}^{(q^1)_2} dq^1 \int_{(q^2)_1}^{(q^2)_2} dq^2 \sqrt{g}\sqrt{g^{33}}\, A_3 \tag{M6.17}$$

where $A_3 = \mathbf{e}_3 \cdot \mathbf{A}$ is the normal component of the vector **A** with respect to the surface element. The integral over the tangential component of **A** is found by taking the vector of the dyadic $\mathbb{S}$

$$\mathbb{S}_{\times} = \int_{(q^1)_1}^{(q^1)_2} dq^1 \int_{(q^2)_1}^{(q^2)_2} dq^2 \sqrt{g}\sqrt{g^{33}}\, \mathbf{e}_3 \times \mathbf{A} \tag{M6.18}$$

Let us consider a surface in space that is enclosing the volume V. By assuming that the field vector **A** is known everywhere within the entire domain of integration, with the help of (M6.12) we obtain the volume integral as

$$\boxed{\mathbf{V} = \int_V d\tau\, \mathbf{A} = \int_{(q^1)_1}^{(q^1)_2} dq^1 \int_{(q^2)_1}^{(q^2)_2} dq^2 \int_{(q^3)_1}^{(q^3)_2} dq^3 \sqrt{g}\mathbf{A}} \tag{M6.19}$$

As before, the boundary coordinates $(q^1)_1$, $(q^1)_2$ are fixed. The boundary coordinates $(q^3)_1$, $(q^3)_2$ must be given functions of q^1 and q^2 and the boundary coordinates $(q^2)_1$, $(q^2)_2$ must be given functions of q^1. The integration proceeds analogously to the volume integration in the Cartesian coordinate system.

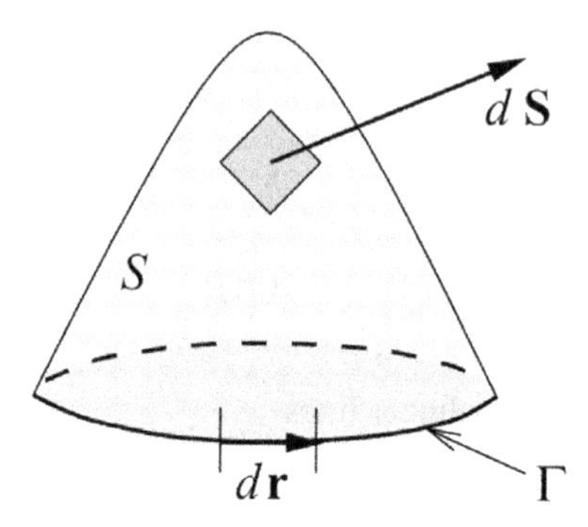

Fig. M6.6 Explanation of the terms occurring in equation (M6.20). $d\mathbf{S}$ is a directed surface element on the spatial surface S which is bordered by the line Γ; $d\mathbf{r}$ is a line element on Γ.

M6.3 Integral theorems

M6.3.1 Stokes' integral theorem

The Stokes integral theorem relating a surface and a line integral is based on the following general expression:

$$\boxed{\int_S d\mathbf{S} \times \nabla(\cdots) = \oint_\Gamma d\mathbf{r}(\cdots)} \tag{M6.20}$$

where the term $(\cdots)$ indicates that this formula may be applied to tensors of arbitrary rank, that is scalars, vectors, dyadics etc. The derivation of this important equation will be omitted here since it is given in many mathematical textbooks on vector analysis. Figure M6.6 depicts the parameters occurring in (M6.20). If the surface S is closed then the border line vanishes and the right hand side of (M6.20) is zero.

Of great importance are the first scalar and the vector of (M6.20). The first scalar results in the Stokes integral theorem

$$\boxed{\int_S d\mathbf{S} \times \nabla \cdot \mathbf{A} = \int_S \nabla \times \mathbf{A} \cdot d\mathbf{S} = \oint_\Gamma d\mathbf{r} \cdot \mathbf{A} = \oint_\Gamma A_\mathrm{T}\, dr} \tag{M6.21}$$

where A_T is the tangential component of $\mathbf{A}$ along the border line Γ. (M6.21) is a fundamental formula not only in fluid dynamics but also in many other branches of science. Using Grassmann's rule (M1.44), we find for the vector of the dyadic

$$\boxed{\int_S (d\mathbf{S} \times \nabla) \times \mathbf{A} = \int_S [\nabla\mathbf{A} - (\nabla \cdot \mathbf{A})\mathbb{E}] \cdot d\mathbf{S} = \oint_\Gamma d\mathbf{r} \times \mathbf{A}} \tag{M6.22}$$

where we have used the property of the conjugate dyadics.

In the following we present some important applications of the Stokes integral theorem. In the first example we apply (M6.21) to a very small surface element $\Delta\mathbf{S} = \Delta S\,\mathbf{e}_3$ so that $\nabla \times \mathbf{A}$ may be viewed as a constant vector and the integral on the left-hand side of (M6.21) may be replaced by $\nabla \times \mathbf{A}\cdot\mathbf{e}_3\,\Delta S$. This yields the *coordinate-free definition of the rotation*,

$$\boxed{\nabla \times \mathbf{A}\cdot\mathbf{e}_3 = \frac{1}{\Delta S}\oint_\Gamma d\mathbf{r}\cdot\mathbf{A}} \tag{M6.23}$$

In the next application we consider the horizontal flow in the Cartesian coordinate system. In this case we write the differentials $d\mathbf{S} = \mathbf{k}\,dx\,dy$ and $d\mathbf{r} = \mathbf{i}\,dx + \mathbf{j}\,dy$. Replacing the vector $\mathbf{A}$ in (M6.21) by the horizontal velocity $\mathbf{v}_\mathrm{h} = \mathbf{i}u + \mathbf{j}v$ yields

$$\int_S \nabla_\mathrm{h} \times \mathbf{v}_\mathrm{h}\cdot\mathbf{k}\,dx\,dy = \oint_\Gamma \mathbf{v}_\mathrm{h}\cdot d\mathbf{r} \tag{M6.24}$$

or

$$\boxed{\int_S\left(\frac{\partial v}{\partial x} - \frac{\partial u}{\partial y}\right)dx\,dy = \int_S \zeta\,dx\,dy = \oint_\Gamma (u\,dx + v\,dy)} \tag{M6.25}$$

The term ζ is known as the *vorticity*, which turns out to be a very useful quantity in dynamic meteorology. Equation (M6.25) relates an integral over the vorticity with the circulation integral.

Let us close this section by giving another important example involving the Stokes theorem. Setting in (M6.21) $\mathbf{A} = \lambda\,\nabla\mu$, where λ and μ are scalar field variables such as pressure and density, the left-hand side of this equation can be written as

$$\int_S d\mathbf{S}\times\nabla\cdot(\lambda\,\nabla\mu) = \int_S d\mathbf{S}\cdot\nabla\times(\lambda\,\nabla\mu) = \int_S d\mathbf{S}\cdot\nabla\lambda\times\nabla\mu \tag{M6.26}$$

Next we replace the directed surface element $d\mathbf{S}$ by means of (M6.4). By interpreting the functions λ and μ as the coordinate lines on the surface S, surrounded by the line Γ, we obtain

$$\begin{aligned}\int_S d\mathbf{S}\cdot\nabla\lambda\times\nabla\mu &= \int_S \sqrt{g}\mathbf{q}^3\cdot\nabla\lambda\times\nabla\mu\,d\lambda\,d\mu\\ &= \int_S \sqrt{g}[\mathbf{q}^3,\mathbf{q}^1,\mathbf{q}^2]\,d\lambda\,d\mu = \int_S d\lambda\,d\mu\end{aligned} \tag{M6.27}$$

with $\nabla\lambda = \mathbf{q}^1$, $\nabla\mu = \mathbf{q}^2$. The right-hand side of (M6.21) may now be written as

$$\oint_\Gamma d\mathbf{r}\cdot\lambda\,\nabla\mu = \oint_\Gamma \lambda\,d_\mathrm{g}\mu \quad \text{or}$$

$$\oint_\Gamma d\mathbf{r}\cdot\lambda\,\nabla\mu = \oint_\Gamma d\mathbf{r}\cdot[\nabla(\lambda\mu) - \mu\,\nabla\lambda] = \oint_\Gamma d_\mathrm{g}(\lambda\mu) - \oint_\Gamma \mu\,d_\mathrm{g}\lambda = -\oint_\Gamma \mu\,d_\mathrm{g}\lambda \tag{M6.28}$$

Since in general λ and μ are time-dependent functions, we have employed the geometric differential d_g defined in (M3.7), indicating that the operations are performed at t = constant. From (M6.27) and (M6.28) we obtain the final result

$$\boxed{\int_S d\lambda\, d\mu = \oint_\Gamma \lambda\, d_g\mu = -\oint_\Gamma \mu\, d_g\lambda} \tag{M6.29}$$

M6.3.2 Gauss' divergence theorem

Let us consider a closed surface S enclosing a volume V. Without proof we accept the general integral theorem

$$\boxed{\int_V d\tau\, \nabla(\cdots) = \oint_S d\mathbf{S}\,(\cdots)} \tag{M6.30}$$

relating a volume integral and an integral over a closed surface. In the following we will consider various applications of equation (M6.30).

The divergence theorem due to Gauss is derived in many textbooks by employing the Cartesian coordinate system. It is of great importance in fluid dynamics and in many other branches of physics. The theorem is obtained by applying (M6.30) to the field vector $\mathbf{A}$ and by taking the first scalar

$$\boxed{\int_V d\tau\, \nabla\cdot\mathbf{A} = \oint_S d\mathbf{S}\cdot\mathbf{A} = \oint_S A_{\rm N}\, dS} \tag{M6.31}$$

where $A_{\rm N}$ is the projection of $\mathbf{A}$ in the direction of $d\mathbf{S}$. The vector of (M6.30) yields

$$\boxed{\int_V d\tau\, \nabla\times\mathbf{A} = \oint_S d\mathbf{S}\times\mathbf{A}} \tag{M6.32}$$

Let us briefly consider some important examples. On shrinking the volume V in (M6.31) to the differential volume element ΔV in which $\nabla\cdot\mathbf{A}$ is considered constant, this equation reduces to the *coordinate-free definition of the divergence* of the vector $\mathbf{A}$,

$$\boxed{\nabla\cdot\mathbf{A} = \frac{1}{\Delta V}\oint_S d\mathbf{S}\cdot\mathbf{A}} \tag{M6.33}$$

In our later studies we will often investigate two-dimensional flow fields in the Cartesian coordinate system. In these studies we will require the application of the two-dimensional version of the Gauss divergence theorem. Instead of deriving this

Table M6.1. *From the three-dimensional divergence theorem of Gauss to the two-dimensional version in Cartesian coordinates*

Volume V	$\Longrightarrow$	area S^* within Γ
Volume element $d\tau$	$\Longrightarrow$	area element $dx\,dy$
Closed surface S	$\Longrightarrow$	closed line Γ
Surface element $d\mathbf{S}$	$\Longrightarrow$	line element $d\mathbf{r}_{\rm N}$ perpendicular to Γ

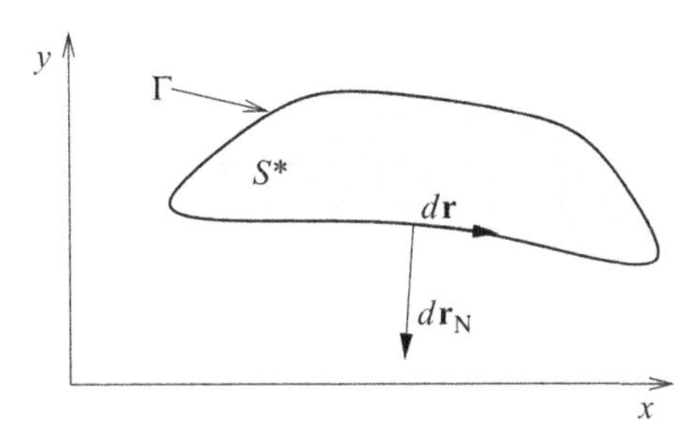

Fig. M6.7 Explanation of the variables of the two-dimensional Gauss divergence theorem (M6.34). $d\mathbf{r}_{\rm N}$ is a directed line element perpendicular to the curve Γ.

theorem we simply employ the replacements summarized in Table M6.1. With the help of this table we recognize that equation (M6.31) reduces to

$$\boxed{\int_{S^*} \nabla_{\rm h} \cdot \mathbf{A}_{\rm h}\, dx\, dy = \oint_\Gamma d\mathbf{r}_{\rm N} \cdot \mathbf{A}} \tag{M6.34}$$

which is the two-dimensional divergence theorem of Gauss. The terms appearing in this equation are displayed in Figure M6.7.

In a simple but important application of (M6.34) we replace the horizontal field vector $\mathbf{A}_{\rm h}$ by the horizontal velocity $\mathbf{v}_{\rm h}$. We immediately obtain with $d\mathbf{r}_{\rm N} = d\mathbf{r} \times \mathbf{k}$

$$\int_{S^*} \nabla_{\rm h} \cdot \mathbf{v}_{\rm h}\, dx\, dy = \oint_\Gamma (\mathbf{i}u + \mathbf{j}v) \cdot (\mathbf{i}\, dy - \mathbf{j}\, dx) \tag{M6.35a}$$

or

$$\boxed{\int_{S^*} \left(\frac{\partial u}{\partial x} + \frac{\partial v}{\partial y}\right) dx\, dy = \oint_\Gamma (u\, dy - v\, dx)} \tag{M6.35b}$$

M6.4 Fluid lines, surfaces, and volumes

Let us consider a fluid in three-dimensional space whose motion is controlled by the velocity field $\mathbf{v}(\mathbf{r}, t)$[1]. A volume $V(t)$ is called a fluid volume if all surface elements $d\mathbf{S}$ of the surface S enclosing $V(t)$ are moving with the momentary velocity $\mathbf{v}(\mathbf{r}, t)$ existing at $d\mathbf{S}$. Since the velocity depends on the position $\mathbf{r}$ and on time t, the volume is bound to deform, thus changing size and shape.

In a first application of the concept of fluid volumes let us replace the field vector $\mathbf{A}$ in (M6.31) by the velocity vector $\mathbf{v}$ so that this equation now reads

$$\int_V \nabla \cdot \mathbf{v}(\mathbf{r}, t)\, d\tau = \oint_S d\mathbf{S} \cdot \mathbf{v}(\mathbf{r}, t) \tag{M6.36}$$

During the time increment Δt each surface element $d\mathbf{S}(\mathbf{r}, t)$ of the volume is covering the distance $\Delta \mathbf{r} = \mathbf{v}(\mathbf{r}, t)\, \Delta t$. Thus the volume element of the difference volume is given by

$$\Delta V = V(t + \Delta t) - V(t) = \oint_S d\mathbf{S} \cdot \Delta \mathbf{r} \tag{M6.37}$$

Using the mean-value theorem for integrals, (M6.36) can now be written as

$$\overline{\nabla \cdot \mathbf{v}} = \frac{1}{V} \lim_{\Delta t \to 0} \frac{1}{\Delta t} \oint_S d\mathbf{S} \cdot \Delta \mathbf{r} = \frac{1}{V} \lim_{\Delta t \to 0} \frac{\Delta V}{\Delta t} = \frac{1}{V} \frac{dV}{dt} \tag{M6.38a}$$

If the volume is small enough that $\nabla \cdot \mathbf{v}$ is constant within V we obtain

$$\boxed{\nabla \cdot \mathbf{v} = \frac{1}{V} \frac{dV}{dt}} \tag{M6.38b}$$

This important expression will be applied soon.

Now we assume that $V(t)$ consists of particles that are also moving with $\mathbf{v}(\mathbf{r}, t)$. Obviously, at all times all particles that are located on the surface of the volume remain on the surface because their velocity and the velocity of the corresponding surface element $d\mathbf{S}$ are identical. Thus no particle can leave the surface of the volume, so the number of particles within $V(t)$ remains constant.

Suppose that it would be possible to dye a small volume of a moving fluid without changing its density. As we follow the motion of the dyed fluid we observe a change of the colored part of the medium. While the original volume has changed its shape, it is always made up of the same particles. Therefore, a fluid volume is also called a *material volume*.

[1] Here and in the following the velocity refers to the absolute system, that is $\mathbf{v} = \mathbf{v}_A$. For simplicity the suffix A has been omitted.

In analogy to the fluid or material volume we define the *fluid or material surface* and the *fluid or material line*. As in the case of the material volume, a material surface and a material line always consist of the same particles.

In order to efficiently describe the properties of a fluid system, we distinguish between *intensive* and *extensive variables*. An intensive variable is a quantity whose value is independent of the mass of the system. Examples are pressure and temperature. The volume, on the other hand, is proportional to its mass M and is, therefore, an example of an extensive variable. Additional examples are the kinetic and the internal energy of the system. External variables will be denoted by Ψ. The density of Ψ is defined by $\widehat{\psi} = \Psi/V$. The *specific value* of Ψ is $\psi = \Psi/M$ so that $\widehat{\psi} = \rho\psi$, where $\rho = M/V$ is the mass density of the fluid.

Of particular interest is not the property Ψ itself but rather the change with time of this quantity, which will now be considered. In general, any extended part of the atmospheric continuum will be inhomogeneous so that the property Ψ of a sizable volume $V(t)$ must be expressed by an integral over the density $\widehat{\psi} = \rho\psi$. Since the total mass of the fluid volume is constant, the change with time of Ψ may be easily calculated as

$$\begin{aligned}\frac{d\Psi}{dt} &= \frac{d}{dt}\int_{V(t)} \rho\psi\, d\tau = \frac{d}{dt}\int_{M=\text{constant}} \psi\, dM \\ &= \int_{M=\text{constant}} \frac{d\psi}{dt}\, dM = \int_{V(t)} \rho\,\frac{d\psi}{dt}\, d\tau\end{aligned} \tag{M6.39}$$

Let us now consider a multicomponent system in which the component k has mass M^k and is moving with the velocity $\mathbf{v}_k(\mathbf{r}, t)$. Hence the total mass of the system is given by

$$M = \sum_{k=0}^{N} M^k \tag{M6.40}$$

where N is the number of components. Each volume element is assumed to move with the momentary *barycentric velocity* $\mathbf{v}(\mathbf{r}, t)$ existing at its position $\mathbf{r}$. If $\rho^k = M^k/V$ is the density of the particle group k and $\rho(\mathbf{r}, t)$ is the total density, the barycentric velocity is defined by

$$\rho(\mathbf{r}, t)\mathbf{v}(\mathbf{r}, t) = \sum_{k=0}^{N} \rho^k(\mathbf{r}, t)\mathbf{v}_k(\mathbf{r}, t) \quad \text{with} \quad \rho(\mathbf{r}, t) = \sum_{k=0}^{N} \rho^k(\mathbf{r}, t) \tag{M6.41}$$

Since each surface element $d\mathbf{S}$ of the volume is moving with the barycentric velocity $\mathbf{v}(\mathbf{r}, t)$, for the particle group k we observe a mass flux $\mathbf{J}^k$ through $d\mathbf{S}$, which is known as the *diffusion flux*:

$$\mathbf{J}^k = \rho^k(\mathbf{v}_k - \mathbf{v}) \tag{M6.42}$$

The difference between the partial velocity $\mathbf{v}_k$ and the barycentric velocity $\mathbf{v}$ is known as the *diffusion velocity* $\mathbf{v}_{k,\text{dif}} = \mathbf{v}_k - \mathbf{v}$.

From (M6.41) and (M6.42) it is easily seen that the total mass flux through $d\mathbf{S}$, that is the sum of all diffusion fluxes, vanishes since

$$\sum_{k=0}^{N} \mathbf{J}^k = 0 \tag{M6.43}$$

We conclude that the diffusion of the partial masses M^k through the surface of the fluid volume may change the mass composition. However, due to (M6.43), at all times the volume conserves the total mass M.

M6.5 Time differentiation of fluid integrals

Often the analytic treatment of fluid-dynamic problems requires the time differentiation of so-called *fluid integrals*. A fluid integral is expressed as an integral over a fluid line, surface, or volume.

M6.5.1 Time differentiation of fluid line integrals

First we derive a formula for the total time differentiation of the fluid line integral:

$$\frac{d}{dt}\left(\int_{L(t)} d\mathbf{r} \cdot \mathbf{A}\right) = \int_{L(t)} d\mathbf{r} \cdot \frac{\partial \mathbf{A}}{\partial t} + \frac{d}{dt}\int_{L(t)} d\mathbf{r} \cdot \mathbf{A} \tag{M6.44}$$

According to the general differentiation rules for products, the total change of the expression within the parentheses is composed of two terms. The first integral on the right-hand side refers to the change with time of the field vector $\mathbf{A}$ for a line fixed in space at time t while the second integral refers to the displacement and the deformation of the line during the time increment Δt while the vector field $\mathbf{A}$ itself is considered fixed in time.

In order to evaluate the latter integral, let us consider the section $(1, 2)$ of a line of fluid particles at time t as shown in Figure M6.8. After the small time increment Δt the particle at position 1 will have moved to $1'$ while the particle at position 2 has moved to $2'$. During the time increment each particle on the line increment $(1, 2)$ is moving the distance $\mathbf{v}\,\Delta t$. Now we apply Stokes' integral theorem (M6.21) to the second integral on the right-hand side of (M6.44). According to Figure M6.8 we integrate over the area $d\mathbf{S} = d\mathbf{r} \times \mathbf{v}\,\Delta t$ which is surrounded by the closed curve Γ connecting the points $(1, 2, 2', 1', 1)$. Substitution of these expressions into

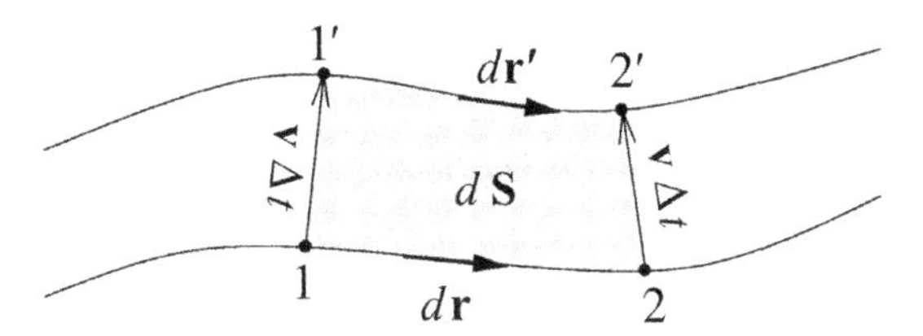

Fig. M6.8 Description of the time differentiation of a fluid line integral using Stokes' integral theorem.

(M6.21) yields

$$
\begin{aligned}
\Delta t \int_1^2 (d\mathbf{r} \times \mathbf{v}) \cdot \nabla \times \mathbf{A} &= \int_1^2 d\mathbf{r} \cdot \mathbf{A} + \Delta t\,(\mathbf{v} \cdot \mathbf{A})_2 - \int_{1'}^{2'} d\mathbf{r}' \cdot \mathbf{A} - \Delta t\,(\mathbf{v} \cdot \mathbf{A})_1 \\
&= \int_1^2 d\mathbf{r} \cdot \mathbf{A} - \int_{1'}^{2'} d\mathbf{r}' \cdot \mathbf{A} + \Delta t \int_1^2 d_{\rm g}(\mathbf{v} \cdot \mathbf{A}) \\
&= \int_1^2 d\mathbf{r} \cdot \mathbf{A} - \int_{1'}^{2'} d\mathbf{r}' \cdot \mathbf{A} + \Delta t \int_1^2 d\mathbf{r} \cdot \nabla(\mathbf{v} \cdot \mathbf{A})
\end{aligned}
\tag{M6.45}
$$

where $d_{\rm g}$ is defined by (M3.22). On dividing this equation by Δt and rearranging terms, we find in the limit $\Delta t \rightarrow 0$

$$
\lim_{\Delta t \rightarrow 0} \frac{\int_{1'}^{2'} d\mathbf{r}' \cdot \mathbf{A} - \int_1^2 d\mathbf{r} \cdot \mathbf{A}}{\Delta t} = \frac{d}{dt} \int_{L(t)} d\mathbf{r} \cdot \mathbf{A} = \int_1^2 d\mathbf{r} \cdot [(\nabla \times \mathbf{A}) \times \mathbf{v} + \nabla(\mathbf{v} \cdot \mathbf{A})]
\tag{M6.46}
$$

where we have employed the rules associated with the scalar triple product. Substituting this result into (M6.44) yields for the total change with time of the fluid line integral

$$
\frac{d}{dt}\left(\int_{L(t)} d\mathbf{r} \cdot \mathbf{A}\right) = \int_{L(t)} d\mathbf{r} \cdot \left[\frac{\partial \mathbf{A}}{\partial t} + (\nabla \times \mathbf{A}) \times \mathbf{v} + \nabla(\mathbf{v} \cdot \mathbf{A})\right] = \int_{L(t)} d\mathbf{r} \cdot \frac{D_1 \mathbf{A}}{Dt}
\tag{M6.47}
$$

Here we have introduced the following differential operator for the time differentiation of fluid line integrals:

$$
\boxed{\frac{D_1 \mathbf{A}}{Dt} = \frac{\partial \mathbf{A}}{\partial t} + (\nabla \times \mathbf{A}) \times \mathbf{v} + \nabla(\mathbf{v} \cdot \mathbf{A}) = \frac{d\mathbf{A}}{dt} + (\nabla \mathbf{v}) \cdot \mathbf{A}}
\tag{M6.48}
$$

A special situation occurs for a closed path $L(t)$. In this case the geometric differential $d_{\rm g}(\mathbf{v} \cdot \mathbf{A})$ vanishes upon integration so that (M6.47) reduces to

$$
\frac{d}{dt}\left(\oint_{L(t)} d\mathbf{r} \cdot \mathbf{A}\right) = \oint_{L(t)} d\mathbf{r} \cdot \left(\frac{d\mathbf{A}}{dt} - \nabla \mathbf{A} \cdot \mathbf{v}\right)
\tag{M6.49}
$$

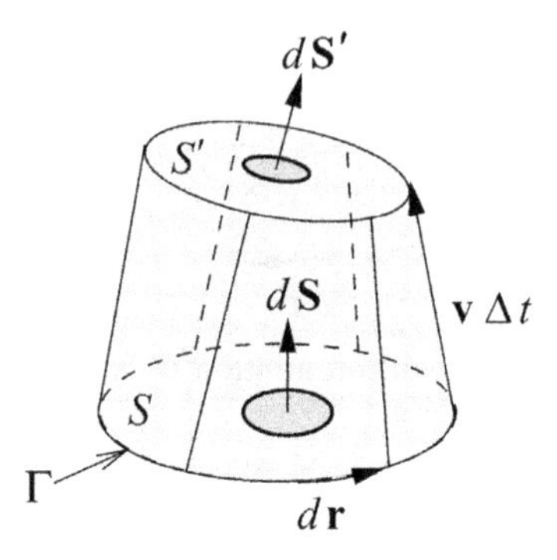

Fig. M6.9 Time differentiation of a fluid surface integral by applying the divergence theorem of Gauss.

M6.5.2 Time differentiation of fluid surface integrals

As in the case of the line integral, the total change of the surface integral consists of two parts,

$$\frac{d}{dt}\left(\int_{S(t)} d\mathbf{S}\cdot\mathbf{A}\right) = \int_{S(t)} d\mathbf{S}\cdot\frac{\partial\mathbf{A}}{\partial t} + \frac{d}{dt}\int_{S(t)} d\mathbf{S}\cdot\mathbf{A} \tag{M6.50}$$

The first integral on the right-hand side describes the change with time of the field vector $\mathbf{A}$ if the surface remains fixed in space at time t while the second integral describes the displacement and the deformation of the surface while the vector field $\mathbf{A}$ remains fixed in time.

In order to evaluate the latter integral we consider Figure M6.9. We first evaluate the integral over the closed surface of the small volume and then apply the Gauss divergence theorem (M6.31). The surface integral consists of three terms. The first two terms are the contributions of the upper and lower surfaces while the third part is the integration over the side surface element which is directed towards $\Delta t\, d\mathbf{r}\times\mathbf{v}$. Thus, we obtain

$$\int_{S'} d\mathbf{S}'\cdot\mathbf{A} - \int_S d\mathbf{S}\cdot\mathbf{A} + \Delta t\oint_\Gamma (d\mathbf{r}\times\mathbf{v})\cdot\mathbf{A} = \int_V \nabla\cdot\mathbf{A}\, d\tau = \Delta t\int_S (\nabla\cdot\mathbf{A})\mathbf{v}\cdot d\mathbf{S} \tag{M6.51}$$

The integration over the surface requires that all surface-element vectors must point to the outside of the volume, which explains the negative sign of the integral over S. The closed line integral on the left-hand side gives the contribution of the complete side surface. This integral will be evaluated by applying the Stokes integral theorem. On the right-hand side of (M6.51) we have replaced the volume element $d\tau$ by $\Delta t\,\mathbf{v}\cdot d\mathbf{S}$.

Dividing (M6.51) by Δt and rearranging terms, in the limit $\Delta t \to 0$ we readily find

$$\lim_{\Delta t \to 0} \frac{\int_{S'} d\mathbf{S}' \cdot \mathbf{A} - \int_S d\mathbf{S} \cdot \mathbf{A}}{\Delta t} = \frac{d}{dt} \int_{S(t)} d\mathbf{S} \cdot \mathbf{A} = \int_S d\mathbf{S} \cdot [\nabla \times (\mathbf{A} \times \mathbf{v}) + \mathbf{v}(\nabla \cdot \mathbf{A})] \tag{M6.52}$$

Hence, equation (M6.50) may be written as

$$\begin{aligned} \frac{d}{dt}\left(\int_{S(t)} d\mathbf{S} \cdot \mathbf{A}\right) &= \int_{S(t)} d\mathbf{S} \cdot \left[\frac{\partial \mathbf{A}}{\partial t} + \nabla \times (\mathbf{A} \times \mathbf{v}) + \mathbf{v}(\nabla \cdot \mathbf{A})\right] \\ &= \int_{S(t)} d\mathbf{S} \cdot \left(\frac{d\mathbf{A}}{dt} + \mathbf{A}\nabla \cdot \mathbf{v} - \overset{\curvearrowleft}{\mathbf{v}\nabla} \cdot \mathbf{A}\right) = \int_{S(t)} d\mathbf{S} \cdot \frac{D_2\mathbf{A}}{Dt} \end{aligned} \tag{M6.53}$$

Some care must be taken to evaluate the term $\nabla \times (\mathbf{A} \times \mathbf{v})$. First we must differentiate this expression and then we apply Grassmann's rule. In (M6.53) we have introduced the differential operator for the time differentiation of a fluid surface,

$$\boxed{\frac{D_2\mathbf{A}}{Dt} = \frac{\partial \mathbf{A}}{\partial t} + \nabla \times (\mathbf{A} \times \mathbf{v}) + \mathbf{v}(\nabla \cdot \mathbf{A}) = \frac{d\mathbf{A}}{dt} + \mathbf{A}\nabla \cdot \mathbf{v} - \overset{\curvearrowleft}{\mathbf{v}\nabla} \cdot \mathbf{A}} \tag{M6.54}$$

It should be noted that (M6.48) and (M6.54) may be applied to all extensive quantities of degree 1 and more. The last term of (M6.54) may also be written as $\mathbf{A} \cdot \nabla\mathbf{v}$. However, for extensive functions of degree 2 and higher the corresponding replacement is not possible since for the scalar product the commutative law is no longer valid.

M6.5.3 Time differentiation of fluid volume integrals

Finally we consider the volume integral over the scalar field function ψ:

$$\frac{d}{dt}\left(\int_{V(t)} \psi \, d\tau\right) = \int_{V(t)} \frac{\partial \psi}{\partial t} d\tau + \frac{d}{dt} \int_{V(t)} \psi \, d\tau \tag{M6.55}$$

As in the previous two situations we obtain two integrals due to the differentiation of a product. The second integral on the right-hand side will now be evaluated according to

$$\begin{aligned} \frac{d}{dt} \int_{V(t)} \psi \, d\tau &= \lim_{\Delta t \to 0} \frac{\int_{V'} \psi \, d\tau - \int_V \psi \, d\tau}{\Delta t} \\ &= \lim_{\Delta t \to 0} \frac{1}{\Delta t} \int_{\Delta V} \psi \, d\tau = \lim_{\Delta t \to 0} \frac{1}{\Delta t} \oint_{S(t)} d\mathbf{S} \cdot \mathbf{v} \, \Delta t \, \psi \\ &= \oint_{S(t)} d\mathbf{S} \cdot (\mathbf{v}\psi) = \int_{V(t)} \nabla \cdot (\mathbf{v}\psi) \, d\tau \end{aligned} \tag{M6.56}$$

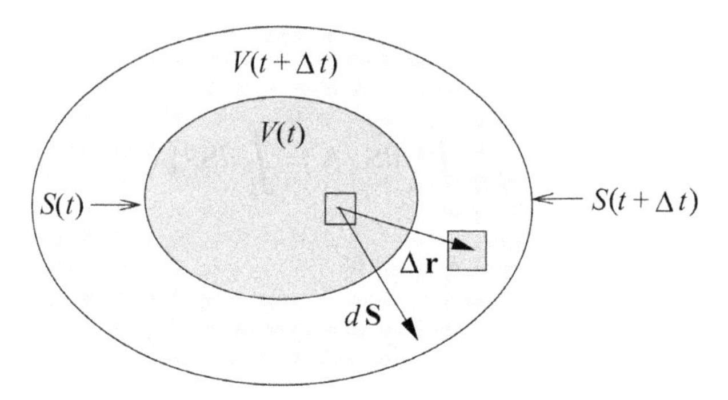

Fig. M6.10 Time differentiation of a fluid volume integral.

where ΔV is equal to the difference volume $V' - V = V(t + \Delta t) - V(t)$. Some details pertaining to these formulas are shown in Figure M6.10. Inspection of this figure shows that the volume element $d\tau$ can be replaced by $d\mathbf{S} \cdot \mathbf{v}\,\Delta t$. The final step in the derivation makes use of the Gauss divergence theorem. Thus, the total change with time of the fluid volume integral is given by

$$\frac{d}{dt}\left(\int_{V(t)} \psi \, d\tau\right) = \int_{V(t)} \left[\frac{\partial \psi}{\partial t} d\tau + \nabla \cdot (\mathbf{v}\psi)\right] d\tau = \int_{V(t)} \frac{D_3\psi}{Dt} d\tau \tag{M6.57}$$

Whenever a fluid volume integral needs to be differentiated with respect to time, we must apply the operator

$$\boxed{\frac{D_3\psi}{Dt} = \frac{\partial \psi}{\partial t} + \nabla \cdot (\mathbf{v}\psi) = \frac{d\psi}{dt} + \psi \, \nabla \cdot \mathbf{v}} \tag{M6.58}$$

This operator will be used numerous times in our studies. For reasons of simplicity most of the time we omit the suffix 3. Similarly to the operators D_1/Dt and D_2/Dt, the operator D_3/Dt may be applied to extensive functions of any degree.

There is a shorter way to obtain equation (M6.58), which is attributed to Lagrange:

$$\begin{aligned}\frac{d}{dt}\left(\int_{V(t)} \psi \, d\tau\right) &= \int_{V(t)} \frac{d\psi}{dt} d\tau + \int_{V(t)} \psi \left(\frac{1}{d\tau}\frac{d}{dt}(d\tau)\right) d\tau \\ &= \int_{V(t)} \frac{d\psi}{dt} d\tau + \int_{V(t)} \psi \, \nabla \cdot \mathbf{v} \, d\tau\end{aligned} \tag{M6.59}$$

where use has been made of (M6.38b) by setting $V = d\tau$.

M6.6 The general form of the budget equation

Theoretical considerations often require the use of budget equations to study the behavior of certain physical quantities such as mass, momentum, and energy in its various forms. The purpose of this section is to derive the general form of a balance or budget equation for these quantities.

Usually any extended part of the atmospheric continuum will be inhomogeneous. To handle an inhomogeneous section of the atmosphere we mentally isolate a sizable fluid volume V that is surrounded by an imaginary surface S. Anything within S belongs to the system; the surroundings or the outside world is found exterior to S. Processes taking place at the surface S itself represent the exchange of the system with the surroundings. These processes are

(i) *mass fluxes* penetrating the surface,
(ii) work contributions by or on the system resulting from surface forces, and
(iii) *heat* and *radiative fluxes* through the surface.

The amount of Ψ contained in the volume V is given by

$$\Psi(\mathbf{r}, t) = \int_V \widehat{\psi}(\mathbf{r}, \mathbf{r}', t)\, d\tau = \int_V \rho(\mathbf{r}, \mathbf{r}', t)\psi(\mathbf{r}, \mathbf{r}', t)\, d\tau \tag{M6.60}$$

where $\mathbf{r}$ is the position vector from a suitable reference point to the volume V. The vector $\mathbf{r}'$ is directed from the endpoint of $\mathbf{r}$ to the volume element $d\tau$. Changes in Ψ caused by the interaction of the system with the exterior will be denoted by $d_\mathrm{e}\Psi/dt$ while interior changes of Ψ will be written as $d_\mathrm{i}\Psi/dt$. The budget equation for Ψ is then given by

$$\frac{d\Psi}{dt} = \frac{d_\mathrm{e}\Psi}{dt} + \frac{d_\mathrm{i}\Psi}{dt} \tag{M6.61}$$

The change with time $d_\mathrm{e}\Psi/dt$ may be expressed by an integral over the surface S bounding V:

$$\frac{d_\mathrm{e}\Psi}{dt} = -\oint_S \mathbf{F}_\psi(\mathbf{r}, \mathbf{r}', t) \cdot d\mathbf{S} = -\int_V \nabla \cdot \mathbf{F}_\psi(\mathbf{r}, \mathbf{r}', t)\, d\tau \tag{M6.62}$$

$\mathbf{F}_\psi$ is a flux vector giving the amount of Ψ per unit surface area and unit time streaming into V or out of it. The negative sign is chosen so that $d_\mathrm{e}\Psi/dt$ is positive when $\mathbf{F}_\psi$ is directed into V. The conversion from the surface to the volume integral is done with the help of the divergence theorem.

If Q_ψ represents the production of Ψ per unit volume and time within V, which is also known as the *source strength*, then we may write

$$\frac{d_\mathrm{i}\Psi}{dt} = \int_V Q_\psi\, d\tau \tag{M6.63}$$

Q_ψ is positive in cases of production of Ψ (sources) and negative in cases of destruction (sinks).

Utilizing the budget operator (M6.57), the time differentiation of (M6.60) yields

$$\frac{d\Psi}{dt} = \frac{d}{dt}\left(\int_V \rho\psi\, d\tau\right) = \int_V \frac{D_3}{Dt}(\rho\psi)\, d\tau \tag{M6.64}$$

Combination of (M6.61)–(M6.64) then results in the integral form of the budget equation

$$\int_V \frac{D_3}{Dt}(\rho\psi)\, d\tau = \int_V \left[\frac{\partial}{\partial t}(\rho\psi) + \nabla\cdot(\rho\psi\mathbf{v})\right] d\tau = \int_V \left[-\nabla\cdot\mathbf{F}_\psi + Q_\psi\right] d\tau \tag{M6.65}$$

This expression describes the following physical processes.

(i) Local time change of Ψ: $\int_V \frac{\partial}{\partial t}(\rho\psi)\, d\tau$

(ii) *Convective flux* of Ψ through S: $\int_V \nabla\cdot(\rho\psi\mathbf{v})\, d\tau = \oint_S \rho\psi\mathbf{v}\cdot d\mathbf{S}$

(iii) *Nonconvective flux* of Ψ through S: $-\int_V \nabla\cdot\mathbf{F}_\psi\, d\tau = -\oint_S \mathbf{F}_\psi\cdot d\mathbf{S}$

(iv) Production or destruction of Ψ: $\int_V Q_\psi\, d\tau$

Since equation (M6.65) is valid for an arbitrary volume V, we immediately obtain the general differential form of the budget equation as

$$\boxed{\frac{D_3}{Dt}(\rho\psi) = \frac{\partial}{\partial t}(\rho\psi) + \nabla\cdot(\rho\psi\mathbf{v}) = -\nabla\cdot\mathbf{F}_\psi + Q_\psi} \tag{M6.66}$$

A very simple but extremely important example of a budget equation is the *continuity equation* describing the conservation of mass of the material volume. Identifying $\Psi = M =$ constant in (M6.60), we have $\psi = 1$ and $Q_\psi = 0$ since mass cannot be created or destroyed. Furthermore, according to (M6.43) the total mass flux through each surface element $d\mathbf{S}$ vanishes so that (M6.66) reduces to

$$\boxed{\frac{D_3\rho}{Dt} = \frac{\partial\rho}{\partial t} + \nabla\cdot(\rho\mathbf{v}) = \frac{d\rho}{dt} + \rho\,\nabla\cdot\mathbf{v} = 0} \tag{M6.67}$$

The continuity equation is a fundamental part of all prognostic meteorological systems.

By expanding the budget operator and utilizing the continuity equation, we obtain

$$\boxed{\frac{D_3}{Dt}(\rho\psi) = \psi\left(\frac{\partial\rho}{\partial t} + \nabla\cdot(\rho\mathbf{v})\right) + \rho\left(\frac{\partial\psi}{\partial t} + \mathbf{v}\cdot\nabla\psi\right) = \rho\,\frac{d\psi}{dt}} \tag{M6.68}$$

This identity is called the *interchange rule*. Thus, the budget equation (M6.66) may be written in the form

$$\boxed{\frac{D_3}{Dt}(\rho\psi) = \rho\,\frac{d\psi}{dt} = -\nabla\cdot\mathbf{F}_\psi + Q_\psi} \tag{M6.69}$$

Division of (M6.61) by the constant total mass M gives

$$\frac{d\psi}{dt} = \frac{d_\text{e}\psi}{dt} + \frac{d_\text{i}\psi}{dt} \tag{M6.70}$$

Substitution of this equation into (M6.69) yields expressions for the *external* and *internal changes* of ψ:

$$\boxed{\rho\frac{d_\text{e}\psi}{dt} = -\nabla\cdot\mathbf{F}_\psi, \qquad \rho\,\frac{d_\text{i}\psi}{dt} = Q_\psi} \tag{M6.71}$$

These two equations are of great importance in thermodynamics.

It is worthwhile to reconsider equation (M6.69) using an argument due to Van Mieghem (1973). Suppose we add the arbitrary vector $\mathbf{X}$ to $\mathbf{F}_\psi$ so that $\nabla\cdot\mathbf{X} = \beta$ and add β to the production term Q_ψ on the right-hand side of this equation. This mathematical operation does not change the budget equation in any way. Consequently, there is no unique definition of either $\mathbf{F}_\psi$ or Q_ψ.

Finally we wish to give two applications of (M6.69) from thermodynamics that will also be useful in our studies.

M6.6.1 The budget equation for the partial masses of atmospheric air

For meteorological applications discussed in this book we denote the partial masses M^k of atmospheric air in the following way: $k = 0$: dry air, $k = 1$: water vapor, $k = 2$: liquid water, and $k = 3$: ice. Hence the summation index N occurring in equations (M6.40)–(M6.43) is restricted to $N = 3$. The index k appears sometimes as a subscript and sometimes as a superscript. Liquid water and ice are treated as bulk water phases; microphysical drop or ice-particle distributions are not considered in this context.

Let ψ represent the mass concentration $m^k = M^k/M$. In this case the term $\mathbf{F}_\psi$ represents the diffusion flux $\mathbf{J}^k$ while the source term Q_ψ is realized by the phase-transition rate I^k describing condensation and evaporation processes of liquid water and ice. Thus (M6.69) assumes the form

$$\boxed{\frac{D_3}{Dt}(\rho m^k) = \rho\,\frac{dm^k}{dt} = -\nabla\cdot\mathbf{J}^k + I^k} \tag{M6.72}$$

This is the *budget equation for the mass concentrations*. For dry air no phase transitions are possible, so $I^0 = 0$. Moreover, since $\sum_{k=0}^{3} m^k = 1$, it can be shown that $\sum_{k=0}^{3} I^k = 0$.

M6.6.2 The first law of thermodynamics

If ψ stands for the internal energy e, then equation (M6.69) represents a form of the *first law of thermodynamics*. In this case we have $\mathbf{F}_\psi = \mathbf{J}^{\mathrm{h}} + \mathbf{F}_{\mathrm{R}}$, where $\mathbf{J}^{\mathrm{h}}$ is the heat flux (*enthalpy flux*) and $\mathbf{F}_{\mathrm{R}}$ is the radiative flux. The source term is given by $Q_\psi = -p\,\nabla\cdot\mathbf{v} + \epsilon$, where $\epsilon = \mathbb{J}\cdot\cdot\nabla\mathbf{v}$ is the *energy dissipation* and $\mathbb{J}$ is the *viscous stress tensor*. Hence we may write

$$\boxed{\frac{D_3}{Dt}(\rho e) = \rho\,\frac{de}{dt} = -\nabla\cdot(\mathbf{J}^{\mathrm{h}} + \mathbf{F}_{\mathrm{R}}) - p\,\nabla\cdot\mathbf{v} + \epsilon} \tag{M6.73}$$

M6.7 Gauss' theorem and the Dirac delta function

Let $\mathbf{r}$, as usual, represent the position vector and apply the Gauss divergence theorem to the function

$$\frac{\mathbf{r}}{r^3} = -\nabla\frac{1}{r} \quad \text{with} \quad r = \sqrt{x^2+y^2+z^2} = |\mathbf{r}| \tag{M6.74}$$

The result is

$$\begin{gathered}\int_V \nabla\cdot\left(\frac{\mathbf{r}}{r^3}\right)d\tau = \int_S \frac{\mathbf{r}}{r^3}\cdot d\mathbf{S} = 0 \\ \text{since} \qquad \nabla\cdot\left(\frac{\mathbf{r}}{r^3}\right) = \frac{3}{r^3} + \mathbf{r}\cdot\mathbf{e}_r\,\frac{\partial}{\partial r}\left(\frac{1}{r^3}\right) = 0\end{gathered} \tag{M6.75}$$

provided that r differs from zero at all points on and within the surface S. This means that (M6.75) is valid only if the origin from which $\mathbf{r}$ is drawn does not lie within the volume V or on its boundary. Since the divergence theorem requires that the functions to which it is applied have continuous first partial derivatives throughout the volume of integration, it cannot be applied to $\mathbf{r}/r^3$ if the origin of $\mathbf{r}$ is within S. To handle this situation, we modify the region of integration by constructing a sphere of radius ϵ having the origin as its center; see Figure M6.11.

Within the region V' between S and S' the function $\mathbf{r}/r^3$ satisfies the condition of the divergence theorem so that (M6.75) is valid, so that

$$\int_S \frac{\mathbf{r}}{r^3}\cdot d\mathbf{S} + \int_{S'} \frac{\mathbf{r}}{r^3}\cdot d\mathbf{S} = 0 \tag{M6.76}$$

According to Figure M6.11, in the last integral of (M6.76) we may make the replacements $r = \epsilon, \mathbf{r} = -\epsilon\mathbf{n}$ so that $\mathbf{r}\cdot d\mathbf{S} = -\epsilon\mathbf{n}\cdot\mathbf{n}\,dS'' = -\epsilon\,dS''$ and we obtain

$$\int_S \frac{\mathbf{r}}{r^3}\cdot d\mathbf{S} = \frac{1}{\epsilon^2}\int_{S'} dS'' = 4\pi \tag{M6.77}$$

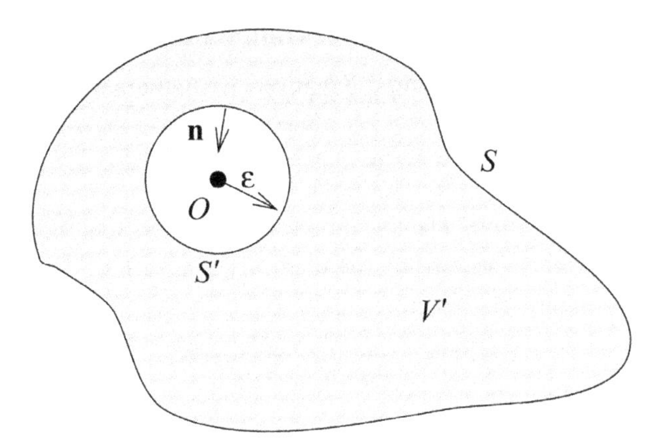

Fig. M6.11 Exclusion of a singular point from a three-dimensional region by an auxiliary spherical surface S'.

If S is a closed regular two-sided surface, then combining (M6.75) and (M6.77) yields

$$\int_S \frac{\mathbf{r}}{r^3} \cdot d\mathbf{S} = \begin{cases} 4\pi & \text{if } O \text{ is inside } S \\ 0 & \text{if } O \text{ is outside } S \end{cases} \tag{M6.78}$$

This is known as Gauss' theorem. We refer to Wylie (1966), where further details may be found.

The one-dimensional Dirac delta function has the properties that

$$\begin{aligned} \delta(x-a) &= 0 \quad \text{for} \quad x \neq a \\ \int_{\Delta x} \delta(x-a)\,dx &= \begin{cases} 1 & \text{if } a \in \Delta x \\ 0 & \text{if } a \notin \Delta x \end{cases} \\ \int_{\Delta x} f(x)\delta(x-a)\,dx &= f(a) \end{aligned} \tag{M6.79}$$

where $f(x)$ is any well-behaved function and a is included in the region of integration. Generalizing, we have

$$\begin{aligned} \delta(\mathbf{r}-\mathbf{r}') &= 0 \quad \text{if} \quad \mathbf{r} \neq \mathbf{r}' \\ \delta(\mathbf{r}-\mathbf{r}') &= \delta(\mathbf{r}'-\mathbf{r}) = \delta(x-x')\delta(y-y')\delta(z-z') \\ \int_V \delta(\mathbf{r}-\mathbf{r}')\,d\tau &= \begin{cases} 1 & \text{if } \mathbf{r}' \in V \\ 0 & \text{if } \mathbf{r}' \notin V \end{cases} \\ \int_V f(\mathbf{r})\delta(\mathbf{r}-\mathbf{r}')\,d\tau &= \begin{cases} f(\mathbf{r}') & \text{if } \mathbf{r}' \in V \\ 0 & \text{if } \mathbf{r}' \notin V \end{cases} \end{aligned} \tag{M6.80}$$

An excellent reference, for example, is Arfken (1970).

M6.8 Solution of Poisson's differential equation

One of the most important differential equations in mathematical physics is Poisson's equation,

$$\nabla^2\phi = Cf(\mathbf{r}) \tag{M6.81}$$

with C = constant and where $f(\mathbf{r})$ is some function to be specified later. We will show that this equation is satisfied by the integral

$$\phi(\mathbf{r}) = -\frac{C}{4\pi}\int_{V'} \frac{f(\mathbf{r}')}{|\mathbf{r}-\mathbf{r}'|}\,d\tau' \tag{M6.82}$$

To verify the validity of this integral we substitute (M6.82) into (M6.81) and carry out the differentiation with respect to $\mathbf{r}$. Formally we may write

$$\nabla^2\phi(\mathbf{r}) = -\frac{C}{4\pi}\int_{V'} f(\mathbf{r}')\,\nabla^2\left(\frac{1}{|\mathbf{r}-\mathbf{r}'|}\right)d\tau' \tag{M6.83}$$

It is convenient and permissible to translate the origin to $\mathbf{r}'$ and consider $\nabla^2\,|1/\mathbf{r}| = \nabla^2(1/r)$. Now we have exactly the situation leading to the development of Gauss' theorem (M6.78). Using this theorem together with (M6.74) and (M6.75), we obtain

$$-\int_V \nabla^2\left(\frac{1}{r}\right)d\tau = \int_V \nabla\cdot\left(\frac{\mathbf{r}}{r^3}\right)d\tau = \begin{cases} 4\pi & \text{if the origin of } \mathbf{r}\in V \\ 0 & \text{if the origin of } \mathbf{r}\notin V \end{cases} \tag{M6.84}$$

This may be conveniently expressed by means of the Dirac delta function as

$$\nabla^2\left(\frac{1}{r}\right) = -4\pi\,\delta(\mathbf{r}-0) \tag{M6.85}$$

We must modify (M6.85) since we have displaced the origin to $\mathbf{r}'$. The term 4π in Gauss' theorem appears if and only if the volume includes the point $\mathbf{r}'$. Therefore, we replace (M6.85) by

$$\nabla^2\left(\frac{1}{|\mathbf{r}-\mathbf{r}'|}\right) = -4\pi\,\delta(\mathbf{r}-\mathbf{r}') \quad \text{with} \quad \mathbf{r}'\neq 0 \tag{M6.86}$$

Substitution of (M6.86) into (M6.83) leads to the original differential equation (M6.81)

$$\nabla^2\phi(\mathbf{r}) = -\frac{C}{4\pi}\int_{V'} f(\mathbf{r}')\big[-4\pi\,\delta(\mathbf{r}-\mathbf{r}')\big]\,d\tau' = Cf(\mathbf{r}) \tag{M6.87}$$

M6.9 Appendix: Remarks on Euclidian and Riemannian spaces

A point in a generalized n-dimensional space is specified by the coordinates $q^1, q^2, \ldots, q^n$. The coordinates of any two neighboring points in this space differ by the differentials $dq^1, dq^2, \ldots, dq^n$. One speaks of a *Riemannian space* if the distance between these two points is defined by the metric fundamental form

$$d\mathbf{r}^2 = g_{mn}\, dq^m\, dq^n \tag{M6.88}$$

If it is possible to transform the entire space so that $d\mathbf{r}^2$ can be expressed by the Cartesian coordinates $x^1, x^2, \ldots, x^n$ with

$$d\mathbf{r}^2 = \delta_{mn}\, dx^m\, dx^n \tag{M6.89}$$

then the space is called a *Euclidian space*. For the common three-dimensional Euclidian space we may write

$$d\mathbf{r}^2 = dx^2 + dy^2 + dz^2 \tag{M6.90}$$

The basic difference between the Riemannian and the Euclidian space is that the Euclidian space is considered flat whereas the Riemannian space is curved. A curved surface embedded in a three-dimensional Euclidian space is the only Riemannian space which is perceptible to us. In general, the transformation from (M6.88) to the form (M6.89) is not possible.

The metric fundamental quantities of a certain space contain all the required information that is necessary in order to find out whether we are dealing with a flat or a curved space. If it turns out that the g_{ij} are constant then the space is flat. Necessarily, in a curved space the g_{ij} are not constants, but depend on the coordinates q^i. However, from the simple fact that the g_{ij} are functions of the coordinates q^i it cannot be concluded that the space is curved. For example, let us consider polar coordinates, in which the distance in space is defined by the metric fundamental form

$$d\mathbf{r}^2 = dr^2 + r^2\, d\alpha^2 \quad \text{with} \quad g_{11} = 1, \quad g_{22} = r^2 \tag{M6.91}$$

so that one of the g_{ij} is a function of the generalized coordinate $q^2 = r$. We leave it to the reader to show that (M6.91) can be transformed into the Cartesian form (M6.89). In this particular case it is quite obvious that the space is flat. Another example is provided by constructing on a plane sheet of paper a Cartesian grid. Rolling the paper to form a cylinder of radius R does not change the distances on the surface, showing that it is possible to transform the fundamental form

$$d\mathbf{r}^2 = dR^2 + R^2\, d\alpha^2 + dz^2 \quad \text{with} \quad g_{11} = 1, \quad g_{22} = R^2, \quad g_{33} = 1 \tag{M6.92}$$

into the Cartesian form (M6.89). A very inportant curved space is the surface of a sphere, which is characterized by the radius r, the latitude φ, the longitude λ, and the metric fundamental form

$$
\begin{aligned}
d\mathbf{r}^2 &= r^2 \cos^2\varphi \, d\lambda^2 + r^2 \, d\varphi^2 + dr^2 \\
\text{with} \quad g_{11} &= r^2\cos^2\varphi, \quad g_{22} = r^2, \quad g_{33} = 1
\end{aligned}
\tag{M6.93}
$$

By inspection or trial-and-error analysis it is very difficult to determine whether a given coordinate-dependent metric tensor $g_{ij}(q^k)$ can be transformed into Cartesian coordinates. Fortunately, there is a systematic method to determine whether the space is flat or curved by calculating the *Riemann–Christoffel tensor* or the *curvature tensor*.

Let us consider the expression $(\nabla_i \nabla_j - \nabla_j \nabla_i)A_k$, where ∇_i is the covariant derivative. By using the methods we have studied previously, omitting details, we can show that

$$
\begin{aligned}
&\nabla_i(\nabla_j A_k) - \nabla_j(\nabla_i A_k) = A_n R^n_{kij} \\
&\quad \text{with} \quad R^l_{kij} = \Gamma^m_{ki}\Gamma^l_{mj} - \frac{\partial}{\partial q^i}\Gamma^l_{kj} - \Gamma^m_{kj}\Gamma^l_{mi} + \frac{\partial}{\partial q^j}\Gamma^l_{ki}
\end{aligned}
\tag{M6.94}
$$

If the curvature tensor $R^l_{kij} = 0$ then the space is flat or uncurved. If $R^l_{kij} \neq 0$ then the space is curved as in the case of a spherical surface. This fact shows that no Cartesian coordinates exist for the sphere.

In atmospheric dynamics we often simplify the metric tensor by assuming that the radius extending from the center of the earth does not change with height throughout the meteorologically relevant part of the atmosphere. The simplification applies only to those terms of $d\mathbf{r}^2$ for which the radius appears in undifferentiated form. This type of Riemannian space is no longer perceptible to us.

Let us briefly consider the parallel transport of a vector since this concept can be used to determine the curvature tensor. By definition, a vector is transported parallelly if its direction and length do not change. Thus in the plane or, more generally, in the Euclidian space the parallel transport does not change the vector. The reason for this is that the basis vectors of the Euclidian space are constant and do not have to be differentiated whenever the vector is differentiated. The parallel displacement of a vector can be used to define the Euclidian space. For such a space the pararallel displacement along an arbitrary closed curve transports the vector to its original position without changing its length and direction.

As an illustration we consider polar coordinates as shown in Figure M6.12. In this two-dimensional space we transport the vector parallelly from point P to Q. While the vector itself remains constant the components ($x = r\cos\alpha$, $y = r\sin\alpha$) of the vector change. If we transport the vector around the circle back to P we obtain the original vector.

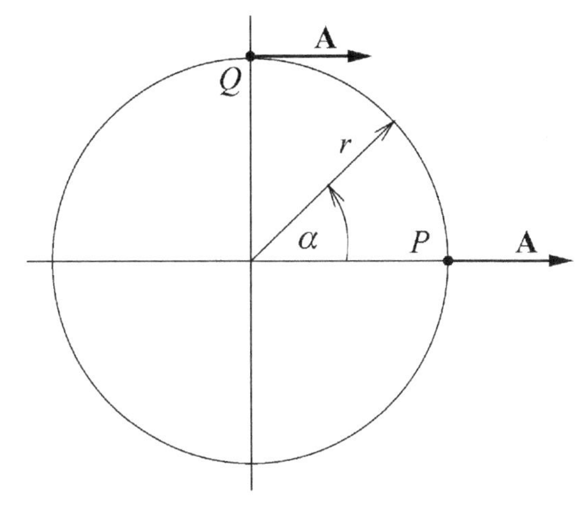

Fig. M6.12 Parallel transport of a vector in the two-dimensional Euclidian space.

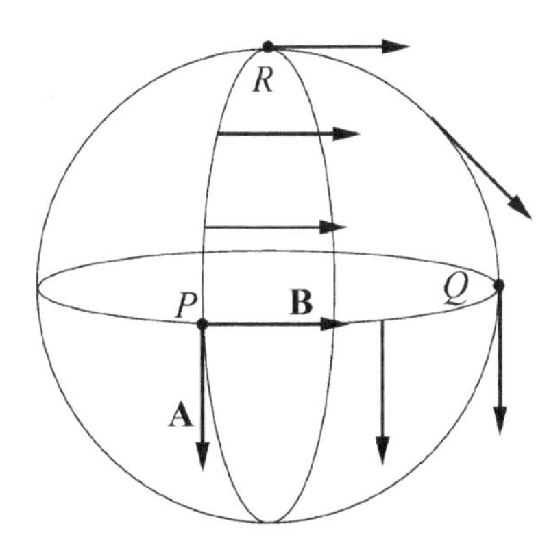

Fig. M6.13 The path dependence of surface parallelism.

A different situation occurs if the vector is transported along a curved surface since surface parallelism depends on the path. The definition "parallel transport of the surface vectors" implies two operations, namely a parallel shift in space, followed by a projection onto the tangent plane. We will demonstrate this situation in Figure M6.13 for a sphere. Let us transport the vector **A** along a great circle from point P to Q. Both points are located on the equator. From point Q the vector is displaced to the pole, point R, and then back to point P. During the transport the angle between the vector and the great circles (geodesic circles) remains constant. The displacement of the vector along the closed path results in a vector **B**, which is perpendicular to the original vector **A**. Thus surface parallelism depends on the path. Had we transported **A** from P to Q and then back to P, of course, we would have obtained the original vector.

M6.10 Problems

M6.1: Show that the integral $\oint_S d\mathbf{S} = 0$.

M6.2: In equation (M6.49) set $\mathbf{A} = (1/\rho)\nabla p$ to verify the following statement:

$$\frac{d}{dt}\left[\oint_{L(t)} d\mathbf{r}\cdot\left(\frac{1}{\rho}\nabla p\right)\right] = \oint_{L(t)}\left[(d_g p)\frac{d}{dt}\left(\frac{1}{\rho}\right) - (d_g\rho)\frac{dp}{dt}\right]$$

where ρ and p are scalar field functions.

M6.3: Use the integral theorems to verify the following statements:

$$\nabla\times\frac{d\mathbf{v}}{dt} = \frac{d}{dt}(\nabla\times\mathbf{v}) + (\nabla\times\mathbf{v})\nabla\cdot\mathbf{v} - (\nabla\times\mathbf{v})\cdot\nabla\mathbf{v}$$

$$\nabla\cdot\frac{d\mathbf{v}}{dt} = \frac{d}{dt}(\nabla\cdot\mathbf{v}) + \nabla\mathbf{v}\cdot\cdot\nabla\mathbf{v}$$

M6.4: Consider the vector field $\mathbf{A} = A_x(x, y)\mathbf{i} + A_y(x, y)\mathbf{j}$ in the (x, y)-plane.
(a) Find the closed line integral $\oint_\Gamma d\mathbf{r}\times\mathbf{A}$. The closed line Γ is a rectangle defined by the corner points $(0, 0)$, $(L, 0)$, (L, M), $(0, M)$.
(b) Apply the result of (a) to the following situation:

$$A_x(x = 0, y) = \cos y, \qquad A_x(x = L, y) = \sin y$$
$$A_y(x, y = 0) = x, \qquad A_y(x, y = M) = -1/L$$

M6.5: The vector field $\mathbf{A} = C\mathbf{r}/r$, with $C =$ constant and $r = |\mathbf{r}|$, is given.
(a) Calculate the line integral $\int_{(1)}^{(2)} d\mathbf{r}\cdot\mathbf{A}$ for an arbitrary path from $(1)\rightarrow(2)$.
(b) Calculate the surface integral $\oint_S d\mathbf{S}\cdot\mathbf{A}$ for a spherical surface of radius a about the origin.

M6.6: For the vector field $\mathbf{v}_\Omega = \boldsymbol{\Omega}\times\mathbf{r}$, $\boldsymbol{\Omega} =$ constant, calculate the line integral $\oint \mathbf{v}_\Omega\cdot d\mathbf{r}$
(a) for a circle of radius a about the axis of rotation, and
(b) for an arbitrary closed curve.
Use Stokes' integral theorem.

M6.7: Show that

$$\int_V \nabla\psi\cdot\nabla\times(\mathbf{A}\cdot\nabla\mathbf{A})\,d\tau = 0$$

On the surface of the volume the vector $\mathbf{A}$ is of the form $\mathbf{A} = \mathbf{B}\times\mathbf{n}$, where $\mathbf{n}$ is the unit vector normal to the surface. In addition to this, assume that $\mathbf{A}\cdot\nabla\psi = 0$, where ψ is a scalar field function. Hint: Use Lamb's transformation (M3.75).

M7

Introduction to the concepts of nonlinear dynamics

By necessity, this introduction is brief and far from complete and may, therefore, be reviewed in a relatively short time.

M7.1 One-dimensional flow

M7.1.1 Fixed points and stability

It is very instructive to discuss a one-dimensional or first-order dynamic system described by the equation $\dot{x} = f(x)$. Since $x(t)$ is a real-valued function of time t, we may consider $\dot{x}$ to be a velocity repesenting the flow along the x-axis. The function $f(x)$ is assumed to be smooth and real-valued. A plot of $f(x)$ may look as shown in Figure M7.1. We imagine a fluid flowing along the x-axis. This imaginary fluid is called the *phase fluid* while the x-axis represents the one-dimensional *phase space*.

The sign of $f(x)$ determines the sign of the one-dimensional velocity $\dot{x}$. The flow is to the right where $f(x) > 0$ and to the left where $f(x) < 0$. The solution of $\dot{x} = f(x)$ is found by considering an imaginary fluid particle, the *phase point*, whose initial position is at $x(t_0) = x_0$. We now observe how this particle is carried along by the flow. As time increases, the phase point moves along the x-axis according to some function $x(t)$, which is called the *trajectory* of the fluid particle. The phase portrait is controlled by the *fixed points* x^*, also known as *equilibrium* or *critical points*, which are found from $f(x^*) = 0$. Fixed points correspond to stagnation points of the flow.

In Figure M7.1 the point P_s is a stable fixed point since the local flow is directed from two sides toward this point. The point P_u is an unstable fixed point since the flow is away from it. Interpreting the original differential equation, fixed points are *equilibrium solutions*. Sometimes they are also called steady, constant, or rest solutions (the fluid is stagnant or at rest). The reason for this terminology is that, if $x = x^*$ initially, then $x(t) = x^*$ for all times. The definition of *stable equilibrium*

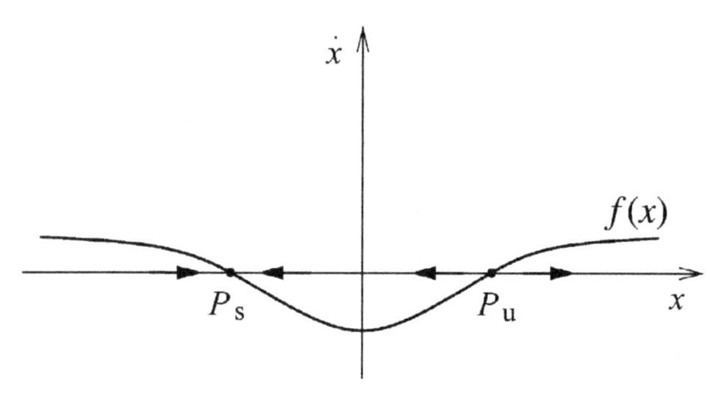

Fig. M7.1 A phase portrait of one-dimensional flow along the x-axis.

is based on considering small perturbations. In Figure M7.1 all small disturbances to P_s will decay. Large disturbances that send the phase point x to the right of P_u will not decay but will be repelled out to $+\infty$. Thus, we say that the fixed point P_s is *locally stable* only. The concept of global stability will be discussed later.

It should be noted that f does not explicitly depend on time. In this case one speaks of an *autonomous system*. If f depends explicitly on t then the equation is *nonautonomous* and usually much more difficult to handle.

Now we practice this way of geometric thinking with a simple example: Find the fixed points for $\dot{x} = x^2 - 4$ and classify their stability. Solution: In this case $f(x) = x^2 - 4$. Setting $f(x^*) = 0$ and solving for x^* yields $x_1^* = 2, x_2^* = -2$. To determine the stability we plot $f(x)$ similarly to Figure M7.1 and obtain a parabola. Thus, we easily see that $x^* = 2$ is unstable whereas $x^* = -2$ is stable.

Now we discuss the concept of *linear stability*. Let $\eta(t)$ represent a small perturbation away from the fixed point x^*,

$$\eta(t) = x(t) - x^* \tag{M7.1}$$

To recognize whether a disturbance grows or decays we need to derive a differential equation for $\eta(t)$. Differentiation of η with respect to time results in

$$\dot{\eta} = \dot{x} = f(x) = f(x^* + \eta) \tag{M7.2}$$

Now we carry out a Taylor expansion of $f(x)$ and discontinue the series after the linear term since only small perturbations are admitted:

$$f(x^* + \eta) = f(x^*) + \eta f'(x^*) + \mathcal{O}(\eta^2) \tag{M7.3}$$

Suppose that $f'(x^*) \neq 0$, then the $\mathcal{O}(\eta^2)$ terms may be ignored. Since $f(x^*) = 0$ we obtain

$$\dot{\eta} = \eta f'(x^*) \tag{M7.4}$$

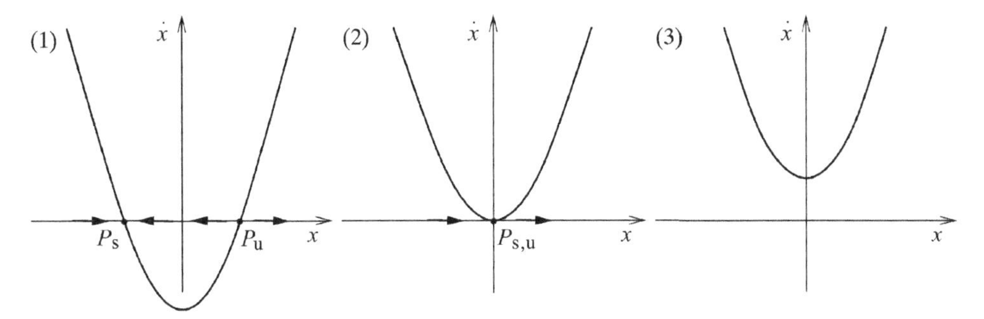

Fig. M7.2 Saddle-node bifurcation. Curve (1): $r < 0$, stable fixed point P_s, unstable fixed point P_u. Curve (2): $r = 0$, half-stable fixed point $P_{s,u}$. Curve (3): $r > 0$, no fixed point.

The linearization about x^* shows that $\eta(t)$ grows exponentially in time if $f'(x^*) > 0$ and decays if $f'(x^*) < 0$. If $f'(x^*) = 0$, the $\mathcal{O}(\eta^2)$ terms cannot be neglected and a nonlinear analysis is needed in order to determine the stability. In the previous example $f'(x^* = 2) = 4$ so that the fixed point $x^* = 2$ is unstable. Since $f'(x^* = -2) = -4$, the fixed point $x^* = -2$ is stable.

Consider now the problem $\dot{x} = x^3$. The fixed point is $x^* = 0$. In this example the linear stability analysis fails since $f'(x^* = 0) = 0$. A plot of $f(x) = x^3$ shows, however, that the origin is an unstable fixed point.

M7.1.2 Bifurcation

More instructive than the above examples is the dependence of x on a parameter since now qualitative changes in the dynamics of systems may occur as the parameter is varied. Whenever this change occurs we speak of bifurcation. The parameter values at which bifurcations occur are called *bifurcation points*. We will now briefly discuss various types of bifurcations.

M7.1.2.1 Saddle-node bifurcation

A *saddle-node bifurcation* is the basic mechanism by which fixed points are created or destroyed. Consider the equation $\dot{x} = r + x^2$, where r is a parameter. Examples of this type may be found in various textbooks or may be constructed. The result of the analysis is shown in Figure M7.2. If $r < 0$ we have one stable and one unstable fixed point (curve 1). As the parameter r approaches 0 from below, the parabola moves up and the two fixed points move toward each other until they collide into a half-stable fixed point $P_{s,u}$ at $x^* = 0$ (curve 2). The half-stable fixed point vanishes with $r > 0$ as shown by curve 3.

There are other possibilities for depicting bifurcations. A popular way is to select r as the abscissa and x as the ordinate. Setting $\dot{x} = 0$, the curve $r = -x^2$ consists

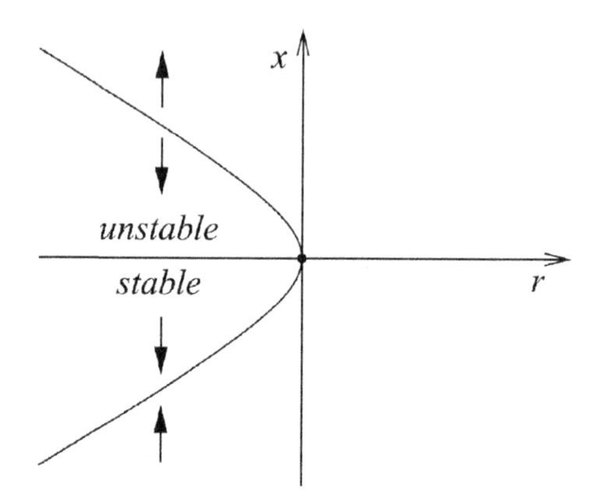

Fig. M7.3 An alternate bifurcation diagram for the saddle-node bifurcation.

of all fixed points of the system; see Figure M7.3. In agreement with Figure M7.2, there is no fixed point for $r > 0$. The arrows in Figure M7.3 indicate the direction of movement toward the fixed-point curve or away from it, thus describing regions of stability and instability, respectively. The origin itself is called a *turning point*.

M7.1.2.2 Transcritical bifurcation

There are situations in which a fixed point must exist for all parameter values of r and cannot be destroyed. However, such a fixed point may change its stability characteristics as r is varied. The *transcritical bifurcation* provides the standard example. The normal form of this type of bifurcation is $\dot{x} = x(r - x)$. There exists a fixed point $x^* = 0$ that is independent of r. Various situations are shown in Figure M7.4. For $r < 0$, $x^* = r$ is an unstable fixed point P_u whereas $x^* = 0$ is a stable fixed point P_s. As $r \to 0$, the unstable fixed point approaches the origin, colliding with it when $r = 0$. Now $x^* = 0$ is a half-stable fixed point $P_{s,u}$. Finally, for $r > 0$ the origin becomes unstable whereas $x^* = r$ is a stable fixed point.

In this case of transcritical bifurcation the two fixed points have not disappeared after the collision with the origin. In fact, an exchange of stability has taken place between the two fixed points. This kind of behavior is in contrast to the saddle-node bifurcation, whereby fixed points are created and destroyed.

M7.1.2.3 Pitchfork bifurcation

This type of bifurcation results from physical problems having symmetry properties. There are two types of *pitchfork bifurcations*.

M7.1.2.3.1. Supercritical pitchfork bifurcation The normal form is given by $\dot{x} = rx - x^3$. This equation is invariant if the variables x and $-x$ are interchanged. The plot of this function is shown in Figure M7.5. If $r \leq 0$ the origin is the only fixed

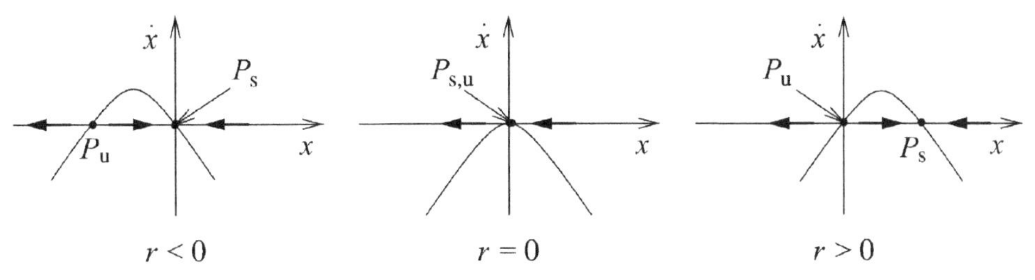

Fig. M7.4 Transcritical bifurcation.

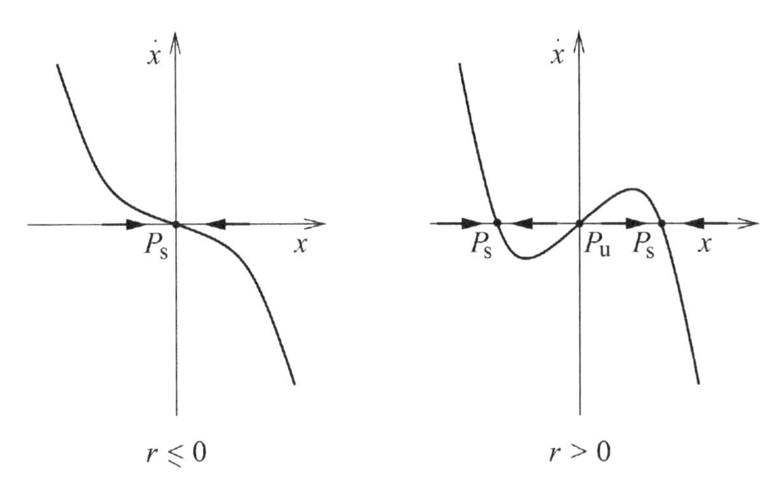

Fig. M7.5 Supercritical pitchfork bifurcation for $\dot{x} = rx - x^3$.

point which is stable. The two curves are quite similar, so only one curve is shown. The stability decreases as r approaches zero. For r = 0 the origin is still stable but more weakly so. If $r > 0$ there are three fixed points. The fixed point at the origin is unstable; the remaining two fixed points are stable.

The curious name pitchfork derives from the bifurcation diagram shown in Figure M7.6. We set $\dot{x} = 0$ and plot $x^2 = r$, yielding a parabola for $r > 0$. From the corresponding curve of Figure M7.5 we recognize that the upper and lower branches refer to stability. If $x = 0$ but $r > 0$, we find instability along the positive part of the abscissa. For $x = 0$ but $r < 0$, from curve 1 of Figure M7.5 we find stability along the negative part of the abscissa.

M7.1.2.3.2 Subcritical pitchfork bifurcation A simple example is given by $\dot{x} = rx + x^3$, see Figure M7.7. For $r \geq 0$ the origin is the only fixed point and it is unstable. Nonzero fixed points exist only for $r < 0$, which are unstable while now the origin is stable. The corresponding bifurcation diagram is depicted in Figure M7.8. Comparison with Figure M7.6 shows that the bifurcation diagrams of the subcritical and supercritical pitchforks are inverted with respect to each other.

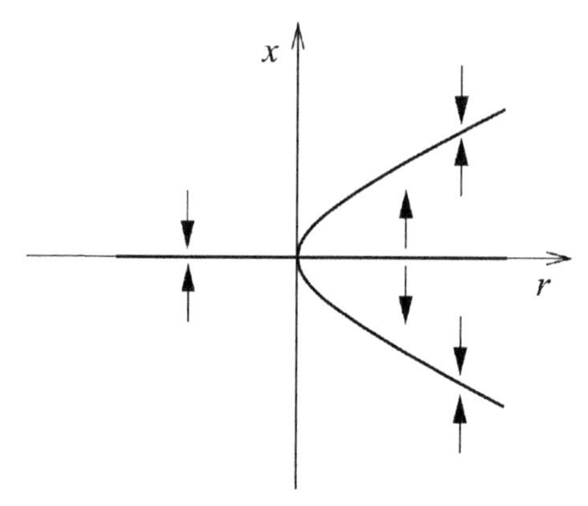

Fig. M7.6 A bifurcation diagram for $\dot{x} = rx - x^3$.

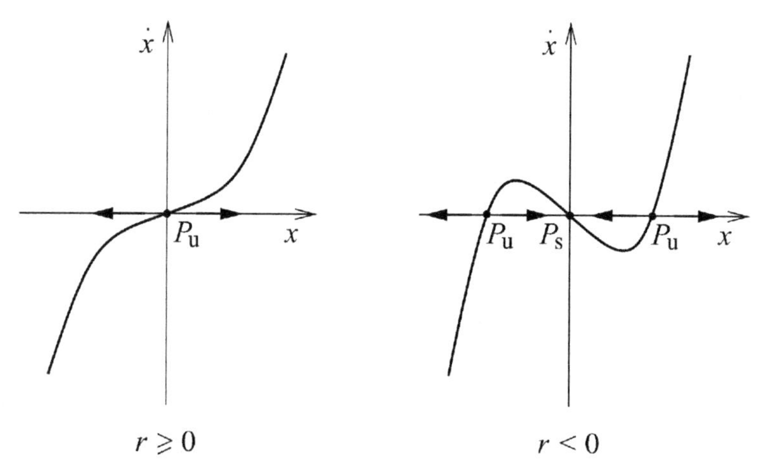

Fig. M7.7 Subcritical pitchfork bifurcation $\dot{x} = rx + x^3$.

M7.2 Two-dimensional flow

M7.2.1 Linear stability analysis

Let us consider the two-dimensional linear system

$$\begin{aligned}\dot{x} &= ax + by \\ \dot{y} &= cx + dy\end{aligned} \quad \text{or} \quad \dot{\mathbf{x}} = A\mathbf{x} \quad \text{with} \quad x = \begin{pmatrix} x \\ y \end{pmatrix}, \qquad A = \begin{pmatrix} a & b \\ c & d \end{pmatrix} \tag{M7.5}$$

where a, b, c, d are constants. The solutions to (M7.5) can be viewed as trajectories moving on the (x, y)-phase plane. Much can be learned from a very simple example,

$$\dot{x} = ax, \qquad \dot{y} = -y \tag{M7.6}$$

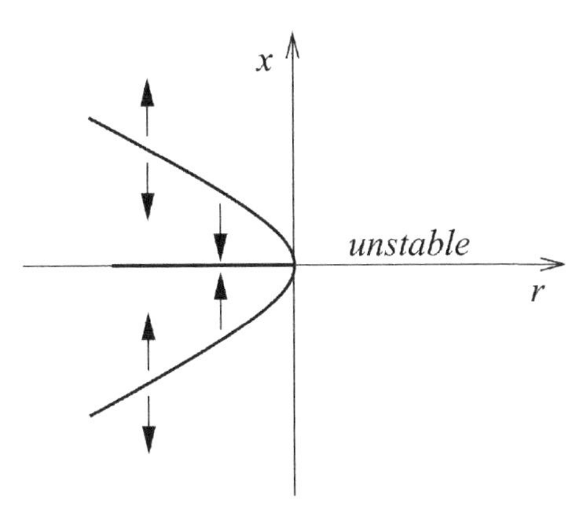

Fig. M7.8 A bifurcation diagram for $\dot{x} = rx + x^3$.

Here the system is uncoupled so that the solution can be found separately for each equation:

$$x(t) = x_0 \exp(at), \qquad y(t) = y_0 \exp(-t) \tag{M7.7}$$

The initial values are (x_0, y_0). The sign of the constant a determines the type of flow, which in this case can easily be depicted; see Figure M7.9. For $a < 0$ every point (x_0, y_0) approaches the origin as $t \to \infty$. The direction of approach depends on the size of a relative to -1. Figure M7.9(a) shows the situation for $a < -1$. The fixed point ($x^* = 0$, $y^* = 0$) is known as a *stable node*. If $a = -1$ the approaches on the abscissa and the ordinate are equally fast so that all trajectories are straight lines through the origin as shown in part (b) of the figure. This also follows from (M7.7), which in this case may be written as $y(t)/x(t) = y_0/x_0 =$ constant. This arrangement is called a *symmetric node* or a *star*.

When $a = 0$ a dramatic change in the flow pattern takes place. Now $x(t) = x_0$ so that there is an infinite line of fixed points (x^*, $y^* = 0$) along the x-axis (Figure M7.9(c)). These stable fixed points are approached from above and below by vertical trajectories depending on the sign of y_0. Finally, for $a > 0$, we obtain a situation in which $(x^*, y^*) = (0, 0)$. This is known as a *saddle point*. With the exception of the trajectories on the y-axis all trajectories are heading out for $+\infty$ or $-\infty$ along the x-axis. In forward time, these trajectories are asymptotic to the x-axis; in backward time ($t \to -\infty$) they are asymptotic to the y-axis. The y-axis is known as the *stable manifold* of the saddle point. More precisely, this is the set of initial conditions (x_0, y_0) such that $[x(t), y(t)] \to (0, 0)$ as $t \to \infty$. Analogously, the unstable manifold of the saddle point is the set of initial conditions (x_0, y_0) such that $[x(t), y(t)] \to (0, 0)$ as $t \to -\infty$. In this example the unstable manifold is the x-axis. It is seen that a typical trajectory approaches the unstable manifold for $t \to \infty$ and the stable manifold for $t \to -\infty$.

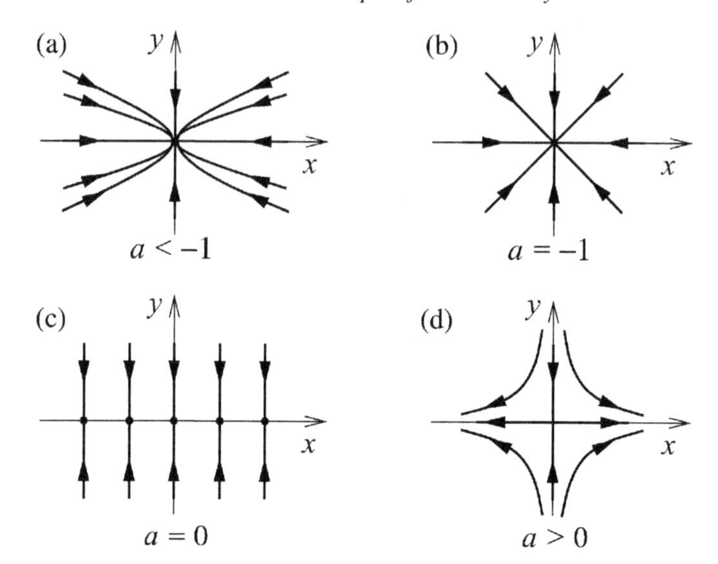

Fig. M7.9 Flow patterns of the differential-equation system (M7.7) for various values of the parameter a.

At this point it is opportune to introduce some important terminologies. Referring to Figures M7.9(a) and (b), $(x^*, y^*) = (0, 0)$ is classified as an *attracting fixed point* since all trajectories starting near this point approach the origin, i.e. $[x(t), y(t)] \to (0, 0)$ as $t \to \infty$. If all path directions are reversed this point is known as a *repellor*. If all trajectories that start sufficiently close to (0, 0) remain close to it for all time, not just as $t \to \infty$, then the fixed point is called *Liapunov stable*. In parts (a), (b), and (c) of Figure M7.9 the origin is Liapunov stable. From Figure M7.9(c) it can be seen that a fixed point can be Liapunov stable but not attracting. Here the fixed point is called *neutrally stable*. In Figure M7.9(d) the fixed point is neither attracting nor Liapunov stable, so $(x^*, y^*) = (0, 0)$ is unstable.

M7.2.2 Classification of linear systems

We will now consider the general solution to the linear system by seeking trajectories of the form

$$\mathbf{x}(t) = \exp(\lambda t)\,\mathbf{b} \tag{M7.8}$$

where the time-independent vector $\mathbf{b} \neq 0$ must be determined. Substitution of (M7.8) into (M7.5) yields

$$A\mathbf{b} = \lambda\mathbf{b} \quad \text{or} \quad (A - E\lambda)\mathbf{b} = 0 \tag{M7.9}$$

Here matrix notation has been utilized and E is the identity matrix. As usual, the eigenvalues $\lambda_{1,2}$ can be found by solving the equation for the determinant of the matrix $(A - \lambda E)$:

$$\begin{vmatrix} a-\lambda & b \\ c & d-\lambda \end{vmatrix} = 0 \Longrightarrow \lambda_{1,2} = \tfrac{1}{2}\left(\tau \pm \sqrt{\tau^2 - 4\Delta}\right) \tag{M7.10}$$

where $\tau = a+d$ is the trace and $\Delta = ad-bc$ is the determinant of A. The solution to (M7.5) may now be written as

$$\mathbf{x}(t) = C_1 \exp(\lambda_1 t)\,\mathbf{b}_1 + C_2 \exp(\lambda_2 t)\,\mathbf{b}_2 \tag{M7.11}$$

In general, the constants C_1, C_2 and the eigenvectors $\mathbf{b}_1$, $\mathbf{b}_2$ of the eigenvalues λ_1, λ_2 are complex quantities.

As an example, let us consider the phase-plane analysis of the simple harmonic oscillator

$$m\ddot{x} + kx = 0 \tag{M7.12}$$

where m is the mass and k is Hooke's constant. By setting $\dot{x} = y$, this second-order differential equation may be written as a system of two first-order differential equations:

$$\dot{x} = y, \quad \dot{y} = -\omega^2 x \quad \text{with} \quad \omega^2 = k/m \tag{M7.13}$$

From this system we immediately recognize that the fixed point is located at $(x^*, y^*) = (0, 0)$. On dividing $\dot{y}$ by $\dot{x}$, we find the equation of the trajectory which upon integration gives the equation of an ellipse:

$$\frac{dy}{dx} = -\omega^2 \frac{x}{y} \Longrightarrow y^2 + \omega^2 x^2 = C \tag{M7.14}$$

The situation is depicted in Figure M7.10. The trajectories form closed lines around the origin, which is therefore known as a *center*. The direction of flow around the center is best recognized by placing an imaginary particle or phase point at a convenient point such as $(x = 0, y > 0)$. Thus, from (M7.13) it follows that $\dot{x} > 0$, so the flow is clockwise. Finally, with the help of (M7.5), the eigenvalues of the system may be easily found as $\lambda_{1,2} = \pm i\omega$, showing that a center is characterized by purely imaginary eigenvalues.

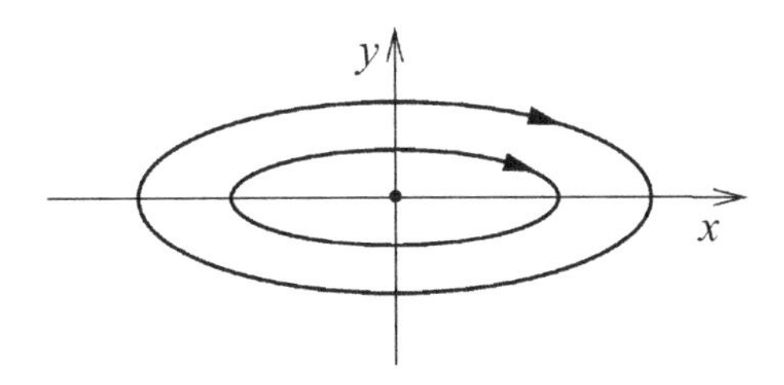

Fig. M7.10 The center for the simple harmonic oscillator.

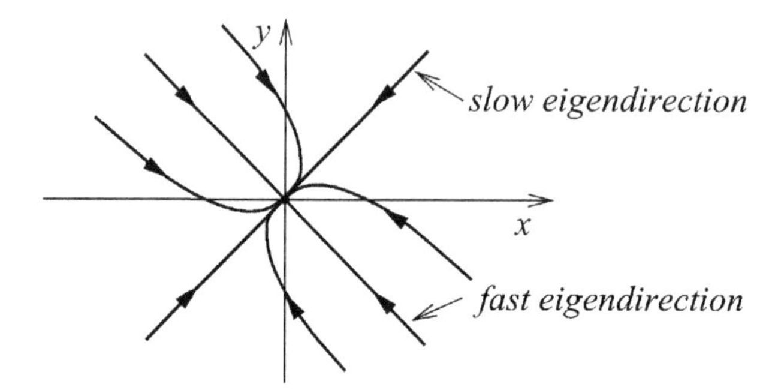

Fig. M7.11 A phase portrait of slow and fast eigendirections.

Before we summarize all important results, let us consider two more simple but important examples. Suppose that, for a given problem, the two eigenvalues are negative, say $\lambda_1 < \lambda_2 < 0$. Thus, both eigensolutions decrease exponentially so that the fixed point is a stable node. The eigenvector resulting from the smaller eigenvalue $|\lambda_2|$ is known as the *slow eigendirection*, while the eigenvector due to the larger value $|\lambda_1|$ is called the *fast eigendirection*. Trajectories typically approach the origin tangentially to the slow eigendirection for $t \rightarrow \infty$. In backward time, $t \rightarrow -\infty$, the trajectories become parallel to the fast eigendirection. The phase portrait may have the appearance of Figure M7.11 representing a stable node. Reversing the directions of the trajectories results in a typical portrait of an unstable node.

Let us briefly return to the phase portrait of the harmonic oscillator; see Figure M7.10. Since nearby trajectories are neither attracted to nor repelled from the fixed point, the center is classified as *neutrally stable*. If the harmonic oscillator were slightly damped, the equation of motion (M7.12) would be modified to read

$$m\ddot{x} + dm\,\dot{x} + kx = 0 \tag{M7.15}$$

where $d > 0$ is the damping constant. Now the trajectories cannot close because the oscillator loses some energy during each cycle. Weak damping is characterized by $\tau^2 - 4\Delta < 0$. The resulting phase portrait is a *stable spiral*. Since $\tau = -d < 0$ the eigenvalues are complex conjugates with a negative real part $\tau/2 = -d/2$. Owing to Euler's formula the solution contains the term $\exp(-\tau t/2)$ yielding exponentially decaying oscillations. Thus the fixed point is a stable spiral. If $d < 0$ in (M7.15), the fixed point is an *unstable spiral*; see Figure M7.12.

It may happen that the eigenvalues of the matrix A are degenerate, i.e. $\lambda_1 = \lambda_2$. In this case there exist two possibilities: Either there are two independent eigenvectors spanning the two-dimensional phase plane or only one eigenvector exists, so that the eigenspace is one-dimensional, and the fixed point is a *degenerate node*.

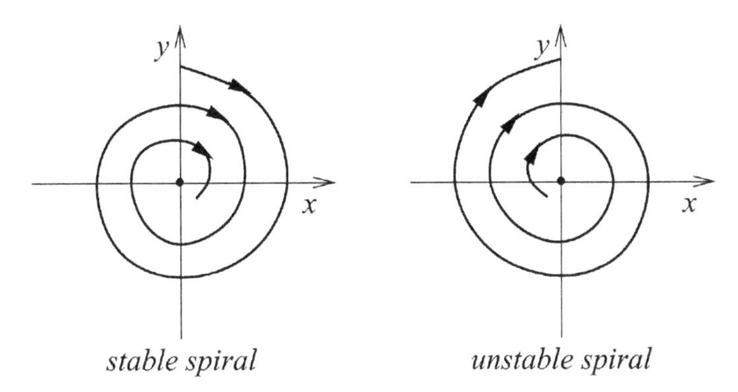

Fig. M7.12 Phase portraits for the stable and the unstable spiral.

From the previous examples we recognize that the type of the eigenvalue determines the stabililty behavior of the fixed point. The essential information is collected in the table of Figure M7.13. This table can be used to construct the stability diagram shown in the figure. In this diagram the ordinate is the trace τ and the abscissa is the determinant Δ of the matrix A of the two-dimensional linear system. Saddle points, nodes, and spirals occur in the large open regions of the (Δ, τ)-plane. They are the most important fixed points. Inspection of Figure M7.13 shows that centers, stars, and degenerate nodes are borderline cases. Of these special cases, centers are by far the most important since they occur in energy-conserving frictionless mechanical systems.

M7.2.3 Two-dimensional nonlinear systems

Before proceeding we state without proof that different trajectories do not intersect. This information may be extracted from the existence and uniqueness theorem of differential equations. If trajectories were to intersect, then there would be two solutions starting from the same point, the crossing point. This would violate the uniqueness part of the theorem. Because of the fact that trajectories do not intersect, we may expect that phase portraits have a "well-groomed" look to them.

By linearizing two-dimensional systems, it is often possible to obtain an approximate phase portrait near the fixed points. The general form of the nonlinear system is given by

$$\dot{x} = f_1(x, y), \qquad \dot{y} = f_2(x, y) \quad \text{or} \quad \dot{\mathbf{x}} = \mathbf{f}(\mathbf{x}) \tag{M7.16}$$

We assume the existence of a fixed point (x^*, y^*) so that $f_1(x^*, y^*) = 0$ and $f_2(x^*, y^*) = 0$. Let $u = x - x^*$ and $v = y - y^*$ represent the components of a small perturbation from the fixed point. To see whether the perturbation grows or is damped out, we need to obtain differential equations for $\dot{u} = \dot{x}$ and $\dot{v} = \dot{y}$.

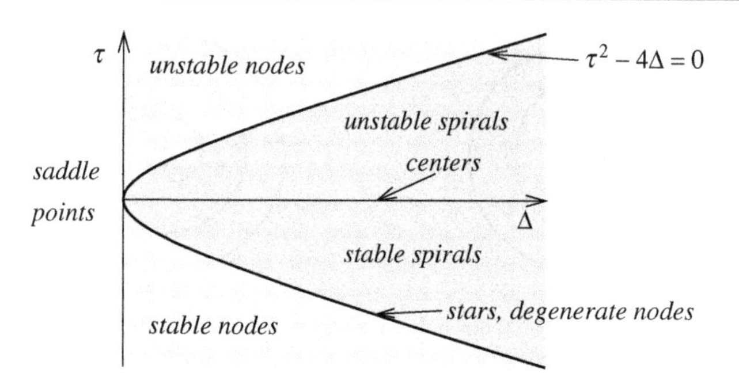

$\lambda_{1,2}$	Δ	τ	$\tau^2 - 4\Delta$	Fixed points
Real, distinct, opposite signs	<0	0	>0	(Unstable) saddle points
Real, distinct, >0	>0	>0	>0	Unstable nodes
Real, distinct, <0	>0	<0	>0	Stable nodes
Complex conjugate, real parts <0	>0	<0	<0	Stable spirals
Complex conjugate, real parts >0	>0	>0	<0	Unstable spirals
Purely imaginary	>0	0	<0	(Stable) centers
Real, $\lambda_1 = \lambda_2 < 0$	>0	<0	0	Stable nodes, star-shaped
Real, $\lambda_1 = \lambda_2 > 0$	>0	>0	0	Unstable nodes, star-shaped

Fig. M7.13 General classification of fixed points of the two-dimensional linear system $\Delta = \lambda_1\lambda_2$, $\tau = \lambda_1 + \lambda_2$.

Expanding f_1 and f_2 in a two-dimensional Taylor series we find

$$f_i(x^* + u, y^* + v) = f_i(x^*, y^*) + u\,\frac{\partial f_i}{\partial x}\bigg|_{x^*,y^*} + v\,\frac{\partial f_i}{\partial y}\bigg|_{x^*,y^*} + \mathcal{O}(u^2, v^2, uv), \quad i = 1, 2 \tag{M7.17}$$

The series will be discontinued after the linear terms since the quadratic terms $\mathcal{O}(u^2, v^2, uv)$ are very small and will be ignored. Because $f_1(x^*, y^*)$ and $f_2(x^*, y^*)$ are zero, the linearized system may be written as

$$\begin{pmatrix} \dot{u} \\ \dot{v} \end{pmatrix} = A \begin{pmatrix} u \\ v \end{pmatrix}, \qquad A = \begin{pmatrix} \dfrac{\partial f_1}{\partial x} & \dfrac{\partial f_1}{\partial y} \\ \dfrac{\partial f_2}{\partial x} & \dfrac{\partial f_2}{\partial y} \end{pmatrix}_{x^*,y^*} \tag{M7.18}$$

where A is the *Jacobian matrix* at the fixed point. The dynamics can now be analyzed by means of the procedures discussed above.

The question of whether the linearization of the nonlinear system gives the correct qualitative picture of the phase portrait near the fixed point (x^*, y^*) arises. This is indeed the case as long as the fixed point is not located on one of the border lines shown in Figure M7.13.

We will demonstrate the linearization procedure by finding and classifying the fixed points of the following example

$$\dot{x} = -x + x^3, \qquad \dot{y} = -2y \tag{M7.19}$$

The fixed points occur where $\dot{x} = 0$, $\dot{y} = 0$ simultaneously. The system is uncoupled so that the fixed points are easily found: $(x^*, y^*) = (0, 0)$, $(1, 0)$, $(-1, 0)$. The Jacobian matrix is given by

$$A = \begin{pmatrix} -1 + 3x^2 & 0 \\ 0 & -2 \end{pmatrix}_{x^*, y^*} \tag{M7.20}$$

Thus, for the three fixed points we find

$$(x^* = 0,\ y^* = 0): \qquad A = \begin{pmatrix} -1 & 0 \\ 0 & -2 \end{pmatrix}, \qquad \tau = -3, \quad \Delta = 2$$

$$(x^* = \pm 1,\ y^* = 0): \qquad A = \begin{pmatrix} 2 & 0 \\ 0 & -2 \end{pmatrix}, \qquad \tau = 0, \quad \Delta = -4 \tag{M7.21}$$

They are shown in Figure M7.14. On consulting the stability diagram of Figure M7.13 we may conclude that the fixed points represent the stable node P_s at the origin and two unstable saddle points P_u. Since we are not treating borderline cases we can be sure that our result is correct.

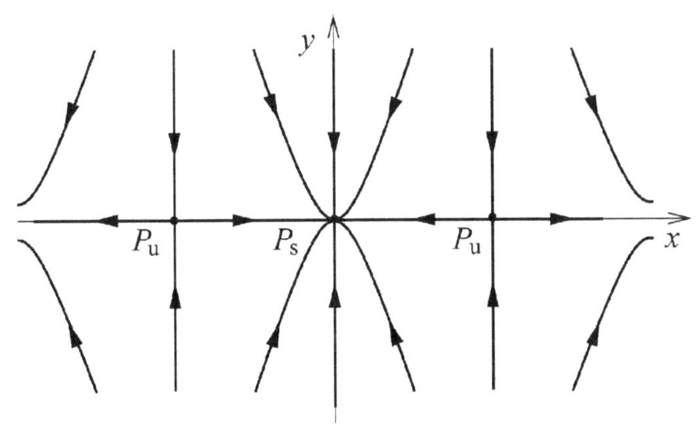

Fig. M7.14 The stability diagram for the fixed points of the example (M7.19).

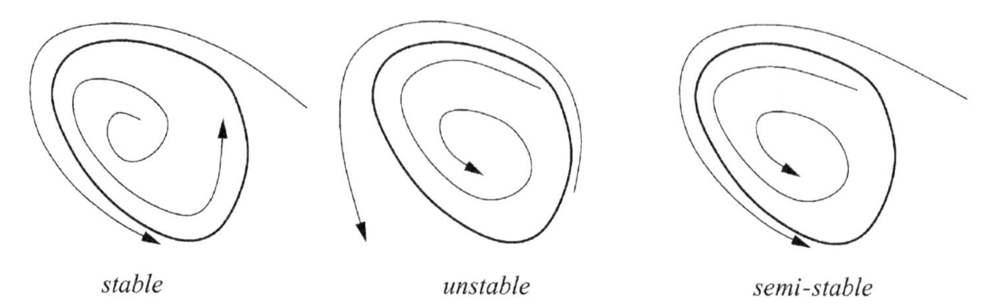

Fig. M7.15 Stable, unstable, and semi-stable limit cycle.

M7.2.4 Limit cycles

Let us now consider a limit cycle, which is a different type of a fixed point. By definition, a limit cycle is an isolated closed trajectory. In this context isolated implies that neighboring trajectories are not closed. They spiral either toward or away from the limit cycle, as shown in Figure M7.15. In the stable case the limit cycle is approached by all neighboring trajectories. Thus, the limit cycle is attracting. In the unstable situation the neighboring trajectories are repelled by the limit cycle. Finally, the rare case of a semi-stable limit cycle is a combination of the first two possibilities.

Let us consider the simple example

$$\dot{r} = r(1 - r^2), \qquad \dot{\theta} = 1 \tag{M7.22}$$

where the dynamic system refers to polar coordinates (r, θ) with $r > 0$. The radial and angular equations are uncoupled so they may be handled independently. We treat the flow as a vector field on the line r. The fixed points are $r^* = 0$ and $r^* = 1$. They are, respectively, unstable and stable. We observe that the equation of a unit circle $r^2 = x^2 + y^2 = 1$ represents a closed trajectory in the phase plane; see Figure M7.16. Since we are dealing with constant angular motion, we should expect that all trajectories spiral toward the limit cycle $r = 1$. We will soon return to this problem.

M7.2.5 Hopf bifurcation

Some bifurcations generate limit cycles or other periodic solutions. One such type is known as the Hopf bifurcation. Let us assume that a two-dimensional system has a stable fixed point. Now the question of in which ways the fixed point could lose stability if the stability parameter μ is varied arises. The position of the eigenvalues of the Jacobian matrix in the complex plane will provide the answer. Since the fixed point is stable by assumption, the two eigenvalues $\lambda_{1,2} = \lambda_r \pm i\lambda_i$ have negative

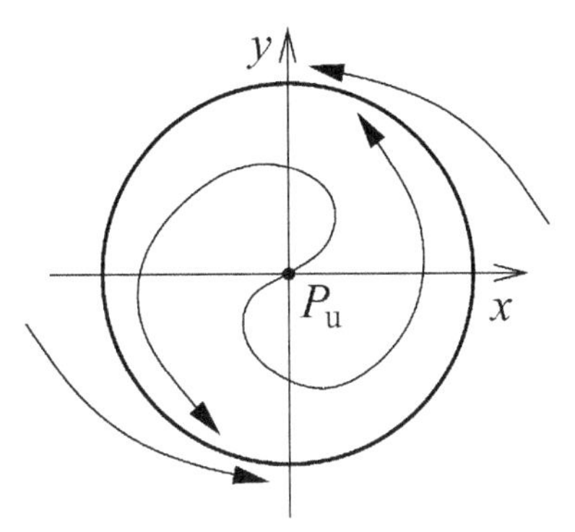

Fig. M7.16 The limit cycle corresponding to equation (M7.22).

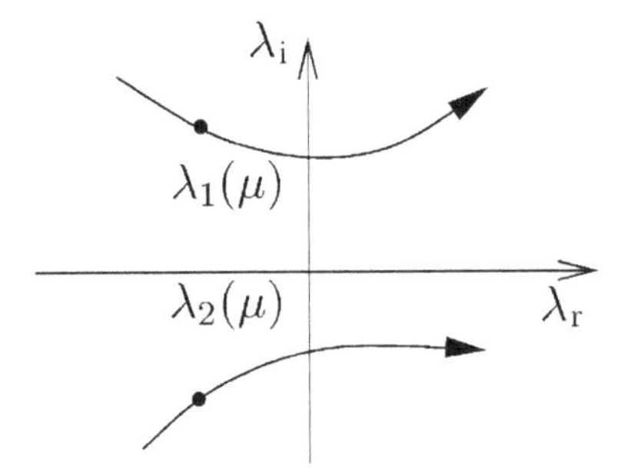

Fig. M7.17 Stability behaviors of the eigenvalues $\lambda_{1,2}$ as functions of the stability parameter μ.

real parts and must therefore lie in the left-hand half of the complex plane. In order to destabilize the fixed point, one or both of the eigenvalues must move into the right-hand half of the complex plane as μ varies; see Figure M7.17.

As an example let us consider the second-order nonlinear system

$$\frac{dx}{dt} = -\omega y + (\mu - x^2 - y^2)x, \qquad \frac{dy}{dt} = \omega x + (\mu - x^2 - y^2)y \tag{M7.23}$$

where μ and ω are real constants. The only fixed point is $(x^*, y^*) = (0, 0)$ so that the Jacobian matrix at the fixed point is expressed by

$$A = \begin{pmatrix} -3x^2 + \mu - y^2 & -\omega - 2xy \\ \omega - 2xy & -3y^2 + \mu \end{pmatrix}_{x^*=0,\ y^*=0} = \begin{pmatrix} \mu & -\omega \\ \omega & \mu \end{pmatrix} \tag{M7.24}$$

The eigenvalues to this matrix are complex:

$$\lambda_{1,2} = \mu \pm i\omega \tag{M7.25}$$

Thus we find

$$\tau = \lambda_1 + \lambda_2 = 2\mu, \qquad \Delta = \lambda_1\lambda_2 = \mu^2 + \omega^2 > 0, \qquad \tau^2 - 4\Delta < 0 \quad \text{(M7.26)}$$

From the stability diagram of Figure M7.13 we recognize that the fixed points are spirals. If $\mu < 0$ the spiral (often also called a *focus*) is stable, whereas for $\mu > 0$ the spiral is unstable.

We are going to confirm this conclusion since the system (M7.23) can be solved analytically by employing polar coordinates

$$x = r\cos\theta, \quad y = r\sin\theta \implies x + iy = r\exp(i\theta) \qquad \text{(M7.27)}$$

from which it follows that

$$\begin{aligned}\frac{d}{dt}\left[r\exp(i\theta)\right] &= \frac{dx}{dt} + i\frac{dy}{dt} = \left(\frac{dr}{dt} + ir\frac{d\theta}{dt}\right)\exp(i\theta) \\ &= \left[\omega r i + (\mu - r^2)r\right]\exp(i\theta)\end{aligned} \qquad \text{(M7.28)}$$

Here use of (M7.23) and (M7.27) has been made. On separating the real and imaginary parts we find

$$\frac{dr}{dt} = r(\mu - r^2), \qquad \frac{d\theta}{dt} = \omega \qquad \text{(M7.29)}$$

where ω is the frequency of the infinitesimal oscillations. By setting $\omega = 1$ and $\mu = 1$ we revert to the problem (M7.22) of the previous section which led to the introduction of the limit cycle.

The system (M7.29) is decoupled so that it is not so difficult to obtain an analytic solution. In general, it is impossible to find analytic solutions to nonlinear differential equations. Dividing (M7.29) by r^3 and setting $n = r^{-2}$ yields

$$\frac{dn}{dt} + 2\mu n = 2 \qquad \text{(M7.30)}$$

Multiplication of this equation by the integrating factor $\exp(2\mu t)$ and integration gives

$$\int_0^t \frac{d}{dt}\left[\exp(2\mu t)\,n(t)\right]dt = 2\int_0^t \exp(2\mu t)\,dt \qquad \text{(M7.31)}$$

from which it follows that

$$n(t) = \frac{1}{\mu} + C\exp(-2\mu t) \quad \text{with} \quad C = n(0) - \frac{1}{\mu}, \qquad n(0) = \frac{1}{r_0^2} \qquad \text{(M7.32)}$$

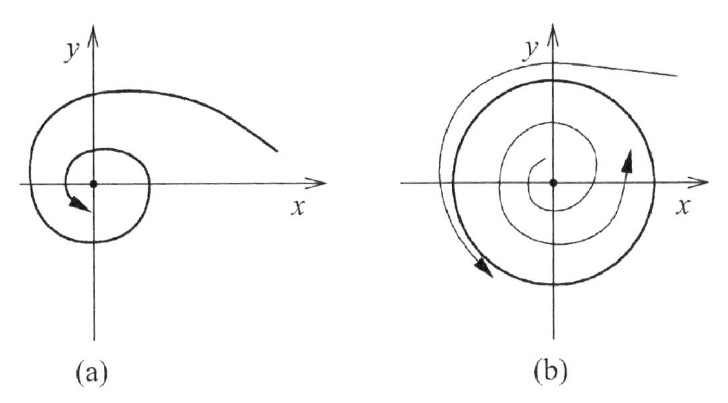

Fig. M7.18 Hopf bifurcation. (a) For $\mu \leq 0$ the origin is a stable spiral. (b) For $\mu > 0$ the origin is an unstable spiral. The circular limit cycle at $r^2 = \mu$ is stable.

Replacing n by r^{-2} and $n(0)$ by r_0^{-2} gives the solution of the differential system (M7.29):

$$r^2(t) = \begin{cases} \dfrac{\mu r_0^2}{r_0^2 + (\mu - r_0^2)\exp(-2\mu t)} & \mu \neq 0 \\ \dfrac{r_0^2}{1 + 2r_0^2 t} & \mu = 0 \end{cases} \tag{M7.33}$$
$$\theta(t) = \omega t + \theta_0$$

We will now present the phase portrait in the (x, y)-plane. As t increases from zero to infinity, we find that, for $\mu \leq 0$, a stable spiral is generated, as shown in Figure M7.18(a). This verifies our previous conclusion, which was derived from the linear stability analysis. In other words, all solutions $\mathbf{x} = [x(t), y(t)]$ tend to zero as t approaches infinity. Thus, each trajectory or orbit spirals into the origin. The sense of the rotation depends on the sign of ω. We observe that, for $\mu = 0$, the linear stability analysis wrongly predicts a center at the origin.

For $\mu > 0$ the origin becomes an unstable focus, see Figure M7.18(b). A new stable periodic solution arises as μ increases through zero and becomes positive. This is the limit cycle $r^2 = \mu$. Two trajectories starting inside and outside of the limit cycle are shown in Figure M7.18(b). Since $x^2 + y^2 = \mu$, the solution is given by

$$x^2 + y^2 = \mu \quad \text{or} \quad x = \sqrt{\mu}\cos(t + \theta_0), \qquad y = \sqrt{\mu}\sin(t + \theta_0) \tag{M7.34}$$

This is an example of a Hopf bifurcation. Hopf (1942) showed that this type of bifurcation occurs quite generally for systems ($n > 2$) of nonlinear differential equations. For further details see, for example, Drazin (1992). In the previous idealized example the limit cycle turned out to be circular. Hopf bifurcations encountered in practice usually result in limit cycles of elliptic shape.

In this example, a stable spiral has changed into an unstable spiral, which is surrounded by a limit cycle since the real part μ of the eigenvalue has crossed the imaginary axis from left to right as μ increased from negative to positive values. This particular type of bifurcation is called the *supercritical Hopf bifurcation*.

The so-called *subcritical Hopf bifurcation* has an entirely different character. After the bifurcation has occurred, the trajectories must jump to a distant attractor, which may be a fixed point, a limit cycle, infinity, or a chaotic attractor if three and higher dimensions are considered. We will study this situation in connection with the Lorenz equations.

M7.2.6 The Liapunov function

Let us consider a system $\dot{\mathbf{x}} = \mathbf{f}(\mathbf{x})$ having a fixed point at $\mathbf{x}^*$. Suppose that we can find a continuously differentiable real-valued function $V(\mathbf{x})$ with the following properties:

$$V(\mathbf{x}) > 0, \qquad \dot{V}(\mathbf{x}) < 0 \quad \forall \quad \mathbf{x} \neq \mathbf{x}^*, \qquad V(\mathbf{x}^*) = 0 \tag{M7.35}$$

This positive definite function is known as the Liapunov function. If this function exists, then the system does not admit closed orbits. The condition $\dot{V}(\mathbf{x}) < 0$ implies that all trajectories flow "downhill" toward $\mathbf{x}^*$. Unfortunately, there is no systematic way to construct such functions.

M7.2.7 Fractal dimensions

Fractal dimensions are characteristic of strange attractors. Therefore, it will be necessary to briefly introduce this concept. A one-dimensional figure, such as a straight line or a curve, can be covered by N one-dimensional boxes of side length ϵ. If L is the length of the line then $N\epsilon = L$, so we may write

$$N(\epsilon) = \left(\frac{L}{\epsilon}\right)^1 \tag{M7.36}$$

Similarly, a square and a three-dimensional cube of side lengths L can be covered by

$$N(\epsilon) = \left(\frac{L}{\epsilon}\right)^2, \qquad N(\epsilon) = \left(\frac{L}{\epsilon}\right)^3 \tag{M7.37}$$

Generalizing, for a d-dimensional box we obtain

$$N(\epsilon) = \left(\frac{L}{\epsilon}\right)^d \tag{M7.38}$$

On taking logarithms we find

$$d = \frac{\ln[N(\epsilon)]}{\ln L - \ln \epsilon} \tag{M7.39}$$

In the limit of small ϵ, the term $\ln L$ can be ignored in comparison with the second term in the denominator of (M7.39). This results in the so-called *capacity dimension* d_c, which is given by

$$d_c = -\lim_{\epsilon \to 0} \frac{\ln[N(\epsilon)]}{\ln \epsilon} \tag{M7.40}$$

It is easy to see that the capacity dimension of a point is zero.

There are other definitions of fractional dimensions. The most important of these is the *Hausdorff dimension* HD, which permits the d-dimensional boxes to vary in size. Thus, the capacity dimension is a special case of the Hausdorff dimension. The inequality

$$HD \leq d_c \tag{M7.41}$$

is valid.

In large parts this chapter follows the excellent textbook on nonlinear dynamics and chaos by Strogatz (1994).

Part 2

Dynamics of the atmosphere

1

The laws of atmospheric motion

1.1 The equation of absolute motion

The foundation to all of atmospheric dynamics is the description of motion in the absolute reference frame. This is a Cartesian coordinate system that is fixed with respect to the "fixed" stars. For all practical purposes we may regard this system as an inertial coordinate system for an earthbound observer. In this reference frame we may apply *Newton's second law of motion* stating that the change of momentum $\mathbf{M}$ with time of an arbitrary body equals the sum of the *real forces* acting on the body. Real forces must be distinguished from *fictitious forces*, which are not due to interactions of a particle with other bodies. Fictitious forces result from the particular type of the coordinate system which is used to describe the motion of the particle.

For a volume $V(t)$ in the absolute frame we may then write

$$\frac{d\mathbf{M}}{dt} = \frac{d}{dt}\int_{V(t)} \rho \mathbf{v}_{\mathrm{A}}\, d\tau = \sum_i \mathbf{F}_i + \sum_i \mathbf{P}_i \tag{1.1}$$

where $d\tau$ is a volume element, ρ is the density of the medium, and $\mathbf{v}_{\mathrm{A}}$ is the *absolute velocity*. The forces appearing on the right-hand side of this equation include *mass* or *volume forces* $\mathbf{F}_i$ and *surface forces* $\mathbf{P}_i$ of the system. We do not need to include molecular-type forces between mass elements occurring in the interior part of the system since their net effect adds up to zero. If the $\mathbf{f}_i$ represent the forces per unit mass and $\mathbf{p}_i$ the surface forces per unit surface area, we may write

$$\sum_i \mathbf{F}_i = \int_{V(t)} \sum_i \rho \mathbf{f}_i\, d\tau, \qquad \sum_i \mathbf{P}_i = \oint_{S(t)} \sum_i \mathbf{p}_i\, dS \tag{1.2}$$

Here, $S(t)$ is the surface enclosing the volume $V(t)$ and dS is a surface element. Combination of these two formulas gives

$$\frac{d}{dt}\int_{V(t)} \rho \mathbf{v}_{\mathrm{A}}\, d\tau = \int_{V(t)} \frac{D_3}{Dt}(\rho \mathbf{v}_{\mathrm{A}})\, d\tau = \int_{V(t)} \sum_i \rho \mathbf{f}_i\, d\tau + \oint_{S(t)} \sum_i \mathbf{p}_i\, dS \tag{1.3}$$

The left-hand side of this equation refers to the rate of change with time of a fluid volume whose surface elements move with the respective absolute velocity $\mathbf{v}_A$. The budget operator D_3/Dt is given by (M6.58). In the following the index 3 will be omitted for brevity.

The only mass force we need to consider is the *gravitational attraction* of the earth. We are not going to deal with tidal forces so that we are justified in ignoring the gravitational pull resulting from celestial bodies. To keep things simple, we consider the entire mass of the earth M_E to be concentrated at its center. According to *Newton's law of attraction*, the *gravitational force* $\mathbf{f}_1 = \mathbf{f}_a$ acting on unit mass is given by

$$\mathbf{f}_a = -\frac{\gamma M_E}{r^2}\mathbf{r}_0 = -\frac{\gamma M_E}{r^3}\mathbf{r} \tag{1.4}$$

where γ is the *gravitational constant* and $\mathbf{r} = r\mathbf{r}_0$ the position vector extending from the center of the earth to the fluid volume element of unit mass. The force $\mathbf{f}_a$ may also be expressed in terms of a potential. To recognize this, we take the curl of $\mathbf{f}_a$ and obtain

$$\nabla \times \mathbf{f}_a = -\gamma M_E\left(\frac{1}{r^3}\nabla \times \mathbf{r} - \frac{3}{r^4}\mathbf{r}_0 \times \mathbf{r}\right) = 0 \tag{1.5}$$

By the rules of vector analysis, $\mathbf{f}_a$ may now be replaced by the gradient of a scalar function,

$$\mathbf{f}_a = -\nabla\phi_a, \qquad \phi_a = -\frac{\gamma M_E}{r} + \text{constant} \tag{1.6}$$

The minus sign is conventional. The term ϕ_a is called the *gravitational potential*. We will subsequently assume that ϕ_a depends on position only and not on time so that

$$\phi_a = \phi_a(\mathbf{r}), \qquad \left(\frac{\partial \phi_a}{\partial t}\right)_{x^i} = 0 \tag{1.7}$$

Next, we consider the surface forces acting on a surface element. These result from the *pressure force* $\mathbf{p}_1 = -p\mathbf{n}$ acting in the opposite direction $-\mathbf{n}$ of $d\mathbf{S}$, and from the normal and tangential *viscous forces* $\mathbf{p}_2 = \mathbf{n}\cdot\mathbb{J}$ which depend on the state of motion of the medium. The quantity $\mathbb{J}$ is known as the *viscous stress tensor*, which is assumed to be symmetric. At this point the stress tensor is introduced only formally, but it will be discussed in more detail in Chapter 5. The surface forces are then given by

$$\sum_i \mathbf{p}_i\, dS = \mathbf{n}\, dS\cdot(-p\mathbb{E} + \mathbb{J}) = d\mathbf{S}\cdot(-p\mathbb{E} + \mathbb{J}) \tag{1.8}$$

where $\mathbb{E}$ is the unit dyadic.

Substituting (1.8) and (1.6) into (1.3) we obtain

$$\int_{V(t)} \left(\frac{D}{Dt}(\rho \mathbf{v}_{\mathrm{A}}) + \rho \nabla \phi_{\mathrm{a}} \right) d\tau + \oint_{S(t)} d\mathbf{S} \cdot (p\mathbb{E} - \mathbb{J}) = 0 \tag{1.9}$$

Application of Gauss' divergence theorem (M6.31) and use of the interchange rule (M6.68) gives

$$\int_{V(t)} \left(\rho \frac{d\mathbf{v}_{\mathrm{A}}}{dt} + \rho \nabla \phi_{\mathrm{a}} + \nabla p - \nabla \cdot \mathbb{J} \right) d\tau = 0 \tag{1.10}$$

Since the volume of integration is completely arbitrary, we obtain the differential form

$$\boxed{\rho \frac{d\mathbf{v}_{\mathrm{A}}}{dt} = -\rho \nabla \phi_{\mathrm{a}} - \nabla p + \nabla \cdot \mathbb{J}} \tag{1.11}$$

This important equation is known as the *equation of absolute motion*. It is fundamental to all of atmospheric dynamics.

In order to formulate the last term of (1.11) we use the following analytic expression for the stress tensor $\mathbb{J}$:

$$\mathbb{J} = \mu(\nabla \mathbf{v}_{\mathrm{A}} + \mathbf{v}_{\mathrm{A}} \overset{\curvearrowleft}{\nabla}) - \lambda \nabla \cdot \mathbf{v}_{\mathrm{A}} \mathbb{E}, \qquad \lambda = \tfrac{2}{3}\mu - l^{11} \tag{1.12}$$

A detailed derivation of this equation may be found, for example, in TH. In (1.12) we have neglected the effects of transitions between the different phases of water on the stress tensor. We have also used *Lamé's coefficients of viscosity* λ and μ, which will be treated as constants. Substitution of (1.12) into (1.11) results in the famous *Navier–Stokes equation*

$$\boxed{\rho \frac{d\mathbf{v}_{\mathrm{A}}}{dt} = -\rho \nabla \phi_{\mathrm{a}} - \nabla p + \mu \nabla^2 \mathbf{v}_{\mathrm{A}} + (\mu - \lambda) \nabla(\nabla \cdot \mathbf{v}_{\mathrm{A}})} \tag{1.13}$$

If we ignore the last two terms involving Lamé's viscosity coefficients then we obtain the *Euler equation*

$$\boxed{\rho \frac{d\mathbf{v}_{\mathrm{A}}}{dt} = -\rho \nabla \phi_{\mathrm{a}} - \nabla p} \tag{1.14}$$

1.2 The energy budget in the absolute reference system

In the absolute reference system the *attractive potential* represents the *potential energy* of unit mass. Application of the budget operator yields

$$\frac{D}{Dt}(\rho\phi_{\rm a}) = \rho\,\frac{d\phi_{\rm a}}{dt} = \rho\,\frac{\partial\phi_{\rm a}}{\partial t} + \rho\mathbf{v}_{\rm A}\cdot\nabla\phi_{\rm a} \tag{1.15}$$

Owing to the condition of stationarity (1.7) the budget equation for the gravitational attraction reduces to

$$\frac{D}{Dt}(\rho\phi_{\rm a}) = \rho\mathbf{v}_{\rm A}\cdot\nabla\phi_{\rm a} \tag{1.16}$$

Next we need to derive the budget equation for the *kinetic energy* $K_{\rm A} = \rho\mathbf{v}_{\rm A}^2/2$ in the absolute reference frame. Scalar multiplication of (1.11) by $\mathbf{v}_{\rm A}$ and application of the identities

$$\begin{aligned}\nabla\cdot(\mathbf{v}_{\rm A}\cdot\mathbb{J}) &= \mathbb{J}\cdot\cdot\nabla\mathbf{v}_{\rm A} + \mathbf{v}_{\rm A}\cdot(\nabla\cdot\mathbb{J})\\ \nabla\cdot(p\mathbf{v}_{\rm A}) &= p\,\nabla\cdot\mathbf{v}_{\rm A} + \mathbf{v}_{\rm A}\cdot\nabla p = \nabla\cdot(p\mathbf{v}_{\rm A}\cdot\mathbb{E})\end{aligned} \tag{1.17}$$

yields the desired budget equation

$$\frac{D}{Dt}\left(\rho\frac{\mathbf{v}_{\rm A}^2}{2}\right) + \nabla\cdot[\mathbf{v}_{\rm A}\cdot(p\mathbb{E}-\mathbb{J})] = -\rho\mathbf{v}_{\rm A}\cdot\nabla\phi_{\rm a} + p\,\nabla\cdot\mathbf{v}_{\rm A} - \mathbb{J}\cdot\cdot\nabla\mathbf{v}_{\rm A} \tag{1.18}$$

Equations (1.16) and (1.18), together with the budget equation for the *internal energy* e (M6.73), constitute the *energy budget for the absolute system*:

$$\boxed{\begin{aligned}\frac{D}{Dt}(\rho\phi_{\rm a}) &= \rho\mathbf{v}_{\rm A}\cdot\nabla\phi_{\rm a}\\ \frac{D}{Dt}\left(\rho\frac{\mathbf{v}_{\rm A}^2}{2}\right) + \nabla\cdot[\mathbf{v}_{\rm A}\cdot(p\mathbb{E}-\mathbb{J})] &= -\rho\mathbf{v}_{\rm A}\cdot\nabla\phi_{\rm a} + p\,\nabla\cdot\mathbf{v}_{\rm A} - \mathbb{J}\cdot\cdot\nabla\mathbf{v}_{\rm A}\\ \frac{D}{Dt}(\rho e) + \nabla\cdot(\mathbf{J}^{\rm h} + \mathbf{F}_{\rm R}) &= -p\,\nabla\cdot\mathbf{v}_{\rm A} + \mathbb{J}\cdot\cdot\nabla\mathbf{v}_{\rm A}\end{aligned}} \tag{1.19}$$

The term $\mathbf{J}^{\rm h}$ occurring in the budget equation for e describes the sensible heat flux and $\mathbf{F}_{\rm R}$ is the radiative flux. This equation is derived in various textbook on thermodynamics; see for example TH.

It can be seen that the sum of the source terms on the right-hand side of (1.19) of the entire system vanishes. This simply means that energy is transformed and exchanged between various parts of the system, but it is neither created nor destroyed. In other words, each source occurring in one equation is compensated by a sink (minus sign) in one of the other equations.

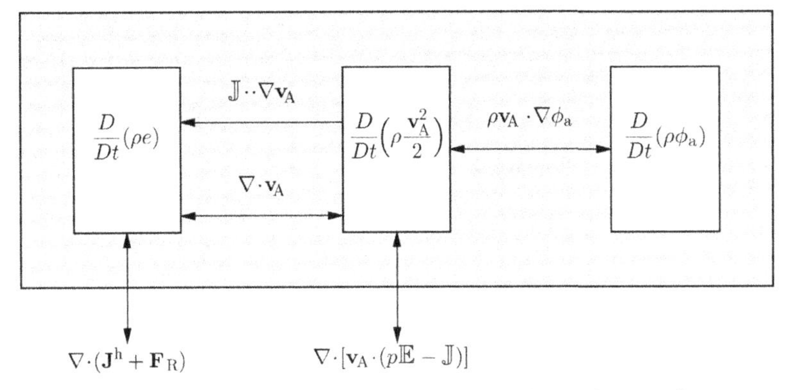

Fig. 1.1 The energy-flux diagram in the absolute reference frame.

For visual purposes it is customary to arrange the energy fluxes (1.19) in the form of a circuit diagram as shown in Figure 1.1. The expressions above the arrows between the three energy boxes describe the energy transitions. These transitions may have either a positive or a negative sign with the exception of the positive definite *energy dissipation* $\mathbb{J}\cdot\cdot\nabla\mathbf{v}_{\mathrm{A}}$ which flows in one direction only. This term is also known as the *Rayleigh dissipation function*. The arrows connecting the system with its surroundings are the divergence terms of the budget equations. If the system is closed energetically then all divergence terms vanish. In this case the *total energy* $\epsilon_{\mathrm{t}} = (\Phi_{\mathrm{a}} + \mathbf{v}_{\mathrm{A}}^2/2 + e)$ is conserved.

1.3 The geographical coordinate system

All meteorological observations are performed on the rotating earth. Therefore, our goal must be to describe the motion of the air from the point of view of an observer participating in the rotational motion. Any motion viewed from a station on the rotating earth is known as *relative motion*. Since the equation of absolute motion (1.13) refers to an inertial system, we must find a relation between the accelerations of absolute and relative motion.

Let us consider a rotating coordinate system whose origin is placed at the earth's center. The vertical axis of this system coincides with the earth's axis so that the coordinate system is rotating with the constant *angular velocity* $\boldsymbol{\Omega}$ of the earth from west to east. Moreover, we assume that the center of the earth and the center of the inertial coordinate system coincide and that the vertical axis of the inertial system is also along the rotational axis of the earth and is pointing to the pole-star. If a point P in this rotating coordinate system remains fixed in the sense that its longitude λ, latitude φ, and the distance r extending from the center of the earth to the point do not change with time, we speak of *rigid rotation* and a rigidly rotating coordinate system. This situation is depicted in Figure 1.2.

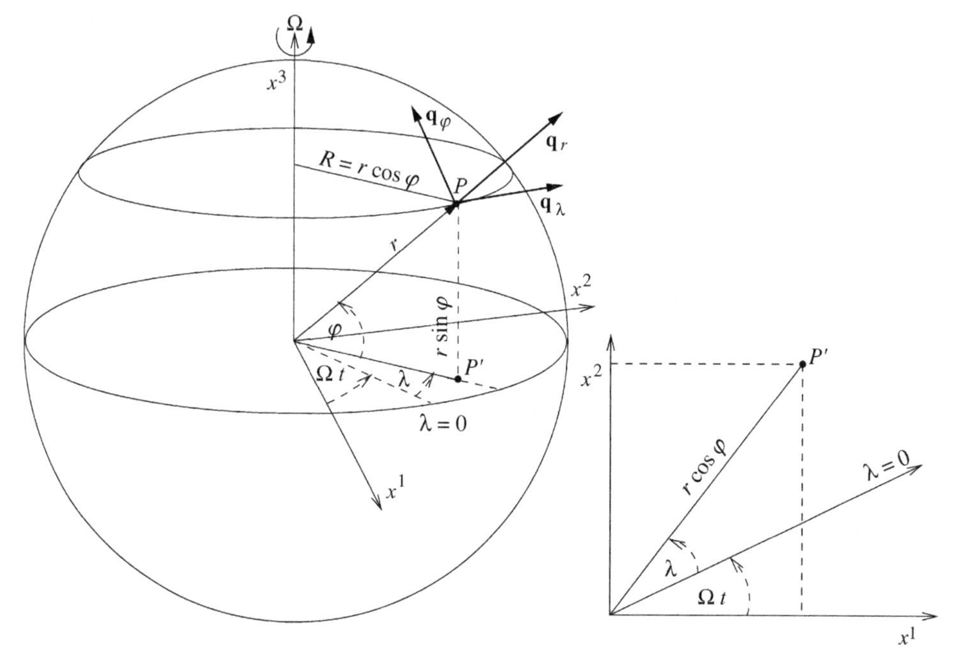

Fig. 1.2 The geographical coordinate system with generalized coordinates $q^i = \lambda, \varphi, r$ representing the longitude λ, the latitude φ, and the distance r of a point P. Also shown are the corresponding basis vectors $\mathbf{q}_\lambda$, $\mathbf{q}_\varphi$, $\mathbf{q}_r$. The axes x^1 and x^2 of the inertial system lie in the equatorial plane of the earth.

We can see immediately that the *rotational velocity* $\mathbf{v}_\Omega$ of this point is directed along the basis vector $\mathbf{q}_\lambda$ and is given by

$$\mathbf{v}_\Omega = \boldsymbol{\Omega} \times \mathbf{r} \tag{1.20}$$

The formal derivation of (1.20) will be given later. The angular velocity $\boldsymbol{\Omega}$ is determined by the period of rotation of the earth with respect to the fixed stars. This period of rotation is called the *sidereal day* (sidereal is an adjective derived from the Latin word for star). Since the earth moves around the sun, the sidereal day differs in length from the solar day which is the period of rotation with respect to the sun. In one year the earth rotates $365\frac{1}{4}$ times with respect to the sun, but $366\frac{1}{4}$ times with respect to the stars so that one year $= 365\frac{1}{4}$ solar days $= 366\frac{1}{4}$ sidereal days. Therefore, we have

$$|\boldsymbol{\Omega}| = \Omega = \frac{2\pi}{\text{sidereal day}} = \frac{366.25}{365.25}\frac{2\pi}{\text{solar day}} = 7.292 \times 10^{-5}\ \text{s}^{-1} \tag{1.21}$$

In our later studies it will be of advantage to introduce pressure or some other suitable variable of state as the vertical coordinate instead of height in order to effectively describe the motion of the air. Let us consider, for example, a point of fixed longitude and latitude on a pressure surface. Owing to heating processes, pressure surfaces will deform so that this point will experience a change in height with respect to the ground and, therefore, with respect to the center of the absolute coordinate system. Of course, this displacement is extremely small relative to the total distance from the center of the earth to the point. The vertical motion of the point, however, cannot be ignored in all situations. Summing it up, this point is not only moving with the rotational velocity of the earth, but also participating in the deformational motion of the pressure surface. This *deformation velocity* $\mathbf{v}_\mathrm{D}$ must be added to $\mathbf{v}_\Omega$ to give the *velocity of the point* $\mathbf{v}_P$ with respect to the absolute system,

$$\mathbf{v}_P = \mathbf{v}_\Omega + \mathbf{v}_\mathrm{D} \tag{1.22}$$

As the next step in our investigation it will be necessary to represent the individual time derivative both in the absolute and in the relative coordinate systems. The individual time derivative itself describes the change of an air parcel in such a way that it is independent of any coordinate system. Therefore, the individual derivative d/dt is called an invariant operator. However, the constituent parts of this operator depend on the coordinate system used to describe the motion and, therefore, they are not invariants. A detailed derivation of the individual time derivative is given in Section M4.2.

It will be recalled that, in general curvilinear coordinates, the position vector is defined only infinitesimally. Exceptions are curvilinear systems for which $\partial\mathbf{e}_i/\partial r = 0$. An example of this exception is the geographical coordinate system. It should be recalled that, in rotating coordinate systems, the basis vectors are functions of time also. The position vector $\mathbf{r}$ in the Cartesian system may also be expressed in terms of the generalized coordinates q^j using the transformation $x^i = x^i(q^j, t)$. Thus, we may write

$$\mathbf{r} = x^n \mathbf{i}_n, \qquad \mathbf{r} = \mathbf{r}(q^j, t) \tag{1.23}$$

In the absolute system we use Cartesian coordinates to represent $d\mathbf{r}/dt$, whereas for the relative system we are going to employ contravariant measure numbers. Application of the invariant operator d/dt to the position vector in the two coordinate systems then gives

$$\begin{aligned}
&x^i \text{ system:} && \frac{d\mathbf{r}}{dt} = \left(\frac{\partial \mathbf{r}}{\partial t}\right)_{x^i} + \frac{\partial \mathbf{r}}{\partial x^n}\frac{dx^n}{dt} = \left(\frac{\partial x^n}{\partial t}\right)_{x^i}\mathbf{i}_n + \dot{x}^n \mathbf{i}_n = \dot{x}^n \mathbf{i}_n = \mathbf{v}_\mathrm{A} \\
&q^i \text{ system:} && \frac{d\mathbf{r}}{dt} = \left(\frac{\partial \mathbf{r}}{\partial t}\right)_{q^i} + \frac{\partial \mathbf{r}}{\partial q^n}\frac{dq^n}{dt} = \left(\frac{\partial \mathbf{r}}{\partial t}\right)_{q^i} + \dot{q}^n \mathbf{q}_n \\
&\text{with} && \frac{\partial \mathbf{r}}{\partial x^i} = \mathbf{i}_i, \qquad \frac{\partial \mathbf{r}}{\partial q^j} = \mathbf{q}_j
\end{aligned} \tag{1.24}$$

In the absolute or the Cartesian system the individual derivative of the position vector is equivalent to the absolute velocity. The local time derivative vanishes since x^i is held constant. In the relative system the partial derivative of $\mathbf{r}$ with respect to time, holding q^i constant, describes the velocity $\mathbf{v}_P$ of the point P as registered in the absolute system. The components of $\mathbf{v}_P$ in the absolute or inertial system are denoted by $\underset{x}{W}^i$. The vector $\mathbf{v}$ describes the velocity of a parcel of air relative to the earth. Therefore, we have

$$\mathbf{v} = \dot{q}^n \mathbf{q}_n, \qquad \mathbf{v}_P = \left(\frac{\partial \mathbf{r}}{\partial t}\right)_{q^i} = \left(\frac{\partial x^n}{\partial t}\right)_{q^i} \mathbf{i}_n = \underset{x}{W}^n \mathbf{i}_n \tag{1.25}$$

From (1.24) and (1.25) we now obtain

$$\boxed{\mathbf{v}_\mathrm{A} = \mathbf{v}_P + \mathbf{v}} \tag{1.26}$$

which is known as the *addition theorem of the velocities*, see also Section M4.2. Hence, the absolute velocity $\mathbf{v}_\mathrm{A}$ of a parcel of air is the sum of the velocity $\mathbf{v}_P$ of the point relative to the absolute system plus the *relative velocity* of the parcel at the point as registered in the moving system. Taking the individual time derivative of (1.26) results in the relation between the accelerations of absolute and relative motion, as was mentioned at the beginning of this section.

1.3.1 Operations involving the rotational velocity $\mathbf{v}_\Omega$

In this section we will present some important relations that will be useful for our later studies.

1.3.1.1 The divergence of $\mathbf{v}_\Omega$ *in the geographical coordinate system*

We will now introduce the coordinates of the geographical system spoken of with the help of Figure 1.2. This system is performing a rigid rotation so that $\mathbf{v}_\mathrm{D} = 0$. The system is rotating with constant angular velocity about the x^3-axis so that

$$\boldsymbol{\Omega} = \mathbf{i}_3 \Omega \tag{1.27}$$

We wish to specify the coordinates of the point P whose projection onto the equatorial plane is denoted by P'. During time t, the longitude $\lambda = 0$ passing through Greenwich moves the angular distance Ωt as measured from the x^1-axis of the absolute system. The angular distance of the point P' then is given by $\Omega t + \lambda$ so that the coordinates of the point P in the absolute system may be expressed in terms of the geographical coordinates (λ, φ, r) by means of

$$\boxed{\begin{aligned} x^1 &= r\cos\varphi\cos(\lambda + \Omega t) \\ x^2 &= r\cos\varphi\sin(\lambda + \Omega t) \\ x^3 &= r\sin\varphi \end{aligned}} \tag{1.28}$$

We now apply (1.25) to find the rotational velocity $\mathbf{v}_\Omega$. In the present situation $\mathbf{v}_\mathrm{D} = 0$ so that

$$\begin{aligned} \mathbf{v}_P = \mathbf{v}_\Omega &= \left(\frac{\partial \mathbf{r}}{\partial t}\right)_{q^i} = \mathbf{i}_n\left(\frac{\partial x^n}{\partial t}\right)_{q^i} \\ &= \underset{x}{W}{}^n\mathbf{i}_n = -\Omega x^2\mathbf{i}_1 + \Omega x^1\mathbf{i}_2 = \boldsymbol{\Omega}\times\mathbf{r} \\ \text{with}\quad \underset{x}{W}{}^1 &= \underset{x}{W}{}_1 = -\Omega x^2, \qquad \underset{x}{W}{}^2 = \underset{x}{W}{}_2 = \Omega x^1 \end{aligned} \tag{1.29}$$

This verifies the validity of (1.20). Recall that in the Cartesian system there is no difference between covariant and contravariant coordinates.

We now wish to find the metric fundamental quantities g_{ij} of the orthogonal geographical coordinate system. For such a system we may apply the relation (M3.15):

$$g_{ij} = \frac{\partial x^n}{\partial q^i}\frac{\partial x^n}{\partial q^j} \tag{1.30}$$

On substituting (1.28) into this expression, after a few easy steps we find

$$\boxed{\begin{gathered} g_{11} = r^2\cos^2\varphi, \qquad g_{22} = r^2, \qquad g_{33} = 1, \qquad g_{ij} = 0 \quad \text{for} \quad i \neq j \implies \\ \sqrt{\underset{q}{g}} = \sqrt{|g_{ij}|} = r^2\cos\varphi \end{gathered}} \tag{1.31}$$

The divergence of the velocity of the point $\mathbf{v}_P = \mathbf{v}_\Omega$ is expressed by the general relation (M4.36)

$$\nabla\cdot\mathbf{v}_P = \frac{1}{\sqrt{\underset{q}{g}}}\left(\frac{\partial\sqrt{\underset{q}{g}}}{\partial t}\right)_{q^i} = \frac{1}{r^2\cos\varphi}\frac{\partial}{\partial t}(r^2\cos\varphi)_{\lambda,\varphi,r} = 0 \tag{1.32}$$

Since the coordinates φ and r are held constant in the time differentiation, it is found that the divergence of $\mathbf{v}_\Omega$ in the rigidly rotating geographical system is zero, as expected. If the deformation velocity differs from zero, we find that the divergence of $\mathbf{v}_\mathrm{D}$ does not vanish. This situation will be treated later. The validity of (1.32) can also be verified by using the definition (1.20), i.e.

$$\nabla\cdot\mathbf{v}_\Omega = \nabla\cdot(\boldsymbol{\Omega}\times\mathbf{r}) = -\boldsymbol{\Omega}\cdot(\nabla\times\mathbf{r}) = 0 \tag{1.33}$$

1.3.1.2 Rotation and the vector gradient of $\mathbf{v}_\Omega$

Using the Grassmann rule (M1.44) we find immediately

$$\nabla \times \mathbf{v}_\Omega = \nabla \times (\boldsymbol{\Omega} \times \mathbf{r}) = (\nabla \cdot \mathbf{r})\boldsymbol{\Omega} - \boldsymbol{\Omega} \cdot (\nabla \mathbf{r}) = 3\boldsymbol{\Omega} - \boldsymbol{\Omega} \cdot \mathbb{E} = 2\boldsymbol{\Omega} \tag{1.34}$$

showing that the rotation of $\mathbf{v}_\Omega$ is a constant vector whose direction is parallel to the rotational axis of the earth.

We will now take the gradient of $\mathbf{v}_\Omega$. Recalling that $\nabla \mathbf{r} = \mathbb{E}$ is a symmetric tensor, that is $\nabla \mathbf{r} = \overset{\curvearrowleft}{\mathbf{r}\nabla}$, we find immediately

$$\begin{aligned} \nabla \mathbf{v}_\Omega &= \nabla(\boldsymbol{\Omega} \times \mathbf{r}) = -\nabla(\mathbf{r} \times \boldsymbol{\Omega}) = -\mathbb{E} \times \boldsymbol{\Omega} = -\boldsymbol{\Omega} \times \mathbb{E} \\ \overset{\curvearrowleft}{\mathbf{v}_\Omega \nabla} &= \overset{\curvearrowleft}{(\boldsymbol{\Omega} \times \mathbf{r})\nabla} = \boldsymbol{\Omega} \times \mathbb{E} = -\nabla \mathbf{v}_\Omega \end{aligned} \tag{1.35}$$

where use of (M2.67) has been made. Therefore, the gradient of $\mathbf{v}_\Omega$ is an anti-symmetric tensor. By taking the scalar product of an arbitrary vector $\mathbf{A}$ with the gradient of $\mathbf{v}_\Omega$, we obtain the very useful expressions

$$\begin{aligned} \mathbf{A} \cdot \nabla \mathbf{v}_\Omega &= -\mathbf{A} \cdot (\mathbb{E} \times \boldsymbol{\Omega}) = \boldsymbol{\Omega} \times \mathbf{A} \\ \nabla \mathbf{v}_\Omega \cdot \mathbf{A} &= -(\boldsymbol{\Omega} \times \mathbb{E}) \cdot \mathbf{A} = \mathbf{A} \times \boldsymbol{\Omega} \end{aligned} \tag{1.36}$$

If $\mathbf{A}$ represents the velocities $\mathbf{v}_\Omega$ and $\mathbf{v}$ we find

$$\begin{aligned} &\text{(a)} \quad \mathbf{v}_\Omega \cdot \nabla \mathbf{v}_\Omega = \boldsymbol{\Omega} \times \mathbf{v}_\Omega = \boldsymbol{\Omega} \times (\boldsymbol{\Omega} \times \mathbf{r}) \\ &\text{(b)} \quad \mathbf{v} \cdot \nabla \mathbf{v}_\Omega = \boldsymbol{\Omega} \times \mathbf{v} \end{aligned} \tag{1.37}$$

These expressions will be needed later.

1.3.2 The centrifugal potential

For meteorological purposes the rotational vector of the earth may be treated as a constant vector. On applying the Grassmann rule to the curl of the vector product $\boldsymbol{\Omega} \times \mathbf{v}_\Omega$ we find with the help of (1.33) and (1.36)

$$\nabla \times (\boldsymbol{\Omega} \times \mathbf{v}_\Omega) = (\nabla \cdot \mathbf{v}_\Omega)\boldsymbol{\Omega} - \boldsymbol{\Omega} \cdot \nabla \mathbf{v}_\Omega = 0 \tag{1.38}$$

From vector analysis we know that an arbitrary vector whose curl is vanishing can be replaced by the gradient of a scalar field function. Therefore, we may write

$$\nabla \phi_z = \boldsymbol{\Omega} \times \mathbf{v}_\Omega \tag{1.39}$$

Since we are dealing with a rotating coordinate system, it is customary to call the scalar field function ϕ_z the centrifugal potential, which will be determined soon.

By taking the scalar product of $\mathbf{v}_\Omega$ and $\nabla\phi_z$ we find that the angle between these two vectors is 90°:

$$\mathbf{v}_\Omega \cdot \nabla\phi_z = \mathbf{v}_\Omega \cdot (\boldsymbol{\Omega} \times \mathbf{v}_\Omega) = 0 \tag{1.40}$$

Owing to the gravitational attraction, equipotential surfaces ϕ_a = constant are of spherical shape. Therefore, we must also have

$$\mathbf{v}_\Omega \cdot \nabla\phi_\mathrm{a} = 0 \tag{1.41}$$

showing that $\mathbf{v}_\Omega$ and $\nabla\phi_\mathrm{a}$ are also orthogonal to each other. This may also be easily verified with the help of (1.20).

In order to find the centrifugal potential itself, we apply (1.37) to (1.39) and utilize the fact that the tensor $\nabla\mathbf{v}_\Omega$ is antisymmetric. Thus we obtain

$$\nabla\phi_z = \boldsymbol{\Omega} \times \mathbf{v}_\Omega = \mathbf{v}_\Omega \cdot \nabla\mathbf{v}_\Omega = -\mathbf{v}_\Omega \cdot \mathbf{v}_\Omega \overset{\curvearrowleft}{\nabla} = -\nabla\left(\frac{\mathbf{v}_\Omega^2}{2}\right) \tag{1.42}$$

By comparison we find, to within an abitrary additive constant, which we set equal to zero, the required expression for the centrifugal potential:

$$\boxed{\phi_z = -\frac{\mathbf{v}_\Omega^2}{2} = -\frac{(\boldsymbol{\Omega} \times \mathbf{r})^2}{2} = -\frac{\Omega^2 R^2}{2}} \tag{1.43}$$

R is the distance from the point under consideration to the earth's axis; see Figure 1.2.

Finally, the invariant individual time derivative of $\mathbf{v}_\Omega$ can be written either in the absolute or in the relative system:

$$\frac{d\mathbf{v}_\Omega}{dt} = \left(\frac{\partial \mathbf{v}_\Omega}{\partial t}\right)_{x^i} + \mathbf{v}_\mathrm{A} \cdot \nabla\mathbf{v}_\Omega = \left(\frac{\partial \mathbf{v}_\Omega}{\partial t}\right)_{q^i} + \mathbf{v} \cdot \nabla\mathbf{v}_\Omega \tag{1.44}$$

In the Cartesian system the local time derivative of $\mathbf{v}_\Omega$ vanishes because

$$\left(\frac{\partial \mathbf{v}_\Omega}{\partial t}\right)_{x^i} = \boldsymbol{\Omega} \times \left(\frac{\partial \mathbf{r}}{\partial t}\right)_{x^i} = \boldsymbol{\Omega} \times \mathbf{i}_n \left(\frac{\partial x^n}{\partial t}\right)_{x^i} = 0 \tag{1.45}$$

In the general q^i system we obain

$$\left(\frac{\partial \mathbf{v}_\Omega}{\partial t}\right)_{q^i} = \boldsymbol{\Omega} \times \left(\frac{\partial \mathbf{r}}{\partial t}\right)_{q^i} = \boldsymbol{\Omega} \times (\mathbf{v}_\Omega + \mathbf{v}_\mathrm{D}) = \nabla\phi_z + \boldsymbol{\Omega} \times \mathbf{v}_\mathrm{D} \tag{1.46}$$

where we have used (1.25) and (1.39). For the rigidly rotating coordinate system with $\mathbf{v}_\mathrm{D} = 0$ this expression reduces to

$$\boxed{\left(\frac{\partial \mathbf{v}_\Omega}{\partial t}\right)_{q^i} = \nabla\phi_z} \tag{1.47}$$

1.3.3 The budget operator

Before we apply the budget operator to the velocity $\mathbf{v}$ of the rigidly rotating system, we will give a general expression for an abitrary system including deformation. We refer to Section M6.5. According to (M6.66), omitting the index 3 for brevity, the budget operator is given by

$$\begin{aligned}\frac{D}{Dt}(\rho\psi) &= \frac{\partial}{\partial t}(\rho\psi)_{x^i} + \nabla\cdot(\rho\psi\,\mathbf{v}_{\rm A}) \\ &= \frac{\partial}{\partial t}(\rho\psi)_{x^i} + \mathbf{v}_{\rm A}\cdot\nabla(\rho\psi) + \rho\psi\,\nabla\cdot\mathbf{v}_{\rm A} \\ &= \frac{d}{dt}(\rho\psi) + \rho\psi\,\nabla\cdot\mathbf{v} + \rho\psi\,\nabla\cdot\mathbf{v}_{\rm D}\end{aligned} \tag{1.48}$$

since $\mathbf{v}_{\rm A} = \mathbf{v} + \mathbf{v}_\Omega + \mathbf{v}_{\rm D}$ and $\nabla\cdot\mathbf{v}_\Omega = 0$. Now we write the total time derivative in the q^i-coordinate system and find

$$\boxed{\frac{D}{Dt}(\rho\psi) = \frac{\partial}{\partial t}(\rho\psi)_{q^i} + \nabla\cdot(\rho\mathbf{v}\psi) + \rho\psi\,\nabla\cdot\mathbf{v}_{\rm D}} \tag{1.49}$$

Setting in (1.48) $\psi = 1$ yields the *general form of the continuity equation*:

$$\boxed{\frac{d\rho}{dt} + \rho\,\nabla\cdot\mathbf{v} + \rho\,\nabla\cdot\mathbf{v}_{\rm D} = 0} \tag{1.50}$$

If the deformation velocity $\mathbf{v}_{\rm D}$ is zero, (1.50) reduces to the continuity equation for the rigidly rotating coordinate system.

Before we derive the equation of motion for the relative system, we will introduce a special notation to avoid notational ambiguities. Suppose that we wish to differentiate locally the vector $\mathbf{A} = A^n\mathbf{q}_n$ with respect to time, yielding

$$\left(\frac{\partial\mathbf{A}}{\partial t}\right)_{q^i} = \left(\frac{\partial A^n}{\partial t}\right)_{q^i}\mathbf{q}_n + A^n\left(\frac{\partial\mathbf{q}_n}{\partial t}\right)_{q^i} \tag{1.51}$$

The first part of the product differentiation on the right-hand side of (1.51) describes the differentiation of the measure numbers A^k with the basis vector $\mathbf{q}_k$ held constant. It might be more convenient to put the vector $\mathbf{A}$ itself into the first operator on the right-hand side instead of its measure numbers. This can be done quite validly since $\mathbf{q}_i$ is held constant. However, in order to avoid an ambiguity in our notation, we place a vertical line $\Big|_{\mathbf{q}_i}$ on the local time-derivative operator to indicate that $\mathbf{q}_i$ is not to be differentiated with respect to time. Thus, the following notation is introduced:

$$\left.\frac{\partial\mathbf{A}}{\partial t}\right|_{\mathbf{q}_i} = \left(\frac{\partial A^n}{\partial t}\right)_{q^i}\mathbf{q}_n \tag{1.52}$$

The second part of the time differentiation in (1.51) involves the basis vector itself:

$$\left(\frac{\partial \mathbf{q}_k}{\partial t}\right)_{q^i} = \frac{\partial}{\partial t}\left(\frac{\partial \mathbf{r}}{\partial q^k}\right)_{q^i} = \frac{\partial}{\partial q^k}\left(\frac{\partial \mathbf{r}}{\partial t}\right)_{q^i} = \frac{\partial \mathbf{v}_P}{\partial q^k} \tag{1.53}$$

where we have substituted the velocity $\mathbf{v}_P$ as defined in (1.25). The local time derivative of $\mathbf{A}$ is then given by

$$\left(\frac{\partial \mathbf{A}}{\partial t}\right)_{q^i} = \left.\frac{\partial \mathbf{A}}{\partial t}\right|_{\mathbf{q}_i} + A^n \frac{\partial \mathbf{v}_P}{\partial q^n} = \left.\frac{\partial \mathbf{A}}{\partial t}\right|_{\mathbf{q}_i} + A^n \mathbf{q}_n \cdot \mathbf{q}^m \frac{\partial \mathbf{v}_P}{\partial q^m} = \left.\frac{\partial \mathbf{A}}{\partial t}\right|_{\mathbf{q}_i} + \mathbf{A}\cdot\nabla\mathbf{v}_P \tag{1.54}$$

With this expression we obtain for the individual time derivative of $\mathbf{A}$

$$\boxed{\begin{aligned} \frac{d\mathbf{A}}{dt} &= \left.\frac{d\mathbf{A}}{dt}\right|_{\mathbf{q}_i} + \mathbf{A}\cdot\nabla\mathbf{v}_P \\ \text{with} \quad \left.\frac{d\mathbf{A}}{dt}\right|_{\mathbf{q}_i} &= \left.\frac{\partial \mathbf{A}}{\partial t}\right|_{\mathbf{q}_i} + \mathbf{v}\cdot\nabla\mathbf{A} \end{aligned}} \tag{1.55}$$

While the vertical line excludes the differentiation of the basis vector $\mathbf{q}_i$ with respect to time, it does not exclude the differentiation with respect to the spatial coordinates. Moreover, it should be clearly recognized that the time dependency of the basis vectors is not lost by any means since it is included in the gradient of $\mathbf{v}_P$.

We proceed similarly with the budget operator. Application of (1.48) to the vector $\mathbf{A}$ results in

$$\boxed{\begin{aligned} \frac{D}{Dt}(\rho\mathbf{A}) &= \left.\frac{D}{Dt}(\rho\mathbf{A})\right|_{\mathbf{q}_i} + \rho\mathbf{A}\cdot\nabla\mathbf{v}_P \\ \text{with} \quad \left.\frac{D}{Dt}(\rho\mathbf{A})\right|_{\mathbf{q}_i} &= \rho\left.\frac{d\mathbf{A}}{dt}\right|_{\mathbf{q}_i} = \left.\frac{\partial}{\partial t}(\rho\mathbf{A})\right|_{\mathbf{q}_i} + \nabla\cdot(\rho\mathbf{v}\mathbf{A}) + \rho\mathbf{A}\,\nabla\cdot\mathbf{v}_{\mathrm{D}} \end{aligned}} \tag{1.56}$$

We would like to point out that equations (1.54)–(1.56) are general and include the deformation velocity.

In the special case of rigid rotation, the deformation velocity is zero so that (1.36) may be applied. Whenever the expression $\mathbf{A}\cdot\nabla\mathbf{v}_P$ appears it may then be replaced by

$$\mathbf{A}\cdot\nabla\mathbf{v}_P = \mathbf{A}\cdot\nabla\mathbf{v}_\Omega = \boldsymbol{\Omega}\times\mathbf{A} \quad \text{for} \quad \mathbf{v}_{\mathrm{D}} = 0 \tag{1.57}$$

1.4 The equation of relative motion

The starting point of the derivation is the equation of absolute motion (1.13). Introducing the addition theorem (1.26), (1.13) may be written as

$$\rho \frac{d\mathbf{v}}{dt} = -\rho \frac{d\mathbf{v}_\Omega}{dt} - \rho \frac{d\mathbf{v}_\mathrm{D}}{\mathrm{d}t} - \rho \nabla\phi_\mathrm{a} - \nabla p + \mu \nabla^2 \mathbf{v}_\mathrm{A} + (\mu - \lambda) \nabla(\nabla \cdot \mathbf{v}_\mathrm{A}) \quad (1.58)$$

We now replace two of the individual derivatives. Application of (1.46) to (1.44) and using (1.37b) gives

$$\frac{d\mathbf{v}_\Omega}{dt} = \nabla\phi_z + \mathbf{\Omega} \times \mathbf{v}_\mathrm{D} + \mathbf{\Omega} \times \mathbf{v} \quad (1.59a)$$

With the help of (1.55) and (1.37b) we obtain

$$\frac{d\mathbf{v}}{dt} = \left.\frac{d\mathbf{v}}{dt}\right|_{\mathbf{q}_i} + \mathbf{\Omega} \times \mathbf{v} + \mathbf{v} \cdot \nabla\mathbf{v}_\mathrm{D} \quad (1.59b)$$

Substitution of (1.59a) and (1.59b) into (1.58) gives the *equation of relative motion*:

$$\boxed{\begin{aligned} \rho \left.\frac{d\mathbf{v}}{dt}\right|_{\mathbf{q}_i} &= -\,2\rho\mathbf{\Omega} \times \mathbf{v} - \rho\mathbf{\Omega} \times \mathbf{v}_\mathrm{D} - \rho\mathbf{v} \cdot \nabla\mathbf{v}_\mathrm{D} - \rho \frac{d\mathbf{v}_\mathrm{D}}{\mathrm{d}t} \\ &\quad - \rho \nabla\phi - \nabla p + \mu \nabla^2 \mathbf{v}_\mathrm{A} + (\mu - \lambda) \nabla(\nabla \cdot \mathbf{v}_\mathrm{A}) \end{aligned}} \quad (1.60)$$

where the *centrifugal* and the *attractive potential* have been combined to give the *geopotential*, that is

$$\boxed{\phi = \phi_\mathrm{a} + \phi_z} \quad (1.61)$$

From (1.6) we know that surfaces of constant gravitational potential are spherical surfaces. The gravitational potential increases with increasing distance from the center of the earth so that $-\nabla\phi_\mathrm{a}$ is pointing toward the center of the earth. According to (1.43) the centrifugal potential ϕ_z = constant is represented by cylindrical surfaces whose negative gradient $-\nabla\phi_z$ is pointing away from the earth's axis. Owing to the rotational symmetry, the two surfaces may be easily added graphically in a cross-section containing the earth's axis. The resulting surfaces of geopotential ϕ = constant are shown in Figure 1.3.

Near the earth's surface, i.e. in the part of the atmosphere relevant to weather, surfaces of constant geopotential may be viewed as rotational ellipsoids. In most of our studies we will even assume that the ellipsoidal surfaces may be replaced by spherical surfaces. Moreover, we assume that the geopotential is a function of position only and set $\partial\phi/\partial t = 0$.

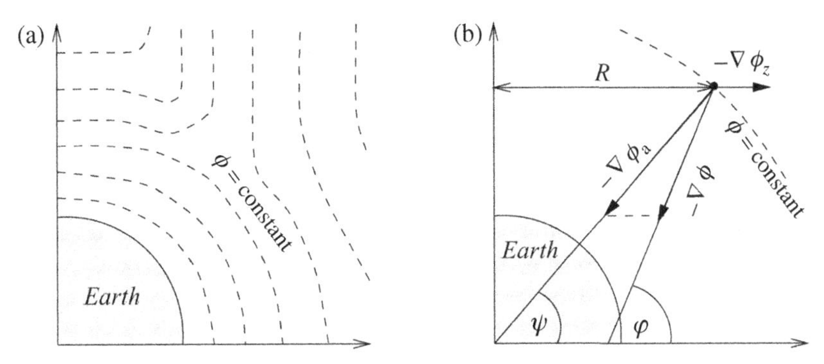

Fig. 1.3 (a) A cross-section of geopotential surfaces. (b) Directions of the negative gradients $-\nabla\phi_a$ and $-\nabla\phi_z$. The angles φ and ψ represent the geographical and the geocentric latitudes.

Equation (1.60) is the general form of the equation of motion in the rotating geographical coordinate system. If the deformation velocity is zero we obtain

$$\boxed{\rho \left.\frac{d\mathbf{v}}{dt}\right|_{\mathbf{q}_i} = -2\rho\boldsymbol{\Omega}\times\mathbf{v} - \rho\,\nabla\phi - \nabla p + \mu\,\nabla^2\mathbf{v} + (\mu-\lambda)\,\nabla(\nabla\cdot\mathbf{v})} \tag{1.62}$$

since $\nabla^2\mathbf{v}_\Omega = 0$. Equation (1.62) applies to the rigidly rotating geographical coordinate system. This is the form of the equation of motion given in most textbooks. It should be clearly understood that the left-hand side of (1.60) represents an artificial acceleration. The vertical line, as discussed, implies that the basis vectors in $\mathbf{v}$ are not to be differentiated with respect to time.

The term $-2\rho\boldsymbol{\Omega}\times\mathbf{v}$ is the *Coriolis force*, which is a fictitious force resulting from the rotation of the coordinate system. An air parcel moving with the relative velocity $\mathbf{v}$ will be deflected to the right in the northern hemisphere and to the left in the southern hemisphere. Of course, the Coriolis force does not perform any work.

1.5 The energy budget of the general relative system

At the end of Section 1.2 we considered the energy budget with reference to the absolute system. Now we wish to derive the energy budget for the general relative system which includes not only the effects of rotation but also that of deformation. As stated before, the total energy is given as the sum of the potential energy due to gravitational attraction, the kinetic energy, and the *internal energy*. In the relative system the relevant variables will be the geopotential instead of the gravitational potential and the relative velocity instead of the absolute velocity. The budget equation for the internal energy e will be included again, with a minor change in the mathematical form that does not change its physical content. With $\mathbf{v}_{\mathrm{A}} = \mathbf{v} + \mathbf{v}_P$ and $\mathbf{v}_P = \mathbf{v}_\Omega + \mathbf{v}_{\mathrm{D}}$ we may write for the *total energy*

$$\epsilon_{\mathrm{t}} = \phi_{\mathrm{a}} + \frac{\mathbf{v}_{\mathrm{A}}^2}{2} + e = \phi + \frac{\mathbf{v}^2}{2} + e + \left(\frac{\mathbf{v}_{\mathrm{A}}^2}{2} - \frac{\mathbf{v}^2}{2} - \phi_z\right) = \phi + \frac{\mathbf{v}^2}{2} + e + \chi \quad (1.63)$$

where we have introduced the centrifugal potential according to (1.61). The expression in parentheses is abbreviated by χ and may be rewritten as

$$\chi = \frac{1}{2}(\mathbf{v} + \mathbf{v}_P)^2 - \frac{\mathbf{v}^2}{2} + \frac{\mathbf{v}_\Omega^2}{2} = \mathbf{v} \cdot \mathbf{v}_P + \frac{\mathbf{v}_P^2}{2} + \frac{\mathbf{v}_\Omega^2}{2} \quad (1.64)$$

where we have replaced ϕ_z by means of (1.43). To describe the energy budget we must obtain budget equations for each term in (1.63). We begin by adding the budget equation of the gravitational attraction

$$\rho \frac{d\phi_{\mathrm{a}}}{dt} = \rho \mathbf{v}_{\mathrm{A}} \cdot \nabla\phi_{\mathrm{a}} = \rho(\mathbf{v} + \mathbf{v}_{\mathrm{D}}) \cdot \nabla\phi_{\mathrm{a}} \quad \text{since} \quad \mathbf{v}_\Omega \cdot \nabla\phi_{\mathrm{a}} = 0 \quad (1.65a)$$

to the budget equation of the centrifugal potential

$$\rho \frac{d\phi_z}{dt} = \rho \mathbf{v}_{\mathrm{A}} \cdot \nabla\phi_z = \rho(\mathbf{v} + \mathbf{v}_{\mathrm{D}}) \cdot \nabla\phi_z \quad \text{since} \quad \mathbf{v}_\Omega \cdot \nabla\phi_z = 0 \quad (1.65b)$$

which must have the same mathematical structure as (1.65a). This gives the budget equation for the geopotential:

$$\rho \frac{d\phi}{dt} = \rho(\mathbf{v} + \mathbf{v}_{\mathrm{D}}) \cdot \nabla\phi \quad (1.66)$$

The next step is the derivation of the budget equation for the *kinetic energy* of the relative system. On substituting (1.59a) into (1.58) we first obtain

$$\rho \frac{d\mathbf{v}}{dt} = -\rho \nabla\phi - \rho\boldsymbol{\Omega} \times (\mathbf{v} + \mathbf{v}_{\mathrm{D}}) - \rho \frac{d\mathbf{v}_{\mathrm{D}}}{dt} - \nabla \cdot (p\mathbb{E} - \mathbb{J}) \quad (1.67)$$

where we have used the definition of the viscous stress tensor $\mathbb{J}$ according to (1.12) to simplify the notation. Scalar multiplication of (1.67) by $\mathbf{v}$ gives the required budget equation for the kinetic energy:

$$\frac{D}{Dt}\left(\rho\frac{\mathbf{v}^2}{2}\right) + \nabla\cdot[\mathbf{v}\cdot(p\mathbb{E} - \mathbb{J})] = -\rho\mathbf{v}\cdot\nabla\phi - \rho\mathbf{v}\cdot\left(\boldsymbol{\Omega} \times \mathbf{v}_{\mathrm{D}} + \frac{d\mathbf{v}_{\mathrm{D}}}{\mathrm{dt}}\right) + (p\mathbb{E} - \mathbb{J})\cdot\cdot\nabla\mathbf{v} \quad (1.68)$$

The budget equation for χ is simply found by subtracting equations (1.65b) and

(1.68) from the second equation of (1.19). The complete *energy budget for the general relative system* is listed as

$$
\begin{array}{ll}
\text{(a)} & \dfrac{D}{Dt}(\rho\phi) = \rho(\mathbf{v} + \mathbf{v}_{\mathrm{D}}) \cdot \nabla\phi \\
\text{(b)} & \dfrac{D}{Dt}\left(\rho\dfrac{\mathbf{v}^2}{2}\right) + \nabla \cdot \left[\mathbf{v} \cdot (p\mathbb{E} - \mathbb{J})\right] = -\rho\mathbf{v} \cdot \nabla\phi + (p\mathbb{E} - \mathbb{J}) \cdot\cdot \nabla\mathbf{v} \\
& \qquad - \rho\mathbf{v} \cdot \left(\mathbf{\Omega} \times \mathbf{v}_{\mathrm{D}} + \dfrac{d\mathbf{v}_{\mathrm{D}}}{dt}\right) \\
\text{(c)} & \dfrac{D}{Dt}(\rho e) + \nabla \cdot (\mathbf{J}^{\mathrm{h}} + \mathbf{F}_{\mathrm{R}}) = -(p\mathbb{E} - \mathbb{J}) \cdot\cdot \nabla(\mathbf{v} + \mathbf{v}_{\mathrm{D}}) \\
\text{(d)} & \dfrac{D}{Dt}(\rho\chi) + \nabla \cdot [(\mathbf{v}_{\Omega} + \mathbf{v}_{\mathrm{D}}) \cdot (p\mathbb{E} - \mathbb{J})] = -\rho\mathbf{v}_{\mathrm{D}} \cdot \nabla\phi + (p\mathbb{E} - \mathbb{J}) \cdot\cdot \nabla\mathbf{v}_{\mathrm{D}} \\
& \qquad + \rho\mathbf{v} \cdot \left(\mathbf{\Omega} \times \mathbf{v}_{\mathrm{D}} + \dfrac{d\mathbf{v}_{\mathrm{D}}}{dt}\right)
\end{array}
\tag{1.69}
$$

In obtaining (1.69c) and (1.69d) we have used the rule that the double scalar product of the symmetric dyadic $(p\mathbb{E} - \mathbb{J})$ and the antisymmetric dyadic $\nabla\mathbf{v}_{\Omega}$ vanishes.

Inspection of (1.69) shows that the sum of source terms on the right-hand sides vanishes as required. Had we considered the budget of the geopotential together with the budget equations of the kinetic energy and the internal energy by themselves, then the sum of the source terms would not have vanished. To correct this deficiency the budget equation of the quantity χ had to be introduced. If desired, an energy flux diagram for the budget (1.69) in the form of Figure 1.1 could be constructed.

Finally, it is also noteworthy that, in the absence of deformational effects ($\mathbf{v}_{\mathrm{D}} = 0$), the budget (1.69) assumes a simplified form describing energetic processes in the relative frame of the rigidly rotating coordinate system. Moreover, the right-hand side of (1.69d) vanishes completely so that the budget of χ is free from sources. Therefore, (1.69d) is no longer a part of the budget system, and we obtain the simplified version of (1.69):

$$
\begin{array}{ll}
\text{(a)} & \dfrac{D}{Dt}(\rho\phi) = \rho\mathbf{v} \cdot \nabla\phi \\
\text{(b)} & \dfrac{D}{Dt}\left(\rho\dfrac{\mathbf{v}^2}{2}\right) + \nabla \cdot \left[\mathbf{v} \cdot (p\mathbb{E} - \mathbb{J})\right] = -\rho\mathbf{v} \cdot \nabla\phi + (p\mathbb{E} - \mathbb{J}) \cdot\cdot \nabla\mathbf{v} \\
\text{(c)} & \dfrac{D}{Dt}(\rho e) + \nabla \cdot (\mathbf{J}^{\mathrm{h}} + \mathbf{F}_{\mathrm{R}}) = -(p\mathbb{E} - \mathbb{J}) \cdot\cdot \nabla\mathbf{v}
\end{array}
\tag{1.70}
$$

By comparing (1.70) with the energy budget of the absolute system (1.19), the similarity of the expressions becomes apparent. In (1.70) the attractive potential $\phi_{\rm a}$ has been replaced by the geopotential ϕ and the absolute velocity $\mathbf{v}_{\rm A}$ by the relative velocity $\mathbf{v}$. In passing we would like to remark that the complete system (1.69) is needed in order to describe the energetic processes of the general circulation.

1.6 The decomposition of the equation of motion

Let us consider the equation of motion (1.62), which is repeated for convenience using the expansion (1.55). For reasons of brevity the final two terms in (1.62) have been rewritten as the divergence of the stress dyadic:

$$\underset{1}{\left.\frac{\partial \mathbf{v}}{\partial t}\right|_{\mathbf{q}_i}} + \underset{2}{\mathbf{v}\cdot\nabla\mathbf{v}} + \underset{3}{\frac{1}{\rho}\nabla p} + \underset{4}{\nabla\phi} + \underset{5}{2\boldsymbol{\Omega}\times\mathbf{v}} - \underset{6}{\frac{1}{\rho}\nabla\cdot\mathbb{J}} = 0 \tag{1.71}$$

The physical meaning of each term will now briefly be explained. Term 1 describes the local change of the velocity whereas the nonlinear term 2 represents the advection of the velocity. Term 3 is most easily comprehended and is usually called the *pressure gradient force*. Term 4 combines the absolute gravitational force and the centrifugal force into a single force often called the *apparent* or *relative gravity*. It is this gravity which is actually observed on the earth. Any surface on which ϕ is constant is called a *level surface* or *equipotential surface*. There is no component of the apparent gravity along such surfaces. Motion along level surfaces is usually referred to as horizontal motion. Multiplying term 5 in (1.71) by -1 results in the Coriolis force, which has already been discussed, whereas term 6 represents frictional effects.

For prognostic purposes it is necessary to decompose the vector equation (1.71) into three equations for the components of the wind field in each direction. There are various ways to obtain the component equations. In order to resolve (1.71) we assume that surfaces of constant geopotential are spherical. The first step is to obtain the metric fundamental quantities g_{ij}. It is best to employ the basic definition

$$\begin{aligned}
&\text{(a)}\quad d\mathbf{r}\cdot d\mathbf{r} = dq^m\,\mathbf{q}_m\cdot dq^n\,\mathbf{q}_n = g_{mn}\,dq^m\,dq^n\\
&\text{(b)}\quad d\mathbf{r} = r\cos\varphi\,d\lambda\,\mathbf{e}_\lambda + r\,d\varphi\,\mathbf{e}_\varphi + dr\,\mathbf{e}_r,\qquad \mathbf{q}_i = \mathbf{e}_i\sqrt{g_{ii}},\qquad \mathbf{e}^i = \mathbf{e}_i
\end{aligned} \tag{1.72}$$

The increment $d\mathbf{r}$ stated in (1.72b) can be easily found from inspection of Figure 1.2. The orthogonal system of the spherical earth is completely described by only three fundamental quantities:

$$g_{11} = r^2\cos^2\varphi,\qquad g_{22} = r^2,\qquad g_{33} = 1,\qquad g_{ij} = 0\quad \text{for}\quad i\neq j \tag{1.73}$$

in agreement with (1.31).

Next we need to specifiy the general gradient operator:

$$\nabla = \mathbf{q}^n \frac{\partial}{\partial q^n} = \mathbf{e}_1\sqrt{g^{11}}\,\frac{\partial}{\partial q^1} + \mathbf{e}_2\sqrt{g^{22}}\,\frac{\partial}{\partial q^2} + \mathbf{e}_3\sqrt{g^{33}}\,\frac{\partial}{\partial q^3} \tag{1.74}$$

which, in the coordinate system being considered, is easily converted to the form

$$\begin{gathered}\nabla = \mathbf{e}_\lambda \frac{1}{r\cos\varphi}\frac{\partial}{\partial\lambda} + \mathbf{e}_\varphi \frac{1}{r}\frac{\partial}{\partial\varphi} + \mathbf{e}_r\frac{\partial}{\partial r} = \mathbf{e}_\lambda\frac{\partial}{\partial\overset{*}{\lambda}} + \mathbf{e}_\varphi\frac{\partial}{\partial\overset{*}{\varphi}} + \mathbf{e}_r\frac{\partial}{\partial z}\\ \text{with}\quad \frac{\partial}{\partial\overset{*}{\lambda}} = \frac{1}{r\cos\varphi}\frac{\partial}{\partial\lambda},\qquad \frac{\partial}{\partial\overset{*}{\varphi}} = \frac{1}{r}\frac{\partial}{\partial\varphi},\qquad \frac{\partial}{\partial z} = \frac{\partial}{\partial r}\end{gathered} \tag{1.75}$$

The relative velocity is simply found by dividing (1.72b) by dt, which results in

$$\begin{gathered}\mathbf{v} = \frac{d\mathbf{r}}{dt} = r\cos\varphi\,\dot{\lambda}\mathbf{e}_\lambda + r\dot{\varphi}\mathbf{e}_\varphi + \dot{r}\mathbf{e}_r = u\mathbf{e}_\lambda + v\mathbf{e}_\varphi + w\mathbf{e}_r\\ \text{with}\quad \overset{*}{u}{}^i = \sqrt{g_{ii}}\dot{q}^i,\qquad \overset{*}{u}{}^1 = u,\qquad \overset{*}{u}{}^2 = v,\qquad \overset{*}{u}{}^3 = w\end{gathered} \tag{1.76}$$

In (1.76) the contravariant velocities $\dot{\lambda}, \dot{\varphi}, \dot{r}$ have been converted into physical measure numbers $\overset{*}{u}{}^i$. The *local velocity dyadic* is then given by

$$\nabla\mathbf{v} = \left(\mathbf{e}_\lambda \frac{1}{r\cos\varphi}\frac{\partial}{\partial\lambda} + \mathbf{e}_\varphi\frac{1}{r}\frac{\partial}{\partial\varphi} + \mathbf{e}_r\frac{\partial}{\partial r}\right)(u\mathbf{e}_\lambda + v\mathbf{e}_\varphi + w\mathbf{e}_r) \tag{1.77}$$

By employing the formulas (M4.45) for the partial derivatives of the unit vectors in the spherical system, we find the following relationships:

$$\begin{aligned}\frac{\partial\mathbf{v}}{\partial\lambda} &= \frac{\partial u}{\partial\lambda}\mathbf{e}_\lambda + u(\sin\varphi\,\mathbf{e}_\varphi - \cos\varphi\,\mathbf{e}_r) + \frac{\partial v}{\partial\lambda}\mathbf{e}_\varphi - v\sin\varphi\,\mathbf{e}_\lambda + \frac{\partial w}{\partial\lambda}\mathbf{e}_r + w\cos\varphi\,\mathbf{e}_\lambda\\ \frac{\partial\mathbf{v}}{\partial\varphi} &= \frac{\partial u}{\partial\varphi}\mathbf{e}_\lambda + \frac{\partial v}{\partial\varphi}\mathbf{e}_\varphi - v\mathbf{e}_r + \frac{\partial w}{\partial\varphi}\mathbf{e}_r + w\mathbf{e}_\varphi\\ \frac{\partial\mathbf{v}}{\partial r} &= \frac{\partial u}{\partial r}\mathbf{e}_\lambda + \frac{\partial v}{\partial r}\mathbf{e}_\varphi + \frac{\partial w}{\partial r}\mathbf{e}_r\end{aligned} \tag{1.78}$$

Using the above formulas it is almost trivial to find the three components of the advection term. The results are

$$\begin{aligned}\mathbf{e}_\lambda\cdot(\mathbf{v}\cdot\nabla\mathbf{v}) &= \frac{u}{r\cos\varphi}\frac{\partial u}{\partial\lambda} + \frac{v}{r}\frac{\partial u}{\partial\varphi} + w\frac{\partial u}{\partial r} + \frac{uw}{r} - \frac{uv}{r}\tan\varphi\\ \mathbf{e}_\varphi\cdot(\mathbf{v}\cdot\nabla\mathbf{v}) &= \frac{u}{r\cos\varphi}\frac{\partial v}{\partial\lambda} + \frac{v}{r}\frac{\partial v}{\partial\varphi} + w\frac{\partial v}{\partial r} + \frac{vw}{r} + \frac{u^2}{r}\tan\varphi\\ \mathbf{e}_r\cdot(\mathbf{v}\cdot\nabla\mathbf{v}) &= \frac{u}{r\cos\varphi}\frac{\partial w}{\partial\lambda} + \frac{v}{r}\frac{\partial w}{\partial\varphi} + w\frac{\partial w}{\partial r} - \frac{1}{r}(u^2+v^2)\end{aligned} \tag{1.79}$$

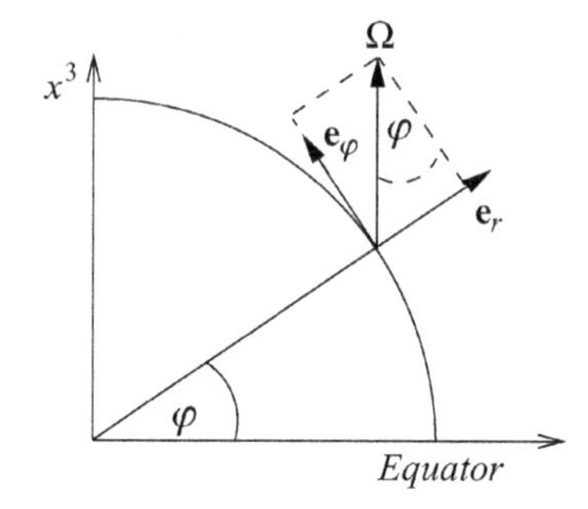

Fig. 1.4 Decomposition of the angular velocity vector.

In order to find the components of the Coriolis force, we need to decompose the angular velocity vector which is oriented perpendicular to the equatorial plane; see Figure 1.2. Therefore, the angular velocity has no component parallel to the equatorial plane, as is shown in Figure 1.4. The result is

$$\boldsymbol{\Omega} = \Omega \cos\varphi\, \mathbf{e}_\varphi + \Omega \sin\varphi\, \mathbf{e}_r = \frac{l}{2}\mathbf{e}_\varphi + \frac{f}{2}\mathbf{e}_r \tag{1.80}$$

with

$$l = 2\Omega \cos\varphi, \qquad f = 2\Omega \sin\varphi \tag{1.81}$$

so that

$$2\boldsymbol{\Omega} \times \mathbf{v} = (lw - fv)\mathbf{e}_\lambda + fu\mathbf{e}_\varphi - lu\mathbf{e}_r \tag{1.82}$$

The terms f and l are the so-called *Coriolis parameters.*

According to our previous discussion, the relation between the *acceleration of gravity* and the geopotential is given by

$$g = -\nabla\phi = -\nabla(\phi_\mathrm{a} + \phi_z) \tag{1.83a}$$

Approximating surfaces of ϕ = constant by spherical surfaces, this equation reduces to

$$\nabla\phi = \mathbf{e}_r g \tag{1.83b}$$

According to (1.43) the centrifugal potential depends on the distance R from the earth's axis. Therefore, g depends on the geographical latitude as well as on the vertical distance from the earth's surface; see Figure 1.2. For most meteorological purposes the height dependence of g may be safely ignored so that $g = g(\varphi)$. However, the latitudinal dependence of g is also relatively weak, yielding values in the range of $9.780 \text{ m s}^{-2} \le g \le 9.832 \text{ m s}^{-2}$ at the equator and the North pole, respectively. Therefore, the φ-dependence of g is usually not explicitly considered in the equation of motion. Instead of this, g is assigned a constant value of 9.81 m s^{-2}. A more complete discussion of this subject may be found, for example, in TH.

Above the planetary boundary layer, which extends to about 1 km in height, frictional effects may often be ignored. For this situation the component form of the equation of motion in spherical coordinates is easily obtained from (1.71) by utilizing (1.79), (1.82), and (1.83) and is given by

$$\frac{\partial u}{\partial t} + \left(u \frac{\partial u}{\partial \overset{*}{\lambda}} + v \frac{\partial u}{\partial \overset{*}{\varphi}} \right) + w \frac{\partial u}{\partial z} + \frac{uw}{r} - \frac{uv}{r} \tan \varphi + lw - fv + \frac{1}{\rho} \frac{\partial p}{\partial \overset{*}{\lambda}} = 0$$

1 2 3 4 5 6 7 8

$$\frac{\partial v}{\partial t} + \left(u \frac{\partial v}{\partial \overset{*}{\lambda}} + v \frac{\partial v}{\partial \overset{*}{\varphi}} \right) + w \frac{\partial v}{\partial z} + \frac{vw}{r} + \frac{u^2}{r} \tan \varphi + fu + \frac{1}{\rho} \frac{\partial p}{\partial \overset{*}{\varphi}} = 0$$

1 2 3 4 5 7 8

$$\frac{\partial w}{\partial t} + \left(u \frac{\partial w}{\partial \overset{*}{\lambda}} + v \frac{\partial w}{\partial \overset{*}{\varphi}} \right) + w \frac{\partial w}{\partial z} - \frac{1}{r}(u^2 + v^2) - lu + \frac{1}{\rho} \frac{\partial p}{\partial z} + g = 0$$

1 2 3 4 6 8 9

(1.84)

Let us briefly discuss the various terms appearing in (1.84), which are numbered for ease of reference. These terms represent either *real* or *fictitious forces*. In each equation term 1 is the local rate of change with time of the velocity component. Terms 2 and 3 denote horizontal and vertical advection, respectively. Fictitious or apparent forces do not result from the interaction of an air parcel with other bodies, but stem from the choice of the rotating coordinate system. Terms 4 and 5 are such apparent forces per unit mass. They are also known as *metric accelerations*, which result from the curvature of the coordinate lines. The metric acceleration or *metric force per unit mass* is perpendicular to the relative velocity as follows from

$$\left[\left(\frac{uw}{r} - \frac{uv}{r} \tan \varphi \right) \mathbf{e}_\lambda + \left(\frac{vw}{r} + \frac{u^2}{r} \tan \varphi \right) \mathbf{e}_\varphi - \frac{u^2 + v^2}{r} \mathbf{e}_r \right] \cdot (u\mathbf{e}_\lambda + v\mathbf{e}_\varphi + w\mathbf{e}_r) = 0 \tag{1.85}$$

so these forces do not perform any work. Terms 6 and 7 are the Coriolis terms which result from the rotation of the coordinate system. By proceeding as in (1.85) we can again verify that the Coriolis force does not perform any work either. Term 8 is the pressure-gradient force and term 9 denotes the acceleration due to gravity.

Equations (1.84) are so general that they describe all scales of motion including local circulations as well as large-scale synoptic systems. For the present, let us consider the motion of dry air only for simplicity. Whenever we introduce moisture with associated phase changes, the situation becomes very involved.

Let us now count the number of dependent variables of the atmospheric system. These are the three velocity components u, v, w, the temperature T, the air density ρ, and pressure p. In order to evaluate these, we must have six equations at our disposal. These are the three component equations of motion for u, v, w, the first law of thermodynamics for T, the continuity equation for ρ, and the ideal-gas law for p. We have just as many equations as unknowns, so we say that this system is closed. We call this system the *molecular system* or the *nonturbulent system*. In contrast, the so-called *microturbulent system*, which we have not yet discussed, is not closed, so there are more unknown quantities than equations. This necessitates the introduction of closure assumptions.

If we compare the numerical values of the various terms appearing in the system (1.84), we find that they may differ by various orders of magnitude. For a particular situation to be studied, it seems reasonable to omit the insignificant terms. There exists a systematic method for deciding how to eliminate these. This method is known as scale analysis and will be described in the next chapter.

1.7 Problems

1.1: Show that

$$\frac{D}{Dt}(\rho\mathbf{v}_\Omega) = \rho\mathbf{\Omega}\times\mathbf{v} - \rho\,\nabla\left(\frac{\mathbf{v}_\Omega^2}{2}\right)$$

$$\frac{d}{dt}(\nabla\psi) = \nabla\,\frac{d\psi}{dt} - \nabla\mathbf{v}\cdot\nabla\psi$$

where ψ is an arbitrary scalar field function.

1.2:

(a) Show that

$$\oint d\mathbf{r}\cdot(\mathbf{\Omega}\times\mathbf{v}) = \frac{d}{dt}\int_S d\mathbf{S}\cdot\mathbf{\Omega}$$

(b) By utilizing this equation, show that, for frictionless motion, equation (1.62) can be written in the form

$$\frac{dC}{dt} = -2\Omega\,\frac{dS'}{dt} - \oint\frac{1}{\rho}\,d_{\mathrm{g}}p \quad \text{with} \quad C = \oint d\mathbf{r}\cdot\mathbf{v}$$

where S' is the projection of the material surface $S(t)$ on the equatorial plane.

1.3: In the absolute system the frictional tensor $\mathbb{J}(\mathbf{v}_\mathrm{A})$ is given by (1.12).

(a) Show in a coordinate-free manner that, for the rigidly rotating earth, we may write

$$\nabla \cdot \mathbb{J}(\mathbf{v}_{\mathrm{A}}) = \nabla \cdot \mathbb{J}(\mathbf{v})$$

(b) Calculate the influence of the frictional force on the velocity profiles

$$\mathbf{v}_1(z) = C_1 \ln\left(\frac{z}{z_0}\right)\mathbf{i} \quad \text{and} \quad \mathbf{v}_2(z) = C_2 z \mathbf{i}$$

where C_1, C_2, and z_0 are constants.

1.4: An incompressible fluid is streaming through a pipe of arbitrary but constant cross-section. Along the axis of the cylinder which is pointing in the x-direction, the fluid velocity is $u = |\mathbf{v}|$ everywhere so that u depends on the coordinates y and z only.
(a) Show that the continuity equation is satisfied.
(b) Find the Navier–Stokes equation for u. Ignore gravity and any convective motion.
(c) Find a solution for u if the pipe is a circular cylinder of radius R_0. The boundary condition is $u = 0$ at $R = R_0$. Use cylindrical coordinates.
(d) Find the amount Q of fluid streaming through the cross-section of the cylinder per unit time.

1.5: The continuity equation for relative motion can be written in the form

$$\frac{d}{dt}\left(\rho \sqrt{g}_{q}\right) + \rho \sqrt{g}_{q} \frac{\partial \dot{q}^n}{\partial q^n} = 0$$

Show that this equation is identical with (1.50).

1.6: Draw and discuss the energy-transformation diagram corresponding to (1.69).

1.7: Show the validity of the following equation:

$$\frac{D}{Dt}(\rho\phi) = \left(\frac{\partial \rho\phi}{\partial t}\right)_{x^i} + \nabla \cdot (\rho \mathbf{v}_{\mathrm{A}} \phi) = \frac{1}{\sqrt{g}_{q}} \left\{ \left[\frac{\partial}{\partial t}\left(\sqrt{g}_{q} \rho\phi\right)\right]_{q^i} + \frac{\partial}{\partial q^n}\left(\sqrt{g}_{q} \rho\phi\dot{q}^n\right)\right\}$$

1.8: Use equations (M4.42) and (M4.45) to verify the following relations for the unit vectors ($\mathbf{e}_\lambda$, $\mathbf{e}_\varphi$, $\mathbf{e}_r$) of the geographical coordinate system:

$$\frac{\partial \mathbf{e}_\lambda}{\partial \lambda} = \mathbf{e}_\varphi \sin\varphi - \mathbf{e}_r \cos\varphi, \qquad \frac{\partial \mathbf{e}_\lambda}{\partial \varphi} = 0, \qquad \frac{\partial \mathbf{e}_\lambda}{\partial r} = 0$$

$$\frac{\partial \mathbf{e}_\varphi}{\partial \lambda} = -\mathbf{e}_\lambda \sin\varphi, \qquad \frac{\partial \mathbf{e}_\varphi}{\partial \varphi} = -\mathbf{e}_r, \qquad \frac{\partial \mathbf{e}_\varphi}{\partial r} = 0$$

$$\frac{\partial \mathbf{e}_r}{\partial \lambda} = \mathbf{e}_\lambda \cos\varphi, \qquad \frac{\partial \mathbf{e}_r}{\partial \varphi} = \mathbf{e}_\varphi, \qquad \frac{\partial \mathbf{e}_r}{\partial r} = 0$$

$$\frac{\partial \mathbf{e}_\lambda}{\partial t} = \mathbf{\Omega} \times \mathbf{e}_\lambda = \Omega \frac{\partial \mathbf{e}_\lambda}{\partial \lambda}, \qquad \frac{\partial \mathbf{e}_\varphi}{\partial t} = \mathbf{\Omega} \times \mathbf{e}_\varphi = \Omega \frac{\partial \mathbf{e}_\varphi}{\partial \lambda}, \qquad \frac{\partial \mathbf{e}_r}{\partial t} = \mathbf{\Omega} \times \mathbf{e}_r = \Omega \frac{\partial \mathbf{e}_r}{\partial \lambda}$$

where $\mathbf{\Omega} = \Omega(\cos\varphi\, \mathbf{e}_\varphi + \sin\varphi\, \mathbf{e}_r)$.

2

Scale analysis

Scale analysis is a systematic method of comparing the magnitudes of the various terms in the hydrodynamical equations describing the atmospheric motion. This theory is instrumental in the design of consistent dynamic–mathematical models for dynamic analysis and numerical weather prediction. Charney (1948) introduced this technique to large-scale dynamics and showed that it is not necessary to use the complete scalar set of Navier–Stokes equations to describe the synoptic and planetary-scale motion. Among others, mainly Burger (1958) and Phillips (1963) used and extended this method. For additional details and a more complete bibliography see Haltiner and Williams (1980). In this chapter we follow Pichler's (1997) excellent introduction to scale analysis.

2.1 An outline of the method

Scale analysis makes it possible to objectively estimate the magnitudes of the various terms in an equation describing a physical system. The basic idea is to formulate a simplified equation by ignoring certain terms in a consistent manner without changing the basic physics. Let us consider an equation of the form

$$\psi_1 + \psi_2 + \cdots + \psi_i + \cdots + \psi_n = 0 \tag{2.1}$$

which may be a part of a more general system. The task ahead is to estimate the magnitudes of the individual terms in (2.1). Let us define the *magnitude* of each term by the symbol

$$\underset{\mathrm{m}}{[\,\psi_i\,]} = \text{magnitude of } \psi_i \tag{2.2}$$

From (2.1) and (2.2) we form a dimensionless expression

$$\langle\, \psi_i \,\rangle = \frac{\psi_i}{\underset{\mathrm{m}}{[\,\psi_i\,]}} \tag{2.3}$$

which by necessity is of magnitude 1. In order to avoid confusion in the notation we have added the suffix m within the bracket in (2.2) to remind the reader that this bracket refers to the magnitude of the term. With (2.3) equation (2.1) can be written as

$$[\underset{\mathrm{m}}{\psi_1}]\langle\psi_1\rangle+[\underset{\mathrm{m}}{\psi_2}]\langle\psi_2\rangle+\cdots+[\underset{\mathrm{m}}{\psi_i}]\langle\psi_i\rangle+\cdots+[\underset{\mathrm{m}}{\psi_n}]\langle\psi_n\rangle=0 \tag{2.4}$$

In order to consistently compare the magnitude of the various terms we introduce dimensionless *characteristic numbers* defined by

$$\psi_{r,i}=[\underset{\mathrm{m}}{\psi_r}]/[\underset{\mathrm{m}}{\psi_i}] \tag{2.5}$$

The importance of one particular term, say term i in (2.1), will now be investigated. We divide equation (2.4) by the magnitude of term i and find, using the definition of the characteristic numbers, the expression

$$\psi_{1,i}\langle\psi_1\rangle+\psi_{2,i}\langle\psi_2\rangle+\cdots+\langle\psi_i\rangle\cdots+\psi_{n,i}\langle\psi_n\rangle=0 \tag{2.6}$$

If, for example, all characteristic numbers are much larger than 1, which is the number multiplying term i, then term i has no significance in relation to the remaining terms and may be ignored. If, on the other hand, all characteristic numbers are much smaller than 1, then term i plays a dominant role and must be considered in the physical treatment under all circumstances.

We will now apply this method to the Navier–Stokes equation (1.71) which excludes the deformational velocity $\mathbf{v}_{\mathrm{D}}$. In the form (2.4) the Navier–Stokes equation can then be written as

$$\begin{aligned}&\Bigg[\underset{\mathrm{m}}{\frac{\partial\mathbf{v}}{\partial t}\bigg|_{\mathbf{q}_i}}\Bigg]\Bigg\langle\frac{\partial\mathbf{v}}{\partial t}\bigg|_{\mathbf{q}_i}\Bigg\rangle+[\mathbf{v}\underset{\mathrm{m}}{\cdot}\nabla\mathbf{v}]\langle\mathbf{v}\cdot\nabla\mathbf{v}\rangle+\Bigg[\underset{\mathrm{m}}{\frac{1}{\rho}\nabla p}\Bigg]\Bigg\langle\frac{1}{\rho}\nabla p\Bigg\rangle\\&+[\underset{\mathrm{m}}{\nabla\phi}]\langle\nabla\phi\rangle+2[\boldsymbol{\Omega}\underset{\mathrm{m}}{\times}\mathbf{v}]\langle\boldsymbol{\Omega}\times\mathbf{v}\rangle-\Bigg[\underset{\mathrm{m}}{\frac{1}{\rho}\nabla\cdot\mathbb{J}}\Bigg]\Bigg\langle\frac{1}{\rho}\nabla\cdot\mathbb{J}\Bigg\rangle=0\end{aligned} \tag{2.7}$$

This is a very convenient form in which to introduce various characteristic numbers that have proven to be very useful in the study of fluids and gases. Of particular interest is the relation of the magnitude of the individual terms to the inertial force per unit mass $[\mathbf{v}\underset{\mathrm{m}}{\cdot}\nabla\mathbf{v}]$. We proceed by dividing the inertial force by the magnitudes of each of the remaining terms. Thus, we obtain five dimensionless numbers:

$$
\begin{aligned}
&\textit{the Strouhal number:} && St = \frac{\underset{\mathrm{m}}{[\,\mathbf{v}\cdot\nabla\mathbf{v}\,]}}{\underset{\mathrm{m}}{\left[\left.\dfrac{\partial\mathbf{v}}{\partial t}\right|_{\mathbf{q}_i}\right]}}\\
&\textit{the Euler number:} && Eu = \frac{\underset{\mathrm{m}}{[\,\mathbf{v}\cdot\nabla\mathbf{v}\,]}}{\underset{\mathrm{m}}{\left[\dfrac{1}{\rho}\nabla p\right]}}\\
&\textit{the Froude number:} && Fr = \frac{\underset{\mathrm{m}}{[\,\mathbf{v}\cdot\nabla\mathbf{v}\,]}}{\underset{\mathrm{m}}{[\,\nabla\phi\,]}}\\
&\textit{the Rossby number:} && Ro = \frac{\underset{\mathrm{m}}{[\,\mathbf{v}\cdot\nabla\mathbf{v}\,]}}{2\underset{\mathrm{m}}{[\,\boldsymbol{\Omega}\times\mathbf{v}\,]}}\\
&\textit{the Reynolds number:} && Re = \frac{\underset{\mathrm{m}}{[\,\mathbf{v}\cdot\nabla\mathbf{v}\,]}}{\underset{\mathrm{m}}{\left[\dfrac{1}{\rho}\nabla\cdot\mathbb{J}\right]}}
\end{aligned}
\tag{2.8}
$$

Each of these numbers expresses the ratio of the magnitude of the inertial force to one of the remaining forces appearing in the Navier–Stokes equation. These numbers are the reciprocals of the characteristic numbers defined by (2.5). By inserting the numbers in (2.8) into (2.7) we obtain the dimensionless form of the Navier–Stokes equation:

$$
\begin{aligned}
&\frac{1}{St}\left\langle\left.\frac{\partial\mathbf{v}}{\partial t}\right|_{\mathbf{q}_i}\right\rangle + \langle\,\mathbf{v}\cdot\nabla\mathbf{v}\,\rangle + \frac{1}{Eu}\left\langle\frac{1}{\rho}\nabla p\right\rangle + \frac{1}{Fr}\langle\,\nabla\phi\,\rangle\\
&\quad + \frac{1}{Ro}\langle\,\boldsymbol{\Omega}\times\mathbf{v}\,\rangle - \frac{1}{Re}\left\langle\frac{1}{\rho}\nabla\cdot\mathbb{J}\right\rangle = 0
\end{aligned}
\tag{2.9}
$$

The characteristic number of the second term in (2.9) equals 1. If, for example, the remaining characteristic numbers are much larger than 1, then the inertial force may be ignored in comparison with the other forces appearing in (2.9). This means that the nonlinear equation (2.9) in this special case reduces to a linear partial differential equation. Inspection of (2.9) shows that the frictional term is important only if the Reynolds number is very small.

2.2 Practical formulation of the dimensionless flow numbers

Equation (2.8) is a collection of various important flow numbers. For practical applications these flow numbers must be expressed in terms of easily accessible variables characterizing the flow field. The basic variables required are time, the wind speed, the angular velocity of the earth, the acceleration due to gravity, pressure, density, and temperature. The scale of motion is characterized by the horizontal (L_1, L_2) and vertical (L_3) length scales of the phenomena to be investigated. The synoptic scale of motion refers to long waves, low- and high-pressure systems so that $L_1, L_2 \geq 10^6$ m. The vertical scale L_3 includes the weather-effective part of the atmosphere and may be approximated by $L_3 \approx 10^4$ m.

According to equation (2.3) we introduce the length scale $\mathbf{x}$ by means of

$$\mathbf{x} = [\underset{\mathrm{m}}{\mathbf{x}}]\langle \mathbf{x} \rangle = S\langle \mathbf{x} \rangle \tag{2.10}$$

Utilizing this expression, the nabla operator may be written as

$$\nabla = [\underset{\mathrm{m}}{\nabla}]\langle \nabla \rangle = \frac{1}{S}\langle \nabla \rangle \tag{2.11}$$

The time scale will be expressed in terms of the local characteristic time interval T_1 describing the nonstationary motion. If the magnitude of the phase velocity of the wave (pressure system) is denoted by C, then we replace the time t by means of

$$t = [\underset{\mathrm{m}}{t}]\langle t \rangle = T_1\langle t \rangle = \frac{S}{C}\langle t \rangle \tag{2.12}$$

Likewise, we write for the remaining variables

$$\begin{gathered}\frac{1}{\rho} = \left[\underset{\mathrm{m}}{\frac{1}{\rho}}\right]\left\langle \frac{1}{\rho}\right\rangle = A\left\langle \frac{1}{\rho}\right\rangle, \quad p = [\underset{\mathrm{m}}{p}]\langle p \rangle = P\langle p \rangle, \quad T = [\underset{\mathrm{m}}{T}]\langle T \rangle = T_0\langle T \rangle \\ \mathbf{v} = [\underset{\mathrm{m}}{\mathbf{v}}]\langle \mathbf{v} \rangle = V\langle \mathbf{v} \rangle, \qquad \boldsymbol{\Omega} = \Omega\langle \boldsymbol{\Omega} \rangle, \quad \nabla\phi = G\langle \nabla\phi \rangle, \quad [\underset{\mathrm{m}}{\mathbb{J}}] = \frac{\mu_0 V}{S}\end{gathered} \tag{2.13}$$

The characteristic magnitudes of pressure, density, and temperature are related by the *ideal-gas law*

$$AP = R_0 T_0 \tag{2.14}$$

where R_0 is the gas constant of dry air. Effects of moisture on temperature are considered unimportant and are omitted. With the help of the definitions (2.13) the magnitudes of all terms occurring in the Navier–Stokes equation (2.7) can now be written as

$$\begin{gathered}\left[\underset{\mathrm{m}}{\left.\frac{\partial \mathbf{v}}{\partial t}\right|_{\mathbf{q}_i}}\right] = \frac{V}{T_1}, \qquad [\underset{\mathrm{m}}{\mathbf{v}\cdot\nabla\mathbf{v}}] = \frac{V^2}{S}, \qquad \left[\underset{\mathrm{m}}{\frac{1}{\rho}\nabla p}\right] = \frac{AP}{S} = \frac{R_0 T_0}{S} \\ [\underset{\mathrm{m}}{\nabla\phi}] = G, \qquad [\underset{\mathrm{m}}{2\boldsymbol{\Omega}\times\mathbf{v}}] = 2\Omega V, \qquad \left[\underset{\mathrm{m}}{\frac{1}{\rho}\nabla\cdot\mathbb{J}}\right] = \frac{A\mu_0 V}{S^2}\end{gathered} \tag{2.15}$$

The dimensionless flow numbers (2.8) assume the practical forms

$$
\begin{aligned}
St &= \frac{VT_1}{S} = \frac{V}{C}, & Ro &= \frac{V}{2\Omega S}, & Eu &= \frac{V^2}{R_0 T_0} \\
Fr &= \frac{V^2}{GS}, & Re &= \frac{VS}{A\mu_0} = \frac{VS}{\nu_0}
\end{aligned}
\tag{2.16}
$$

The equation of motion (1.71) in scale-analytic form can now be written as

$$
\begin{aligned}
&\frac{V}{T_1}\left\langle \left.\frac{\partial \mathbf{v}}{\partial t}\right|_{\mathbf{q}_i} \right\rangle + \frac{V^2}{S}\langle \mathbf{v}\cdot\nabla\mathbf{v}\rangle + \frac{AP}{S}\left\langle \frac{1}{\rho}\nabla p\right\rangle \\
&\quad - G\langle\nabla\phi\rangle + 2\Omega V\langle \boldsymbol{\Omega}\times\mathbf{v}\rangle - \frac{A\mu_0 V}{S^2}\left\langle \frac{1}{\rho}\nabla\cdot\mathbb{J}\right\rangle = 0
\end{aligned}
\tag{2.17}
$$

Dividing this equation by the magnitude of the inertial force and using the practical forms of the flow numbers (2.16) again yields (2.9).

The scale-analytic form of the Navier–Stokes equation may be used to estimate the relative importance of each term. For many applications, however, it is of advantage not to subject the vector equation (2.9) directly to a scale analysis but to use the equivalent scalar equations. How to proceed for large-scale frictionless flow will be explained in the next section. The effect of friction will be treated in a later chapter when we have acquired the necessary background. In passing, we would like to remark that the method of scale analysis is quite general and may be applied not only to the equation of motion but also to other equations.

2.3 Scale analysis of large-scale frictionless motion

As stated before, we wish to apply the method of scale analysis to the scalar form of the equation of motion. We describe the motion in spherical coordinates in the form (1.84). In order to proceed efficiently, we apply equation (2.3) to the latitude and longitude, height, time, and other pertinent variables, yielding

$$
\begin{aligned}
&\text{(a)} \quad \overset{*}{\lambda} = L\left\langle \overset{*}{\lambda}\right\rangle, && \overset{*}{\varphi} = L\left\langle \overset{*}{\varphi}\right\rangle, && z = D\langle z\rangle \\
&\text{(b)} \quad u = U\langle u\rangle, && v = U\langle v\rangle, && w = W\langle w\rangle \\
&\text{(c)} \quad t = T_1\langle t\rangle = \frac{L}{C}\langle t\rangle \\
&\text{(d)} \quad g = G\langle g\rangle, && p = P\langle p\rangle \\
&\text{(e)} \quad H_p = H_0\left\langle H_p\right\rangle, && R_0 T = gH_p = GH_0\left\langle gH_p\right\rangle \\
&\text{(f)} \quad f = f_0\langle f\rangle, && l = l_0\langle l\rangle \\
&\text{(g)} \quad r = a
\end{aligned}
\tag{2.18}
$$

The stars labeling latitude and longitude, as defined by (1.75), are to remind us that we are using physical measure numbers. From observations it is known that the large scales of motion in longitudinal and latitudinal directions are of the same order of magnitude L whereas the vertical scale D is quite different, see (2.18a). The same is true for the magnitudes of the horizontal (U) and vertical (W) components of the wind velocity. Equations (2.18c) and (2.18d) give the scaled forms of time, of the acceleration due to gravity, and of the pressure p. It is also customary to introduce the *pressure scale height* H_p as given by (2.18e). The Coriolis parameters f and l are scaled in (2.18f) whereby the suffix 0 refers to the mean latitude of a geographical latitude belt for which the motion is considered. Finally, the radius r may be replaced by the mean radius a of the earth.

According to observations in the atmosphere the following typical numerical values are used for the quantities occurring in (2.18)[1]

$$\begin{aligned}
&U \sim 10\ \mathrm{m\,s^{-1}}, \qquad W \le 0.1\ \mathrm{m\,s^{-1}}, \qquad C \le U\\
&L \sim 10^6\ \mathrm{m}, \qquad D = H_0 \sim 10^4\ \mathrm{m}\\
&G \sim 10\ \mathrm{m\,s^{-2}}, \qquad a \sim 10^7\ \mathrm{m}\\
&l_0 \sim 10^{-4}\ \mathrm{s^{-1}}, \qquad f_0 \sim \begin{cases} 10^{-4}\ \mathrm{s^{-1}} & \text{for } 25^\circ \le \varphi_0 \le 80^\circ \\ 10^{-5}\ \mathrm{s^{-1}} & \text{for } \varphi_0 < 25^\circ \end{cases}
\end{aligned} \tag{2.19}$$

It is seen that the characteristic vertical velocity is much smaller than the characteristic horizontal velocity so that $W \ll U$. Furthermore, the phase velocity C of synoptic systems such as ridges and troughs satisfies the inequality $C \le U$ so that, according to (2.16), the Strouhal number is $St \ge 1$. Therefore, the local time scale $T_\mathrm{l} = S/C$ is larger than or equal to the so-called *convective time scale* $T_\mathrm{c} = S/U$, i.e. $T_\mathrm{l} \ge T_\mathrm{c}$. The latitudinal dependence of the Coriolis parameter f_0 is accounted for by using different values for the two geographical latitude bands $25^\circ \le \varphi_0 \le 80^\circ$ representing the broad range of mid- and high latitudes and $\varphi_0 < 25^\circ$ for the low latitudes.

Using (2.18) the pressure-gradient force terms appearing in (1.84) may be written as

$$\begin{aligned}
\frac{1}{\rho}\frac{\partial p}{\partial \overset{*}{\lambda}} &= \frac{gH_p}{p}\frac{\partial p}{\partial \overset{*}{\lambda}} = \frac{GH_0}{L}\frac{\Delta p_\mathrm{h}}{P}\left\langle \frac{gH_p}{p}\frac{\partial p}{\partial \overset{*}{\lambda}} \right\rangle\\
\frac{1}{\rho}\frac{\partial p}{\partial \overset{*}{\varphi}} &= \frac{gH_p}{p}\frac{\partial p}{\partial \overset{*}{\varphi}} = \frac{GH_0}{L}\frac{\Delta p_\mathrm{h}}{P}\left\langle \frac{gH_p}{p}\frac{\partial p}{\partial \overset{*}{\varphi}} \right\rangle\\
\frac{1}{\rho}\frac{\partial p}{\partial z} &= \frac{gH_p}{p}\frac{\partial p}{\partial z} = \frac{GH_0}{D}\frac{\Delta p_\mathrm{v}}{P}\left\langle \frac{gH_p}{p}\frac{\partial p}{\partial z} \right\rangle
\end{aligned} \tag{2.20}$$

[1] If dynamic processes of the boundary layer, where friction cannot be ignored, are being investigated, then we must use $D = 10^3$ m.

In these expressions the magnitudes of the spatial pressure changes $\Delta p_{\mathrm{h,v}}$ have been introduced, whereby the subscripts h and v have been added in order to distinguish between horizontal and vertical pressure changes. However, as the expressions suggest, it is profitable to approximate the magnitudes of the relative pressure changes $\Delta p_{\mathrm{h,v}}/P$. From observations we find the following typical values:

$$\frac{\Delta p_{\mathrm{h}}}{P} \sim \begin{cases} 10^{-2} & \text{for } 25^{\circ} \leq \varphi_0 \leq 80^{\circ} \\ 10^{-3} & \text{for } \varphi_0 < 25^{\circ} \end{cases}, \qquad \frac{\Delta p_{\mathrm{v}}}{P} \sim 1 \tag{2.21}$$

The vertical pressure change refers to the entire vertical extent H_0 of the atmosphere.

Observations indicate that the magnitudes of the changes in velocity along the horizontal and vertical length scales are similar and equal to the order of magnitude of the velocity:

$$\Delta u_{\mathrm{h}} \sim U, \qquad \Delta u_{\mathrm{v}} \sim U, \qquad \Delta v_{\mathrm{h}} \sim U, \qquad \Delta v_{\mathrm{v}} \sim U \tag{2.22}$$

Now we have finished all preparatory work for the scale analysis of the equation of motion in the scalar form. By substituting (2.20) for the pressure-gradient terms into (1.84) and using the scale-analytic expressions given in the previous equations, we obtain without difficulty

$$\begin{array}{lcccc} & \dfrac{CU}{L}\left\langle \dfrac{\partial u}{\partial t}\right\rangle + & \dfrac{U^2}{L}\left(\left\langle u\,\dfrac{\partial u}{\partial \overset{*}{\lambda}}\right\rangle + \left\langle v\,\dfrac{\partial u}{\partial \overset{*}{\varphi}}\right\rangle\right) + & \dfrac{UW}{H_0}\left\langle w\,\dfrac{\partial u}{\partial z}\right\rangle + & \dfrac{UW}{a}\langle uw\rangle \\ \text{(i)} & \sim 10^{-4} & \sim 10^{-4} & \sim 10^{-4} & \sim 10^{-7} \\ \text{(ii)} & \sim 10^{-4} & \sim 10^{-4} & \sim 10^{-4} & \sim 10^{-7} \\ & 1 & 2 & 3 & 4 \end{array}$$

$$\begin{array}{lcccc} & -\dfrac{U^2}{a}\tan\varphi\langle uv\rangle + & l_0 W\langle lw\rangle - & f_0 U\langle fv\rangle + & \dfrac{GH_0}{L}\dfrac{\Delta p_{\mathrm{h}}}{P}\left\langle \dfrac{gH_p}{p}\dfrac{\partial p}{\partial \overset{*}{\lambda}}\right\rangle = 0 \\ \text{(i)} & \sim 10^{-5} & \sim 10^{-5} & \sim 10^{-3} & \sim 10^{-3} \\ \text{(ii)} & \sim 10^{-5} & \sim 10^{-5} & \sim 10^{-4} & \sim 10^{-4} \\ & 5 & 6 & 7 & 8 \end{array}$$

(2.23a)

$$\frac{CU}{L}\left\langle \frac{\partial v}{\partial t} \right\rangle + \frac{U^2}{L}\left(\left\langle u\frac{\partial v}{\partial \overset{*}{\lambda}} \right\rangle + \left\langle v\frac{\partial v}{\partial \overset{*}{\varphi}} \right\rangle\right) + \frac{UW}{H_0}\left\langle w\frac{\partial v}{\partial z} \right\rangle + \frac{UW}{a}\langle vw \rangle$$

	1	2	3	4
(i)	$\sim 10^{-4}$	$\sim 10^{-4}$	$\sim 10^{-4}$	$\sim 10^{-7}$
(ii)	$\sim 10^{-4}$	$\sim 10^{-4}$	$\sim 10^{-4}$	$\sim 10^{-7}$

$$+\frac{U^2}{a}\langle u^2 \rangle \tan\varphi + f_0 U \langle fu \rangle + \frac{GH_0}{L}\frac{\Delta p_{\mathrm{h}}}{P}\left\langle \frac{gH_p}{p}\frac{\partial p}{\partial \overset{*}{\varphi}} \right\rangle = 0 \tag{2.23b}$$

	5	7	8
(i)	$\sim 10^{-5}$	$\sim 10^{-3}$	$\sim 10^{-3}$
(ii)	$\sim 10^{-5}$	$\sim 10^{-4}$	$\sim 10^{-4}$

$$\frac{CW}{L}\left\langle \frac{\partial w}{\partial t} \right\rangle + \frac{UW}{L}\left(\left\langle u\frac{\partial w}{\partial \overset{*}{\lambda}} \right\rangle + \left\langle v\frac{\partial w}{\partial \overset{*}{\varphi}} \right\rangle\right) + \frac{W^2}{H_0}\left\langle w\frac{\partial w}{\partial z} \right\rangle$$

	1	2	3
(i)	$\sim 10^{-6}$	$\sim 10^{-6}$	$\sim 10^{-6}$
(ii)	$\sim 10^{-6}$	$\sim 10^{-6}$	$\sim 10^{-6}$

$$-\frac{U^2}{a}(\langle u^2 \rangle + \langle v^2 \rangle) - l_0 U \langle lu \rangle + G\frac{\Delta p_{\mathrm{v}}}{P}\left\langle \frac{gH_p}{p}\frac{\partial p}{\partial z} \right\rangle + G\langle g \rangle = 0 \tag{2.23c}$$

	4	6	8	9
(i)	$\sim 10^{-5}$	$\sim 10^{-3}$	$\sim 10^{1}$	$\sim 10^{1}$
(ii)	$\sim 10^{-5}$	$\sim 10^{-3}$	$\sim 10^{1}$	$\sim 10^{1}$

Below each term we have written the approximate magnitudes for the two latitude bands (i) $25^\circ \leq \varphi_0 \leq 80^\circ$ and (ii) $\varphi_0 < 25^\circ$. For ease of identification the terms have also been numbered. Each term is the product of a dimensional factor representing force per unit mass and of dimensionless quantities $\langle \cdots \rangle$ of magnitude 1. Comparison of individual terms then shows which terms may be safely omitted in large-scale frictionless motion.

Let us first consider the prognostic equation (2.23a) for the u-component of the wind field. Term 1 representing the local rate of change with time of the velocity and terms 2 and 3 denoting horizontal and vertical advection are all of the same order of magnitude 10^{-4}. Terms 4 and 5 both involve the radius a of the earth. Since term 4 is at least two orders of magnitude smaller than the remaining terms, it may be safely ignored. Terms 6 and 7 are due to the earth's rotation and represent the Coriolis force. The main effect of the Coriolis force is included in term 7, so term 6 may be omitted. One might be tempted to also ignore term 5, which is of the same magnitude as term 6, but these two terms represent entirely different forces. Term 5 is a *metric acceleration* and should be retained in comparison with the metric acceleration term 4. Metric accelerations are apparent accelerations resulting from the particular shape of the coordinate system being considered. They do not perform any work. The estimated magnitude of the pressure-gradient term shows that this term must be retained under all circumstances. The same type of argument may be applied to equation (2.23b) describing the equation for the v-component of the wind field. Only the metric term 4 will be ignored due to its small magnitude.

A somewhat different situation arises in (2.23c). Inspection shows that the vertical pressure gradient and the gravitational effect represented by terms 8 and 9 are of the same order of magnitude and seven orders of magnitude larger than the individual acceleration represented by the terms 1–3. Moreover, terms 4 and 6 may also be ignored in comparison with terms 8 and 9, so this equation degenerates to two terms only. Nevertheless, we will momentarily retain terms 1–3 adding up to the individual vertical acceleration. The analysis of some small-scale circulations can be carried out only if the vertical acceleration is accounted for.

The information gained by the above scale analysis may now be utilized to obtain the approximate form of the equation of motion (1.84) for the components u, v, w of the wind field in the geographical coordinate system for the description of large-scale frictionless flow fields

$$\begin{aligned}
&\frac{\partial u}{\partial t} + \frac{u}{a\cos\varphi}\frac{\partial u}{\partial \lambda} + \frac{v}{a}\frac{\partial u}{\partial \varphi} + w\frac{\partial u}{\partial z} - \frac{uv}{a}\tan\varphi - fv = -\frac{1}{\rho a\cos\varphi}\frac{\partial p}{\partial \lambda}\\
&\frac{\partial v}{\partial t} + \frac{u}{a\cos\varphi}\frac{\partial v}{\partial \lambda} + \frac{v}{a}\frac{\partial v}{\partial \varphi} + w\frac{\partial v}{\partial z} + \frac{u^2}{a}\tan\varphi + fu = -\frac{1}{\rho a}\frac{\partial p}{\partial \varphi}\\
&\frac{\partial w}{\partial t} + \frac{u}{a\cos\varphi}\frac{\partial w}{\partial \lambda} + \frac{v}{a}\frac{\partial w}{\partial \varphi} + w\frac{\partial w}{\partial z} = -\frac{1}{\rho}\frac{\partial p}{\partial z} - g
\end{aligned} \tag{2.24}$$

Recalling the definition of the individual derivative

$$\frac{d}{dt} = \frac{\partial}{\partial t} + \frac{u}{a\cos\varphi}\frac{\partial}{\partial \lambda} + \frac{v}{a}\frac{\partial}{\partial \varphi} + w\frac{\partial}{\partial z} \quad \text{with} \quad u = a\cos\varphi\,\dot{\lambda}, \quad v = a\dot{\varphi}, \quad w = \dot{r} \tag{2.25}$$

the three components of the equation of motion in their approximate form may be rewritten as

$$
\boxed{
\begin{aligned}
&\text{(a)} \quad \frac{du}{dt} - \frac{uv}{a}\tan\varphi - fv = -\frac{1}{\rho a \cos\varphi}\frac{\partial p}{\partial \lambda} \\
&\text{(b)} \quad \frac{dv}{dt} + \frac{u^2}{a}\tan\varphi + fu = -\frac{1}{\rho a}\frac{\partial p}{\partial \varphi} \\
&\text{(c)} \quad \frac{dw}{dt} = -\frac{1}{\rho}\frac{\partial p}{\partial z} - g
\end{aligned}
}
\tag{2.26}
$$

Owing to the small magnitudes of terms 1 to 3 in (2.23c) for large-scale motion, which combine to give the vertical acceleration, we need to retain only terms 8 and 9. For this reason (2.26c) degenerates to

$$
\boxed{\frac{1}{\rho}\frac{\partial p}{\partial z} = -g}
\tag{2.27}
$$

This equation is known as the *hydrostatic approximation*. The large-scale or the synoptic-scale motion is quasistatic even in the presence of horizontal gradients of the thermodynamic variables (p, T, ρ). We must clearly understand that quasistatic motion does not imply that there is no vertical motion since w is still contained in the advection term of (2.24). In fact, we have not set $dw/dt = 0$ but used scale analysis to show that the absolute value of dw/dt is much smaller than the absolute values of the vertical pressure gradient and the acceleration due to gravity. If the third equation of motion (2.26c) is replaced by (2.27) it is no longer available for the prognostic determination of w, which must be obtained in some other way. The large-scale motion is then characterized by equations (2.26a) and (2.26b) together with equation (2.27).

We will now multiply (2.26a) and (2.26b) by the unit vectors $\mathbf{e}_\lambda$ and $\mathbf{e}_\varphi$, respectively, and then add these two equations representing the horizontal motion. The two terms containing $\tan\varphi$ may be combined to give a vector expression representing the metric acceleration. For brevity, we also combine the two acceleration terms $(du/dt, dv/dt)$ with the metric acceleration, yielding

$$
\left(\left.\frac{d\mathbf{v}_\text{h}}{dt}\right|_{\mathbf{e}_i}\right)_\text{h} = \mathbf{e}_\lambda \frac{du}{dt} + \mathbf{e}_\varphi \frac{dv}{dt} + \frac{u}{a}\tan\varphi\,(\mathbf{e}_r \times \mathbf{v}_\text{h})
\tag{2.28a}
$$

with $\mathbf{v}_\text{h} = \mathbf{e}_\lambda u + \mathbf{e}_\varphi v$. This is the horizontal part of the total acceleration of the horizontal wind given by

$$
\left.\frac{d\mathbf{v}_\text{h}}{dt}\right|_{\mathbf{e}_i} = \left(\left.\frac{d\mathbf{v}_\text{h}}{dt}\right|_{\mathbf{e}_i}\right)_\text{h} - \frac{u^2+v^2}{a}\mathbf{e}_r
\tag{2.28b}
$$

The last term of this equation appears in (2.23c). Recall that the vertical line in the differential operator on the left-hand side symbolizes that the unit vectors are not to be differentiated with respect to time. Likewise, the two terms containing the Coriolis parameter may be combined to give $\mathbf{C}_\mathrm{h}$, which is an approximate expression for the Coriolis force. Therefore, the equation of horizontal motion may be written in simplified form as

$$\boxed{\left(\frac{d\mathbf{v}_\mathrm{h}}{dt}\bigg|_{\mathbf{e}_i}\right)_\mathrm{h} - \mathbf{C}_\mathrm{h} = -\frac{1}{\rho}\nabla_\mathrm{h} p \quad \text{with} \quad \mathbf{C}_\mathrm{h} = -f\mathbf{e}_r \times \mathbf{v}_\mathrm{h}} \tag{2.29}$$

From (2.29) it can be seen that the Coriolis force $\mathbf{C}_\mathrm{h}$ is oriented normal to $\mathbf{v}_\mathrm{h}$. In the northern (southern) hemisphere the Coriolis force is pointing to the right (left) if the observer is looking in the direction of the horizontal wind. This approximation is also valid at low latitudes with the exception of the equatorial belt. Even at the latitude of $\varphi = \pm 5^\circ$ the term lw appearing in (2.23a) is one order of magnitude smaller than fv.

2.4 The geostrophic wind and the Euler wind

We wish to consider the magnitudes of the various terms of (2.29). Inspection of (2.23a) and (2.23b) shows that, at mid- and high latitudes, the individual derivatives $du/dt, dv/dt$ and the metric acceleration are at least one order of magnitude smaller than the components of $\mathbf{C}_\mathrm{h}$ and the horizontal pressure gradient. This is also apparent from the scaling of (2.29):

$$\frac{U^2}{L}\left\langle\left(\frac{d\mathbf{v}_\mathrm{h}}{dt}\bigg|_{\mathbf{e}_i}\right)_\mathrm{h}\right\rangle + f_0 U\langle f\mathbf{e}_r \times \mathbf{v}_\mathrm{h}\rangle = -\frac{GH_0}{L}\frac{\Delta p_\mathrm{h}}{P}\left\langle\frac{gH_p}{p}\nabla_\mathrm{h} p\right\rangle \tag{2.30}$$

where the upper limit $C = U$ has been used. By utilizing in (2.16) the magnitudes listed in (2.19) we obtain for the Rossby and Froude numbers

$$Ro = \frac{U}{f_0 L}, \qquad Fr = \frac{U^2}{GH_0} \tag{2.31}$$

Substitution of these expressions into (2.30) yields

$$Ro\left\langle\left(\frac{d\mathbf{v}_\mathrm{h}}{dt}\bigg|_{\mathbf{e}_i}\right)_\mathrm{h}\right\rangle + \langle f\mathbf{e}_r \times \mathbf{v}_\mathrm{h}\rangle = -\frac{Ro}{Fr}\frac{\Delta p_\mathrm{h}}{P}\left\langle\frac{gH_p}{p}\nabla_\mathrm{h} p\right\rangle \tag{2.32}$$

For the large-scale synoptic flow regimes at the mid- and high latitudes we obtain from (2.19) and (2.21)

$$25^\circ \le \varphi_0 \le 80^\circ: \qquad Ro \sim 10^{-1}, \qquad \frac{Ro}{Fr}\frac{\Delta p_\mathrm{h}}{P} \sim 1 \tag{2.33}$$

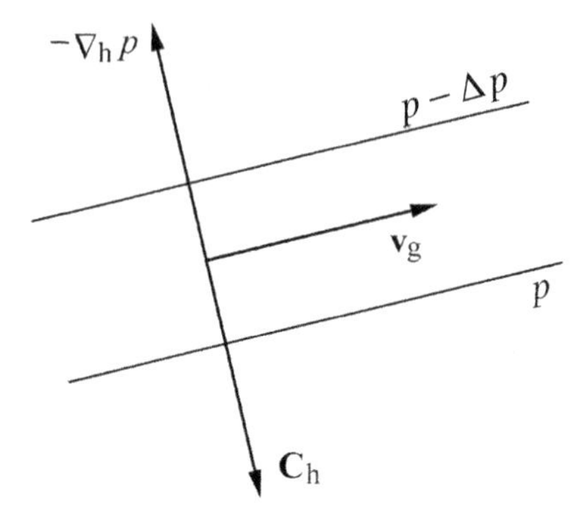

Fig. 2.1 The geostrophic wind in the northern hemisphere.

In the limiting case (L very large so that $Ro \to 0$, but $(u, v) \neq 0$), we find from (2.32) an exact balance between the Coriolis force and the pressure-gradient force. The wind resulting from this balance is known as the geostrophic wind $\mathbf{v}_g$ and may be easily obtained from (2.29), which reduces to

$$f\mathbf{e}_r \times \mathbf{v}_g = -\frac{1}{\rho}\nabla_h p \tag{2.34}$$

On solving for $\mathbf{v}_g$ we find

$$\boxed{\mathbf{v}_g = \frac{1}{\rho f}\mathbf{e}_r \times \nabla_h p} \tag{2.35}$$

The isobars run parallel to the geostrophic wind vector so that $\mathbf{v}_g$ is normal to the pressure-gradient force. According to (2.34) the directions of the Coriolis force and the pressure-gradient force are opposite, as shown in Figure 2.1 for the northern hemisphere. Above the atmospheric boundary layer where frictional effects are very small, the geostrophic wind deviates very little from the actually observed wind. However, this small deviation is very important since it is responsible for atmospheric developments. Owing to its smallness, the geostrophic deviation in the free atmosphere can hardly be determined from routine measurements.

Let us briefly consider the tropical latitudes. In this situation we obtain from (2.19) and (2.21) (see also Charney (1963))

$$\varphi_0 < 25^\circ: \qquad Ro \sim 1, \qquad \frac{Ro}{Fr}\frac{\Delta p_h}{P} \sim 1 \tag{2.36}$$

Approaching the equator either from the south or from the north, the Rossby number becomes very large so that the Coriolis term in (2.32) may be ignored. This reduces

(2.29) to the horizontal form of the *Euler equation*:

$$\boxed{\left(\left.\frac{d\mathbf{v}_{\mathrm{h}}}{dt}\right|_{\mathbf{e}_i}\right)_{\mathrm{h}} = -\frac{1}{\rho}\nabla_h p} \tag{2.37}$$

which is the equilibrium condition at $\varphi = 0^\circ$ and very near to the equator.

2.5 The equation of motion on a tangential plane

Let us briefly review the scale-analyzed equation of motion (2.26) of the rotating spherical-coordinate system. The singularites at the poles ($\varphi = \pm 90^\circ$) arising from $\cos\varphi$ and $\tan\phi$ are troublesome. Later, when we introduce the so-called stereographic coordinate system, these singularities disappear. At this point it will be very convenient for us to introduce a rotating rectangular coordinate system in which the x-axis is pointing toward the east, the y-axis toward the north, and the z-axis toward the local zenith. This type of coordinate system is sufficient for the study of various meteorological problems. We will reduce equation (2.26) by "brute strength" to obtain the desired result.

We imagine a plane that is tangential to the spherical earth at a selected point. The rotational speed of the plane is $\Omega \sin\varphi = f/2$, as follows from inspection of Figure 1.4. By omitting from (2.26) the two terms containing $\tan\varphi$ and by replacing the increments of the physical coordinates (see equation (1.75)) $\delta\lambda^* = a\cos\varphi\,\delta\lambda$ and $\delta\varphi^* = a\,\delta\varphi$ in the pressure-gradient force by δx and δy, respectively, we obtain the equation of motion for the tangential plane:

$$\boxed{\begin{aligned} \frac{du}{dt} - fv &= -\frac{1}{\rho}\frac{\partial p}{\partial x} \\ \frac{dv}{dt} + fu &= -\frac{1}{\rho}\frac{\partial p}{\partial y} \\ \frac{dw}{dt} &= -\frac{1}{\rho}\frac{\partial p}{\partial z} - g \end{aligned}} \tag{2.38}$$

We shall use this form of the equation of motion from time to time.

2.6 Problems

2.1: Check whether the magnitudes in (2.23) are correct. Use the numerical values stated in the text.

2.2: Verify equation (2.28b).

2.3: Let M_λ, M_φ, and M_r represent the metric accelerations.
(a) With the help of (2.23) write down proper expressions for M_λ, M_φ, and M_r.
(b) Show that the metric acceleration does not perform any work.

2.4: The equation of motion can be written as

$$\left.\frac{\partial \rho \mathbf{v}}{\partial t}\right|_{\mathbf{q}_i} = -\nabla\cdot(\rho\mathbf{v}\mathbf{v}) - \nabla\cdot(p\mathbb{E} - \mathbb{J}) - \rho\,\nabla\phi + 2\rho\mathbf{v}\times\Omega$$

$$\text{with}\quad \mathbb{J} = \mu(\nabla\mathbf{v} + \overset{\curvearrowleft}{\mathbf{v}\nabla} - \tfrac{2}{3}\nabla\cdot\mathbf{v}\mathbb{E})$$
$$2\boldsymbol{\Omega} = 2\Omega\cos\varphi\,\mathbf{i}_2 + 2\Omega\sin\varphi\,\mathbf{i}_3 = f_2\mathbf{i}_2 + f_3\mathbf{i}_3$$

Show that the component form of this equation in Cartesian coordinates is given by

$$\begin{aligned}\frac{\partial\rho\,u_i}{\partial t} &= -\frac{\partial}{\partial x_n}(\rho u_n u_i) + \epsilon_{imn} f_n \rho u_m - \frac{\partial p}{\partial x_i} - \delta_{i3}\rho g \\ &\quad + \frac{\partial}{\partial x_n}\left[\mu\left(\frac{\partial u_i}{\partial x_n} + \frac{\partial u_n}{\partial x_i}\right) - \frac{2}{3}\delta_{in}\frac{\partial u_m}{\partial x_m}\right]\end{aligned}$$

$$\text{with}\quad \mathbf{i}_1\cdot\mathbf{A}\times\mathbf{B} = \epsilon_{imn}A_m B_n$$
$$\epsilon_{ijk} = \begin{cases} 1 & \text{for } (i,j,k) = (1,2,3), (2,3,1), (3,1,2) \\ -1 & \text{for } (i,j,k) = (3,2,1), (2,1,3), (1,3,2) \\ 0 & \text{else}\end{cases}$$

3

The material and the local description of flow

The kinematics of the atmosphere is the mathematical description of atmospheric flow fields without regarding the cause of the motion. Therefore, kinematics stands in contrast to dynamics, in which the governing equations are derived from considerations of forces acting on the fluid particles. There exist two methods describing the atmospheric motion. These are the methods of Lagrange and Euler. We shall begin our discussion with the so-called material description of Lagrange, in which the velocity field is represented as a function of time at the position of the moving particle.

3.1 The description of Lagrange

Suppose that x_0^1, x_0^2, x_0^3 are the initial coordinates of a fluid particle at time t_0 and x^1, x^2, x^3 the coordinates at some later time t. If x^i, $i = 1, 2, 3$, can be expressed as a function of the initial coordinates and the time, we know the history or the *trajectory* of the particle. Formally, this can be stated as

$$x^i = x^i\left(x_0^1, x_0^2, x_0^3, t\right), \qquad i = 1, 2, 3 \tag{3.1}$$

so that, in the Lagrangian system, x_0^i and t are the independent and x^i the dependent variables. Equation (3.1) is the formal parameter representation of the trajectory of a particle whose initial position is x_0^i. Changing the initial coordinates simply means that we have selected a different fluid particle. It is evident that (3.1) is the solution of a system of prognostic equations for the trajectory of a particle as given by

$$\frac{dx^i}{dt} = f_i(x^1, x^2, x^3, t), \qquad i = 1, 2, 3 \tag{3.2}$$

A very simple example is $dx^1/dt = u = \text{constant} \Longrightarrow x^1(x_0^1, t) = x_0^1 + u(t - t_0)$.

Lagrange's method is characterized by the introduction of the so-called *enumeration coordinates* a^i, $i = 1, 2, 3$, which are simply the coordinates of the particle at the initial time $t = t_0$, i.e. $a^i = x_0^i$, as depicted in Figure 3.1.

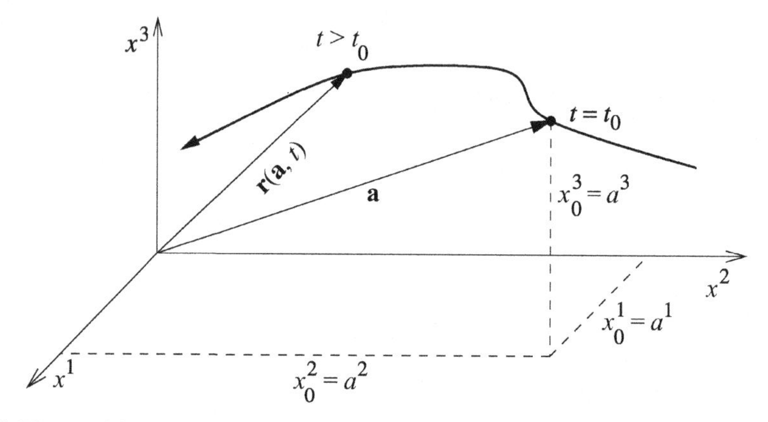

Fig. 3.1 The position vector **r** of the trajectory of a particle. Initially the particle is at position **a**.

For simplicity the selected initial coordinates are assumed to be Cartesian, but, in general, any coordinate system could be used. The formal representation of the trajectory may then be written as

$$x^i = x^i(a^1, a^2, a^3, t) \quad \text{or} \quad \mathbf{r} = \mathbf{r}(\mathbf{a}, t) \tag{3.3}$$

This representation also serves as the transformation equation between the coordinates x^i and a^i,

$$a^i = a^i(x^1, x^2, x^3, t) \quad \text{or} \quad \mathbf{a} = \mathbf{a}(\mathbf{r}, t) \tag{3.4}$$

A unique transformation is possible only if the functional determinant of the transformation differs from zero, i.e.

$$\left| \frac{\partial(x^1, x^2, x^3)}{\partial(a^1, a^2, a^3)} \right| \neq 0 \tag{3.5}$$

In general, the a^i-coordinate system, as stated in (3.4), is curvilinear, nonorthogonal, and time-dependent, as displayed in Figure 3.2, where the initial coordinate system is assumed to be the rectangular Cartesian system. For $t > t_0$ the original orthogonal unit vectors will transform into nonorthogonal non-normalized basis vectors.

It should be kept in mind that, in general, the displacement of individual particles along their trajectories results in a deformation of the surfaces a^i = constant. Whereas in the original coordinate system the distance between any two particles, for example two particles on the x^1-axis with the distance δx^1 between them, changes continually with time, the increment δa^1 between two arbitrary particles remains unchanged since the enumeration coordinates of the particles are fixed at all times.

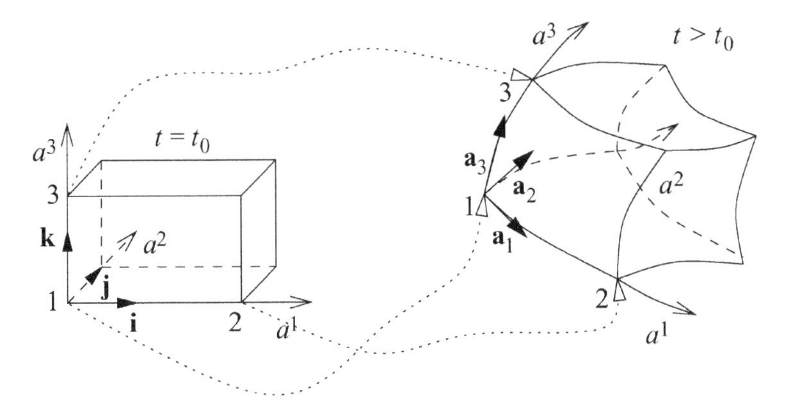

Fig. 3.2 Conceptional displacement of a material volume.

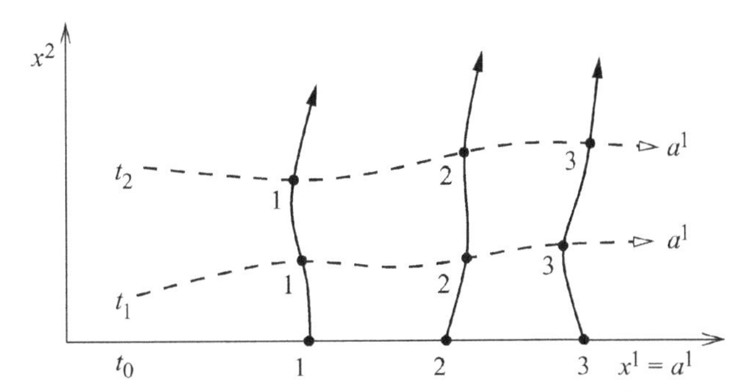

Fig. 3.3 Displacement of three particles aligned on the x^1-axis at $t = t_0$.

To clarify the idea, consider the following simplified two-dimensional coordinate system of Figure 3.3, where three particles in their original positions are shown. At time $t = t_0$ the coordinate axes x^1 and a^1 are identical and the distance between the particles is $\delta a^1 = 1$. The distance δx^1 of the three particles moving along their individual trajectories changes with time. Since the particles have fixed enumeration coordinates they remain neighboring particles on the a^1-axis at all times so that the a^1-coordinate line must be drawn as shown.

3.2 Lagrange's version of the continuity equation

3.2.1 Preliminaries

Consider an increment $d\mathbf{s}^i$ along the coordinate line a^i as given by

$$d\mathbf{s}^i = \frac{\partial \mathbf{r}}{\partial a^i} da^i = \mathbf{a}_i \, da^i \tag{3.6}$$

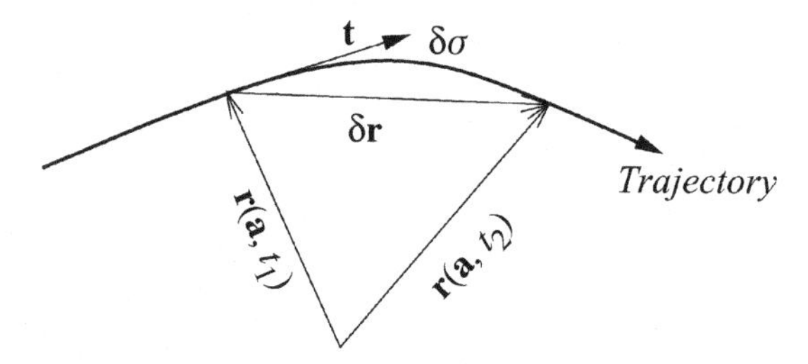

Fig. 3.4 The tangential unit vector along the trajectory.

where the basis vector $\mathbf{a}_i$ is shown in Figure 3.2. If the unit tangential vector $\mathbf{e}_\mathrm{T} = \mathbf{a}_i/\sqrt{g_{ii}}$ to the coordinate line a^i is introduced as well as the physical coordinate $\overset{*}{a}{}^i$ by

$$d\mathbf{s}^i = \mathbf{e}_\mathrm{T}\sqrt{g_{ii}}\,da^i = \mathbf{e}_\mathrm{T}\,d\overset{*}{a}{}^i \tag{3.7}$$

then $ds^i = |d\mathbf{s}^i|$ becomes the arclength which is identical to an increment on the physical coordinate line $\overset{*}{a}{}^i$. The unit vector along the trajectory of a particular particle is given by

$$\mathbf{t} = \left(\frac{\partial \mathbf{r}}{\partial \sigma}\right)_{a^i} \tag{3.8}$$

where $\delta\sigma$ represents the arclength. On the trajectory the enumeration coordinates a^i, $i = 1, 2, 3$, are constant; see Figure 3.3.

For the velocity vector of the particle we may write

$$\mathbf{v} = |\mathbf{v}|\,\mathbf{t}, \qquad |\mathbf{v}| = \frac{d\sigma}{dt} \tag{3.9}$$

The acceleration of an individual particle in the a^i-coordinate system can be obtained from the Eulerian development

$$\frac{d\mathbf{v}}{dt} = \left(\frac{\partial \mathbf{v}}{\partial t}\right)_{a^i} + \dot{a}^n\,\frac{\partial \mathbf{v}}{\partial a^n} \tag{3.10}$$

The important point is that the enumeration coordinates do not change with time, so $da^i/dt = 0$. Therefore, the acceleration in the Lagrangian system is given by

$$\frac{d\mathbf{v}}{dt} = \left(\frac{\partial \mathbf{v}}{\partial t}\right)_{a^i} \tag{3.11}$$

Since the Euler expansion is valid for any field function ψ, $\boldsymbol{\Psi}$, $\mathbb{\Psi}$, we have, in general, for the Lagrangian system

$$\frac{d}{dt}\begin{pmatrix}\psi\\ \boldsymbol{\Psi}\\ \mathbb{\Psi}\end{pmatrix} = \frac{\partial}{\partial t}\begin{pmatrix}\psi\\ \boldsymbol{\Psi}\\ \mathbb{\Psi}\end{pmatrix}_{a^i} \tag{3.12}$$

3.2.2 The mass-conservation equation in the Lagrangian form

We are now ready to formulate the mass-conservation or the continuity equation in the Lagrangian form. Starting with the general coordinate-free form of the continuity equation (M6.67) it is easy to show that, in the q^i system, this equation may be written as

$$\frac{d}{dt}\left(\rho\sqrt{\underset{q}{g}}\right) + \rho\sqrt{\underset{q}{g}}\,\frac{\partial \dot{q}^n}{\partial q^n} = 0 \tag{3.13}$$

Setting here $q^i = a^i$, we obtain

$$\frac{d}{dt}\left(\rho\sqrt{\underset{a}{g}}\right) = \left[\frac{\partial}{\partial t}\left(\rho\sqrt{\underset{a}{g}}\right)\right]_{a^i} = 0 \tag{3.14}$$

since $da^i/dt = \dot{a}^i = 0$. This conservative condition is analogous, for example, to isentropic motion $d\theta/dt = 0$, where θ is constant along the trajectory. Integration of (3.14) yields the Lagrangian form of the continuity equation

$$\left(\rho\sqrt{\underset{a}{g}}\right)_t = \left(\rho\sqrt{\underset{a}{g}}\right)_{t=t_0} = \text{constant} \tag{3.15}$$

This equation can be easily interpreted by realizing that the density ρ and $\sqrt{\underset{a}{g}}$ generally change with time. The quantity $\sqrt{\underset{a}{g}}$ corresponds to the scalar triple product $[\mathbf{a}_1, \mathbf{a}_2, \mathbf{a}_3]$ which represents the volume of a parallelepiped. If the volume expands (contracts) the density must decrease (increase), which is the principle of conservation of mass.

Finally, we consider the special but interesting case that, at $t = t_0$, the Lagrangian and the Cartesian system are identical so that $\sqrt{\underset{x}{g}} = \sqrt{\underset{a}{g}}\big|_{t_0} = 1$. In this case (3.15) reduces to

$$\left(\rho\sqrt{\underset{a}{g}}\right)_t = \rho(t = t_0) = \text{constant} \tag{3.16}$$

In case of incompressibility we have $\rho = \rho(t = t_0)$ and we obtain $\sqrt{\underset{a}{g}} = 1$.

3.3 An example of the use of Lagrangian coordinates

3.3.1 General remarks

In order to appreciate more fully the method of Lagrangian coordinates, we will work out an example and show how to find the approximate numerical solution to a one-dimensional hyperbolic system involving the thermo-hydrodynamic differential equations. We refer to Chapter 12 of Richtmeyer and Morton (1967). The fluid system is assumed to be frictionless, and Coriolis effects are ignored. For simplicity we disregard any subgrid heat and mass fluxes as well as heat sources and the gravitational force. The thermodynamic properties of the fluid will be expressed in the form

$$p = p(e, \alpha) \tag{3.17}$$

where p is the air pressure, e the specific internal energy, and $\alpha = 1/\rho$ the specific volume.

If we consider an ideal gas, for which the internal energy depends on temperature only, then (3.17) reduces to the ideal-gas law. Later the *Courant–Friedrichs–Lewy stability criterion* of the numerical solution, which involves the isentropic *speed of sound* c, will be discussed briefly. This quantity is defined by

$$c^2 = \frac{dp}{d\rho} \quad \text{or} \quad c = \alpha\sqrt{-\frac{dp}{d\alpha}} \tag{3.18}$$

On expanding (3.17) and replacing de with the help of the first law of thermodynamics, we obtain

$$dp = \left(\frac{\partial p}{\partial e}\right)_\alpha de + \left(\frac{\partial p}{\partial \alpha}\right)_e d\alpha, \qquad de = -p\, d\alpha \tag{3.19}$$

and

$$\frac{dp}{d\alpha} = -p\left(\frac{\partial p}{\partial e}\right)_\alpha + \left(\frac{\partial p}{\partial \alpha}\right)_e \tag{3.20}$$

Therefore, the speed of sound is

$$c = \alpha\sqrt{p\left(\frac{\partial p}{\partial e}\right)_\alpha - \left(\frac{\partial p}{\partial \alpha}\right)_e} \tag{3.21}$$

We use the following notation for the finite-difference equations of the numerical scheme. Let $\psi(x, t)$ represent an arbitrary function of the spatial variable x and time t, then ψ_j^n stands for the finite-difference approximation $\psi(j\,\Delta x, n\,\Delta t)$ with Δx, Δt the discrete distances in the space-time grid. Since central-difference approximations will be used, j and n will assume integer as well as half-integer values.

3.3.2 The thermo-hydrodynamic equations

In our flow problem the Lagrangian coordinates a^i of a fluid particle will be represented by the Cartesian coordinates x^i at time $t = t_0$. In the one-dimensional case, which is considered here, we have

$$a = x(t_0) \tag{3.22}$$

Therefore, the transformation equation between the Cartesian and the Lagrangian coordinates at the arbitrary time t is given by

$$x = x(a, t) \tag{3.23}$$

This is the formal one-dimensional parameter representation of the trajectory where the value of a, i.e. the Cartesian coordinate at $t = t_0$, is constant along the trajectory. Using the assumptions stated above, we obtain in Cartesian coordinates the following thermo-hydrodynamic system:

$$\boxed{\begin{aligned} &\text{Equation of motion:} && \frac{du}{dt} = -\alpha \frac{\partial p}{\partial x} \\ &\text{Trajectory:} && \frac{dx}{dt} = u \\ &\text{Continuity equation:} && \frac{d\alpha}{dt} = \alpha \frac{\partial u}{\partial x} \\ &\text{Energy equation:} && \frac{de}{dt} = -p \frac{d\alpha}{dt} \\ &\text{Equation of state:} && p = p(e, \alpha) \end{aligned}} \tag{3.24}$$

The first equation is identical with (2.38a) if we set $f = 0$. Therefore, the equation of motion refers to the absolute coordinate system.

The equations (3.24) must now be transformed into the Lagrangian coordinates. First of all we adapt (3.15) to the present problem. From the general definition (see (M4.21))

$$\sqrt{g}_a = \sqrt{g}_x \left| \frac{\partial(x^1, x^2, x^3)}{\partial(a^1, a^2, a^3)} \right| \tag{3.25a}$$

we find

$$\sqrt{g}_a = \frac{\partial x^1}{\partial a^1} = \frac{\partial x}{\partial a} \quad \text{with} \quad \sqrt{g}_x = 1 \tag{3.25b}$$

Using (3.11) and multiplying both sides of the equation of motion by $\partial x/\partial a$, we obtain first

$$\frac{\partial x}{\partial a}\left(\frac{\partial u}{\partial t}\right)_a = -\alpha \frac{\partial p}{\partial a} \tag{3.26}$$

From (3.12) we find the equation of the trajectory as

$$\left(\frac{\partial x}{\partial t}\right)_a = u \tag{3.27}$$

The continuity equation is given by (3.16), which in our case simplifies to

$$\alpha = \alpha_0 \frac{\partial x}{\partial a} \tag{3.28}$$

With the help of the continuity equation in Lagrangian coordinates, (3.26) may be written in the form

$$\left(\frac{\partial u}{\partial t}\right)_a = -\alpha_0 \frac{\partial p}{\partial a} \tag{3.29}$$

The quantity $\alpha_0 = \alpha(a, t_0)$ represents the specific volume at time $t = t_0$ of the medium being considered. In this example we take $\alpha_0 =$ constant, for simplicity.

The energy equation transforms likewise. With the help of (3.12) we get from the fourth equation of (3.24)

$$\left(\frac{\partial e}{\partial t}\right)_a = -p\left(\frac{\partial \alpha}{\partial t}\right)_a \tag{3.30}$$

For ease of reference the system of equations to be solved will be collected in (3.31):

$$\begin{array}{rr}
\text{Equation of motion:} & \left(\dfrac{\partial u}{\partial t}\right)_a = -\alpha_0 \dfrac{\partial p}{\partial a} \\
\text{Trajectory:} & \left(\dfrac{\partial x}{\partial t}\right)_a = u \\
\text{Continuity equation:} & \alpha = \alpha_0 \dfrac{\partial x}{\partial a} \\
\text{Energy equation:} & \left(\dfrac{\partial e}{\partial t}\right)_a = -p\left(\dfrac{\partial \alpha}{\partial t}\right)_a \\
\text{Equation of state:} & p = p(e, \alpha)
\end{array} \tag{3.31}$$

3.3.3 Difference approximations

The central-difference scheme that we wish to use for the numerical approximation of equation (3.31) is shown in Figure 3.5 together with the various quantities to be calculated.

The arrows show the time step. If we succeed in calculating the various state quantities at times $(n + \frac{1}{2})\,\Delta t$ and $(n + 1)\,\Delta t$ from $(n - \frac{1}{2})\,\Delta t$ and $n\,\Delta t$, then the method can be used for an arbitrary time integration using Lagrangian coordinates. We will now discretize the equations (3.31).

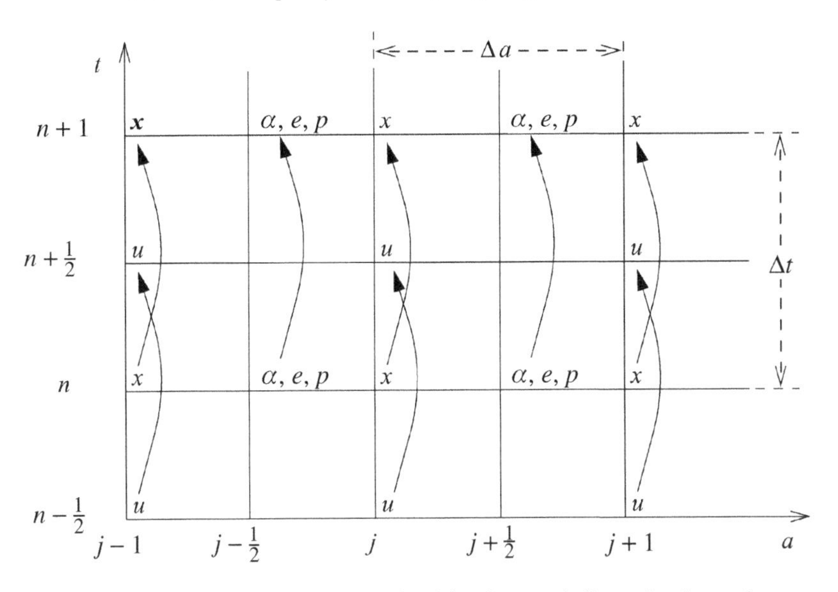

Fig. 3.5 The rectangular network of grid points and discretization scheme.

Equation of motion:

$$u_j^{n+1/2} = u_j^{n-1/2} - \alpha_0 \frac{\Delta t}{\Delta a}\left(p_{j+1/2}^n - p_{j-1/2}^n\right) \tag{3.32a}$$

Equation (3.32a) is of the explicit form. The quantity $u_j^{n+1/2} = u(a_j, (n + \frac{1}{2})\,\Delta t)$ is the velocity at time $t = (n + \frac{1}{2})\,\Delta t$ of the trajectory T_j.

Trajectory:

$$x_j^{n+1} = x_j^n + \Delta t\, u_j^{n+1/2} \tag{3.32b}$$

Since $u_j^{n+1/2}$ is known from (3.32a), this equation is explicit also. The calculated quantity x_j^{n+1} represents the moving fluid particle at time $t = (n + 1)\,\Delta t$ of the trajectory T_j.

Continuity equation:

$$\alpha_{j+1/2}^{n+1} = \frac{\alpha_0}{\Delta a}\left(x_{j+1}^{n+1} - x_j^{n+1}\right) \tag{3.32c}$$

The values of x_j^{n+1} and x_{j+1}^{n+1} are considered known so that (3.32c) is explicit also. The quantity $\alpha_{j+1/2}^{n+1}$ is the specific volume at time $t = (n + 1)\,\Delta t$ of the trajectory $T_{j+1/2}$.

Energy equation:

$$e_{j+1/2}^{n+1} = e_{j+1/2}^n - \frac{1}{2}\left(p_{j+1/2}^n + p_{j+1/2}^{n+1}\right)\left(\alpha_{j+1/2}^{n+1} - \alpha_{j+1/2}^n\right) \tag{3.32d}$$

Using the equation of state

$$p_{j+1/2}^{n+1} = p\left(e_{j+1/2}^{n+1}, \alpha_{j+1/2}^{n+1}\right) \tag{3.32e}$$

and replacing the term $p_{j+1/2}^{n+1}$ results in the only implicit difference equation

$$e_{j+1/2}^{n+1} = e_{j+1/2}^{n} - \tfrac{1}{2}\left[p_{j+1/2}^{n} + p\left(e_{j+1/2}^{n+1}, \alpha_{j+1/2}^{n+1}\right)\right]\left(\alpha_{j+1/2}^{n+1} - \alpha_{j+1/2}^{n}\right) \tag{3.32f}$$

Since $\alpha_{j+1/2}^{n+1}$ is known from (3.32c), only $e_{j+1/2}^{n+1}$ needs to be determined, but it also occurs on the right-hand side of the equation. Therefore, we must proceed iteratively to find $e_{j+1/2}^{n+1}$, which is the specific internal energy at time $t = (n+1)\,\Delta t$ of the trajectory $T_{j+1/2}$.

It should be noted that the difference equations must be solved in the given order. The required state quantities can be determined explicitly, except for the specific internal energy, which must be found iteratively for each $n\,\Delta t$ and at each grid point $(j + \frac{1}{2})$. Instead of (3.32f) we could also use a simpler explicit version of the difference equation, but this would decrease the numerical reliability of the scheme. In the finite-difference scheme adopted all other finite-difference equations use central differences so that, for each point in the (a, t)-plane, the partial derivatives are approximated to second-order accuracy $\mathcal{O}(\Delta t^2)$ and $\mathcal{O}(\Delta a^2)$.

Some schematic model results are shown in Figure 3.6 for a section of the space-time grid beginning with time $t = t_0$. The trajectories of the various particles are labelled according to their positions at time $t = t_0$. Therefore, the trajectory T_j traces the path of the particle whose Lagrangian coordinate is a_j as described by $x_j = x(a_j, t)$, where a_j is the value x_j of the trajectory at time $t = t_0$.

3.3.4 Initial values and boundary conditions

The initial time $t = t_0$ is taken at $n = 0$. At this time, at all gridpoints j the initial values of the trajectories $x_j = a_j$ must be known according to (3.22). At all points between $j-1, j, j+1$, i.e. at $j - \frac{1}{2}, j + \frac{1}{2}$ etc., initial values of α, e, p must be available also. Additionally, for time $t = -\Delta t/2$ all velocities $u(x_j, -\Delta t/2)$, $j = 0, 1, \ldots, J$ must be given.

An interpolation scheme can be used to find the still-missing initial values $u_j^{-1/2} = u(a_j, -\Delta t/2) \neq u(x_j, -\Delta t/2)$ on the trajectory T_j to start the calculations.

3.3.4.1 *Approximate determination of* $\widetilde{u}_j^{1/2}$ *and* $\widetilde{x}_j^1$

Using (3.32a) and (3.32b) and the given gridpoint value $u(x_j, -\Delta t/2)$ we initially estimate the values (indicated by the tilde)

$$\begin{aligned} \widetilde{u}_j^{1/2} &= u(x_j, -\Delta t/2) - \alpha_0 \frac{\Delta t}{\Delta a}\left(p_{j+1/2}^0 - p_{j-1/2}^0\right) \\ \widetilde{x}_j^1 &= x_j^0 + \Delta t\,\widetilde{u}_j^{1/2} \end{aligned} \tag{3.33}$$

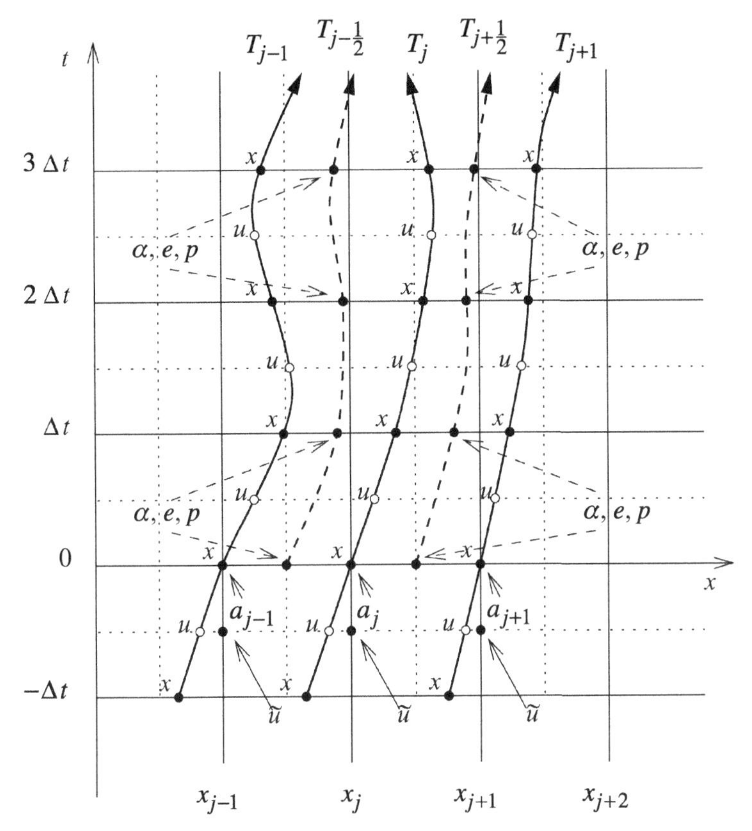

Fig. 3.6 Schematic results for the state variables u, α, e, p, and $x(a, t)$. T_j and $T_{j\pm1}$ are the calculated trajectories of particles $a_j, a_{j\pm1}$ while $T_{j\pm1/2}$ are the interpolated trajectories of particles $a_{j\pm1/2}$.

3.3.4.2 The interpolated velocity $u_j^{-1/2}$ on the trajectory

By means of a straight-line interpolation in the backward direction the trajectory point T_j at time $-\Delta t/2$ is approximated as

$$\widetilde{x}_j^{-1/2} = x(a_j, -\Delta t/2) = x_j^0 - \tfrac{1}{2}\left(\widetilde{x}_j^1 - x_j^0\right) \tag{3.34}$$

Now the required starting value $u_j^{-1/2}$ can be found from interpolation so that

$$u_j^{-1/2} = u(\widetilde{x}_j^{-1/2}, -\Delta t/2) \tag{3.35}$$

Now the so-far-missing values of the velocity u are known and the procedure (3.32a)–(3.32f) can be used to find the solution, provided that the boundary conditions are known.

Table 3.1. *Arrangements of variables for Euler and Lagrange schemes*

Scheme	Independent variables	Dependent variables	Examples
Euler	x^1, x^2, x^3, t	$a^1 = a^1(x^1, x^2, x^3, t)$ $a^2 = a^2(x^1, x^2, x^3, t)$ $a^3 = a^3(x^1, x^2, x^3, t)$	$\mathbf{v}(x^i, t)$ $\frac{d}{dt}[\mathbf{v}(x^i, t)]$
Lagrange	a^1, a^2, a^3, t	$x^1 = x^1(a^1, a^2, a^3, t)$ $x^2 = x^2(a^1, a^2, a^3, t)$ $x^3 = x^3(a^1, a^2, a^3, t)$	$\mathbf{v}(a^i, t)$ $\frac{\partial}{\partial t}[\mathbf{v}(a^i, t)]$

Typical boundary conditions at $j = J$ may be specified as follows:

rigid wall: $u_J^{n\pm 1/2} = 0$ for all n,

free surface: $p^n_{J+1/2} = -p^n_{J-1/2}$ for all n.

The latter boundary condition has the effect that the interpolated value of p vanishes at $j = J$.

3.3.5 The numerical stability condition

The numerical solution of the present problem requires that the Courant–Friedrichs–Lewy stability criterion be obeyed. The Cartesian form of this criterion is given by

$$c\frac{\Delta t}{\Delta x} \leq 1 \tag{3.36}$$

Using (3.28) we find

$$\Delta t \leq \frac{\alpha\,\Delta a}{\alpha_0 c} \tag{3.37}$$

where the isentropic velocity of sound c is determined by (3.21).

3.4 The local description of Euler

The description of fluid motion according to Euler requires knowledge of the velocity field at fixed points within the fluid. If measurements of the velocity are carried out simultaneously at many points then we obtain a spatial picture of the flow.

The descriptions of the fluid according to the methods of Lagrange and Euler require different variables. The reciprocal arrangement of the dependent and independent variables is listed in Table 3.1.

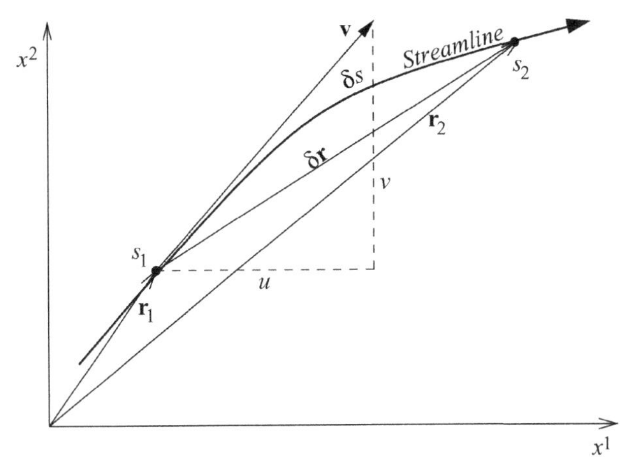

Fig. 3.7 A two-dimensional flow field and representative streamlines, $t = t_0$.

Figure 3.7, as an example of the Eulerian method, shows a two-dimensional velocity vector field for a fixed time $t = t_0$. This snapshot of the flow field will be used to introduce the important concept of the *streamline*. A vector line that is tangential everywhere to the instantaneous wind vector is called a streamline. The direction of the wind vector is fixed by the unit tangential vector **t** which is tangential to the streamline as well as to the trajectory, as stated by the following equation:

$$\mathbf{t} = \underset{\text{Euler}}{\left(\frac{\partial \mathbf{r}}{\partial s}\right)_t} = \underset{\text{Lagrange}}{\left(\frac{\partial \mathbf{r}}{\partial \sigma}\right)_{a^i}} \tag{3.38}$$

It should be noted that different arclengths are used for the Euler (streamline) and the Lagrange (trajectory) representations.

By definition, for fixed $t = t_0$, the velocity vector is tangential to every increment $\Delta\mathbf{r} \to 0$ of the streamline, so that

$$d\mathbf{r} \times \mathbf{v} = 0 \quad \text{with} \quad d\mathbf{r} = \mathbf{i}_n\, dx^n, \quad \mathbf{v} = \mathbf{i}_1 u + \mathbf{i}_2 v + \mathbf{i}_3 w \tag{3.39}$$

From this condition the component forms can be written down immediately,

$$\begin{aligned}
&\text{(a)} \quad \frac{dx^3}{dx^2} = \frac{w(x^i, t_0)}{v(x^i, t_0)} \\
&\text{(b)} \quad \frac{dx^3}{dx^1} = \frac{w(x^i, t_0)}{u(x^i, t_0)} \\
&\text{(c)} \quad \frac{dx^2}{dx^1} = \frac{v(x^i, t_0)}{u(x^i, t_0)}
\end{aligned} \tag{3.40}$$

Only two of these are independent since, for example, division of (3.40b) by (3.40a) gives (3.40c). The integration of two of these differential equations gives the parameter represention of the streamline,

$$F_i(x^j, t_0) = C_i, \qquad i = 1, 2, \quad j = 1, 2, 3 \tag{3.41}$$

A simple example will clarify the idea for the case of a horizontal streamline. We have

$$\frac{dx^2}{dx^1} = \frac{dy}{dx} = \frac{v}{u} = A\cos[k(x - ct_0)] \tag{3.42}$$

where k is the wavenumber and c the phase velocity. Integration of (3.42) results in

$$y - \frac{A}{k}\sin[k(x - ct_0)] = C \tag{3.43}$$

Another example will be given shortly.

In contrast to the streamline which refers to the fixed time $t = t_0$, the trajectory, according to Euler, exhibits an explicit time dependency. The differential equations specifying the velocity vector $\mathbf{v}$ of the trajectory are given by

$$\frac{dx^i}{dt} = v^i(x^j, t), \qquad v^1 = u, \; v^2 = v, \; v^3 = w \tag{3.44}$$

The solution is formally given by

$$x^i = x^i(x_0^j, t) \tag{3.45}$$

where the integration constants x_0^j are the initial coordinates of the particle of concern at time $t = t_0$.

In general, the streamlines and trajectories are different; only for the steady state do they coincide. The following simple example will demonstrate this. Consider the motion of a fluid in a vertical plane with

$$u = \frac{dx}{dt} = x + t, \qquad w = \frac{dz}{dt} = -z + t \tag{3.46}$$

where we have used the more familiar coordinates (x, z) instead of (x^1, x^3). Now find

(a) the family of streamlines and the particular streamline passing through the point $P(x, z) = (-1, -1)$ at $t = t_0 = 0$, and
(b) the trajectory of the particle passing through the same point at $t = t_0$.

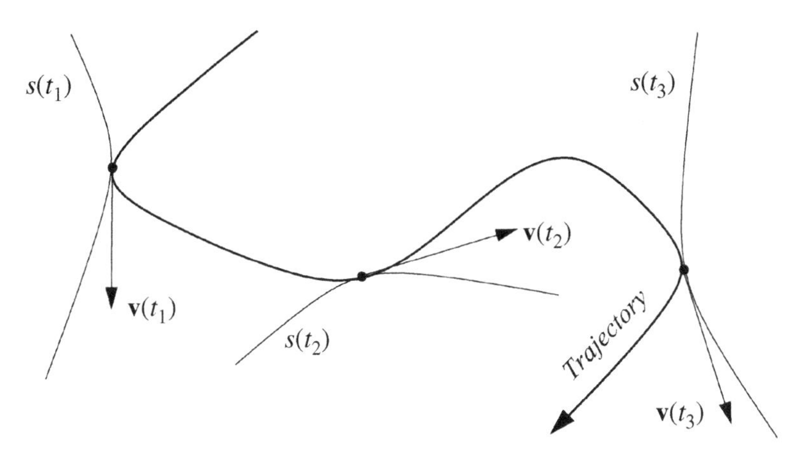

Fig. 3.8 The envelope of a system of successive streamlines.

Solution:
(i) The flow is two-dimensional and nonstationary since u and w contain t explicitly. From (3.46) we obtain

$$\frac{dx}{x+t} = \frac{dz}{-z+t} \tag{3.47}$$

The integration is carried out for $t = t_0 = \text{constant}$, giving

$$\ln(x+t) = -\ln(-z+t) + \ln C \implies (x+t)(-z+t) = C \tag{3.48}$$

which is a family of hyperbolas. For $t = t_0$ at $(x, z) = (-1, -1)$ we have $C = -1$ so that the particular streamline passing through the given point is given by

$$xz = 1 \tag{3.49}$$

(ii) First of all we note that the differential equations are decoupled. The solution is easily carried out by standard methods, with the result

$$x = C_1 \exp t - t - 1, \qquad z = C_2 \exp(-t) + t - 1 \tag{3.50}$$

from which it follows that $C_1 = C_2 = 0$. Elimination of t gives

$$x + z = -2 \tag{3.51}$$

which is the equation of a straight line, showing that the trajectory and the streamlines do not coincide.

Finally, Figure 3.8 shows that the trajectory of a particle is the envelope of a system of successive streamlines.

Occasionally, the concept of a *streak line*, which is a line connecting all the particles that have passed a given geometric point, is used. A plume of smoke from a chimney may be viewed as a streak line.

3.5 Transformation from the Eulerian to the Lagrangian system

We consider an arbitrary field function ψ in the Cartesian Eulerian x^i system and the Lagrangian a^i system, i.e.

$$\begin{aligned} \text{Euler:} \quad & \psi = \psi(x^i, t) \\ \text{Lagrange:} \quad & \psi = \psi(a^i, t) \end{aligned} \tag{3.52}$$

The gradient operator in these systems is given by

$$\begin{aligned} \text{Euler:} \quad & \nabla\psi = \mathbf{i}^n \frac{\partial\psi}{\partial x^n} \\ \text{Lagrange:} \quad & \nabla\psi = \mathbf{a}^n \frac{\partial\psi}{\partial a^n} \end{aligned} \tag{3.53}$$

The transformation rules of the partial derivatives as derived in (M4.24) may be written as

$$\begin{aligned} \frac{\partial\psi}{\partial a^i} &= \underset{a}{\nabla}_i\psi = \frac{\partial x^n}{\partial a^i}\frac{\partial\psi}{\partial x^n} = \frac{\partial x^n}{\partial a^i}\underset{x}{\nabla}_n\psi \\ \frac{\partial\psi}{\partial x^i} &= \underset{x}{\nabla}_i\psi = \frac{\partial a^n}{\partial x^i}\frac{\partial\psi}{\partial a^n} = \frac{\partial a^n}{\partial x^i}\underset{a}{\nabla}_n\psi \end{aligned} \tag{3.54}$$

It may be helpful to rewrite the transformation relations by using matrix notation. If $\widetilde{T}^{i\cdot}_{\cdot j}$ represents the transpose of the transformation matrix $T^{i\cdot}_{\cdot j} = (\partial x^i/\partial a^j)$, where i labels the row and j the column, we may write instead of (3.54)

$$\begin{pmatrix} \dfrac{\partial}{\partial a^1} \\ \dfrac{\partial}{\partial a^2} \\ \dfrac{\partial}{\partial a^3} \end{pmatrix} = \left(\frac{\partial x^j}{\partial a^i}\right)\begin{pmatrix} \dfrac{\partial}{\partial x^1} \\ \dfrac{\partial}{\partial x^2} \\ \dfrac{\partial}{\partial x^3} \end{pmatrix} = \widetilde{T}^{i\cdot}_{\cdot j}\begin{pmatrix} \dfrac{\partial}{\partial x^1} \\ \dfrac{\partial}{\partial x^2} \\ \dfrac{\partial}{\partial x^3} \end{pmatrix} \tag{3.55}$$

The inverse relation of (3.55), denoted by the overbar, is given by

$$\begin{pmatrix} \dfrac{\partial}{\partial x^1} \\ \dfrac{\partial}{\partial x^2} \\ \dfrac{\partial}{\partial x^3} \end{pmatrix} = \left(\frac{\partial a^j}{\partial x^i}\right)\begin{pmatrix} \dfrac{\partial}{\partial a^1} \\ \dfrac{\partial}{\partial a^2} \\ \dfrac{\partial}{\partial a^3} \end{pmatrix} = \overline{\widetilde{T}}^{i\cdot}_{\cdot j}\begin{pmatrix} \dfrac{\partial}{\partial a^1} \\ \dfrac{\partial}{\partial a^2} \\ \dfrac{\partial}{\partial a^3} \end{pmatrix} \tag{3.56}$$

Since (3.55) and (3.56) are inverse relations, we must have

$$\begin{aligned} \left(\widetilde{T}^{i\cdot}_{\cdot j}\right)\left(\overline{\widetilde{T}}^{i\cdot}_{\cdot j}\right) &= \left(\frac{\partial x^j}{\partial a^i}\right)\left(\frac{\partial a^j}{\partial x^i}\right) = \left(\frac{\partial x^n}{\partial a^i}\frac{\partial a^j}{\partial x^n}\right) = \left(\delta^j_i\right) \\ \left(T^{i\cdot}_{\cdot j}\right)\left(\overline{T}^{i\cdot}_{\cdot j}\right) &= \left(\frac{\partial x^i}{\partial a^j}\right)\left(\frac{\partial a^i}{\partial x^j}\right) = \left(\frac{\partial x^i}{\partial a^n}\frac{\partial a^n}{\partial x^j}\right) = \left(\delta^i_j\right) \end{aligned} \tag{3.57}$$

For completeness we state the individual time derivatives in both systems:

$$\begin{aligned} \text{Euler:} \quad \frac{d\psi}{dt} &= \left(\frac{\partial \psi}{\partial t}\right)_{x^i} + \dot{x}^n \frac{\partial \psi}{\partial x^n} \\ \text{Lagrange:} \quad \frac{d\psi}{dt} &= \left(\frac{\partial \psi}{\partial t}\right)_{a^i} \quad \text{since} \quad \dot{a}^i = 0 \end{aligned} \tag{3.58}$$

The individual or material derivative represents the total change of ψ as viewed by an observer following the fluid particle. The first expression of (3.58) is sometimes called the *Lagrangian derivative* as expressed in terms of the Eulerian coordinates. The local derivative $(\partial\psi/\partial t)_{x^i}$ is occasionally called the *Eulerian derivative* expressing the change of ψ at any point fixed in space. The term $\dot{x}^n\, \partial\psi/\partial x^n$ has the meaning that, in time-independent flows, the fluid properties of ψ depend on the spatial coordinates only. For further details see, for example, Currie (1974).

We conclude this section by restating the transformation relations for the basis vectors. These are given by the rules derived in (M4.8) as

$$\begin{aligned} \mathbf{a}_i &= \frac{\partial x^n}{\partial a^i}\,\mathbf{i}_n, & \mathbf{i}_i &= \frac{\partial a^n}{\partial x^i}\,\mathbf{a}_n \\ \mathbf{a}^i &= \frac{\partial a^i}{\partial x^n}\,\mathbf{i}^n, & \mathbf{i}^i &= \frac{\partial x^i}{\partial a^n}\,\mathbf{a}^n \\ \mathbf{i}^k &= \mathbf{i}_k, & x^k &= x_k \end{aligned} \tag{3.59}$$

since in the orthogonal Cartesian system there is no difference between covariant and contravariant basis vectors and measure numbers.

The equation of relative motion expressed in terms of the Lagrangian enumeration coordinates will be presented later when the necessary background is available.

3.6 Problems

3.1: Starting with the continuity equation in the general coordinate-free form

$$\frac{D\rho}{Dt} = \frac{d\rho}{dt} + \rho\,\nabla\cdot\mathbf{v}_{\mathrm{A}} = 0 \quad \text{with} \quad \mathbf{v}_{\mathrm{A}} = \mathbf{v} + \mathbf{v}_P$$

prove the validity of equation (3.13).

3.2: Consider a two-dimensional flow field described by

$$u = x(1+2t), \qquad v = y$$

(a) Find the equation of the streamline passing through the point $(x, y) = (1, 1)$ at time $t = 0$.

(b) Find the equation of the trajectory.

3.3: A particular two-dimensional flow is defined by the velocity components $u = A + Bt$, $v = C$. A, B, and C are constants.

(a) Show that the streamlines are straight lines.

(b) Show that the trajectories are parabolas.

4

Atmospheric flow fields

4.1 The velocity dyadic

In this chapter we will recognize that the velocity dyadic is of great importance in the kinematics of atmospheric motion. First of all we will discuss some general properties in three dimensions. This will be followed by two-dimensional considerations since the large-scale atmospheric motion may be considered quasi-horizontal. We will restrict the discussion to Cartesian coordinates.

4.1.1 The three-dimensional velocity dyadic

As is well known from tensor analysis (see Chapter M2), any dyadic may be written as the sum of the symmetric and antisymmetric parts

$$
\begin{aligned}
&\nabla\mathbf{v} = \mathbb{V}^{\mathrm{s}} + \mathbb{V}^{\mathrm{a}} = \frac{\nabla\mathbf{v} + \overset{\curvearrowleft}{\mathbf{v}\nabla}}{2} + \frac{\nabla\mathbf{v} - \overset{\curvearrowleft}{\mathbf{v}\nabla}}{2} \\
\text{with}\quad &\mathbb{V}^{\mathrm{s}} = \mathbb{D} = \frac{\nabla\mathbf{v} + \overset{\curvearrowleft}{\mathbf{v}\nabla}}{2} \quad \text{deformation dyadic} \\
&\mathbb{V}^{\mathrm{a}} = \mathfrak{R} = \frac{\nabla\mathbf{v} - \overset{\curvearrowleft}{\mathbf{v}\nabla}}{2} \quad \text{rotation dyadic}
\end{aligned}
\tag{4.1}
$$

The definitions are general and may be applied to any vector $\mathbf{A}$. Therefore, at this point it is not necessary to specify whether $\mathbf{v}$ refers to relative or absolute motion. The reason why the symmetric and antisymmetric parts are given the designations *deformation dyadic* and *rotation dyadic* will become obvious shortly. As has already been practiced in Section M2.4.1, we will decompose the symmetric deformation dyadic and find

$$
\begin{aligned}
\mathbb{D} &= \mathbb{D}_{\mathrm{ai}} + \mathbb{D}_{\mathrm{i}} \\
\text{with} \quad \mathbb{D}_{\mathrm{ai}} &= \frac{\nabla \mathbf{v} + \overset{\curvearrowleft}{\mathbf{v}\nabla}}{2} - \frac{\nabla \cdot \mathbf{v}}{3}\mathbb{E} \\
\mathbb{D}_{\mathrm{i}} &= \frac{\nabla \cdot \mathbf{v}}{3}\mathbb{E}
\end{aligned}
\tag{4.2}
$$

where the suffices ai and i stand for anisotropic and isotropic. Splitting the deformation dyadic is mainly for mathematical convenience since certain operations to be applied to this dyadic cause some parts to vanish.

The reason why the antisymmetric part of the local velocity dyadic is associated with the rotational part of the flow field can best be demonstrated by recalling the vector identity (M2.98), which is restated here for a special case

$$
\mathbb{E} \times (\mathbf{B} \times \mathbf{C}) = \mathbb{E} \cdot (\mathbf{CB} - \mathbf{BC})
\tag{4.3}
$$

On replacing the unspecified vectors $\mathbf{B}$ and $\mathbf{C}$ by the gradient operator and the velocity $\mathbf{v}$, respectively, we obtain

$$
\frac{1}{2}\mathbb{E} \times (\nabla \times \mathbf{v}) = -\mathbb{E} \cdot \left(\frac{\nabla \mathbf{v} - \overset{\curvearrowleft}{\mathbf{v}\nabla}}{2} \right) = -\mathbb{E} \cdot \mathbb{\Omega} = -\mathbb{\Omega}
\tag{4.4}
$$

thus justifying the name rotation dyadic. Using the above definitions, the velocity dyadic can now be written as

$$
\nabla \mathbf{v} = \begin{pmatrix}
\dfrac{\partial u}{\partial x}\mathbf{ii} & +\dfrac{\partial v}{\partial x}\mathbf{ij} & +\dfrac{\partial w}{\partial x}\mathbf{ik} \\
+\dfrac{\partial u}{\partial y}\mathbf{ji} & +\dfrac{\partial v}{\partial y}\mathbf{jj} & +\dfrac{\partial w}{\partial y}\mathbf{jk} \\
+\dfrac{\partial u}{\partial z}\mathbf{ki} & +\dfrac{\partial v}{\partial z}\mathbf{kj} & +\dfrac{\partial w}{\partial z}\mathbf{kk}
\end{pmatrix}
$$

$$
= \frac{\nabla \cdot \mathbf{v}}{3}\mathbb{E} + \left(\frac{\nabla \mathbf{v} + \overset{\curvearrowleft}{\mathbf{v}\nabla}}{2} - \frac{\nabla \cdot \mathbf{v}}{3}\mathbb{E} \right) - \frac{1}{2}\mathbb{E} \times (\nabla \times \mathbf{v}) = \mathbb{D}_{\mathrm{i}} + \mathbb{D}_{\mathrm{ai}} + \mathbb{\Omega}
$$

$$\mathbb{D}_{\rm i} = \frac{1}{3}\begin{pmatrix} \left(\frac{\partial u}{\partial x}+\frac{\partial v}{\partial y}+\frac{\partial w}{\partial z}\right)\mathbf{ii} & +0 & +0 \\ +0 & +\left(\frac{\partial u}{\partial x}+\frac{\partial v}{\partial y}+\frac{\partial w}{\partial z}\right)\mathbf{jj} & +0 \\ +0 & +0 & +\left(\frac{\partial u}{\partial x}+\frac{\partial v}{\partial y}+\frac{\partial w}{\partial z}\right)\mathbf{kk} \end{pmatrix}$$

$$\mathbb{D}_{\rm ai} = \begin{pmatrix} \frac{1}{3}\left(2\frac{\partial u}{\partial x}-\frac{\partial v}{\partial y}-\frac{\partial w}{\partial z}\right)\mathbf{ii} & +\frac{1}{2}\left(\frac{\partial v}{\partial x}+\frac{\partial u}{\partial y}\right)\mathbf{ij} & +\frac{1}{2}\left(\frac{\partial w}{\partial x}+\frac{\partial u}{\partial z}\right)\mathbf{ik} \\ +\frac{1}{2}\left(\frac{\partial v}{\partial x}+\frac{\partial u}{\partial y}\right)\mathbf{ji} & +\frac{1}{3}\left(2\frac{\partial v}{\partial y}-\frac{\partial u}{\partial x}-\frac{\partial w}{\partial z}\right)\mathbf{jj} & +\frac{1}{2}\left(\frac{\partial w}{\partial y}+\frac{\partial v}{\partial z}\right)\mathbf{jk} \\ +\frac{1}{2}\left(\frac{\partial w}{\partial x}+\frac{\partial u}{\partial z}\right)\mathbf{ki} & +\frac{1}{2}\left(\frac{\partial w}{\partial y}+\frac{\partial v}{\partial z}\right)\mathbf{kj} & +\frac{1}{3}\left(2\frac{\partial w}{\partial z}-\frac{\partial u}{\partial x}-\frac{\partial v}{\partial y}\right)\mathbf{kk} \end{pmatrix}$$

$$\mathfrak{R} = \frac{1}{2}\begin{pmatrix} 0 & +\left(\frac{\partial v}{\partial x}-\frac{\partial u}{\partial y}\right)\mathbf{ij} & +\left(\frac{\partial w}{\partial x}-\frac{\partial u}{\partial z}\right)\mathbf{ik} \\ -\left(\frac{\partial v}{\partial x}-\frac{\partial u}{\partial y}\right)\mathbf{ji} & +0 & +\left(\frac{\partial w}{\partial y}-\frac{\partial v}{\partial z}\right)\mathbf{jk} \\ -\left(\frac{\partial w}{\partial x}-\frac{\partial u}{\partial z}\right)\mathbf{ki} & -\left(\frac{\partial w}{\partial y}-\frac{\partial v}{\partial z}\right)\mathbf{kj} & +0 \end{pmatrix} \tag{4.5}$$

For ease of reference we have explicitly stated all parts.

Before we reduce the previous formulas to the two-dimensional case, we wish to introduce an important definition known as the *kinematic vorticity*, $\zeta_{\rm kin}$,

$$\boxed{\zeta_{\rm kin} = \frac{\sqrt{\mathfrak{R}\cdot\cdot\widetilde{\mathfrak{R}}}}{\sqrt{\mathbb{D}\cdot\cdot\widetilde{\mathbb{D}}}} = \frac{\sqrt{\frac{1}{2}\mathfrak{R}_\times\cdot\mathfrak{R}_\times}}{\sqrt{\mathbb{D}\cdot\cdot\mathbb{D}}} = \frac{|\nabla\times\mathbf{v}|}{\sqrt{2\mathbb{D}\cdot\cdot\mathbb{D}}}} \tag{4.6}$$

with $\mathfrak{R} = -\widetilde{\mathfrak{R}}$, $\mathbb{D} = \widetilde{\mathbb{D}}$ and $\mathfrak{R}\cdot\cdot\widetilde{\mathfrak{R}} = \mathfrak{R}_\times\cdot\mathfrak{R}_\times/2 = (\nabla\times\mathbf{v})^2/2$. $\zeta_{\rm kin}$ is a measure of the rigidity of the motion since it involves the deformation dyadic $\mathbb{D}$. Obviously we may define three limiting cases, given by

$$\zeta_{\rm kin} = \begin{cases} 0 & |\nabla\times\mathbf{v}| = 0 \quad \text{vortex-free deformation flow} \\ 1 & |\nabla\times\mathbf{v}| = \sqrt{2\mathbb{D}\cdot\cdot\mathbb{D}} \quad \text{rotation equals deformation} \\ \infty & \sqrt{2\mathbb{D}\cdot\cdot\mathbb{D}} = 0 \quad \text{vortex flow without deformation} \end{cases} \tag{4.7}$$

which characterize the flow field.

4.1.2 The two-dimensional velocity dyadic

Inspection of equation (4.5) shows that the horizontal velocity dyadic simplifies to

$$\nabla_{\rm h}\mathbf{v}_{\rm h} = \begin{pmatrix} \dfrac{\partial u}{\partial x}\mathbf{ii} & +\dfrac{\partial v}{\partial x}\mathbf{ij} \\ +\dfrac{\partial u}{\partial y}\mathbf{ji} & +\dfrac{\partial v}{\partial y}\mathbf{jj} \end{pmatrix} = \mathbb{D}_{\rm i} + \mathbb{D}_{\rm ai} + \mathfrak{R}$$

$$\text{with}\quad \mathbf{v}_{\rm h} = u\mathbf{i} + v\mathbf{j}, \qquad \nabla_{\rm h} = \mathbf{i}\frac{\partial}{\partial x} + \mathbf{j}\frac{\partial}{\partial y}$$

$$\mathbb{D}_{\rm i} = \frac{1}{2}\begin{pmatrix} \left(\dfrac{\partial u}{\partial x} + \dfrac{\partial v}{\partial y}\right)\mathbf{ii} & +0 \\ +0 & +\left(\dfrac{\partial u}{\partial x} + \dfrac{\partial v}{\partial y}\right)\mathbf{jj} \end{pmatrix}$$

$$\mathbb{D}_{\rm ai} = \frac{1}{2}\begin{pmatrix} \left(\dfrac{\partial u}{\partial x} - \dfrac{\partial v}{\partial y}\right)\mathbf{ii} & +\left(\dfrac{\partial u}{\partial y} + \dfrac{\partial v}{\partial x}\right)\mathbf{ij} \\ +\left(\dfrac{\partial u}{\partial y} + \dfrac{\partial v}{\partial x}\right)\mathbf{ji} & -\left(\dfrac{\partial u}{\partial x} - \dfrac{\partial v}{\partial y}\right)\mathbf{jj} \end{pmatrix} \tag{4.8}$$

$$\mathfrak{R} = \frac{1}{2}\begin{pmatrix} 0 & +\left(\dfrac{\partial v}{\partial x} - \dfrac{\partial u}{\partial y}\right)\mathbf{ij} \\ -\left(\dfrac{\partial v}{\partial x} - \dfrac{\partial u}{\partial y}\right)\mathbf{ji} & +0 \end{pmatrix}$$

Note that the factor $\frac{1}{2}$ instead of $\frac{1}{3}$ appears in the deformation dyadic $\mathbb{D}_{\rm i}$ since the first scalar of the two-dimensional unit dyadic is 2. Various combinations of partial derivatives appear, which will be abbreviated as

$$\begin{aligned}
&\text{(a)}\quad D = \nabla_{\rm h}\cdot\mathbf{v}_{\rm h} = \frac{\partial u}{\partial x} + \frac{\partial v}{\partial y} \\
&\text{(b)}\quad \zeta = \mathbf{k}\cdot\nabla_{\rm h}\times\mathbf{v}_{\rm h} = \frac{\partial v}{\partial x} - \frac{\partial u}{\partial y} \implies \nabla_{\rm h}\times\mathbf{v}_{\rm h} = \mathbf{k}\zeta \\
&\text{(c)}\quad A = \frac{\partial u}{\partial x} - \frac{\partial v}{\partial y}, \qquad B = \frac{\partial u}{\partial y} + \frac{\partial v}{\partial x} \\
&\text{(d)}\quad def = \sqrt{A^2 + B^2} = \sqrt{2\mathbb{D}_{\rm ai}\cdot\cdot\mathbb{D}_{\rm ai}} \\
&\text{(e)}\quad Def = \sqrt{A^2 + B^2 + D^2} = \sqrt{2\mathbb{D}\cdot\cdot\mathbb{D}}
\end{aligned} \tag{4.9}$$

The symbols D and ζ denote the two-dimensional divergences of the horizontal velocity field and the vorticity, respectively. The quantities A and B of (4.9c) are known as the principal parts of the deformation field. Some geometric interpretations will be given at the end of this chapter. It should be noted that the quantities

D, ζ, etc. were introduced without reference to the rotating earth. If the rotation of the earth needs to be accounted for, we would have to allow for certain modifications. For example, the vorticity ζ would have to be replaced by $\zeta + f$. Details will be given in a later chapter.

Recall that any dyadic is an extensive quantity that is independent of a particular coordinate system. If the general multiplication in dyadic products is replaced by the vectorial or the scalar multiplication new quantities are produced, which are again independent of the coordinate system. Such quantities are known as invariants. D and ζ are invariants whereas A and B are not. Furthermore, the velocity deformation fields def and Def, given in (4.9d) and (4.9e), are also invariants because of the double scalar products involved in these terms. The various invariants of the local two-dimensional velocity dyadic are listed next:

$$\begin{aligned}
\text{First scalar:} \quad & (\nabla_{\rm h}\mathbf{v}_{\rm h})_{\rm I} = \nabla_{\rm h} \cdot \mathbf{v}_{\rm h} = D \\
\text{Vector:} \quad & (\nabla_{\rm h}\mathbf{v}_{\rm h})_{\times} = \nabla_{\rm h} \times \mathbf{v}_{\rm h} = \mathbf{k}\zeta \\
\text{Second scalar:} \quad & (\nabla_{\rm h}\mathbf{v}_{\rm h})_{\rm II} = J(u, v) = \frac{D^2}{4} + \frac{\zeta^2}{4} - \frac{def^2}{4} \\
\text{Third scalar:} \quad & (\nabla_{\rm h}\mathbf{v}_{\rm h})_{\rm III} = J(u, v)
\end{aligned} \tag{4.10}$$

Note that, in the two-dimensional case, the second and the third scalar are identical.

In analogy to the three-dimensional case (4.6) we define the two-dimensional kinematic vorticity

$$\zeta_{\rm kin} = \frac{\sqrt{\mathfrak{R}\cdot\cdot\widetilde{\mathfrak{R}}}}{\sqrt{\mathbb{D}\cdot\cdot\mathbb{D}}} = \frac{\zeta}{Def} \tag{4.11}$$

Thus, $\zeta_{\rm kin}$ is the ratio of two invariants. We shall not discuss this concept in more detail.

4.2 The deformation of the continuum

4.2.1 The representation of the wind field

We consider the structure of the wind field at a fixed time in the infinitesimal surroundings of a point P as shown in Figure 4.1. In the immediate neighborhood at point P' we may write

$$\begin{aligned}
\mathbf{v}(P') &= \mathbf{v}(P) + \delta\mathbf{r} \cdot \nabla\mathbf{v}(P) \\
&= \mathbf{v}(P) + \delta\mathbf{r} \cdot \mathbb{D}_{\rm i} + \delta\mathbf{r} \cdot \mathbb{D}_{\rm ai} + \delta\mathbf{r} \cdot \mathfrak{R} \\
\text{with} \quad \delta\mathbf{r} \cdot \mathfrak{R} &= \tfrac{1}{2}(\nabla \times \mathbf{v}) \times \delta\mathbf{r} = \boldsymbol{\Omega} \times \delta\mathbf{r}
\end{aligned} \tag{4.12}$$

where we have discontinued the Taylor expansion after the linear term. In (4.12) we have introduced the local velocity dyadic $\nabla\mathbf{v}$ which is considered a constant

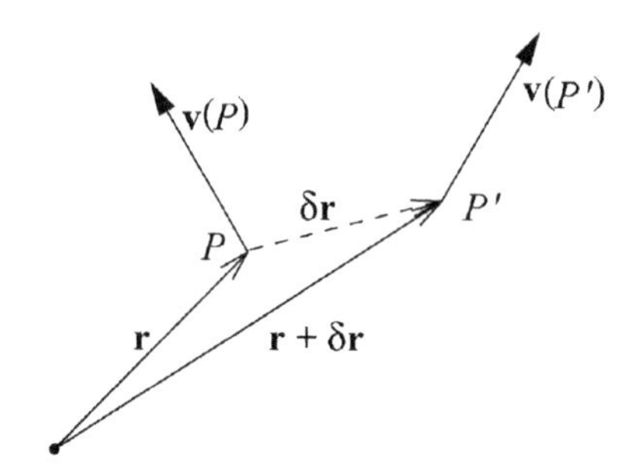

Fig. 4.1 A representation of the wind field.

quantity in the immediate neighborhood of P. The velocity $\mathbf{v}(P')$ consists of two parts. $\mathbf{v}(P)$ represents the joint translation of the two points while $\delta\mathbf{r} \cdot \nabla\mathbf{v}(P)$ is the velocity of point P' relative to P. The vector $\boldsymbol{\Omega}$ represents the angular velocity of a rigid rotation of P' about an axis through P.

By carrying out the scalar multiplication in (4.12) and using the definitions (4.9), we immediately find the component form of the two-dimensional wind field:

$$\begin{aligned} u &= u_0 + \frac{A+D}{2}x + \frac{B-\zeta}{2}y \\ v &= v_0 + \frac{B+\zeta}{2}x + \frac{D-A}{2}y \end{aligned} \tag{4.13}$$

where the suffix 0 represents the velocity components at point P. Moreover, we have replaced δx and δy by x and y.

We have previously stated that $A^2 + B^2$ is an invariant. By axis rotation this allows us to take $A > 0$ and $B = 0$ so that $\partial u/\partial y = -\partial v/\partial x$. Thus (4.13) reduces to

$$\begin{aligned} u &= u_0 + \tfrac{1}{2}(Ax + Dx - \zeta y) = u_0 + u_1 + u_2 + u_3 \\ v &= v_0 + \tfrac{1}{2}(-Ay + Dy + \zeta x) = v_0 + v_1 + v_2 + v_3 \end{aligned} \tag{4.14}$$

The various velocity components appearing in this equation denote the following types of linear flow field near the point P: (u_0, v_0): translation, (u_1, v_1): deformation, (u_2, v_2): divergence, (u_3, v_3): rotation. These flow types are taken from Petterssen (1956) or Panchev (1985) and are depicted in Figure 4.2.

The x- and y-components of the deformation field are known as *dilatation* and *contraction*, respectively. We will now determine the coordinates (x_c, y_c) of the so-called *kinematic center*, which is a point where the velocity vanishes. By setting $u = v = 0$ in (4.14) and then solving for the coordinates $x = x_c$ and $y = y_c$ of the system we find

$$x_c = -2\frac{u_0(D-A) + v_0\zeta}{D^2 - A^2 + \zeta^2}, \qquad y_c = -2\frac{v_0(D+A) - u_0\zeta}{D^2 - A^2 + \zeta^2} \tag{4.15}$$

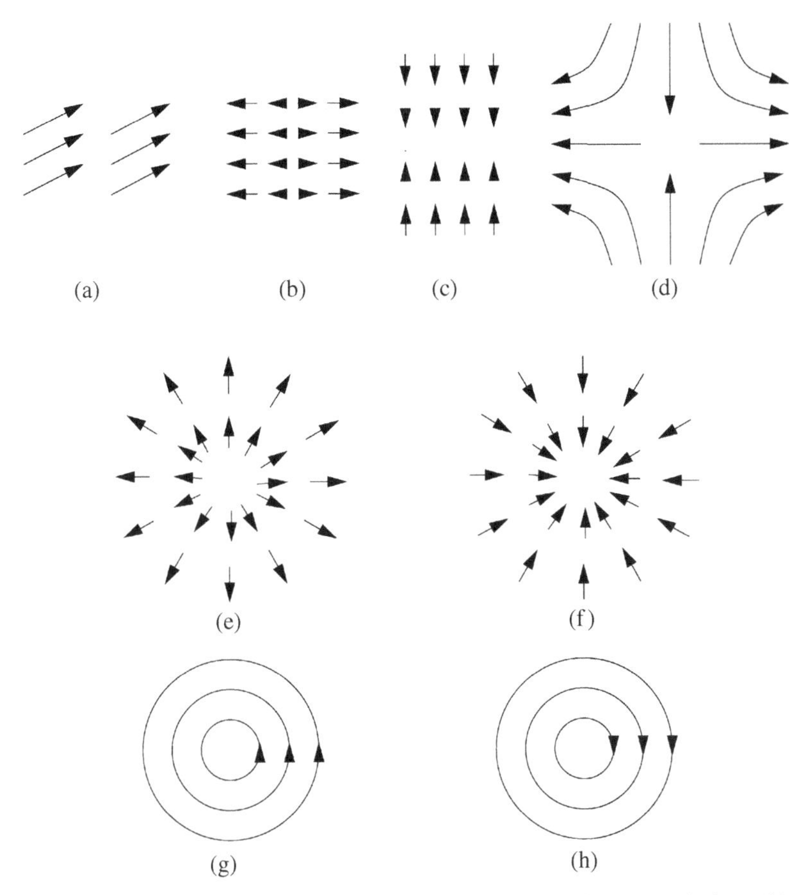

Fig. 4.2 Component motions of the linear flow field: (a) uniform translation, (b) and (c) deformation, (d) total deformation, (e) divergence, (f) convergence, (g) positive rotation–idealized northern-hemispheric low-pressure system, and (h) negative rotation–idealized northern-hemispheric high-pressure system.

The condition for the existence of a kinematic center may therefore be written as

$$D^2 - A^2 + \zeta^2 \neq 0 \tag{4.16}$$

On translating the origin of the coordinate system to the kinematic center, equation (4.14) reduces to

$$u = \frac{D + A}{2}x - \frac{\zeta}{2}y, \qquad v = \frac{\zeta}{2}x + \frac{D - A}{2}y \tag{4.17}$$

The question of whether there are straight streamlines through the center arises quite naturally. If the streamlines are to be straight, then the equation of the streamline

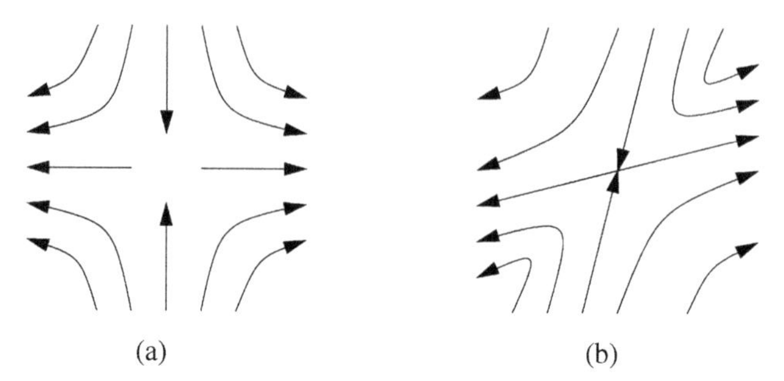

Fig. 4.3 Hyperbolic streamline patterns describing deformation: (a) pure deformation, and (b) with added rotation.

can be written as

$$\frac{dy}{dx} = \frac{v}{u} = \frac{y}{x} = \tan\vartheta \tag{4.18}$$

where ϑ is the angle between the x-axis and the streamline.

On substituting from (4.17) we obtain

$$\tan\vartheta = \frac{A \pm \sqrt{A^2 - \zeta^2}}{\zeta} \tag{4.19}$$

First of all we note that the angle is independent of the divergence D. Analyzing equation (4.19) leads to the following conclusions.

(i) If $A = \zeta = 0$ we have pure divergence. In this unrealistic case an infinite number of straight streamlines will pass through the center, see Figures 4.2(e) and 4.2(f).

(ii) If $A > \zeta$ then the deformative component exceeds the rotational part. In this case there are two roots so that two streamlines pass through the center dividing the plane into four sectors. In each sector the streamlines are of the hyperbolic type, as shown in Figure 4.3, where two idealized cases are presented.

(iii) If $A^2 < \zeta^2$ then there are no real solutions for β. Consequently, there will be no straight streamlines passing through the center. Typical situations for this case are shown in Figure 4.4.

4.2.2 Flow patterns and stability

We conclude this section by looking at the flow pattern from the modern point of view involving elementary stability theory. For more details see also Section M7.2. We proceed by writing equation (4.17) in the form

$$\dot{x} = ax + by = X(x, y), \qquad \dot{y} = cx + dy = Y(x, y) \tag{4.20}$$

The parameters a, b, c, and d can be identified by comparison with (4.17). Beginning at an arbitrary initial point, the solution $x(t), y(t)$ of (4.20) traces out a

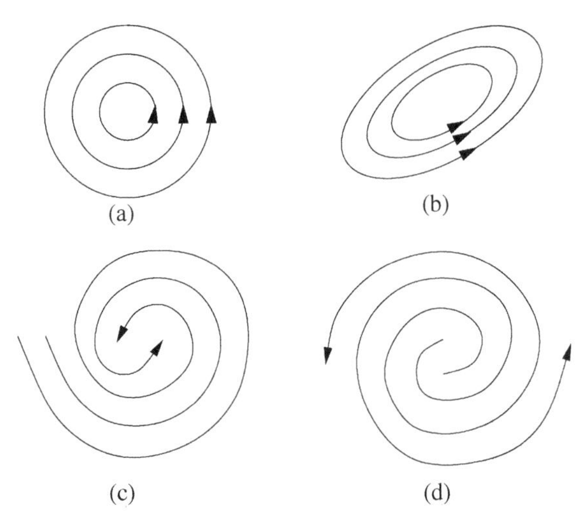

Fig. 4.4 Flow patterns without straight streamlines: (a) pure rotation, (b) rotation and deformation, (c) convergence superimposed on (b), and (d) divergence superimposed on (b); (a) and (b) are also called centers whereas (c) and (d) are known as spirals.

directed curve in the (x, y)-plane which is known as the *phase path*. From (4.20) we may easily form the streamline equation, which can always be solved:

$$\frac{dy}{dx} = \frac{cx + dy}{ax + by} \tag{4.21}$$

Let us consider the fixed point (x^*, y^*) satisfying the relations

$$X(x^*, y^*) = 0, \qquad Y(x^*, y^*) = 0 \tag{4.22}$$

This point determines the qualitative behavior of the solution. It is convenient to write (4.20) in matrix form:

$$\dot{\mathbf{x}} = F\mathbf{x}, \qquad \mathbf{x} = \begin{pmatrix} x \\ y \end{pmatrix}, \qquad F = \begin{pmatrix} a & b \\ c & d \end{pmatrix} \tag{4.23}$$

According to (M7.11), the general solution to (4.20) is given by

$$\mathbf{x}(t) = C_1\mathbf{x}_1 \exp(\lambda_1 t) + C_2\mathbf{x}_2 \exp(\lambda_2 t) \tag{4.24}$$

where the λ_i are the eigenvalues or the characteristic values of the matrix F and the $\mathbf{x}_i$ are the corresponding eigenvectors. The eigenvalues are found from the characteristic equation

$$\begin{vmatrix} a - \lambda & b \\ c & d - \lambda \end{vmatrix} = 0 \tag{4.25}$$

The expanded form can be written as

$$\begin{aligned}
&\lambda^2 - \tau\lambda + \Delta = 0 \\
\text{with} \quad &\tau = \text{trace}(F) = a + d = \lambda_1 + \lambda_2 \\
&\Delta = |\, F \,| = ad - bc = \lambda_1\lambda_2 \\
&\lambda_{1,2} = \frac{\tau \pm \sqrt{\tau^2 - 4\Delta}}{2}
\end{aligned} \tag{4.26}$$

Let us consider one more example. For the matrix

$$F = \begin{pmatrix} 1 & 1 \\ 4 & -2 \end{pmatrix} \tag{4.27a}$$

we find the eigenvalues and eigenvectors

$$\begin{aligned}
\lambda_1 &= 2, \qquad \mathbf{x}_1 = \begin{pmatrix} x_1 \\ y_1 \end{pmatrix} = \begin{pmatrix} 1 \\ 1 \end{pmatrix} \\
\lambda_2 &= -3, \qquad \mathbf{x}_2 = \begin{pmatrix} x_2 \\ y_2 \end{pmatrix} = \begin{pmatrix} 1 \\ -4 \end{pmatrix}
\end{aligned} \tag{4.27b}$$

so that the formal solution (4.24) is given by

$$\mathbf{x} = C_1 \begin{pmatrix} 1 \\ 1 \end{pmatrix} \exp(2t) + C_2 \begin{pmatrix} 1 \\ -4 \end{pmatrix} \exp(-3t) \tag{4.28}$$

The eigenvectors may also be multiplied by a constant factor, which could be absorbed by the integration constants. The eigensolution corresponding to λ_1 grows exponentially while the second eigensolution decays, meaning that the origin is a saddle point. The stable manifold is the line spanned by the eigenvector multiplying $\exp(-3t)$ while the unstable manifold is the line spanned by the eigenvector multiplying $\exp(2t)$. To get a better impression of the flow field we have included additional trajectories by continuity, as is qualitatively shown in Figure 4.5. The similarity to Figure 4.3(b) is apparent.

The fixed points may be classified according to the simple scheme shown in Figure M7.13. Since $\Delta = -6$ the solution is a saddle point.

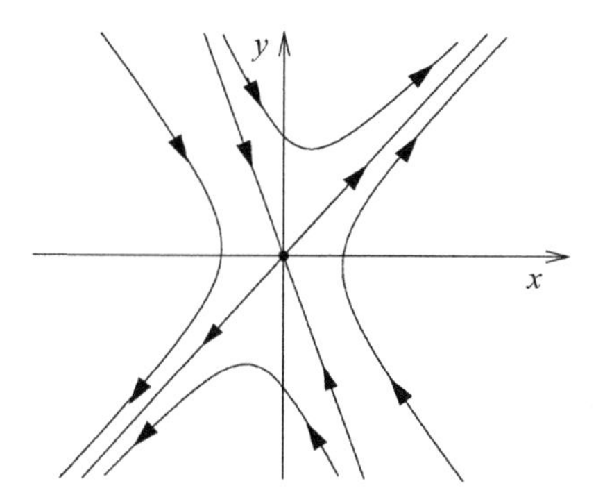

Fig. 4.5 Trajectories of the example given by equation (4.28).

It may be recognized that the information content of equation (4.19) and the information expressed by the two-dimensional eigenvectors are equivalent. To show this, we identify the parameters a, b, c, and d by comparison of (4.17) and (4.20). The result is

$$a = \frac{D+A}{2}, \qquad b = -\frac{\zeta}{2}, \qquad c = \frac{\zeta}{2}, \qquad d = \frac{D-A}{2} \tag{4.29}$$

With this identification we easily find the eigenvalues and the eigenvectors of the flow system (4.17) as

$$\lambda_{1,2} = \frac{D \pm \sqrt{A^2 - \zeta^2}}{2}, \qquad \mathbf{x}_{1,2} = \begin{pmatrix} 1 \\ \dfrac{\lambda_{1,2} - a}{b} \end{pmatrix} \tag{4.30}$$

On taking the ratio of the components of the eigenvectors we again obtain (4.19),

$$\begin{aligned} \frac{y_1}{x_1} &= \frac{\lambda_1 - a}{b} = \frac{A - \sqrt{A^2 - \zeta^2}}{\zeta} = \tan \vartheta_1 \\ \frac{y_2}{x_2} &= \frac{\lambda_2 - a}{b} = \frac{A + \sqrt{A^2 - \zeta^2}}{\zeta} = \tan \vartheta_2 \end{aligned} \tag{4.31}$$

We should not have expected anything else since the two ways of characterizing the flow are equivalent.

4.3 Individual changes with time of geometric fluid configurations

We will now derive expressions for time changes of fluid elements (moving lines, surfaces, volumes). Let us consider a small line element $\delta\mathbf{r}$. Replacing in (M6.47) the vector $\mathbf{A}$ by the unit dyadic $\mathbb{E}$ we find immediately

$$\begin{aligned} \frac{d}{dt}\left(\int_{L(t)} d\mathbf{r} \cdot \mathbb{E}\right) &= \frac{d}{dt}\left(\int_{L(t)} d\mathbf{r}\right) = \frac{d}{dt}(\delta\mathbf{r}) = \int_{L(t)} d\mathbf{r} \cdot (\nabla\mathbf{v} \cdot \mathbb{E}) \\ &= \int_{L(t)} d\mathbf{r} \cdot \nabla\mathbf{v} = \delta\mathbf{r} \cdot \nabla\mathbf{v} \quad \text{with} \quad \int_{L(t)} d\mathbf{r} = \delta\mathbf{r} \end{aligned} \tag{4.32}$$

Similarly we proceed in case of a surface and a volume. The corresponding expressions are obtained by replacing $\mathbf{A}$ in (M6.53) by $\mathbb{E}$ and by setting $\psi = 1$ in (M6.57). The results are collected in

$$\begin{aligned} &\text{(a)} \quad \frac{d}{dt}(\delta\mathbf{r}) = \delta\mathbf{r} \cdot \nabla\mathbf{v}, && \int_{L(t)} d\mathbf{r} = \delta\mathbf{r} \\ &\text{(b)} \quad \frac{d}{dt}(\delta\mathbf{S}) = \delta\mathbf{S}\, \nabla \cdot \mathbf{v} - \nabla\mathbf{v} \cdot \delta\mathbf{S}, && \int_{\Delta S(t)} d\mathbf{S} = \delta\mathbf{S} \\ &\text{(c)} \quad \frac{d}{dt}(\delta\tau) = \delta\tau\, \nabla \cdot \mathbf{v}, && \int_{\Delta V(t)} d\tau = \delta\tau \end{aligned} \tag{4.33}$$

Here the terms $\nabla\mathbf{v}$ and $\nabla\cdot\mathbf{v}$ have been treated as constants. We should keep in mind that these expressions are only approximate. However, for most practical purposes they are sufficiently accurate. If the initial configurations of $\delta\mathbf{r}$, $\delta\mathbf{S}$, and $\delta\tau$ are known, then the deformations of the fluid elements can be calculated by means of time integration.

In the next section we discuss various geometric properties resulting from the application of the local dyadic. Using as an example the liquid or material line element in the flow field, we find from (4.33a) and (4.5)

$$\frac{d}{dt}(\delta\mathbf{r}) = \delta\mathbf{r}\cdot\nabla\mathbf{v} = \delta\mathbf{r}\cdot\mathbb{D} - \frac{1}{2}\,\delta\mathbf{r}\cdot\mathbb{E}\times(\nabla\times\mathbf{v}) = \delta\mathbf{r}\cdot\mathbb{D}_{\mathrm{i}} + \delta\mathbf{r}\cdot\mathbb{D}_{\mathrm{ai}} + \frac{1}{2}(\nabla\times\mathbf{v})\times\delta\mathbf{r} \tag{4.34}$$

where $\delta\mathbf{r} = \delta r\,\mathbf{e}$ and $\mathbf{e}$ is a unit vector in the direction of $\delta\mathbf{r}$. Several special cases that can easily be comprehended arise.

4.3.1 The relative change of the material line element

The following manipulation of the relative change δr is obvious:

$$\frac{1}{\delta r}\frac{d}{dt}(\delta r) = \frac{1}{(\delta r)^2}\frac{d}{dt}\frac{(\delta r)^2}{2} = \frac{1}{(\delta r)^2}\frac{d}{dt}\frac{(\delta\mathbf{r})^2}{2} = \frac{1}{(\delta r)^2}\left(\delta\mathbf{r}\cdot\frac{d}{dt}(\delta\mathbf{r})\right) \tag{4.35}$$

On substituting from (4.33a) we obtain immediately for the relative change

$$\frac{1}{\delta r}\frac{d}{dt}(\delta r) = \frac{1}{(\delta r)^2}(\delta\mathbf{r}\cdot\nabla\mathbf{v}\cdot\delta\mathbf{r}) = \mathbf{e}\cdot\nabla\mathbf{v}\cdot\mathbf{e} \tag{4.36}$$

or

$$\frac{1}{\delta r}\frac{d}{dt}(\delta r) = \mathbf{e}\cdot\mathbb{D}\cdot\mathbf{e} \quad \text{since} \quad \mathbf{e}\cdot\frac{(\nabla\times\mathbf{v})}{2}\times\mathbf{e} = 0 \tag{4.37}$$

Therefore, the relative change with time of the material line element depends only on the symmetric part of the local dyadic, which is equivalent to the deformation dyadic. This means that the antisymmetric part of the local velocity dyadic does not contribute to the relative change.

4.3.2 The directional change of the material line element

If we wish to determine the directional change of the line element we must also consider the change with time of the unit vector, or

$$\frac{d}{dt}(\delta\mathbf{r}) = \delta r\,\frac{d\mathbf{e}}{dt} + \mathbf{e}\,\frac{d}{dt}(\delta r) \tag{4.38}$$

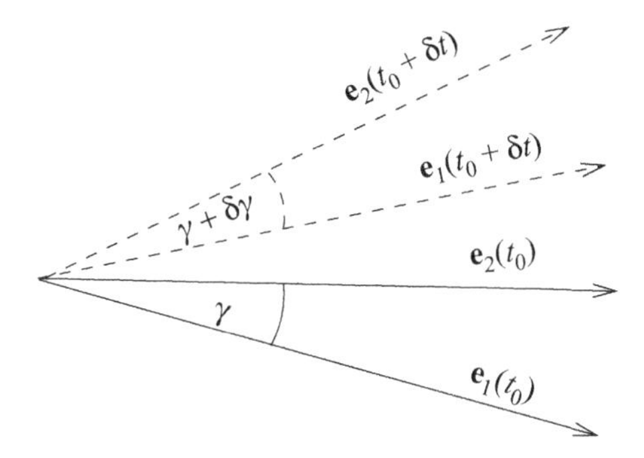

Fig. 4.6 The change in direction between two material fluid elements.

so that

$$\frac{d\mathbf{e}}{dt} = \frac{1}{\delta r}\frac{d}{dt}(\delta\mathbf{r}) - \frac{\mathbf{e}}{\delta r}\frac{d}{dt}(\delta r) \tag{4.39}$$

By use of equations (4.34) and (4.37) we find without difficulty

$$\frac{d\mathbf{e}}{dt} = \mathbf{e}\cdot(\mathbb{D}_{\rm i} + \mathbb{D}_{\rm ai}) + \frac{1}{2}(\nabla\times\mathbf{v})\times\mathbf{e} - \mathbf{e}(\mathbf{e}\cdot\mathbb{D}_{\rm i}\cdot\mathbf{e} + \mathbf{e}\cdot\mathbb{D}_{\rm ai}\cdot\mathbf{e}) \tag{4.40}$$

The manipulations with the isotropic part of the deformation dyadic $\mathbb{D}_{\rm i}$

$$\mathbf{e}\cdot\mathbb{D}_{\rm i} = \mathbf{e}\cdot\frac{\nabla\cdot\mathbf{v}}{3}\mathbf{e}\mathbf{e} = \frac{\nabla\cdot\mathbf{v}}{3}\mathbf{e}, \qquad \mathbf{e}\cdot\mathbb{D}_{\rm i}\cdot\mathbf{e} = \frac{\nabla\cdot\mathbf{v}}{3} \tag{4.41}$$

lead to the desired result

$$\frac{d\mathbf{e}}{dt} = \mathbf{e}\cdot\mathbb{D}_{\rm ai} - \mathbf{e}(\mathbf{e}\cdot\mathbb{D}_{\rm ai}\cdot\mathbf{e}) + \frac{1}{2}(\nabla\times\mathbf{v})\times\mathbf{e} \tag{4.42}$$

giving the change with time of the direction of the material fluid element. Recall that $d\mathbf{e}/dt$ is perpendicular to the unit vector $\mathbf{e}$ itself. Equation (4.42) shows that the change in direction with time depends only on the anisotropic part of the deformation dyadic and on the rotation vector.

An immediate application of (4.42) lies in the possibility of calculating the change of the angle γ between two line elements, see Figure 4.6. On applying (4.42) to the two directions we find at once

$$\begin{aligned} &\text{(a)} \quad \frac{d\mathbf{e}_1}{dt} = \mathbf{e}_1\cdot\mathbb{D}_{\rm ai} - \mathbf{e}_1(\mathbf{e}_1\cdot\mathbb{D}_{\rm ai}\cdot\mathbf{e}_1) + \frac{1}{2}(\nabla\times\mathbf{v})\times\mathbf{e}_1 \\ &\text{(b)} \quad \frac{d\mathbf{e}_2}{dt} = \mathbf{e}_2\cdot\mathbb{D}_{\rm ai} - \mathbf{e}_2(\mathbf{e}_2\cdot\mathbb{D}_{\rm ai}\cdot\mathbf{e}_2) + \frac{1}{2}(\nabla\times\mathbf{v})\times\mathbf{e}_2 \end{aligned} \tag{4.43}$$

Scalar multiplication of parts (a) and (b) by $\mathbf{e}_2$ and $\mathbf{e}_1$, and then adding the results, gives

$$\begin{aligned} \frac{d}{dt}(\mathbf{e}_1\cdot\mathbf{e}_2) &= \frac{d}{dt}(\cos\gamma) = -\sin\gamma\,\frac{d\gamma}{dt} \\ &= 2\mathbf{e}_1\cdot\mathbb{D}_{\rm ai}\cdot\mathbf{e}_2 - (\mathbf{e}_1\cdot\mathbb{D}_{\rm ai}\cdot\mathbf{e}_1 + \mathbf{e}_2\cdot\mathbb{D}_{\rm ai}\cdot\mathbf{e}_2)\cos\gamma \end{aligned} \tag{4.44}$$

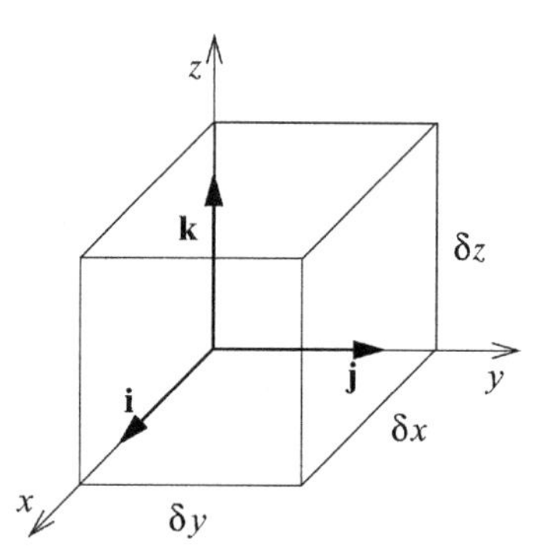

Fig. 4.7 A three-dimensional box with sides δx, δy, and δz.

since the rotational parts cancel out. Inspection of this equation shows that the angular change is caused solely by the anisotropic part of the deformation tensor.

4.3.3 The change in volume of a rectangular fluid box

In this case the application of equation (4.37) is particularly simple due to the rectangular system. We consider the relative changes of the three sides of the rectangular box shown in Figure 4.7.

On splitting the deformational dyadic into the isotropic and anisotropic parts we find from (4.36)

$$
\begin{aligned}
\frac{1}{\delta x}\frac{d}{dt}(\delta x) &= \mathbf{i}\cdot\mathbb{D}_{\mathrm{i}}\cdot\mathbf{i}+\mathbf{i}\cdot\mathbb{D}_{\mathrm{ai}}\cdot\mathbf{i} = \frac{\nabla\cdot\mathbf{v}}{3}+(D_{\mathrm{ai}})_{11} \\
\frac{1}{\delta y}\frac{d}{dt}(\delta y) &= \frac{\nabla\cdot\mathbf{v}}{3}+(D_{\mathrm{ai}})_{22} \\
\frac{1}{\delta z}\frac{d}{dt}(\delta z) &= \frac{\nabla\cdot\mathbf{v}}{3}+(D_{\mathrm{ai}})_{33}
\end{aligned}
\tag{4.45}
$$

where we have used the explicit representations of $\mathbb{D}_{\mathrm{i}}$ and $\mathbb{D}_{\mathrm{ai}}$ listed in (4.5). The D_{ai} are the measure numbers of the dyadic $\mathbb{D}_{\mathrm{ai}}$. On forming the trace of the matrix of $\mathbb{D}_{\mathrm{ai}}$, which is the same as summing the diagonal elements, we find immediately

$$
(D_{\mathrm{ai}})_{11}+(D_{\mathrm{ai}})_{22}+(D_{\mathrm{ai}})_{33} = (D_{\mathrm{ai}})_{\mathrm{I}} = 0 \tag{4.46}
$$

so that the total deformation amounts to the three-dimensional velocity divergence or

$$
\frac{1}{\delta x}\frac{d}{dt}(\delta x)+\frac{1}{\delta y}\frac{d}{dt}(\delta y)+\frac{1}{\delta z}\frac{d}{dt}(\delta z) = \frac{1}{\delta\tau}\frac{d}{dt}(\delta\tau) = \nabla\cdot\mathbf{v} = (\mathbb{D}_{\mathrm{i}})_{\mathrm{I}} \tag{4.47}
$$

For the special case that the divergence vanishes we find from (4.45) that the relative change in volume is zero:

$$\frac{1}{\delta\tau}\frac{d}{dt}(\delta\tau) = (D_{\text{ai}})_{11} + (D_{\text{ai}})_{22} + (D_{\text{ai}})_{33} = 0 \tag{4.48}$$

Therefore, a volume-true deformation (the relative change vanishes) of the three-dimensional box is described by three measure numbers, which are located on the diagonal of the dyadic $\mathbb{D}_{\text{ai}}$. We consider the brief example

$$\begin{gathered} (D_{\text{ai}})_{11} = 0 \implies (D_{\text{ai}})_{22} = -(D_{\text{ai}})_{33} \quad \text{or} \quad \frac{1}{\delta x}\frac{d}{dt}(\delta x) = 0 \\ \text{if} \quad (D_{\text{ai}})_{22} < 0 \implies (D_{\text{ai}})_{33} > 0 \end{gathered} \tag{4.49}$$

This means that a decrease in length in the y-direction is accompanied by an increase in length in the z-direction. These changes take place in such a way that the total volume remains constant, which is called a *volume-true change*. Moreover, the volume remains rectangular.

Let us now briefly discuss the situation in which the form of the box changes. Since the box originally was rectangular, equation (4.44) simplifies because the cosine term vanishes. For the angular changes we find from (4.44) using (4.5) the simple relations

$$\begin{aligned} \frac{d}{dt}(\gamma_{xy}) &= -2\mathbf{i}\cdot\mathbb{D}_{\text{ai}}\cdot\mathbf{j} = -2(D_{\text{ai}})_{12} \\ \frac{d}{dt}(\gamma_{xz}) &= -2\mathbf{i}\cdot\mathbb{D}_{\text{ai}}\cdot\mathbf{k} = -2(D_{\text{ai}})_{13} \\ \frac{d}{dt}(\gamma_{yz}) &= -2\mathbf{j}\cdot\mathbb{D}_{\text{ai}}\cdot\mathbf{k} = -2(D_{\text{ai}})_{23} \end{aligned} \tag{4.50}$$

We recognize that the off-diagonal elements of the coefficient matrix of D_{ai} are responsible for the volume-true changes of the three angles.

Many additional examples could be given, such as the relative change in volume of a fluid sphere. As should be expected, the change in volume is again equal to the divergence.

A brief summary of our results may be helpful. With reference to (4.12) we observe

$$\mathbf{v}(P') = \mathbf{v}(P) + \delta\mathbf{r}\cdot\mathbb{D}_{\text{i}} + \delta\mathbf{r}\cdot\mathbb{D}_{\text{ai}} + \tfrac{1}{2}(\nabla\times\mathbf{v})\times\delta\mathbf{r} \tag{4.51}$$

so that the general motion of a fluid element consists of

(1) *Rigid translation*: $\mathbf{v}(P)$.
(2) *Dilatation*: $\mathbf{v}_{\text{DIL}} = \delta\mathbf{r}\cdot\mathbb{D}_{\text{i}} = \delta\mathbf{r}\cdot\frac{1}{3}\nabla\cdot\mathbf{v}\mathbb{E}$, form-invariant change in volume.
(3) *Distortion*: $\mathbf{v}_{\text{DIS}} = \delta\mathbf{r}\cdot\mathbb{D}_{\text{ai}}$.

(a) Owing to the $(D_{\text{ai}})_{ii}$ elements a deformation whereby angles and volume do not change occurs.

$$\frac{1}{\delta x}\frac{d}{dt}(\delta x) = \mathbf{i} \cdot \mathbb{D} \cdot \mathbf{i} = \frac{1}{2}(D + A)$$
$$\frac{1}{\delta y}\frac{d}{dt}(\delta y) = \mathbf{j} \cdot \mathbb{D} \cdot \mathbf{j} = \frac{1}{2}(D - A)$$

Fig. 4.8(a) Length changes of the sides of a rectangle.

$$\frac{1}{\delta x}\frac{d}{dt}(\delta x) + \frac{1}{\delta y}\frac{d}{dt}(\delta y) = \frac{1}{\delta x\,\delta y}\frac{d}{dt}(\delta x\,\delta y) = D$$

Fig. 4.8(b) A change in area of a rectangle.

$$\boldsymbol{\Omega} = \frac{1}{2}\nabla_{\rm h} \times \mathbf{v}_{\rm h}$$

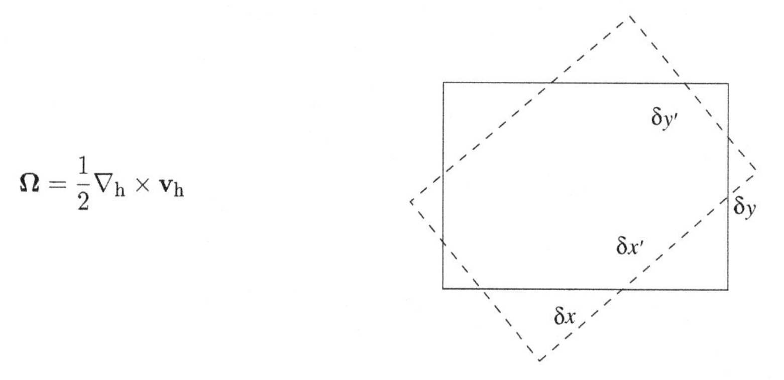

Fig. 4.8(c) Rigid rotation of a rectangle, $D = 0$.

(b) Owing to the $(D_{\rm ai})_{ij}$ $(i \neq j)$ elements a distortion occurs but the volume remains unchanged.

(4) Rotation: $\mathbf{v}_{\rm ROT} = \frac{1}{2}(\nabla \times \mathbf{v}) \times \delta\mathbf{r}$, rigid rotation.

4.3.4 Two-dimensional examples

In order to demonstrate more completely the geometric meaning of the principal deformation quantities A and B, see equation (4.9), we consider several special areal expansions. The results are collected in Figures 4.8(a)–(e). Part (a) follows directly from (4.37) on identifying the unit vector $\mathbf{e}$ with the Cartesian unit vectors. Verification is easily accomplished by expanding the two-dimensional deformation

$$\frac{d\gamma}{dt} = -2\mathbf{i} \cdot \mathbb{D}_{\text{ai}} \cdot \mathbf{j} = -2(D_{\text{ai}})_{12} = -B$$

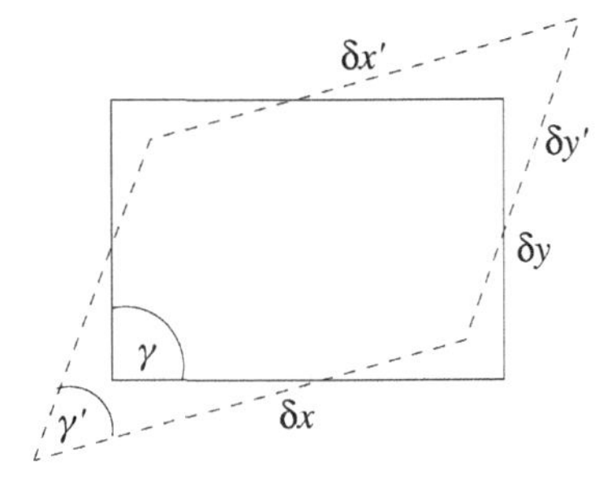

Fig. 4.8(d) A change in angle between two sides of a rectangle without a change in area, $D = 0$.

$$\frac{1}{\delta x}\frac{d}{dt}(\delta x) = (D_{\text{ai}})_{11} = \frac{A}{2}$$
$$\frac{1}{\delta y}\frac{d}{dt}(\delta y) = (D_{\text{ai}})_{22} = -\frac{A}{2}$$

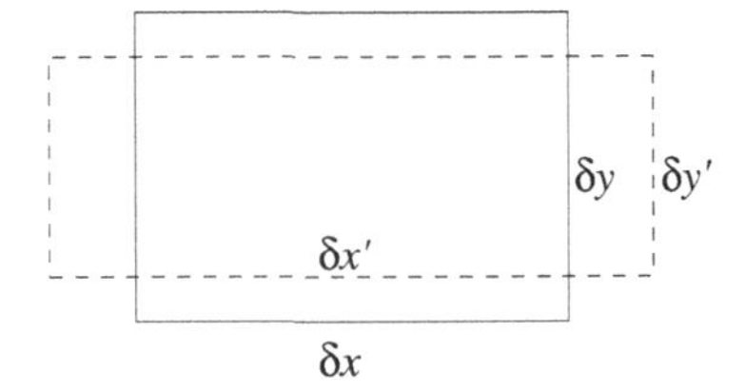

Fig. 4.8(e) A change in form of a rectangle, with no change in angle and area, $D = 0$.

dyadic according to (4.8). Part (b) is verified with the help of (4.47) by omitting the height term. Part (c) shows a rigid rotation with the angular velocity $\mathbf{\Omega}$ due to the antisymmetric dyadic $\mathfrak{R}$, see (4.12). Part (d) follows from (4.44) since the cosine term vanishes. Part (e) results from (4.48) on specializing to two dimensions.

4.4 Problems

4.1: Show that $\mathfrak{R} \cdot\cdot\, \mathfrak{R} = (\nabla \times \mathbf{v})^2/2$

4.2: Verify equations (4.9d) and (4.9e).

4.3: Consider the initial (x, y)-coordinate system. Rotate this system by a fixed angle θ to obtain the rotated (x', y')-coordinate system. Differentiation of (x, y) and (x', y') with respect to time gives a relation between the velocity components (u, v) and (u', v'). Show that the divergence and the vorticity are invariant under the rotation, i.e. $D = D'$, $\zeta = \zeta'$.

4.4: Derive (4.33b) and (4.33c).

4.5: Assume (a) a line element in the (x, y)-plane, (b) a surface element in the (x, y)-plane, and (c) a volume element in (x, y, z)-space. Verify equation (4.33).

5

The Navier–Stokes stress tensor

In this chapter we are going to derive the connection between the Navier–Stokes stress tensor $\mathbb{J}$ and the deformation dyadic (tensor) $\mathbb{D}$. Once more, the paramount importance of the local velocity dyadic becomes apparent. First of all, it will be necessary to introduce the general stress tensor $\mathbb{T}$ which includes $\mathbb{J}$.

5.1 The general stress tensor

The description of deformable media requires the definition of *internal* and *external forces*. Internal forces are molecular-type forces between mass elements, which may be excluded from our considerations. Owing to Newton's principle *actio* = *reactio* the net effect of these forces adds up to zero when we integrate over a specified volume of the continuum. There exist two types of external forces.

5.1.1 Volume forces

These forces are proportional to the mass. As is customary, we define these with respect to the unit mass and denote them by the symbol $\mathbf{f}_i$. In general, we distinguish between *attractive* and *inertial forces*.

(a) Attractive forces due to the presence of the earth and other celestial bodies. The gravitational pulls due to the sun and the moon are accounted for only if tidal effects are considered. These are real forces since they are caused by the interaction of an atmospheric particle with other bodies.

(b) Inertial forces such as the Coriolis force and the centrifugal force. These are fictitious forces and stem from the rotating coordinate system used to describe the motion of the particle.

5.1.2 Surface forces

These forces act in directions normal and tangential to a surface. They will be defined with respect to unit area and denoted by $\mathbf{p}_i$. In the atmosphere we have to deal with two types of surface forces.

(a) The *pressure force* $\mathbf{p}_1(p)$ results from the action of the all-directional atmospheric pressure. It is always acting in the direction opposite to the normal of a surface element of the fluid volume to which it is applied. If $\mathbf{n}$ is the unit normal defining the direction of the surface, then we must have

$$\mathbf{p}_1 = -p\mathbf{n} \tag{5.1}$$

On identifying $\mathbf{n}$ in succession by the Cartesian unit vectors $\mathbf{i}, \mathbf{j}$, and $\mathbf{k}$ we recognize the local isotropy of the pressure field since in each case we find $|\mathbf{p}_1(p)| = p$.

(b) The *frictional stress force* $\mathbf{p}_2(\mathbf{v}_\text{A})$ is a type of surface force that depends on the motion of the fluid (gas) and on the orientation of the surface to which it is applied. In contrast to $\mathbf{p}_1(p)$ the frictional stress is not limited to the perpendicular direction, but acts also tangentially to the surface of the fluid volume. $\mathbf{p}_2(\mathbf{v}_\text{A})$ may be represented by the linear vector function

$$\mathbf{p}_2 = \mathbf{n} \cdot \mathbb{J} \tag{5.2}$$

where $\mathbb{J}$ is the viscous stress tensor or dyadic which was introduced previously. Summing up, we find for the surface force

$$\begin{aligned} &\mathbf{p}_1 + \mathbf{p}_2 = -p\mathbf{n} \cdot \mathbb{E} + \mathbf{n} \cdot \mathbb{J} = \mathbf{n} \cdot \mathbb{T} \\ \text{with} \quad &\mathbb{T} = -p\mathbb{E} + \mathbb{J} = -p\mathbf{i}_n\mathbf{i}_n + \tau_{mn}\mathbf{i}_m\mathbf{i}_n \end{aligned} \tag{5.3}$$

where $\mathbb{T}$ is the general stress tensor which also includes the effect of pressure. The nine possible elements of $\mathbb{J}$ acting on the fluid volume element are illustrated in Figure 5.1.

The row subscript i in the matrix (τ_{ij}) refers to the surface element on which the stress is acting. The column index j refers to the direction of the stress. If $i = j$ then we are dealing with normal stresses; otherwise $(i \neq j)$ with tangential stresses. The *viscous stress vector* $\mathbf{J}_i$ for surfaces $i = 1, 2, 3$ is given by

$$\mathbf{J}_i = \mathbf{i}_i \cdot \mathbb{J} = \tau_{i1}\mathbf{i}_1 + \tau_{i2}\mathbf{i}_2 + \tau_{i3}\mathbf{i}_3 \tag{5.4}$$

Before we focus our attention on equilibrium conditions in the stress field, we need to restate the integral form of the equation of absolute motion,

$$\begin{aligned} &\frac{d}{dt}\left(\int_{V(t)} \rho\mathbf{v}_\text{A}\,d\tau\right) = \int_{V(t)} \rho\frac{d\mathbf{v}_\text{A}}{dt}\,d\tau = \int_{V(t)} \rho\mathbf{f}_\text{a}\,d\tau + \oint_{S(t)} d\mathbf{S} \cdot \mathbb{T} \\ \text{with} \quad &\oint_{S(t)} d\mathbf{S} \cdot \mathbb{T} = \int_{V(t)} \nabla \cdot \mathbb{T}\,d\tau = \int_{V(t)} (-\nabla p + \nabla \cdot \mathbb{J})\,d\tau \end{aligned} \tag{5.5}$$

This form of the equation of motion will now be used to show that the general and the viscous stress tensors $\mathbb{T}$ and $\mathbb{J}$ are symmetric.

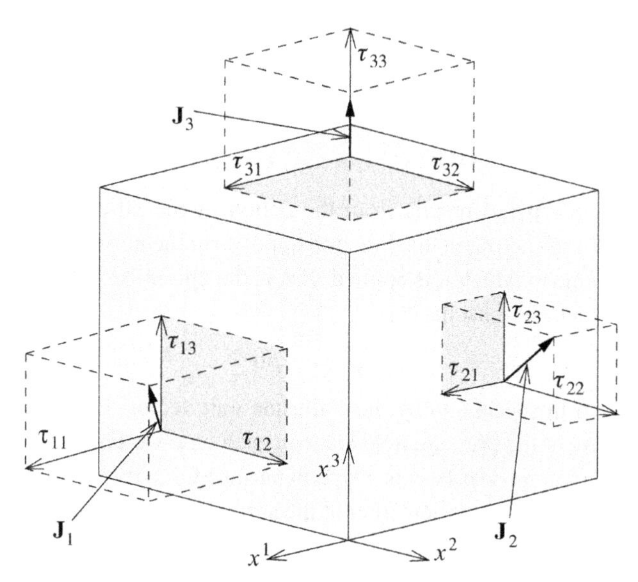

Fig. 5.1 Illustration of the viscous stresses acting on a volume element.

5.2 Equilibrium conditions in the stress field

We wish to prove the following statement:

Stresses on infinitesimally small fluid volumes must be in equilibrium.

The proof is very brief. We divide (5.5) by the surface S enclosing V and then implement the following limiting process:

$$\lim_{S\to 0}\left[\frac{1}{S}\left(-\int_V \rho\frac{d\mathbf{v}_{\mathrm{A}}}{dt}\,d\tau+\int_V \rho\mathbf{f}_{\mathrm{a}}\,d\tau+\oint_S d\mathbf{S}\cdot\mathbb{T}\right)\right]=0 \tag{5.6}$$

For easier understanding of the limiting process, let us momentarily think of the volume as a small cube whose sides have length Δl. The volume integrals extend over the volume $(\Delta l)^3$ while the surface S is proportional to $(\Delta l)^2$. By choosing Δl arbitrarily small, we find that, in the limit $\Delta l \to 0$, the numerators of the first two terms go to zero faster than do the denominators. The same type of argument holds for volumes of any shape, which can always be decomposed into numerous elementary cubes. Since the first two integrals go to zero, we can argue that the third integral must go to zero also. Therefore, using (5.3), we may also write

$$\lim_{S\to 0}\oint_S p\,d\mathbf{S}=0, \qquad \lim_{S\to 0}\oint_S d\mathbf{S}\cdot\mathbb{J}=0 \tag{5.7}$$

from which we conclude that the total force resulting from all surface forces acting on the small fluid volume must vanish.

5.3 Symmetry of the stress tensor

The proof follows from a law of mechanics stating that the individual time derivative of the angular momentum of the system equals the sum of the moments of the external forces. First, we observe that

$$\frac{d}{dt}\left(\int_{V(t)} \mathbf{r} \times \rho \mathbf{v}_{\mathrm{A}}\, d\tau\right) = \int_{V(t)} \mathbf{r} \times \rho \frac{d\mathbf{v}_{\mathrm{A}}}{dt}\, d\tau \tag{5.8}$$

where we have used the differentiation rule for fluid volumes. On performing the vectorial multiplication with the position vector under the integral sign, we find from (5.5)

$$\int_{V(t)} \mathbf{r} \times \rho \frac{d\mathbf{v}_{\mathrm{A}}}{dt}\, d\tau - \int_{V(t)} \mathbf{r} \times \rho \mathbf{f}_{\mathrm{a}}\, d\tau - \oint_{S(t)} \mathbf{r} \times (d\mathbf{S} \cdot \mathbb{T}) = 0 \tag{5.9}$$

Next we use Gauss' divergence theorem, which is also applicable to dyadics. The third term can then be written as

$$\begin{aligned}
-\oint_{S(t)} \mathbf{r} \times (d\mathbf{S} \cdot \mathbb{T}) &= \int_{V(t)} \nabla \cdot (\mathbb{T} \times \mathbf{r})\, d\tau = \int_{V(t)} \left(\nabla \cdot \overset{\downarrow}{\mathbb{T}} \times \mathbf{r} + \nabla \cdot \mathbb{T} \times \overset{\downarrow}{\mathbf{r}}\right) d\tau \\
&= \int_{V(t)} (-\mathbf{r} \times \nabla \cdot \mathbb{T} - \mathbb{T}_{\times})\, d\tau \\
\text{since} \quad \nabla \cdot \mathbb{T} \times \overset{\downarrow}{\mathbf{r}} &= (\widetilde{\mathbb{T}} \cdot \nabla) \times \mathbf{r} = \left[(\widetilde{\mathbb{T}} \cdot \nabla)\mathbf{r}\right]_{\times} = (\widetilde{\mathbb{T}} \cdot \mathbb{E})_{\times} = -\mathbb{T}_{\times}
\end{aligned} \tag{5.10}$$

where $\mathbb{T}_{\times}$ is the vector of the dyadic $\mathbb{T}$ (see also Section M2.4.2). Substitution of (5.10) into (5.9) yields

$$\int_{V(t)} \mathbf{r} \times \left(\rho \frac{d\mathbf{v}_{\mathrm{A}}}{dt} - \rho \mathbf{f}_{\mathrm{a}} - \nabla \cdot \mathbb{T}\right) d\tau - \int_{V(t)} \mathbb{T}_{\times}\, d\tau = 0 \tag{5.11}$$

The expression in parentheses is actually the equation of absolute motion and must vanish. Therefore

$$\mathbb{T}_{\times} = 0 \tag{5.12}$$

We already know that the vector of a symmetric dyadic is zero, whereas the vector of an antisymmetric dyadic differs from zero (see Section M2.4.2). Thus, we conclude that the general stress tensor $\mathbb{T}$ is symmetric so that the viscous stress tensor $\mathbb{J}$ must also be symmetric:

$$\mathbb{J} = \widetilde{\mathbb{J}} \Longrightarrow \mathbf{n} \cdot \mathbb{J} = \widetilde{\mathbb{J}} \cdot \mathbf{n} = \mathbb{J} \cdot \mathbf{n} \tag{5.13}$$

This justifies all previous mathematical operations with the stress tensor whenever we assumed that $\mathbb{J}$ is symmetric.

5.4 The frictional stress tensor and the deformation dyadic

It stands to reason that the motion-dependent frictional force $\mathbf{p}_2(\mathbf{v}_\mathrm{A})$ at an arbitrary point P within the fluid medium cannot result from rigid rotation or from translation of an entire region surrounding P. The frictional stress is caused only by deformative velocities in the immediate surrounding of P. Therefore, we should expect a relation between $\mathbb{J}$ and the deformation dyadic $\mathbb{D}$ defined in the previous chapter. Since the atmosphere may be considered an isotropic medium, we may deduce that the principal axes of the tensors $\mathbb{J}$ and $\mathbb{D}$ coincide. By means of a suitable rotation of the coordinate system, the so-called principal-axis transformation (see also Section M2.3), it is possible to reduce the dyadics $\mathbb{J}$ and $\mathbb{D}$ to the simple normal forms

$$\begin{aligned} \mathbb{J} &= \tau_1 \mathbf{e}_1 \mathbf{e}_1 + \tau_2 \mathbf{e}_2 \mathbf{e}_2 + \tau_3 \mathbf{e}_3 \mathbf{e}_3 \\ \mathbb{D} &= d_1 \mathbf{e}_1 \mathbf{e}_1 + d_2 \mathbf{e}_2 \mathbf{e}_2 + d_3 \mathbf{e}_3 \mathbf{e}_3 \end{aligned} \tag{5.14}$$

The quantities τ_i and d_i represent the principal stresses and the relative changes in length along the orthogonal unit vectors $\mathbf{e}_i$ defining the directions of the principal axes. Moreover, the τ_i and d_i are the eigenvalues of the matrices representing $\mathbb{J}$ and $\mathbb{D}$. Since the matrices are symmetric, the resulting eigenvalues must be real. Next we need to apply *Hooke's law* in the generalized form stating that changes in length and cross-contractions are proportional to the stresses. Owing to the isotropy of the medium we may write

$$\begin{aligned} &\text{(a)} \quad \tau_1 = \nu d_1 - \lambda(d_2 + d_3) = 2\mu d_1 - \lambda(d_1 + d_2 + d_3) \\ &\text{(b)} \quad \tau_2 = \nu d_2 - \lambda(d_1 + d_3) = 2\mu d_2 - \lambda(d_1 + d_2 + d_3) \\ &\text{(c)} \quad \tau_3 = \nu d_3 - \lambda(d_1 + d_2) = 2\mu d_3 - \lambda(d_1 + d_2 + d_3) \end{aligned} \tag{5.15}$$

where $\mu = (\nu + \lambda)/2$ is known as the molecular coefficient of the deformation viscosity. The quantities λ and μ are known as *Lamé's coefficients*, and will be treated here as constants. On multiplying (5.15a) by $\mathbf{e}_1\mathbf{e}_1$, (5.15b) by $\mathbf{e}_2\mathbf{e}_2$, and (5.15c) by $\mathbf{e}_3\mathbf{e}_3$, and adding the resulting equations, we find

$$\mathbb{J} = 2\mu\mathbb{D} - \lambda\mathbb{D}_\mathrm{I}\mathbb{E} \tag{5.16}$$

On splitting $\mathbb{D}$ into its isotropic and anisotropic parts and reverting to the general form (4.2), we find

$$\mathbb{J} = 2\mu\mathbb{D}_{\mathrm{ai}} + \left(\frac{2\mu}{3} - \lambda\right)\nabla\cdot\mathbf{v}_{\mathrm{A}}\mathbb{E} \tag{5.17}$$

We assume that λ and μ do not explicitly depend on the spatial coordinates, i.e.

$$\nabla\mu = 0, \qquad \nabla\lambda = 0 \tag{5.18}$$

However, μ exhibits some dependency on the temperature. Values of μ can be found in various handbooks. The coefficient $(2\mu/3 - \lambda)$ is known as the *coefficient of volume viscosity* since it multiplies the divergence term in (5.17) describing the relative changes in volume. In atmospheric systems this coefficient may be neglected. This statement can be motivated with the help of statistical thermodynamics. There it is shown that this coefficient is zero for monatomic molecules so that, to an acceptable approximation, we may write

$$\mathbb{J} = 2\mu\mathbb{D}_{\mathrm{ai}} \tag{5.19}$$

We conclude this chapter with a simple example. Consider the symmetric stress tensor $\mathbb{J}$ defined by (5.19):

$$\mathbb{J} = \mu\left(\frac{\nabla\mathbf{v}_{\mathrm{A}} + \overset{\curvearrowleft}{\mathbf{v}_{\mathrm{A}}\nabla}}{2}\right) - \frac{2}{3}\mu\,\nabla\cdot\mathbf{v}_{\mathrm{A}}\mathbb{E} \tag{5.20}$$

We recognize that $\mathbb{J}$ is known at a certain point P if $\nabla\mathbf{v}_{\mathrm{A}}$ is known there. Assuming that the absolute velocity $\mathbf{v}_{\mathrm{A}}$ is given by

$$\mathbf{v}_{\mathrm{A}} = u(z)\mathbf{i} + v(x)\mathbf{j} + w(y)\mathbf{k} \tag{5.21}$$

we find $\nabla\cdot\mathbf{v}_{\mathrm{A}} = 0$ so that

$$\mathbb{J} = \mu\left(\frac{\partial u}{\partial z}(\mathbf{ki} + \mathbf{ik}) + \frac{\partial v}{\partial x}(\mathbf{ij} + \mathbf{ji}) + \frac{\partial w}{\partial y}(\mathbf{jk} + \mathbf{kj})\right) \tag{5.22}$$

For the viscous surface force $\mathbf{p}_2(\mathbf{v}_{\mathrm{A}})$ defined by (5.2), we find for different orientations of the surface unit normal

$$\begin{aligned}
\mathbf{n} = \mathbf{i}: &\qquad \mathbf{p}_2(\mathbf{v}_{\mathrm{A}}) = \mathbf{i}\cdot\mathbb{J} = \mu\left(\frac{\partial u}{\partial z}\mathbf{k} + \frac{\partial v}{\partial x}\mathbf{j}\right)\\
\mathbf{n} = \mathbf{j}: &\qquad \mathbf{p}_2(\mathbf{v}_{\mathrm{A}}) = \mathbf{j}\cdot\mathbb{J} = \mu\left(\frac{\partial v}{\partial x}\mathbf{i} + \frac{\partial w}{\partial y}\mathbf{k}\right)\\
\mathbf{n} = \mathbf{k}: &\qquad \mathbf{p}_2(\mathbf{v}_{\mathrm{A}}) = \mathbf{k}\cdot\mathbb{J} = \mu\left(\frac{\partial u}{\partial z}\mathbf{i} + \frac{\partial w}{\partial y}\mathbf{j}\right)
\end{aligned} \tag{5.23}$$

For $\mathbf{n} = \mathbf{i}$ the surface force at P has the orientation shown in Figure 5.2.

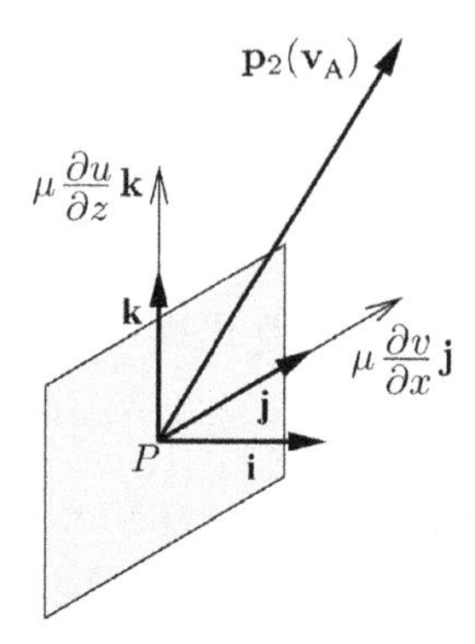

Fig. 5.2 The viscous force for $\mathbf{n} = \mathbf{i}$. $\mathbf{p}_2(\mathbf{v}_A)$ at point P results from the addition of the two vectors in the $\mathbf{j}$- and $\mathbf{k}$-directions.

5.5 Problems

5.1: Assuming the validity of (5.19) and μ = constant, find an expression for the term $\nabla \cdot \mathbb{J}$.

5.2: Show that

$$\begin{aligned} &\mathbf{v}_A \cdot \nabla \cdot \mathbb{J} = \nabla \cdot (\mathbf{v}_A \cdot \mathbb{J}) - \nabla \mathbf{v}_A \cdot\cdot \mathbb{J} \\ \text{with}\quad &\text{(a) } \mathbf{v}_A \cdot \mathbb{J} = \mu[\nabla \mathbf{v}_A^2 + (\nabla \times \mathbf{v}_A) \times \mathbf{v}_A - \tfrac{2}{3}(\nabla \cdot \mathbf{v}_A)\mathbf{v}_A] \\ &\text{(b) } \nabla \mathbf{v}_A \cdot\cdot \mathbb{J} = 2\mu \mathbb{D}_{ai} \cdot\cdot \mathbb{D}_{ai} = \rho\epsilon \end{aligned}$$

The quantity ϵ is the Rayleigh dissipation function, which is a source term for the internal energy and a sink term for the kinetic energy.

5.3: Assume that we have incompressible two-dimensional and steady flow on an inclined plane of infinite extent. The x-axis lies in the plane and the z-direction is perpendicular to it. Thus, we have $v = 0$, $w = 0$, $\partial/\partial y = 0$, $\partial/\partial t = 0$.

(a) Find an expression for the profile of the velocity u in the x-direction using the boundary conditions $u(z = 0) = 0$ and $(\partial u/\partial z)_{z=H} = 0$. Ignore Coriolis effects. Hint: Return to equation (1.60) with $\mathbf{v}_D = 0$.
(b) What is the orientation of the isobars?
(c) Find the amount of fluid q per unit time and unit width flowing between $z = 0$ and $z = H$.
(d) Find the energy dissipation $\rho\epsilon$ within the channel.

5.4: Assume that we have incompressible two-dimensional and steady flow between two parallel plates. The lower plate is at rest. Ignore Coriolis effects.

(a) Suppose that the upper plate at $z = H$ is moving with the velocity $u(z = H)$ in the x-direction. Starting with equation (1.60), show that a linear velocity profile will form. This flow is known as the *Couette flow*. The lower boundary condition is $u(z = 0) = 0$.

(b) In contrast to part (a), assume now that both plates are at rest, but a pressure gradient $-(1/\rho)(\partial p/\partial x)$ is admitted. This is known as *Poiseuille flow*. The boundary conditions are $u(z = 0) = 0$, $u(z = H) = 0$.

5.5: Assume that the flow field is proportional to the position vector, that is $\mathbf{v}_{\mathrm{A}} = C\mathbf{r}$. The constant of proportionality C carries the units per second. Find the stress tensor $\mathbb{J}$, the force due to viscous stress $\mathbf{p}_2(\mathbf{v}_{\mathrm{A}})$, and the tensor ellipsoid.

6

The Helmholtz theorem

6.1 The three-dimensional Helmholtz theorem

If within a region of interest the divergence $\nabla \cdot \mathbf{v}$ and the rotation $\nabla \times \mathbf{v}$ of the velocity vector $\mathbf{v}$ are given and if these vanish sufficiently fast on approaching infinity, then $\mathbf{v}$ is uniquely specified. More generally we may state:

> *Every vector field* $\mathbf{v}$ *whose divergence and rotation possess potentials can be written as the sum of a divergence-free vector field plus another vector field that is irrotational.*

We will now proceed with the proof by assuming that the vector $\mathbf{v}$ can be written as the sum of two vectors such that the rotational part is divergence-free and the divergent part is irrotational, or

$$\begin{aligned} &\mathbf{v} = \mathbf{v}_{\mathrm{ROT}} + \mathbf{v}_{\mathrm{DIV}} \\ \text{with}\quad &\nabla \cdot \mathbf{v}_{\mathrm{ROT}} = 0, \qquad \nabla \times \mathbf{v}_{\mathrm{ROT}} \neq 0 \\ &\nabla \cdot \mathbf{v}_{\mathrm{DIV}} \neq 0, \qquad \nabla \times \mathbf{v}_{\mathrm{DIV}} = 0 \end{aligned} \tag{6.1}$$

From the rules of vector analysis we recognize that (6.1) is satisfied if

$$\mathbf{v}_{\mathrm{ROT}} = \nabla \times \mathbf{A}, \qquad \mathbf{v}_{\mathrm{DIV}} = -\nabla \chi \tag{6.2}$$

where $\mathbf{A}$ and χ are known as the *vector potential* and the *scalar potential of* $\mathbf{v}$, respectively. The minus sign in the second equation is conventional. Often the scalar velocity potential is simply called the *velocity potential.*

The potentials $\mathbf{A}$ and χ are unknown quantities. For (6.2) to hold it is also necessary that the divergence of the vector potential vanishes. The proof will be given later. As a consequence of (6.2) we may write

$$\mathbf{v} = \nabla \times \mathbf{A} - \nabla \chi \tag{6.3}$$

To show that it is possible to split the vector $\mathbf{v}$ in this manner we take the divergence and then the curl of (6.3) and find

$$\nabla^2 \chi = -\nabla \cdot \mathbf{v}, \qquad \nabla^2 \mathbf{A} = -\nabla \times \mathbf{v} \tag{6.4}$$

where we have made use of the assumption $\nabla \cdot \mathbf{A} = 0$. According to (M6.82) the formal solutions of these two Poisson equations are given by

$$\chi(\mathbf{r}) = \frac{1}{4\pi} \int_{V'} \frac{\nabla \cdot \mathbf{v}'}{|\mathbf{r} - \mathbf{r}'|} d\tau', \qquad \mathbf{A}(\mathbf{r}) = \frac{1}{4\pi} \int_{V'} \frac{\nabla \times \mathbf{v}'}{|\mathbf{r} - \mathbf{r}'|} d\tau' \tag{6.5}$$

where we have replaced the unspecified function $f(\mathbf{r}')$. With the help of (6.5), provided that the divergence and the curl of the vector $\mathbf{v}$ are known, we may calculate $\mathbf{A}$ and χ and then the vector $\mathbf{v}$ itself. The proof will be complete if we can show that the divergence of the vector potential $\nabla \cdot \mathbf{A}$ vanishes.

It is immediately obvious that

$$\nabla \left(\frac{1}{\left| \overset{\downarrow}{\mathbf{r}} - \mathbf{r}' \right|} \right) = -\nabla \left(\frac{1}{\left| \mathbf{r} - \overset{\downarrow}{\mathbf{r}'} \right|} \right) \tag{6.6}$$

so that

$$\nabla \cdot \mathbf{A}(\mathbf{r}) = -\frac{1}{4\pi} \int_{V'} \nabla \times \mathbf{v}' \cdot \nabla \left(\frac{1}{\left| \mathbf{r} - \overset{\downarrow}{\mathbf{r}'} \right|} \right) d\tau' \tag{6.7}$$

Applying the vector rule $\mathbf{B} \cdot \nabla C = \nabla \cdot (\mathbf{B}C) - (\nabla \cdot \mathbf{B})C$, setting in (6.7) $\mathbf{B} = \nabla \times \mathbf{v}'$ and $C = 1/\left| \mathbf{r} - \overset{\downarrow}{\mathbf{r}'} \right|$, we obtain

$$\begin{aligned} \nabla \cdot \mathbf{A}(\mathbf{r}) &= -\frac{1}{4\pi} \int_{V'} \nabla \cdot \left(\frac{\nabla \times \mathbf{v}'}{|\mathbf{r} - \mathbf{r}'|} \right) d\tau' + \frac{1}{4\pi} \int_{V'} \frac{1}{|\mathbf{r} - \mathbf{r}'|} \nabla \cdot \nabla \times \mathbf{v}' \, d\tau' \\ &= -\frac{1}{4\pi} \oint_{S_\infty} \left(\frac{\nabla \times \mathbf{v}'}{|\mathbf{r} - \mathbf{r}'|} \right) \cdot d\mathbf{S} = 0 \end{aligned} \tag{6.8}$$

since $\nabla \cdot \nabla \times \mathbf{v}'$ vanishes. As the last step we have applied the divergence theorem. We have introduced a sufficiently large surface S_∞ since it was assumed that $\nabla \times \mathbf{v}'$ vanishes on approaching infinity.

From these results we conclude that, for irrotational stationary *potential flow* $\nabla \times \mathbf{v} = 0$, the velocity is given by

$$\mathbf{v} = -\nabla \chi \implies \nabla \cdot \mathbf{v} = -\nabla^2 \chi \tag{6.9}$$

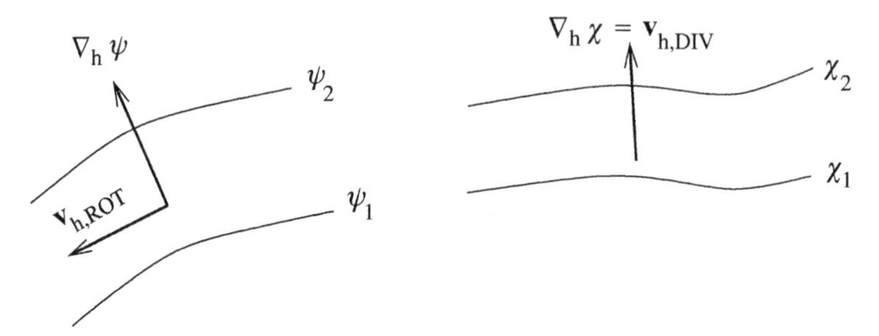

Fig. 6.1 The stream function ψ and velocity potential χ.

As we have seen, if the divergence of the flow field is known, the velocity potential can be calculated with the help of (6.5) if the region of integration is infinitely large. In this case the boundary is undefined. In practical meteorology, in general, the region has a finite boundary and equation (6.9) must be treated as an elliptic boundary-value problem by specifying the velocity potential (the *Dirichlet problem*) or its normal gradient (the *Neumann problem*) on the boundary. The solution is usually found by numerical methods. It is also possible in the case of a finite boundary to proceed analytically. This leads to the introduction of *Green's function*, which often is difficult or even impossible to obtain since it depends on the shape of the boundary. We will not broaden this topic since potential flow does not occur in large-scale dynamics.

6.2 The two-dimensional Helmholtz theorem

The three-dimensional Helmholtz theorem has its two-dimensional counterpart. Since the large-scale motion can often be approximated as two-dimensional flow it is useful to consider the consequences of Helmholtz's theorem. Instead of (6.1) we now have

$$\begin{aligned} &\mathbf{v}_h = \mathbf{v}_{h,ROT} + \mathbf{v}_{h,DIV} \\ &\nabla_h \cdot \mathbf{v}_{h,ROT} = 0, \qquad \mathbf{k} \cdot \nabla_h \times \mathbf{v}_{h,ROT} \neq 0 \\ &\nabla_h \cdot \mathbf{v}_{h,DIV} \neq 0, \qquad \mathbf{k} \cdot \nabla_h \times \mathbf{v}_{h,DIV} = 0 \end{aligned} \tag{6.10}$$

From the rules of vector analysis we immediately recognize that

$$\begin{aligned} \nabla_h \cdot \mathbf{v}_{h,ROT} = 0 \quad &\text{if} \quad \mathbf{v}_{h,ROT} = \mathbf{k} \times \nabla_h \psi \\ \mathbf{k} \cdot \nabla_h \times \mathbf{v}_{h,DIV} = 0 \quad &\text{if} \quad \mathbf{v}_{h,DIV} = \nabla_h \chi \end{aligned} \tag{6.11}$$

This leads to the definition of the *stream function* ψ and the two-dimensional velocity potential χ; see Figure 6.1.

The stream function should not be confused with the streamline, which can be constructed for any type of flow, whereas the stream function results from the first condition of (6.11). Using the relations (6.11) involving the stream function and the velocity potential we may rewrite (6.10) as

$$\mathbf{v}_\mathrm{h} = \mathbf{k} \times \nabla_\mathrm{h}\psi + \nabla_\mathrm{h}\chi \tag{6.12a}$$

or in components

$$u = -\frac{\partial \psi}{\partial y} + \frac{\partial \chi}{\partial x}, \qquad v = \frac{\partial \psi}{\partial x} + \frac{\partial \chi}{\partial y} \tag{6.12b}$$

Taking the vertical component of the curl and the divergence of $\mathbf{v}_\mathrm{h}$, observing that the curl of the gradient of the velocity potential vanishes, we obtain

$$\begin{aligned} \mathbf{k} \cdot \nabla_\mathrm{h} \times \mathbf{v}_\mathrm{h} &= \nabla_\mathrm{h}^2 \psi = \zeta = \frac{\partial v}{\partial x} - \frac{\partial u}{\partial y} \\ \nabla_\mathrm{h} \cdot \mathbf{v}_\mathrm{h} &= \nabla_\mathrm{h}^2 \chi = D = \frac{\partial u}{\partial x} + \frac{\partial v}{\partial y} \end{aligned} \tag{6.13}$$

By providing proper boundary conditions the Poisson equations for ψ and χ can be solved. Moreover, solutions analogous to (6.5) can be written down. By suitable differentiation of ζ and D in (6.13) with respect to x and y and adding (subtracting) the resulting equations, we immediately find two additional Poisson equations for the horizontal velocity components u and v:

$$\nabla_\mathrm{h}^2 u = -\frac{\partial \zeta}{\partial y} + \frac{\partial D}{\partial x}, \qquad \nabla_\mathrm{h}^2 v = \frac{\partial \zeta}{\partial x} + \frac{\partial D}{\partial y} \tag{6.14}$$

Equations (6.12)–(6.14) show the inter-relations among the horizontal velocity components u, v, the stream function ψ, the velocity potential χ, the vorticity ζ, and the divergence D. Finally, we would like to remark that it is possible to generalize the stream-function method to three dimensions; see Sievers (1995).

6.3 Problems

6.1: Consider the three-dimensional velocity field defined by

$$\mathbf{v} = -\nabla \times \mathbf{\Psi} + \nabla \phi$$

Apply the condition $\nabla \cdot \mathbf{\Psi} = 0$.
(a) Find the relations of $(\nabla \mathbf{v})_\mathrm{I}$, $(\nabla \mathbf{v})_\times$, and $(\nabla \mathbf{v})''$.
(b) Express the vortex equation

$$\frac{d}{dt}\nabla \times \mathbf{v} = -(\nabla \times \mathbf{v})(\nabla \cdot \mathbf{v}) + (\nabla \times \mathbf{v}) \cdot \nabla \mathbf{v} - \nabla\left(\frac{1}{\rho}\right) \times \nabla p$$

in terms of $\mathbf{\Psi}$ and ϕ.

7

Kinematics of two-dimensional flow

7.1 Atmospheric flow fields

In the previous chapter we have shown that the horizontal wind vector may be decomposed into one part $\mathbf{v}_{h,ROT}$, which is divergence-free, and a second part $\mathbf{v}_{h,DIV}$, which is irrotational. On the basis of this decomposition we will briefly discuss various combinations of the flow field associated either with $\nabla_h \cdot \mathbf{v}_h = 0$ and $\mathbf{k} \cdot \nabla_h \times \mathbf{v}_h \neq 0$ or with the case in which the signs $=$ and $\neq$ are interchanged. First of all, we consider the more general situation:

$$\text{(i)} \qquad \nabla_h \cdot \mathbf{v}_h \neq 0 \quad \text{and} \quad \mathbf{k} \cdot \nabla_h \times \mathbf{v}_h \neq 0$$

In an atmospheric flow field whose characteristic length varies from roughly 50 to 2000 km, both parts, $\mathbf{v}_{h,ROT}$ and $\mathbf{v}_{h,DIV}$, must be accounted for. In order to gain qualitative insight into the structure of the fields of the stream function ψ and the velocity potential χ, we assume a normal mode solution of the form

$$\begin{pmatrix} \psi \\ \chi \end{pmatrix} \propto \exp[i(k_x + k_y + \omega t)] \quad \text{with} \quad k_{x,y} = \frac{2\pi}{L_{x,y}}, \qquad \omega = c\sqrt{k_x^2 + k_y^2} \tag{7.1}$$

The quantity c is the phase velocity of the wave and k_x and k_y are the wavenumbers corresponding to wavelengths L_x and L_y as shown in Figure 7.1.

We substitute (7.1) into the Poisson equations (6.13). Because of

$$\nabla_h^2 \begin{pmatrix} \psi \\ \chi \end{pmatrix} \propto -(k_x^2 + k_y^2) \begin{pmatrix} \psi \\ \chi \end{pmatrix} \tag{7.2}$$

we obtain the qualitative statement

$$-(k_x^2 + k_y^2) \begin{pmatrix} \psi \\ \chi \end{pmatrix} \propto \begin{pmatrix} \zeta \\ D \end{pmatrix} \tag{7.3}$$

showing that the field of the stream function is characterized by the vorticity field whereas the field of the velocity potential has the structure of the divergence field.

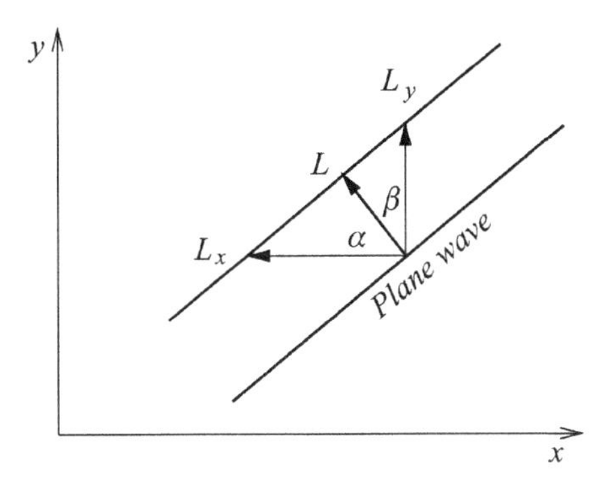

Fig. 7.1 Propagation of a plane wave.

$$\text{(ii)} \qquad \nabla_{\text{h}} \cdot \mathbf{v}_{\text{h}} = 0 \quad \text{and} \quad \mathbf{k} \cdot \nabla_{\text{h}} \times \mathbf{v}_{\text{h}} \neq 0$$

Now we will discuss some special situations. If the horizontal flow field is free from divergence, Poisson's equation for χ reduces to

$$\nabla_{\text{h}}^2 \chi = 0 \tag{7.4}$$

which is *Laplace's equation* for the velocity potential. A function with continuous second derivatives satisfying Laplace's equation is known as a *harmonic function*. We may easily prove that a harmonic function within the region being considered is zero if it is zero on its boundary. On multiplying (7.4) by χ and integrating over the surface of interest, we find

$$\int_S \chi \nabla_{\text{h}}^2 \chi \, dS' = \int_S \nabla_{\text{h}} \cdot (\chi \nabla_{\text{h}} \chi) \, dS' - \int_S (\nabla_{\text{h}} \chi)^2 \, dS' = 0 \tag{7.5}$$

Application of Gauss' two-dimensional integration theorem (M6.34) to the first term on the right-hand side of (7.5) shows that this term is zero if we assume that $\chi = 0$ on the boundary Γ of the region, or

$$\int_S \nabla_{\text{h}} \cdot (\chi \, \nabla_{\text{h}} \chi) \, dS' = \oint_L \chi \, \nabla_{\text{h}} \chi \cdot d\mathbf{s}_\perp = 0 \tag{7.6}$$

where $d\mathbf{s}_\perp$ takes the places of $d\mathbf{r}_{\text{N}}$. The second term of (7.5) is positive definite so that $\nabla_{\text{h}} \chi = 0$ or $\chi =$ constant within the entire region of integration. Since χ is zero on the boundary, the constant must be zero also. Therefore, the general statement (6.12) reduces to

$$\mathbf{v}_{\text{h}} = \mathbf{k} \times \nabla_{\text{h}} \psi \quad \text{or} \quad u = -\frac{\partial \psi}{\partial y}, \qquad v = \frac{\partial \psi}{\partial x} \tag{7.7}$$

so that the velocity may be described solely in terms of the stream function ψ. For large-scale atmospheric flow fields exceeding a characteristic length scale of about 2000 km, the divergent part of the velocity may often be disregarded. The flow then follows the lines representing the constant stream function.

$$\text{(iii)} \qquad \nabla_h \cdot \mathbf{v}_h \neq 0 \quad \text{and} \quad \mathbf{k} \cdot \nabla_h \times \mathbf{v}_h = 0$$

In the case that the rotation vanishes, we find from (6.13) the Laplace equation

$$\nabla_h^2 \psi = 0 \tag{7.8}$$

Repeating the argument of case (ii), we may conclude that $\psi = 0$. This is the situation of purely *potential flow* as expressed by

$$\mathbf{v}_h = \nabla_h \chi, \qquad \nabla_h^2 \chi = D \tag{7.9}$$

This type of flow is applicable to small-scale atmospheric motion such as mountain-valley winds and land–sea breezes whose characteristic length scale is usually less than 50 km. The essential difference from large-scale motion is that the Coriolis effect is of minor importance. In the absence of centripetal forces the motion follows the negative pressure gradient from high to low pressure but along the positive gradient of the velocity potential. In the so-called *fine structure of the pressure field* a minimum value of p is associated with a maximum of χ and vice versa.

$$\text{(iv)} \qquad \nabla_h \cdot \mathbf{v}_h = 0 \quad \text{and} \quad \mathbf{k} \cdot \nabla_h \times \mathbf{v}_h = 0$$

In this case it is possible to equate the horizontal velocities according to (7.7) and (7.9) to give

$$\nabla_h \chi = \mathbf{k} \times \nabla_h \psi \tag{7.10}$$

or, in component form,

$$u = \frac{\partial \chi}{\partial x} = -\frac{\partial \psi}{\partial y}, \qquad v = \frac{\partial \chi}{\partial y} = \frac{\partial \psi}{\partial x} \tag{7.11}$$

These are the famous *Cauchy–Riemann equations* from complex-variable theory, which are rarely satisfied for atmospheric conditions. Nevertheless, these conditions provide a perfect example of purely *deformational flow*.

As an exercise we will obtain the distributions of the stream function and the velocity potential for the simple flow field described by

$$u = \lambda x, \qquad v = -\lambda y \tag{7.12}$$

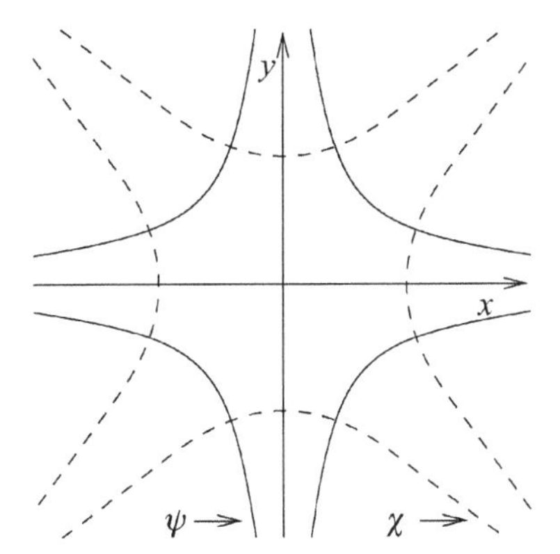

Fig. 7.2 Representative lines of the stream function ψ and lines of constant velocity potential χ.

Application of the Cauchy–Riemann equations and integration immediately results in

$$\begin{aligned} \frac{\partial \psi}{\partial y} &= -\lambda x \implies \psi = -\lambda x y + f(x) \\ \frac{\partial \psi}{\partial x} &= -\lambda y \implies \psi = -\lambda y x + f(y) \\ & \qquad\qquad\quad \psi = -\lambda x y + C_1 \end{aligned} \tag{7.13}$$

Since we are dealing with partial differential equations, integration yields the functions $f(x)$ and $f(y)$ instead of integration constants. However, comparison of the first two equations of (7.13) shows that $f(x) = f(y)$ so that both functions must be equal to a constant. Analogously to ψ, the velocity potential χ is obtained by using the remaining pair of Cauchy–Riemann equations:

$$\begin{aligned} \frac{\partial \chi}{\partial x} &= \lambda x & &\implies \chi = \frac{\lambda x^2}{2} + f(y) \\ \frac{\partial \chi}{\partial y} &= -\lambda y = \frac{df}{dy} & &\implies f(y) = -\frac{\lambda y^2}{2} + C_2 \\ & & & \qquad \chi = \frac{\lambda}{2}(x^2 - y^2) + C_2 \end{aligned} \tag{7.14}$$

We will now show that, in this situation, the isolines of the stream function and the velocity potential are orthogonal since the scalar product of the gradients of $\nabla_h \psi$ and $\nabla_h \chi$ is zero,

$$\nabla_h \psi \cdot \nabla_h \chi = \nabla_h \psi \cdot \mathbf{k} \times \nabla_h \psi = 0 \tag{7.15}$$

A sketch of the flow field is given in Figure 7.2.

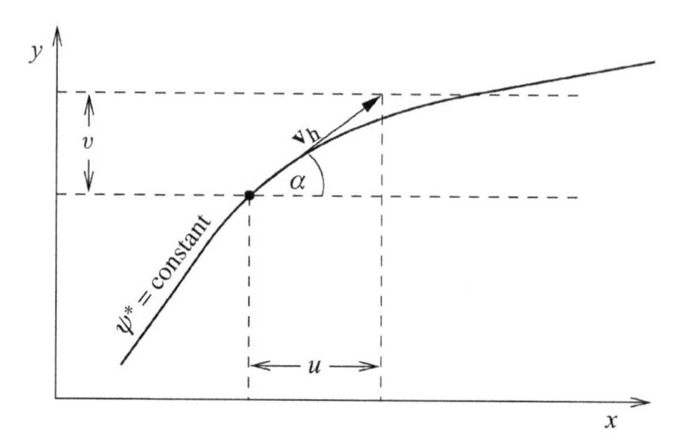

Fig. 7.3 The horizontal streamline $\psi^* =$ constant.

From the definitions stated in (4.9) one obtains

$$\zeta = 0, \qquad D = 0, \qquad def = \sqrt{A^2 + B^2} = 2\lambda \tag{7.16}$$

This verifies the statement that we are dealing with a purely deformational flow field.

7.2 Two-dimensional streamlines and normals

7.2.1 Two-dimensional streamlines

For the horizontal flow the equation of streamlines (3.39) reduces to

$$\mathbf{v}_\mathrm{h} \times d\mathbf{r}_\mathrm{s} = 0 \Longrightarrow \frac{dy}{dx} = \frac{v(x, y, t_0)}{u(x, y, t_0)} \quad \text{or} \quad -u\,dy + v\,dx = 0 \tag{7.17}$$

where $d\mathbf{r}_\mathrm{s}$ is a line element along the streamline. From analysis we know that (7.17) is an exact differential equation if there exists a function $\psi^*(x, y)$ such that

$$d\psi^* = \frac{\partial \psi^*}{\partial y}\,dy + \frac{\partial \psi^*}{\partial x}\,dx = -u\,dy + v\,dx = 0 \Longrightarrow \psi^*(x, y) = \text{constant} \tag{7.18}$$

See Figure 7.3. The condition of exactness, also known as *the integrability condition*, is given by

$$\frac{\partial^2 \psi^*}{\partial x\,\partial y} = \frac{\partial^2 \psi^*}{\partial y\,\partial x} \Longrightarrow \frac{\partial u}{\partial x} + \frac{\partial v}{\partial y} = \nabla_\mathrm{h} \cdot \mathbf{v}_\mathrm{h} = 0 \tag{7.19}$$

showing that the divergence of the flow field is zero.

From (7.18) we have

$$u = -\frac{\partial \psi^*}{\partial y}, \qquad v = \frac{\partial \psi^*}{\partial x} \quad \text{or} \quad \mathbf{v}_\mathrm{h} = \mathbf{k} \times \nabla_\mathrm{h} \psi^* \tag{7.20}$$

Since the divergence $D = 0$, the streamline is identical to the stream function, see (6.11), which is defined by

$$\mathbf{v}_{\mathrm{h,ROT}} = \mathbf{k} \times \nabla_{\mathrm{h}} \psi \tag{7.21}$$

If the divergence differs from zero, the integrability condition (7.19) is not satisfied, that is

$$\frac{\partial u}{\partial x} + \frac{\partial v}{\partial y} \neq 0 \tag{7.22}$$

Now the streamlines and lines of the constant stream functions are no longer identical. In order to find ψ^*, we multiply (7.17) by a suitable integrating factor $\mu(x, y)$, causing this equation to become exact,

$$-u\mu\, dy + v\mu\, dx = 0 \tag{7.23}$$

In this case the integrability condition is given by

$$-\frac{\partial}{\partial x}(u\mu) = \frac{\partial}{\partial y}(v\mu) \tag{7.24}$$

It may be very difficult to find such an integrating factor. If it has been found, we have to solve

$$u\mu = -\frac{\partial \psi^*}{\partial y}, \qquad v\mu = \frac{\partial \psi^*}{\partial x} \tag{7.25}$$

to find the streamline ψ^*, which is no longer identical to the stream function ψ.

We will now consider a very simple example satisfying condition (7.19) so that $D = 0$. If the velocity components are given by

$$u = U = \text{constant} > 0, \qquad v = -kA \sin(kx) \tag{7.26}$$

then the equation of the streamline reads

$$-U\, dy - kA \sin(kx)\, dx = 0 \tag{7.27}$$

Application of (7.18) results in

$$\begin{aligned}
&\text{(a)} \quad \frac{\partial \psi^*}{\partial y} = -U \implies \psi^* = -Uy + f(x) \\
&\text{(b)} \quad \frac{\partial \psi^*}{\partial x} = -kA \sin(kx) = \frac{df}{dx} \implies f(x) = A \cos(kx) + \text{constant} \\
&\text{(c)} \quad \psi^* = A \cos(kx) - Uy + \psi_0^* \implies y = \frac{1}{U}\left[\psi_0^* - \psi^* + A \cos(kx)\right]
\end{aligned} \tag{7.28}$$

where the integration constant has been denoted by ψ_0^*. The solution is a flow field representing a *stationary Rossby wave*; see Figure 7.4.

If the wave is displaced with the phase velocity c in the eastward direction then the last equation of (7.28) must be modified to read

$$y = \text{constant} + \frac{A}{U} \cos[k(x - ct)] \tag{7.29}$$

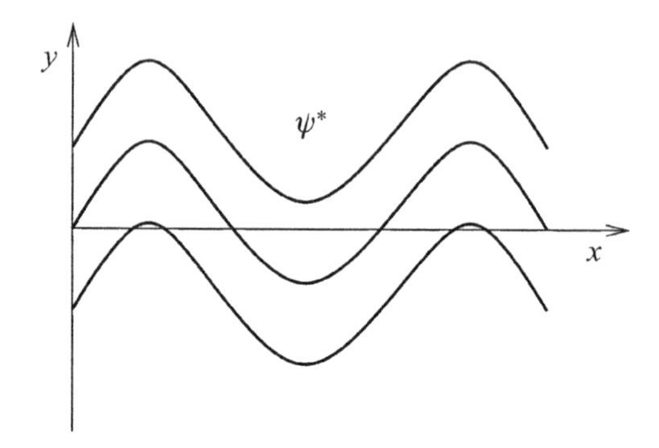

Fig. 7.4 Streamlines representing the stationary flow field.

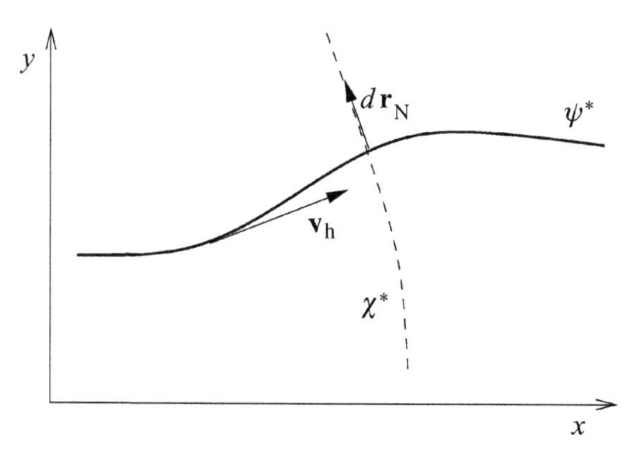

Fig. 7.5 Normals χ^* to the streamlines ψ^*.

7.2.2 Construction of normals

At every point of the horizontal flow field at t = constant streamlines ψ^* and normals χ^* form an orthogonal system. Since the instantaneous velocity vector is tangential to the streamlines, the differential equation of the normals must be given by

$$\mathbf{v}_h \cdot d\mathbf{r}_N = 0 \quad \text{or} \quad u\,dx + v\,dy = 0 \tag{7.30}$$

where $d\mathbf{r}_N$ is a line element along the normal; see Figure 7.5.

Equation (7.30) is a total differential of the normal χ^* if the integrability condition

$$\frac{\partial v}{\partial x} = \frac{\partial u}{\partial y} \tag{7.31}$$

is satisfied or if the vorticity ζ is zero. In this case we may write

$$d\chi^* = \frac{\partial \chi^*}{\partial x}dx + \frac{\partial \chi^*}{\partial y}dy = u\,dx + v\,dy = 0 \tag{7.32}$$

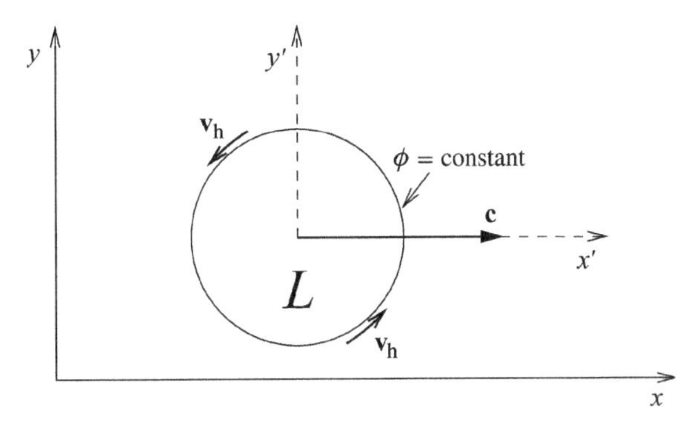

Fig. 7.6 The drifting coordinate system.

from which it follows that

$$u = \frac{\partial \chi^*}{\partial x}, \qquad v = \frac{\partial \chi^*}{\partial y} \quad \text{or} \quad \mathbf{v}_\text{h} = \nabla_\text{h}\chi^* \tag{7.33}$$

From this equation we may find analytic expressions for the normals in the same way as that in which we have found stream functions. A comparison with (7.9) shows that χ^* is identical with the velocity potential. If the integrability condition (7.31) is not satisfied we may introduce an integrating factor and proceed analogously to finding streamlines; see equations (7.22)–(7.25). In this case the normal χ^* will not coincide with the velocity potential χ.

7.3 Streamlines in a drifting coordinate system

We consider a horizontal pressure disturbance ϕ, which is displaced with phase velocity $\mathbf{c}$ in the eastward direction. The air particles themselves move with the horizontal velocities $\mathbf{v}_\text{h}$ and $\mathbf{v}_\text{h}'$ with respect to the stationary (x, y)- and the moving (x', y')-coordinate system, see Figure 7.6.

On replacing in the addition theorem of the velocities (M4.32) the velocity $\mathbf{v}_P$ of the moving system by $\mathbf{c}$ and the relative velocity $\mathbf{v}$ by $\mathbf{v}_\text{h}'$, we find

$$\mathbf{v}_\text{h} = \mathbf{v}_\text{h}' + \mathbf{c} \tag{7.34}$$

Application of Euler's development gives for the resting system the local change with time of the field function

$$\left(\frac{\partial \phi}{\partial t}\right)_{x,y} = \frac{d\phi}{dt} - \mathbf{v}_\text{h} \cdot \nabla_\text{h}\phi \tag{7.35}$$

In the moving (x', y') system we find

$$\left(\frac{\partial \phi}{\partial t}\right)_{x',y'} = \frac{d\phi}{dt} - \mathbf{v}_\text{h}' \cdot \nabla_\text{h}\phi \tag{7.36}$$

Since the individual change with time $d\phi/dt$ is independent of the coordinate system, it may be eliminated by subtracting (7.35) from (7.36), yielding

$$\left(\frac{\partial \phi}{\partial t}\right)_{x,y} = \left(\frac{\partial \phi}{\partial t}\right)_{x'y'} - \mathbf{c} \cdot \nabla_{\mathrm{h}} \phi \tag{7.37}$$

This relation expresses the local change with time of the function ϕ in both systems, which involves, as should be expected, the displacement velocity $\mathbf{c}$. The local change with time $(\partial\phi/\partial t)_{x',y'}$ in the moving system represents the development of the pressure system. In case of stationarity in the moving system, there is no development and $(\partial\phi/\partial t)_{x',y'} = 0$.

Next we consider a system of streamlines moving with the phase velocity $\mathbf{c}$ in the direction of the positive x-axis. We assume that the velocity components are given by

$$u = U = \text{constant}, \qquad v = -kA \sin[k(x - ct)] \tag{7.38}$$

Substitution of (7.38) into the streamline equation (7.17), integrating for fixed time $t = t_0$, gives the streamline representation

$$y = \frac{A}{U} \cos[k(x - ct_0)] + \text{constant} \tag{7.39}$$

The question of the shape of the streamlines in the moving system now arises. The coordinate relation between the two systems is taken from Figure 7.6 but now we assume that the y- and the y'-axis coincide, or

$$x' = x - ct, \qquad y' = y \tag{7.40}$$

From this it follows immediately that

$$u' = \frac{dx'}{dt} = \frac{dx}{dt} - c = U - c, \qquad v' = \frac{dy'}{dt} = v = -kA \sin(kx') \tag{7.41}$$

Inspection of (7.41) shows that the velocity field (u', v') in the (x', y') system is time-independent or stationary. The differential equation of the streamline in the moving system is then given by

$$\frac{dy'}{dx'} = \frac{v'}{u'} = \frac{v}{U - c} = \frac{-kA \sin(kx')}{U - c} \tag{7.42}$$

Integration of (7.42) gives the equation of the streamline of the moving system:

$$y' = \frac{A}{U - c} \cos[k(x - ct)] + \text{constant} \tag{7.43}$$

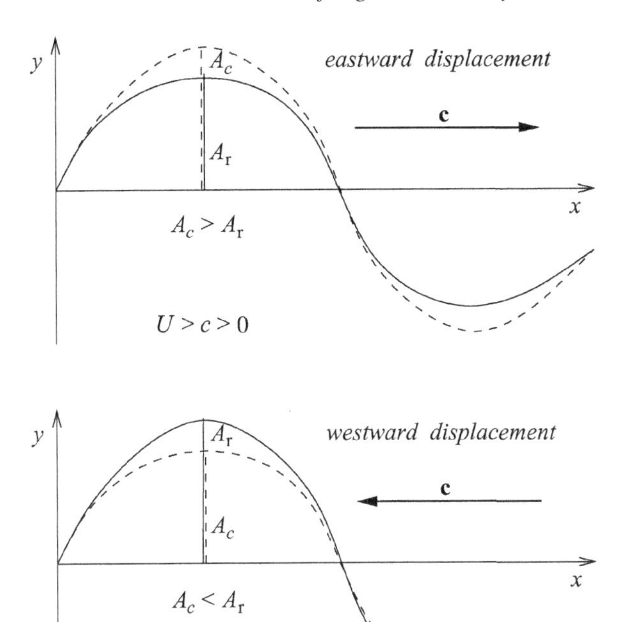

Fig. 7.7 Eastward and westward displacement of the streamline system.

From (7.39) and (7.43) we find the relationship between the amplitudes of the system at rest ($A_r = A/U$) and the moving system [$A_c = A/(U - c)$],

$$A_c = A_r \frac{U}{U - c} \tag{7.44}$$

Since the wavelength is the same for the streamlines of both systems, we may arrange the streamlines as shown in Figure 7.7.

On solving (7.44) for the phase velocity c we, find

$$c = U\left(1 - \frac{A_r}{A_c}\right) \tag{7.45}$$

In case of an eastward displacement of the streamline system we have $A_c > A_r$, and $A_c < A_r$ for the displacement in the opposite direction.

We conclude this section by giving an example pertaining to the moving coordinate system. We shall assume that the potential temperature is a conservative quantity during the motion so that $\dot{\Theta} = 0$. Furthermore, we shall assume stationarity in the moving system. Therefore, the development term $(\partial\Theta/\partial t)_{x',y'} = 0$ and, analogously to (7.36), we find

$$\mathbf{v}_h' \cdot \nabla_h \theta = 0 \tag{7.46}$$

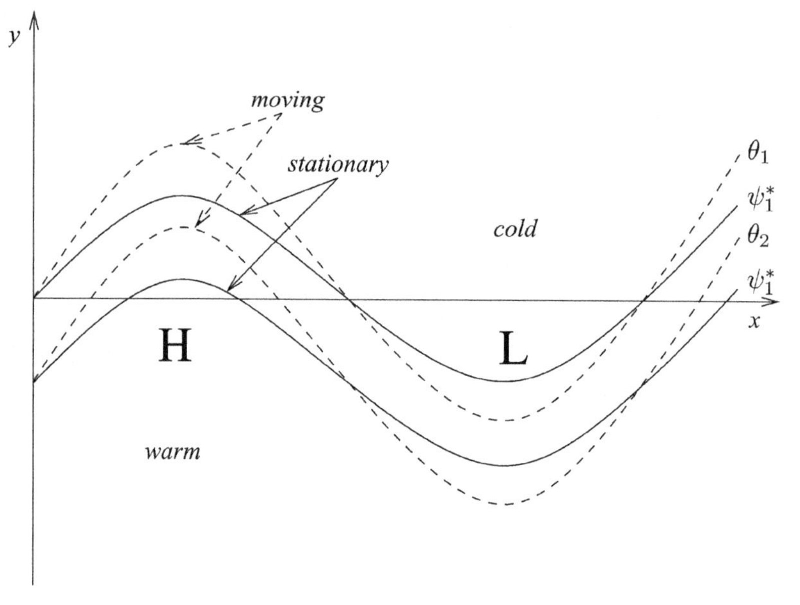

Fig. 7.8 Interpretation of relative streamlines, $\psi_2^* > \psi_1^*$.

This means that lines of constant potential temperature are arranged parallel to the horizontal velocity $\mathbf{v}_{\mathrm{h}}'$ that is parallel to the streamlines $(\psi^*)'$ of the moving system. In the system at rest, assuming geostrophic conditions, the streamlines $\psi^* = \phi/f$ correspond to lines of constant geopotential on an isobaric surface $p =$ constant. In the tropospheric westward drift most frequently $U > c > 0$. According to (7.44) the amplitudes of the streamlines in the moving system exceed the amplitudes in the system at rest so that $A_c > A_{\mathrm{r}}$, as shown in Figure 7.8.

By relabeling the relative streamlines $(\psi^*)'$ as lines of constant potential temperature we find that moving troughs are colder than moving ridges.

7.4 Problems

7.1: Consider two-dimensional potential flow in a polar coordinate system.
(a) Find the components v_r and v_t where r and t refer to the radial and the tangential direction.
(b) Find the continuity equation for incompressible flow in the polar coordinate system.

7.2: The circulation is defined by

$$C = \oint_L \mathbf{v} \cdot d\mathbf{r}$$

where L is the length of an arbitrary arc. Find the circulation in terms of potential flow.

7.3: Suppose that the potential flow is expressed by $\chi = \theta$, where θ is the angle shown in the figure.

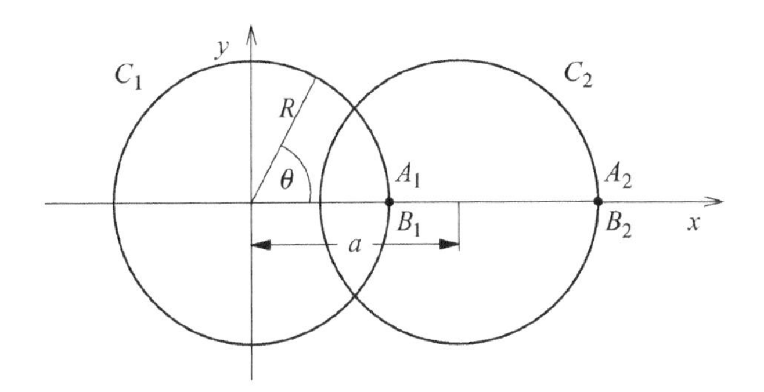

(a) Calculate the circulation along the circle C_1.
(b) Calculate the circulation along the circle C_2 defined by $(x-a)^2+y^2 = R^2$, $a > R$.

7.4: The velocity potential is assumed to be given by $\chi = ax + by$, where a and b are real numbers.
(a) Find the total velocity of the flow field.
(b) Find the stream function of the flow field.

7.5: Assuming that the velocity potential is in the form $\chi = \ln\left(\sqrt{x^2+y^2}\right)$, show that the stream function is given by $\psi = -\arctan(x/y)$ if we ignore an arbitrary constant.

8

Natural coordinates

8.1 Introduction

In this chapter we consider horizontal and frictionless flow in a Cartesian coordinate system on a plane tangential to the earth's surface as explained in Section 2.5. For convenience we repeat the equations required for the horizontal wind field:

$$\frac{du}{dt} = fv - \frac{1}{\rho}\frac{\partial p}{\partial x}, \qquad \frac{dv}{dt} = -fu - \frac{1}{\rho}\frac{\partial p}{\partial y} \tag{8.1}$$

The Cartesian system will then be transformed into the *natural coordinate system*. The orientation of this system is described by the unit vector $\mathbf{e}_s$ defining the direction of the horizontal wind vector $\mathbf{v}_{\rm h} = |\mathbf{v}_{\rm h}|\, \mathbf{e}_s$ and the orthogonal unit vector $\mathbf{e}_n$ in the direction of lines normal to the *streamline* (normals). To complete the coordinate system we define the unit vector $\mathbf{e}_z$ in the direction normal to the tangent plane.

Figure 8.1 refers to a section of the horizontal flow field and shows the *trajectory* T of a particle at the positions $P_T(t_0)$ and $P_T(t_1)$ at times t_0 and t_1. At these locations the horizontal velocity is given by $\mathbf{v}_{\rm h}(t_0)$ and $\mathbf{v}_{\rm h}(t_1)$. As described in Section 3.4, the velocity vector is jointly tangential to the streamline and to the trajectory. While the trajectory defines a sequence of particle positions, both streamlines and their normals refer to fixed time. The infinitesimal arclengths Δs_T, Δs and Δn of the trajectory T, the streamline S, and the normal N together with the *contingency angle* α between $\mathbf{e}_s$ and the x-axis define the corresponding radii of curvature as given by

$$R_T = \frac{ds_T}{d\alpha}, \qquad R_s = \left(\frac{\partial s}{\partial \alpha}\right)_t, \qquad R_n = \left(\frac{\partial n}{\partial \alpha}\right)_t \tag{8.2}$$

Inspection of Figure 8.1 also shows that $dR_s = -dn$ and $dR_n = ds$ as Δn and Δs shrink to zero. Since the flow is assumed to be horizontal, the natural coordinate system is particularly simple and the contingency angles play the dominating role. Since the coordinate lines s = constant and n = constant are defined only

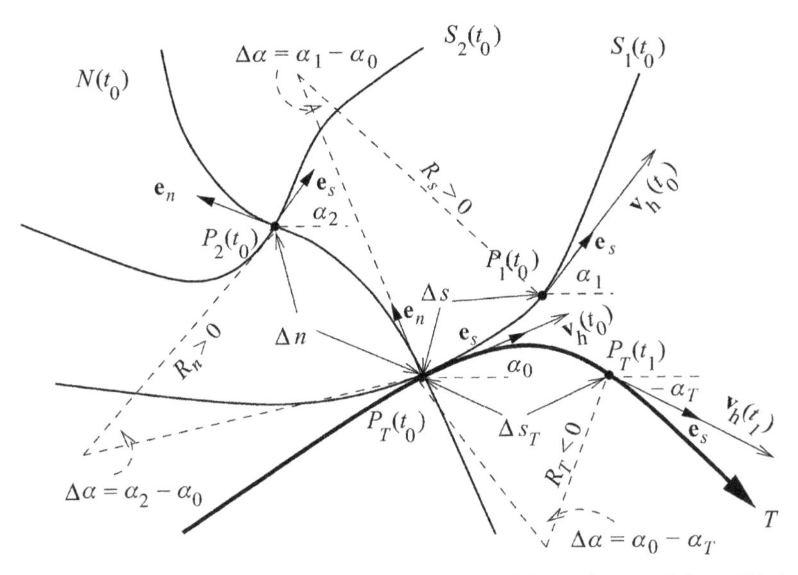

Fig. 8.1 The trajectory T, streamlines $S_1(t_0)$, $S_2(t_0)$, and normal line $N(t_0)$.

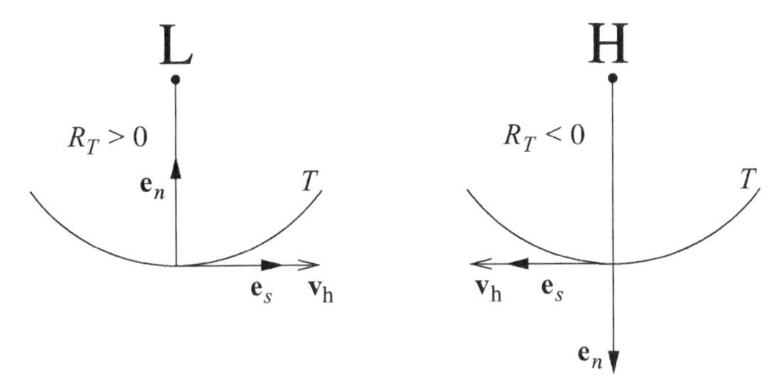

Fig. 8.2 Sign conventions for the radius of curvature of the trajectory, for the northern hemisphere.

piecewise, the intregability condition is not satisfied, so the order of the partial derivatives with respect to the independent coordinates cannot be interchanged. In the next section we will demonstrate this in detail.

Since the curvature may assume either sign, we introduce the generally accepted convention shown in Figure 8.2. If $\mathbf{e}_s$ is pointing in the counterclockwise direction (rotation in the positive sense) the radius of curvature of the trajectory R_T is taken as positive. If the motion is in the clockwise direction R_T is negative.

8.2 Differential definitions of the coordinate lines

The natural coordinate system is defined by the orthogonal vector basis $(\mathbf{e}_s, \mathbf{e}_n, \mathbf{e}_z)$, where $\mathbf{e}_s$ defines the direction of motion of a particle so that $\mathbf{e}_s \times \mathbf{e}_n = \mathbf{e}_z$. Therefore, the horizontal velocity is given by

$$\mathbf{v}_\mathrm{h} = \mathbf{i}u + \mathbf{j}v = \mathbf{i}\,|\mathbf{v}_\mathrm{h}|\cos\alpha + \mathbf{j}\,|\mathbf{v}_\mathrm{h}|\sin\alpha = \mathbf{e}_s\,|\mathbf{v}_\mathrm{h}| = \mathbf{e}_s \overset{*}{v}_\mathrm{h} \tag{8.3}$$

where $\overset{*}{v}_\mathrm{h}$ is the physical measure number of the horizontal velocity. Differentials of the coordinate lines are defined by

$$d\overset{*}{q}^1 = ds = R_s\, d\alpha, \qquad d\overset{*}{q}^2 = dn = R_n\, d\alpha, \qquad d\overset{*}{q}^3 = dz \tag{8.4}$$

Here $d\overset{*}{q}^1$ and $d\overset{*}{q}^2$ are the elements of arclength of the streamline and of the normal, while $d\overset{*}{q}^3$ is the line element along $\mathbf{e}_z$.

Before proceeding, we need to state the transformation relationships between the Cartesian and the natural coordinate system. At time $t =$ constant these simply correspond to the rotation of the Cartesian system about the vertical axis by the angle α counterclockwise. Let the orthogonal transformation matrix be denoted by

$$(T_{ij}) = \begin{pmatrix} \cos\alpha & -\sin\alpha & 0 \\ \sin\alpha & \cos\alpha & 0 \\ 0 & 0 & 1 \end{pmatrix} \tag{8.5}$$

Then

$$\begin{pmatrix} \mathrm{i} \\ \mathrm{j} \\ \mathrm{k} \end{pmatrix} = (T_{ij}) \begin{pmatrix} \mathbf{e}_s \\ \mathbf{e}_n \\ \mathbf{e}_z \end{pmatrix}, \qquad \begin{pmatrix} \mathbf{e}_s \\ \mathbf{e}_n \\ \mathbf{e}_z \end{pmatrix} = (\widetilde{T_{ij}}) \begin{pmatrix} \mathbf{i} \\ \mathbf{j} \\ \mathbf{k} \end{pmatrix} \tag{8.6}$$

[1]The variation of the second expression of (8.6) is given by

$$\delta \begin{pmatrix} \mathbf{e}_s \\ \mathbf{e}_n \\ \mathbf{e}_z \end{pmatrix} = (\delta\widetilde{T_{ij}})(T_{ij}) \begin{pmatrix} \mathbf{e}_s \\ \mathbf{e}_n \\ \mathbf{e}_z \end{pmatrix} = \begin{pmatrix} 0 & \delta\alpha & 0 \\ -\delta\alpha & 0 & 0 \\ 0 & 0 & 0 \end{pmatrix} \begin{pmatrix} \mathbf{e}_s \\ \mathbf{e}_n \\ \mathbf{e}_z \end{pmatrix} = \delta\alpha \begin{pmatrix} \mathbf{e}_n \\ -\mathbf{e}_s \\ 0 \end{pmatrix} \tag{8.7}$$

or, in component form, by

$$\delta\mathbf{e}_s = \mathbf{e}_n\,\delta\alpha, \qquad \delta\mathbf{e}_n = -\mathbf{e}_s\,\delta\alpha, \qquad \delta\mathbf{e}_z = 0 \tag{8.8}$$

[1] The tilde denotes the transpose of (T_{ij}).

The operator δ takes the place of all differential operators such as d/dt, $\partial/\partial t$, d/ds, and $\partial/\partial s$. However, the budget operator D/Dt is not included in this group. Therefore, we may write

$$
\begin{array}{llll}
\text{(a)} & \dfrac{d\mathbf{e}_s}{dt} = \mathbf{e}_n \dfrac{d\alpha}{dt}, & \dfrac{d\mathbf{e}_n}{dt} = -\mathbf{e}_s \dfrac{d\alpha}{dt} & \\
\text{(b)} & \dfrac{d\mathbf{e}_s}{ds_T} = \mathbf{e}_n \dfrac{d\alpha}{ds_T} = \dfrac{\mathbf{e}_n}{R_T}, & \dfrac{d\mathbf{e}_n}{ds_T} = -\mathbf{e}_s \dfrac{d\alpha}{ds_T} = -\dfrac{\mathbf{e}_s}{R_T} & \\
\text{(c)} & \left(\dfrac{\partial \mathbf{e}_s}{\partial t}\right)_{s,n,z} = \mathbf{e}_n \left(\dfrac{\partial \alpha}{\partial t}\right)_{s,n,z}, & \left(\dfrac{\partial \mathbf{e}_n}{\partial t}\right)_{s,n,z} = -\mathbf{e}_s \left(\dfrac{\partial \alpha}{\partial t}\right)_{s,n,z} & (8.9) \\
\text{(d)} & \dfrac{\partial \mathbf{e}_s}{\partial n} = \mathbf{e}_n \dfrac{\partial \alpha}{\partial n} = \dfrac{\mathbf{e}_n}{R_n}, & \dfrac{\partial \mathbf{e}_n}{\partial n} = -\mathbf{e}_s \dfrac{\partial \alpha}{\partial n} = -\dfrac{\mathbf{e}_s}{R_n} & \\
\text{(e)} & \dfrac{\partial \mathbf{e}_s}{\partial s} = \mathbf{e}_n \dfrac{\partial \alpha}{\partial s} = \dfrac{\mathbf{e}_n}{R_s}, & \dfrac{\partial \mathbf{e}_n}{\partial s} = -\mathbf{e}_s \dfrac{\partial \alpha}{\partial s} = -\dfrac{\mathbf{e}_s}{R_s} &
\end{array}
$$

For convenience, in parts (d) and (e) we have not explicitly stated the variables to be held constant. These are the famous *Frenet–Serret formulas* for the two-dimensional situation. A detailed discussion of the three-dimensional case, for example, may be found in Lass (1950). We conclude this section by stating the gradient operator in the natural coordinate system:

$$
\nabla\psi = \mathbf{q}^n \frac{\partial \psi}{\partial q^n} = \mathbf{e}_s \frac{\partial \psi}{\partial s} + \mathbf{e}_n \frac{\partial \psi}{\partial n} + \mathbf{e}_z \frac{\partial \psi}{\partial z} \tag{8.10}
$$

where ψ is some arbitrary but well-defined field function. Application of (8.3) to (8.10) results in the advection term

$$
\mathbf{v}_{\mathrm{h}} \cdot \nabla\psi = \mathring{v}_{\mathrm{h}} \frac{\partial \psi}{\partial s} \tag{8.11}
$$

which will be needed later.

As we have stated above, the order of the partial derivatives with respect to the independent variables s and n cannot be interchanged. We will prove this now. Since the partial derivatives transform in the same way as the corresponding unit vectors, we may write from (8.6)

$$
\begin{aligned}
&\text{(a)} \quad \begin{pmatrix} \dfrac{\partial}{\partial s} \\ \dfrac{\partial}{\partial n} \\ \dfrac{\partial}{\partial z} \end{pmatrix} = \left(\widetilde{T_{ij}}\right) \begin{pmatrix} \dfrac{\partial}{\partial x} \\ \dfrac{\partial}{\partial y} \\ \dfrac{\partial}{\partial z} \end{pmatrix} \\
&\text{(b)} \quad \begin{pmatrix} \dfrac{\partial}{\partial x} \\ \dfrac{\partial}{\partial y} \\ \dfrac{\partial}{\partial z} \end{pmatrix} = \left(T_{ij}\right) \begin{pmatrix} \dfrac{\partial}{\partial s} \\ \dfrac{\partial}{\partial n} \\ \dfrac{\partial}{\partial z} \end{pmatrix}
\end{aligned}
\tag{8.12}
$$

Application of (8.12a) to the arbitrary field function ψ yields

$$
\frac{\partial \psi}{\partial s} = \cos\alpha \, \frac{\partial \psi}{\partial x} + \sin\alpha \, \frac{\partial \psi}{\partial y} \tag{8.13}
$$

Taking the partial derivative with respect to time, we obtain

$$
\frac{\partial^2 \psi}{\partial t \, \partial s} = -\sin\alpha \, \frac{\partial \alpha}{\partial t} \frac{\partial \psi}{\partial x} + \cos\alpha \, \frac{\partial^2 \psi}{\partial t \, \partial x} + \cos\alpha \, \frac{\partial \alpha}{\partial t} \frac{\partial \psi}{\partial y} + \sin\alpha \, \frac{\partial^2 \psi}{\partial t \, \partial y} \tag{8.14a}
$$

The order of partial differentiation with respect to (t, x) and (t, y) may be interchanged so that

$$
\frac{\partial^2 \psi}{\partial t \, \partial s} = -\sin\alpha \, \frac{\partial \alpha}{\partial t} \frac{\partial \psi}{\partial x} + \cos\alpha \, \frac{\partial^2 \psi}{\partial x \, \partial t} + \cos\alpha \, \frac{\partial \alpha}{\partial t} \frac{\partial \psi}{\partial y} + \sin\alpha \, \frac{\partial^2 \psi}{\partial y \, \partial t} \tag{8.14b}
$$

Using the above transformation rule (8.12b), we find from (8.14b)

$$
\frac{\partial^2 \psi}{\partial t \, \partial s} = \frac{\partial \alpha}{\partial t} \frac{\partial \psi}{\partial n} + \frac{\partial^2 \psi}{\partial s \, \partial t} \tag{8.14c}
$$

which is clearly showing that the mixed partial derivatives are not identical. For the remaining partial derivatives of the natural coordinates analogous expressions may be derived. All transformation rules are summarized in

$$
\boxed{
\begin{aligned}
&\frac{\partial^2\psi}{\partial t\,\partial s} = \frac{\partial\alpha}{\partial t}\frac{\partial\psi}{\partial n} + \frac{\partial^2\psi}{\partial s\,\partial t}\\
&\frac{\partial^2\psi}{\partial z\,\partial s} = \frac{\partial\alpha}{\partial z}\frac{\partial\psi}{\partial n} + \frac{\partial^2\psi}{\partial s\,\partial z}\\
&\frac{\partial^2\psi}{\partial t\,\partial n} = -\frac{\partial\alpha}{\partial t}\frac{\partial\psi}{\partial s} + \frac{\partial^2\psi}{\partial n\,\partial t}\\
&\frac{\partial^2\psi}{\partial z\,\partial n} = -\frac{\partial\alpha}{\partial z}\frac{\partial\psi}{\partial s} + \frac{\partial^2\psi}{\partial n\,\partial z}\\
&\frac{\partial^2\psi}{\partial n\,\partial s} = \frac{\partial\alpha}{\partial s}\frac{\partial\psi}{\partial s} + \frac{\partial\alpha}{\partial n}\frac{\partial\psi}{\partial n} + \frac{\partial^2\psi}{\partial s\,\partial n}\\
&\frac{\partial^2\psi}{\partial z\,\partial t} = \frac{\partial^2\psi}{\partial t\,\partial z}
\end{aligned}
}
\tag{8.15}
$$

The derivation of the second from last expression is a little tricky. We take the partial derivative of the field function ψ first with respect to s and then with respect to n. Next we reverse the order of differentiation and then combine the resulting equations.

8.3 Metric relationships

The metric coefficients of the orthogonal natural coordinate system are obtained from the fundamental metric form

$$
\begin{aligned}
(d\mathbf{r})^2 &= g_{nn}(dq^n)^2 = (ds)^2 + (dn)^2 + (dz)^2\\
&= R_s^2(d\alpha)^2 + R_n^2(d\alpha)^2 + (dz)^2
\end{aligned}
\tag{8.16}
$$

Comparison of coefficients gives immediately

$$
g_{11} = R_s^2, \qquad g_{22} = R_n^2, \qquad g_{33} = 1 \tag{8.17}
$$

The functional determinant then follows directly as

$$
\sqrt{g} = \sqrt{|g_{ij}|} = R_s R_n \tag{8.18}
$$

We know from Chapter M1 that, in orthonormal systems such as the natural coordinate system of the tangential plane, there is no difference between covariant and contravariant unit vectors and the corresponding physical measure numbers. The covariant and contravariant basis vectors, however, differ and can be easily formulated with the help of (8.17).

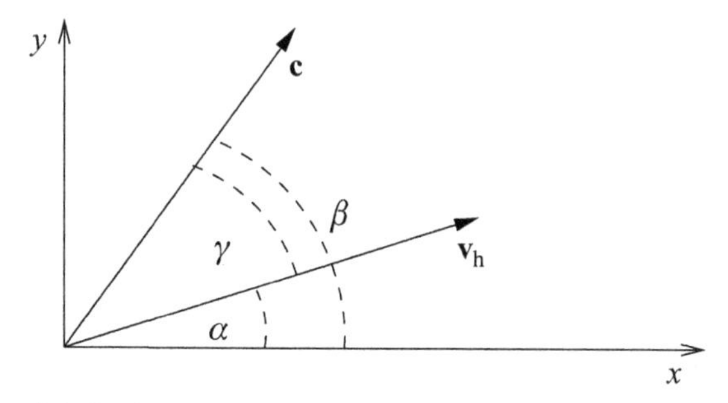

Fig. 8.3 Definitions of angles appearing in equation (8.23).

8.4 Blaton's equation

Blaton's equation gives a relation between the radii of curvature of the streamline and the trajectory. The Euler development of the contingency angle α gives

$$\begin{aligned} \frac{d\alpha}{dt} &= \left(\frac{\partial \alpha}{\partial t}\right)_{n,s,z} + \mathbf{v}_h \cdot \nabla \alpha \\ &= \left(\frac{\partial \alpha}{\partial t}\right)_{n,s,z} + \overset{*}{v}_h \frac{\partial \alpha}{\partial s} = \left(\frac{\partial \alpha}{\partial t}\right)_{n,s,z} + \frac{\overset{*}{v}_h}{R_s} \end{aligned} \tag{8.19}$$

where use has been made of equations (8.2) and (8.11). By writing $d\alpha/dt$ as

$$\frac{d\alpha}{dt} = \frac{d\alpha}{ds_T}\frac{ds_T}{dt} = \frac{\overset{*}{v}_h}{R_T} \tag{8.20}$$

we find the desired result

$$\boxed{\left(\frac{\partial \alpha}{\partial t}\right)_{n,s,z} = \overset{*}{v}_h (K_T - K_s)} \tag{8.21}$$

The quantities $K_T = 1/R_T$ and $K_s = 1/R_s$ are the curvatures of the trajectory and the streamline. In case of *directional stationarity* Blaton's equation (8.21) reduces to $K_T = K_s$.

It might be instructive to apply (8.21) to the case of a nondeveloping field moving with the phase speed $\mathbf{c}$. From (7.37) and with reference to Figure 8.3 we obtain

$$\left(\frac{\partial \alpha}{\partial t}\right)_{x,y} = \left(\frac{\partial \alpha}{\partial t}\right)_{x',y'} - \mathbf{c} \cdot \nabla \alpha = -c \cos \beta \frac{\partial \alpha}{\partial x} - c \sin \beta \frac{\partial \alpha}{\partial y} \tag{8.22a}$$

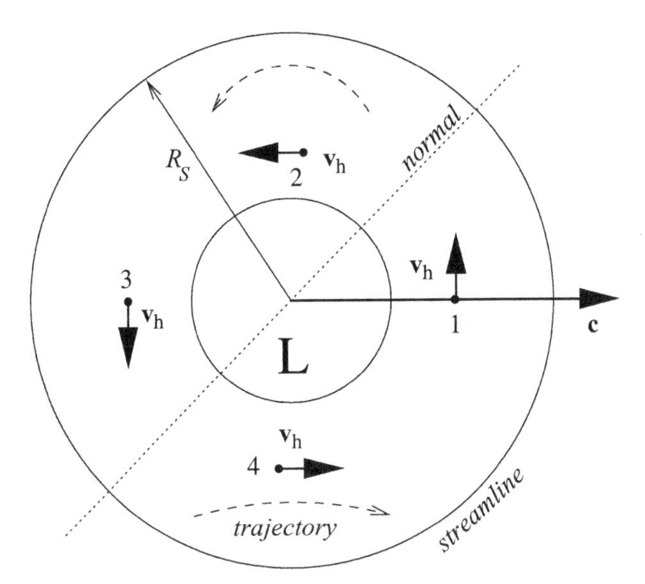

Fig. 8.4 Curvatures of the trajectory at various points within a speedily moving circular cyclone.

The local time derivative at constant x', y' vanishes due to the assumed stationarity. Using (8.12b) to replace $\partial\alpha/\partial x$ and $\partial\alpha/\partial y$ in terms of $\partial\alpha/\partial s = K_s$ and $\partial\alpha/\partial n = K_n$ we obtain

$$\left(\frac{\partial\alpha}{\partial t}\right)_{n,s} = -c(K_s \cos\gamma + K_n \sin\gamma) \tag{8.22b}$$

where $K_n = 1/R_n$ is the curvature of the normal. The relation between α, β and the angle γ, as defined in Figure 8.3. is found from the addition theorems of the trigonometric functions. Combining (8.21) and (8.22b) finally gives

$$\boxed{K_T = K_s - \frac{c}{\overset{*}{v}_{\rm h}}(K_s \cos\gamma + K_n \sin\gamma)} \tag{8.23}$$

We will conclude this section by applying (8.23) to the situation of a speedily moving cyclone for which $c/\overset{*}{v}_{\rm h} > 1$. Figure 8.4 demonstrates the situation.

Obviously, the curvature K_n of the normal is zero. The curvature K_s of the low-pressure system L is positive by convention. The curvature of the trajectory K_T and its sign may then be easily found from (8.23). At points 1 ($\gamma = 90°$) and 3 ($\gamma = 270°$) the curvatures of the trajectory and the streamline coincide, i.e. $K_T = K_s$. At point 2 ($\gamma = 180°$) the curvature of the trajectory K_T exceeds the

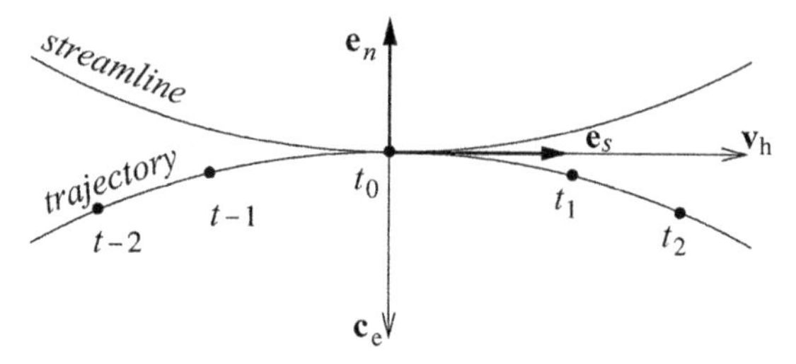

Fig. 8.5 Tangential and centripetal acceleration.

cyclonic curvature of the streamline K_s, whereas at point 4 ($\gamma = 0°$) the curvature of the trajectory is anticyclonic, that is $K_T < 0$. For an additional discussion of this topic see, for example, Petterssen (1956).

8.5 Individual and local time derivatives of the velocity

The individual derivative of (8.3) gives

$$\frac{d\mathbf{v}_h}{dt} = \mathbf{e}_s \frac{d\overset{*}{v}_h}{dt} + \frac{d\mathbf{e}_s}{dt}\overset{*}{v}_h \tag{8.24}$$

From (8.9) and (8.2) we find

$$\frac{d\mathbf{e}_s}{dt} = \mathbf{e}_n \frac{d\alpha}{ds_T}\frac{ds_T}{dt} = \mathbf{e}_n \frac{\overset{*}{v}_h}{R_T} \tag{8.25}$$

Substitution of (8.25) into (8.24) results in

$$\frac{d\mathbf{v}_h}{dt} = \mathbf{e}_s \frac{d\overset{*}{v}_h}{dt} + \mathbf{e}_n \frac{\overset{*}{v}{}_h^2}{R_T} \tag{8.26}$$

The first term on the right-hand side of (8.26) has the direction of $\mathbf{v}_h$ and is known as the *tangential acceleration*. The second term has a direction perpendicular to $\mathbf{v}_h$ and represents the *centripetal acceleration* $\mathbf{c}_e$ which is acting on an air particle whenever the trajectory is not straight-line flow; see Figure 8.5.

We proceed similarly with the partial time derivative of the velocity. From (8.9c) and with the help of Blaton's equation (8.21) we first obtain

$$\left(\frac{\partial \mathbf{e}_s}{\partial t}\right)_{n,s,z} = \mathbf{e}_n \left(\frac{\partial \alpha}{\partial t}\right)_{n,s,z} = \mathbf{e}_n \overset{*}{v}_h (K_T - K_s) \tag{8.27}$$

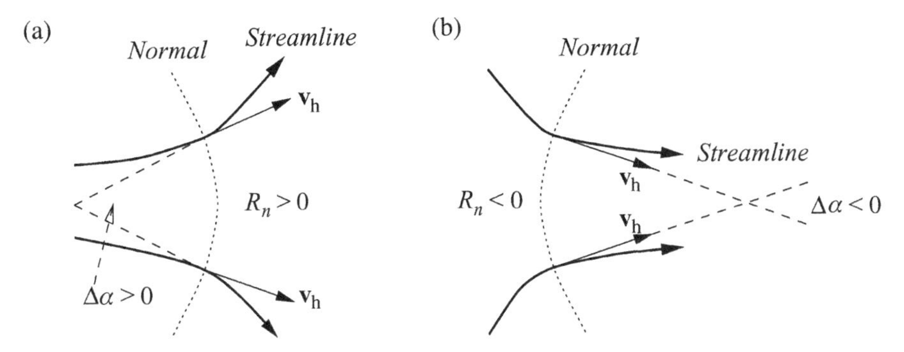

Fig. 8.6 Regions of (a) diffluence and (b) confluence.

Using this equation, we find from (8.3) for the local time change in the natural coordinate system the expression

$$\left(\frac{\partial \mathbf{v}_h}{\partial t}\right)_{n,s,z} = \mathbf{e}_s \left(\frac{\partial \overset{*}{v}_h}{\partial t}\right)_{n,s,z} + \mathbf{e}_n \overset{*}{v}_h^2 (K_T - K_s) \tag{8.28}$$

which is needed whenever the Euler expansion is required.

8.6 Differential invariants

8.6.1 The horizontal divergence of the velocity

The starting point of the derivation is the development of $\nabla \cdot \mathbf{v}_h$ in the natural coordinate system:

$$\nabla \cdot \mathbf{v}_h = \left(\mathbf{e}_s \frac{\partial}{\partial s} + \mathbf{e}_n \frac{\partial}{\partial n}\right) \cdot (\mathbf{e}_s \overset{*}{v}_h) \tag{8.29}$$

Since this system is orthonormal, one immediately obtains

$$\nabla \cdot \mathbf{v}_h = \mathbf{e}_n \cdot \frac{\partial \mathbf{e}_s}{\partial n} \overset{*}{v}_h + \frac{\partial \overset{*}{v}_h}{\partial s} = \frac{\overset{*}{v}_h}{R_n} + \frac{\partial \overset{*}{v}_h}{\partial s} \tag{8.30}$$

where (8.9d) and (8.9e) have been used to evaluate the partial derivatives. The following abbreviations are introduced:

$$D_d = \frac{\overset{*}{v}_h}{R_n}, \qquad D_v = \frac{\partial \overset{*}{v}_h}{\partial s} \implies \nabla \cdot \mathbf{v}_h = D_d + D_v \tag{8.31}$$

The part D_d is known as the *divergence due to directional change* while D_v refers to the *velocity divergence*. The physical interpretation of (8.31) follows from Figure 8.6 showing regions of *diffluence* and *confluence* as well as the sign convention for the radius of curvature of the normal. In the region of diffluence the curvature is positive in the mathematical sense so that $R_n > 0$, whereas in the region of confluence the curvature of the normal is negative or $R_n < 0$.

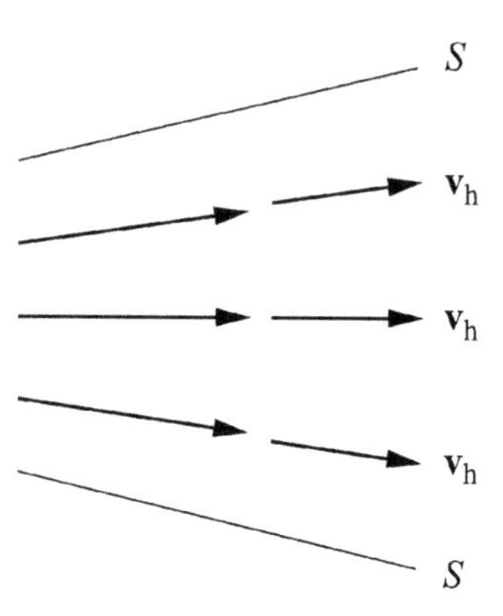

Fig. 8.7 A region of diffluence in the atmosphere, showing both parts of the divergence with $D_\mathrm{d} > 0$ and $D_\mathrm{v} < 0$.

In large-scale atmospheric motion the horizontal divergence is very small and of the order of $|\nabla \cdot \mathbf{v}_\mathrm{h}| = 10^{-5}$–$10^{-6}\ \mathrm{s}^{-1}$. This results from the fact that $|\nabla \cdot \mathbf{v}_\mathrm{h}|$, in general, is composed of the two parts D_d and D_v which nearly compensate each other. Therefore, it is very difficult to measure the horizontal velocity divergence. The idea is demonstrated in Figure 8.7.

8.6.2 Vorticity or the vertical component of $\nabla \times \mathbf{v}_\mathrm{h}$

The concept of vorticity is very important in meteorology since the vorticity is a measure of rotation. Here we will only briefly dwell on this subject, but in later chapters we will exploit it fully since it is closely related to atmospheric circulation. In the natural coordinate system the components of $\nabla \times \mathbf{v}_\mathrm{h}$ are given by

$$\nabla \times \mathbf{v}_\mathrm{h} = \left(\mathbf{e}_s \frac{\partial}{\partial s} + \mathbf{e}_n \frac{\partial}{\partial n}\right) \times (\mathbf{e}_s \overset{*}{v}_\mathrm{h}) = \mathbf{e}_s \times \frac{\partial \mathbf{e}_s}{\partial s} \overset{*}{v}_\mathrm{h} + \mathbf{e}_n \times \mathbf{e}_s \frac{\partial \overset{*}{v}_\mathrm{h}}{\partial n} \tag{8.32}$$

where use has been made of equations (8.9d) and (8.9e). The vertical component of this expression is then

$$\zeta = \mathbf{e}_z \cdot \nabla \times \mathbf{v}_\mathrm{h} = \frac{\overset{*}{v}_\mathrm{h}}{R_s} - \frac{\partial \overset{*}{v}_\mathrm{h}}{\partial n} \tag{8.33}$$

where again (8.9d) has been utilized. The vorticity ζ consists of two parts, caused by the curvature of the streamlines ζ_c and the velocity shear ζ_s.

$$\zeta = \zeta_\mathrm{c} + \zeta_\mathrm{s} = \frac{\overset{*}{v}_\mathrm{h}}{R_s} - \frac{\partial \overset{*}{v}_\mathrm{h}}{\partial n} \tag{8.34}$$

As has already been mentioned, in the northern hemisphere the radius of curvature of the streamlines R_s is defined to be positive for cyclonic flow so that $R_s > 0$,

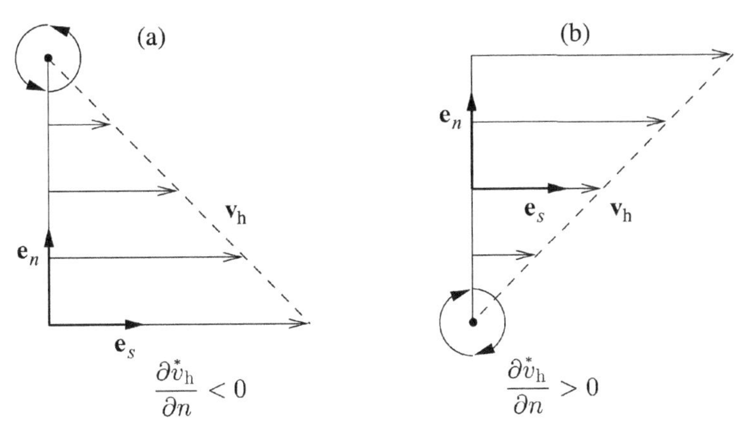

Fig. 8.8 The sign convention for the shear vorticity in northern-hemispheric flow: (a) cyclonic wind shear, and (b) anticyclonic wind shear.

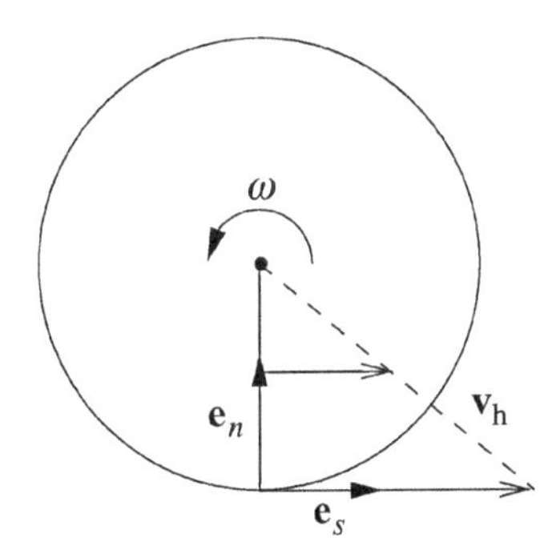

Fig. 8.9 The vorticity of rigid rotation of a disk.

whereas for anticyclonic flow $R_s < 0$; see Figure 8.2. The physical meaning and the sign convention for the shear vorticity are shown in Figure 8.8.

Finally, we give a very brief example relating to the vorticity of rigid rotation of a circular disk with $\mathring{v}_h = \omega R_s$; see Figure 8.9. The simple calculation is outlined in equation (8.35), which does not require any additional comment:

$$\zeta_c = \frac{\omega R_s}{R_s} = \omega, \qquad \zeta_s = -\frac{\partial \mathring{v}_h}{\partial n} = \frac{\partial \mathring{v}_h}{\partial R_s} = \omega \implies \zeta = 2\omega \tag{8.35}$$

In large-scale atmospheric motion the order of magnitude of the vorticity $|\zeta|$ is $\approx 10^{-4}$–10^{-5} s^{-1}.

8.6.3 The Jacobian operator and the Laplacian

Since we are not going to use these operators, we will not derive them; but we will state them for reference. They can be derived from the general formulas which were given previously. Checking the validity of the formulas

$$
\begin{aligned}
J(u,v) &= \frac{\partial u}{\partial x}\frac{\partial v}{\partial y} - \frac{\partial u}{\partial y}\frac{\partial v}{\partial x} = D_{\mathrm{v}} D_{\mathrm{d}} + \zeta_{\mathrm{c}}\zeta_{\mathrm{s}} \\
\nabla^2\psi &= \frac{\partial^2\psi}{\partial s^2} + \frac{\partial^2\psi}{\partial n^2} + \frac{1}{R_n}\frac{\partial\psi}{\partial s} - \frac{1}{R_s}\frac{\partial\psi}{\partial n}
\end{aligned}
\tag{8.36}
$$

will be left for the exercises.

8.7 The equation of motion for frictionless horizontal flow

After having taken some interesting but necessary detours, we are now ready to find the equation of motion for frictionless horizontal flow in the natural coordinate system. We proceed by transforming each term of the equation of motion (2.38) in the Cartesian system

$$
\begin{aligned}
\frac{d\mathbf{v}_{\mathrm{h}}}{dt} &= -\frac{1}{\rho}\nabla p - \nabla\phi - f\mathbf{e}_z \times \mathbf{v}_{\mathrm{h}} \\
&= \mathbf{e}_s \frac{d\overset{*}{v}_{\mathrm{h}}}{dt} + \mathbf{e}_n \frac{\overset{*}{v}{}_{\mathrm{h}}^2}{R_T}
\end{aligned}
\tag{8.37}
$$

The last equation follows from (8.26). The individual derivative of the first term on the right-hand side can be expanded with the help of (8.11) to give

$$
\frac{d\overset{*}{v}_{\mathrm{h}}}{dt} = \left(\frac{\partial \overset{*}{v}_{\mathrm{h}}}{\partial t}\right)_{n,s,z} + \overset{*}{v}_{\mathrm{h}} \frac{\partial \overset{*}{v}_{\mathrm{h}}}{\partial s} = \left(\frac{\partial \overset{*}{v}_{\mathrm{h}}}{\partial t}\right)_{n,s,z} + \frac{\partial}{\partial s}\left(\frac{\overset{*}{v}{}_{\mathrm{h}}^2}{2}\right)_{n,z,t}
\tag{8.38}
$$

The local derivative is explicitly given by (8.28).

The gradient of an arbitrary field function ψ is specified by (8.10) and can be directly applied to the pressure term and to the geopotential. Assuming that the geopotential is approximated by $\phi = gz$, we find

$$
\nabla\phi = g\mathbf{e}_z
\tag{8.39}
$$

The Coriolis term can be written as

$$
-f\mathbf{e}_z \times \mathbf{v}_{\mathrm{h}} = -f\mathbf{e}_z \times \overset{*}{v}_{\mathrm{h}}\mathbf{e}_s = -\mathbf{e}_n f \overset{*}{v}_{\mathrm{h}}
\tag{8.40}
$$

Using these transformations, we find the desired equation of motion in the natural coordinate system:

$$\boxed{\mathbf{e}_s \frac{d\overset{*}{v}_\mathrm{h}}{dt} + \mathbf{e}_n \frac{\overset{*}{v}_\mathrm{h}^2}{R_T} = -\frac{1}{\rho}\nabla p - g\mathbf{e}_z - \mathbf{e}_n f \overset{*}{v}_\mathrm{h}} \tag{8.41}$$

The s-component of this equation yields a prognostic equation for the magnitude of the velocity:

$$\left(\frac{\partial \overset{*}{v}_\mathrm{h}}{\partial t}\right)_{n,s,z} + \frac{\partial}{\partial s}\left(\frac{\overset{*}{v}_\mathrm{h}^2}{2}\right)_{n,z,t} = -\frac{1}{\rho}\left(\frac{\partial p}{\partial s}\right)_{n,z,t} \tag{8.42}$$

The n-component of (8.41) results in a diagnostic relation involving the balance of the centripetal force, the pressure gradient force, and the Coriolis force:

$$\frac{\overset{*}{v}_\mathrm{h}^2}{R_T} = -\frac{1}{\rho}\frac{\partial p}{\partial n} - f\overset{*}{v}_\mathrm{h} \tag{8.43}$$

while the vertical component of (8.41) is the hydrostatic equation

$$0 = -\frac{1}{\rho}\frac{\partial p}{\partial z} - g \tag{8.44}$$

8.8 The gradient wind relation

If the flow is solely governed by (8.43), we obtain the gradient wind relations. With the abbreviation $v_\mathrm{h} = \overset{*}{v}_\mathrm{h} > 0$ equation (8.43) changes into

$$\frac{v_\mathrm{h}^2}{R_T} + f v_\mathrm{h} + \frac{1}{\rho}\frac{\partial p}{\partial n} = 0 \tag{8.45}$$

Note that $\partial p/\partial n < 0$. First of all, we observe that, if $R_T \longrightarrow \infty$, then we obtain the geostrophic flow, which has already been discussed in an earlier chapter:

$$v_\mathrm{g} = -\frac{1}{\rho f}\frac{\partial p}{\partial n} > 0, \qquad R_T \longrightarrow \infty \tag{8.46}$$

Solving the quadratic equation (8.43) results in the gradient wind equations for curved *cyclonic* or *anticyclonic flow*:

$$\begin{aligned} \text{cyclonic flow:} \quad v_\mathrm{h} &= -\frac{f R_T}{2} + \sqrt{\frac{f^2 R_T^2}{4} - \frac{R_T}{\rho}\frac{\partial p}{\partial n}}, \qquad R_T > 0 \\ \text{anticyclonic flow:} \quad v_\mathrm{h} &= -\frac{f R_T}{2} - \sqrt{\frac{f^2 R_T^2}{4} - \frac{R_T}{\rho}\frac{\partial p}{\partial n}}, \qquad R_T < 0 \end{aligned} \tag{8.47}$$

Note that $\mathbf{v}_\mathrm{h}$ is always positive in the natural coordinate system. The type of flow is determined by the sign convention which we introduced previously. To prevent the wind speed from assuming complex values, we must assure that the argument of the root does not become negative for $R_T < 0$. Therefore, the wind speed around the high-pressure system is limited to the maximum value

$$v_\mathrm{h}(\max) = -\frac{f R_T}{2}, \qquad R_T < 0 \tag{8.48}$$

If the pressure gradient force approximately balances the centripetal force then we obtain the so-called *cyclostrophic wind* given by

$$v_\mathrm{cycl} = \sqrt{-\frac{R_T}{\rho}\frac{\partial p}{\partial n}}, \qquad R_T > 0 \tag{8.49}$$

This type of flow is important near the centers of tropical cyclones at low latitudes, where the centripetal force may outweigh the Coriolis force by as much as 25 to 1 as remarked by Byers (1959), so that the term $f v_\mathrm{h}$ in (8.45) may be neglected. Similar discussions on gradient flow may be found in various textbooks, e.g. Hess (1959).

8.9 Problems

8.1: Let α represent the angle between the x-axis and the horizontal wind vector $\mathbf{v}_\mathrm{h}$. Curves along which α = constant and $|\mathbf{v}_\mathrm{h}|$ = constant are called *isogons* and *isotachs*, respectively. Consider the auxiliary vectors $\mathbf{G}$ and $\mathbf{H}$ shown in the figure.

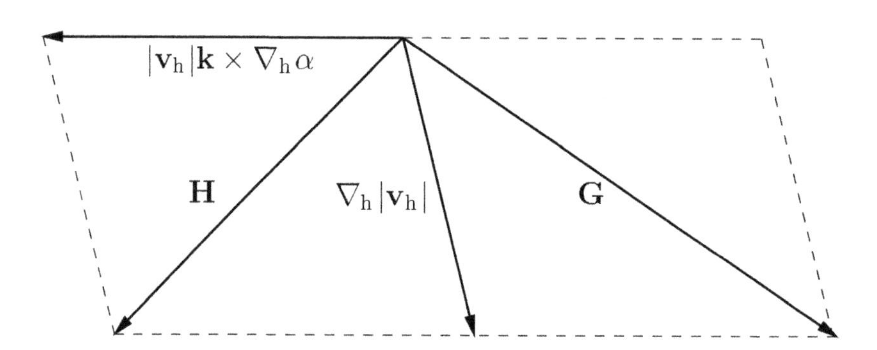

Use natural coordinates to show that
(a) the divergence is given by $D = \mathbf{G} \cdot \mathbf{e}_s$,
(b) the vorticity is given by $\zeta = -\mathbf{G} \cdot \mathbf{e}_n$,
(c) $D^2 + \zeta^2 = \mathbf{G}^2$.

(d) Let $A = \partial u/\partial x - \partial v/\partial y$ and $B = \partial u/\partial y + \partial v/\partial x$. Show that

$$\begin{aligned} A &= \mathbf{H} \cdot [(\cos(2\alpha)\,\mathbf{e}_s - \sin(2\alpha)\,\mathbf{e}_n)] \\ B &= \mathbf{H} \cdot [(\sin(2\alpha)\,\mathbf{e}_s + \cos(2\alpha)\,\mathbf{e}_n)] \\ \mathbf{H}^2 &= \mathbf{H} \cdot \mathbf{H} = A^2 + B^2 = def^2 \end{aligned}$$

(e) Show that

$$\begin{aligned} \mathbf{G} &= D\mathbf{e}_s - \zeta\mathbf{e}_n \\ H &= [A\cos(2\alpha) + B\sin(2\alpha)]\mathbf{e}_s + [-A\sin(2\alpha) + B\cos(2\alpha)]\mathbf{e}_n \end{aligned}$$

(f) Use the answer to part (e) to demonstrate that the horizontal wind vector $\mathbf{v}_{\mathrm{h}}$ can be expressed in terms of the kinematic fields.

Hint: Consider the arbitrary vector $\mathbf{A} = A_s\mathbf{e}_s + A_n\mathbf{e}_n$. Show that $\mathbf{e}_s = (A_n\mathbf{A} \times \mathbf{k} + A_s\mathbf{A})/(A_n^2 + A_s^2)$. Replace $\mathbf{A}$ by $\mathbf{G}$ to show that

$$\mathbf{v}_{\mathrm{h}} = \frac{|\mathbf{v}_{\mathrm{h}}|}{D^2 + \zeta^2}[D(\nabla_{\mathrm{h}}\,|\mathbf{v}_{\mathrm{h}}| - |\mathbf{v}_{\mathrm{h}}|\,\mathbf{k} \times \nabla_{\mathrm{h}}\alpha) + \zeta(\mathbf{k} \times \nabla_{\mathrm{h}}\,|\mathbf{v}_{\mathrm{h}}| + |\mathbf{v}_{\mathrm{h}}|\,\nabla_{\mathrm{h}}\alpha)]$$

8.2: Suppose that $\mathbf{v}_{\mathrm{h}} \cdot \nabla_{\mathrm{h}}\mathbf{v}_{\mathrm{h}} = 0$. Show that this statement is equivalent to

$$D\nabla_{\mathrm{h}}\,|\mathbf{v}_{\mathrm{h}}| = -|\mathbf{v}_{\mathrm{h}}|\,\zeta\,\nabla_{\mathrm{h}}\alpha$$

Hint: Find two expressions for $\mathbf{v}_{\mathrm{h}} \cdot \nabla_{\mathrm{h}}\mathbf{v}_{\mathrm{h}}$. Obtain one of them with the help of Lamb's development, which is also valid for the two-dimensional case.

8.3: Verify equation (8.36).

9

Boundary surfaces and boundary conditions

9.1 Introduction

The continuity equation and the equation of motion are applicable only to fluid regions in which the physical variables change in a continuous fashion. Only in these regions is it possible to form the required derivatives of the variables as they appear in the various terms of the prognostic and diagnostic equations. However, there exist external boundary surfaces at which the fluid is constrained by a wall or bounded by a vacuum, where the field functions or their nth derivative experience discontinuous changes. Such surfaces are called *discontinuity surfaces* (DSs). At external as well as internal boundary surfaces the continuity equation and the equation of motion must be replaced by the so-called kinematic and dynamic boundary-surface conditions.

It is customary to classify the DS according to its order. A boundary surface is said to be of nth order if the lowest discontinuous derivative of the field function being considered is of nth order. Let the symbol $\{\psi\}$ represent the jump experienced by the field function ψ at the DS as shown in Figure 9.1, so that

$$\psi^{(2)} - \psi^{(1)} = \{\psi\} \tag{9.1}$$

A boundary surface of nth-order discontinuity is then defined by

$$\{\psi\} = 0, \qquad \left\{\frac{\partial \psi}{\partial s}\right\} = 0, \qquad \left\{\frac{\partial^2 \psi}{\partial s^2}\right\} = 0, \qquad \ldots, \qquad \left\{\frac{\partial^n \psi}{\partial s^n}\right\} \neq 0 \tag{9.2}$$

Actual discontinuities do not form in the atmosphere but there are narrow zones of transition between two air masses, which, in large-scale motion, may be viewed as discontinuities. Consider, for example, an idealized warm front or a cold front that is a DS of order zero in terms of temperature, density, and wind, that is $\{T\} \neq 0$, $\{\rho\} \neq 0$, $\{\mathbf{v}\} \neq 0$; see Figure 9.2.

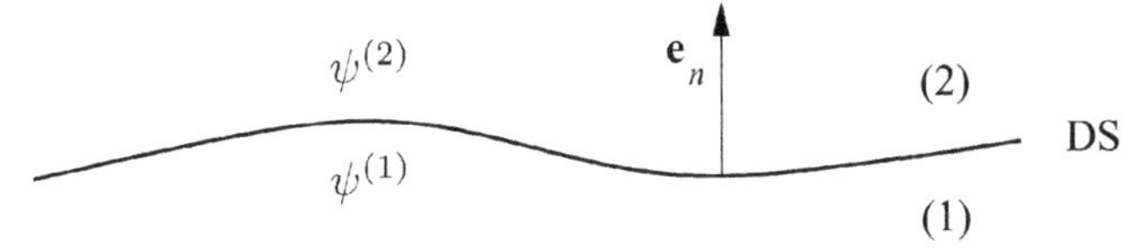

Fig. 9.1 A jump of the field function at a discontinuity surface.

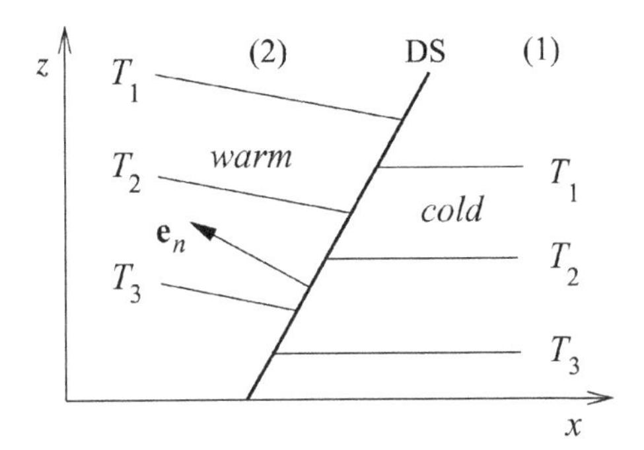

Fig. 9.2 A discontinuity surface DS of order zero.

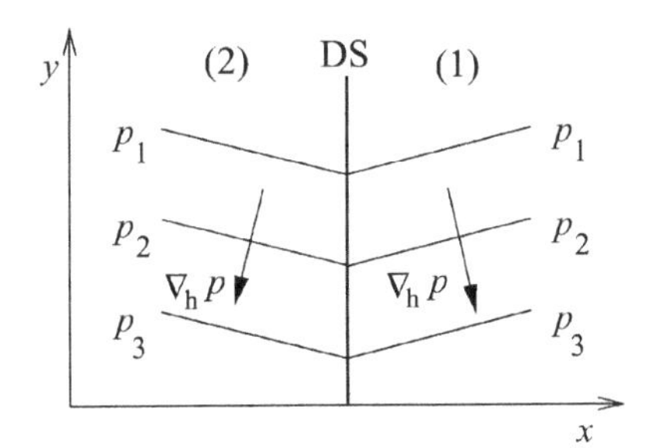

Fig. 9.3 A first-order DS relative to pressure.

A discontinuity surface of order one is defined by the condition that the first spatial derivative experiences a jump. The tropopause, for example, is a first-order DS relative to temperature, that is $\{T\} = 0$ and $\{\partial T/\partial z\} \neq 0$. Frontal surfaces are DSs of first order relative to pressure with $\{p\} = 0$ and $\{\nabla_{\mathrm{h}} p\} \neq 0$; see Figure 9.3. Before we proceed with our discussion, it is imperative to define various differential operators applicable to the DS.

9.2 Differential operations at discontinuity surfaces

We consider an arbitrary extensive field function ψ, which is assumed to be discontinuous at the DS so that we are dealing with a zeroth-order DS. Utilizing the

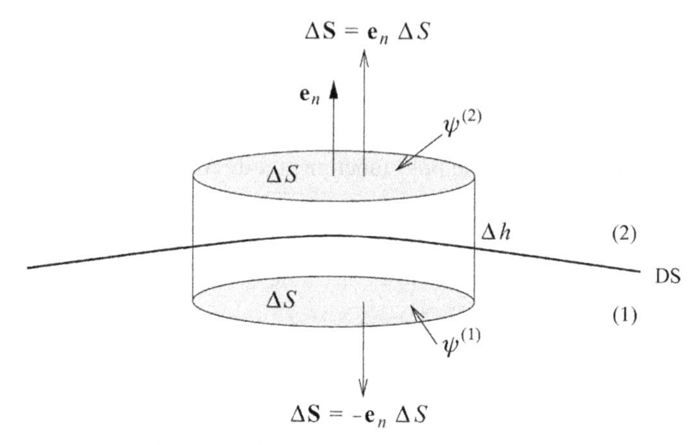

Fig. 9.4 The definition of the surface gradient.

general integral theorem (M6.30), in a region where ψ changes in a continuous fashion, the gradient of ψ may be written as

$$\nabla\psi = \lim_{\Delta\tau\to 0} \frac{1}{\Delta\tau} \oint_S \psi \, d\mathbf{S} \tag{9.3}$$

where $\Delta\tau$ is an infinitesimally small volume with surface S. In order to account for the discontinuity of the field function at the DS, we introduce the *surface gradient* in analogy to (9.3) by means of

$$\| \, \psi = \lim_{\Delta S\to 0} \frac{1}{\Delta S} \oint_S \psi \, d\mathbf{S} \tag{9.4}$$

where the surface integral is taken over the upper and lower surfaces of the small cylindrical volume of infinitesimal thickness Δh; see Figure 9.4.

The contributions by the sides $\Delta h \to 0$ of the pillbox to the surface integral are considered negligible. Integration of (9.4) results in

$$\| \, \psi = \mathbf{e}_n(\psi^{(2)} - \psi^{(1)}) = \mathbf{e}_n\{\psi\} \tag{9.5}$$

It should be noted that the superscripts (1) and (2) simply mean that the function ψ is taken directly at the corresponding sides of the DS. The operator $\| \, \psi$ at the DS replaces Hamilton's nabla operator, which is valid only in the continuous region of the fluid. Therefore, (9.5) is often called the *surface Hamilton operator*.

For the arbitrary vector $\mathbf{\Psi}$ we formally define the *surface divergence* for the zeroth-order DS by

$$\boxed{\| \cdot \mathbf{\Psi} = \mathbf{e}_n \cdot \{\mathbf{\Psi}\}} \tag{9.6}$$

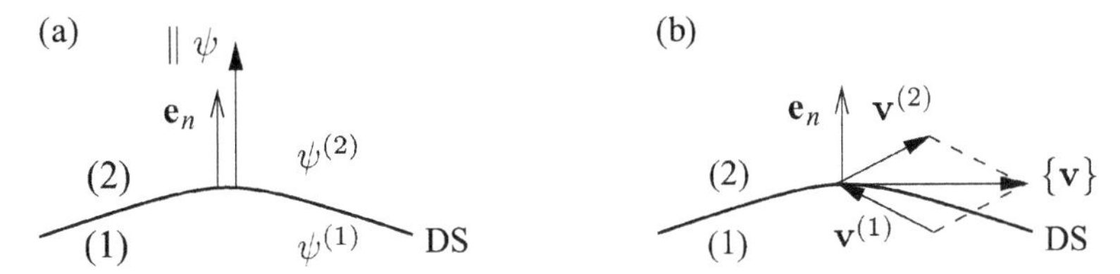

Fig. 9.5 Surface Hamilton operators for (a) the scalar field function ψ and (b) the vectorial field function $\mathbf{v}$.

The *surface rotation* may be written as

$$\boxed{\| \times \mathbf{\Psi} = \mathbf{e}_n \times \{\mathbf{\Psi}\}} \tag{9.7}$$

Some brief examples will clarify the concept. Figure 9.5(a) shows the gradient of the scalar field function ψ. If ψ represents the velocity vector $\mathbf{v}$ then the surface gradient, the divergence, and the rotation are defined by

$$\| \, \mathbf{v} = \mathbf{e}_n\{\mathbf{v}\}, \qquad \| \cdot \mathbf{v} = \mathbf{e}_n \cdot \{\mathbf{v}\}, \qquad \| \times \mathbf{v} = \mathbf{e}_n \times \{\mathbf{v}\} \tag{9.8}$$

We will now consider the following situations.

(i) $\mathbf{v}$ is source-free on the DS

Then we have

$$\| \cdot \mathbf{v} = \mathbf{e}_n \cdot \{\mathbf{v}\} = 0 \tag{9.9}$$

Since by assumption $\{\mathbf{v}\} \neq 0$, we must conclude that $\mathbf{e}_n$ is perpendicular to $\{\mathbf{v}\}$, meaning that the velocity jump is located in the plane tangential to the DS; see Figure 9.5(b). We may also write

$$\mathbf{e}_n \cdot \mathbf{v}^{(1)} = \mathbf{e}_n \cdot \mathbf{v}^{(2)} \tag{9.10}$$

showing that the normal components of the velocity vector $\mathbf{v}$ are continuous on the DS.

(ii) $\mathbf{v}$ is irrotational on the DS

Then we have

$$\| \times \mathbf{v} = \mathbf{e}_n \times \{\mathbf{v}\} = 0 \tag{9.11}$$

From this it follows that $\{\mathbf{v}\}$ is parallel or antiparallel to $\mathbf{e}_n$, that is perpendicular to the DS. By writing (9.11) in the form

$$\mathbf{e}_n \times \mathbf{v}^{(1)} = \mathbf{e}_n \times \mathbf{v}^{(2)} \tag{9.12}$$

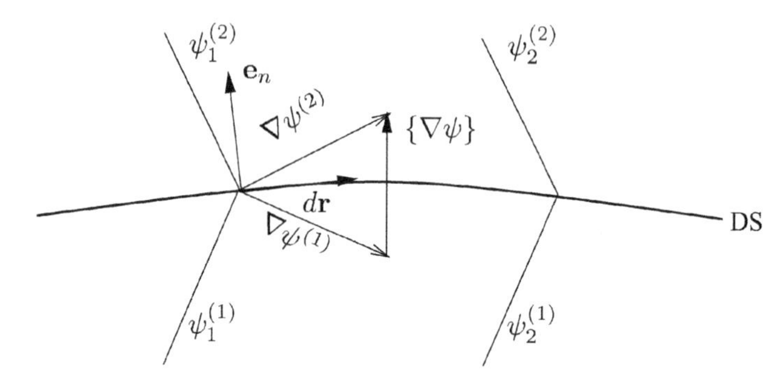

Fig. 9.6 Relevant quantities for the proof of (9.14).

we find that the tangential components of the velocity vector are continuous on the DS.

$$\text{(iii)} \qquad \| \cdot \mathbf{v} = 0 \text{ and } \| \times \mathbf{v} = 0.$$

The conditions $\| \cdot \mathbf{v} = 0$ and $\| \times \mathbf{v} = 0$ cannot be satisfied simultaneously on a zeroth-order DS since this would require $\{\mathbf{v}\} = 0$, which contradicts our assumption.

This section will be concluded by proving that, on a first-order DS, as defined by

$$\{\psi\} = 0, \qquad \{\nabla\psi\} \neq 0 \tag{9.13}$$

the surface rotation of the gradient must vanish:

$$\| \times \nabla\psi = \mathbf{e}_n \times \{\nabla\psi\} = 0 \tag{9.14}$$

For the proof consider Figure 9.6, where the relevant quantities are shown. The geometric change $d_g\psi$ on both sides of the DS can be represented with the help of the displacement vector $d\mathbf{r}$ which is located along the DS. The gradients of the field function on both sides of the DS are also shown in Figure 9.6, as is the jump of the gradient $\{\nabla\psi\}$. The geometric changes are given by

$$d_g\psi^{(i)} = d\mathbf{r} \cdot \nabla\psi^{(i)}, \qquad i = 1, 2 \tag{9.15}$$

Since the field function ψ is continuous on the DS, the geometric changes are the same on both sides, so we obtain

$$\{d_g\psi\} = d\mathbf{r} \cdot \{\nabla\psi\} = 0 \quad \text{or} \quad \mathbf{e}_n \times \{\nabla\psi\} = 0, \qquad \{\nabla\psi\} \neq 0 \tag{9.16}$$

Therefore, the jump $\{\nabla\psi\}$ is orthogonal to the DS as shown in Figure 9.6. This completes the proof.

9.3 Particle invariance at boundary surfaces, displacement velocities

The assumption of particle invariance at a boundary surface implies that the DS is composed of the same group of particles for as long as it exists. If a particle is a part of the DS it has to remain in the DS; it cannot penetrate the surface. Thus, the DS is a material or fluid surface: see also Section M6.4. Otherwise the particle would experience an infinitely large variation of its scalar value, say $\Delta T > 0$, so that $\Delta T/\Delta h \to \infty$. In order to realize the assumption of particle invariance, the normal velocity on both sides of the DS must be the same, i.e. $\mathbf{v}^{(1)} \cdot \mathbf{e}_n = \mathbf{v}^{(2)} \cdot \mathbf{e}_n$. Let us consider a DS defined by

$$z = z_{\mathrm{DS}}(x, y, t) \tag{9.17}$$

The function

$$F(x, y, z, t) = z - z_{\mathrm{DS}}(x, y, t) = 0 \tag{9.18}$$

may be considered to be the defining equation of the DS. Since all particles of the DS must remain within the DS, we may write the *condition of particle invariance* as

$$\boxed{\begin{gathered} F = 0, \qquad \frac{dF}{dt} = 0 \Longrightarrow \\ \left(\frac{dF}{dt}\right)^{(i)} = \frac{\partial F}{\partial t} + \mathbf{v}^{(i)} \cdot \nabla F = \frac{\partial F}{\partial t} + \mathbf{v}^{(i)} \cdot \mathbf{e}_n \, |\nabla F| = 0, \qquad i = 1, 2 \end{gathered}} \tag{9.19}$$

The last expression follows from the fact that ∇F is perpendicular to the DS. By evaluating (9.19) for $i = 1$ and $i = 2$ and subtracting one of the results from the other we again obtain (9.10).

The displacement velocity $\mathbf{c}$ of the DS and the unit normal $\mathbf{e}_n$ to the DS are considered positive if they are pointing from side (1) to (2); see Figure 9.7. In order to keep the mathematical analysis as simple as possible, we have arranged the coordinate system in such a way that the trace of the DS (front) is parallel to the y-axis. The unit vector $\mathbf{i}$ which is perpendicular to the trace of the DS is pointing in the direction of the rising boundary surface. The angle α defines the inclination of the DS.

The relation between the unit vector $\mathbf{e}_n$ and the Cartesian vectors $\mathbf{i}$ and $\mathbf{k}$ is easily found from

$$\mathbf{e}_n = \mathbf{e}_n \cdot \mathbb{E} = \mathbf{e}_n \cdot \mathbf{ii} + \mathbf{e}_n \cdot \mathbf{jj} + \mathbf{e}_n \cdot \mathbf{kk} = -\sin\alpha\,\mathbf{i} + \cos\alpha\,\mathbf{k} \tag{9.20}$$

Owing to the particle invariance, the displacement velocity $\mathbf{c} = c\mathbf{e}_n$ of the DS must be equal to the normal component of the wind velocity

$$\mathbf{c} = (\mathbf{v}^{(i)} \cdot \mathbf{e}_n)\mathbf{e}_n = (-u^{(i)} \sin\alpha + w^{(i)} \cos\alpha)\mathbf{e}_n = c\mathbf{e}_n, \qquad i = 1, 2 \tag{9.21}$$

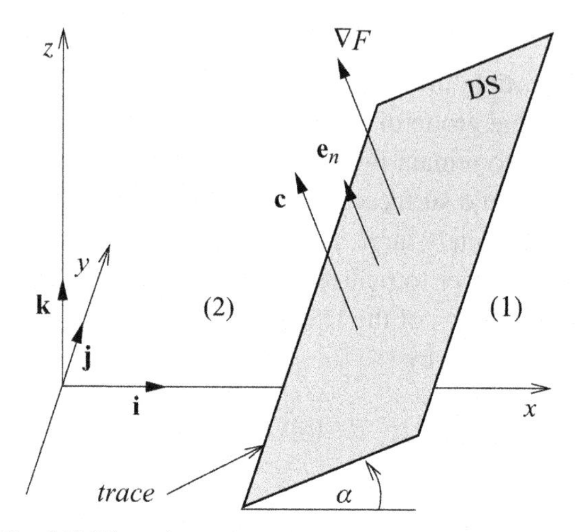

Fig. 9.7 The orientation of the discontinuity surface.

Moreover, the displacement velocity $\mathbf{c}$ can also be expressed by means of the particle invariance. From (9.19) we find

$$\mathbf{v}^{(i)} \cdot \mathbf{e}_n = -\frac{\partial F}{\partial t} |\nabla F|^{-1}, \qquad i = 1, 2 \tag{9.22}$$

so that the displacement velocity is also given by

$$\mathbf{c} = -\frac{\partial F}{\partial t} |\nabla F|^{-1} \mathbf{e}_n = -\frac{\partial F}{\partial t} (\mathbf{e}_n \cdot \nabla F)^{-1} \mathbf{e}_n \tag{9.23}$$

Of greater interest than the displacement velocity $\mathbf{c}$ itself is the horizontal displacement velocity $\mathbf{c}_\mathrm{h}$ shown in Figure 9.8. From the figure and from (9.23) we find for the horizontal displacement the relation

$$\mathbf{c}_\mathrm{h} = c_\mathrm{h}\mathbf{i} = -\frac{c}{\sin\alpha}\mathbf{i} = \frac{c}{\mathbf{i} \cdot \mathbf{e}_n}\mathbf{i} = -\frac{\partial F}{\partial t} (\mathbf{i} \cdot \nabla F)^{-1} \mathbf{i} \tag{9.24}$$

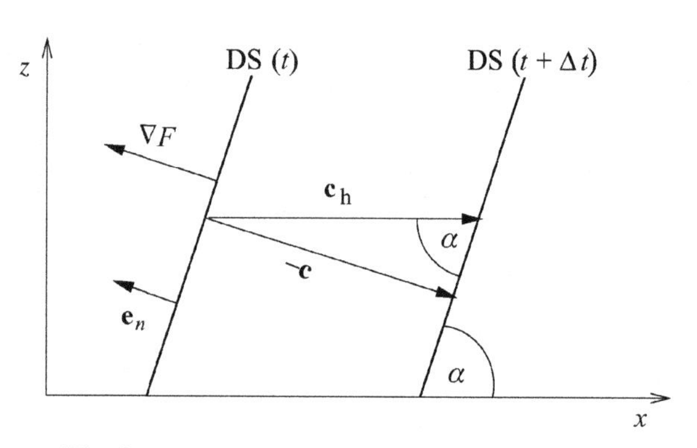

Fig. 9.8 The horizontal displacement velocity.

where the measure number $c_{\rm h}$ can have either sign. Owing to the special orientation of the coordinate system, the same result also follows on expanding the equation of particle invariance:

$$F = 0, \qquad dF = 0 = \frac{\partial F}{\partial t}\,dt + \frac{\partial F}{\partial x}\,dx + \frac{\partial F}{\partial z}\,dz \tag{9.25}$$

Application of this equation for z = constant yields the horizontal displacement velocity

$$F = 0, \quad z = \text{constant:} \qquad \left(\frac{dx}{dt}\right)_{\rm DS} = c_{\rm h} = -\frac{\partial F}{\partial t}\left(\frac{\partial F}{\partial x}\right)^{-1} = -\frac{\partial F}{\partial t}\,\frac{1}{(\mathbf{i}\cdot\nabla F)} \tag{9.26}$$

which is identical with (9.24). From (9.25) we also find the inclination of the DS:

$$F = 0, \quad t = \text{constant:} \qquad \left(\frac{dz}{dx}\right)_{\rm DS} = -\frac{\partial F}{\partial x}\left(\frac{\partial F}{\partial z}\right)^{-1} = -\frac{\mathbf{i}\cdot\nabla F}{\mathbf{k}\cdot\nabla F} = \tan\alpha \tag{9.27}$$

On substituting (9.21) into (9.24) we find the useful expression

$$\mathbf{c}_{\rm h} = -\frac{c}{\sin\alpha}\mathbf{i} = \frac{u^{(i)}\sin\alpha - w^{(i)}\cos\alpha}{\sin\alpha}\mathbf{i} = (u^{(i)} - w^{(i)}\cot\alpha)\mathbf{i}, \qquad i = 1, 2 \tag{9.28}$$

This equation states that, for equal conditions regarding u and w, a steep DS ($\cot\alpha$ is small) moves faster than does a DS with a shallow inclination. A slope of a frontal surface of 1:50 is considered steep whereas 1:300 is regarded as slight. By expressing (9.28) for $i = 1$ and $i = 2$, upon subtraction of one equation from the other we find another expression for the inclination of the DS:

$$\tan\alpha = \frac{\{w\}}{\{u\}} = \frac{w^{(2)} - w^{(1)}}{u^{(2)} - u^{(1)}} \tag{9.29}$$

involving the discontinuity jump of the velocity components.

9.4 The kinematic boundary-surface condition

There are various types of boundary surface. It is necessary to make a distinction between outer or external and internal boundary surfaces. An external boundary surface, for example, is the surface of the earth for which a lower boundary condition must be formulated. An internal boundary surface is an imagined separation boundary between two fluids of differing densities.

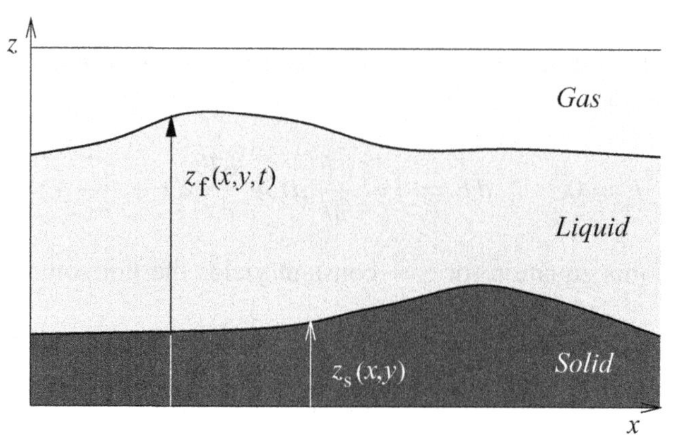

Fig. 9.9 External boundary surfaces.

9.4.1 External boundary surfaces

Let us consider an ideal frictionless fluid whose so-called *free surface* z_f is separating the fluid from a vacuum or a gas-filled space. To a good approximation this is realized by the boundary separating the ocean and the atmosphere. While the free surface is time-dependent, a time-independent surface such as the surface of the earth is considered to be a rigid wall z_s, see Figure 9.9. Therefore, the boundary equations may be written as

$$F(x, y, z, t) = z - z_f(x, y, t) = 0, \qquad F(x, y, z) = z - z_s(x, y) = 0 \quad (9.30)$$

As has already been mentioned, any boundary surface, as long as it exists, is considered to be particle-invariant, meaning that the boundary surface is always made up of the same group of particles. From (9.30) follows the so-called kinematic boundary-surface condition. For the nonstationary free surface we may write

$$\text{Free surface:} \quad F = 0, \qquad \frac{\partial F}{\partial t} \neq 0, \qquad \frac{dF}{dt} = \frac{\partial F}{\partial t} + \mathbf{v} \cdot \mathbf{e}_n |\nabla F| = 0 \quad (9.31)$$

where the velocity refers to the fluid medium. With the help of (9.23) this formula can be rewritten as

$$F = 0, \qquad \frac{\partial F}{\partial t} |\nabla F|^{-1} + \mathbf{v} \cdot \mathbf{e}_n = -c + \mathbf{e}_n \cdot \mathbf{v} = -\mathbf{e}_n \cdot \mathbf{e}_n c + \mathbf{e}_n \cdot \mathbf{v} = 0 \quad (9.32)$$

or by means of

$$F = 0, \qquad \mathbf{e}_n \cdot (\mathbf{v} - \mathbf{c}) = 0 \quad (9.33)$$

Since the unit vector $\mathbf{e}_n$ is oriented normal to the DS, the vector $\mathbf{v} - \mathbf{c}$, representing the velocity difference, must lie in the DS.

For a rigid surface we find analogously

$$\text{Rigid surface:} \quad F = 0. \qquad \frac{\partial F}{\partial t} = 0, \qquad \frac{dF}{dt} = 0, \qquad \mathbf{v} \cdot \mathbf{e}_n = 0 \tag{9.34}$$

so that the velocity vector itself lies in the DS.

In a viscous fluid not only does the normal component of the velocity vanish at a rigid wall as implied by (9.34) but also the tangential component must vanish. In this more realistic case the kinematic boundary-surface condition must be written as

$$F = 0, \qquad \mathbf{v} = 0 \tag{9.35}$$

9.4.2 Internal boundary surfaces

Let us consider an internal DS between two frictionless fluids such as a temperature DS of order zero. We again start our analysis from the condition of particle invariance (9.19) which may also be written as

$$\frac{\partial F}{\partial t} |\nabla F|^{-1} \mathbf{e}_n \cdot \mathbf{e}_n + \mathbf{v}^{(i)} \cdot \mathbf{e}_n = 0, \qquad i = 1, 2 \tag{9.36a}$$

Scalar multiplication of (9.23) by $\mathbf{e}_n$ and substitution of the result into (9.36a) gives

$$\mathbf{e}_n \cdot (\mathbf{v}^{(i)} - \mathbf{c}) = (-u^{(i)} \sin\alpha + w^{(i)} \cos\alpha) - c = 0, \qquad i = 1, 2 \tag{9.36b}$$

where we have used the orientation of the coordinate system shown in Figure 9.7. Setting in succession in (9.36a) $i = 1, 2$ and subtracting one of the results from the other yields the kinematic boundary-surface condition for the internal DS:

$$\boxed{\begin{aligned} & F = 0, \qquad \frac{\partial F}{\partial t} \neq 0: \\ &\text{(a)} \quad \{\mathbf{v}\} \cdot \nabla F = 0 \\ &\text{(b)} \quad \{\mathbf{v}\} \cdot \mathbf{e}_n = \| \cdot \mathbf{v} = 0 \\ &\text{(c)} \quad w^{(2)} - w^{(1)} = (u^{(2)} - u^{(1)}) \tan\alpha \quad \text{or} \quad \{w\}/\{u\} = \tan\alpha \end{aligned}} \tag{9.37}$$

Equations (9.37b) and (9.37c) follow directly from (9.37a) since $|\nabla F| \neq 0$. In (9.37b) we have also used the definition (9.8) for the surface divergence. Again, we have obtained equation (9.29) giving the inclination of the discontinuity surface.

Equations (9.37a) and (9.37b) require that the velocity-jump vector $\{\mathbf{v}\}$ at the zeroth-order DS must be tangential to the DS while the normal component of $\mathbf{v}$

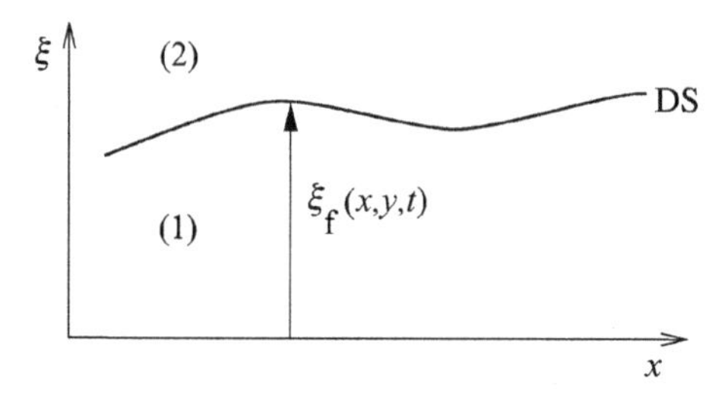

Fig. 9.10 The generalized height of a discontinuity surface.

must be identical on both sides of the DS. The component form (9.37c) of the kinematic boundary-surface condition gives a relation between the slope $\tan\alpha$ and the velocity jumps of the components of the wind velocity.

Equation (9.37) is valid for a nonstationary ($\partial F/\partial t \neq 0$) internal boundary surface. For a stationary boundary surface ($\partial F/\partial t = 0$) the kinematic boundary-surface condition reduces to

$$\boxed{\begin{array}{cccc} & F = 0, & \dfrac{\partial F}{\partial t} = 0: & \\ \mathbf{v}^{(i)} \cdot \nabla F = 0, & \mathbf{v}^{(i)} \cdot \mathbf{e}_n = 0, & \tan\alpha = w^{(i)}/u^{(i)}, & i = 1, 2 \end{array}} \tag{9.38}$$

9.4.3 The generalized vertical velocity at boundary surfaces

The condition of particle invariance can be used to derive the vertical velocity at outer and internal boundary surfaces. We will first introduce the *generalized vertical coordinate* which will also be of importance in our future work when we consider the atmospheric motion in arbitrary coordinate systems; see Figure 9.10.

As specific examples of the generalized vertical coordinates we consider the atmospheric pressure p and the height z of a pressure surface which generally depend on the horizontal coordinates x and y. Therefore, we may write

$$\begin{aligned} p = p_{\mathrm{f}}(x, y, t) &\Longrightarrow p - p_{\mathrm{f}}(x, y, t) = F_1(x, y, p, t) = 0 \\ z = z_{\mathrm{f}}(x, y, t) &\Longrightarrow z - z_{\mathrm{f}}(x, y, t) = F_2(x, y, z, t) = 0 \end{aligned} \tag{9.39}$$

or, in general,

$$F = \xi - \xi_{\mathrm{f}}(x, y, t) = 0, \qquad \frac{dF}{dt} = 0 \tag{9.40}$$

Individual differentiation with respect to time gives

$$F = 0, \qquad \dot{\xi}^{(i)} = \frac{\partial \xi_{\mathrm{f}}}{\partial t} + \mathbf{v}_{\mathrm{h}}^{(i)} \cdot \nabla_{\mathrm{h}} \xi_{\mathrm{f}}, \qquad i = 1, 2 \tag{9.41}$$

which is the generalized velocity at the boundary surface. If we are dealing with an outer stationary or nonstationary boundary surface then (9.41) refers to the interior fluid. Writing (9.41) down for both sides of the DS ($i = 1, 2$) and subtraction of one of the results from the other gives

$$F = 0, \qquad \{\dot{\xi}\} = \{\mathbf{v}_\mathrm{h}\} \cdot \nabla_\mathrm{h}\xi_\mathrm{f} \tag{9.42}$$

This equation is equivalent to the kinematic boundary-surface condition since it reduces to (9.37a) if the y-axis is taken along the trace of the discontinuity surface. Let us now consider various cases of (9.41) and (9.42), which are collected in the following equations.

$$\text{(I)} \qquad \xi = z, \qquad \dot{\xi} = w$$

The horizontal surface of the earth: $z_\mathrm{s} = \text{constant}$

$$\frac{\partial z_\mathrm{s}}{\partial t} = 0, \qquad \nabla_\mathrm{h} z_\mathrm{s} = 0, \qquad w_\mathrm{s} = 0 \tag{9.43a}$$

Earth's surface with topography: $z_\mathrm{s} = z_\mathrm{s}(x, y)$

$$\frac{\partial z_\mathrm{s}}{\partial t} = 0, \qquad \nabla_\mathrm{h} z_\mathrm{s} \neq 0, \qquad w_\mathrm{s} = \mathbf{v}_\mathrm{s} \cdot \nabla_\mathrm{h} z_\mathrm{s} \tag{9.43b}$$

A free nonstationary surface: $\xi_\mathrm{f} = H$

$$w_H = \frac{\partial H}{\partial t} + \mathbf{v}_\mathrm{h} \cdot \nabla_\mathrm{h} H \tag{9.43c}$$

A nonstationary internal DS:

$$w_\mathrm{f}^{(2)} - w_\mathrm{f}^{(1)} = (\mathbf{v}_\mathrm{h}^{(2)} - \mathbf{v}_\mathrm{h}^{(1)}) \cdot \nabla_\mathrm{h} z_\mathrm{f} \tag{9.43d}$$

A stationary internal DS:

$$w_\mathrm{f}^{(i)} = \mathbf{v}_\mathrm{h}^{(i)} \cdot \nabla_\mathrm{h} z_\mathrm{f}, \qquad i = 1, 2 \tag{9.43e}$$

$$\text{(II)} \qquad \xi = -p, \qquad \dot{\xi} = -\dot{p} = -\omega$$

A nonstationary internal DS:

$$\omega_\mathrm{f}^{(2)} - \omega_\mathrm{f}^{(1)} = (\mathbf{v}^{(2)} - \mathbf{v}^{(1)}) \cdot \nabla_\mathrm{h} p_\mathrm{f} \tag{9.43f}$$

A stationary internal DS:

$$\omega_\mathrm{f}^{(i)} = \mathbf{v}_\mathrm{h}^{(i)} \cdot \nabla_\mathrm{h} p_\mathrm{f}, \qquad i = 1, 2 \tag{9.43g}$$

These expressions do not require additional explanations.

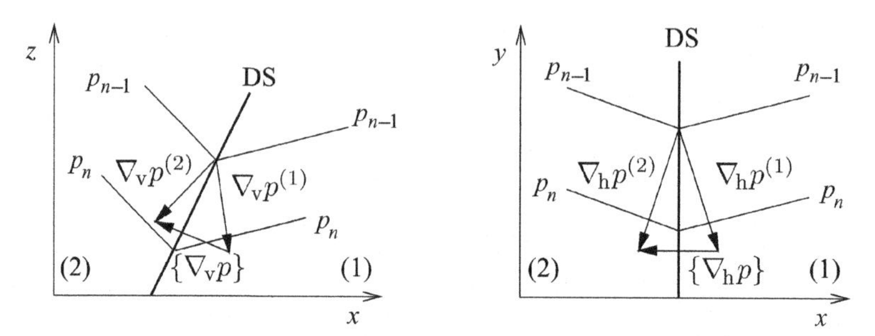

Fig. 9.11 The pressure-gradient jump at a vertical (left) and a horizontal (right) cross-section of a pressure-discontinuity surface. $\nabla_v = \mathbf{i}\,\partial/\partial x + \mathbf{k}\,\partial/\partial z$.

9.5 The dynamic boundary-surface condition

At a boundary surface the stress tensor $\mathbb{T}$ as defined in (5.3) must be continuous, otherwise infinitely large pressure gradients would result. This physical condition is known as the dynamic boundary-surface condition, which can be written as

$$\begin{aligned} &\text{(a)} \quad \{\mathbb{T}\} = 0 \\ &\text{(b)} \quad \{p\} = 0 \\ &\text{(c)} \quad p = 0, \qquad \frac{dp}{dt} = 0 \end{aligned} \tag{9.44}$$

Equation (9.44a) is the general condition for viscous fluids whereas (9.44b) refers to frictionless fluids. This condition states that the pressure boundary surface must be a DS of first order at least. If we consider the upper boundary of the atmosphere as a free material surface then condition (9.44c) must apply. With reference to equations (9.14) and (9.16) we may write

$$\| \times \nabla p = \mathbf{e}_n \times \{\nabla p\} = 0, \qquad d\mathbf{r} \cdot \{\nabla p\} = 0 \tag{9.45}$$

where $d\mathbf{r}$ lies in the DS. Equation (9.45) shows that, in a frictionless fluid, the jump of the pressure gradient at a first-order DS relative to p must be perpendicular to the DS. The pressure itself, however, is continuous as stated by (9.44b). The situation is demonstrated in Figure 9.11.

Finally, we may combine the kinematic and the dynamic boundary-surface conditions to give the so-called *mixed boundary-surface condition* for frictionless fluids by setting in (9.30)

$$F(x, y, z, t) = p^{(2)}(x, y, z, t) - p^{(1)}(x, y, z, t) = \{p\} = 0 \tag{9.46}$$

The result is

$$\left\{\frac{\partial p}{\partial t}\right\} + \mathbf{v}^{(i)} \cdot \{\nabla p\} = 0, \qquad i = 1, 2 \tag{9.47}$$

On writing this expression down for both sides of the boundary surface and subtracting one of the results from the other we obtain

$$\boxed{\{\mathbf{v}\} \cdot \{\nabla p\} = 0} \tag{9.48}$$

Since $\{\nabla p\} \neq 0$ by assumption, we find that, in a frictionless fluid, the velocity jump must be perpendicular to the jump of the pressure gradient.

9.6 The zeroth-order discontinuity surface

9.6.1 The inclination of the zeroth-order DS

We wish to treat briefly the inclination of a zeroth-order DS in a frictionless fluid for various conditions. In our example we consider a boundary surface separating two air masses of different densities. Using the ideal-gas law and recalling that $\{p\} = 0$ at the DS, it is easy to show that this corresponds to a zeroth-order DS of the virtual temperature T_v or

$$\{\rho\} = -\{T_v\}p/(R_0 T_v^{(1)} T_v^{(2)}) \tag{9.49}$$

It stands to reason that, for air masses at rest, the DS must be horizontal, with the warmer lighter air on top of the colder denser air. There can be no equilibrium between two air masses of different densities seperated by a vertical DS. If the air masses are in motion the colder air will form a wedge under the warmer air. It is important to realize that both the kinematic and the dynamic boundary-surface conditions must be satisfied at the DS. We repeat the dynamic boundary condition (9.45) and write

$$\begin{gathered}\{\nabla p\} \cdot d\mathbf{r} = \{\nabla_v p\} \cdot d\mathbf{r} = \{\nabla_v p\} \cdot \mathbf{i}\, dx + \left\{\frac{\partial p}{\partial z}\right\} dz = 0 \\ \text{with} \quad \nabla_v = \mathbf{i}\frac{\partial}{\partial x} + \mathbf{k}\frac{\partial}{\partial z}\end{gathered} \tag{9.50}$$

Again we have used the special arrangement of the coordinate system. From this equation we obtain the inclination of the DS, which is given by

$$\tan\alpha = \left(\frac{dz}{dx}\right)_{\text{DS}} = \mathbf{i} \cdot \nabla_v z_{\text{DS}} = -\mathbf{i} \cdot \{\nabla_v p\}\left\{\frac{\partial p}{\partial z}\right\}^{-1} \tag{9.51a}$$

or by

$$\nabla_v z_{\text{DS}} = \left(\nabla_v p^{(1)} - \nabla_v p^{(2)}\right)\left(\frac{\partial p^{(2)}}{\partial z} - \frac{\partial p^{(1)}}{\partial z}\right)^{-1} \tag{9.51b}$$

Assuming the validity of the hydrostastic equation for the present situation, we may rewrite (9.51b) and obtain an equivalent expression for the slope of the DS:

$$\nabla_{\mathrm{v}} z_{\mathrm{DS}} = \frac{\{\nabla_{\mathrm{v}} p\}}{g\{\rho\}} = \frac{\nabla_{\mathrm{v}} p^{(2)} - \nabla_{\mathrm{v}} p^{(1)}}{g(\rho^{(2)} - \rho^{(1)})} \quad \text{or} \quad \tan\alpha = \frac{\mathbf{i}\cdot\{\nabla_{\mathrm{v}} p\}}{g\{\rho\}} \tag{9.52}$$

Owing to the choice of the coordinate system, see Figure 9.7, the slope is always greater than zero, or $\tan\alpha > 0$. Furthermore, we assume that we have stable atmospheric conditions in the sense that the denser colder air is located below the lighter warmer air. We note that the denominator in (9.52) is smaller than zero, so the numerator must be negative in order to make the fraction positive. We will use this information by considering the surface divergence of ∇p:

$$\| \cdot \nabla p = \mathbf{e}_n \cdot \{\nabla p\} = (-\sin\alpha\,\mathbf{i} + \cos\alpha\,\mathbf{k}) \cdot \left\{\nabla_{\mathrm{h}} p + \mathbf{k}\frac{\partial p}{\partial z}\right\} \tag{9.53}$$

where we have replaced the unit normal $\mathbf{e}_n$ by means of (9.20). On carrying out the scalar multiplication, we obtain the equivalent expression

$$\begin{gathered} \| \cdot \nabla p = -\mathbf{i}\cdot\{\nabla_{\mathrm{h}} p\}\sin\alpha + \left\{\frac{\partial p}{\partial z}\right\}\cos\alpha > 0 \\ \text{since} \quad \mathbf{i}\cdot\{\nabla_{\mathrm{v}} p\} = \mathbf{i}\cdot\{\nabla_{\mathrm{h}} p\} < 0 \quad \text{and} \quad \left\{\frac{\partial p}{\partial z}\right\} = g\left(\rho^{(1)} - \rho^{(2)}\right) > 0 \end{gathered} \tag{9.54}$$

See Figure 9.11. Since both the horizontal and the vertical parts of (9.54) are larger than zero, we find for reasons of stability that the total surface divergence of the pressure gradient must also be larger than zero. As shown in the left-hand panel of Figure 9.12, in general this is possible only in case of a cyclonic pressure-gradient jump (cyclonic air motion). Otherwise the horizontal part of the surface divergence would not be positive. The anticyclonic pressure-gradient jump would violate the requirement that the first term of (9.54) must be greater than zero; see the right-hand panel of Figure 9.12.

We are now ready to find an expression for the inclination of the DS in the presence of a geostrophic wind field.

9.6.2 A discontinuity surface of zeroth-order in the geostrophic wind field

This type of discontinuity surface is also known as the *Margules boundary surface*. To obtain a suitable expression for the pressure-gradient jump at the DS, we use the approximate form of the equation of motion

$$\rho\frac{d\mathbf{v}_{\mathrm{h}}}{dt} + \rho f\mathbf{k}\times\mathbf{v}_{\mathrm{h}} + \rho g\mathbf{k} = -\nabla_{\mathrm{h}} p - \frac{\partial p}{\partial z}\mathbf{k} = -\nabla p \tag{9.55}$$

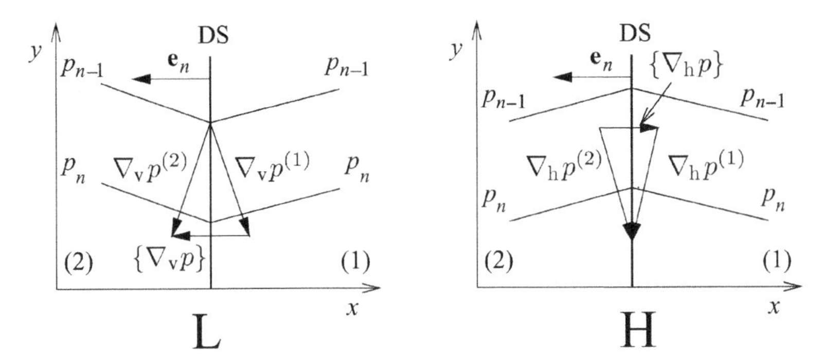

Fig. 9.12 Allowed (left) and forbidden (right) pressure-gradient jumps.

For simplicity we have omitted the vertical bar on the individual time derivative of $\mathbf{v}_{\mathrm{h}}$ expressing that the basis vectors are not to be differentiated with respect to time. This equation may be easily obtained by combining the horizontal equation of motion (2.29) with the hydrostatic approximation (2.27). From (9.55) we obtain for the jump of the pressure gradient

$$\{\nabla p\} = -g\mathbf{k}\{\rho\} + f\{\rho v\}\mathbf{i} - f\{\rho u\}\mathbf{j} - \left\{\rho\,\frac{d\mathbf{v}_{\mathrm{h}}}{dt}\right\} \tag{9.56}$$

The inclination of the DS according to (9.51a) or (9.52) can then be expressed by

$$\tan\alpha = \mathbf{i}\cdot\nabla_{\mathrm{v}}z_{\mathrm{DS}} = \frac{\mathbf{i}\cdot\{\nabla_{\mathrm{v}}p\}}{g\{\rho\}} = \frac{-\{\rho\,du/dt\} + f\{\rho v\}}{g\{\rho\}} \tag{9.57a}$$

$$\text{since}\quad \mathbf{i}\cdot\{\nabla p\} = \mathbf{i}\cdot\{\nabla_{\mathrm{v}}p\}$$

Owing to the special orientation of the coordinate system ($\partial/\partial y = 0$) we also obtain

$$\mathbf{j}\cdot\{\nabla p\} = \mathbf{j}\cdot\{\nabla_{\mathrm{v}}p\} = -\left\{\rho\,\frac{dv}{dt}\right\} - f\{\rho u\} = 0 \tag{9.57b}$$

Now we assume that the acceleration is zero so that du/dt and dv/dt vanish. Since we also ignore viscosity effects, the flow is geostrophic. Therefore, we may write

$$\begin{aligned}
&\text{(a)}\qquad \tan\alpha = \mathbf{i}\cdot\nabla_{\mathrm{v}}z_{\mathrm{DS}} = \frac{\mathbf{i}\cdot\{\nabla_{\mathrm{v}}p\}}{g\{\rho\}} = -\frac{f\mathbf{i}\cdot\mathbf{k}\times\{\rho\mathbf{v}_{\mathrm{g}}\}}{g\{\rho\}} = \frac{f\{\rho v_{\mathrm{g}}\}}{g\{\rho\}}\\
&\text{(b)}\quad \mathbf{j}\cdot\{\nabla_{\mathrm{v}}p\} = -f\mathbf{j}\cdot\mathbf{k}\times\{\rho\mathbf{v}_{\mathrm{g}}\} = -f\{\rho u_{\mathrm{g}}\} = 0 \implies \mathbf{i}\cdot\{\rho\mathbf{v}_{\mathrm{g}}\} = 0
\end{aligned} \tag{9.58}$$

Equation (9.58b) leads to the conclusion that $\{\rho\mathbf{v}_{\mathrm{g}}\}$ lies along the surface of discontinuity. Moreover, it follows that the geostrophic momentum perpendicular to the trace of the DS is equal on both sides of the boundary surface. A slope of the DS of 1:50 is considered steep whereas 1:300 is regarded as shallow.

For purely horizontal geostrophic motion we find from (9.28) and (9.58b)

$$\begin{aligned} c_{\mathrm{h}} &= u_{\mathrm{g}}^{(1)} = u_{\mathrm{g}}^{(2)} = u_{\mathrm{g}} \\ \{\rho u_{\mathrm{g}}\} &= \{\rho\}\, u_{\mathrm{g}} = \{\rho\}\, c_{\mathrm{h}} = 0 \\ \{\rho\} \neq 0 &\Longrightarrow c_{\mathrm{h}} = 0 \quad \text{but} \quad v_{\mathrm{g}}^{(i)} \neq 0 \end{aligned} \tag{9.59}$$

Since the density jump across the DS is assumed to differ from zero, we find that, for the geostrophic case, the displacement along the x-axis $c_x = c_{\mathrm{h}}$ must vanish. However, in case of accelerated horizontal flow, we find from (9.57b) together with (9.28) the following relation:

$$-f\{\rho\}c_{\mathrm{h}} - \left\{\rho\,\frac{dv}{dt}\right\} = 0 \tag{9.60}$$

or

$$c_{\mathrm{h}} = c_x = -\frac{\{\rho\, dv/dt\}}{f\{\rho\}} \tag{9.61}$$

Therefore, for purely horizontal flow a displacement of the front in the x-direction is possible only if the flow is accelerated in the y-direction.

For the geostrophic flow the wind direction must be parallel to the isobars. Since the pressure-gradient jump is cyclonic, the horizontal wind shear along the DS must be cyclonic also. According to the kinematic boundary condition, the velocity jump is located along the DS. From Figure 9.13 it can be seen that $\{\mathbf{v}_{\mathrm{g}}\}$ is directed along the negative y-axis. The direction of $\{\mathbf{v}_{\mathrm{g}}\}$ can also be found from the condition $\tan\alpha > 0$. This will be shown shortly. For reference see also Figure 9.8.

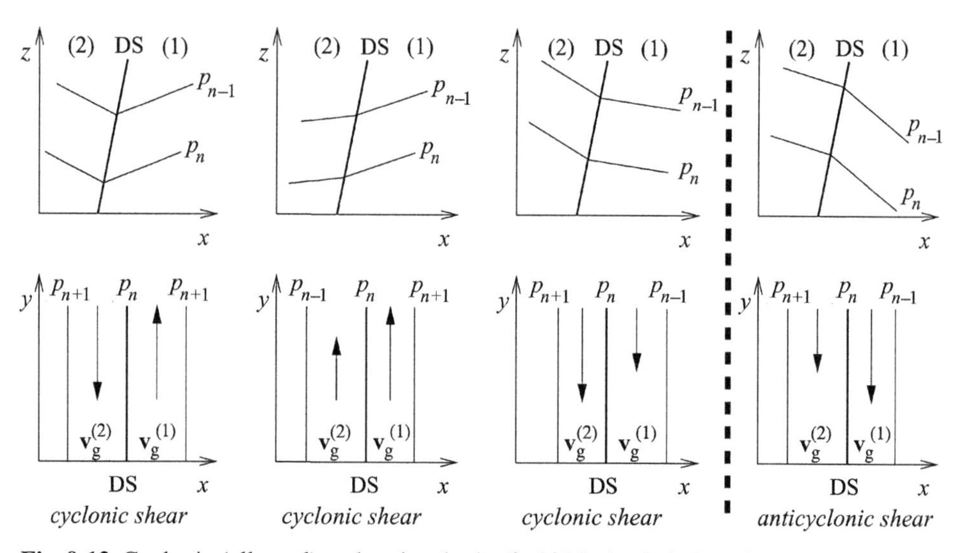

Fig. 9.13 Cyclonic (allowed) and anticyclonic (forbidden) wind shear in geostrophic flow at a discontinuity surface.

Let us briefly return to equation (9.58a) and consider three special cases. (i) If $\{\rho\} = 0$ then $\alpha = 90^\circ$ and no boundary surface exists. (ii) If $\{\rho v_{\rm g}\} = 0$ but $\{\rho\} \neq 0$, then $\alpha = 0$ and the boundary surface is horizontal. (iii) If $v_{\rm g}^{(1)} = v_{\rm g}^{(2)} = v_{\rm g}$ is assumed then the inclination of the boundary surface DS equals the inclination of an isobaric surface. From (9.58a) it follows that

$$\tan\alpha = \mathbf{i}\cdot\nabla_{\rm v} z_{\rm DS} = \frac{f v_{\rm g}}{g} = \frac{1}{\rho g}\frac{\partial p}{\partial x} = \tan\alpha_p \tag{9.62}$$

Let us rewrite (9.58a) with the help of the ideal-gas law and express the density in terms of the virtual temperature. Since the pressure at the boundary surface is continuous, we obtain immediately

$$\nabla_{\rm v} z_{\rm DS} = -\frac{f\mathbf{k}\times\{\rho\mathbf{v}_{\rm g}\}}{g\{\rho\}} = \frac{f\mathbf{k}\times\left(T_{\rm v}^{(1)}\mathbf{v}_{\rm g}^{(2)} - T_{\rm v}^{(2)}\mathbf{v}_{\rm g}^{(1)}\right)}{g\left(T_{\rm v}^{(2)} - T_{\rm v}^{(1)}\right)} \tag{9.63}$$

Expressing the geostrophic wind and the virtual temperature in the two different air masses separated by the DS in terms of their mean values

$$\overline{\mathbf{v}}_{\rm g} = \frac{u_{\rm g}^{(1)} + u_{\rm g}^{(2)}}{2}\mathbf{i} + \frac{v_{\rm g}^{(1)} + v_{\rm g}^{(2)}}{2}\mathbf{j}, \qquad \overline{T}_{\rm v} = \frac{T_{\rm v}^{(1)} + T_{\rm v}^{(2)}}{2} \tag{9.64}$$

and jumps, we find

$$\begin{aligned} \mathbf{v}_{\rm g}^{(2)} &= \overline{\mathbf{v}}_{\rm g} + \frac{\{\mathbf{v}_{\rm g}\}}{2}, & \mathbf{v}_{\rm g}^{(1)} &= \overline{\mathbf{v}}_{\rm g} - \frac{\{\mathbf{v}_{\rm g}\}}{2} \\ T_{\rm v}^{(2)} &= \overline{T}_{\rm v} + \frac{\{T_{\rm v}\}}{2}, & T_{\rm v}^{(1)} &= \overline{T}_{\rm v} - \frac{\{T_{\rm v}\}}{2} \end{aligned} \tag{9.65}$$

On substituting (9.65) into (9.63) we obtain

$$\boxed{\nabla_{\rm v} z_{\rm DS} = \frac{f}{g}\mathbf{k}\times\left(\{\mathbf{v}_{\rm g}\}\frac{\overline{T}_{\rm v}}{\{T_{\rm v}\}} - \overline{\mathbf{v}}_{\rm g}\right)} \tag{9.66}$$

As follows from comparison with (9.62), the second term on the right-hand side represents the inclination of an isobaric surface whose average midlatitude inclination is approximately $\pm 10^{-4}$. For average atmospheric conditions the first term in (9.66) is approximately 30 times larger than the second term. If we ignore the second term we obtain with acceptable accuracy for the inclination of the boundary surface

$$\tan\alpha = \mathbf{i}\cdot\nabla_{\rm v} z_{\rm DS} \approx \frac{f\overline{T}_{\rm v}}{g\{T_{\rm v}\}}\left(\mathbf{i}\cdot\mathbf{k}\times\{\mathbf{v}_{\rm g}\}\right) > 0 \tag{9.67}$$

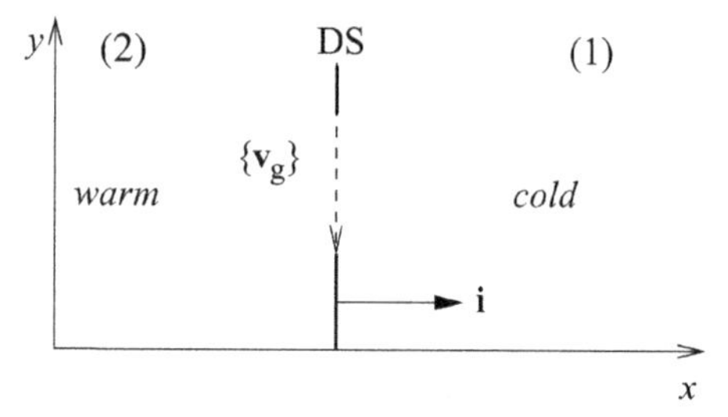

Fig. 9.14 The positions of the warm and the cold air relative to the jump of the geostrophic wind.

This expression admits an important interpretation. According to the kinematic boundary-surface condition the jump $\{\mathbf{v}_g\}$ must lie in the DS and can be interpreted as a horizontal as well as a vertical jump. Since the warm air overlies the wedge of cold air, the temperature jump $\{T_\mathrm{v}\}$ must be positive. A consequence of the condition $\tan\alpha > 0$ is that $\{\mathbf{v}_\mathrm{g}\}$ must have the direction shown in Figure 9.14. Thus, looking in the direction of $\{\mathbf{v}_\mathrm{g}\}$, the warm air must always be situated to the right of the jump $\{\mathbf{v}_\mathrm{g}\}$, as shown in the figure. This result is in agreement with Figure 9.13.

The behavior of $\{\mathbf{v}_\mathrm{g}\}$ with respect to the positions of the warm and the cold air corresponds exactly to the behavior of the *thermal wind* $\mathbf{v}_T$ in the continuous field. The thermal wind is defined as the variation of the geostrophic wind with height. For convenience, we will consider geostrophic motion on an isobaric surface. Thus we have to transform the equation for the geostrophic wind from the (x, y, z)-system to the (x, y, p)-system. This is easily accomplished with the help of (M4.51) by setting there $q^3 = z, \xi = p$, and $\psi = p$. Utilizing the hydrostatic equation together with $\phi = gz$, we obtain $\nabla_{\mathrm{h},z}p = \rho\,\nabla_{\mathrm{h},p}\phi$ so that the equation for the geostrophic wind in the p system is given by

$$\mathbf{v}_\mathrm{g} = \frac{1}{f}\mathbf{k} \times \nabla_{\mathrm{h},p}\phi \tag{9.68}$$

On differentiating this equation with respect to pressure we obtain the differential form of the thermal wind in the p system:

$$\frac{\partial \mathbf{v}_\mathrm{g}}{\partial p} = \frac{1}{f}\mathbf{k} \times \nabla_{\mathrm{h},p}\frac{\partial \phi}{\partial p} = -\frac{1}{f}\mathbf{k} \times \nabla_{\mathrm{h},p}\frac{1}{\rho} = -\frac{R_0}{fp}\mathbf{k} \times \nabla_{\mathrm{h},p}T_\mathrm{v} \tag{9.69}$$

A slight rearrangement of this formula gives

$$\boxed{\mathbf{v}_T = \Delta\mathbf{v}_\mathrm{g} = \frac{\partial \mathbf{v}_\mathrm{g}}{\partial \phi}\Delta\phi = \frac{\partial \mathbf{v}_\mathrm{g}}{\partial p}\frac{\partial p}{\partial \phi}\Delta\phi = \frac{\Delta\phi}{fT_\mathrm{v}}\mathbf{k} \times \nabla_{\mathrm{h},p}T_\mathrm{v}} \tag{9.70}$$

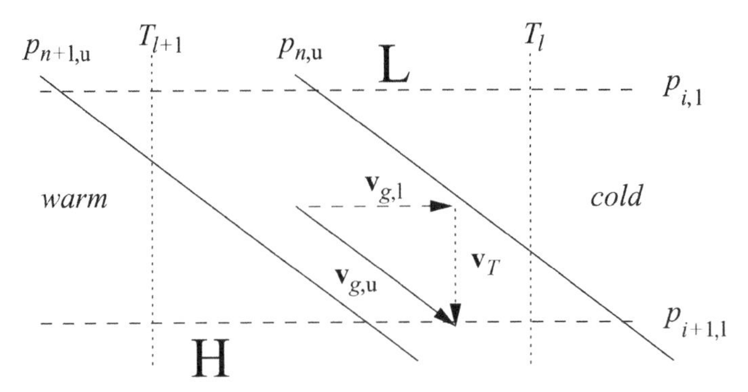

Fig. 9.15 The positions of the warm and cold air relative to the thermal wind, subscripts l and u refer to lower and upper layers.

showing that the warm air is always located to the right of the thermal wind; see Figure 9.15.

It can be recognized from Figure 9.15 that clockwise turning of the geostrophic wind with height (veering) is associated with warm-air advection whereas anti-clockwise turning (backing) with height causes cold-air advection. The rule is reversed in the southern hemisphere.

At the conclusion of this section we wish to point out that mathematical expressions for obtaining general equations for the slope of discontinuity surfaces for accelerated frictionless flow have been worked out. These expressions are not very important for our work and will be omitted.

9.7 An example of a first-order discontinuity surface

An important example of a first-order discontinuity surface is the tropopause, where the temperature is continuous while the vertical temperature gradient is discontinuous; see Figure 9.16. We will only briefly consider this subject and calculate the midlatitude slope of the tropopause. More detailed information is given, for example, by Lowell (1951); see also Haltiner and Martin (1957).

Let us consider two points along the tropopause. The temperature variation along the tropopause is approximately given by

$$d_g T^{(i)} = \left(\frac{\partial T}{\partial x}\right)^{(i)} dx + \left(\frac{\partial T}{\partial z}\right)^{(i)} dz, \qquad i = 1, 2 \tag{9.71}$$

with $d_g T^{(1)} = d_g T^{(2)}$. Since the geometric variation is the same on both sides of the DS, we find upon subtraction

$$\left\{\frac{\partial T}{\partial x}\right\} dx + \left\{\frac{\partial T}{\partial z}\right\} dz = 0 \tag{9.72}$$

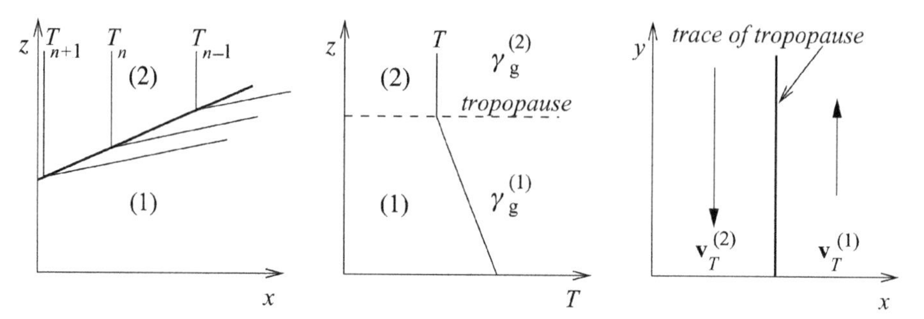

Fig. 9.16 The temperature distribution and the geostrophic wind shear at the tropopause. The x-axis is pointing toward the equator.

The slope of the tropopause is then given by

$$\tan \alpha_T = \frac{dz}{dx} = \frac{\left\{\dfrac{\partial T}{\partial x}\right\}_z}{\{\gamma_g\}} \tag{9.73}$$

where $\{\gamma_g\}$ represents the jump of the lapse rate. We now transform from the z-system to the pressure system by using the transformation rule

$$\left(\frac{\partial T}{\partial x}\right)_z = \left(\frac{\partial T}{\partial x}\right)_p - \frac{\partial T}{\partial z}\left(\frac{\partial z}{\partial x}\right)_p \tag{9.74}$$

This expression is obtained from (M4.51) by setting there $q^1 = x,\ q^3 = z$, and $\zeta = p$. Thus we find

$$\tan \alpha_T = \frac{\left\{\dfrac{\partial T}{\partial x}\right\}_p}{\{\gamma_g\}} + \left(\frac{\partial z}{\partial x}\right)_p \tag{9.75}$$

where the second term represents the slope of the pressure surface in the x-direction. Using the definition of the geostrophic wind in the pressure system (9.68) and rewriting the thermal-wind equation (9.69) with the help of the hydrostatic equation, we find with $T \approx T_v$

$$\begin{aligned} \tan \alpha_T &= \frac{f v_g}{g} + \frac{T_v f}{g} \frac{\{v_T\}}{\{\gamma_g\}} \\ \text{with} \quad \{v_T\} &= \left\{\frac{\partial v_g}{\partial z}\right\} = \frac{g}{f T_v}\left\{\left(\frac{\partial T_v}{\partial x}\right)_p\right\} \end{aligned} \tag{9.76}$$

Here v_T is the differential form of the thermal wind. From the temperature distribution in the vicinity of the tropopause we find the direction of the geostrophic wind shear along the trace of the tropopause. Assuming that $v_T^{(1)} = 4\ \mathrm{m\ s^{-1}\ km^{-1}}$, $v_T^{(2)} = -6\ \mathrm{m\ s^{-1}\ km^{-1}}$, $v_\mathrm{g} = 30\ \mathrm{m\ s^{-1}}$, $T = 240$ K at the tropopause, and $\{\gamma_\mathrm{g}\} = 6\ \mathrm{K\ km^{-1}}$, we find from (9.76) $\tan\alpha_T = 43\ \mathrm{m}\ (10\ \mathrm{km})^{-1}$. This means that, in the midlatitudes, the tropopause rises by about 43 m if we move a distance of 10 km toward the equator.

It should be observed that there exists no entirely satisfactory theory for the formation and the existence of the tropoause, which is not a continuous band rising from about 9 km at the North pole to 18 km at the equator. The tropopause is often fractured in the region of strong jet streams, permitting the exchange of air between the troposphere and the stratosphere.

9.8 Problems

9.1: Let $\xi = p$. With the help of the proper transformation equations and the kinematic boundary-surface condition, find an expression for the generalized vertical velocity $\omega_\mathrm{s} = \dot{p}_\mathrm{s}$ for the earth's surface. This expression should involve the geopotential.

9.2: With the help of (9.63), find an expression for the slope of the discontinuity surface.

9.3: Sketch three figures of the type shown in Figure 9.13. Show the distribution of the isobars, the direction and roughly the magnitude of the geostrophic wind, and the jump of $\mathbf{v}_\mathrm{g}$ for the following three situations:
(a) cold high, warm high,
(b) cold high, warm low,
(c) warm low, warm high.
Place the pressure systems on the appropriate side of the discontinuity surface. Check your figures by applying the conditions that $\tan\alpha > 0$ and $\{\mathbf{v}_\mathrm{g}\}$ is cyclonic.

10
Circulation and vorticity theorems

In this chapter a number of very important circulation and vorticity theorems will be introduced. Instead of following the historical development, they will be deduced from a general baroclinic vortex law in order to demonstrate the close relationships among the various theorems. This particular way of presentation is chosen in order to better appreciate the great beauty of the underlying theory. Of course, the vorticity equation in Cartesian coordinates, for example, could be easily derived from the horizontal equations of motion by differentiating these with respect to the horizontal coordinates and subtracting the resulting formulas from each other. This operational process is easily understood and carried out, but the reader probably fails to appreciate the fairly general character of the entire theory. A basic tool needed in the following derivations is Ertel's form of the continuity equation, which will be presented in the next section. Another important tool in our work is the Weber transformation, which will be discussed in Section 10.2.

10.1 Ertel's form of the continuity equation

The derivation rests on the general formulation of the continuity equation. Using (M3.56) and (M4.36), equation (1.50) can be written in the following form:

$$\begin{gathered}\frac{d}{dt}\left(\rho\sqrt{\underset{q}{g}}\right)+\rho\sqrt{\underset{q}{g}}\,\frac{\partial\dot{q}^n}{\partial q^n}=0\\ \text{with}\quad \sqrt{\underset{q}{g}}=\sqrt{\underset{x}{g}}\left|\frac{\partial(x^1,x^2,x^3)}{\partial(q^1,q^2,q^3)}\right|=\frac{1}{J_x(q^1,q^2,q^3)}\end{gathered}\tag{10.1}$$

This equation involves a transformation from the Cartesian to the general coordinates as discussed in Section M4.1.3. Replacing the density ρ by the specific volume α and carrying out the differentiation gives

$$\frac{d}{dt}\left[\alpha J_x(q^1, q^2, q^3)\right] = \alpha J_x(q^1, q^2, q^3)\frac{\partial \dot{q}^n}{\partial q^n} \tag{10.2}$$

In $\sqrt{g_q}$ the partial derivatives with respect to q^k will now be replaced by using the proper transformation rules (M4.24):

$$\frac{\partial}{\partial x^i} = \frac{\partial q^n}{\partial x^i}\frac{\partial}{\partial q^n} \quad \text{or} \quad \begin{pmatrix} \dfrac{\partial}{\partial x^1} \\ \dfrac{\partial}{\partial x^2} \\ \dfrac{\partial}{\partial x^3} \end{pmatrix} = \left(\frac{\partial q^j}{\partial x^i}\right)\begin{pmatrix} \dfrac{\partial}{\partial q^1} \\ \dfrac{\partial}{\partial q^2} \\ \dfrac{\partial}{\partial q^3} \end{pmatrix} \tag{10.3}$$

By inversion of (10.3) we find the expression

$$\begin{pmatrix} \dfrac{\partial}{\partial q^1} \\ \dfrac{\partial}{\partial q^2} \\ \dfrac{\partial}{\partial q^3} \end{pmatrix} = \left(\frac{\partial x^j}{\partial q^i}\right)\begin{pmatrix} \dfrac{\partial}{\partial x^1} \\ \dfrac{\partial}{\partial x^2} \\ \dfrac{\partial}{\partial x^3} \end{pmatrix} = \frac{\left(M_{i\cdot}^{\cdot j}\right)}{J_x(q^1, q^2, q^3)}\begin{pmatrix} \dfrac{\partial}{\partial x^1} \\ \dfrac{\partial}{\partial x^2} \\ \dfrac{\partial}{\partial x^3} \end{pmatrix} \tag{10.4}$$

where $J_x(q^1, q^2, q^3) = \mathrm{Det}(\partial q^j/\partial x^i)$ and $M_{i\cdot}^{\cdot j}$ are the adjoints of the matrix $(\partial q^j/\partial x^i)$. It should be kept in mind that the value of a determinant is not changed by transposition of rows and columns. The reader may easily convince himself that

$$\begin{pmatrix} M_{1\cdot}^{\cdot 1} & M_{1\cdot}^{\cdot 2} & M_{1\cdot}^{\cdot 3} \\ M_{2\cdot}^{\cdot 1} & M_{2\cdot}^{\cdot 2} & M_{2\cdot}^{\cdot 3} \\ M_{3\cdot}^{\cdot 1} & M_{3\cdot}^{\cdot 2} & M_{3\cdot}^{\cdot 3} \end{pmatrix}\begin{pmatrix} \dfrac{\partial}{\partial x^1} \\ \dfrac{\partial}{\partial x^2} \\ \dfrac{\partial}{\partial x^3} \end{pmatrix} = \begin{pmatrix} J_x(\quad, q^2, q^3) \\ J_x(q^1, \quad, q^3) \\ J_x(q^1, q^2, \quad) \end{pmatrix} \tag{10.5}$$

where, for example,

$$J_x(\quad, q^2, q^3) = \begin{vmatrix} \dfrac{\partial}{\partial x^1} & \dfrac{\partial}{\partial x^2} & \dfrac{\partial}{\partial x^3} \\ \dfrac{\partial q^2}{\partial x^1} & \dfrac{\partial q^2}{\partial x^2} & \dfrac{\partial q^2}{\partial x^3} \\ \dfrac{\partial q^3}{\partial x^1} & \dfrac{\partial q^3}{\partial x^2} & \dfrac{\partial q^3}{\partial x^3} \end{vmatrix} \tag{10.6}$$

By application of (10.4) and using (10.5), we find directly

$$\begin{pmatrix} \dfrac{\partial \dot{q}^1}{\partial q^1} \\ \dfrac{\partial \dot{q}^2}{\partial q^2} \\ \dfrac{\partial \dot{q}^3}{\partial q^3} \end{pmatrix} = \frac{1}{J_x(q^1, q^2, q^3)} \begin{pmatrix} J_x(\dot{q}^1, q^2, q^3) \\ J_x(q^1, \dot{q}^2, q^3) \\ J_x(q^1, q^2, \dot{q}^3) \end{pmatrix} \tag{10.7}$$

Substitution of this expression into (10.2) gives

$$\boxed{\frac{d}{dt}\left[\alpha J_x(q^1, q^2, q^3)\right] = \alpha\left[J_x(\dot{q}^1, q^2, q^3) + J_x(q^1, \dot{q}^2, q^3) + J_x(q^1, q^2, \dot{q}^3)\right]} \tag{10.8}$$

where all spatial derivatives are taken with respect to the Cartesian coordinates. This version of the continuity equation is due to Ertel (1960), and is of great advantage whenever conservative quantities, i.e. invariant field functions, are involved. These are always characterized by the conservation equation

$$\frac{d\psi_i}{dt} = \dot{\psi}_i = 0 \tag{10.9}$$

Two brief examples will demonstrate this.

Example 1 Suppose that the q^k are the Lagrangian enumeration coordinates a^k which are characterized by $\dot{a}^k = 0$. Then it is easily seen that the right-hand side of (10.8) vanishes so that

$$\frac{d}{dt}\left[\alpha J_x(a^1, a^2, a^3)\right] = \frac{d}{dt}\left[\alpha \nabla_x a^1 \cdot \nabla_x a^2 \times \nabla_x a^3\right] = 0 \tag{10.10}$$

Thus $[\alpha \nabla_x a^1 \cdot \nabla_x a^2 \times \nabla_x a^3]$ is an invariant. From (10.1) follows immediately the continuity equation in Lagrangian enumeration coordinates which was discussed previously:

$$\frac{d}{dt}\left(\frac{\alpha}{\sqrt{g_a}}\right) = \frac{\partial}{\partial t}\left(\frac{\alpha}{\sqrt{g_a}}\right)_{a^i} = 0 \quad \text{or} \quad \frac{d}{dt}\left(\rho\sqrt{g_a}\right) = \frac{\partial}{\partial t}\left(\rho\sqrt{g_a}\right)_{a^i} = 0 \tag{10.11}$$

with $(d/dt)(\cdots) = (\partial/\partial t)(\cdots)_{a^i}$ since $\dot{a}^k = 0$. Integration yields

$$\rho\sqrt{g_a} = \text{constant} = \left(\rho\sqrt{g_a}\right)_{t=t_0} \tag{10.12}$$

Example 2 Since the q^k may be expressed as functions of the Cartesian coordinates, we may consider these as independent field functions of x^i. Let ψ be an arbitrary field function, s the specific entropy ($\dot{s} = 0$ for isentropic processes), and P the potential vorticity, which has not yet been discussed. P turns out to be an invariant quantity if certain conditions are met. Substitution of these assignments into (10.8) gives

$$\frac{d}{dt}(\alpha\,\nabla_x\psi\cdot\nabla_x s\times\nabla_x P) = \alpha\left[\nabla_x\left(\frac{d\psi}{dt}\right)\cdot\nabla_x s\times\nabla_x P + \nabla_x\psi\cdot\nabla_x\left(\frac{ds}{dt}\right)\times\nabla_x P + \nabla_x\psi\cdot\nabla_x s\times\nabla_x\left(\frac{dP}{dt}\right)\right] \tag{10.13}$$

For isentropic motion ($\dot{s} = 0$), with, as will be shown later, $\dot{P} = 0$ also, we find

$$\frac{d}{dt}(\alpha\,\nabla\psi\cdot\nabla s\times\nabla P) = \alpha\,\nabla\frac{d\psi}{dt}\cdot\nabla s\times\nabla P \tag{10.14}$$

where we have omitted the reference to Cartesian coordinates. This equation has the form of an interchange relation of operators, which we recognize by setting

$$\delta_1 = \frac{d(\;)}{dt}, \qquad \delta_2 = \alpha\,\nabla s\times\nabla P\cdot\nabla(\;) \tag{10.15}$$

so that

$$\delta_1\delta_2\psi = \delta_2\delta_1\psi \tag{10.16}$$

10.2 The baroclinic Weber transformation

The Weber transformation is an essential tool in the derivation of circulation theorems and hydrodynamic invariants. We begin the somewhat lengthy but straightforward derivation by restating equation (3.54), which gives the transformation between the Lagrangian and the Cartesian coordinates:

$$\frac{\partial\psi}{\partial a^i} = \frac{\partial x^n}{\partial a^i}\frac{\partial\psi}{\partial x^n}, \qquad \frac{\partial\psi}{\partial x^i} = \frac{\partial a^n}{\partial x^i}\frac{\partial\psi}{\partial a^n} \tag{10.17}$$

Next we introduce the specific entropy s of dry air into the equation of absolute motion for frictionless flow by eliminating the pressure-gradient term. Using Gibbs' fundamental equation in the form for the enthalpy h (see TH or any other textbook on thermodynamics)

$$d_g h = T\,d_g s + \alpha\,d_g p \quad \text{with} \quad h = e + p\alpha \tag{10.18}$$

and recalling that $d_g\psi = d\mathbf{r}\cdot\nabla\psi$, we may immediately deduce that

$$\nabla h = T\,\nabla s + \alpha\,\nabla p \tag{10.19}$$

On substituting this equation into the equation of motion

$$\frac{d\mathbf{v}_A}{dt} = \frac{d^2\mathbf{r}}{dt^2} = -\nabla\phi_a - \alpha\,\nabla p \tag{10.20}$$

we obtain

$$\frac{d^2\mathbf{r}}{dt^2} = -\nabla(\phi_a + h) + T\,\nabla s \tag{10.21}$$

In component form, using Cartesian coordinates, (10.21) reads

$$\frac{d^2x^k}{dt^2} = -\frac{\partial}{\partial x^k}(\phi_a + h) + T\,\frac{\partial s}{\partial x^k}, \qquad k = 1, 2, 3 \tag{10.22}$$

This equation may also be written in Lagrangian coordinates by simply replacing d^2/dt^2 by $(\partial^2/\partial t^2)_L$, see Section 3.2.1, where the subscript L denotes the Lagrangian system. Multiplying (10.22) on both sides by $\partial x^k/\partial a^i$ and then summing over k, we have

$$\frac{\partial x^n}{\partial a^i}\left(\frac{\partial^2 x^n}{\partial t^2}\right)_L = -\frac{\partial x^n}{\partial a^i}\,\frac{\partial}{\partial x^n}(\phi_a + h) + T\,\frac{\partial s}{\partial x^n}\,\frac{\partial x^n}{\partial a^i} \tag{10.23}$$

Using (10.17) in (10.23), we obtain

$$\frac{\partial x^n}{\partial a^i}\left(\frac{\partial^2 x^n}{\partial t^2}\right)_L = -\frac{\partial}{\partial a^i}(\phi_a + h) + T\,\frac{\partial s}{\partial a^i} \tag{10.24}$$

We will now integrate this equation assuming isentropic flow ($\dot{s} = 0$). First the left-hand side of (10.24) will be rewritten as

$$\frac{\partial x^n}{\partial a^i}\left(\frac{\partial^2 x^n}{\partial t^2}\right)_L = \left[\frac{\partial}{\partial t}\left(\frac{\partial x^n}{\partial a^i}\,\frac{\partial x^n}{\partial t}\right)\right]_L - \left(\frac{\partial x^n}{\partial t}\right)_L\frac{\partial}{\partial a^i}\left(\frac{\partial x^n}{\partial t}\right)_L \tag{10.25}$$

The second term on the right-hand side is equal to

$$\frac{\partial}{\partial a^i}\left(\frac{1}{2}\frac{\partial x^n}{\partial t}\,\frac{\partial x^n}{\partial t}\right)_L = \frac{\partial}{\partial a^i}\left(\frac{\mathbf{v}_A^2}{2}\right) \tag{10.26}$$

so that (10.24) assumes the form

$$\left[\frac{\partial}{\partial t}\left(\frac{\partial x^n}{\partial a^i}\,\frac{\partial x^n}{\partial t}\right)\right]_L = \frac{\partial L_A}{\partial a^i} + T\,\frac{\partial s}{\partial a^i} \tag{10.27}$$

with

$$L_{\mathrm{A}} = \frac{\mathbf{v}_{\mathrm{A}}^2}{2} - \phi_{\mathrm{a}} - h \tag{10.28}$$

The symbol L_{A} is the *Lagrangian function in the absolute system*. This function will be treated in great detail in later chapters. Since we assumed isentropic conditions, we have

$$\frac{ds}{dt} = \left(\frac{\partial s}{\partial t}\right)_{\mathrm{L}} = 0 \tag{10.29}$$

so that s is constant on the fluid-particle trajectory. Next we introduce the *action integral* of the absolute system:

$$\boxed{W_{\mathrm{A}} = \int_0^t L_{\mathrm{A}}\, dt} \tag{10.30}$$

and similarly

$$\beta = \int_0^t T\, dt \quad \text{or} \quad \frac{d\beta}{dt} = \dot{\beta} = T \tag{10.31}$$

where T is the absolute temperature. Using this notation, we carry out the time integration of (10.27) and find

$$\left(\frac{\partial x^n}{\partial a^i}\frac{\partial x^n}{\partial t}\right)_{\mathrm{L},t} = \left(\frac{\partial x^n}{\partial a^i}\frac{\partial x^n}{\partial t}\right)_{t=0} + \frac{\partial W_{\mathrm{A}}}{\partial a^i} + \beta\,\frac{\partial s}{\partial a^i} \tag{10.32}$$

since the enumeration coordinates do not change with time. It should be recognized that the first term on the right-hand side is invariant in time. In the Eulerian system this invariance may be stated as

$$\frac{d}{dt}\left(\frac{\partial x^n}{\partial a^i}\frac{\partial x^n}{\partial t}\right)_{t=0} = 0 \tag{10.33}$$

At time $t = 0$, as before, we require that the Lagrangian and the Eulerian systems coincide so that

$$\left(\frac{\partial x^n}{\partial a^i}\frac{\partial x^n}{\partial t}\right)_{t=0} = \left(\frac{\partial x^n}{\partial x^i}\frac{\partial x^n}{\partial t}\right)_{t=0} = \left(\delta_i^n\,\frac{\partial x^n}{\partial t}\right)_{t=0} = \left(\frac{\partial x^i}{\partial t}\right)_{t=0} = \left(u_{\mathrm{A}}^i\right)_{t=0} \tag{10.34}$$

Using this identity, we finally get from (10.32) the expression

$$\left(\frac{\partial x^n}{\partial a^i}\frac{\partial x^n}{\partial t}\right)_{\mathrm{L},t} = \left(\frac{\partial x^i}{\partial t}\right)_{t=0} + \frac{\partial W_{\mathrm{A}}}{\partial a^i} + \beta\,\frac{\partial s}{\partial a^i} \tag{10.35}$$

It is desirable to return to the x^i system. This is accomplished by setting $i = r$ in (10.35), multiplying both sides by $(\partial a^r/\partial x^k)$, and then summing over r. The result is

$$\left(\frac{\partial a^r}{\partial x^k}\frac{\partial x^n}{\partial a^r}\frac{\partial x^n}{\partial t}\right)_{\mathrm{L},t} = \frac{\partial a^r}{\partial x^k}\left(\frac{\partial x^r}{\partial t}\right)_{t=0} + \frac{\partial a^r}{\partial x^k}\frac{\partial W_{\mathrm{A}}}{\partial a^r} + \beta\frac{\partial a^r}{\partial x^k}\frac{\partial s}{\partial a^r} \tag{10.36}$$

We now recall (10.17), where we put $i = k$ and $n = r$ for conformity of notation with (10.36), yielding

$$\frac{\partial \psi}{\partial x^k} = \frac{\partial a^r}{\partial x^k}\frac{\partial \psi}{\partial a^r} \tag{10.37}$$

Now we identify $\psi = x^n$, W_{A}, and s in succession, and find

$$\frac{\partial x^n}{\partial x^k} = \frac{\partial a^r}{\partial x^k}\frac{\partial x^n}{\partial a^r} = \delta_k^n, \qquad \frac{\partial W_{\mathrm{A}}}{\partial x^k} = \frac{\partial a^r}{\partial x^k}\frac{\partial W_{\mathrm{A}}}{\partial a^r}, \qquad \frac{\partial s}{\partial x^k} = \frac{\partial a^r}{\partial x^k}\frac{\partial s}{\partial a^r} \tag{10.38}$$

Using these expressions in (10.36) we obtain

$$\left(\frac{\partial x^k}{\partial t}\right)_{\mathrm{L}} = \frac{\partial a^n}{\partial x^k}\left(\frac{\partial x^n}{\partial t}\right)_{t=0} + \frac{\partial W_{\mathrm{A}}}{\partial x^k} + \beta\frac{\partial s}{\partial x^k}, \qquad k = 1, 2, 3 \tag{10.39}$$

Next we write (10.39) down for $k = 1, 2, 3$ and multiply each equation by the unit vectors $\mathbf{i}, \mathbf{j}, \mathbf{k}$, respectively. By adding the results and using the definitions

$$\begin{aligned} &\nabla a^k = \mathbf{i}\frac{\partial a^k}{\partial x^1} + \mathbf{j}\frac{\partial a^k}{\partial x^2} + \mathbf{k}\frac{\partial a^k}{\partial x^3} \quad \text{and} \quad \mathbf{v}_{\mathrm{A}} = \mathbf{i}u_{\mathrm{A}} + \mathbf{j}v_{\mathrm{A}} + \mathbf{k}w_{\mathrm{A}} \\ &\text{with} \quad \left(\frac{\partial x^1}{\partial t}\right)_{\mathrm{L}} = u_{\mathrm{A}}, \qquad \left(\frac{\partial x^2}{\partial t}\right)_{\mathrm{L}} = v_{\mathrm{A}}, \qquad \left(\frac{\partial x^3}{\partial t}\right)_{\mathrm{L}} = w_{\mathrm{A}} \end{aligned} \tag{10.40}$$

we obtain the absolute velocity

$$\boxed{\mathbf{v}_{\mathrm{A}} = u_{\mathrm{A}_0}\nabla a^1 + v_{\mathrm{A}_0}\nabla a^2 + w_{\mathrm{A}_0}\nabla a^3 + \nabla W_{\mathrm{A}} + \beta\nabla s} \tag{10.41}$$

This equation is the general Weber transformation. The suffix 0 refers to $t = 0$. We observe that

$$\dot{u}_{\mathrm{A}_0} = 0, \quad \dot{v}_{\mathrm{A}_0} = 0, \quad \dot{w}_{\mathrm{A}_0} = 0, \qquad \dot{a}^1 = 0, \quad \dot{a}^2 = 0, \quad \dot{a}^3 = 0 \tag{10.42}$$

Now we define the relation

$$\begin{aligned} &\text{(a)} \qquad \mathbf{B}_{\mathrm{A}} = \mathbf{v}_{\mathrm{A}} - \nabla \mathrm{W}_{\mathrm{A}} - \beta\nabla s \implies \\ &\text{(b)} \quad \nabla\times\mathbf{B}_{\mathrm{A}} = \nabla\times\mathbf{v}_{\mathrm{A}} - \nabla\beta\times\nabla s \end{aligned} \tag{10.43}$$

With (10.43a) the Weber transformation can be written more briefly as

$$\boxed{\mathbf{B}_{\mathrm{A}} = u_{\mathrm{A}_0}\nabla a^1 + v_{\mathrm{A}_0}\nabla a^2 + w_{\mathrm{A}_0}\nabla a^3} \tag{10.44}$$

As stated at the beginning of this chapter, the Weber transformation serves as an important tool in some parts of the following sections.

10.3 The baroclinic Ertel–Rossby invariant

A first application of the Weber transformation (10.41) in combination with Ertel's version of the continuity equation (10.8) will now be given. Here we replace the coordinates $q^k(x^i)$ by the conservative field functions ψ^i, $i = 1, 2, 3$. In this case the continuity equation becomes

$$\frac{d}{dt}\big[\alpha J_x(\psi^1, \psi^2, \psi^3)\big] = \frac{d}{dt}\big(\alpha\, \nabla\psi^1 \cdot \nabla\psi^2 \times \nabla\psi^3\big) = 0 \tag{10.45}$$

The next step is to take the curl of $\mathbf{B}_\mathrm{A}$, yielding

$$\nabla \times \mathbf{B}_\mathrm{A} = \nabla u_{\mathrm{A}_0} \times \nabla a^1 + \nabla v_{\mathrm{A}_0} \times \nabla a^2 + \nabla w_{\mathrm{A}_0} \times \nabla a^3 \tag{10.46}$$

Scalar multiplication of this expression by $\alpha \mathbf{B}_\mathrm{A}$ yields for the right-hand side

$$\begin{aligned}\alpha\mathbf{B}_\mathrm{A} \cdot \nabla \times \mathbf{B}_\mathrm{A} = \alpha(&v_{\mathrm{A}_0}\, \nabla a^2 \cdot \nabla u_{\mathrm{A}_0} \times \nabla a^1 + w_{\mathrm{A}_0}\, \nabla a^3 \cdot \nabla u_{\mathrm{A}_0} \times \nabla a^1 \\ &+ u_{\mathrm{A}_0}\, \nabla a^1 \cdot \nabla v_{\mathrm{A}_0} \times \nabla a^2 + w_{\mathrm{A}_0}\, \nabla a^3 \cdot \nabla v_{\mathrm{A}_0} \times \nabla a^2 \\ &+ u_{\mathrm{A}_0}\, \nabla a^1 \cdot \nabla w_{\mathrm{A}_0} \times \nabla a^3 + v_{\mathrm{A}_0}\, \nabla a^2 \cdot \nabla w_{\mathrm{A}_0} \times \nabla a^3)\end{aligned} \tag{10.47}$$

In view of (10.42) and (10.45) we find that d/dt of each term is zero. Thus, for the absolute system, we find

$$\boxed{\frac{d}{dt}(\alpha\mathbf{B}_\mathrm{A} \cdot \nabla \times \mathbf{B}_\mathrm{A}) = 0} \tag{10.48}$$

The invariant field function

$$\boxed{\psi_\mathrm{ER} = \alpha\mathbf{B}_\mathrm{A} \cdot \nabla \times \mathbf{B}_\mathrm{A}} \tag{10.49}$$

is called the *baroclinic Ertel–Rossby invariant*. For additional details see Ertel and Rossby (1949). For relative motion on the rotating earth we replace $\mathbf{v}_\mathrm{A}$ by $\mathbf{v} + \mathbf{v}_\Omega$ with $\mathbf{v}_\Omega = \Omega \times \mathbf{r}$ and obtain with (10.43b)

$$\frac{d}{dt}\{\alpha(\mathbf{v} + \mathbf{v}_\Omega - \nabla W_\mathrm{A} - \beta\, \nabla s) \cdot [\nabla \times (\mathbf{v} + \mathbf{v}_\Omega) - \nabla\beta \times \nabla s]\} = 0 \tag{10.50}$$

where the action integral (10.28) is now given by

$$W_\mathrm{A} = \int_0^t \left(\frac{(\mathbf{v} + \mathbf{v}_\Omega)^2}{2} - \phi - h\right) dt \tag{10.51}$$

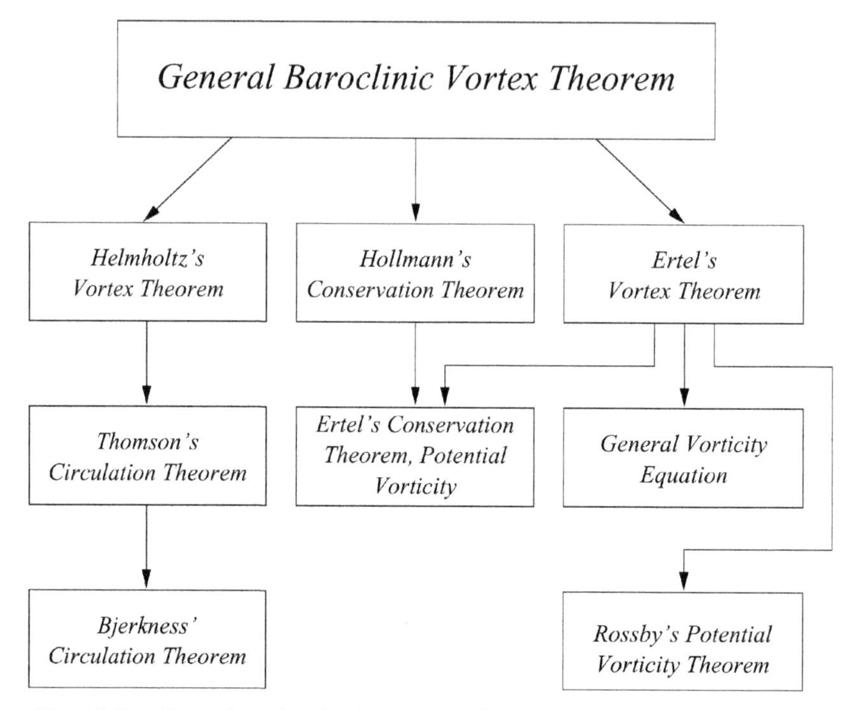

Fig. 10.1 A flow chart for the derivation of vortex and circulation theorems.

10.4 Circulation and vorticity theorems for frictionless baroclinic flow

In the previous section we have given a first example of an invariant. Other invariants will be derived in the next section, together with the equations describing the circulation and vorticity in baroclinic media. There are various ways of obtaining the desired results irrespective of the historic development. At the beginning of our analysis we derive a very general baroclinic vortex law from which all other considerations follow as shown in Figure 10.1.

10.4.1 A general baroclinic vortex theorem

The first step in the derivation of the general baroclinic vortex theorem is taken by performing scalar multiplication of (10.46) by $\alpha \nabla \psi$, where ψ is an arbitrary scalar field function that is not necessarily invariant. We simply obtain

$$\begin{aligned} \alpha \nabla \times \mathbf{B}_{\rm A} \cdot \nabla \psi &= \alpha \nabla u_{\rm A_0} \times \nabla a^1 \cdot \nabla \psi + \alpha \nabla v_{\rm A_0} \times \nabla a^2 \cdot \nabla \psi + \alpha \nabla w_{\rm A_0} \times \nabla a^3 \cdot \nabla \psi \\ &= \alpha J(u_{\rm A_0}, a^1, \psi) + \alpha J(v_{\rm A_0}, a^2, \psi) + \alpha J(w_{\rm A_0}, a^3, \psi) \end{aligned} \tag{10.52}$$

Consider, for example, the first term on the right-hand side of this equation. Since u_{A_0} and a^1 are invariants, we find from Ertel's version of the continuity equation that

$$\frac{d}{dt}\big[\alpha J(u_{A_0}, a^1, \psi)\big] = \alpha J(u_{A_0}, a^1, \dot{\psi}) \tag{10.53}$$

For convenience we have omitted the suffix x on the Jacobian J. Corresponding equations can be written for the second and third terms on the right-hand side of (10.52). Therefore, taking the individual time derivative of (10.52), we find

$$\boxed{\begin{aligned} &\dot{s} = 0: \\ &\frac{d}{dt}(\alpha \nabla \times \mathbf{B}_A \cdot \nabla\psi) = \alpha J(u_{A_0}, a^1, \dot{\psi}) + \alpha J(v_{A_0}, a^2, \dot{\psi}) + \alpha J(w_{A_0}, a^3, \dot{\psi}) \\ &\qquad = \alpha \nabla \times \mathbf{B}_A \cdot \nabla\dot{\psi} \end{aligned}} \tag{10.54}$$

This is the desired general vortex law for the absolute system where the arbitrary field function ψ is at our disposal and may be assigned to various variables. Since the Weber transformation was obtained for the case of isentropic motion, we must not ignore the condition $\dot{s} = 0$ as expressed in (10.54). If additionally the arbitrary field function ψ is invariant, i.e. $\dot{\psi} = 0$, then (10.54) reduces to *Hollmann's conservation law*

$$\boxed{\begin{aligned} &\dot{s} = 0, \quad \dot{\psi} = 0: \\ &\frac{d}{dt}(\alpha \nabla \times \mathbf{B}_A \cdot \nabla\psi) = 0 \end{aligned}} \tag{10.55}$$

The paper by Hollmann (1965) should be consulted for further details. In contrast to (10.48), now we have a one-parametric conservation law since the arbitrary field function ψ is still at our disposal. The only restriction on ψ is that it must be invariant. Transformation of (10.54) and (10.55) into the relative system of the rotating earth gives

$$\begin{aligned} &\dot{s} = 0: \\ &\frac{d}{dt}\{\alpha[\nabla \times (\mathbf{v} + \mathbf{v}_\Omega) - \nabla\beta \times \nabla s] \cdot \nabla\psi\} = \alpha\big[\nabla \times (\mathbf{v} + \mathbf{v}_\Omega) - \nabla\beta \times \nabla s\big] \cdot \nabla\dot{\psi} \\ &\dot{s} = 0, \quad \dot{\psi} = 0: \\ &\frac{d}{dt}\{\alpha[\nabla \times (\mathbf{v} + \mathbf{v}_\Omega) - \nabla\beta \times \nabla s] \cdot \nabla\psi\} = 0 \end{aligned} \tag{10.56}$$

where we have replaced $\mathbf{B}_A$ by means of (10.43a).

10.4.2 Ertel's vortex theorem

The general baroclinic vortex theorem contains Ertel's vortex theorem (Ertel, 1942) as a special case. We show this by introducing (10.43b) into (10.54). This results in the following equation:

$$\frac{d}{dt}(\alpha\,\nabla\times\mathbf{v}_{\mathrm{A}}\cdot\nabla\psi)-\frac{d}{dt}\big[\alpha J(\beta,s,\psi)\big]=\alpha\,\nabla\times\mathbf{v}_{\mathrm{A}}\cdot\nabla\dot{\psi}-\alpha J(\beta,s,\dot{\psi}) \qquad (10.57)$$

Since $\dot{s}=0$ by assumption, we find from Ertel's form of the continuity equation (10.8) for the second term on the left of (10.57) with $\dot{\beta}=T$ (see (10.31)) the expression

$$\frac{d}{dt}\big[\alpha J(\beta,s,\psi)\big]=\alpha J(T,s,\psi)+\alpha J(\beta,s,\dot{\psi}) \qquad (10.58)$$

The first term on the right-hand side can be reformulated with the help of Gibbs' fundamental equation for frictionless dry air in the form (10.19). Taking the curl of this expression, we find

$$\nabla T\times\nabla s=-\nabla\alpha\times\nabla p \qquad (10.59)$$

Using the definition of the Jacobian $J(T,s,\psi)$, we find from (10.59)

$$J(T,s,\psi)=\nabla p\times\nabla\alpha\cdot\nabla\psi \qquad (10.60)$$

so that (10.57) can finally be written as

$$\boxed{\frac{d}{dt}(\alpha\,\nabla\times\mathbf{v}_{\mathrm{A}}\cdot\nabla\psi)-\alpha\,\nabla\times\mathbf{v}_{\mathrm{A}}\cdot\nabla\dot{\psi}=\alpha\,\nabla p\times\nabla\alpha\cdot\nabla\psi=\alpha J(p,\alpha,\psi)} \qquad (10.61)$$

This is Ertel's celebrated vortex theorem from which various other theorems follow, as shown in Figure 10.1. If this theorem is required for the system of the rotating earth, then the velocity $\mathbf{v}_{\mathrm{A}}$ must be replaced by $\mathbf{v}+\mathbf{v}_{\Omega}$. The interested reader may wish to read a paper by Ertel (1954). Moreover, Fortak (1956) has addressed the question of general hydrodynamic vortex laws.

10.4.3 Ertel's conservation theorem, potential vorticity

First of all we observe that the right-hand side of (10.61) vanishes not only when the barotropic condition ($\nabla p\times\nabla\alpha=0$) applies but also if the arbitrary function ψ depends on α and p so that $\psi=\psi(\alpha,p)$. The specific entropy possesses this property, so we may write

$$\nabla s=\left(\frac{\partial s}{\partial p}\right)_{\alpha}\nabla p+\left(\frac{\partial s}{\partial \alpha}\right)_{p}\nabla\alpha \qquad (10.62)$$

For isentropic state changes we find from (10.61) the important conservation law

$$\boxed{\frac{d}{dt}(\alpha\,\nabla\times\mathbf{v}_{\mathrm{A}}\cdot\nabla s)=0} \tag{10.63}$$

The expression

$$\boxed{P_{\mathrm{E}}=\alpha\,\nabla\times\mathbf{v}_{\mathrm{A}}\cdot\nabla s} \tag{10.64}$$

is known as the potential vorticity or *Ertel's vortex invariant*. This conservation law is valid for isentropic motion. There exists another conservation law leading to the formulation of the potential vorticity according to Rossby (1940). This formulation will be discussed in a later section.

It is interesting to remark that Ertel's conservation theorem could have been obtained directly from the general vortex theorem (10.54) by setting $\psi=s$.

10.4.4 The general vorticity theorem

There are various ways to derive the general vorticity theorem. A very elegant way is based on Ertel's vortex theorem which we will use to derive the general vorticity equation in geographical coordinates. Owing to the rigid rotation of the geographical coordinate system with $\mathbf{v}_{\Omega}=\boldsymbol{\Omega}\times\mathbf{r}$ we have

$$\nabla\cdot\mathbf{v}_{\Omega}=0,\qquad \nabla\times\mathbf{v}_{\Omega}=2\boldsymbol{\Omega}=2\Omega\sin\varphi\,\mathbf{e}_r+2\Omega\cos\varphi\,\mathbf{e}_{\varphi}=f\mathbf{e}_r+l\mathbf{e}_{\varphi} \tag{10.65}$$

Now we define the *relative vorticity* ζ and the *absolute vorticity,* η by means of

$$\boxed{\zeta=\mathbf{e}_r\cdot\nabla\times\mathbf{v},\qquad \eta=\mathbf{e}_r\cdot\nabla\times\mathbf{v}_{\mathrm{A}}=\zeta+f} \tag{10.66}$$

On setting $\psi=r$ in Ertel's vortex theorem (10.61) and using the definition of the absolute vorticity, we obtain with $\dot{r}=w$ and $\nabla r=\mathbf{e}_r$

$$\boxed{\frac{d\eta}{dt}=\nabla\times\mathbf{v}_{\mathrm{A}}\cdot\nabla w+\nabla p\times\nabla\alpha\cdot\mathbf{e}_r-\eta\,\nabla\cdot\mathbf{v}_{\mathrm{A}}} \tag{10.67}$$

where use of the continuity equation has been made. We now introduce the geographical coordinates and describe the relative velocity by means of physical measure numbers (u, v, w). This requires several manipulations, which are listed next:

$$
\begin{aligned}
\nabla \cdot \mathbf{v}_\mathrm{A} &= \nabla \cdot \mathbf{v}_{\mathrm{h,A}} + \nabla \cdot (\mathbf{e}_r w) = \nabla_\mathrm{h} \cdot \mathbf{v}_\mathrm{h} + \frac{2w}{r} + \frac{\partial w}{\partial r} \\
\nabla \times \mathbf{v}_\mathrm{A} &= \nabla \times \mathbf{v} + 2\boldsymbol{\Omega} \\
&= (\nabla \times \mathbf{v})_\lambda \mathbf{e}_\lambda + (\nabla \times \mathbf{v})_\varphi \mathbf{e}_\varphi + (\nabla \times \mathbf{v})_r \mathbf{e}_r + l \mathbf{e}_\varphi + f \mathbf{e}_r \\
&= (\nabla \times \mathbf{v})_\lambda \mathbf{e}_\lambda + \left[(\nabla \times \mathbf{v})_\varphi + l\right] \mathbf{e}_\varphi + \eta \mathbf{e}_r \\
\nabla \times \mathbf{v}_\mathrm{A} \cdot \nabla w &= \{(\nabla \times \mathbf{v})_\lambda \mathbf{e}_\lambda + \left[(\nabla \times \mathbf{v})_\varphi + l\right] \mathbf{e}_\varphi\} \cdot \nabla_\mathrm{h} w + \eta \frac{\partial w}{\partial r} \\
&= -\frac{1}{r\cos\varphi}\frac{\partial w}{\partial \lambda}\left(\frac{\partial v}{\partial r} + \frac{v}{r}\right) + \frac{1}{r}\frac{\partial w}{\partial \varphi}\left(\frac{\partial u}{\partial r} + \frac{u}{r} + l\right) + \eta \frac{\partial w}{\partial r} \\
\nabla p \times \nabla \alpha \cdot \mathbf{e}_r &= \left(\nabla_\mathrm{h} p + \mathbf{e}_r \frac{\partial p}{\partial r}\right) \times \left(\nabla_\mathrm{h} \alpha + \mathbf{e}_r \frac{\partial \alpha}{\partial r}\right) \cdot \mathbf{e}_r = \nabla_\mathrm{h} p \times \nabla_\mathrm{h} \alpha \cdot \mathbf{e}_r
\end{aligned}
\tag{10.68}
$$

Here we have identified the horizontal part of the gradient operator by the suffix h. Substitution of these expressions into (10.67) gives the vorticity equation in geographical coordinates:

$$
\begin{aligned}
\frac{d\eta}{dt} &= -\eta\left(\nabla_\mathrm{h} \cdot \mathbf{v}_\mathrm{h} + \underline{\frac{2w}{r}}\right) - \frac{1}{r\cos\varphi}\frac{\partial w}{\partial \lambda}\left(\frac{\partial v}{\partial r} + \underline{\frac{v}{r}}\right) \\
&\quad + \frac{1}{r}\frac{\partial w}{\partial \varphi}\left(\frac{\partial u}{\partial r} + \underline{\frac{u}{r} + l}\right) + \nabla_\mathrm{h} p \times \nabla_\mathrm{h} \alpha \cdot \mathbf{e}_r
\end{aligned}
\tag{10.69}
$$

A scale analysis of the vorticity equation would show that the underlined terms may be omitted. If this is done we obtain the simplified form of the vorticity equation

$$
\boxed{\frac{d\eta}{dt} = -\eta\, \nabla_\mathrm{h} \cdot \mathbf{v}_\mathrm{h} + \nabla_\mathrm{h} w \cdot \mathbf{e}_r \times \frac{\partial \mathbf{v}_\mathrm{h}}{\partial r} + \nabla_\mathrm{h} p \times \nabla_\mathrm{h} \alpha \cdot \mathbf{e}_r}
\tag{10.70}
$$

where we have made use of

$$
\mathbf{e}_r \times \frac{\partial \mathbf{v}_\mathrm{h}}{\partial r} = -\mathbf{e}_\lambda \frac{\partial v}{\partial r} + \mathbf{e}_\varphi \frac{\partial u}{\partial r}
\tag{10.71}
$$

According to (10.70) the change of the absolute vorticity with time is due to three effects.

(i) The *divergence effect* describing the horizontal divergence of the two-dimensional flow (the first term on the right-hand side).

(ii) Production of vorticity due to the interaction of the vertical gradient of the horizontal velocity with the horizontal gradient of the vertical velocity. This *tipping-term effect* will be discussed in some detail in connection with the so-called quasi-geostrophic theory (the second term on the right-hand side).

(iii) The *solenoidal effect*, i.e. the number of solenoids per unit area (the last term on the right-hand side).

It is of some advantage to express the vorticity equation in the p system, in which pressure is the vertical coordinate. In this system the solenoidal term does not appear explicitly. Again, the starting point is Ertel's vortex theorem (10.61), to which we apply the following simplifications.

(i) We omit the vertical component w of the relative velocity from the term $\nabla \times \mathbf{v}_A$.
(ii) We omit the horizontal component l in $2\mathbf{\Omega}$.
(iii) We apply the metric simplification $u/r = v/r = 0$ as in (10.69).
(iv) We apply the hydrostatic approximation.

These simplifications are summarized next, together with the resulting approximation of the curl of the absolute velocity in the geographical coordinate system (λ, φ, r). The deformation velocity is approximated as a one-component vector. It is also assumed that W_D^3 depends only on p so that the curl of $\mathbf{v}_D$ vanishes:

$$
\begin{aligned}
&\mathbf{v}_A = \mathbf{v} + \mathbf{v}_\Omega + \mathbf{v}_D = \mathbf{v}_h + \mathbf{\Omega} \times \mathbf{r} + \mathbf{q}_3 W_D^3 \\
&\nabla \times \mathbf{v}_\Omega = 2\mathbf{\Omega} = f\mathbf{e}_r, \qquad \nabla \times \mathbf{v}_D = 0, \qquad \frac{u}{r} = \frac{v}{r} = 0 \\
&\nabla \times \mathbf{v}_A = \nabla \times \mathbf{v}_h + f\mathbf{e}_r = -\mathbf{e}_\lambda \frac{\partial v}{\partial r} + \mathbf{e}_\varphi \frac{\partial u}{\partial r} + \mathbf{e}_r(\zeta + f) \\
&\qquad = \mathbf{e}_r \times \frac{\partial \mathbf{v}_h}{\partial r} + \mathbf{e}_r \eta = -g\rho \mathbf{e}_r \times \frac{\partial \mathbf{v}_h}{\partial p} + \mathbf{e}_r \eta \\
&\text{where} \quad \zeta = \frac{1}{r^2 \cos\varphi}\left(\frac{\partial}{\partial \lambda}(rv) - \frac{\partial}{\partial \varphi}(ru\cos\varphi)\right)
\end{aligned}
\tag{10.72}
$$

The hydrostatic approximation was introduced as the last step. Next the general field function ψ will be replaced in (10.61) by the pressure coordinate p. By splitting the gradient of p into its horizontal $(\nabla_h p)$ and vertical $(\mathbf{e}_r\, \partial p/\partial r)$ parts, observing (10.72) and the hydrostatic relation, we easily find from vector-analytic operations the expression

$$
\alpha \nabla p \cdot \nabla \times \mathbf{v}_A = -g\, \nabla_h p \cdot \mathbf{e}_r \times \frac{\partial \mathbf{v}_h}{\partial p} - g\eta \tag{10.73}
$$

Therefore, Ertel's theorem can be written as

$$
\frac{d}{dt}\left(\eta + \nabla_h p \cdot \mathbf{e}_r \times \frac{\partial \mathbf{v}_h}{\partial p}\right) = \nabla_h \dot{p} \cdot \mathbf{e}_r \times \frac{\partial \mathbf{v}_h}{\partial p} + \eta \frac{\partial \dot{p}}{\partial p} \tag{10.74}
$$

To complete the transformation to the (λ, φ, p) system we replace the horizontal gradient by the horizontal gradient in the p system:

$$\nabla_{\rm h} = \nabla_{{\rm h},p} + \rho(\nabla_{{\rm h},p}\phi)\frac{\partial}{\partial p} \tag{10.75}$$

The proof of this equation is left as an exercise. Now we use (10.75) to establish a relation between the absolute vorticities in these two systems. On applying the operators in the form $\mathbf{e}_r \cdot \nabla \times \mathbf{v}_{\rm h}$, we find by splitting $\mathbf{v}$ and ∇ into their horizontal and vertical parts, for ζ

$$\begin{aligned}
\zeta = \mathbf{e}_r \cdot \nabla \times \mathbf{v} = \mathbf{e}_r \cdot \nabla_{\rm h} \times \mathbf{v}_{\rm h} &= \mathbf{e}_r \cdot \nabla_{{\rm h},p} \times \mathbf{v}_{\rm h} + \rho\mathbf{e}_r \cdot \nabla_{{\rm h},p}\phi \times \frac{\partial \mathbf{v}_{\rm h}}{\partial p} \\
&= \zeta_p - \rho\, \nabla_{{\rm h},p}\phi \cdot \left(\mathbf{e}_r \times \frac{\partial \mathbf{v}_{\rm h}}{\partial p}\right) \\
\text{or} \quad \eta &= \eta_p - \rho\, \nabla_{{\rm h},p}\phi \cdot \left(\mathbf{e}_r \times \frac{\partial \mathbf{v}_{\rm h}}{\partial p}\right) \\
\text{with} \quad \zeta_p &= \mathbf{e}_r \cdot \nabla_{{\rm h},p} \times \mathbf{v}_{\rm h}
\end{aligned} \tag{10.76}$$

On replacing $\nabla_{\rm h}(\cdots)$ in (10.74) by (10.75) and recognizing that the horizontal pressure gradient in the p system vanishes, we find

$$\boxed{\frac{d\eta_p}{dt} = \left(\mathbf{e}_r \times \frac{\partial \mathbf{v}_{\rm h}}{\partial p}\right) \cdot \nabla_{{\rm h},p}\omega + \eta_p\frac{\partial \omega}{\partial p}} \tag{10.77}$$

The change in pressure $dp/dt = \dot{p}$ has been replaced by the vertical velocity ω in the p system and in the absolute vorticity η_p. This very useful equation and its application will be discussed in some detail in conjunction with the quasi-geostrophic theory. Once again it is pointed out that the solenoidal term does not appear explicitly since we have replaced ψ by p in (10.61). Since a coordinate transformation cannot change the physical content of the vorticity equation (10.70), the solenoidal effect must then be hidden in the remaining terms.

10.4.5 Rossby's formulation of the potential vorticity

Originally Rossby (1940) used the metrically simplified horizontal equations of motion in the θ system to obtain a very useful conservative quantity known as the *Rossby potential vorticity*. To demonstrate once again the central role of Ertel's vortex theorem we will not follow the original derivation but proceed differently. Since Rossby assumed frictionless isentropic dry air motion ($\dot{\theta} = 0$), we put $\psi = \theta$

in (10.61). As a first step in the derivation we combine the equation of potential temperature with the ideal-gas law and find

$$\frac{\nabla\theta}{\theta} = \frac{\nabla\alpha}{\alpha} + \frac{c_v}{c_p}\frac{\nabla p}{p} \tag{10.78}$$

where c_p and c_v are the specific heats at constant pressure and volume, respectively. In view of (10.78) we recognize that the entire right-hand side of (10.61) vanishes. The resulting conservation law

$$\boxed{\frac{d}{dt}(\alpha\,\nabla\times\mathbf{v}_{\mathrm{A}}\cdot\nabla\theta) = 0} \tag{10.79}$$

will now be rewritten in the form proposed by Rossby. As before, we decompose the gradient operator of the scalar function into its horizontal and vertical parts. The conserved quantity, using (10.72), assumes the form

$$\begin{aligned}(\alpha\,\nabla\times\mathbf{v}_{\mathrm{A}}\cdot\nabla\theta) &= \alpha\left(\mathbf{e}_r\times\frac{\partial\mathbf{v}_{\mathrm{h}}}{\partial r} + \mathbf{e}_r\eta\right)\cdot\left(\nabla_{\mathrm{h}}\theta + \mathbf{e}_r\frac{\partial\theta}{\partial r}\right)\\ &= -g\left(\nabla_{\mathrm{h}}\theta\cdot\mathbf{e}_r\times\frac{\partial\mathbf{v}_{\mathrm{h}}}{\partial p} + \eta\frac{\partial\theta}{\partial p}\right)\end{aligned} \tag{10.80}$$

where, once again, we have used the hydrostatic approximation.

Next, we wish to introduce pressure as the vertical coordinate. Using the transformation rules of partial derivatives as discussed in Section M4.2, we readily find

$$\nabla_{\mathrm{h},q^3} = \nabla_{\mathrm{h},\xi} - (\nabla_{\mathrm{h},\xi}q^3)\frac{\partial}{\partial q^3} \tag{10.81}$$

When we apply this equation to the potential temperature we obtain the relation of the horizontal gradients in the two systems using $r = q^3$ and $p = \xi$ as vertical coordinates:

$$\nabla_{\mathrm{h},r}\theta = \nabla_{\mathrm{h},p}\theta + \rho\frac{\partial\theta}{\partial p}\nabla_{\mathrm{h},p}\phi \tag{10.82}$$

Now we replace the horizontal gradients in (10.79) by means of (10.82), using (10.76), and find

$$\frac{d}{dt}\left[\nabla_{\mathrm{h},p}\theta\cdot\left(\mathbf{e}_r\times\frac{\partial\mathbf{v}_{\mathrm{h}}}{\partial p}\right) + \eta_p\frac{\partial\theta}{\partial p}\right] = 0 \tag{10.83}$$

Here the conserved quantity P_{R},

$$\boxed{P_{\mathrm{R}} = \nabla_{\mathrm{h},p}\theta\cdot\left(\mathbf{e}_r\times\frac{\partial\mathbf{v}_{\mathrm{h}}}{\partial p}\right) + \eta_p\frac{\partial\theta}{\partial p}} \tag{10.84}$$

is Rossby's formulation of the potential vorticity in the p system.

Before we discuss this very useful concept in some detail, we wish to transform (10.83) so that the potential temperature rather than the pressure appears as the vertical coordinate. We accomplish this by first setting $q^3 = \theta$ and $\xi = p$ in (10.81). With the help of the hydrostatic equation we find

$$\nabla_{\mathrm{h},\theta} = \nabla_{\mathrm{h},p} - \frac{\partial p}{\partial \theta} \nabla_{\mathrm{h},p}\theta \frac{\partial}{\partial p} \tag{10.85}$$

In order to introduce the vorticity with respect to an isentropic surface, we take the vector product of the operators appearing in (10.85) with the horizontal velocity and then use scalar multiplication by the vertical unit vector of the geographical coordinates. The result is

$$\mathbf{e}_r \cdot \nabla_{\mathrm{h},\theta} \times \mathbf{v}_\mathrm{h} = \zeta_\theta = \mathbf{e}_r \cdot \nabla_{\mathrm{h},p} \times \mathbf{v}_\mathrm{h} - \frac{\partial p}{\partial \theta} \mathbf{e}_r \cdot \nabla_{\mathrm{h},p}\theta \times \frac{\partial \mathbf{v}_\mathrm{h}}{\partial p} \tag{10.86}$$

or

$$\boxed{\begin{aligned} \zeta_\theta &= \zeta_p - \frac{\partial p}{\partial \theta} \mathbf{e}_r \cdot \nabla_{\mathrm{h},p}\theta \times \frac{\partial \mathbf{v}_\mathrm{h}}{\partial p} \\ \eta_\theta &= \eta_p - \frac{\partial p}{\partial \theta} \mathbf{e}_r \cdot \nabla_{\mathrm{h},p}\theta \times \frac{\partial \mathbf{v}_\mathrm{h}}{\partial p} \end{aligned}} \tag{10.87}$$

The equation for the absolute vorticity has been found by adding f to both sides of the first equation. Equation (10.87) relates the two systems with potential temperature and pressure as vertical coordinates. As the final step we eliminate η_p from (10.83) with the help of (10.87) and find the desired expression

$$\boxed{\frac{dP_\mathrm{R}}{dt} = \frac{d}{dt}\left(\eta_\theta \frac{\partial \theta}{\partial p}\right) = 0} \tag{10.88}$$

If the assumptions leading to this equation are satisfied then the potential vorticity is invariant or a conservative quantity along the trajectory of an air parcel. The term $\partial\theta/\partial p$ is a measure of hydrostatc stability.

Before applying this conservation rule to a problem of large-scale motion we will obtain from (10.88) an approximate but useful formula. From thermodynamics and atmospheric statics the following expression can be easily derived:

$$\frac{\partial \theta}{\partial p} = -\frac{\theta}{T} \frac{\gamma_\mathrm{a} - \gamma_\mathrm{g}}{g\rho} \tag{10.89}$$

where $\gamma_\mathrm{a} = g/c_{p,0}$ and γ_g represent the dry adiabatic and the observed lapse rates. Next, we approximate the actual wind shear by the thermal wind in the (x, y, p)-coordinates, see (9.69),

$$\frac{\partial \mathbf{v}_\mathrm{h}}{\partial p} \approx -\frac{R_0}{pf} \mathbf{e}_r \times \nabla_{\mathrm{h},p} T \tag{10.90}$$

Logarithmic differentiation of the defining relation for the potential temperature on isobaric surfaces yields

$$\nabla_{\mathrm{h},p}\theta = \frac{\theta}{T}\nabla_{h,p}T \tag{10.91}$$

With these relations (10.84) can be written as

$$P_{\mathrm{R}} = -(\zeta_p + f)\frac{\theta}{T}\frac{\gamma_{\mathrm{a}} - \gamma_{\mathrm{g}}}{g}\frac{R_0 T}{p} + \frac{\theta}{T}\frac{R_0}{pf}(\nabla_{\mathrm{h},p}T)^2 \tag{10.92}$$

In a last step, expressing the velocity gradients in the Cartesian system on a constant-pressure surface, we finally get

$$\begin{aligned} P_{\mathrm{R}} = \left(\frac{p_0}{p}\right)^{R_0/c_{p,0}} \frac{R_0}{p}\Bigg\{ &-\left[\left(\frac{\partial v}{\partial x}\right)_p - \left(\frac{\partial u}{\partial y}\right)_p + f\right]\frac{(\gamma_{\mathrm{a}} - \gamma_{\mathrm{g}})T}{g} \\ &+ \frac{1}{f}\left[\left(\frac{\partial T}{\partial x}\right)_p^2 + \left(\frac{\partial T}{\partial y}\right)_p^2\right]\Bigg\} \end{aligned} \tag{10.93}$$

All quantities appearing in this approximate relation can now be obtained from a single map, provided that γ_{g} is known.

Equation (10.88) is a powerful constraint on the large-scale motion of the atmosphere, as will be illustrated by considering air flow over a symmetric mountain barrier oriented south–north; see Figure 10.2. Since the flow is assumed to be isentropic some distance above the ground where frictional effects are small, the air is constrained to move between two isentropic surfaces that more or less follow the contour of the ground. On writing (10.88) in finite differences for a fixed value of $\Delta\theta$, we find

$$\left(\frac{\zeta_\theta + f}{\Delta p}\right)_{\mathrm{A}} = \left(\frac{\zeta_\theta + f}{\Delta p}\right)_{\mathrm{B}} = \left(\frac{\zeta_\theta + f}{\Delta p}\right)_{\mathrm{C}} \tag{10.94}$$

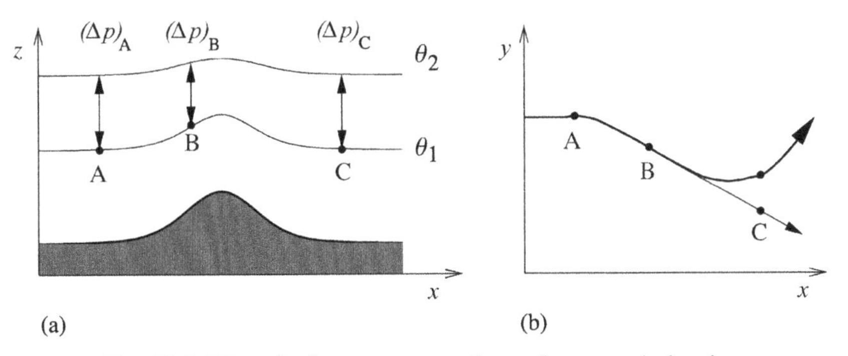

Fig. 10.2 Westerly flow over a south–north mountain barrier.

since the potential vorticity is conserved. From this relation we my explain qualitatively the behavior of the trajectory of an air parcel as it crosses the mountain barrier. Suppose that at position A there is straight-line flow in the eastward direction without any shear of the horizontal wind vector. We write the potential vorticity in Cartesian coordinates

$$\zeta_\theta = \left(\frac{\partial v}{\partial x}\right)_\theta - \left(\frac{\partial u}{\partial y}\right)_\theta \tag{10.95}$$

where the partial derivatives are evaluated along the θ-surface. Thus we immediately recognize that $\zeta_\theta(\mathrm{A}) = 0$. When the air parcel begins to cross the mountain range, the difference in pressure between the isentropic surfaces decreases so that the relative vorticity over the mountain range must be negative in order to conserve absolute vorticity; see part (b) of Figure 10.2. This results in anticyclonic curvature of the trajectory in the northern hemisphere. After the air parcel has crossed the mountain the difference in pressure between the two isentropic surfaces has returned to its original value. Without the effect of the changing Coriolis parameter the air parcel would have reached the position C, where straight line flow would result due to the assumed symmetry of the mountain barrier. This means that the straight westerly flow has changed to a flow from the north-west. So far we have assumed that the Coriolis parameter f remains constant. However, as the air parcel is moving south, the Coriolis parameter is decreasing, thus causing the relative vorticity to become positive so that the resulting trajectory would have a positive curvature at C due to the conservation principle. The formation of the lee-side trough is observed quite regularly to the east of the Rocky Mountains; see Bolin (1950). Further details are given, for example, by Holton (1972) and Pichler (1997).

10.4.6 Helmholtz's baroclinic vortex theorem

The general vortex theorem in the absolute system (10.54) can be used to deduce Helmholtz's famous vortex theorem, which is the starting point for the derivation of some circulation laws to be discussed in detail. On replacing ψ by (x, y, z) successively in (10.54) with $(\nabla x = \mathbf{i}, \nabla y = \mathbf{j}, \nabla z = \mathbf{k})$ and $(\dot{x} = u_\mathrm{A}, \dot{y} = v_\mathrm{A}, \dot{z} = w_\mathrm{A})$ we obtain

$$\begin{aligned}
&\frac{d}{dt}(\alpha\, \nabla \times \mathbf{B}_\mathrm{A} \cdot \mathbf{i}) - \alpha\, \nabla \times \mathbf{B}_\mathrm{A} \cdot \nabla u_\mathrm{A} = 0 \\
&\frac{d}{dt}(\alpha\, \nabla \times \mathbf{B}_\mathrm{A} \cdot \mathbf{j}) - \alpha\, \nabla \times \mathbf{B}_\mathrm{A} \cdot \nabla v_\mathrm{A} = 0 \\
&\frac{d}{dt}(\alpha\, \nabla \times \mathbf{B}_\mathrm{A} \cdot \mathbf{k}) - \alpha\, \nabla \times \mathbf{B}_\mathrm{A} \cdot \nabla w_\mathrm{A} = 0
\end{aligned} \tag{10.96}$$

On performing dyadic multiplication of these equations by the unit vectors $\mathbf{i}$, $\mathbf{j}$, $\mathbf{k}$ and adding the resulting equations the vector $\alpha \nabla \times \mathbf{B}_\mathrm{A}$ is obtained since scalar multiplication of a vector by the unit dyadic gives the vector itself. Hence we obtain

$$\frac{d}{dt}(\alpha \nabla \times \mathbf{B}_\mathrm{A}) - \alpha \nabla \times \mathbf{B}_\mathrm{A} \cdot \nabla \mathbf{v}_\mathrm{A} = 0 \tag{10.97}$$

Carrying out the differentiation and using the continuity equation immediately leads to the baroclinic version of the Helmholtz vortex theorem:

$$\boxed{\frac{d}{dt}(\nabla \times \mathbf{B}_\mathrm{A}) = \nabla \times \mathbf{B}_\mathrm{A} \cdot \nabla \mathbf{v}_\mathrm{A} - \nabla \cdot \mathbf{v}_\mathrm{A}(\nabla \times \mathbf{B}_\mathrm{A})} \tag{10.98}$$

This theorem states, for example, that $\nabla \times \mathbf{B}_\mathrm{A}$ remains zero at all times if it is zero at time $t = 0$. This statement is not immediately obvious but requires a brief mathematical discussion.

First of all we expand $\nabla \times \mathbf{B}_\mathrm{A}$ in a Taylor series about the time $t = 0$:

$$\nabla \times \mathbf{B}_\mathrm{A} = (\nabla \times \mathbf{B}_\mathrm{A})_{t=0} + t\left[\frac{d}{dt}(\nabla \times \mathbf{B}_\mathrm{A})\right]_{t=0} + \frac{t^2}{2!}\left[\frac{d^2}{dt^2}(\nabla \times \mathbf{B}_\mathrm{A})\right]_{t=0} + \cdots \tag{10.99}$$

If $\nabla \times \mathbf{B}_\mathrm{A} = 0$ at $t = 0$ then (10.98) shows that

$$\left[\frac{d}{dt}(\nabla \times \mathbf{B}_\mathrm{A})\right]_{t=0} = 0 \tag{10.100}$$

On differentiating (10.98) with respect to time we find without difficulty, using (10.100), that

$$\left[\frac{d^2}{dt^2}(\nabla \times \mathbf{B}_\mathrm{A})\right]_{t=0} = 0 \tag{10.101}$$

Continuing this procedure verifies the assertion that, if $\nabla \times \mathbf{B}_\mathrm{A} = 0$ at $t = 0$, it remains zero at all times, as now follows from (10.99).

Finally, we wish to state the Helmholtz theorem in a more practical form for the rotating earth. This is accomplished by substituting (10.43b) into (10.98) in order to replace the Weber transformation. The result is

$$\begin{aligned}\frac{d}{dt}\big[\nabla \times (\mathbf{v} + \mathbf{v}_P) - \nabla\beta \times \nabla s\big] &= \big[\nabla \times (\mathbf{v} + \mathbf{v}_P) - \nabla\beta \times \nabla s\big] \cdot \nabla(\mathbf{v} + \mathbf{v}_P) \\ &\quad - \big[\nabla \times (\mathbf{v} + \mathbf{v}_P) - \nabla\beta \times \nabla s\big]\, \nabla \cdot (\mathbf{v} + \mathbf{v}_P)\end{aligned} \tag{10.102}$$

For the relative system with rigid rotation we may use the following identities:

$$\begin{gathered}\mathbf{v}_P = \boldsymbol{\Omega} \times \mathbf{r}, \qquad \nabla \times \mathbf{v}_P = 2\boldsymbol{\Omega}, \qquad \nabla \mathbf{v}_P = -\mathbb{E} \times \boldsymbol{\Omega}, \qquad \nabla \cdot \mathbf{v}_P = 0 \\ \frac{d}{dt}(\nabla \times \mathbf{B}_\mathrm{A}) = \frac{d}{dt}(\nabla \times \mathbf{B}_\mathrm{A})\Big|_{\mathbf{q}_i \neq \mathbf{q}_i(t)} + \nabla \times \mathbf{B}_\mathrm{A} \cdot \nabla \mathbf{v}_P \\ \nabla \times \mathbf{B}_\mathrm{A} \cdot \nabla \mathbf{v}_P = -(\nabla \times \mathbf{B}_\mathrm{A}) \times \boldsymbol{\Omega}\end{gathered} \tag{10.103}$$

By substituting (10.103) into (10.102) we finally find

$$\boxed{\begin{aligned}&\frac{d}{dt}(\nabla\times\mathbf{v}+2\boldsymbol{\Omega}-\nabla\beta\times\nabla s)\Big|_{\mathbf{q}_i\neq\mathbf{q}_i(t)}\\&\quad=(\nabla\times\mathbf{v}+2\boldsymbol{\Omega}-\nabla\beta\times\nabla s)\cdot(\nabla\mathbf{v}-\nabla\cdot\mathbf{v}\mathbb{E})\end{aligned}}\tag{10.104}$$

In this form Helmholtz's theorem is not so easily interpreted in meteorological terms. However, this theorem is the starting point for the derivation of circulation theorems that give much physical insight.

10.4.7 Thomson's and Bjerkness' baroclinic circulation theorems

Helmholtz's baroclinic vortex law (10.98) can be transformed into an integral statement by using a theorem for the motion of material surfaces S, see Section M6.5.2:

$$\frac{d}{dt}\left(\int_S\mathbf{A}\cdot d\mathbf{S}\right)=\int_S\left(\frac{d\mathbf{A}}{dt}+\mathbf{A}\,\nabla\cdot\mathbf{v}_\mathrm{A}-\mathbf{A}\cdot\nabla\mathbf{v}_\mathrm{A}\right)\cdot d\mathbf{S}=\int_S\frac{D_2\mathbf{A}}{Dt}\cdot d\mathbf{S}\tag{10.105}$$

First we set $\mathbf{A}=\nabla\times\mathbf{B}_\mathrm{A}$ in (10.105). Next we integrate (10.98) over the surface S. Comparison of these two equations gives

$$\begin{aligned}\frac{d}{dt}\left(\int_S\nabla\times\mathbf{B}_\mathrm{A}\cdot d\mathbf{S}\right)&=\int_S\left(\frac{d}{dt}(\nabla\times\mathbf{B}_\mathrm{A})+(\nabla\times\mathbf{B}_\mathrm{A})\nabla\cdot\mathbf{v}_\mathrm{A}\right)\cdot d\mathbf{S}\\&\quad-\int_S(\nabla\times\mathbf{B}_\mathrm{A})\cdot\nabla\mathbf{v}_\mathrm{A}\cdot d\mathbf{S}=0\end{aligned}\tag{10.106}$$

Application of Stokes' integral theorem results in the so-called baroclinic version of Thomson's circulation theorem in the absolute system,

$$\boxed{\frac{d}{dt}\left(\int_S\nabla\times\mathbf{B}_\mathrm{A}\cdot d\mathbf{S}\right)=\frac{d}{dt}\left(\oint_\Gamma\mathbf{B}_\mathrm{A}\cdot d\mathbf{r}\right)=0}\tag{10.107}$$

This is not a very explicit form. Therefore, we replace the Weber transformation (10.43a) in this equation and find

$$\begin{aligned}\frac{d}{dt}\left(\oint_\Gamma\mathbf{B}_\mathrm{A}\cdot d\mathbf{r}\right)&=\frac{d}{dt}\left(\oint_\Gamma\mathbf{v}_\mathrm{A}\cdot d\mathbf{r}\right)-\frac{d}{dt}\left(\oint_\Gamma d_\mathrm{g}W_\mathrm{A}\right)-\frac{d}{dt}\left(\oint_\Gamma\beta\,\nabla s\cdot d\mathbf{r}\right)\\&=\frac{d}{dt}\left(\oint_\Gamma\mathbf{v}_\mathrm{A}\cdot d\mathbf{r}\right)-\frac{d}{dt}\left(\oint_\Gamma\beta\,\nabla s\cdot d\mathbf{r}\right)=0\end{aligned}\tag{10.108}$$

since the closed line integral of $d_g W_A$ vanishes. Next we define the *absolute* and the *relative circulation*:

$$\boxed{C_A = \oint_\Gamma \mathbf{v}_A \cdot d\mathbf{r}, \qquad C = \oint_\Gamma \mathbf{v} \cdot d\mathbf{r}} \tag{10.109}$$

Using the definition of the absolute circulation gives a second version of Thomson's circulation theorem:

$$\boxed{\frac{dC_A}{dt} = \frac{d}{dt}\left(\oint_\Gamma \beta\,\nabla s \cdot d\mathbf{r}\right) = \frac{d}{dt}\left(\oint_\Gamma \beta\, d_g s\right)} \tag{10.110}$$

On replacing the absolute velocity in (10.108), assuming rigid rotation of the coordinate system, we obtain Bjerkness' circulation theorem,

$$\boxed{\frac{d}{dt}\left(\oint_\Gamma (\mathbf{v} + \mathbf{v}_\Omega - \beta\,\nabla s) \cdot d\mathbf{r}\right) = 0 \quad \text{or} \quad \frac{dC}{dt} = \frac{d}{dt}\left(\oint_\Gamma (-\mathbf{v}_\Omega + \beta\,\nabla s) \cdot d\mathbf{r}\right)} \tag{10.111}$$

This very useful version of the circulation theorem is easily interpreted and will be discussed in detail in the next section.

10.4.8 Interpretation of Bjerkness' circulation theorem

The interpretation of this theorem is facilitated by rewriting (10.111). With the help of (10.65) we find

$$\frac{d}{dt}\left(\oint_\Gamma \mathbf{v}_\Omega \cdot d\mathbf{r}\right) = \frac{d}{dt}\left(\int_S \nabla \times \mathbf{v}_\Omega \cdot d\mathbf{S}\right) = 2\boldsymbol{\Omega} \cdot \frac{d\mathbf{S}}{dt} \tag{10.112}$$

where use of the Stokes integration theorem has been made. We recognize from Figure 10.3 that

$$S_E = \mathbf{S} \cdot \frac{\boldsymbol{\Omega}}{|\boldsymbol{\Omega}|} \tag{10.113}$$

is the projection of the integration surface onto the equatorial plane. By using the differentiation rule for the closed line integral we find (see problem 10.5)

$$\frac{d}{dt}\left(\oint_\Gamma \beta\,\nabla s \cdot d\mathbf{r}\right) = \oint_\Gamma \left(\frac{d\beta}{dt}\, d_g s - \frac{ds}{dt}\, d_g \beta\right) \tag{10.114}$$

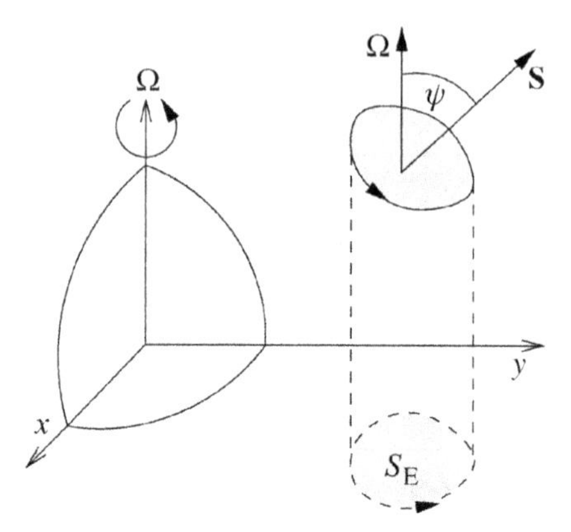

Fig. 10.3 Projection of the surface of integration **S** onto the equatorial plane.

Since isentropic processes were assumed in all derivations leading to the circulation laws, we have to set $ds/dt = 0$. Furthermore, recalling that $d\beta/dt = T$, we find from (10.111)

$$\frac{dC}{dt} + 2\Omega \frac{dS_{\mathrm{E}}}{dt} = \oint_{\Gamma} T \, d_{\mathrm{g}}s \tag{10.115}$$

By integrating equation (10.18) over the closed curve Γ we find

$$\oint_{\Gamma} T \, d_{\mathrm{g}}s = \oint_{\Gamma} d_{\mathrm{g}}h - \oint_{\Gamma} \alpha \, d_{\mathrm{g}}p = -\oint_{\Gamma} d_{\mathrm{g}}(\alpha p) + \oint_{\Gamma} p \, d_{\mathrm{g}}\alpha = \oint_{\Gamma} p \, d_{\mathrm{g}}\alpha \tag{10.116}$$

since closed line integrals of the exact differentials must vanish. With the help of Stokes' integral theorem we finally find the following three versions of Bjerkness' circulation theorem:

$$\begin{aligned} \frac{dC}{dt} + 2\Omega \frac{dS_{\mathrm{E}}}{dt} &= \oint_{\Gamma} T \, d_{\mathrm{g}}s = \oint_{\Gamma} T \, \nabla s \cdot d\mathbf{r} = \int_{S} \nabla T \times \nabla s \cdot d\mathbf{S} \\ &= -\oint_{\Gamma} \alpha \, d_{\mathrm{g}}p = -\oint_{\Gamma} \alpha \, \nabla p \cdot d\mathbf{r} = -\int_{S} \nabla \alpha \times \nabla p \cdot d\mathbf{S} \\ &= \oint_{\Gamma} p \, d_{\mathrm{g}}\alpha = \oint_{\Gamma} p \, \nabla \alpha \cdot d\mathbf{r} = \int_{S} \nabla p \times \nabla \alpha \cdot d\mathbf{S} \end{aligned} \tag{10.117}$$

Whenever the acceleration of the circulation $dC/dt > 0$ we speak of the *direct circulation*; when $dC/dt < 0$ it is called the *indirect circulation*. The vector $\nabla \alpha \times \nabla p$ describing the baroclinicity of the system is called the *baroclinicity vector* **N**. As can be seen from Figure 10.4, two sets of intersecting surfaces $\alpha =$ constant and $p =$ constant divide a particular volume into a continuous family of isosteric–isobaric tubes. These tubes are called *solenoids*. In the vertical cross-section depicted in Figure 10.4 they appear as parallelogram-type figures.

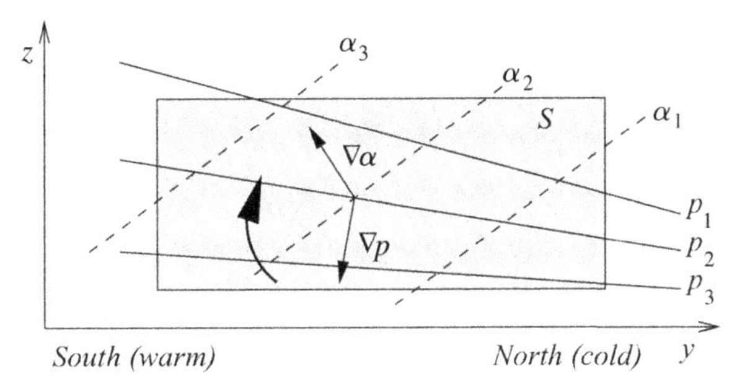

Fig. 10.4 The solenoidal effect.

In (M6.27) the validity of the following equation was shown:

$$\int_S \nabla p \times \nabla \alpha \cdot d\mathbf{S} = \int_S dp\, d\alpha \tag{10.118}$$

The right-hand side of this equation represents the number $N(\alpha, -p)$ of solenoids contained within the integration surface S. In the term $N(\alpha, -p)$ pressure is given a negative sign since p decreases but α increases with height. Comparison of (10.118) with (10.117) shows that the number of solenoids may also be expressed in a temperature and entropy coordinate system.

In the absence of friction we may now write the Bjerkness circulation theorem in the form

$$\boxed{\frac{dC}{dt} = -2\Omega \frac{dS_{\mathrm{E}}}{dt} + N(\alpha, -p)} \tag{10.119}$$

The solenoidal term produces a direct circulation, as will be recognized from Figure 10.4.

We will now explain the meaning of the first term on the right-hand side of (10.119). This term results from the Coriolis effect. If a material surface expands in time its projection onto the equatorial plane increases. As a consequence an existing cyclonic circulation ($C > 0$) weakens whereas an anticyclonic circulation ($C < 0$) intensifies. If the material surface contracts the opposite effects take place. Let us consider a closed material curve along a latitude circle that is displaced toward the north pole. This results in an intensification of the westerlies, whereas a displacement toward the equator has the opposite effect.

Next we consider the displacement of a material surface not enclosing the rotational axis of the earth; see Figure 10.5. This surface is oriented nearly parallel

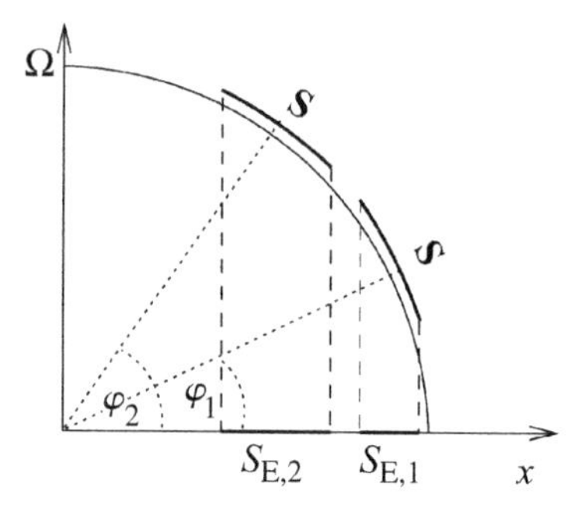

Fig. 10.5 The latitudinal effect.

to the earth's surface. A poleward displacement increases the projection onto the equatorial plane so that a cyclonic circulation along the enclosed surface is weakened whereas an existing anticyclonic circulation is intensified. A displacement toward the equator has the opposite effect. The change of the circulation due to a latitudinal displacement is known as the *latitudinal effect*. The convergence or divergence of the flow field resulting in an contraction or expansion of the material surfaces is known as the *divergence effect*. In summary, expansion (contraction) of the material surface and poleward (southward) motion work in the same direction. The combination of displacement with contraction or expansion may also occur.

On the small spatial scales of land–sea breezes and mountain–valley winds the rotational effect of the earth may be disregarded. The observed circulation pattern is then solely due to the solenoidal effect. On this small scale the isobaric surfaces may be considered horizontal, see Figure 10.6, in which the circulation patterns for the land-and-sea breeze are shown for a calm and cloudless summer situation. During the day the land is heated while the water surface remains relatively cool. The isosteric surfaces then assume the orientation depicted, causing the wind to blow from the sea toward the land. This is shown by the direction of the solenoidal vector. During the night the opposite situation occurs. The land surface cools off and the water remains relatively warm so that the wind blows from the land toward the sea. This type of circulation takes place on a relatively small scale. Even in well developed situations the scales of the motion hardly exceed a few hundred meters in the vertical direction and about 80 km in the horizontal direction if large bodies of water are involved. A similar situation develops for mountain and valley breezes, for which the slopes are heated more rapidly than the air during daytime and cooled more strongly than the air during the night.

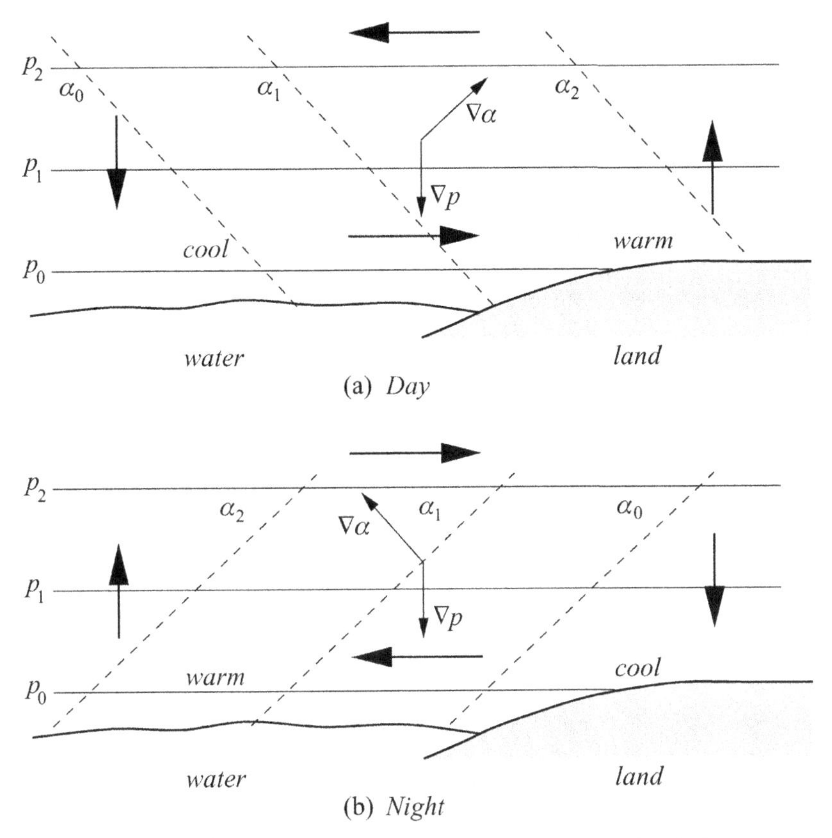

Fig. 10.6 Representation of a land-and-sea breeze circulation.

10.5 Circulation and vorticity theorems for frictionless barotropic flow

10.5.1 The barotropic Ertel–Rossby invariant

A barotropic fluid is characterized by the condition $\mathbf{N} = 0$. This implies that isobaric and isosteric surfaces are parallel so that solenoids cannot form. Whenever we deal with a barotropic fluid we effectively ignore the laws of thermodynamics since the specific volume is a function of pressure only. Therefore, we cannot proceed as before to eliminate the pressure gradient in terms of enthalpy, temperature, and entropy. Instead of this, we leave the equation of absolute motion in the original form, but we replace the pressure gradient force by

$$\alpha \nabla p = \nabla\left(\int \alpha \, dp\right) \tag{10.120}$$

The proof of this formula is left as an exercise. Now the equation of motion of the absolute system reduces to

$$\frac{d\mathbf{v}_{\mathrm{A}}}{\mathrm{dt}} = -\nabla\left(\phi_{\mathrm{a}} + \int \alpha\, dp\right) \tag{10.121}$$

By comparison of (10.121) with the baroclinic form of the equation of motion (10.21) and retracing the steps leading to (10.24), we find the barotropic form of the equation of motion in Lagrangian coordinates:

$$\frac{\partial x^n}{\partial a^i}\left(\frac{\partial^2 x^n}{\partial t^2}\right)_{\mathrm{L}} = -\frac{\partial}{\partial a^i}\left(\phi_{\mathrm{a}} + \int \alpha\, dp\right) \tag{10.122}$$

Inspection of the mathematical steps involved in going from (10.24) to (10.28) reveals that, in the barotropic case, the Lagrangian function is given by

$$L_{\mathrm{A}} = \frac{\mathbf{v}_{\mathrm{A}}^2}{2} - \phi_{\mathrm{a}} - \int \alpha\, dp \tag{10.123}$$

so that the action integral W_{A} now reads

$$W_{\mathrm{A}}^* = \int\left(\frac{\mathbf{v}_{\mathrm{A}}^2}{2} - \phi_{\mathrm{a}} - \int \alpha\, dp\right) dt \tag{10.124}$$

The star denotes the barotropic system. Instead of (10.43), the Weber transformation is now given by

$$\mathbf{B}_{\mathrm{A}}^* = \mathbf{v}_{\mathrm{A}} - \nabla W_{\mathrm{A}}^*, \qquad \nabla \times \mathbf{B}_{\mathrm{A}}^* = \nabla \times \mathbf{v}_{\mathrm{A}} \tag{10.125}$$

The transition from the baroclinic to the barotropic vortex laws may be accomplished by replacing $\mathbf{B}_{\mathrm{A}}$ by $\mathbf{B}_{\mathrm{A}}^*$.

In the barotropic atmosphere the baroclinic Ertel–Rossby conservation law (10.48) of the absolute system reduces to

$$\boxed{\frac{d}{dt}\left[\alpha(\mathbf{v}_{\mathrm{A}} - \nabla W_{\mathrm{A}}^*) \cdot \nabla \times \mathbf{v}_{\mathrm{A}}\right] = 0} \tag{10.126}$$

where

$$\boxed{\psi_{\mathrm{ER}}^* = \alpha(\mathbf{v}_{\mathrm{A}} - \nabla W_{\mathrm{A}}^*) \cdot \nabla \times \mathbf{v}_{\mathrm{A}}} \tag{10.127}$$

is the *barotropic Ertel–Rossby invariant*. If $\mathbf{v}_{\mathrm{A}}$ is replaced by $\mathbf{v} + \mathbf{v}_{\Omega}$ then the theorem refers to relative motion.

10.5.2 Barotropic vortex theorems of Ertel, Helmholtz, and Thomson

Owing to the condition of barotropy $\nabla\alpha \times \nabla p = 0$, the right-hand side of (10.61) vanishes. If the arbitrary field function $\psi = s$ and we assume that changes of state are isentropic $(ds/dt = 0)$, Ertel's vortex theorem reduces to

$$\boxed{\frac{d}{dt}(\alpha\,\nabla \times \mathbf{v}_{\mathrm{A}} \cdot \nabla s) = 0} \tag{10.128}$$

This expression is formally identical to the corresponding conservation law (10.63) of the baroclinic system.

In order to obtain the Helmholtz vortex theorem for the absolute system in a barotropic atmosphere we again replace $\mathbf{B}_{\mathrm{A}}$ by $\mathbf{B}_{\mathrm{A}}^{*}$. Instead of (10.98) we obtain for frictionless motion

$$\boxed{\frac{d}{dt}(\nabla \times \mathbf{v}_{\mathrm{A}}) = \nabla \times \mathbf{v}_{\mathrm{A}} \cdot \nabla \mathbf{v}_{\mathrm{A}} - \nabla \cdot \mathbf{v}_{\mathrm{A}}(\nabla \times \mathbf{v}_{\mathrm{A}})} \tag{10.129}$$

On repeating the arguments of Section 10.4.6 we see that $\nabla \times \mathbf{v}_{\mathrm{A}}$ remains zero if it is zero at time $t = 0$. The transition to the relative system is accomplished by replacing $\mathbf{v}_{\mathrm{A}}$ by $\mathbf{v} + \mathbf{v}_{\Omega}$.

Thomson's barotropic circulation theorem is an integral statement of the Helmholtz barotropic vortex theorem and can be derived in the same manner as its baroclinic counterpart. More easily we find it from Thomson's baroclinic circulation theorem if $\mathbf{B}_{\mathrm{A}}$ is replaced by $\mathbf{B}_{\mathrm{A}}^{*}$ in (10.108). This yields

$$\frac{d}{dt}\left(\oint_{\Gamma} \mathbf{B}_{\mathrm{A}}^{*} \cdot d\mathbf{r}\right) = \frac{d}{dt}\left(\oint_{\Gamma} \mathbf{v}_{\mathrm{A}} \cdot d\mathbf{r}\right) - \frac{d}{dt}\left(\oint_{\Gamma} d_{\mathrm{g}} W_{\mathrm{A}}^{*}\right) = 0 \tag{10.130}$$

The closed line integral of an exact differential is zero, so the last term vanishes. On substituting the definition of the absolute circulation from (10.109) and applying Stokes' integral theorem, we find

$$\boxed{\frac{dC_{\mathrm{A}}}{dt} = \frac{d}{dt}\left(\int_{S} \nabla \times \mathbf{v}_{\mathrm{A}} \cdot d\mathbf{S}\right) = \frac{d}{dt}\left(\oint_{\Gamma} \mathbf{v}_{\mathrm{A}} \cdot d\mathbf{r}\right) = 0} \tag{10.131}$$

The physical content of this equation is that the absolute circulation along the closed material curve Γ is a constant. This theorem has a number of important consequences, which will now be discussed in some detail.

10.5.3 Vortex lines and vortex tubes

Vortex lines may be introduced analogously to streamlines, which are defined by the differential equation (3.39). Thus vortex lines are defined by

$$d\mathbf{r} \times (\nabla \times \mathbf{v}_{\mathrm{A}}) = 0 \tag{10.132}$$

representing curves whose tangents are parallel to the vorticity vector $\nabla \times \mathbf{v}_{\mathrm{A}}$. Before we introduce the idea of a vortex tube we consider the concept of a stream tube. The definition of a streamline may be extended to a stream tube whose side walls are composed of streamlines. For any closed contour in a flow field each point on the contour will have a streamline passing through it. By considering all points on the contour an infinite number of streamlines is obtained, forming a surface known as a *stream tube*. Figure 10.7(a) shows a section of a stream tube defined by a contour enclosing the surface S_1. The corresponding contour a small distance away encloses the surface S_2. If the cross-section of the stream tube is infinitesimally small it is called a *stream filament.*

The definition of a vortex tube is analogous to that of a stream tube. Any point on a closed contour in the flow field will have a vortex line passing through it, thus forming a vortex tube. If the contour encloses the surface area S we obtain the configuration as shown in Figure 10.7(b). A vortex tube whose cross-section is infinitesimally small is known as a *vortex filament.*

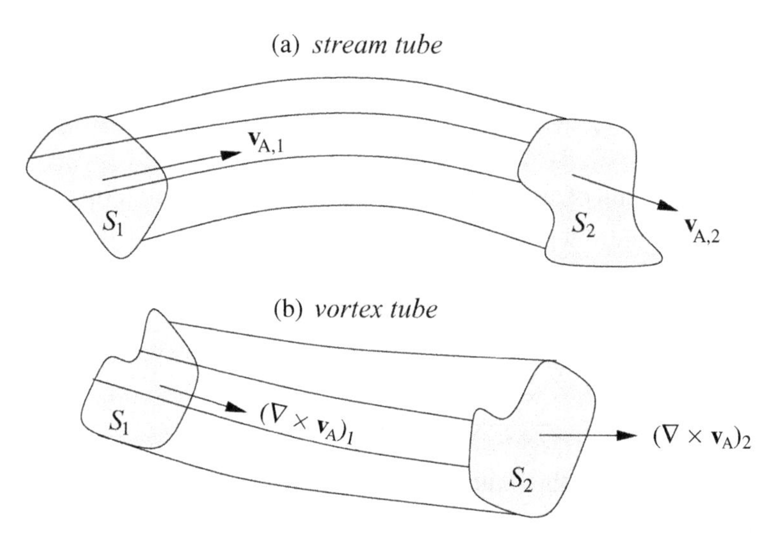

Fig. 10.7 Stream tubes and vortex tubes.

Some very interesting properties of the flow field can be derived from the fact that the divergence of the curl of any vector vanishes. Applying this to the vector field $\mathbf{v}_A$, we have

$$\nabla \cdot (\nabla \times \mathbf{v}_A) = 0 \tag{10.133}$$

Since the vorticity vector is divergence-free there can be no sources and sinks of the vorticity in the fluid itself. This means that vortex lines either form closed loops or must terminate on the boundary of the fluid, which may either be a solid surface or a free surface. Because the vorticity vector is divergence-free, we have an analogy with the flow of an incompressible fluid whose continuity equation is $\nabla \cdot \mathbf{v}_A = 0$, showing that the velocity vector $\mathbf{v}_A$ is divergence-free. Integration of this expression over the closed surface of the stream tube V_t or stream filament gives

$$\int_{V_t} \nabla \cdot \mathbf{v}_A \, d\tau = \int_{S_t} \mathbf{v}_A \cdot d\mathbf{S} = 0 \tag{10.134}$$

The surface-element vector $d\mathbf{S}$ is orthogonal to the side walls of the stream tube or to $\mathbf{v}_A$, so the side walls cannot make a contribution to the surface integral. Therefore, (10.134) reduces to

$$\int_{S_1} \mathbf{v}_A \cdot d\mathbf{S} + \int_{S_2} \mathbf{v}_A \cdot d\mathbf{S} = 0 \tag{10.135}$$

Since $d\mathbf{S}$, by definition, is always pointing in the direction of the outward normal, we may define the flow rate with respect to surfaces S_1 and S_2 by means of

$$\int_{S_1} \mathbf{v}_A \cdot d\mathbf{S} = -F_1, \qquad \int_{S_2} \mathbf{v}_A \cdot d\mathbf{S} = F_2 \tag{10.136}$$

or

$$F_1 = F_2 \tag{10.137}$$

as follows from (10.135). Divergence-free flow of the velocity vector results in equal flow rates for the fluid crossing S_1 and S_2.

We now integrate (10.133) over the volume and obtain in complete analogy to (10.134) the expression

$$\int_{V_t} \nabla \cdot \nabla \times \mathbf{v}_A \, d\tau = \int_{F_t} \nabla \times \mathbf{v}_A \cdot d\mathbf{S} = 0 \tag{10.138}$$

Since $\nabla \times \mathbf{v}_A \cdot d\mathbf{S} = 0$ on the walls of the vortex tube, we have

$$\int_{S_1} \nabla \times \mathbf{v}_A \cdot d\mathbf{S} + \int_{S_2} \nabla \times \mathbf{v}_A \cdot d\mathbf{S} = 0 \tag{10.139}$$

Application of Stokes' integral theorem gives

$$\begin{aligned}\int_{S_1} \nabla \times \mathbf{v}_{\mathrm{A}} \cdot d\mathbf{S} &= \oint_{\Gamma_1} \mathbf{v}_{\mathrm{A}} \cdot d\mathbf{r} = -C_{S_1} \\ \int_{S_2} \nabla \times \mathbf{v}_{\mathrm{A}} \cdot d\mathbf{S} &= \oint_{\Gamma_2} \mathbf{v}_{\mathrm{A}} \cdot d\mathbf{r} = C_{S_2}\end{aligned} \tag{10.140}$$

or, from (10.139),

$$C_{S_1} = C_{S_2} \tag{10.141}$$

This simple statement shows that the absolute circulations around the limiting contours of surfaces S_1 and S_2 are identical. Alternately we may say that the circulation through each cross-sectional area is the same. Comparison of (10.137) and (10.141) shows that, in case of the stream tube, the flow rates at S_1 and S_2 are equal while the absolute circulation along the vortex tube is the same.

Equation (10.139) can also be interpreted in a different manner. If the cross-sectional area S of the vortex tube is sufficiently small, we may define a meaningful average vorticity by means of

$$\int_S \nabla \times \mathbf{v}_{\mathrm{A}} \cdot d\mathbf{S}' = \overline{\nabla \times \mathbf{v}_{\mathrm{A}}} \cdot \int_S d\mathbf{S}' = \left|\overline{\nabla \times \mathbf{v}_{\mathrm{A}}}\right| S \tag{10.142}$$

This expression is known as the *vortex strength*. With reference to (10.139) and (10.140) we find for the flow through cross-sections S_1 and S_2 the expression

$$\left|\overline{\nabla \times \mathbf{v}_{\mathrm{A}}}\right| S_1 = \left|\overline{\nabla \times \mathbf{v}_{\mathrm{A}}}\right| S_2 \tag{10.143}$$

stating that the vortex strength is constant along the vortex tube. From the fact that the vorticity vector is divergence-free it follows that vortex tubes must be closed, terminate on a solid boundary, or terminate on a free surface. Observational evidence for closed vortex tubes is provided by smoke rings, whereas a vortex tube at a free surface flow may have one end at the free surface and the other end at the solid boundary of the fluid.

Additional information on the kinematics of vortex tubes may be obtained by considering a closed curve Γ located entirely on the side wall of the vortex tube which does not enclose the axis of the vortex tube; see Figure 10.8.

Since $\nabla \times \mathbf{v}_{\mathrm{A}}$ is perpendicular to $d\mathbf{S}$ we find for the absolute circulation

$$C_A = \oint_{\Gamma} \mathbf{v}_{\mathrm{A}} \cdot d\mathbf{r} = \int_{S_\Gamma} \nabla \times \mathbf{v}_{\mathrm{A}} \cdot d\mathbf{S} = 0 \tag{10.144}$$

According to Thomson's theorem (10.131) C_{A} is independent of time so that Γ remains on the side walls. We may now think of the entire surface area of the vortex

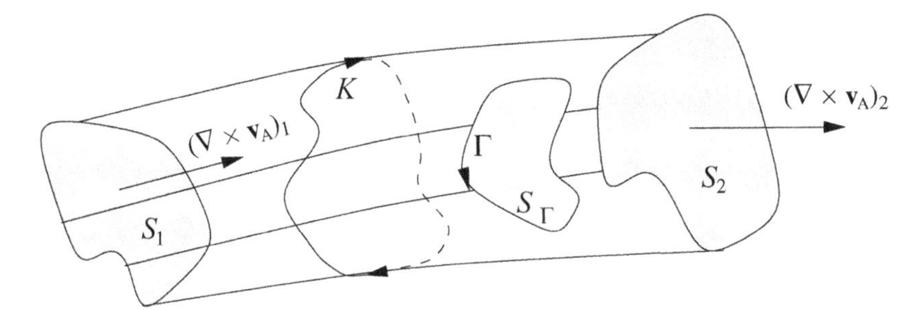

Fig. 10.8 Flow contours on a vortex tube.

tube's side wall as being composed of a number of subsections, each bounded by a certain Γ not enclosing the axis of the vortex. On each one of these boundaries the circulation vanishes, which means that the vortex tube always consists of the same particles.

One more important conclusion may be drawn by considering the curve K of the same figure enclosing the vortex tube. From (10.143) it follows that the circulation along K equals the vortex strength of the tube. The circulation along K is constant in time so that the vortex strength is constant in time also. From this we derive the important statement that, in case of an ideal barotropic medium, vortices cannot be created or destroyed. With this statement we close the train of thought beginning with the Helmholtz vortex theorem from which we derived that $\nabla \times \mathbf{v}_A$ remains zero if it was zero to begin with. A nice and consistent treatment of flow kinematics is given in Currie (1974).

A barotropic fluid is characterized by a lack of solenoids so that the solenoidal vector vanishes. Therefore, the Bjerkness circulation theorem (10.119) reduces to

$$\boxed{\frac{dC}{dt} + 2\Omega \frac{dS_E}{dt} = 0} \tag{10.145}$$

For the physical interpretation of the remaining terms we refer to Section 10.4.8. Without any problems equation (10.145) could also have been derived from Thomson's theorem (10.111) for a baroclinic fluid.

10.5.4 The vorticity theorem for the barotropic atmosphere

The baroclinic vorticity theorem (10.70) simplifies to its barotropic counterpart on considering the following points.

(i) The barotropic condition requires $\nabla\alpha \times \nabla p = 0$.
(ii) There is no twisting or tilting term since the vertical derivatives of the horizontal motion vanish in the barotropic model atmosphere.

Condition (ii) is satisfied if the following prerequisites are valid: (a) barotropy, i.e. $\alpha = \alpha(p)$, (b) the hydrostatic approximation is valid, and (c) $\partial \mathbf{v}_h/\partial p = 0$ for any t. Therefore, equation (10.70) reduces to the *barotropic vorticity theorem*

$$\boxed{\frac{d\eta}{dt} = -\eta\, \nabla_h \cdot \mathbf{v}_h} \tag{10.146}$$

We will discuss this equation in some detail when we deal with the barotropic forecast model. One more simplification is possible by dropping the divergence term in (10.146):

$$\boxed{\frac{d\eta}{dt} = 0, \qquad \eta = \text{constant}} \tag{10.147}$$

In this case the absolute vorticity is conserved. From this conservation theorem we can predict changes of the relative vorticity if low- or high-pressure systems are displaced in the northward or southward direction due to the accompanying changes of the Coriolis parameter. Consider as an example the southward displacement of a low-pressure system in the northern hemisphere. Since f is decreasing the relative vorticity is increasing, implying a strengthening of the system. Equation (10.147) does not, however, give any information about the trajectory of a displaced system, so the conservation of absolute vorticity is merely a qualitative tool.

As a final point of this section we derive Rossby's potential vorticity theorem in the manner suggested by Rossby. In order to do so we need to have recourse to the continuity equation for a barotropic fluid, which will be derived in a later chapter. It is given by

$$\frac{d}{dt}(\phi - \phi_s) = -(\phi - \phi_s)\, \nabla_h \cdot \mathbf{v}_h \tag{10.148}$$

where ϕ_s is the geopotential of the earth's surface. On eliminating the divergence term in (10.146) by (10.148) we obtain the conservation equation

$$\boxed{\frac{d}{dt}\left(\frac{\eta}{\phi - \phi_s}\right) = 0, \qquad \frac{\eta}{\phi - \phi_s} = P_{R,b}} \tag{10.149}$$

where $P_{R,b}$ is known as *Rossby's potential vorticity for a barotropic fluid*. We shall refrain from discussing it at this point.

Our discussion on vortex theorems is far from complete. In an early paper, Fortak (1956) already addressed the question of the general formalism of vortex theorems. As a final remark in this chapter, we would like to point out that research on the existence of conservation laws for special conditions has not ceased. Two examples are papers by Herbert and Pichler (1994) and Egger and Schär (1994), which deserve serious study.

10.6 Problems

10.1: Find equation (10.63) from equation (10.55).

10.2: Verify all parts of equation (10.69).

10.3: Verify all parts of equation (10.72) by using the approximations which were introduced just before this equation.

10.4: By utilizing the proper transformation rules given in Chapter M4 together with the hydrostatic approximation, prove the validity of (10.75).

10.5: Start with equation (10.84) to verify the mathematical steps leading to equation (10.93).

10.6: Verify equation (10.114).

Hint: Rewrite the right-hand side of equation (M6.49) in the form

$$\oint_{L(t)} d\mathbf{r} \cdot \left(\frac{\partial \mathbf{A}}{\partial t} - \mathbf{v} \times \nabla \times \mathbf{A} \right)$$

10.7: Apply the barotropic condition to show that

$$\alpha \nabla p = \nabla P^* \quad \text{with} \quad P^* = \int_{p_0}^{p} \alpha \, dp$$

P^* is a scalar field function with $P^* = 0$ at $p = p_0$.

10.8: Derive equation (10.145) from equation (10.111).

11

Turbulent systems

The basic equations of motion and the budget equations presented so far refer to the molecular system implying laminar flow. The atmosphere, however, does not behave like a laminar fluid since it is turbulent everywhere and at all times. Therefore, the pertinent equations have to be modified in order to handle turbulent flow. Since most meteorological observations represent average values over some time interval and spatial region it will be necessary to average the governing equations of the molecular system. In the next few sections some averaging procedures will be discussed and averaging operators will be introduced. This leads to the introduction of the so-called *microturbulent system*.

11.1 Simple averages and fluctuations

There are several types of averages. Perhaps the best known average is the so-called *ensemble average*, also known as the *simple mean value* or the *expectation value*. For the arbitrary variable f the ensemble average is defined by

$$\boxed{\overline{f} = \frac{1}{N}\sum_{i=1}^{N} f_i} \tag{11.1}$$

where the simple averaging operator is denoted by the overbar. Each of the f_i represents one of N data points. We think of the ensemble as a large number of realizations of a physical experiment carried out under identical external conditions. The essential idea involved in the definition of the ensemble average is that each individual realization f_i is influenced by random errors that cannot be controlled externally. In general, individual measurements will differ due to stochastic disturbances.

We will now give some rules pertaining to the simple mean value. We define an α quantity as a quantity that is the same for each member of the ensemble, meaning

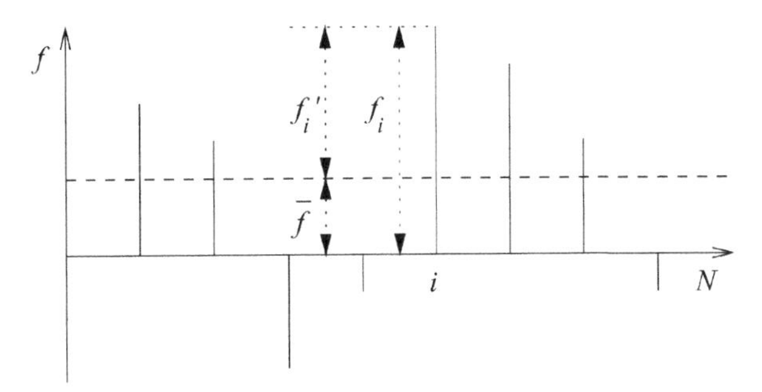

Fig. 11.1 A schematic representation of the fluctuations and the mean value.

that α is independent of the subscript i, so that $\alpha_i = \alpha$. Obviously, the α quantity includes all constants. Moreover, explicit functions of space and time, such as the metric fundamental quantities, will not be averaged in the sense of (11.1). This yields the following averaging rules:

$$\begin{gathered}\overline{\alpha} = \alpha, \qquad \overline{\alpha f} = \alpha \overline{f}, \qquad \overline{\overline{f}} = \overline{f}, \qquad \overline{\overline{f} g} = \overline{f}\,\overline{g} \\ \overline{f + g} = \overline{f} + \overline{g}, \qquad \overline{\alpha_1 f + \alpha_2 g} = \alpha_1 \overline{f} + \alpha_2 \overline{g}\end{gathered} \tag{11.2}$$

In the expression $\overline{\overline{f} g}$ the average value $\overline{f}$ is common to all members of the ensemble. Therefore, it is an α quantity and may be factored out of the sum. Repeated averaging of a quantity that has already been averaged does not change its value. This is known as the *idempotent rule*. The sum rules given in (11.2) demonstrate that $(\overline{\ \ \ })$ is a linear operator.

The value f_i of an ensemble will now be split into two parts. The deviation f_i' from the mean value $\overline{f}$ is known as the *simple fluctuation*, as indicated in

$$f_i = \overline{f} + f_i' \quad \text{or} \quad f_i' = f_i - \overline{f} \tag{11.3}$$

and depicted in Figure 11.1. From this expression it is easily seen that the average value of the fluctuations of the ensemble is zero:

$$\overline{f'} = \frac{1}{N} \sum_{i=1}^{N} f_i' = \frac{1}{N} \sum_{i=1}^{N} (f_i - \overline{f}) = \overline{f} - \overline{\overline{f}} = \overline{f} - \overline{f} = 0 \tag{11.4}$$

Of particular interest is the average of the product of two variables f and g:

$$\overline{fg} = \frac{1}{N} \sum_{i=1}^{N} f_i g_i = \frac{1}{N} \sum_{i=1}^{N} (\overline{f} + f_i')(\overline{g} + g_i') = \overline{f}\,\overline{g} + \overline{f'g'} \tag{11.5}$$

since $\overline{\overline{f}g'} = \overline{f}\,\overline{g'} = 0$ and $\overline{f'\overline{g}} = \overline{f'}\,\overline{g} = 0$. In this expression the so-called *correlation product* $\overline{f'g'}$ has been introduced according to

$$\overline{f'g'} = \frac{1}{N}\sum_{i=1}^{N} f_i' g_i' \tag{11.6}$$

By utilizing the averaging rules (11.2) and (11.4) it is easily seen that the correlation product can be written in various ways:

$$\overline{f'g'} = \overline{f'g} = \overline{fg'} \tag{11.7}$$

Let δ represent a differential operator that is required to be the same for all members of the ensemble so that δ is independent of the subscript i. Let the differential operator be applied to the variable f. Then the mean or average value of this expression is given by

$$\overline{\delta f} = \frac{1}{N}\sum_{i=1}^{N}(\delta f)_i = \frac{1}{N}\delta\sum_{i=1}^{N} f_i = \delta\overline{f} \tag{11.8}$$

Since δ is free from the subscript i we find the interesting rule that the differential operator can be separated from the averaging operator $(\overline{\ \ })$. Examples are $\delta = \partial/\partial x, \partial/\partial t, \nabla$, etc. However, the budget operator $\delta = D/Dt$ as well as the individual time derivative $\delta = d/dt$ do not follow this rule. Explicit expressions for the budget operator and the total time derivative will be worked out later.

11.2 Weighted averages and fluctuations

An important generalization of the simple mean value is the *weighted mean value* or *weighted expectation value* defined by

$$\boxed{\widehat{f} = \frac{\frac{1}{N}\sum_{i=1}^{N}\rho_i f_i}{\frac{1}{N}\sum_{i=1}^{N}\rho_i}} \tag{11.9}$$

The symbol $\widehat{\ \ }$ defines the weighted-averaging operator. The various ρ_i are the weights. From (11.1) and (11.9) it follows that

$$\boxed{\widehat{f} = \frac{\overline{\rho f}}{\overline{\rho}} \implies \overline{\rho f} = \overline{\rho}\widehat{f}} \tag{11.10}$$

which is also known as the *Hesselberg average*. Equations (11.9) and (11.10) will be used many times in the following sections.

Some of the rules given for the simple average are of general character and, therefore, may also be applied to the weighted average. In analogy to (11.2) the weighted average is independent of the subscript i so that it can be handled as an α quantity. Furthermore, the sum rules also apply to the weighted average, yielding

$$\begin{aligned} &\widehat{\alpha} = \alpha, \qquad \widehat{\alpha f} = \alpha \widehat{f}, \qquad \widehat{\widehat{f}} = \widehat{f}, \qquad \widehat{\widehat{f} g} = \widehat{f}\,\widehat{g} \\ &\widehat{(f+g)} = \widehat{f} + \widehat{g}, \qquad \widehat{(\alpha_1 f + \alpha_2 g)} = \alpha_1 \widehat{f} + \alpha_2 \widehat{g} \end{aligned} \tag{11.11}$$

Thus, the weighted average also has the properties of a linear operator. Additional rules are given in

$$\widehat{\overline{f} g} = \overline{f}\,\widehat{g}, \qquad \widehat{\overline{f}} = \overline{f}, \qquad \overline{\widehat{f}} = \widehat{f} \tag{11.12}$$

These are largely self-explanatory.

Let us again consider the differential operator δ applied to the variable f. The weighted average of this expression is given in

$$\widehat{\delta f} = \frac{\sum_{i=1}^{N} \rho_i (\delta f)_i}{\sum_{i=1}^{N} \rho_i} \neq \delta \widehat{f} \quad \text{since} \quad \delta \widehat{f} = \delta \left(\frac{\sum_{i=1}^{N} (\rho_i f_i)}{\sum_{i=1}^{N} \rho_i} \right) \tag{11.13}$$

Thus, in contrast to (11.8), which refers to the simple mean, it is not possible to separate δ from the weighted average.

In analogy to (11.3) we now split the variable f into its weighted mean $\widehat{f}$ and fluctuation f_i'':

$$f_i = \widehat{f} + f_i'' \quad \text{or} \quad f_i'' = f_i - \widehat{f} \tag{11.14}$$

whereby the mean or average value of the weighted fluctuation also vanishes:

$$\widehat{f''} = \frac{\frac{1}{N}\sum_{i=1}^{N} \rho_i f_i''}{\frac{1}{N}\sum_{i=1}^{N} \rho_i} = \frac{\frac{1}{N}\sum_{i=1}^{N} \rho_i (f_i - \widehat{f})}{\frac{1}{N}\sum_{i=1}^{N} \rho_i} = \widehat{f} - \widehat{\widehat{f}} = 0 \tag{11.15}$$

In order to get a clearer picture of the meaning of the weighted average, let us consider Figure 11.2. We observe that each member of the ensemble is characterized by a rectangular area of height f_i and width ρ_i. The rectangular area on the right-hand side of Figure 11.2 represents the average area $F = \overline{\rho f} = \overline{\rho}\,\widehat{f}$. The height of this rectangular area is the weighted value $\widehat{f}$ while the width is given by $\overline{\rho}$.

Mean values of fluctuations vanish only if fluctuations and the type of the average correspond, as demonstrated in

$$\overline{f''} = \overline{f} - \overline{\widehat{f}} = \overline{f} - \widehat{f} \neq 0, \qquad \widehat{f'} = \widehat{f} - \widehat{\overline{f}} = \widehat{f} - \overline{f} \neq 0 \tag{11.16}$$

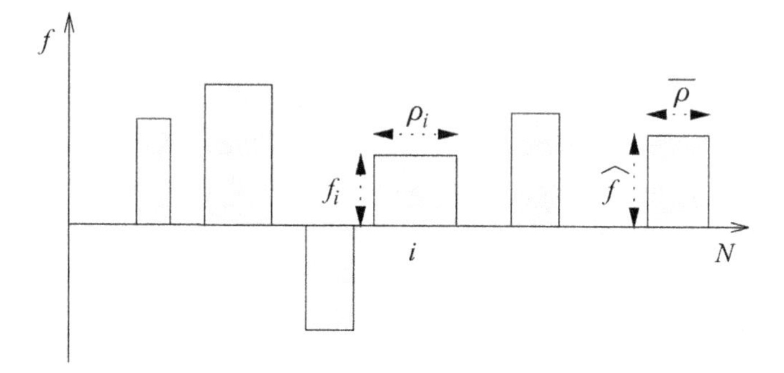

Fig. 11.2 A representation of the weighted average and the weights.

The product rule

$$\widehat{fg} = \widehat{f}\widehat{g} + \widehat{f''g''} \tag{11.17}$$

corresponds fully to (11.5). The correlation $\widehat{f''g''}$ may be written in various ways as shown in

$$\widehat{f''g''} = \widehat{f''g} = \widehat{fg''} \tag{11.18}$$

This property turns out to be very helpful, as will become apparent in later sections. Other important relations following from (11.10) and (11.18) are

$$\overline{\rho f''} = \overline{\rho}\widehat{f''} = 0, \qquad \overline{\rho f''g''} = \overline{\rho f''g} = \overline{\rho f g''} \tag{11.19}$$

11.3 Averaging the individual time derivative and the budget operator

As has already been mentioned, great care must be taken when one is averaging the individual time derivative d/dt and the budget operator D/Dt. We will now show how to proceed. The first step for the individual time derivative is given by

$$\overline{\frac{d\psi}{dt}} = \overline{\frac{\partial \psi}{\partial t}} + \overline{\mathbf{v}_{\mathrm{A}} \cdot \nabla\psi} = \frac{\partial \overline{\psi}}{\partial t} + \widehat{\mathbf{v}}_{\mathrm{A}} \cdot \nabla\overline{\psi} + \overline{\mathbf{v}''_{\mathrm{A}} \cdot \nabla\psi} \tag{11.20}$$

Here the velocity vector $\mathbf{v}_{\mathrm{A}}$ has been split into its weighted average $\widehat{\mathbf{v}}_{\mathrm{A}}$, which is treated as an α quantity, and the fluctuation $\mathbf{v}''_{\mathrm{A}}$. Utilizing a rigidly rotating coordinate system with $\mathbf{v}_{\mathrm{D}} = 0$, we have $\mathbf{v}_{\mathrm{A}} = \mathbf{v} + \mathbf{v}_{\Omega}$ and $\mathbf{v}''_{\mathrm{A}} = \mathbf{v}''$ since $\mathbf{v}_{\Omega}$ is an explicit function of space. Recall that, according to (11.8), $\partial/\partial t$ and ∇ may be separated from the simple average operator. By defining the individual time derivative of the averaged system by means of

$$\frac{\widehat{d} \ldots}{dt} = \frac{\partial \ldots}{\partial t} + \widehat{\mathbf{v}}_{\mathrm{A}} \cdot \nabla \ldots \tag{11.21}$$

we finally obtain for the averaged individual time derivative

$$\boxed{\overline{\frac{d\psi}{dt}} = \frac{\widehat{d}\,\overline{\psi}}{dt} + \overline{\mathbf{v}'' \cdot \nabla\psi}} \tag{11.22}$$

In complete analogy we proceed with the budget operator and obtain

$$\begin{aligned}\overline{\frac{D}{Dt}(\rho\psi)} &= \frac{\partial}{\partial t}(\overline{\rho\psi}) + \nabla\cdot(\widehat{\mathbf{v}}_{\mathrm{A}}\overline{\rho\psi}) + \nabla\cdot(\overline{\mathbf{v}''\rho\psi}) \\ &= \frac{\widehat{D}}{Dt}(\overline{\rho\psi}) + \nabla\cdot\overline{\mathbf{v}''\rho\psi} \\ \text{with}\quad \frac{\widehat{D}\ldots}{Dt} &= \frac{\partial\ldots}{\partial t} + \nabla\cdot(\widehat{\mathbf{v}}_{\mathrm{A}}\ldots)\end{aligned} \tag{11.23}$$

Averaging the molecular form of the continuity equation yields the expression

$$\boxed{\overline{\frac{\partial\rho}{\partial t} + \nabla\cdot(\rho\mathbf{v}_{\mathrm{A}})} = \frac{\partial\overline{\rho}}{\partial t} + \nabla\cdot(\widehat{\mathbf{v}}_{\mathrm{A}}\overline{\rho}) = \frac{\widehat{D}}{Dt}\overline{\rho} = 0} \tag{11.24}$$

where we have used the Hesselberg averaging procedure. Here the advantage of the Hesselberg mean becomes clearly apparent since the averaged form of the continuity equation (11.24) is completely analogous to the molecular form. Had we used the Reynolds mean then the additional term $\overline{\nabla\cdot\mathbf{v}'_{\mathrm{A}}\rho'}$, which is not easily accessible, would have appeared in the averaged continuity equation.

In the exercises we will show that the interchange rule (M6.68) of the molecular system also holds for the averaged system

$$\boxed{\frac{\widehat{D}}{Dt}(\overline{\rho}\psi) = \overline{\rho}\,\frac{\widehat{d}\;\psi}{dt}} \tag{11.25}$$

11.4 Integral means

Obviously, we are unable to control the atmosphere so that it is impossible to produce and observe identical weather systems. Therefore, the ensemble average is of doubtful practical use. For this reason we direct our attention to time and space averages. The time average at a fixed point is defined by means of

$$\overline{\psi}^{t}(\mathbf{r},t) = \frac{1}{\Delta t}\int_{t-\Delta t/2}^{t+\Delta t/2}\psi(\mathbf{r},t,t')\,dt' \tag{11.26}$$

where Δt is a suitable averaging interval about t and ψ is the quantity to be averaged over time at the fixed position $\mathbf{r}$. The integral mean can be obtained for the Cartesian system as well as for the general q^i system by using the proper metric fundamental quantities. To give a specific example, let us consider the hypothetical velocity spectrum shown in Figure 11.3. For turbulent steady flow the average

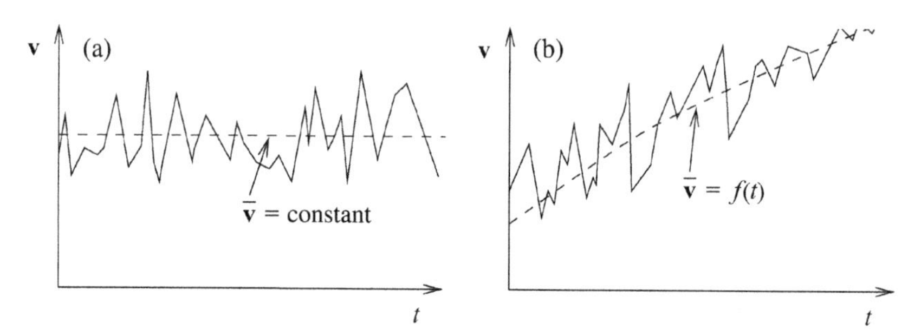

Fig. 11.3 (a) Turbulent steady flow. (b) Turbulent unsteady flow.

is time-independent; see Figure 11.3(a). For extended time intervals the average itself may become a function of time, as shown in part (b) of Figure 11.3. The time-dependent mean is often called the *gliding mean* or the *moving mean.*

According to the choice of the averaging interval, the gliding mean has the tendency to suppress a part of the high-frequency fluctuations. For this reason the gliding mean acts as a *low-pass filter*. In the same sense most measuring devices act as low-pass filters since they are incapable of recording rapid oscillations.

Next we define the *spatial mean for fixed time* by

$$\overline{\psi}^{s}(\mathbf{r}, t) = \frac{1}{V(\mathbf{r})} \int_{V(\mathbf{r})} \psi(\mathbf{r}, \mathbf{r}', t)\, dV' \tag{11.27}$$

There is a problem with this type of mean since we generally cannot expect that the value measured at a certain point is representative of a larger surrounding area. Since the observational grid is rarely dense enough to evaluate (11.27), we are hardly able to use this equation. For our purposes, however, it seems convenient to think of the average as a *space-time average* defined either by

$$\overline{\psi}(\mathbf{r}, t) = \overline{\psi}(\mathbf{r}_{\mathrm{st}}) = \frac{1}{G(\mathbf{r}_{\mathrm{st}})} \int_{G(\mathbf{r}_{\mathrm{st}})} \psi(\mathbf{r}_{\mathrm{st}}, \mathbf{r}'_{\mathrm{st}})\, dG' \tag{11.28}$$

or by

$$\widehat{\psi}(\mathbf{r}, t) = \widehat{\psi}(\mathbf{r}_{\mathrm{st}}) = \frac{\dfrac{1}{G(\mathbf{r}_{\mathrm{st}})} \displaystyle\int_{G(\mathbf{r}_{\mathrm{st}})} \rho(\mathbf{r}_{\mathrm{st}}, \mathbf{r}'_{\mathrm{st}}) \psi(\mathbf{r}_{\mathrm{st}}, \mathbf{r}'_{\mathrm{st}})\, dG'}{\dfrac{1}{G(\mathbf{r}_{\mathrm{st}})} \displaystyle\int_{G(\mathbf{r}_{\mathrm{st}})} \rho(\mathbf{r}_{\mathrm{st}}, \mathbf{r}'_{\mathrm{st}})\, dG'} \tag{11.29}$$

representing the simple and weighted means. Here we have formally introduced the space-time vectors $\mathbf{r}_{\mathrm{st}}$ and $\mathbf{r}'_{\mathrm{st}}$ as shown in Figure 11.4. Actually space-time vectors cannot be made visible, but this figure is a good tool for demonstration purposes. $\mathbf{r}_{\mathrm{st}}$ is the space-time vector of the original system whereas $\mathbf{r}'_{\mathrm{st}}$ is the space-time vector of

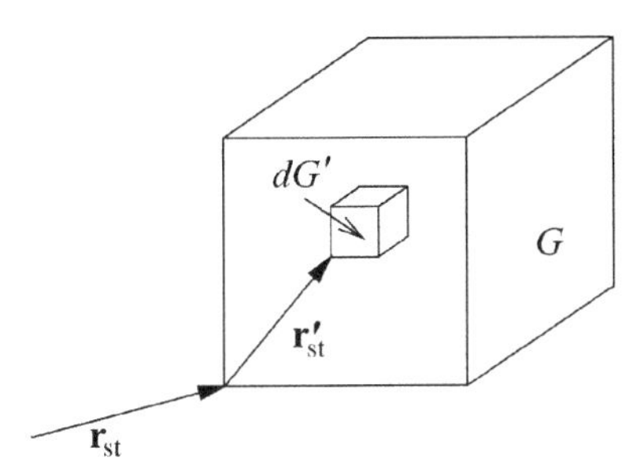

Fig. 11.4 Averaging for fixed **r** and t (i.e. $\mathbf{r}_{st}$ = constant) over the space-time 'volume' G.

the averaging space-time 'volume' G whose origin is taken at the endpoint of $\mathbf{r}_{st}$. For the Cartesian system we have $dG' = d\underset{x}{G'} = dx'\,dy'\,dz'\,dt'$ whereas for the general q^i system $dG' = d\underset{q}{G'} = \sqrt{\underset{q}{g}}'\,dq^{1'}\,dq^{2'}\,dq^{3'}\,dt'$. During the averaging process the space-time volume G and $\mathbf{r}_{st}$ are fixed while the averaging itself is carried out with the help of $\mathbf{r}'_{st}$ so that the coordinates of $\mathbf{r}_{st}$ and $\mathbf{r}'_{st}$ are entirely independent.

Let us now consider the special situation that the statistical parameters characterizing the turbulence are independent of space and statistically not changing with time. In this case we speak of *homogeneous and stationary turbulence*. Now the ensemble, time, and space averages yield the same results. This is known as the *ergodic condition*. To make the turbulence problem more tractable, in our studies we will assume that the ergodic condition applies. Thus, all results obtained with the help of the ensemble average will be considered valid for the other averages as well.

To get a better understanding of the concept of averaging, we will show how the one-dimensional ensemble average can be transformed into the corresponding integral average if the number of realizations N becomes very large. Moreover, we will demonstrate with the help of Figure 11.5 how to usefully interpret the average and what is meant by holding x constant and by the integration over x'. Let us consider the hypothetical spectrum depicted in Figure 11.5(a). First we select the averaging interval Δx which is centered at x_i. Then we rotate the averaging interval by 90° at the point x_i and introduce the x'-axis as shown in part (b) of Figure 11.5. At the point x_i one now has a collection of N realizations x'_j, implying N fluctuations.

On integrating over this newly formed collection of N realizations we find at the point x_i the ensemble average, which transforms to the integral average:

$$\overline{\psi}^{x}(x_i) = \lim_{N\to\infty}\left(\frac{1}{N}\sum_{j=1}^{N}\psi(x_i, x'_j)\right) = \frac{1}{\Delta x}\int_{x_i-\Delta x/2}^{x_i+\Delta x/2}\psi(x_i, x')\,dx' \qquad (11.30)$$

if N becomes very large. The rotation is then carried out at each point of the x-axis so that for each x value one has a mean value $\overline{\psi}(x)$ representing N fluctuations.

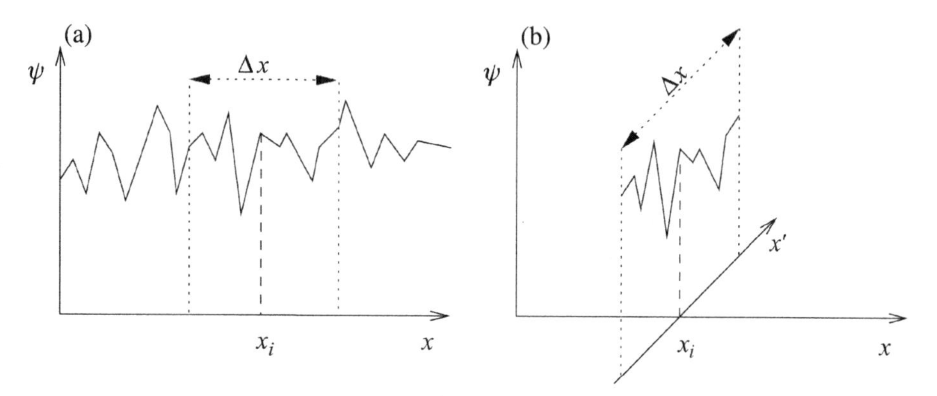

Fig. 11.5 A schematic representation of the averaging procedure for a one-dimensional spectrum.

In another example we are going to show that the derivative of the time average as found from the definition of the ensemble average may also be found from the integral definition. By observing that the average quantity $\overline{\psi}$ depends only on $\mathbf{r}_{\rm st}$ and not on $\mathbf{r}'_{\rm st}$ and that the boundaries of the averaging domain G are held fixed, we obtain according to the Leibniz rule

$$
\begin{aligned}
\overline{\frac{\partial}{\partial t}\psi(\mathbf{r}_{\rm st})} &= \frac{1}{G}\int_G \frac{\partial}{\partial t}\psi(\mathbf{r}_{\rm st}, \mathbf{r}'_{\rm st})\, dG' \\
&= \frac{1}{G}\int_G \frac{\partial}{\partial t}\left[\overline{\psi}(\mathbf{r}_{\rm st}) + \psi'(\mathbf{r}_{\rm st}, \mathbf{r}'_{\rm st})\right] dG' \\
&= \frac{\partial}{\partial t}\left(\frac{1}{G}\int_G \left[\overline{\psi}(\mathbf{r}_{\rm st}) + \psi'(\mathbf{r}_{\rm st}, \mathbf{r}'_{\rm st})\right] dG'\right) \\
&= \frac{\partial \overline{\psi}(\mathbf{r}_{\rm st})}{\partial t} + \frac{\partial}{\partial t}\left(\frac{1}{G}\int_G \psi'(\mathbf{r}_{\rm st}, \mathbf{r}'_{\rm st})\, dG'\right) \\
&= \frac{\partial \overline{\psi}(\mathbf{r}_{\rm st})}{\partial t} + \frac{\partial \overline{\psi'}(\mathbf{r}_{\rm st})}{\partial t} = \frac{\partial \overline{\psi}(\mathbf{r}_{\rm st})}{\partial t}
\end{aligned}
\tag{11.31}
$$

The last integral in (11.31) represents the mean value of the fluctuations and must vanish. Analogous arguments hold for the space derivatives.

11.5 Budget equations of the turbulent system

As we have pointed out several times, each of the important atmospheric prognostic equations of the molecular system can be written in the form of the budget equation. We will now proceed and derive the general form of the budget equation for the microturbulent system, which for simplicity will henceforth be called the turbulent system. This is done by averaging the molecular form (M6.66) of the budget equation. By steps that are now well known we find

$$\frac{\partial}{\partial t}(\overline{\rho}\widehat{\psi}) + \nabla\cdot(\widehat{\mathbf{v}}_{\mathrm{A}}\overline{\rho}\widehat{\psi}) + \nabla\cdot(\overline{\mathbf{v}''\rho\psi''}) + \nabla\cdot\overline{\mathbf{F}}_{\psi} = \overline{Q}_{\psi} \tag{11.32}$$

since $\overline{\mathbf{v}_{\mathrm{A}}\rho\psi} = \widehat{\mathbf{v}}_{\mathrm{A}}\overline{\rho}\widehat{\psi} + \overline{\mathbf{v}''\rho\psi''}$. Instead of one term describing the molecular convective flux or molecular transport term, we have two terms, i.e. $\widehat{\mathbf{v}}_{\mathrm{A}}\overline{\rho}\widehat{\psi}$ representing the mean convective flux in the turbulent system plus the *turbulent flux* $\mathbf{F}_{\psi,\mathrm{t}} = \overline{\mathbf{v}''\rho\psi''}$. Using the operator (11.23) we obtain the budget equation of the turbulent system in the form

$$\frac{\widehat{D}}{Dt}(\overline{\rho}\widehat{\psi}) + \nabla\cdot(\overline{\mathbf{F}}_{\psi} + \mathbf{F}_{\psi,\mathrm{t}}) = \overline{Q}_{\psi} \tag{11.33}$$

Next we are going to apply this equation to various state variables. For reasons of clarity we first collect the prognostic equations of the molecular system. These are the continuity equation for the total mass (M6.67), the continuity for the partial masses or concentrations (M6.72), the first law of thermodynamics (M6.73), and the equation of motion (1.71):

$$\begin{aligned}
\frac{D\rho}{Dt} &= 0\\
\frac{D}{Dt}(\rho m^k) &= -\nabla\cdot\mathbf{J}^k + I^k\\
\frac{D}{Dt}(\rho e) &= -\nabla\cdot(\mathbf{J}^h + \mathbf{F}_{\mathrm{R}}) - p\,\nabla\cdot\mathbf{v}_{\mathrm{A}} + \epsilon\\
\frac{D}{Dt}(\rho\mathbf{v}) &= -\nabla p - \rho\,\nabla\phi - 2\rho\boldsymbol{\Omega}\times\mathbf{v} + \nabla\cdot\mathbb{J}
\end{aligned} \tag{11.34}$$

where $\epsilon = \mathbb{J}\cdot\cdot\nabla\mathbf{v}_{\mathrm{A}}$ is the dissipation of energy. These equations will then be averaged using the averaging procedures we have discussed so thoroughly above. The results are

$$\boxed{\begin{aligned}
&\text{(a)} && \frac{\widehat{D}\overline{\rho}}{Dt} = 0\\
&\text{(b)} && \frac{\widehat{D}}{Dt}(\overline{\rho}\,\widehat{m}^k) = -\nabla\cdot\overline{\mathbf{J}}^k - \nabla\cdot\mathbf{J}^k_{t} + \overline{I}^k\\
&\text{(c)} && \frac{\widehat{D}}{Dt}(\overline{\rho}\,\widehat{e}) = -\nabla\cdot\overline{\mathbf{J}}^h - \nabla\cdot\overline{\mathbf{F}}_{\mathrm{R}} - \nabla\cdot\mathbf{J}^h_{\mathrm{t}}\\
& && \qquad - \overline{p}\,\nabla\cdot\widehat{\mathbf{v}}_{\mathrm{A}} + \overline{\mathbf{v}''\cdot\nabla p} + \overline{\epsilon}\\
&\text{(d)} && \frac{\widehat{D}}{Dt}(\overline{\rho}\,\widehat{\mathbf{v}}) = -\nabla\overline{p} + \nabla\cdot\overline{\mathbb{J}} - \nabla\cdot\mathbb{J}_{\mathrm{t}} - \overline{\rho}\,\nabla\phi + 2\overline{\rho}\Omega\times\widehat{\mathbf{v}}
\end{aligned}} \tag{11.35}$$

One essential part, in contrast to the molecular system, is the appearance of the turbulent flux terms $\mathbf{F}_{\psi,\mathrm{t}}$. These are given by

$$\boxed{\mathbf{J}_\mathrm{t}^k = \overline{\mathbf{v}''\rho m^k}, \qquad \mathbf{J}_\mathrm{t}^h = \overline{\mathbf{v}''\rho h}, \qquad \mathbb{J}_\mathrm{t} = \overline{\mathbf{v}''\rho \mathbf{v}_\mathrm{A}}} \tag{11.36}$$

where $h = e + p\alpha$ is the enthalpy. The reason why the term $\mathbf{J}_\mathrm{t}^e = \overline{\mathbf{v}''\rho e}$ does not appear is left as a problem. For the turbulent diffusion flux $\mathbf{J}_\mathrm{t}^k$ and the mean molecular phase transitions $\overline{I}^k$ the following summation conditions hold:

$$\begin{gathered}\sum_{k=0}^{3} \mathbf{J}_\mathrm{t}^k = \overline{\mathbf{v}''\rho \sum_{k=0}^{3} m^k} = \overline{\rho \mathbf{v}''} = \overline{\rho}\widehat{\mathbf{v}''} = 0 \\ \overline{I}^0 = 0, \qquad \sum_{k=1}^{3} \overline{I}^k = 0\end{gathered} \tag{11.37}$$

The turbulent momentum flux tensor $\mathbb{J}_\mathrm{t}$ and the *Reynolds tensor* $\mathbb{R}$, which is frequently used in the literature, are related by

$$\mathbb{R} = -\overline{\mathbf{v}''\rho \mathbf{v}_\mathrm{A}} = -\overline{\mathbf{v}''\rho \mathbf{v}''} = -\mathbb{J}_\mathrm{t} \tag{11.38}$$

A few remarks about the meaning of the turbulent fluxes may be helpful. Consider a test volume V moving approximately with the average velocity $\widehat{\mathbf{v}}_\mathrm{A}$ so that during time step Δt the volume is displaced the distance $\widehat{\mathbf{v}}_\mathrm{A}\,\Delta t$. Fluid elements, however, are moving with the velocity $\widehat{\mathbf{v}}_\mathrm{A} + \mathbf{v}''$. During this time step the velocity fluctuations cause some fluid to enter the test volume at some parts of the imagined volume boundary while at other sections of the boundary some fluid leaves the test volume. While the total mass is conserved on average, as is guaranteed by the continuity equation, there is no reason to believe that the quantity $\Psi_V = \int_V \rho\psi \, dV'$ is conserved also. In fact Ψ_V may be viewed as the total content of the property ψ within the volume V at a particular time. Since Ψ_V changes with time, there must exist a flux that is penetrating the volume surface of the fluid volume. It now stands to reason that $\mathbf{F}_{\psi,\mathrm{t}} = \overline{\mathbf{v}''\rho\psi}$, as listed in (11.36), may be interpreted as the turbulent flux which is the mean ψ stream through the surface of V which is moving with the velocity $\widehat{\mathbf{v}}_\mathrm{A}$. In a later chapter it will be shown that, within the framework of this discussion, the turbulent flux is directed from regions of higher to lower ψ values so that this flux causes an equalization of the ψ field. Occasionally, there are situations in which the transport appears to flow against the gradient. This is known as the *counter-gradient flow*, which is really not well understood at this time and will be left out of our discussion.

We close this section with a remark on the Hesselberg average. Only specific values of extensive variables $(\widehat{\mathbf{v}}, \widehat{e}, \widehat{h}, \widehat{\alpha})$ and their derivatives with respect to intensive coordinates p and T carry the roof symbol $\widehat{\ }$.

11.6 The energy budget of the turbulent system

Our next step is to show that the total energy budget of the turbulent system is balanced or source-free. In order to proceed efficiently, we again take the energy budget of the molecular system in the absence of deformational effects ($\mathbf{v}_D = 0$). The system consists of the budget equation for the potential energy as expressed by the geopotential, the budget equation for the kinetic energy, and the first law of thermodynamics:

$$\begin{aligned}
&\text{(a)} \quad \frac{D}{Dt}(\rho\phi) = \rho\mathbf{v}\cdot\nabla\phi \\
&\text{(b)} \quad \frac{D}{Dt}\left(\frac{\rho\mathbf{v}^2}{2}\right) + \nabla\cdot\left[\mathbf{v}\cdot(p\mathbb{E} - \mathbb{J})\right] = -\rho\mathbf{v}\cdot\nabla\phi + p\,\nabla\cdot\mathbf{v} - \epsilon \\
&\text{(c)} \quad \frac{D}{Dt}(\rho e) + \nabla\cdot(\mathbf{J}^h + \mathbf{F}_R) = -p\,\nabla\cdot\mathbf{v} + \epsilon
\end{aligned} \tag{11.39}$$

For later convenience we add the enthalpy equation

$$\frac{D}{Dt}(\rho h) + \nabla\cdot(\mathbf{J}^h + \mathbf{F}_R) = \frac{dp}{dt} + \epsilon \tag{11.40}$$

This equation may be easily obtained by substituting into (11.39c) $e = h - p\alpha$ and $\nabla\cdot\mathbf{v} = \rho\, d\alpha/dt$ with $\alpha = 1/\rho$. We recall that quantities depending explicitly on $\mathbf{r}$ and t are not averaged in the microphysical sense. This includes the geopotential $\phi = gz$.

First of all we direct our attention to averaging (11.39b) by using (11.23). The result is

$$\frac{\widehat{D}}{Dt}\left(\frac{\overline{\rho}\widehat{\mathbf{v}^2}}{2}\right) + \nabla\cdot\left(\overline{\mathbf{v}\cdot(p\mathbb{E} - \mathbb{J})} + \overline{\mathbf{v}''\frac{\rho\mathbf{v}^2}{2}}\right) = -\overline{\rho}\,\widehat{\mathbf{v}}\cdot\nabla\phi + \overline{p\,\nabla\cdot\mathbf{v}} - \overline{\epsilon} \tag{11.41}$$

By using the expressions

$$\begin{gathered}
\overline{\mathbf{v}''\frac{\rho\mathbf{v}^2}{2}} = \mathbf{k}_t - \widehat{\mathbf{v}}\cdot\mathbb{R}, \qquad \mathbf{k}_t = \overline{\mathbf{v}''\frac{\rho\mathbf{v}''^2}{2}} \\
\overline{p\,\nabla\cdot\mathbf{v}} = \overline{p}\,\nabla\cdot\widehat{\mathbf{v}} + \nabla\cdot(\overline{\mathbf{v}''p}) - \overline{\mathbf{v}''\cdot\nabla p} \\
\overline{\mathbf{v}\cdot(p\mathbb{E} - \mathbb{J})} = \widehat{\mathbf{v}}\cdot(\overline{p}\mathbb{E} - \overline{\mathbb{J}}) + \overline{\mathbf{v}''p} - \overline{\mathbf{v}''\cdot\mathbb{J}}
\end{gathered} \tag{11.42}$$

we obtain for (11.41) the budget equation for the total kinetic energy:

$$\boxed{\begin{aligned}
&\frac{\widehat{D}}{Dt}\left(\frac{\overline{\rho}\widehat{\mathbf{v}^2}}{2}\right) + \nabla\cdot\left[\widehat{\mathbf{v}}\cdot(\overline{p}\mathbb{E} - \overline{\mathbb{J}} - \mathbb{R}) - \overline{\mathbf{v}''\cdot\mathbb{J}} + \mathbf{k}_t\right] \\
&\quad = -\overline{\rho}\,\widehat{\mathbf{v}}\cdot\nabla\phi + \overline{p}\,\nabla\cdot\widehat{\mathbf{v}} - \overline{\mathbf{v}''\cdot\nabla p} - \overline{\epsilon}
\end{aligned}} \tag{11.43}$$

The vector $\mathbf{k}_t$ describes the turbulent flux of turbulent kinetic energy. Note that the effect of the earth's rotation does not appear in the equation of the kinetic energy (molecular or turbulent) since the Coriolis force does not perform any work.

Our next goal is to find an expression for the turbulent kinetic energy $\widehat{k}$ which is defined by means of

$$\widehat{k} = \frac{\widehat{\mathbf{v}''^2}}{2} = \frac{\widehat{\mathbf{v}^2}}{2} - \frac{\widehat{\mathbf{v}}^2}{2} \tag{11.44}$$

In order to proceed we must derive an expression for the kinetic energy $\widehat{\mathbf{v}}^2/2$ of the mean motion. Scalar multiplication of the equation of mean motion (11.35d) by $\widehat{\mathbf{v}}$ yields

$$\boxed{\frac{\widehat{D}}{Dt}\left(\frac{\overline{\rho}\,\widehat{\mathbf{v}}^2}{2}\right) + \nabla\cdot\left[\widehat{\mathbf{v}}\cdot(\overline{p}\mathbb{E} - \overline{\mathbb{J}} - \mathbb{R})\right] = -\overline{\rho}\,\widehat{\mathbf{v}}\cdot\nabla\phi + \overline{p}\,\nabla\cdot\widehat{\mathbf{v}} - \nabla\widehat{\mathbf{v}}\cdot\cdot(\overline{\mathbb{J}} + \mathbb{R})} \tag{11.45}$$

Now we subtract this expression from (11.43) and obtain the prognostic equation for the turbulent kinetic energy:

$$\boxed{\frac{\widehat{D}}{Dt}(\overline{\rho}\widehat{k}) + \nabla\cdot(\mathbf{k}_t - \overline{\mathbf{v}''\cdot\mathbb{J}}) = \nabla\widehat{\mathbf{v}}\cdot\cdot(\overline{\mathbb{J}} + \mathbb{R}) - \overline{\epsilon} - \overline{\mathbf{v}''\cdot\nabla p}} \tag{11.46}$$

The turbulent budget system is completed by stating the prognostic equations for the potential energy,

$$\boxed{\frac{\widehat{D}}{Dt}(\overline{\rho}\phi) = \overline{\rho}\,\widehat{\mathbf{v}}\cdot\nabla\phi} \tag{11.47}$$

and the internal energy,

$$\boxed{\frac{\widehat{D}}{Dt}(\overline{\rho}\,\widehat{e}) + \nabla\cdot(\overline{\mathbf{J}}^h + \overline{\mathbf{F}}_R + \mathbf{J}_t^h) = -\overline{p}\,\nabla\cdot\widehat{\mathbf{v}} + \overline{\mathbf{v}''\cdot\nabla p} + \overline{\epsilon}} \tag{11.48}$$

The fact that we have introduced $\mathbf{J}_t^h$ instead of $\mathbf{J}_t^e$ is responsible for the appearance of the term $\overline{\mathbf{v}''\cdot\nabla p}$. By adding equations (11.45)–(11.48) we find that the sum of all source terms vanishes. Thus, we conclude that the energy budget of the turbulent system is balanced. The reader may wish to display graphically the energy transformations by repeating the procedure used for the molecular system.

Of special interest in the energy transformations is the interaction of the turbulent kinetic energy and the internal energy by means of the advection-type term $\overline{\mathbf{v}''\cdot\nabla p}$. If this term is positive a loss of turbulent kinetic energy occurs in favor of the internal energy. A loss of turbulent kinetic energy represents a damping effect that occurs for stable atmospheric stratification. If this source term is negative, turbulence will increase, which is indicative of unstable atmospheric stratification. It should be noted that the representation (11.45)–(11.48) is not unique. By some mathematical manipulations it is possible to separate the source terms in a different way but the sum of the sources still adds up to zero.

For later convenience we will also average the enthalpy equation (11.40) of the molecular system, resulting in

$$\frac{\widehat{D}}{Dt}(\overline{\rho}\widehat{h}) + \nabla \cdot (\overline{\mathbf{J}}^{h} + \mathbf{J}_{\mathrm{t}}^{h} + \overline{\mathbf{F}}_{\mathrm{R}}) = \frac{\widehat{d}\,\overline{p}}{dt} + \overline{\mathbf{v}'' \cdot \nabla p} + \overline{\epsilon} \tag{11.49}$$

11.7 Diagnostic and prognostic equations of turbulent systems

We begin with a simple example by considering the ideal-gas law of the molecular system as given by

$$p = R_0 \rho T_{\mathrm{v}} \tag{11.50}$$

where T_{v} is the virtual temperature and R_0 the gas constant of dry air. Taking the Reynolds average to simulate a turbulent system, we find

$$\frac{\overline{p}}{R_0} = \overline{\rho}\,\overline{T_{\mathrm{v}}} + \overline{\rho' T_{\mathrm{v}}'} \tag{11.51}$$

The final term of this equation containing the fluctuations is usually much smaller in magnitude than the remaining terms. Ignoring the fluctuation term, the ideal-gas law in the mean retains the molecular form (11.50),

$$\overline{p} = R_0 \overline{\rho}\,\overline{T_{\mathrm{v}}} \tag{11.52}$$

Another approach is to postulate a general diagnostic formula in order to avoid the appearance of correlations to begin with. Proceeding in this manner, a particular diagnostic equation of the turbulent system is identical in form with the corresponding molecular equation. To set the pattern, we make use of the general diagnostic equation which is discussed more thoroughly in TH:

$$\begin{aligned}
&\text{(a)} \quad \psi = \psi_n m^n, \qquad \psi_k = \psi_k(p, T, m^0, m^1, m^2, m^3) \\
&\text{(b)} \quad d\psi = \left(\frac{\partial \psi}{\partial p}\right)_{T,m^k} dp + \left(\frac{\partial \psi}{\partial T}\right)_{p,m^k} dT + \psi_n\, dm^n
\end{aligned} \tag{11.53}$$

Here we make use of the Einstein summation convention which omits the summation sign in the expression $\psi_n m^n$. The term ψ represents various thermodynamic functions, such as internal energy, enthalpy, entropy, specific volume, and others. ψ_k is known as the *specific partial quantity* of Ψ. Next we define the function

$$\widehat{\psi} = \widetilde{\psi}_n \widehat{m}^n, \qquad \widetilde{\psi}_k = \psi_k(\overline{p}, \overline{T}, \widehat{m}^0, \widehat{m}^1, \widehat{m}^2, \widehat{m}^3) \tag{11.54}$$

which applies to the turbulent system. The tilde above ψ_k does not represent an average value itself but rather represents a function depending on the average state variables. Analogously to (11.53) we obtain for the turbulent system

$$\widehat{d}\,\widehat{\psi} = \left(\frac{\partial \widehat{\psi}}{\partial \overline{p}}\right)_{\overline{T},\widehat{m}^k} \widehat{d}\overline{p} + \left(\frac{\partial \widehat{\psi}}{\partial \overline{T}}\right)_{\overline{p},\widehat{m}^k} \widehat{d}\,\overline{T} + \widetilde{\psi}_n\, \widehat{d}\widehat{m}^n \tag{11.55}$$

The operator $\widehat{d}$ is defined in (11.21). A few examples will clarify the meaning of (11.55).

Example 1 In the first example we give the equation of state consisting of dry air, water vapor, liquid water, and ice

$$\alpha = \frac{1}{\rho} = \alpha_n m^n = \left(\frac{R_0 T}{p}\right) m^0 + \left(\frac{R_1 T}{p}\right) m^1 + \alpha_2 m^2 + \alpha_3 m^3 \tag{11.56}$$

The corresponding statement for the turbulent system is

$$\widehat{\alpha} = \frac{1}{\overline{\rho}} = \widetilde{\alpha}_n \widehat{m}^n = \left(\frac{R_0 \overline{T}}{\overline{p}}\right) \widehat{m}^0 + \left(\frac{R_1 \overline{T}}{\overline{p}}\right) \widehat{m}^1 + \alpha_2 \widehat{m}^2 + \alpha_3 \widehat{m}^3 \tag{11.57}$$

since $\alpha_2, \alpha_3 =$ constant.

Example 2 The differential of the enthalpy of the turbulent system is obtained by setting $\psi = h$ in (11.55) so that

$$\begin{aligned} &\widehat{d}\,\widehat{h} = \widehat{c}_p\, \widehat{d}\,\overline{T} + \widetilde{h}_n\, \widehat{d}\widehat{m}^n \\ &\text{with} \quad \widehat{c}_p = \left(\frac{\partial \widehat{h}}{\partial \overline{T}}\right)_{\widehat{m}^k,\overline{p}} = c_{p,n}\widehat{m}^n, \qquad c_{p,k} = \text{constant} \\ &\widetilde{h}_k = \widetilde{h}_k(\overline{T} = T_0) + c_{p,k}(\overline{T} - T_0) \end{aligned} \tag{11.58}$$

Here we have assumed that the pressure dependency of the enthalpy can be ignored. This assumption is rigorously true for any ideal gas.

Equation (11.58) can be used to obtain the prognostic equation for the temperature of the turbulent system. First we multiply (11.58) by the density and then divide the result by dt. Next we eliminate $\widehat{d}\widehat{m}^k/dt$ with the help of (11.35b), to find

$$
\begin{aligned}
\overline{\rho}\,\frac{\widehat{d}\,\widehat{h}}{dt} &= \overline{\rho}\,\widehat{c}_p\,\frac{\widehat{d}\,\widehat{T}}{dt} + \overline{\rho}\widetilde{h}_n\,\frac{\widehat{d}\,\widehat{m}^n}{dt} \quad \Longrightarrow \\
\frac{\widehat{D}}{Dt}(\overline{\rho}\widehat{h}) &= \widehat{c}_p\,\frac{\widehat{D}}{Dt}(\overline{\rho}\widehat{T}) + \widetilde{h}_n\,\frac{\widehat{D}}{Dt}(\overline{\rho}\widehat{m}^n) \\
&= \widehat{c}_p\,\frac{\widehat{D}}{Dt}(\overline{\rho}\widehat{T}) + \widetilde{h}_n\Big[-\nabla\cdot(\overline{\mathbf{J}}^n + \mathbf{J}^n_{\mathrm{t}}) + \overline{I}^n\Big] \\
&= \widehat{c}_p\,\frac{\widehat{D}}{Dt}(\overline{\rho}\widehat{T}) - \nabla\cdot\Big[\widetilde{h}_n(\overline{\mathbf{J}}^n + \mathbf{J}^n_{\mathrm{t}})\Big] + \widetilde{h}_n\overline{I}^n + (\overline{\mathbf{J}}^n + \mathbf{J}^n_{\mathrm{t}})\cdot\nabla\widetilde{h}_n
\end{aligned}
\tag{11.59}
$$

By substituting this equation into (11.49) we obtain the prognostic equation for the temperature:

$$
\boxed{
\begin{aligned}
\widehat{c}_p\,\frac{\widehat{D}}{Dt}(\overline{\rho}\widehat{T}) + \nabla\cdot(\overline{\mathbf{J}}^h_{\mathrm{s}} + \mathbf{J}^h_{\mathrm{s,t}} + \overline{\mathbf{F}}_{\mathrm{R}}) = \frac{\widehat{d}\,\overline{p}}{dt} + \overline{\mathbf{v}''\cdot\nabla p} + \overline{\epsilon} - \widetilde{h}_n\overline{I}^n \\
- (\overline{\mathbf{J}}^n + \mathbf{J}^n_{\mathrm{t}})\cdot\nabla\widetilde{h}_n
\end{aligned}
}
\tag{11.60}
$$

In this equation the following fluxes appear:

$$
\begin{aligned}
&\textit{the mean molecular sensible enthalpy flux:} && \overline{\mathbf{J}}^h_{\mathrm{s}} = \overline{\mathbf{J}}^h - \overline{\mathbf{J}}^n\widetilde{h}_n \\
&\textit{the turbulent sensible enthalpy flux:} && \mathbf{J}^h_{\mathrm{s,t}} = \mathbf{J}^h_{\mathrm{t}} - \mathbf{J}^n_{\mathrm{t}}\widetilde{h}_n \\
&\textit{the turbulent latent enthalpy flux:} && \mathbf{J}^h_{\mathrm{l,t}} = \mathbf{J}^n_{\mathrm{t}}\widetilde{h}_n
\end{aligned}
\tag{11.61}
$$

In several equations, e.g. (11.35), (11.43), and (11.60), the term $\overline{\mathbf{v}''\cdot\nabla p}$, which is very awkward to handle, appears. Since there is no reason to ignore this term, we will try to replace it by a reasonable approximation. To simplify the mathematical treatment we begin with the molecular system of dry air which is described by

$$
p = R_0\rho T, \qquad \Pi = \left(\frac{p}{p_0}\right)^{k_0}, \qquad \theta = \frac{T}{\Pi}
\tag{11.62}
$$

where Π is the *Exner function*, $k_0 = R_0/c_{p,0}$, and θ is the potential temperature. With the help of these expressions we easily obtain

$$
\overline{\mathbf{v}''\cdot\nabla p} = \overline{\rho c_{p,0}\theta\mathbf{v}''\cdot\nabla\Pi}
\tag{11.63}
$$

We decompose the Exner function as shown in

$$
\Pi = \widetilde{\Pi} + \Pi' \quad \text{with} \quad \widetilde{\Pi} = \left(\frac{\overline{p}}{p_0}\right)^{k_0}
\tag{11.64}
$$

Substituting (11.64) into (11.63) gives

$$\overline{\mathbf{v}'' \cdot \nabla p} = \widetilde{\Pi}\overline{\rho c_{p,0}\theta \mathbf{v}''} \cdot \nabla \ln \widetilde{\Pi} + \widetilde{\Pi}\overline{\rho c_{p,0}\theta \mathbf{v}'' \cdot \frac{\nabla \Pi'}{\widetilde{\Pi}}} \tag{11.65}$$

Before proceeding we wish to remark that, with the help of the entropy-production equation and the so-called *linear Onsager theory*, it is possible to obtain proper parameterized expressions for various fluxes. This subject is thoroughly discussed in various textbooks on thermodynamics. Often one speaks of the *phenomenological theory*. We refer to TH.

The first term on the right-hand side of (11.65) has the desired form required by the linear Onsager theory which will be used later to find proper expressions for various turbulent fluxes. The first factor in each of the expressions on the right-hand side is a flux while the second factor is the thermodynamic force. For reasons that will become apparent later we split the flux:

$$\widetilde{\Pi}\overline{\rho c_{p,0}\theta \mathbf{v}''} = \overline{\rho c_{p,0}\theta \Pi \mathbf{v}''} - \overline{\rho c_{p,0}\theta \Pi' \mathbf{v}''} = \overline{\rho c_{p,0}T' \mathbf{v}''} - \overline{\rho c_{p,0}\theta \Pi' \mathbf{v}''} \tag{11.66}$$

since $\overline{\rho c_{p,0} T \mathbf{v}''} = \overline{\rho c_{p,0} T' \mathbf{v}''}$. In a simplified manner we now extrapolate from the system of dry air to the general multicomponent system by replacing $c_{p,0}$ in (11.66) by $c_p = c_{p,n} m^n$. We proceed analogously with (11.65). The result is

$$\begin{aligned} \overline{\mathbf{v}'' \cdot \nabla p} &= \widetilde{\Pi}\overline{\rho c_p \theta \mathbf{v}''} \cdot \nabla \ln \widetilde{\Pi} + \widetilde{\Pi}\overline{\rho c_p \theta \mathbf{v}'' \cdot \frac{\nabla \Pi'}{\widetilde{\Pi}}} \\ \widetilde{\Pi}\overline{\rho c_p \theta \mathbf{v}''} &= \overline{\rho c_p \theta \Pi \mathbf{v}''} - \overline{\rho c_p \theta \Pi' \mathbf{v}''} = \overline{\rho c_p T' \mathbf{v}''} - \overline{\rho c_p \theta \Pi' \mathbf{v}''} \end{aligned} \tag{11.67}$$

The heat fluxes appearing in the above equations are given the following names:

the turbulent heat flux:	$\mathbf{J}_{\mathrm{t}}^{\theta} = \widetilde{\Pi}\overline{\rho c_p \theta \mathbf{v}''}$
the turbulent Exner flux:	$\mathbf{J}_{\mathrm{t}}^{\Pi} = \overline{\rho c_p \theta \Pi' \mathbf{v}''}$
the turbulent sensible enthalpy flux:	$\mathbf{J}_{\mathrm{s,t}}^{h} = \overline{\rho c_p T' \mathbf{v}''}$

(11.68)

They are related by

$$\mathbf{J}_{\mathrm{t}}^{\theta} = \mathbf{J}_{\mathrm{s,t}}^{h} - \mathbf{J}_{\mathrm{t}}^{\Pi} \tag{11.69}$$

Note that neglecting the pressure fluctuation term Π' results in an identity of the heat flux $\mathbf{J}_{\mathrm{t}}^{\theta}$ and the sensible enthalpy flux $\mathbf{J}_{\mathrm{s,t}}^{h}$. Very often we do not distinguish between these two fluxes since $\left|\mathbf{J}_{\mathrm{t}}^{\Pi}\right|$ is usually very small in comparison with $\left|\mathbf{J}_{\mathrm{t}}^{\theta}\right|$.

Utilizing the expressions derived above, equation (11.65) can finally be written as

$$\overline{\mathbf{v}'' \cdot \nabla p} = \mathbf{J}_{\mathrm{t}}^{\theta} \cdot \nabla \ln \widetilde{\Pi} + \overline{\mathbf{J}_{\mathrm{t}}^{\theta} \cdot \mathbf{A}} \quad \text{with} \quad \mathbf{A} = \frac{\nabla \Pi'}{\widetilde{\Pi}} \tag{11.70}$$

It should be noted that the second term on the right-hand side of (11.70) cannot be evaluated in terms of the phenomenological equations by using the theory in the form presented here. The same is also true for the energy dissipation $\overline{\epsilon}$.

In this section we have obtained an approximate parameterization of the term $\overline{\mathbf{v}'' \cdot \nabla p}$, but we have failed to derive an expression for the dissipation of energy. A different approach in the treatment of the thermodynamics of turbulent systems is to avoid the appearance of $\overline{\mathbf{v}'' \cdot \nabla p}$ altogether. This can be done by identifying some of the variables of the turbulent system with average values of the molecular system. The remaining variables must be defined in a suitable manner. One possible way to proceed is described by the so-called *exclusive system*. The curious name derives from the treatment of the internal energy of the system which excludes, or, which amounts to the same thing, does not include, the turbulent kinetic energy. While this system offers various advantages, it does not provide a method by which one can find the Reynolds tensor $\mathbb{R}$ and the dissipation of energy $\overline{\epsilon}$ without additional assumptions. The least desirable property of the exclusive system is that energy is not strictly conserved. In the more refined *inclusive system* the turbulent kinetic energy is included as part of the internal energy. In this system it is still not possible to obtain $\overline{\epsilon}$ but the system does provide access to $\mathbb{R}$. The theory is difficult and not yet complete. For more details see the papers by Sievers (1982, 1984).

11.8 Production of entropy in the microturbulent system

In order to derive the proper forms of the various turbulent fluxes we employ the entropy-production equation. This equation will now be derived for the microturbulent system. The starting point in the derivation is *Gibbs' fundamental equation* for the molecular system. Again we refer to TH. If s is the entropy and $\mu_k = h_k - T s_k$ the chemical potential, then we have

$$T \frac{ds}{dt} = \frac{de}{dt} + p \frac{d\alpha}{dt} - \mu_n \frac{dm^n}{dt} \tag{11.71}$$

We assume that the Gibbs equation is also valid for the microturbulent system. We proceed by introducing the proper variables and differential operators of the microturbulent system in the form

$$\boxed{\overline{T} \frac{\widehat{d}\,\widehat{s}}{dt} = \frac{\widehat{d}\,\widehat{e}}{dt} + \overline{p} \frac{\widehat{d}\,\widehat{\alpha}}{dt} - \widetilde{\mu}_n \frac{\widehat{d}\,\widehat{m}^n}{dt}} \tag{11.72}$$

We want to point out that this equation does not arise simply from averaging (11.71), but rather must be introduced in the sense of an axiomatic statement. The quantitities $\widehat{s}$ and $\widetilde{\mu}_k$ are functions of the averaged state variables $(\overline{p}, \overline{T}, \widehat{m}^k)$. First we introduce into (11.72) the continuity equation

$$\frac{\widehat{d}\,\widehat{\alpha}}{dt} = \widehat{\alpha}\,\nabla \cdot \widehat{\mathbf{v}}_{\mathrm{A}} \tag{11.73}$$

which follows from (11.35a). On multiplying (11.72) by the air density we find

$$\overline{T}\,\frac{\widehat{D}}{Dt}(\overline{\rho}\,\widehat{s}) = \frac{\widehat{D}}{Dt}(\overline{\rho}\,\widehat{e}) + \overline{p}\,\nabla \cdot \widehat{\mathbf{v}}_{\mathrm{A}} - \widetilde{\mu}_n\,\frac{\widehat{D}}{Dt}(\overline{\rho}\widehat{m}^n) \tag{11.74}$$

Inspection of this equation shows that there are three derivative expressions. Two of these will be eliminated by substituting the budget equations for the partial concentrations (11.35b) and for the internal energy (11.35c) into (11.74), yielding, after some slight rearrangements,

$$\overline{T}\,\frac{\widehat{D}}{Dt}(\overline{\rho}\,\widehat{s}) = -\nabla \cdot (\overline{\mathbf{J}}^h + \mathbf{J}_{\mathrm{t}}^h) + \mathbf{J}_{\mathrm{t}}^\theta \cdot \nabla \ln \widetilde{\Pi} + \overline{\epsilon} - \widetilde{\mu}_n\Big[\overline{I}^n - \nabla \cdot (\overline{\mathbf{J}}^n + \mathbf{J}_{\mathrm{t}}^n)\Big] \tag{11.75}$$

We have omitted the radiative flux, which is obtained by solving the radiative-transfer equation. Moreover, we have used (11.70) to replace $\overline{\mathbf{v}'' \cdot \nabla p}$, assuming that the second term on the right-hand side of this equation may be ignored. Now we divide (11.75) by $\overline{T}$ and write the resulting expression in budget form. We thus obtain

$$\frac{\widehat{D}}{Dt}(\overline{\rho}\,\widehat{s}) + \nabla \cdot \left(\frac{1}{\overline{T}}\Big[\overline{\mathbf{J}}^h + \mathbf{J}_{\mathrm{t}}^h - \widetilde{\mu}_n(\overline{\mathbf{J}}^n + \mathbf{J}_{\mathrm{t}}^n)\Big]\right) = \overline{Q_{\widehat{s}}} \tag{11.76}$$

with

$$\begin{aligned}\overline{Q_{\widehat{s}}} = &-\frac{\widetilde{\mu}_n \overline{I}^n}{\overline{T}} + \frac{\mathbf{J}_{\mathrm{t}}^\theta}{\overline{T}} \cdot \nabla \ln \widetilde{\Pi} + \frac{\overline{\epsilon}}{\overline{T}} - (\overline{\mathbf{J}}^h + \mathbf{J}_{\mathrm{t}}^h) \cdot \frac{\nabla \overline{T}}{\overline{T}^2} \\ &- (\overline{\mathbf{J}}^n + \mathbf{J}_{\mathrm{t}}^n) \cdot \nabla\left(\frac{\widetilde{\mu}_n}{\overline{T}}\right) \geq 0\end{aligned} \tag{11.77}$$

The inequality results from the fact that, according to the second law of thermodynamics, the entropy production is positive definite.

To obtain convenient flux representations we are going to rewrite the entropy production $\overline{Q_{\widehat{s}}}$ by using a well-known thermodynamic relation. For the microturbulent system this identity may be written as

$$\nabla\left(\frac{\widetilde{\mu}_k}{\overline{T}}\right) = \frac{1}{\overline{T}}(\nabla \widetilde{\mu}_k)_{\overline{T}} - \frac{\widetilde{h}_k}{\overline{T}^2}\,\nabla \overline{T} \tag{11.78}$$

For the molecular system the averaging symbols must be omitted. The subscript $\overline{T}$ indicates that the derivative is taken at constant $\overline{T}$. Substitution of this expression into (11.77) results in the entropy-production equation

$$\overline{T}\,\overline{Q}_{\widehat{s}} = -\,\widetilde{\mu}_n \overline{I}^n - \left[(\overline{\mathbf{J}}^h + \mathbf{J}_{\mathrm{t}}^h) - \widetilde{h}_n(\overline{\mathbf{J}}^n + \mathbf{J}_{\mathrm{t}}^n)\right] \cdot \frac{\nabla \overline{T}}{\overline{T}} \\ -\,(\overline{\mathbf{J}}^n + \mathbf{J}_{\mathrm{t}}^n) \cdot (\nabla \widetilde{\mu}_n)_{\overline{T}} + \mathbf{J}_{\mathrm{t}}^{\theta} \cdot \nabla \ln \widetilde{\Pi} + \overline{\epsilon} \geq 0 \tag{11.79}$$

with $\overline{\epsilon} = \overline{\mathbb{J}} \cdot\cdot \nabla \widehat{\mathbf{v}}_{\mathrm{A}} + \overline{\mathbb{J} \cdot\cdot \nabla \mathbf{v}''}$. By introducing the sensible enthalpy fluxes we obtain the final form of the entropy production:

$$\boxed{\overline{T}\,\overline{Q}_{\widehat{s}} = -\,\widetilde{\mu}_n \overline{I}^n - (\overline{\mathbf{J}}_{\mathrm{s}}^h + \mathbf{J}_{\mathrm{s,t}}^h) \cdot \nabla \ln \overline{T} \\ -\,(\overline{\mathbf{J}}^n + \mathbf{J}_{\mathrm{t}}^n) \cdot (\nabla \widetilde{\mu}_n)_{\overline{T}} + \mathbf{J}_{\mathrm{t}}^{\theta} \cdot \nabla \ln \widetilde{\Pi} + \overline{\epsilon} \geq 0} \tag{11.80}$$

We assume that *Curie's principle*, as explained in TH and elsewhere, also applies to turbulent systems. This permits us to split the total entropy production into parts. Each part belongs to a certain tensorial class (scalar, vectorial, and dyadic) and is required to be positive definite. Thus, the scalar part of the entropy production containing the phase-transition fluxes and the energy dissipation is given by

$$-\widetilde{\mu}_n \overline{I}^n + \overline{\epsilon} \geq 0 \tag{11.81}$$

For the vectorial part we may write

$$-(\overline{\mathbf{J}}_{\mathrm{s}}^h + \mathbf{J}_{\mathrm{s,t}}^h) \cdot \nabla \ln \overline{T} - (\overline{\mathbf{J}}^n + \mathbf{J}_{\mathrm{t}}^n) \cdot (\nabla \widetilde{\mu}_n)_{\overline{T}} + \mathbf{J}_{\mathrm{t}}^{\theta} \cdot \nabla \ln \widetilde{\Pi} \geq 0 \tag{11.82}$$

We will now attempt to parameterize the various fluxes. The parameterization of the mean energy dissipation $\overline{\epsilon}$ is best left to the statistical theory of turbulence. Review of (11.35d) and of (11.45) shows that the equation of motion and the equation of the kinetic energy of the mean flow contain the sum of the mean molecular and turbulent momentum fluxes. These fluxes cannot be found with the help (11.80) since they do not appear in this particular version of the entropy-production equation. The treatment of these fluxes would require an entropy-production equation in a different but equivalent form. It turns out, however, that the determination of the dyadic fluxes would be a rather involved mathematical process. For this reason we will treat the dyadic fluxes in an approximate manner, which will be sufficient for our purposes.

In order to determine the fluxes we should involve the full linear Onsager theory. However, the numerical evaluation of the resulting fluxes as part of prognostic models would be prohibitively expensive in computer expenditure now and for some time to come. Therefore, we will make some simplifying assumptions.

11.8.1 Scalar fluxes

Every scalar flux is driven only by its conjugated thermodynamic force, thus ignoring all superpositions. The thermodynamic force for the phase-transitions is the so-called *chemical affinity* $\widetilde{a}_{k1} = \widetilde{\mu}_k - \widetilde{\mu}_1$. To obtain the phase-transition rate, the $\widetilde{a}_{k1}$ must be multiplied by the phenomenological coefficients $l^{(k)}$. The phase-transition fluxes are then given by

$$\begin{aligned} &\overline{I}^0 = 0, \qquad \overline{I}^1 = -\overline{I}^2 - \overline{I}^3, \qquad \overline{I}^2 = -l^{(2)}\widetilde{a}_{21}, \qquad \overline{I}^3 = -l^{(3)}\widetilde{a}_{31} \\ &l^{(2)} \geq 0, \qquad l^{(3)} \geq 0, \qquad \widetilde{a}_{21} = (\widetilde{\mu}_2 - \widetilde{\mu}_1), \qquad \widetilde{a}_{31} = (\widetilde{\mu}_3 - \widetilde{\mu}_1) \end{aligned} \tag{11.83}$$

The chemical affinities $\widetilde{a}_{21}$ and $\widetilde{a}_{31}$ refer to the phase transitions between liquid water and water vapor and between ice and water vapor, respectively. The phenomenological coefficients should be evaluated in terms of average values of the state variables.

11.8.2 Vectorial fluxes

Again we assume that superpositions are excluded. First we rewrite the inequality (11.82). This treatment is motivated by the fact that the potential temperature is nearly conserved in many atmospheric processes so that the turbulent flux $\mathbf{J}_{\mathrm{t}}^{\theta}$ is expected to be driven by the gradient of the potential temperature. We replace $\mathbf{J}_{\mathrm{s,t}}^{h}$ according to (11.69), thus obtaining a different arrangement of fluxes and their thermodynamic forces:

$$-(\overline{\mathbf{J}}_{\mathrm{s}}^{h} + \mathbf{J}_{\mathrm{t}}^{\Pi}) \cdot \nabla \ln \overline{T} - (\overline{\mathbf{J}}^{n} + \mathbf{J}_{\mathrm{t}}^{n}) \cdot (\nabla \widetilde{\mu}_n)_{\overline{T}} - \mathbf{J}_{\mathrm{t}}^{\theta} \cdot \nabla \ln \widetilde{\theta} \geq 0 \tag{11.84}$$

Now the flux $\mathbf{J}_{\mathrm{t}}^{\theta}$ is driven by the gradient of the potential temperature as desired.

In order to guarantee that this inequality is satisfied, each flux must be proportional to the thermodynamic force driving the flux. To make sure that each measure number of the flux depends on each measure number of the thermodynamic forces, we introduce dyadic phenomenological coefficients. The coefficient matrix representing the dyadic coefficient, according to the linear Onsager theory, must be symmetric. The vectorial fluxes are then given by the following set of equations:

$$\begin{aligned} &\text{(a)} \qquad \mathbf{J}_{\mathrm{t}}^{\theta} = -\mathbb{B}^{\theta} \cdot \nabla \ln \widetilde{\theta} = -\overline{\rho}\, \widehat{c}_p \mathbb{K}^{\theta} \cdot \nabla \widetilde{\theta}, \qquad \mathbb{B}^{\theta} = \overline{\rho}\, \widehat{c}_p \widetilde{\theta} \mathbb{K}^{\theta} \\ &\text{(b)} \qquad \overline{\mathbf{J}}_{\mathrm{s}}^{h} + \mathbf{J}_{\mathrm{t}}^{\Pi} = -\mathbb{B}^{T} \cdot \nabla \ln \overline{T} = -\overline{\rho}\, \widehat{c}_p \mathbb{K}^{T} \cdot \nabla \overline{T}, \qquad \mathbb{B}^{T} = -\overline{\rho}\, \widehat{c}_p \overline{T} \mathbb{K}^{T} \\ &\text{(c)} \qquad \overline{\mathbf{J}}^{k} + \mathbf{J}_{\mathrm{t}}^{k} = -\mathbb{B}^{k} \cdot \nabla(\widetilde{\mu}_k - \widetilde{\mu}_0)_{\overline{T}}, \qquad k = 1, 2, 3 \end{aligned} \tag{11.85}$$

In order to evaluate (11.85c) we use the identity

$$\nabla(\widetilde{\mu}_k - \widetilde{\mu}_0)_{\overline{T}} = \frac{\partial}{\partial p}(\widetilde{\mu}_k - \widetilde{\mu}_0)\, \nabla \overline{p} + \frac{\partial}{\partial \widehat{m}^n}(\widetilde{\mu}_k - \widetilde{\mu}_0)\, \nabla \widehat{m}^n \tag{11.86}$$

The various $\mathbb{K}$ appearing in (11.85) are known as the *exchange dyadics*. It is very difficult to evaluate the turbulent flux $\mathbf{J}_{\mathrm{t}}^{\Pi}$ in (11.85b), which is often considered to be negligibly small. Since the mean molecular sensible heat flux is also of a very small magnitude in comparison with $\mathbf{J}_{\mathrm{t}}^{\theta}$, equation (11.85b) is often totally ignored.

It should be clearly understood that the fluxes appearing in (11.85) cannot be considered as fully parameterized since we did not show how the elements of the matrices representing the various $\mathbb{B}$ can be calculated. In fact, the determination of these elements is very difficult and we will have to be satisfied with several approximations.

Finally, we are going to assume that the tensor ellipsoids characterizing the state of atmospheric turbulence are rotational ellipsoids about the z-axis. In this case each symmetric dyadic in the x, y, z system is described by only two measure numbers. In the general case we need nine coefficients. With the assumption of rotational symmetry, the phenomenological equations assume a simplified form. Since we are going to use the Cartesian system, we must also use the corresponding unit vectors $(\mathbf{i}_1, \mathbf{i}_2, \mathbf{i}_3)$.

11.8.3 The scalar phenomenological equations

The scalar phase-transition fluxes given by equation (11.83) are not affected by the above assumption and retain their validity.

11.8.4 The vectorial phenomenological equations

The symmetric coefficient dyadics $\mathbb{B}$ are represented by two measure numbers only. Omitting any superscripts, we may write this dyadic as

$$\mathbb{B} = B_{\mathrm{h}}(\mathbf{i}_1\mathbf{i}_1 + \mathbf{i}_2\mathbf{i}_2) + B_{\mathrm{v}}\mathbf{i}_3\mathbf{i}_3 \tag{11.87}$$

where B_{h} and B_{v} represent the horizontal and vertical measure numbers. With this simplified representation of the exchange dyadic the vectorial fluxes assume the simplified forms

$$\begin{aligned}
&\text{(a)} \qquad \mathbf{J}_{\mathrm{t}}^{\theta} = -\overline{\rho}\,\widehat{c}_p\big[K_{\mathrm{h}}^{\theta}(\mathbf{i}_1\mathbf{i}_1 + \mathbf{i}_2\mathbf{i}_2) + K_{\mathrm{v}}^{\theta}\mathbf{i}_3\mathbf{i}_3\big]\cdot\nabla\widetilde{\theta} \\
&\text{(b)} \quad \overline{\mathbf{J}}_{\mathrm{s}}^{h} + \mathbf{J}_{\mathrm{t}}^{\Pi} = -\overline{\rho}\,\widehat{c}_p\big[K_{\mathrm{h}}^{T}(\mathbf{i}_1\mathbf{i}_1 + \mathbf{i}_2\mathbf{i}_2) + K_{\mathrm{v}}^{T}\mathbf{i}_3\mathbf{i}_3\big]\cdot\nabla\overline{T} \\
&\text{(c)} \quad \overline{\mathbf{J}}^{k} + \mathbf{J}_{\mathrm{t}}^{k} = -\overline{\rho}\big[K_{\mathrm{h}}^{k}(\mathbf{i}_1\mathbf{i}_1 + \mathbf{i}_2\mathbf{i}_2) + K_{\mathrm{v}}^{k}\mathbf{i}_3\mathbf{i}_3\big]\cdot\nabla(\widetilde{\mu}_k - \widetilde{\mu}_0)_{\overline{T}}
\end{aligned} \tag{11.88}$$

The factors K_{h} and K_{v} are known as *exchange coefficients*.

11.8.5 The dyadic fluxes

As has previously been stated, the sum of the fluxes $\overline{\mathbb{J}} + \mathbb{R}$ cannot be dealt with in terms of the entropy production (11.80) since they do not appear in this equation. We will now give a form for these dyadic fluxes that is acceptable for many situations by assuming that the molecular and turbulent fluxes can be stated in analogous forms. For the molecular system we have

$$\mathbb{J} = \mu(\nabla \mathbf{v}_A + \overset{\curvearrowleft}{\mathbf{v}_A \nabla}) - \left(\tfrac{2}{3}\mu - l^{11}\right) \nabla \cdot \mathbf{v}_A \mathbb{E} \tag{11.89}$$

In the following we will assume that the divergence term may be neglected. It seems logical to parameterize the fluxes for the turbulent system by replacing the velocity $\mathbf{v}_A$ by the average $\widehat{\mathbf{v}}_A$. In this case we may first write

$$\nabla \widehat{\mathbf{v}}_A + \overset{\curvearrowleft}{\widehat{\mathbf{v}}_A \nabla} = \left(\nabla_h + \mathbf{i}_3 \frac{\partial}{\partial z}\right)(\widehat{\mathbf{v}}_{A,h} + \mathbf{i}_3 \widehat{w}_A) + (\widehat{\mathbf{v}}_{A,h} + \mathbf{i}_3 \widehat{w}_A)\left(\nabla_h + \mathbf{i}_3 \frac{\partial}{\partial z}\right) \tag{11.90}$$

Assuming additionaly that $\widehat{w}_A = 0$, we find

$$\nabla \widehat{\mathbf{v}}_A + \overset{\curvearrowleft}{\widehat{\mathbf{v}}_A \nabla} = \nabla_h \widehat{\mathbf{v}}_{A,h} + \overset{\curvearrowleft}{\widehat{\mathbf{v}}_{A,h} \nabla_h} + \mathbf{i}_3 \frac{\partial \widehat{\mathbf{v}}_{A,h}}{\partial z} + \frac{\partial \widehat{\mathbf{v}}_{A,h}}{\partial z} \mathbf{i}_3 \tag{11.91}$$

By using only two turbulence coefficients, $K_h^v \geq 0$ and $K_v^v \geq 0$, we find the approximate expression

$$\overline{\mathbb{J}} + \mathbb{R} = \overline{\rho} K_h^v (\nabla_h \widehat{\mathbf{v}}_{A,h} + \overset{\curvearrowleft}{\widehat{\mathbf{v}}_{A,h} \nabla_h}) + \overline{\rho} K_v^v \left(\mathbf{i}_3 \frac{\partial \widehat{\mathbf{v}}_{A,h}}{\partial z} + \frac{\partial \widehat{\mathbf{v}}_{A,h}}{\partial z} \mathbf{i}_3\right) \tag{11.92}$$

It should be clearly understood that the steps leading to (11.91) are not part of a rigorous derivation. The equations obtained in this section will be used in a later chapter when we are dealing with the physics of the atmospheric boundary layer.

11.9 Problems

11.1: Verify the validity of equations (11.7), (11.18), and (11.19).

11.2: Show that the interchange rule (11.25) applies to the microturbulent system.

11.3: Verify the validity of

$$-p \nabla \cdot \mathbf{v}_A = -\overline{p} \nabla \cdot \widehat{\mathbf{v}}_A - \overline{\nabla \cdot (p\mathbf{v}'')} + \overline{\mathbf{v}'' \cdot \nabla p}$$

11.4: Verify the first and the last equation of (11.42).

11.5: Perform the necessary operations to obtain equation (11.45) from (11.35d).

11.6: Show that

$$\sum_{k=0}^{3} \overline{I}^{k} = 0$$

11.7: Show that

$$\frac{\widehat{D}}{Dt}(\overline{\rho}\widehat{k}) = -\nabla \cdot \mathbf{k}_{\mathrm{t}} + \overline{\rho \mathbf{v}'' \cdot \left[\frac{d\mathbf{v}''}{dt} + \nabla\left(\frac{\mathbf{v}''^2}{2}\right)\right]}$$

$$\text{with} \quad \widehat{k} = \widehat{\frac{\mathbf{v}''^2}{2}} \quad \text{and} \quad \mathbf{k}_{\mathrm{t}} = \overline{\rho \mathbf{v}'' \frac{\mathbf{v}''^2}{2}}$$

11.8: Inspection of equation (11.33) seems to indicate that the divergence of the term $\mathbf{J}_{\mathrm{t}}^{e} = \overline{\rho e \mathbf{v}''}$ should appear in (11.48) instead of the turbulent enthalpy flux $\mathbf{J}_{\mathrm{t}}^{h}$. By developing the expression $-\overline{p\,\nabla \cdot \mathbf{v}_{\mathrm{A}}}$ show that the appearence of the turbulent enthalpy flux is correct.

12

An excursion into spectral turbulence theory

The phenomenological theory discussed in the previous chapter did not permit the parameterization of the energy dissipation. In this chapter spectral turbulence theory will be presented to the extent that we appreciate the connections among the turbulent exchange coefficient, the energy dissipation, and the turbulent kinetic energy. In the spectral representation we think of the longer waves as the averaged quantities and the short waves as the turbulent fluctuations. Since the system of atmospheric prediction equations is very complicated we will be compelled to apply some simplifications.

12.1 Fourier representation of the continuity equation and the equation of motion

Before we begin with the actual transformation it may be useful to briefly review some basic concepts. For this reason let us consider the function $a(x)$ which has been defined on the interval L only. In order to represent the function by a Fourier series, we extend it by assuming spatial periodicity. Using Cartesian coordinates we obtain a plot as exemplified in Figure 12.1. The period L is taken to be large enough that averaged quantities within L may vary, i.e. the averaging interval $\Delta x \ll L$.

Certain conditions must be imposed on $a(x)$ in order to make the expansion valid. The function $a(x)$ must be a bounded periodic function that in any one period has at most a finite number of local maxima and minima and a finite number of points of discontinuity. If these conditions are met then the Fourier expansion of $a(x)$ converges to the function $a(x)$ at all points where the function is continuous and to the average of the right-hand and left-hand limits at each point where $a(x)$ is discontinuous. These are the *Dirichlet conditions* which are usually met in

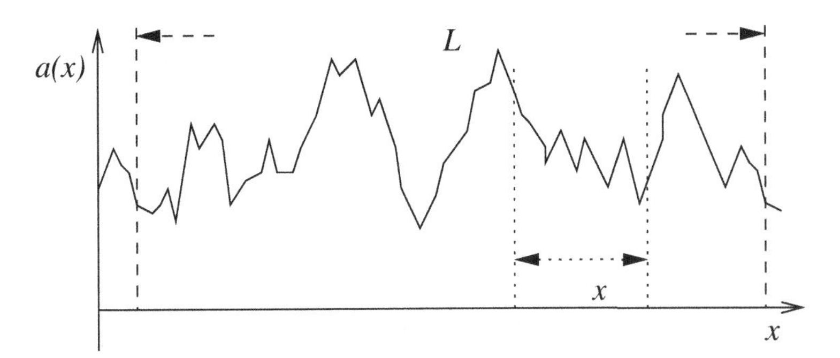

Fig. 12.1 The function $a(x)$ and its periodic extension. L is the period and Δx the averaging interval.

meteorological analysis. The Fourier expansion of $a(x)$ is given by

$$a(x) = \sum_{l=-\infty}^{\infty} A(l) \exp\left(i \frac{2\pi}{L} l x\right), \qquad l = \ldots, -2, -1, 0, 1, 2, \ldots \tag{12.1}$$

or, using the summation convention, by

$$a(x) = A(n) \exp\left(i \frac{2\pi}{L} n x\right), \qquad n = \ldots, -2, -1, 0, 1, 2, \ldots \tag{12.2}$$

In order to obtain the amplitude $A(k)$, we multiply (12.2) by the factor $\exp[-i(2\pi/L)kx]$ and integrate the resulting expression over the expansion interval L. This results in

$$\int_0^L a(x) \exp\left(-i \frac{2\pi}{L} k x\right) dx = A(n) \int_0^L \exp\left(i \frac{2\pi}{L} (n-k) x\right) dx = \delta_n^k L A(n) \tag{12.3}$$

so that the amplitude and its conjugate are given by

$$\begin{aligned} A(k) &= \frac{1}{L} \int_0^L a(x) \exp\left(-i \frac{2\pi}{L} k x\right) dx, \\ \widetilde{A}(k) &= \frac{1}{L} \int_0^L a(x) \exp\left(i \frac{2\pi}{L} k x\right) dx = A(-k) \end{aligned} \tag{12.4}$$

A schematic representation of the first three waves is shown in Figure 12.2, together with the corresponding amplitudes. It will be recognized that a large value of l refers to short waves whereas a small value signifies longer waves.

We will now obtain a useful mathematical expression that will be helpful in our work. We first multiply the functions $a(x)$ and $b(x)$ and their expansions, yielding

$$\begin{aligned} a(x)b(x) &= A(n) \exp\left(i \frac{2\pi}{L} n x\right) B(m) \exp\left(i \frac{2\pi}{L} m x\right) \\ &= A(n)B(m) \exp\left(i \frac{2\pi}{L} (n+m) x\right) \end{aligned} \tag{12.5a}$$

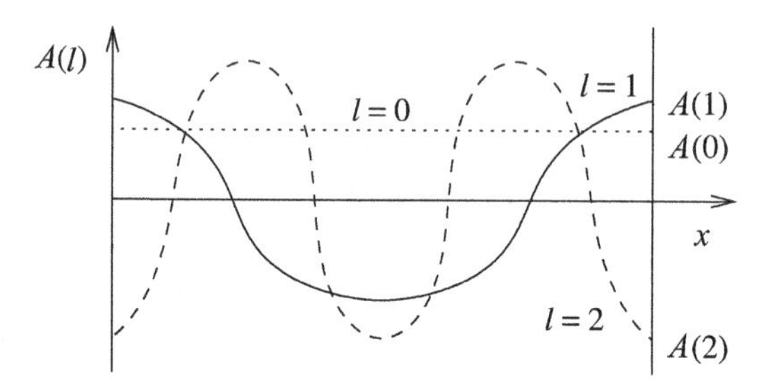

Fig. 12.2 A schematic representation of the first three waves in the interval L and the corresponding amplitudes.

On integrating this expression over the expansion interval we immediately find

$$\int_0^L a(x)b(x)\,dx = A(n)B(m)\int_0^L \exp\left(i\frac{2\pi}{L}(n+m)x\right)dx = A(n)B(m)\delta^0_{n+m}L \tag{12.5b}$$

From the Kronecker symbol it follows that $m = -n$, so that

$$A(n)B(-n) = \frac{1}{L}\int_0^L a(x)b(x)\,dx \tag{12.5c}$$

For the special case $A = B$ we obtain

$$A(n)A(-n) = \frac{1}{L}\int_0^L a^2(x)\,dx \tag{12.6}$$

which is known as *Parseval's identity for Fourier series*; summation over n is implied. Obviously this expression is real. For example, let the function $a(x)$ represent the velocity component $u(x)$. In this case we obtain

$$\frac{1}{L}\int_0^L \frac{u^2(x)}{2}\,dx = \frac{\overline{u^2}}{2} = \frac{U(n)U(-n)}{2} = E_u \tag{12.7}$$

representing the average value of the kinetic energy per unit mass.

We will now generalize the previous treatment to three dimensions and formally admit the time dependency of the function a. Now the interval of expansion changes to the volume of expansion as shown in Figure 12.3. In analogy to the one-dimensional case, periodicity is now required for the volume.

The formal expansion is then given by

$$a(x^1, x^2, x^3, t) = \sum_{l_1=-\infty}^{\infty}\sum_{l_2=-\infty}^{\infty}\sum_{l_3=-\infty}^{\infty} A(l_1, l_2, l_3, t)\exp\left[i2\pi\left(\frac{l_1}{L_1}x^1 + \frac{l_2}{L_2}x^2 + \frac{l_3}{L_3}x^3\right)\right] \tag{12.8}$$

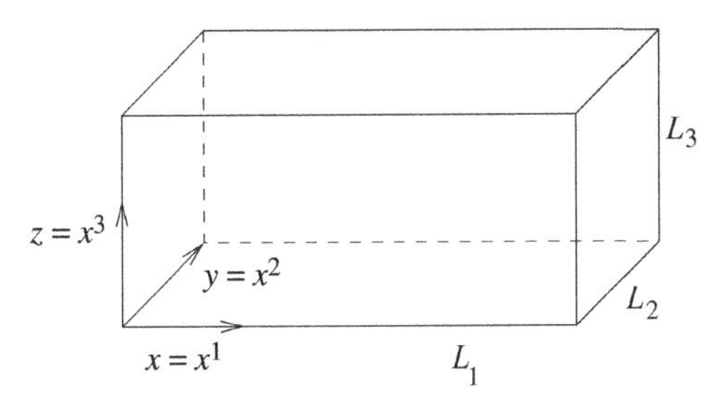

Fig. 12.3 The expansion volume $V = L_1 L_2 L_3$ for the function $a(x^1, x^2, x^3, t)$.

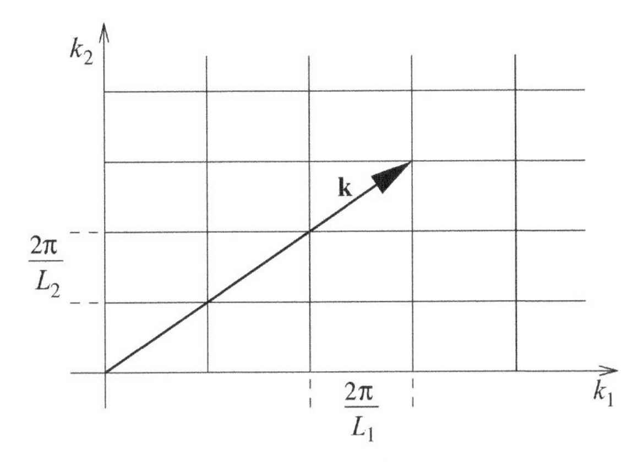

Fig. 12.4 A schematic representation of the wavenumber vector in the plane.

We now introduce the abbreviations

$$k_i = \frac{2\pi l_i}{L_i}, \qquad \mathbf{r} = \mathbf{i}_1 x^1 + \mathbf{i}_2 x^2 + \mathbf{i}_3 x^3, \qquad \mathbf{k} = \mathbf{i}^1 k_1 + \mathbf{i}^2 k_2 + \mathbf{i}^3 k_3 \tag{12.9a}$$

where the third equation represents the wavenumber vector. The exponent in (12.8) can then be written as the scalar product:

$$2\pi\left(\frac{l_1}{L_1}x^1 + \frac{l_2}{L_2}x^2 + \frac{l_3}{L_3}x^3\right) = k_1 x^1 + k_2 x^2 + k_3 x^3 = \mathbf{k}\cdot\mathbf{r} \tag{12.9b}$$

To simplify the notation we formally introduce the summation *wavenumber vector* **k**

$$\sum_{\mathbf{k}} = \sum_{l_1=-\infty}^{\infty} \sum_{l_2=-\infty}^{\infty} \sum_{l_3=-\infty}^{\infty} \tag{12.9c}$$

and obtain the representation

$$a(\mathbf{r}, t) = \sum_{\mathbf{k}} A(\mathbf{k}, t) \exp(i\mathbf{k}\cdot\mathbf{r}) = A(\mathbf{n}, t) \exp(i\mathbf{n}\cdot\mathbf{r}) \tag{12.9d}$$

Figure 12.4 shows a schematic representation of the wavenumber vector in the plane.

As stated at the beginning of this chapter, we need to simplify the physical–mathematical system in order to make the analysis more tractable. First of all we postulate a dry constant-density atmosphere, that is $\alpha = 1/\rho =$ constant, resulting in a simplified continuity equation. Expanding the velocity vector according to (12.9d) we may write

$$\mathbf{v}(\mathbf{r}, t) = \mathbf{V}(\mathbf{n}, t) \exp(i\mathbf{n} \cdot \mathbf{r}) \tag{12.10}$$

so that the continuity equation can be expressed as

$$\nabla \cdot \mathbf{v} = \mathbf{V}(\mathbf{n}, t) \cdot \nabla \exp(i\mathbf{n} \cdot \mathbf{r}) = 0 \tag{12.11}$$

The gradient operator acting on the exponential part with wavenumber vector $\mathbf{k}$ results in

$$\nabla \exp(i\mathbf{k} \cdot \mathbf{r}) = \exp(i\mathbf{k} \cdot \mathbf{r})\, \nabla(i\mathbf{k} \cdot \mathbf{r}) = i \exp(i\mathbf{k} \cdot \mathbf{r})\, \nabla\mathbf{r} \cdot \mathbf{k} = i\mathbf{k} \exp(i\mathbf{k} \cdot \mathbf{r}) \tag{12.12}$$

It is now convenient to write the Laplacian of the exponential part as

$$\nabla^2 \exp(i\mathbf{k} \cdot \mathbf{r}) = \nabla \cdot \nabla \exp(i\mathbf{k} \cdot \mathbf{r}) = -\mathbf{k}^2 \exp(i\mathbf{k} \cdot \mathbf{r}) \tag{12.13}$$

since this expression will be needed soon. Using (12.12), the continuity equation assumes the form

$$\mathbf{V}(\mathbf{n}, t) \cdot \mathbf{n} i \exp(i\mathbf{n} \cdot \mathbf{r}) = 0 \tag{12.14}$$

Since this expression must hold for every wavenumber vector $\mathbf{k}$, the expression

$$\mathbf{V}(\mathbf{k}, t) \cdot \mathbf{k} = 0, \qquad \mathbf{k} \perp \mathbf{V}(\mathbf{k}, t) \tag{12.15}$$

is also true, thus proving that the velocity amplitude $\mathbf{V}(\mathbf{k}, t)$ is perpendicular to the wavenumber vector $\mathbf{k}$.

Our next task is to write the equation of motion in the spectral form. By ignoring the Coriolis force the equation of motion may be written as

$$\frac{\partial \mathbf{v}}{\partial t} + \nabla \cdot (\mathbf{v}\mathbf{v}) = -\nabla \Pi + \frac{1}{\rho} \nabla \cdot \mathbb{J} \tag{12.16}$$

Here the abbreviation $\Pi = \alpha p + \phi$ has been used. Previously the same symbol represented the Exner function, but confusion is unlikely to arise. Since the divergence of the velocity is zero the stress tensor $\mathbb{J}$ reduces to the simplified form

$$\mathbb{J} = \mu(\nabla\mathbf{v} + \overset{\curvearrowleft}{\mathbf{v}\nabla}) \Longrightarrow \frac{1}{\rho} \nabla \cdot \mathbb{J} = \nu \nabla^2 \mathbf{v}, \qquad \nu = \frac{\mu}{\rho} \tag{12.17}$$

On substituting (12.10) into the equation of motion and the analogous expression for Π, we obtain

$$\frac{\partial \mathbf{V}(\mathbf{n},t)}{\partial t}\exp(i\mathbf{n}\cdot\mathbf{r}) + \nabla\cdot\left[\mathbf{V}(\mathbf{m},t)\mathbf{V}(\mathbf{q},t)\exp(i(\mathbf{m}+\mathbf{q})\cdot\mathbf{r})\right]$$
$$= -\nabla\Pi(\mathbf{n},t)\exp(i\mathbf{n}\cdot\mathbf{r}) + \nu\,\nabla^2\left[\mathbf{V}(\mathbf{n},t)\exp(i\mathbf{n}\cdot\mathbf{r})\right] \tag{12.18a}$$

By carrying out the differential operations involving the gradient operator and the Laplacian, using (12.12) and (12.13), we find without difficulty the expression

$$\frac{\partial \mathbf{V}(\mathbf{n},t)}{\partial t}\exp(i\mathbf{n}\cdot\mathbf{r}) + i(\mathbf{m}+\mathbf{q})\cdot\mathbf{V}(\mathbf{m},t)\mathbf{V}(\mathbf{q},t)\exp[i(\mathbf{m}+\mathbf{q})\cdot r]$$
$$= -i\Pi(\mathbf{n},t)\mathbf{n}\exp(i\mathbf{n}\cdot\mathbf{r}) - \nu\mathbf{n}^2\mathbf{V}(\mathbf{n},t)\exp(i\mathbf{n}\cdot\mathbf{r}) \tag{12.18b}$$

where $\mathbf{m}$, $\mathbf{n}$, and $\mathbf{q}$ are summation vectors.

The differential equation (12.18b) must be valid for the entire range of $\mathbf{n}$ including the term $\mathbf{k}$. From the double sum over $\mathbf{m}$ and $\mathbf{q}$ we factor out the terms with

$$\mathbf{m}+\mathbf{q}=\mathbf{k} \quad \text{or} \quad \mathbf{q}=\mathbf{k}-\mathbf{m} \tag{12.19}$$

The double sum over $\mathbf{m}$ and $\mathbf{q}$ now becomes a simple sum over $\mathbf{m}$ for each $\mathbf{k}$. The result is given by

$$\frac{\partial \mathbf{V}(\mathbf{k},t)}{\partial t} + i\mathbf{k}\cdot\mathbf{V}(\mathbf{m},t)\mathbf{V}(\mathbf{k}-\mathbf{m},t) = -i\Pi(\mathbf{k},t)\mathbf{k} - \nu\mathbf{k}^2\mathbf{V}(\mathbf{k},t) \tag{12.20}$$

By slightly rewriting this formula we obtain

$$\left(\frac{\partial}{\partial t} + \nu\mathbf{k}^2\right)\mathbf{V}(\mathbf{k},t) = -i\Pi(\mathbf{k},t)\mathbf{k} - i\mathbf{k}\cdot\mathbf{V}(\mathbf{m},t)\mathbf{V}(\mathbf{k}-\mathbf{m},t) \tag{12.21}$$

which is the more common spectral form of the equation of motion for the complex amplitude vector $\mathbf{V}(\mathbf{k},t)$. Summation over all $\mathbf{m}$ is implied. The nonlinear advection term in the original equation of motion is the reason for the appearance of the product of the two amplitude vectors.

12.2 The budget equation for the amplitude of the kinetic energy

First of all we generalize the expression (12.7) which is the u-component of the kinetic energy. The average value of the total kinetic energy is then given by

$$\frac{1}{V}\int_0^V \frac{\mathbf{v}^2}{2}\,dV' = \frac{\overline{\mathbf{v}^2}}{2} = \frac{\mathbf{V}(\mathbf{n},t)\cdot\mathbf{V}(-\mathbf{n},t)}{2} = E(t) \tag{12.22}$$

where the summation over all amplitude vectors must be carried out. For each individual component $\mathbf{k}$ we may write

$$E(\mathbf{k},t) = \frac{\mathbf{V}(\mathbf{k},t)\cdot\mathbf{V}(-\mathbf{k},t)}{2} \tag{12.23}$$

As before, the conjugated velocity amplitude vector is obtained by formally replacing $\mathbf{k}$ by $-\mathbf{k}$, that is $\widetilde{\mathbf{V}}(\mathbf{k},t) = \mathbf{V}(-\mathbf{k},t)$. Summing over all $\mathbf{k}$ gives the total kinetic energy

$$E(t) = \sum_{\mathbf{k}=-\infty}^{\infty} E(\mathbf{k},t) \tag{12.24}$$

thus repeating (12.22). In order to find a prognostic equation for $E(\mathbf{k},t)$, we first replace the wavenumber vector $\mathbf{k}$ in (12.21) by $-\mathbf{k}$ to give

$$\left(\frac{\partial}{\partial t} + \nu\mathbf{k}^2\right)\mathbf{V}(-\mathbf{k},t) = -i\Pi(-\mathbf{k},t)(-\mathbf{k}) + i\mathbf{k}\cdot\mathbf{V}(\mathbf{m},t)\mathbf{V}(-\mathbf{k}-\mathbf{m},t) \tag{12.25}$$

Next we carry out two scalar multiplications by multiplying (12.21) by $\mathbf{V}(-\mathbf{k},t)$ and (12.25) by $\mathbf{V}(\mathbf{k},t)$. By observing the orthogonality relation stated in (12.15), the terms multiplying the function Π disappear. The result is shown in

$$\begin{aligned}
&\mathbf{V}(-\mathbf{k},t)\cdot\frac{\partial\mathbf{V}(\mathbf{k},t)}{\partial t} + \nu\mathbf{k}^2\mathbf{V}(-\mathbf{k},t)\cdot\mathbf{V}(\mathbf{k},t)\\
&\quad = -i\mathbf{k}\cdot\mathbf{V}(\mathbf{m},t)\mathbf{V}(\mathbf{k}-\mathbf{m},t)\cdot\mathbf{V}(-\mathbf{k},t)\\
&\mathbf{V}(\mathbf{k},t)\cdot\frac{\partial\mathbf{V}(-\mathbf{k},t)}{\partial t} + \nu\mathbf{k}^2\mathbf{V}(\mathbf{k},t)\cdot\mathbf{V}(-\mathbf{k},t)\\
&\quad = i\mathbf{k}\cdot\mathbf{V}(\mathbf{m},t)\mathbf{V}(-\mathbf{k}-\mathbf{m},t)\cdot\mathbf{V}(\mathbf{k},t)
\end{aligned} \tag{12.26}$$

Adding these two equations and using the definition (12.23) gives the budget equation for the amplitude of the kinetic energy per unit mass

$$\begin{aligned}
\frac{\partial E(\mathbf{k},t)}{\partial t} + 2\nu\mathbf{k}^2 E(\mathbf{k},t) &= -\frac{i}{2}\mathbf{k}\cdot\mathbf{V}(\mathbf{m},t)\mathbf{V}(\mathbf{k}-\mathbf{m},t)\cdot\mathbf{V}(-\mathbf{k},t)\\
&\quad + \frac{i}{2}\mathbf{k}\cdot\mathbf{V}(\mathbf{n},t)\mathbf{V}(-\mathbf{k}-\mathbf{n},t)\cdot\mathbf{V}(\mathbf{k},t)
\end{aligned} \tag{12.27a}$$

Summation over $\mathbf{m}$ and $\mathbf{n}$ is implied. From (12.24) it follows that $E(t)$ is a real number. At first glance the right-hand side, however, appears to be a complex quantity, but all imaginary terms will cancel out in the summation over $\mathbf{m}$ and $\mathbf{n}$ from minus to plus infinity. In order to get a more concise form of (12.27a), we substitute into the first term on the right-hand side $\mathbf{k}' = \mathbf{k} - \mathbf{m}$ and in the last term we use $-\mathbf{k}' = -\mathbf{k} - \mathbf{n}$. Thus, we obtain

$$\begin{aligned}
\frac{\partial E(\mathbf{k},t)}{\partial t} + 2\nu\mathbf{k}^2 E(\mathbf{k},t) &= -\sum_{\mathbf{k}'}\frac{i}{2}\mathbf{k}\cdot\mathbf{V}(\mathbf{k}-\mathbf{k}',t)\mathbf{V}(\mathbf{k}',t)\cdot\mathbf{V}(-\mathbf{k},t)\\
&\quad + \sum_{\mathbf{k}'}\frac{i}{2}\mathbf{k}\cdot\mathbf{V}(\mathbf{k}'-\mathbf{k},t)\mathbf{V}(-\mathbf{k}',t)\cdot\mathbf{V}(\mathbf{k},t)
\end{aligned} \tag{12.27b}$$

where now, in contrast to (12.27a), the summation over $\mathbf{k}'$ has been written down explicitly. Introducing the abbreviation

$$\sum_{\mathbf{k}'} W(\mathbf{k},\mathbf{k}',t) = -\sum_{\mathbf{k}'} \frac{i}{2}\mathbf{k}\cdot\mathbf{V}(\mathbf{k}-\mathbf{k}',t)\mathbf{V}(\mathbf{k}',t)\cdot\mathbf{V}(-\mathbf{k},t) + \sum_{\mathbf{k}'} \frac{i}{2}\mathbf{k}\cdot\mathbf{V}(\mathbf{k}'-\mathbf{k},t)\mathbf{V}(-\mathbf{k}',t)\cdot\mathbf{V}(\mathbf{k},t) \tag{12.28}$$

leads to the final form of the budget equation for the spectral kinetic energy:

$$\boxed{\frac{\partial E(\mathbf{k},t)}{\partial t} + 2\nu\mathbf{k}^2 E(\mathbf{k},t) = \sum_{\mathbf{k}'} W(\mathbf{k},\mathbf{k}',t)} \tag{12.29}$$

Often the expression $W(\mathbf{k},\mathbf{k}',t)$ is called the *energy-transfer function*. This function is antisymmetric, that is $W(\mathbf{k},\mathbf{k}',t) = -W(\mathbf{k}',\mathbf{k},t)$, as follows very easily from inspection of (12.28). The interpretation of (12.29) is not difficult. The first term represents the tendency of E to either increase or descrease in time. The second term on the left-hand side describes the energy dissipation giving that part of the kinetic energy which is transformed into internal energy. It should be recalled that the dissipation of short waves greatly exceeds the dissipation of long waves. The term on the right-hand side of (12.29) represents the exchange of energy of cell $\mathbf{k}$ with neighboring and also with more distant cells. It will be recognized that the antisymmetric property of the energy-transfer function is closely related to the principle of conservation of energy. Whatever cell $\mathbf{k}$ gains from cell $\mathbf{k}'$ is a loss for cell $\mathbf{k}'$ in favor of cell $\mathbf{k}$. Moreover, the right-hand side of (12.29) reflects the nonlinearity of the equation of motion.

12.3 Isotropic conditions, the transition to the continuous wavenumber space

The solution of (12.29) is very complicated because of the dependency of the transfer function on the entire velocity spectrum $\mathbf{V}(\mathbf{k},t)$. Therefore, it seems necessary to introduce simplifying assumptions. A great deal of simplification is obtained by introducing the so-called *isotropic condition* so that the directional dependencies of $\mathbf{V}$ and E on the wavenumber vector $\mathbf{k}$ are suppressed, that is

$$\mathbf{V}(\mathbf{k},t) \longrightarrow \mathbf{V}(k,t), \qquad E(\mathbf{k},t) \longrightarrow E(k,t), \qquad |\mathbf{k}| = k \tag{12.30}$$

The velocity vector and the kinetic energy now depend only on the scalar wavenumber k and on the time t. In order to simplify the notation the time dependencies of all variables will henceforth be omitted, e.g. $E(k,t)$ will simply be written as $E(k)$.

Certainly, the assumption (12.30) results in the desired analytic simplification, but much of the physical significance is lost since turbulent processes generally depend on direction. For the isotropic case the spectral energy-budget equation assumes the form

$$\frac{\partial E(k)}{\partial t} + 2\nu k^2 E(k) = \sum_{k'} W(k, k') \tag{12.31}$$

We think of the discrete amplitudes $E(k)$ of the energy spectrum as representing a certain volume element in k-space. If we divide $E(k)$ by a sufficiently small volume element ΔV_k in this space, we obtain a continuous function $\rho(k)$ that is known as the *spectral energy density*,

$$\rho(k) = \frac{E(k)}{\Delta V_k} \tag{12.32a}$$

The energy contained within a thin spherical shell of radius k and thickness Δk is then given by

$$\rho(k)\,\Delta V_k = 4\pi k^2 \rho(k)\,\Delta k = E(k) = \epsilon(k)\,\Delta k \tag{12.32b}$$

with $\Delta V_k = 4\pi k^2\,\Delta k$. Here we have also introduced the energy $\epsilon(k)$ per unit wavenumber, given by $\epsilon(k) = 4\pi k^2 \rho(k)$. On dividing (12.31) by ΔV_k we obtain the *budget equation for the energy density:*

$$\boxed{\frac{\partial \rho(k)}{\partial t} + 2\nu k^2 \rho(k) = \frac{1}{\Delta V_k} \sum_{k'} W(k, k')} \tag{12.33}$$

Multiplication of (12.33) by $4\pi k^2$ results in a budget equation for the energy per unit wavenumber:

$$\frac{\partial \epsilon(k)}{\partial t} + 2\nu k^2 \epsilon(k) = \frac{4\pi k^2}{\Delta V_k} \sum_{k'} W(k, k') \tag{12.34}$$

In order to treat the interaction of wavenumber k with all other wavenumbers k' we replace the summation over k' by an integral over the volume $dV_k' = 4\pi (k')^2\,dk'$. By introducing the function

$$T(k, k') = \frac{4\pi k^2}{\Delta V_k} W(k, k') \frac{4\pi (k')^2}{\Delta V_k'} \tag{12.35}$$

into (12.34), we finally obtain the compact form of the budget equation for the spectral energy per unit wavenumber:

$$\boxed{\frac{\partial \epsilon(k)}{\partial t} + 2\nu k^2 \epsilon(k) = \int_0^\infty T(k, k')\,dk'} \tag{12.36}$$

It will be observed that $T(k, k')$ is also antisymmetric since $W(k, k')$ has this property. In this prognostic equation the function $T(k, k')$ describes the exchange of energy in k-space between shell k and all other shells since the integration ranges from $k' = 0$ to $k' = \infty$.

Our goal is to get some information about the properties of $T(k, k')$. This information could be obtained by solving the spectral equation of motion (12.25) for each amplitude $\mathbf{V}(k)$ to compute $T(k, k')$. In order to avoid such complex numerical calculations, various closure hypotheses have been introduced in order to obtain $T(k, k')$ alone from knowledge of the energy spectrum $\epsilon(k)$. For a detailed discussion see, for example, Hinze (1959) and Rotta (1972), where an extensive literature on isotropic turbulence can be found. In the following brief description we shall only introduce the approach of Heisenberg, who considers the case of stationary or at least quasi-stationary turbulence. He assumed that the spectrum $\epsilon(k)$ can be divided into two regions that are separated by the wavenumber $k = k^*$.

In the region of small wavenumbers (large wavelengths, region I) where $k < k^*$, turbulence is strongly influenced by external parameters. Examples would be the geometric characteristics of the flow domain and the type of turbulence generation. For this region it is not possible to draw general conclusions about the energy spectrum and the direction of the energy transport.

In the region of large wavenumbers (small wavelengths, region II) where $k > k^*$, external influences are not important or do not exist. In this wavenumber range the spectrum should be characterized by universal laws. In this *universality region*, the energy transfer is always directed from smaller to larger wavenumbers. This means that an *energy cascade* is taking place, whereby turbulent energy is directed from smaller to larger wavenumbers, and from these to still larger wavenumbers, until the energy is finally dissipated at the largest wavenumbers.

Heisenberg postulated that, in this energy transfer, the action of large-wavenumber eddies upon small-wavenumber eddies is much like the appearance of an additional viscosity. For this turbulence viscosity within the fluid, in analogy to the frictional term $2\nu k^2 \rho(k)$, he assumed the relationship

$$T(k, k') = -2\kappa_{\rm H} k^2 \epsilon(k) g(k', \epsilon(k')) \tag{12.37}$$

The function $g(k', \epsilon(k'))$ is thought to be a universal function. The constant of proportionality $\kappa_{\rm H}$ is known as *Heisenberg's constant*, which is a pure number, $\kappa_{\rm H} = 0.5 \pm 0.03$.

The form of the universal function g follows from a dimensional analysis that is carried out next. In the mks system the units of energy density per unit mass and unit wavenumber are $[\epsilon(k)] = \mathrm{m}^3\,\mathrm{s}^{-2}$, the units of the time derivative are therefore $\left[\partial\epsilon(k)/\partial t\right] = \mathrm{m}^3\,\mathrm{s}^{-3}$. From (12.36) we find the units of $T(k, k')$ as $\left[T(k, k')\right] =$

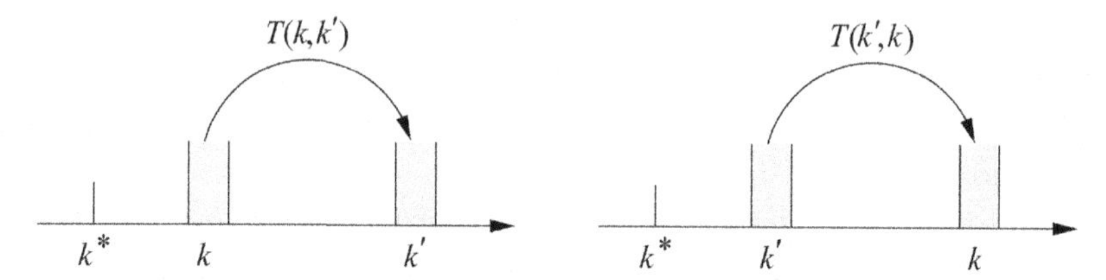

Fig. 12.5 A schematic view of energy transfer in region II.

$m^4\,s^{-3}$ so that the units of g follow immediately from (12.37), $\left[g(k', \epsilon(k'))\right] = m^3\,s^{-1}$. Since g depends only on k' and $\epsilon(k')$, we may write

$$g(k', \epsilon(k')) = (k')^{\alpha} \epsilon(k')^{\beta} \tag{12.38a}$$

where the exponents α and β must be determined. By comparison with the dimension of g we find $\left[(k')^{\alpha}\epsilon(k')^{\beta}\right] = m^3\,s^{-1}$. This leads to the values $\alpha = -\frac{3}{2}$ and $\beta = \frac{1}{2}$. Utilizing these values, for g we finally get the form

$$g(k') = (k')^{-3/2}\epsilon(k')^{1/2} = \sqrt{\frac{\epsilon(k')}{(k')^3}} \tag{12.38b}$$

Therefore, according to (12.37), the transfer function $T(k, k')$ is given by

$$T(k, k') = -2\kappa_{\mathrm{H}} k^2 \epsilon(k) \sqrt{\frac{\epsilon(k')}{(k')^3}} \tag{12.39}$$

We will now briefly discuss the transfer function $T(k, k')$ by considering the energy cascade shown schematically in Figure 12.5. The outflow from cell k to cell $k' > k$ is directly given by (12.39). Using the fact that the transfer function is antisymmetric, i.e. $T(k, k') = -T(k', k)$, the flow from cell k' to cell $k > k'$ can be written down directly. This results in

$$T(k, k') = \begin{cases} -2\kappa_{\mathrm{H}} k^2 \epsilon(k) \sqrt{\dfrac{\epsilon(k')}{(k')^3}} & k' > k: \quad \text{outflow} \\ 2\kappa_{\mathrm{H}} (k')^2 \epsilon(k') \sqrt{\dfrac{\epsilon(k)}{(k)^3}} & k' < k: \quad \text{inflow} \end{cases} \tag{12.40}$$

12.4 The Heisenberg spectrum

The task ahead is to calculate the energy spectrum for the region $k > k^*$ by assuming stationary conditions. For such conditions it will be necessary to compensate for the steady loss of turbulence energy by dissipation by gaining an equal amount of energy in order to maintain stationarity. Therefore, it will be assumed that production of energy $\sigma(k)$ takes place in the region $k < k^*$. Eventually, the energy produced in the production region is transported to the short waves where it will be dissipated. Instead of equation (12.36) we now write the energy budget as

$$\left(\frac{\partial}{\partial t} + 2\nu k^2\right)\epsilon(k) = \int_0^\infty T(k, k')\, dk' + \sigma(k) \tag{12.41}$$

with $\sigma(k) = 0$ for $k > k^*$. This is the spectral energy-balance equation with production, keeping in mind that no production is allowed to take place in the universality range. In order to obtain the budget equation for an entire spectral interval, we integrate (12.41) over a wavenumber range to the wavenumber $k > k^*$ as indicated by

$$\frac{\partial}{\partial t}\int_0^k \epsilon(k')\, dk' + 2\nu \int_0^k (k')^2\epsilon(k')\, dk' = \int_0^k dk' \int_0^\infty T(k', k'')\, dk'' + \int_0^k \sigma(k')\, dk' \tag{12.42}$$

For stationary conditions the production and dissipation of energy must be balanced:

$$\begin{aligned} &\int_0^k \sigma(k')\, dk' = 2\nu \int_0^\infty (k')^2\epsilon(k')\, dk' = \overline{\epsilon}_{\rm M} \\ \text{with}\quad &\int_0^k \sigma(k')\, dk' = \int_0^\infty \sigma(k')\, dk' \end{aligned} \tag{12.43}$$

since $\sigma(k') = 0$ for $k' > k^*$. Here $\overline{\epsilon}_{\rm M}$ is the dissipation of energy per unit mass. Setting the tendency term in (12.42) equal to zero and using (12.43), we may write

$$\begin{aligned} 2\nu \int_0^k (k')^2\epsilon(k')\, dk' &= \int_0^k dk' \int_0^\infty T(k', k'')\, dk'' + \overline{\epsilon}_{\rm M} \\ &= \int_0^k dk' \int_k^\infty T(k', k'')\, dk'' + \overline{\epsilon}_{\rm M} \\ \text{since}\quad T(k', k'') = -T(k'', k') &\Longrightarrow \int_0^k dk' \int_0^k T(k', k'')\, dk'' = 0 \end{aligned} \tag{12.44}$$

On substituting the transfer function $T(k, k')$ according to (12.39) into this equation, the spectral energy equation for stationary conditions can be written as

$$\left(\nu + \kappa_{\rm H} \int_k^\infty \sqrt{\frac{\epsilon(k'')}{(k'')^3}}\, dk''\right) 2 \int_0^k (k')^2\epsilon(k')\, dk' = \overline{\epsilon}_{\rm M} \tag{12.45a}$$

The integral expression within the large parentheses,

$$K = \kappa_{\rm H} \int_k^\infty \sqrt{\frac{\epsilon(k'')}{(k'')^3}}\, dk'' \tag{12.45b}$$

is known as the *Heisenberg exchange coefficient* K, which depends on ϵ and on the wavenumber k. We observe that, for large k (small wavelengths), the exchange coefficient is small, tending to zero as k tends to infinity. In this case the entire energy loss is then due to molecular dissipation. If $\epsilon(k'')$ is zero, then the exchange coefficient is zero as well. In order to give a quantitative treatment of the exchange coefficient, we must have information on ϵ. Later we will try to express $\epsilon(k)$ as a function of K.

We are now going to calculate the energy spectrum from equation (12.45). For this we define the integral multiplying the bracket in (12.45) by

$$y(k) = 2\int_0^k (k')^2\epsilon(k')\,dk' \tag{12.46a}$$

so that (12.45) can be written as

$$\nu + \kappa_{\mathrm{H}}\int_k^\infty \sqrt{\frac{y'(k')}{2(k')^5}}\,dk' = \frac{\overline{\epsilon}_{\mathrm{M}}}{y(k)} \tag{12.46b}$$

Here $y' = dy/dk = 2k^2\epsilon(k)$, which follows from the differentiation of (12.46a) with respect to the variable upper integration limit k. Differentiating (12.46b) with respect to k and squaring the result gives

$$\frac{\kappa_{\mathrm{H}}^2}{2k^5} = \frac{\overline{\epsilon}_{\mathrm{M}}^2}{y^4(k)}y' \tag{12.46c}$$

This expression is integrated by separating the variables, yielding

$$\frac{\kappa_{\mathrm{H}}^2}{8k^4} + C = \frac{\overline{\epsilon}_{\mathrm{M}}^2}{3y^3(k)} \tag{12.46d}$$

The constant of integration C is found from the definition of $\overline{\epsilon}_{\mathrm{M}}$ given by (12.43):

$$\overline{\epsilon}_{\mathrm{M}} = 2\nu\int_0^\infty (k')^2\epsilon(k')\,dk' = \nu\lim_{k\to\infty} y(k) \implies y(\infty) = \frac{\overline{\epsilon}_{\mathrm{M}}}{\nu} \tag{12.46e}$$

where (12.46a) has been used. Therefore, for $k \to \infty$ we obtain from (12.46d) $C = \overline{\epsilon}_{\mathrm{M}}/[3y(\infty)] = \nu^3/(3\overline{\epsilon}_{\mathrm{M}})$. Substituting this expression into (12.46d) and solving for y gives

$$y(k) = \frac{\overline{\epsilon}_{\mathrm{M}}^{2/3}}{3^{1/3}}\left(\frac{\kappa_{\mathrm{H}}^2}{8k^4} + \frac{\nu^3}{3\overline{\epsilon}_{\mathrm{M}}}\right)^{-1/3} \tag{12.46f}$$

Differentiating this equation with respect to k with $dy/dk = y' = 2k^2\epsilon(k)$ and then solving for $\epsilon(k)$ gives

$$\epsilon(k) = \frac{\overline{\epsilon}_{\mathrm{M}}^{2/3}}{3^{1/3}}\left(\frac{2}{3}\right)\left(\frac{\kappa_{\mathrm{H}}^2}{8k^4} + \frac{\nu^3}{3\overline{\epsilon}_{\mathrm{M}}}\right)^{-4/3}\frac{\kappa_{\mathrm{H}}^2}{8k^7}, \qquad k > k' \tag{12.46g}$$

This can be rewritten in the form

$$\boxed{\epsilon(k) = \left(\frac{8}{9\kappa_{\mathrm{H}}}\right)^{2/3} \overline{\epsilon}_{\mathrm{M}}^{2/3} k^{-5/3} \left(1 + \frac{8\nu^3 k^4}{3\overline{\epsilon}_{\mathrm{M}}\kappa_{\mathrm{H}}^2}\right)^{-4/3}} \tag{12.47}$$

First of all we observe from (12.47) that the spectrum may be separated into two wavenumber regions by the wavenumber k_{c}, given by

$$k_{\mathrm{c}} = \left(\frac{\overline{\epsilon}_{\mathrm{M}}}{\nu^3}\right)^{1/4} \tag{12.48}$$

since for $k > k_{\mathrm{c}}$ the second expression within the second set of parentheses exceeds the number 1 by at least one order of magnitude. Ignoring the number 1 within the parantheses, we obtain the section of the spectrum to the far right. This wing is known as the *dissipation range*, where the energy spectrum decays with the minus seventh power of k:

$$\boxed{k > k_{\mathrm{c}}: \qquad \epsilon(k) = \left(\frac{\kappa_{\mathrm{H}}\overline{\epsilon}_{\mathrm{M}}}{2\nu^2}\right)^2 k^{-7} \Longrightarrow \epsilon(k) \sim k^{-7}} \tag{12.49}$$

Now we consider the region defined by $k^* < k \ll k_{\mathrm{c}}$. By omitting now from (12.47) the second expression within the parentheses we obtain

$$\boxed{k^* < k \ll k_{\mathrm{c}}: \qquad \epsilon(k) = \left(\frac{8}{9\kappa_{\mathrm{H}}^2}\right)^{2/3} \overline{\epsilon}_{\mathrm{M}}^{2/3} k^{-5/3} \Longrightarrow \epsilon(k) \sim k^{-5/3}} \tag{12.50}$$

Hence, it is seen that, in the long-wave spectral range, $\epsilon(k)$ is proportional to $k^{-5/3}$. This phenomenon is called the $k^{-5/3}$ *law*. The spectral range for which the $k^{-5/3}$ law is valid is known as the *transfer region* or the *inertial subrange* since in this region the viscosity may be neglected. The physical picture is that production and dissipation occur in different spectral regions so that the eddies in the inertial subrange receive their energy initially from the larger eddies and transfer their energy to the smaller eddies. The transfer takes place at such a rate that production is exactly balanced by dissipation. Observations show that, in the production range ($k < k^*$), the left wing of the energy spectrum falls off to the fourth power. For the whole spectrum the distribution of $\epsilon(k)$ is schematically shown in Figure 12.6.

Often equation (12.47) is written in a somewhat different way by introducing the *Kolmogorov constant* κ_{K}, the characteristic length l_{c}, and the characteristic time t_{c} defined by

$$\kappa_{\mathrm{K}} = \left(\frac{8}{9\kappa_{\mathrm{H}}}\right)^{2/3} = 1.44, \qquad l_{\mathrm{c}} = \left(\frac{\nu^3}{\overline{\epsilon}_{\mathrm{M}}}\right)^{1/4}, \qquad t_{\mathrm{c}} = \left(\frac{\nu}{\overline{\epsilon}_{\mathrm{M}}}\right)^{1/2} \tag{12.51}$$

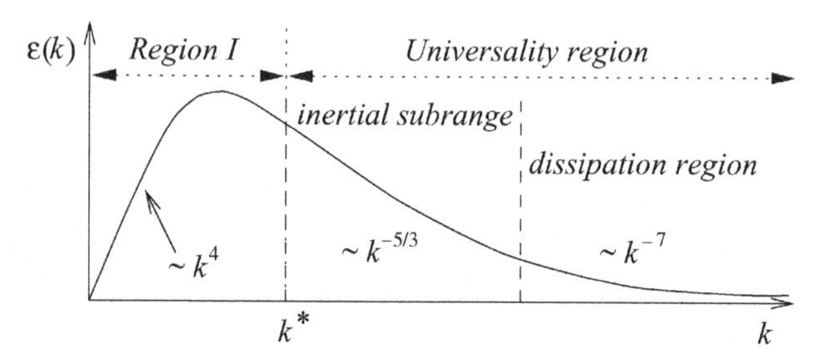

Fig. 12.6 The Heisenberg spectrum.

with $\kappa_{\rm H} = 0.51$. The Kolmogorov constant is named after the scientist who pioneered research into turbulence spectra. Substituting $\kappa_{\rm K}$ and $l_{\rm c}$ into (12.47), the energy spectrum may be written as

$$\epsilon(k) = \kappa_{\rm K}\overline{\epsilon}_{\rm M}^{2/3} k^{-5/3}\left(1 + \frac{8(kl_{\rm c})^4}{3\kappa_{\rm H}^2}\right)^{-4/3}, \qquad k > k^* \tag{12.52}$$

This new notation, of course, does not change the physical content of the energy spectrum. Now the long-wave range is isolated by means of $kl_{\rm c} \ll 1$ and the short -wave range by $kl_{\rm c} \gg 1$.

12.5 Relations for the Heisenberg exchange coefficient

The task ahead is to evaluate the integral (12.45b) defining the Heisenberg exchange coefficient. This will be accomplished by first substituting the energy expression for the long-wave approximation (12.50) for the stationary case into (12.45b), yielding

$$K = \kappa_{\rm H}\int_k^\infty \sqrt{\frac{\epsilon(k')}{(k')^3}}\,dk' = \kappa_{\rm H}\int_k^\infty \left(\frac{\kappa_{\rm K}\overline{\epsilon}_{\rm M}^{2/3}(k')^{-5/3}}{(k')^3}\right)^{1/2} dk' = \frac{3}{4}\kappa_{\rm K}^{1/2}\kappa_{\rm H}\overline{\epsilon}_{\rm M}^{1/3} k^{-4/3} \tag{12.53}$$

We now assume that the wavenumber k separates the short-wave region from the long-wave region. Then we can express the average kinetic energy $\widehat{\mathbf{v}}^2/2$ of the longer waves by

$$\frac{\widehat{\mathbf{v}}^2}{2} = \int_0^k \epsilon(k')\,dk' \tag{12.54a}$$

and the turbulent kinetic energy $\widehat{k}$ of the shorter waves by

$$\widehat{k} = \int_k^\infty \epsilon(k')\,dk' \tag{12.54b}$$

We substitute the spectral energy approximation for the inertial subrange (12.50) into (12.54b) and obtain

$$\widehat{k} = \kappa_{\rm K}\int_k^\infty \overline{\epsilon}_{\rm M}^{2/3}(k')^{-5/3}\,dk' = \frac{3}{2}\kappa_{\rm K}\overline{\epsilon}_{\rm M}^{2/3} k^{-2/3} \tag{12.55}$$

We have implicitly assumed that the $k^{-5/3}$ law holds for the entire range $k' \geq k$; the error we make thereby is likely to be small. Now we solve (12.55) for $\overline{\epsilon}_{\mathrm{M}}$ and obtain

$$\overline{\epsilon}_{\mathrm{M}} = ak\widehat{k}^{3/2} \quad \text{with} \quad a = \left(\frac{3\kappa_{\mathrm{K}}}{2}\right)^{-3/2} \tag{12.56}$$

Substituting this expression into (12.53) gives

$$K = bk^{-1}\widehat{k}^{1/2} \quad \text{with} \quad b = \frac{3}{4}\left(\frac{2}{3}\right)^{-1/2} \kappa_{\mathrm{H}} \tag{12.57}$$

Now we have obtained the desired relations among $\overline{\epsilon}_{\mathrm{M}}$, $\widehat{k}$, and K. Finally, the wavenumber k will be replaced by a characteristic length l by means of

$$k = \frac{2\pi}{L} = \frac{1}{l} \tag{12.58}$$

From (12.57), using the definition (12.58), it follows that

$$\widehat{k} = \left(\frac{K}{bl}\right)^2 \tag{12.59a}$$

so that (12.56) can be written as

$$\overline{\epsilon}_{\mathrm{M}} = \frac{a}{l}\left(\frac{K}{bl}\right)^3 = \frac{aK^3}{b^3 l^4} \tag{12.59b}$$

By combining (12.59b) with (12.56) and (12.57), we obtain

$$K = ab\frac{\widehat{k}^2}{\overline{\epsilon}_{\mathrm{M}}} \tag{12.60}$$

Hence, it is seen that the exchange coefficient itself is proportional to the square of the turbulent kinetic energy and inversely proportional to the dissipation of energy $\overline{\epsilon}_{\mathrm{M}}$. We are going to consider the Heisenberg exchange coefficient as a typical exchange coefficient that is valid not only for the turbulent momentum flux but also for the transport of other turbulent fluxes. In the next chapter we will refine this concept.

12.6 A prognostic equation for the exchange coefficient

Before continuing our discussion, we wish to briefly elucidate the action of the turbulent flux $\overline{\mathbf{v}''\rho\psi}$. As an example, let us consider a simple one-dimensional

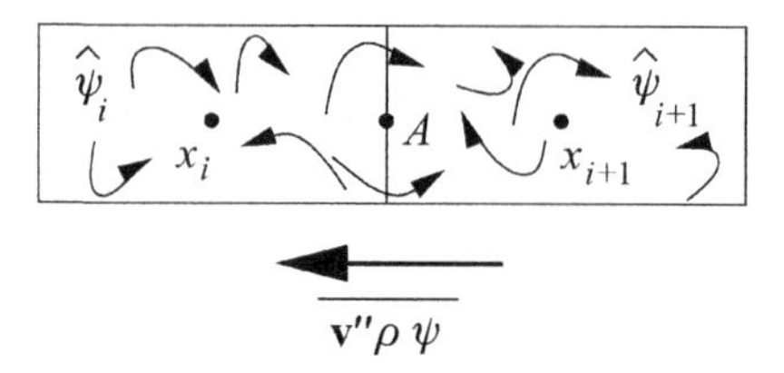

Fig. 12.7 A schematic view of the action of a turbulent flux.

numerical grid as it may be used for the evaluation of the atmospheric equations. The process of turbulent mixing is shown schematically in Figure 12.7. At each central grid point x_i the value of a physical quantity $\widehat{\psi}_i$ represents an average for the entire grid cell. These central values are thought to be known without discussing how they are obtained. Within each of these cells we expect irregular turbulent motion to take place. In order to obtain information about the turbulent flux at point A between two grid cells x_i and x_{i+1}, we do not need (and usually cannot obtain) information about each individual eddy or fluctuation. It is sufficient to know the average value $\overline{\mathbf{v}''\rho\psi}$. For definiteness let us assume that the value of $\widehat{\psi}$ increases from left to right so that $\widehat{\psi}_{i+1} > \widehat{\psi}_i$. The turbulent eddies perform a completely disordered or stochastic motion, but each of these will be associated with a particular value of ψ. From experience we know that, on average, a directed transport of the quantity ψ will take place from higher toward lower ψ values. For the turbulent flux we may therefore write

$$\overline{\mathbf{v}''\rho\psi} \sim -\nabla\widehat{\psi}, \qquad \overline{\mathbf{v}''\rho\psi} = -\rho K\,\nabla\widehat{\psi} \tag{12.61}$$

The proportionality factor is the exchange coefficient K, which depends on the turbulent kinetic energy of the system.

With these ideas in mind, we are ready to obtain a prognostic equation for the exchange coefficient K. The starting point for the derivation is the prognostic equation (11.46) for the turbulent kinetic energy, which is repeated here for convenience:

$$\frac{\widehat{D}}{Dt}(\overline{\rho}\widehat{k}) + \nabla\cdot(\mathbf{k}_{\rm t} - \overline{\mathbf{v}''\cdot\mathbb{J}}) = \nabla\widehat{\mathbf{v}}\cdot\cdot(\overline{\mathbb{J}} + \mathbb{R}) - \overline{\epsilon} - \overline{\mathbf{v}''\cdot\nabla p} \tag{12.62}$$

In this equation we replace $\overline{\mathbf{v}''\cdot\nabla p}$ by a simplified form of (11.70):

$$\overline{\mathbf{v}''\cdot\nabla p} = \mathbf{J}_{\rm t}^{\theta}\cdot\nabla\ln\widetilde{\Pi}, \qquad \mathbf{k}_{\rm t} = \tfrac{1}{2}\overline{\mathbf{v}''\rho(\mathbf{v}'')^2} \tag{12.63}$$

For reference the definition of the turbulent kinetic energy flux $\mathbf{k}_{\rm t}$, see equation (11.42), is included as part of this equation. By replacing the turbulent kinetic energy $\widehat{k}$ in terms of the exchange coefficient K according to (12.59a) and $\overline{\epsilon}$ by (12.59b), we obtain a prognostic equation for the exchange coefficient:

$$\frac{\widehat{D}}{Dt}\left[\overline{\rho}\left(\frac{K}{bl}\right)^2\right] + \nabla\cdot(\mathbf{k}_{\mathrm{t}} - \overline{\mathbf{v}''\cdot\mathbb{J}}) = \nabla\widehat{\mathbf{v}}\cdot\cdot(\overline{\mathbb{J}} + \mathbb{R}) - \mathbf{J}_{\mathrm{t}}^{\theta}\cdot\ln\widetilde{\Pi} - \frac{\overline{\rho}a}{b^3l^4}K^3 \qquad (12.64)$$

In the final term the air density $\overline{\rho}$ has been inserted since we divided (12.16) by the density. In this equation $\widehat{\mathbf{v}}$, $\overline{\rho}$, and $\widetilde{\Pi}$ are considered to be known quantities. The remaining variables are still unknown and must be replaced by suitable parameterizations.

The entire analysis of this chapter is based on the assumption that the atmosphere is characterized by $\rho =$ constant so that the velocity divergence is zero. In this case the averaged molecular stress tensor $\mathbb{J}$ can be written as

$$\overline{\mathbb{J}} = \mu(\nabla\widehat{\mathbf{v}} + \overset{\curvearrowleft}{\widehat{\mathbf{v}}\nabla}) \qquad (12.65)$$

In analogy to this, assuming isotropic conditions, we may write for the Reynolds stress tensor

$$\mathbb{R} = \overline{\rho}K(\nabla\widehat{\mathbf{v}} + \overset{\curvearrowleft}{\widehat{\mathbf{v}}\nabla}) \qquad (12.66)$$

where K is the turbulent exchange coefficient. Since the turbulent exchange coefficient K is much larger than the dynamic viscosity multiplied by the air density, we may ignore the contribution of the mean molecular stress. Therefore, we parameterize the first term on the right-hand side of (12.64) as

$$\nabla\widehat{\mathbf{v}}\cdot\cdot(\overline{\mathbb{J}} + \mathbb{R}) = \nabla\widehat{\mathbf{v}}\cdot\cdot\overline{\rho}K(\nabla\widehat{\mathbf{v}} + \overset{\curvearrowleft}{\widehat{\mathbf{v}}\nabla}) \qquad (12.67)$$

We now consider the vectorial expression in the divergence term of equation (12.64). Assuming that the second term may be ignored, we may write in analogy to (12.61) the parameterized form

$$(\mathbf{k}_{\mathrm{t}} - \overline{\mathbf{v}''\cdot\mathbb{J}}) = -\overline{\rho}K\,\nabla\left(\frac{\widehat{\mathbf{v}''^2}}{2}\right) = -\overline{\rho}K\,\nabla\widehat{k} = -\overline{\rho}K\,\nabla\left(\frac{K}{bl}\right)^2 \qquad (12.68)$$

The negative sign shows that the flow is from larger to smaller values of the transported quantity. Finally, we need to parameterize the turbulent heat flux which was defined by equation (11.88). For the isotropic case the exchange tensor may be replaced by the scalar exchange coefficient. On the basis of the brief discussion we have given above and from dimensional requirements, we find the parameterized form

$$\mathbf{J}_{\mathrm{t}}^{\theta} = -\overline{\rho}\widehat{c}_pK\,\nabla\widetilde{\theta} \qquad (12.69)$$

While in (11.88) the two scalar coefficients K_{h}^{θ} and K_{v}^{θ} are used to parameterize $\mathbf{J}_{\mathrm{t}}^{\theta}$, this simplified treatment admits only the exchange coefficient K. On substituting

(12.67)–(12.69) into (12.64), the prognostic equation for the exchange coefficient will be written as

$$\frac{\widehat{D}}{Dt}\left[\overline{\rho}\left(\frac{K}{bl}\right)^2\right] - \nabla\cdot\left[\overline{\rho}K\,\nabla\left(\frac{K}{bl}\right)^2\right] = \nabla\widehat{\mathbf{v}}\cdot\cdot\overline{\rho}K(\nabla\widehat{\mathbf{v}}+\widehat{\mathbf{v}}\overleftarrow{\nabla}) + \overline{\rho}\widehat{c}_pK\,\nabla\widetilde{\theta}\cdot\nabla\ln\widetilde{\Pi} - \frac{\overline{\rho}a}{b^3l^4}K^3 \tag{12.70}$$

Equation (12.70) is rather complex and its evaluation requires a great deal of numerical effort. From this equation, however, we may derive an entire hierarchy of simplifications, which were often applied when computer capabilities were insufficient. Numerical investigations of (12.70) have shown that, for many situations, the tendency term may be neglected. For this case, after expanding the budget operator and disregarding the local time derivative, we obtain

$$\nabla\cdot\left[\overline{\rho}\,\widehat{\mathbf{v}}\left(\frac{K}{bl}\right)^2\right] - \nabla\cdot\left[\overline{\rho}K\,\nabla\left(\frac{K}{bl}\right)^2\right] = \nabla\widehat{\mathbf{v}}\cdot\cdot\overline{\rho}K(\nabla\widehat{\mathbf{v}}+\widehat{\mathbf{v}}\overleftarrow{\nabla}) + \overline{\rho}\,\widehat{c}_pK\,\nabla\widetilde{\theta}\cdot\nabla\ln\widetilde{\Pi} - \frac{\overline{\rho}a}{b^3l^4}K^3 \tag{12.71}$$

This is a partial differential equation of the elliptic type representing a boundary value problem. We select a rectangular parallelepiped as the integration domain, for which it is customary to set $K = 0$ at the upper boundary $z = H$. At the lower boundary, assuming neutral conditions, we set K proportional to the frictional velocity u^*, $K = \kappa z_0 u^*$. The constant $\kappa = 0.4$ is known as the *Von Karman constant* and z_0 is the roughness height. The precise definition of u^* and the justification of this form of the exchange coefficient at the surface will be offered in the following chapter.

The next simplification is to assume that we have horizontal homogeneity, so that the gradient operator reduces to

$$\nabla_{\rm h}\ldots = 0, \qquad \nabla\ldots = \mathbf{k}\,\frac{\partial\ldots}{\partial z} \tag{12.72a}$$

By assuming additionally that $\widehat{w} = 0$ we obtain $\widehat{\mathbf{v}} = \widehat{\mathbf{v}}_{\rm h}$, from which it follows that

$$\mathbf{k}\cdot\widehat{\mathbf{v}}_{\rm h} = 0, \qquad \nabla\widehat{\mathbf{v}}\cdot\cdot(\nabla\widehat{\mathbf{v}}+\widehat{\mathbf{v}}\overleftarrow{\nabla}) = \left(\frac{\partial\widehat{u}}{\partial z}\right)^2 + \left(\frac{\partial\widehat{v}}{\partial z}\right)^2 = \left(\frac{\partial\,\widehat{\mathbf{v}}_h}{\partial z}\right)^2 \tag{12.72b}$$

For these conditions the stationary K equation is expressed by

$$\frac{\partial}{\partial z}\left[\overline{\rho}K\frac{\partial}{\partial z}\left(\frac{K}{bl}\right)\right]+\overline{\rho}K\left(\frac{\partial\widehat{\mathbf{v}}_{\rm h}}{\partial z}\right)^2+\overline{\rho}c_{p,0}K\frac{\partial\widetilde{\theta}}{\partial z}\frac{\partial\ln\widetilde{\Pi}}{\partial z}-\overline{\rho}\frac{a}{b^3l^4}K^3=0 \qquad (12.73)$$

where we have made the highly satisfactory assumption that $\widehat{c}_p = c_{p,0}$. We will now eliminate the Exner function $\widetilde{\Pi}$ as shown next:

$$\ln\widetilde{\Pi}=\frac{R_0}{c_{p,0}}\ln\overline{p}+\text{constant}\implies\frac{\partial\ln\widetilde{\Pi}}{\partial z}=\frac{R_0}{c_{p,0}\overline{p}}\frac{\partial\overline{p}}{\partial z}=-\frac{g}{c_{p,0}\widetilde{T}} \qquad (12.74)$$

This results in the desired form

$$\frac{\partial}{\partial z}\left[\overline{\rho}K\frac{\partial}{\partial z}\left(\frac{K}{bl}\right)\right]+\overline{\rho}K\left(\frac{\partial\widehat{\mathbf{v}}_{\rm h}}{\partial z}\right)^2-\frac{\overline{\rho}gK}{\widetilde{T}}\frac{\partial\widetilde{\theta}}{\partial z}-\overline{\rho}\frac{a}{b^3l^4}K^3=0 \qquad (12.75)$$

The final step in the hierarchy of approximations is to ignore the divergence term. Solving for K^2 results in

$$K^2=\frac{b^3l^4}{a}\left(\frac{\partial\widehat{\mathbf{v}}_{\rm h}}{\partial z}\right)^2\left[1-\frac{g}{\widetilde{T}}\frac{\partial\widetilde{\theta}}{\partial z}\left(\frac{\partial\widehat{\mathbf{v}}_{\rm h}}{\partial z}\right)^{-2}\right] \qquad (12.76)$$

The fraction in the square bracket is known as the *Richardson number* Ri,

$$\boxed{Ri=\frac{g}{\widetilde{T}}\frac{\partial\widetilde{\theta}}{\partial z}\left(\frac{\partial\widehat{\mathbf{v}}_{\rm h}}{\partial z}\right)^{-2}} \qquad (12.77)$$

Ri is a measure of the atmospheric stability. A detailed discussion follows in the next chapter. Inserting Ri into (28.76) gives

$$\boxed{K=\sqrt{\frac{b^3}{a}}l^2\left|\frac{\partial\widehat{\mathbf{v}}}{\partial z}\right|\sqrt{1-Ri}=(l')^2\left|\frac{\partial\widehat{\mathbf{v}}}{\partial z}\right|\sqrt{1-Ri},\qquad Ri\le 1} \qquad (12.78)$$

showing that the exchange coefficient is a function of the wind shear and the thermal stability of the atmosphere. We have also combined the product of the constant $\sqrt{(b^3/a)}$ with the square of the characteristic length l^2 to give a new characteristic length $(l')^2$. The quantity l' is some sort of mixing length, which will be discussed more fully in the following chapter.[1] The resulting formula (12.78) is often used as the definition of the exchange coefficient or *Austausch coefficient* (the German word *Austausch* means exchange). The concept of the mixing length will be discussed more fully in the following chapter.

[1] For neutral conditions Blackadar postulated the following form of the mixing length: $l' = kz/(1+kz/\lambda)$, where $k=0.4$ is the Von Karman constant and λ an asymptotic value for l' that is reached at great heights.

12.7 Concluding remarks on closure procedures

There are several excellent books on boundary-layer theory dealing in great detail with various closure schemes. A very illuminating account is given, for example, by Stull (1989), where the reader can find an extensive bibliography on this subject as well as many observational results. Here we can give only a few brief statements on this topic.

There are *local closure schemes* and *nonlocal closure schemes*. The closure technique described in Section 12.6 is based on the turbulent-kinetic-energy equation. This type of closure technique belongs to the group of local closure schemes. They are called local since an unknown quantity at a point in space is parameterized by values and gradients of known quantities at or near the same point. If nonlocal closure techniques are used, the unknown quantity at one point in space is parameterized by using values of known quantities at many points. The idea behind this concept is that larger eddies transport fluid over larger distances before the smaller eddies have a chance to cause mixing. This is the so-called *transilient turbulence theory* described in some detail by Stull (1989). There is a second nonlocal scheme called *spectral diffusivity theory*. This theory has its origin in the spectral theory which we have previously discussed.

Furthermore, it is customary to distinguish between first- and higher-order closure schemes. To convey the idea of higher-order closure let us consider the prognostic equation for the mean velocity $\widehat{\mathbf{v}}$; see equation (11.35d). This equation includes the divergence of the Reynolds tensor which is essentially the double correlation $\overline{\mathbf{v}''\mathbf{v}''}$ of the velocity fluctuations. The idea behind this closure principle is to derive a differential equation for $\overline{\mathbf{v}''\mathbf{v}''}$. The mathematical steps involved are not particularly difficult but rather lengthy so we will restrict ourselves to a brief verbal description to demonstrate the principle. For simplicity, we assume that the density ρ is a constant. By subtracting the equation of mean motion from the molecular form of the Navier–Stokes equation, we obtain a differential equation for the velocity fluctuation $d\mathbf{v}''/dt$. On the right-hand side of this equation, among other terms, there still appears the divergence of the Reynolds tensor. Dyadic multiplication of $d\mathbf{v}''/dt$ first from the right and then from the left gives two differential equations, i.e. $(d\mathbf{v}''/dt)\mathbf{v}'' = \cdots$ and $\mathbf{v}''\,d\mathbf{v}''/dt = \cdots$. Averaging these two equations and adding the results gives the desired differential equation for the Reynolds tensor $d(\mathbf{v}''\mathbf{v}'')/dt$. Having derived this differential equation does not complete the closure problem by any means since this rather complex differential equation now contains the divergence of the unknown triple correlation $\overline{\mathbf{v}''\mathbf{v}''\mathbf{v}''}$. If we try to derive a differential equation for the triple correlation, we end up with the appearence of a still higher unknown correlation $\overline{\mathbf{v}''\mathbf{v}''\mathbf{v}''\mathbf{v}''}$. In order to avoid the escalation of this problem, it is customary to introduce *closure assumptions*. An extensive literature is available on this topic. We refer to Mellor and Yamada (1974), who propose a second-order closure parameterization for $\overline{\mathbf{v}''\mathbf{v}''\mathbf{v}''}$.

Even second-order closure schemes are rather complicated. In order to avoid such complications, quite early in the development of the boundary-layer theory, the *first-order closure scheme* or *K theory* was introduced to handle the turbulence problem. Suppose that we decompose the Reynolds tensor in a suitable manner. Among other components we would obtain the correlation

$$\overline{\mathbf{v}''w''} = -K\,\frac{\partial \widehat{\mathbf{v}}_{\mathrm{h}}}{\partial z} \tag{12.79}$$

The first-order closure scheme simply requires that the mean value of the double correlation is proportional to the vertical gradient of the horizontal mean velocity. The coefficient of proportionality is the exchange coefficient which must be specified or predicted in some manner. The form of equation (12.79) explains why the K theory is also called the *gradient transport theory*. Let us briefly return to Section 12.6, where a differential equation for the turbulent exchange coefficient was derived. This type of approach is more realistic than the first-order closure scheme, in which the exchange coefficient is usually specified, but it is not as refined as the second-order closure scheme. For this reason the approach presented in Section 12.6 is called a *1.5-order closure scheme*, of which many variants are described in the literature.

Among other approaches to solving the turbulence problems, so-called *large-eddy simulation* (LES) is becoming more prominent as electronic computers become increasingly more powerful. The concept of LES is based on the idea that some filter operation, for example averaging, is applied to the equations of the turbulent atmosphere to separate the large and small scales of turbulent flow. The form of the atmospheric equations modeling the flow remains unchanged, with the turbulent-stress term replacing the molecular-viscosity term. For this method the numerical grid is chosen fine enough to resolve the larger eddies containing the bulk of the turbulent energy. All the information regarding the small unresolved scales is extracted from a subgrid model. The larger resolved eddies are sensitive to the geometry of the flow and the thermal stratification. The much-less-sensitive unresolved small-scale eddies can be efficiently parameterized by a relatively simple first-order closure scheme. This type of approach has successfully been applied to a variety of turbulent-flow problems. Comprehensive reviews have been given by Rodi (1993) and by Mason (1999), for example.

12.8 Problems

12.1: In order to better understand the transition from (12.18b) to (12.20), perform simplified summations. Consider the special case that the vectors **m** and **q** are scalars. Perform the summation $(m + q)V(m)V(q)\exp(m + q)$ with $m, q = 1, 2, \ldots, 5$ and factor out the sum multiplying exp(6). Compare your result with the sum $6V(m)V(6 - m)$.

12.2: Show explicitly that the transfer function $W(\mathbf{k}, \mathbf{k}', t)$ is antisymmetric.

12.3: Verify that α and β of equation (12.38a) are given by $\alpha = -\frac{3}{2}$ and $\beta = \frac{1}{2}$.

13

The atmospheric boundary layer

13.1 Introduction

The vertical structure of the atmospheric boundary layer is depicted in Figure 13.1. The lowest atmospheric layer is known as the *laminar sublayer* and has a thickness of only a few millimeters. It is difficult to verify the existence of this layer because of its small vertical extent. Within the laminar sublayer all physical processes such as the transport of momentum and heat are regulated by molecular motion. In most boundary layer models the existence of this layer is not explicitly treated. It stands to reason that there also exists some type of a transitional layer between the laminar sublayer and the so-called *Prandtl layer* where turbulence is fully developed.

The lower boundary of the Prandtl or surface layer is the *roughness height* z_0 where the mean wind is assumed to vanish. The vertical extent of the Prandtl layer is regulated by the thermal stratification of the air and may vary from about 20 to 100 m. In this layer all turbulent fluxes are approximately constant with height. The influence of the Coriolis force may be ignored this close to the earth's surface, so the turning of the wind within the Prandtl layer may be ignored. The wind speed, however, increases very strongly in this layer, reaching a value of more than half the wind speed at the top of the boundary layer.

Above the Prandtl layer we find the so-called *Ekman layer*, reaching a height exceeding 1000 m depending on the stability of the air. Turbulent fluxes in this layer decrease to zero at the top of the Ekman layer. Above the Ekman layer the air flow is more or less nonturbulent. The influence of the Coriolis force in this layer causes the turning of the wind vector. For this reason the Ekman layer is often called the *spiral layer* since the nomogram of the wind vector is a spiral. The entire region from the earth's surface to the top of the Ekman layer is called the *planetary boundary layer*.

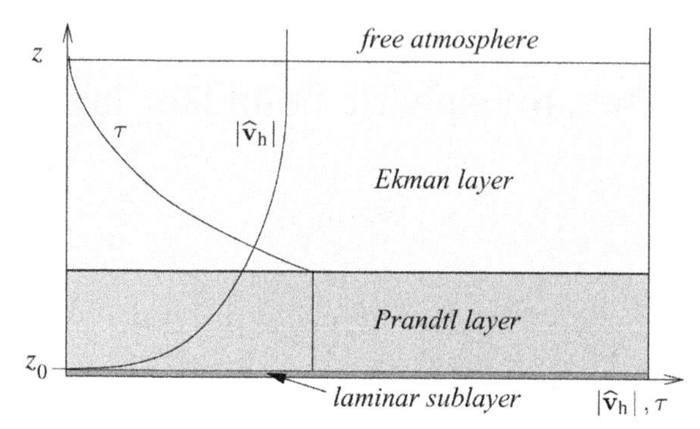

Fig. 13.1 Subdivision of the planetary boundary layer, showing the vertical distribution of the horizontal wind $\widehat{\mathbf{v}}_{\mathrm{h}}$ and the stress τ within the boundary layer.

13.2 Prandtl-layer theory

It is customary to use the Cartesian coordinate system to discuss the Prandtl-layer theory. To begin with we will summarize the hypothetical Prandtl-layer conditions:

$$
\boxed{\begin{array}{lllll}
\text{(a)} & \widehat{\mathbf{v}}(z=z_0)=0, & \widehat{\mathbf{v}}_{\mathrm{h}}(z=z_0)=0, & \widehat{w}(z=z_0)=0 & \\
\text{(b)} & \boldsymbol{\Omega}=0 & & & \\
\text{(c)} & \widehat{m}^2=0, & \widehat{m}^3=0, & \overline{I}^2=0, & \overline{I}^3=0 \\
\text{(d)} & \nabla\cdot\overline{\mathbf{F}}_{\mathrm{R}}=0 & & & \\
\text{(e)} & J_{\mathrm{t}}^{\theta}=\mathbf{i}_3\cdot\mathbf{J}_{\mathrm{t}}^{\theta}=\text{constant} & & & \\
\text{(f)} & \dfrac{\partial\ldots}{\partial t}=0, & \nabla_{\mathrm{h}}\ldots=0, & \nabla=\mathbf{i}_3\,\dfrac{\partial}{\partial z} &
\end{array}} \tag{13.1}
$$

The lower boundary of this layer is the roughness height z_0 where the wind speed vanishes, (13.1a). This condition implies a flat surface of the earth. Since the Coriolis force is of negligible importance, we ignore this force by formally setting the angular velocity of the earth equal to zero in the equation of motion, (13.1b). Condensation of water vapor is not allowed to take place in the Prandtl-layer theory, so fog does not form. This means that only dry air and water vapor are admitted. Therefore, phase transition fluxes I^k do not appear in the Prandtl-layer equations, (13.1c). The divergence of the radiative flux is ignored, (13.1d). The vertical component of the turbulent heat flux J_{t}^{θ} does not vary with height, (13.1e). The most essential part of the Prandtl-layer theory is the hypothesis of stationarity and of horizontal homogeneity of all averaged variables of state, (13.1f). In order to investigate the formation and dissipation of ground fog, the Prandtl-layer theory must be relaxed and radiative processes are then of paramount importance.

13.2.1 The modified budget equations

We are now ready to state the budget equations in their modified forms as they apply to the Prandtl layer.

13.2.1.1 The continuity equation

Owing to stationarity and horizontal homogeneity, the continuity equation (11.35a) reduces to

$$\frac{\widehat{D}\overline{\rho}}{Dt} = \frac{\partial}{\partial z}(\overline{\rho}\widehat{w}) = 0 \quad \text{or} \quad \overline{\rho}\widehat{w} = \text{constant} \tag{13.2}$$

It is customary to treat $\overline{\rho}$ as a constant. Since $\widehat{w}(z_0) = 0$ we find within the Prandtl layer

$$\widehat{w}(z) = 0 \tag{13.3}$$

so that the velocity divergence vanishes:

$$\nabla \cdot (\overline{\rho}\,\widehat{\mathbf{v}}) = \mathbf{i}_3 \cdot \frac{\partial}{\partial z}(\overline{\rho}\,\widehat{\mathbf{v}}_{\rm h}) = 0, \qquad \nabla \cdot \widehat{\mathbf{v}} = 0 \tag{13.4}$$

As a consequence of these equations and due to the stationarity, the averaged budget operator and the individual time derivative also vanish:

$$\frac{\widehat{D}}{Dt}(\overline{\rho}\psi) = 0, \qquad \frac{\widehat{d}\,\psi}{dt} = 0 \tag{13.5}$$

13.2.1.2 The continuity equation for the concentrations

The continuity equation for the concentration of water vapor (11.35b) reduces to the divergence expression

$$\nabla \cdot \left(\overline{\mathbf{J}}^1 + \mathbf{J}_{\rm t}^1\right) = \frac{\partial}{\partial z}\mathbf{i}_3 \cdot \left(\overline{\mathbf{J}}^1 + \mathbf{J}_{\rm t}^1\right) = 0 \Longrightarrow \mathbf{i}_3 \cdot \left(\overline{\mathbf{J}}^1 + \mathbf{J}_{\rm t}^1\right) = \overline{J}^1 + J_{\rm t}^1 = \text{constant} \tag{13.6}$$

From the requirement of horizontal homogeneity it follows immediately that the sum of the vertical components of the diffusive water-vapor fluxes is constant with height. The diffusion fluxes for dry air can be directly obtained from

$$\overline{J}^0 + J_{\rm t}^0 = -\left(\overline{J}^1 + J_{\rm t}^1\right) \tag{13.7}$$

13.2.1.3 The equation of motion

From (11.35d), together with the various hypothetical Prandtl-layer statements, we immediately obtain

$$\mathbf{i}_3\,\frac{\partial\overline{p}}{\partial z}-\mathbf{i}_3\cdot\frac{\partial}{\partial z}(\overline{\mathbb{J}}+\mathbb{R})=-\mathbf{i}_3\overline{\rho}g \tag{13.8}$$

since $\nabla_{\rm h}\phi=0$ and $\mathbf{i}_3\,\partial\phi/\partial z=\mathbf{i}_3 g$. We will now define the *stress vector of the boundary layer* as

$$\mathbf{T}=\mathbf{i}_3\cdot(\overline{\mathbb{J}}+\mathbb{R}) \tag{13.9}$$

Next, we require some information on the turbulence state of the Prandtl layer. We assume that we have horizontal isotropy of turbulence as discussed in Section 11.8. From (11.92) it follows that

$$\mathbf{T}=\overline{\rho}K_{\rm v}^{v}\,\frac{\partial\widehat{\mathbf{v}}_{\rm h}}{\partial z} \tag{13.10}$$

This is a horizontally directed vector that is parallel to $\partial\widehat{\mathbf{v}}_{\rm h}/\partial z$. Equation (13.8) now splits into a horizontal and a vertical part:

$$\boxed{\frac{\partial\overline{p}}{\partial z}=-\overline{\rho}g,\qquad \frac{\partial\mathbf{T}}{\partial z}=0\implies\mathbf{T}=\text{constant}} \tag{13.11}$$

The first equation is the hydrostatic equation for the averaged pressure and density. From the second equation it follows that the stress vector is constant with height within the Prandtl layer. Direct application of the hypothetical Prandtl-layer conditions to the kinetic-energy equation of mean motion (11.45) also yields $\mathbf{T}=$ constant.

Since the horizontal velocity vanishes at z_0, the stress vector must also be parallel to the velocity vector itself. This is most easily seen by writing the vertical derivative as a finite difference. Since $\mathbf{T}$, $\partial\widehat{\mathbf{v}}_{\rm h}/\partial z$, and $\widehat{\mathbf{v}}_{\rm h}$ all have the same direction, it is of advantage to rotate the coordinate system about the z-axis so that one of the horizontal axes is pointing in the direction of the three vectors. It is customary to select the x-axis so that we may write

$$|\widehat{\mathbf{v}}_{\rm h}|=\widehat{u},\quad \widehat{v}=0,\quad |\mathbf{T}|=\tau=\overline{\rho}K_{\rm v}^{v}\,\frac{\partial\widehat{u}}{\partial z},\quad \frac{\partial\tau}{\partial z}=0,\quad \mathbf{T}\cdot\frac{\partial\widehat{\mathbf{v}}_{\rm h}}{\partial z}=\tau\,\frac{\partial\widehat{u}}{\partial z} \tag{13.12}$$

The vertical distribution of τ is shown in Figure 13.1.

13.2.1.4 The budget equation of the turbulent kinetic energy

We will now direct our attention to the budget of the turbulent kinetic energy as stated in (11.46). The budget operator vanishes according to (13.5). Owing to horizontal homogeneity the divergence part degenerates to

$$\nabla \cdot \left(\overline{\mathbf{k}_{\mathrm{t}} - \mathbf{v}'' \cdot \mathbb{J}}\right) = \frac{\partial E}{\partial z} \quad \text{with} \quad E = \mathbf{i}_3 \cdot \left(\overline{\mathbf{k}_{\mathrm{t}} - \mathbf{v}'' \cdot \mathbb{J}}\right), \qquad \mathbf{k}_{\mathrm{t}} = \overline{\rho \mathbf{v}'' \frac{\mathbf{v}''^2}{2}} \tag{13.13}$$

The double scalar product on the right-hand side of (11.46) may be rewritten as

$$\nabla \widehat{\mathbf{v}} \cdot\cdot \left(\overline{\mathbb{J}} + \mathbb{R}\right) = \mathbf{i}_3 \frac{\partial \widehat{\mathbf{v}}_{\mathrm{h}}}{\partial z} \cdot\cdot \left(\overline{\mathbb{J}} + \mathbb{R}\right) = \mathbf{i}_3 \cdot \left(\overline{\mathbb{J}} + \mathbb{R}\right) \cdot \frac{\partial \widehat{\mathbf{v}}_{\mathrm{h}}}{\partial z} = \mathbf{T} \cdot \frac{\partial \widehat{\mathbf{v}}_{\mathrm{h}}}{\partial z} \tag{13.14}$$

since $\overline{\mathbb{J}}$ and $\mathbb{R}$ are symmetric tensors. The power term $\overline{\mathbf{v}'' \cdot \nabla p}$ appearing in (11.46) will be replaced by (11.70) with the approximation $\overline{\mathbf{J}^{\theta} \cdot \mathbf{A}} = 0$. Utilizing $R_0 \overline{\rho}/\overline{p} = 1/\overline{T}$, the term multiplying the heat flux $\mathbf{J}_{\mathrm{t}}^{\theta}$ in (11.70) is approximated as

$$\nabla \ln \widetilde{\Pi} = \frac{\partial \ln \widetilde{\Pi}}{\partial z} \mathbf{i}_3 = -\frac{g}{c_{p,0} \overline{T}} \mathbf{i}_3 \tag{13.15}$$

Here and in the following $\widehat{c}_p$ will be approximated by $c_{p,0}$. With these approximations the budget of the turbulent kinetic energy can finally be written as

$$\boxed{\frac{\partial E}{\partial z} = \mathbf{T} \cdot \frac{\partial \widehat{\mathbf{v}}_{\mathrm{h}}}{\partial z} - \overline{\epsilon} - \mathbf{J}_{\mathrm{t}}^{\theta} \cdot \nabla \ln \widetilde{\Pi} = \tau \frac{\partial \widehat{u}}{\partial z} - \overline{\epsilon} + J_{\mathrm{t}}^{\theta} \frac{g}{c_{p,0} \overline{T}}} \tag{13.16}$$

with $\overline{\epsilon} = \overline{\mathbb{J} \cdot\cdot \nabla \mathbf{v}} \geq 0$

13.2.1.5 The budget equation of the internal energy

Application of the hypothetical Prandtl-layer conditions to (11.48) gives the first version of the budget equation for the internal energy

$$\frac{\partial}{\partial z}\left(\overline{J}^{h} + J_{\mathrm{t}}^{h}\right) = -J_{\mathrm{t}}^{\theta} \frac{g}{c_{p,0} \overline{T}} + \overline{\epsilon} \tag{13.17}$$

Next we introduce the sensible enthalpy fluxes

$$\overline{J}_{\mathrm{s}}^{h} = \overline{J}^{h} - \overline{J}^{n} \widetilde{h}_n, \qquad J_{\mathrm{s,t}}^{h} = J_{\mathrm{t}}^{h} - J_{\mathrm{t}}^{n} \widetilde{h}_n \tag{13.18}$$

Utilizing the latter equation, the vertical component of the vector relation (11.69) can be written as

$$J_{\mathrm{t}}^{h} = J_{\mathrm{t}}^{\theta} + J_{\mathrm{t}}^{\Pi} + J_{\mathrm{t}}^{n} \widetilde{h}_n \tag{13.19}$$

In the absence of ice we may write for the atmospheric system $J_{\rm t}^n\widetilde{h}_n = J_{\rm t}^0\widetilde{h}_0 + J_{\rm t}^1\widetilde{h}_1 + J_{\rm t}^2\widetilde{h}_2$. An analogous expression holds also for $\overline{J}^n\widetilde{h}_n$. The enthalpies $\widetilde{h}_k$ depend on temperature, and, for the condensed phase, even on pressure. As we have seen, the Prandtl-layer theory is based on numerous hypothetical conditions. Thus we feel justified in applying the additional assumption that the $\widetilde{h}_k$ may be treated as constants. Since the definition of any thermodynamic potential includes an arbitrary constant that is at our disposal, we set $\widetilde{h}_0 = 0$ and $\widetilde{h}_2 = 0$. Thus the latent heat $l_{21} = \widetilde{h}_1 - \widetilde{h}_2 = \widetilde{h}_1$ is a constant also and we may write

$$\overline{J}^n\widetilde{h}_n = l_{21}\overline{J}^1, \qquad J_{\rm t}^n\widetilde{h}_n = l_{21}J_{\rm t}^1 \tag{13.20}$$

Instead of (13.17) we obtain for the budget equation for the internal energy

$$\frac{\partial}{\partial z}\left(\overline{J}^h + J_{\rm t}^h\right) = \frac{\partial}{\partial z}\left[\overline{J}_{\rm s}^h + J_{\rm t}^\theta + J_{\rm t}^\Pi + (\overline{J}^n + J_{\rm t}^n)\widetilde{h}_n\right] = -J_{\rm t}^\theta\frac{g}{c_{p,0}\overline{\overline{T}}} + \overline{\epsilon} \tag{13.21}$$

Owing to (13.6) and (13.20) we may write

$$(\overline{J}^n + J_{\rm t}^n)\widetilde{h}_n = \left(\overline{J}^1 + J_{\rm t}^1\right)l_{21}, \qquad \frac{\partial}{\partial z}\left[(\overline{J}^1 + J_{\rm t}^1)l_{21}\right] = 0 \tag{13.22}$$

According to (13.1e) the heat flux $J_{\rm t}^\theta$ is constant with height so that (13.21) finally reduces to

$$\boxed{\frac{\partial}{\partial z}\left(\overline{J}_{\rm s}^h + J_{\rm t}^\Pi\right) = -J_{\rm t}^\theta\frac{g}{c_{p,0}\overline{\overline{T}}} + \overline{\epsilon}} \tag{13.23}$$

It is customary to introduce the following abbreviating symbols:

$$\begin{aligned} & W = \overline{J}_{\rm s}^h + J_{\rm t}^\Pi \\ \text{Heat flux:}\quad & H = J_{\rm t}^\theta \qquad \text{with} \quad \frac{\partial H}{\partial z} = 0 \\ \text{Moisture flux:}\quad & Q = \overline{J}^1 + J_{\rm t}^1 \qquad \text{with} \quad \frac{\partial Q}{\partial z} = 0 \end{aligned} \tag{13.24}$$

No special name is given to W. Utilizing these definitions the energy budget of the Prandtl layer is summarized as

$$\boxed{\begin{aligned} &\text{Kinetic energy } \frac{\widehat{\mathbf{v}}^2}{2}: && -\frac{\partial}{\partial z}(\tau\widehat{u}) = -\tau\frac{\partial\widehat{u}}{\partial z} \\ &\text{Turbulent kinetic energy } \widehat{k}: && \frac{\partial E}{\partial z} = \tau\frac{\partial\widehat{u}}{\partial z} - \overline{\epsilon} + \frac{gH}{c_{p,0}\overline{\overline{T}}} \\ &\text{Internal energy } \widehat{e}: && \frac{\partial W}{\partial z} = \overline{\epsilon} - \frac{gH}{c_{p,0}\overline{\overline{T}}} \end{aligned}} \tag{13.25}$$

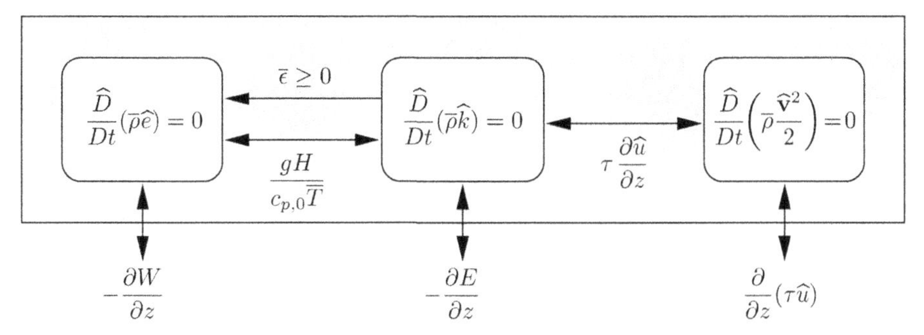

Fig. 13.2 The energy balance within the Prandtl layer.

The budget equation of the kinetic energy of mean motion follows directly from (13.11) where we have shown that $\mathbf{T}$ = constant. As required, the sum of the sources, i.e. the sum of all right-hand-side terms, equals zero. The energy budget can be usefully displayed graphically as shown in Figure 13.2. We think of a spatially fixed unit volume V. The three boxes within V represent changes with time per unit volume of the mean values of the internal energy, the turbulent kinetic energy, and the kinetic energy of the mean motion. The mean change in total energy contained in V is given by the sum of the three boxes. The arrows piercing the outer line marking V denote the interaction with the outside world, i.e. with neighboring boxes. The arrows connecting the energy boxes represent energy transformations within V. With one exception all internal transformations may go in two directions. Only the dissipation of energy is positive definite and can go in one direction only, from the turbulent kinetic energy to the internal energy.

Now we are going to give explicit expressions for H and Q. Scalar multiplication of (11.88a) by $\mathbf{i}_3$ gives the desired relation for H. We assume that Q may be expressed analogously. Together with the corresponding equation (13.12) for τ, we obtain the phenomenological equations of the Prandtl layer as

$$\boxed{\tau = \overline{\rho} K^{v} \frac{\partial \widehat{u}}{\partial z}, \qquad H = -c_{p,0}\overline{\rho} K^{\theta} \frac{\partial \widetilde{\theta}}{\partial z}, \qquad Q = -\overline{\rho} K^{q} \frac{\partial \widehat{q}}{\partial z}} \tag{13.26}$$

where we have suppressed the subscript v in the exchange coefficients since in the Prandtl layer only vertical fluxes exist. Furthermore, we have introduced the variable $\widehat{q} = \widehat{m}^1$ usually denoting the specific humidity in the system of moist air. While H directly follows from (11.88a), the quantity Q is only an approximation, as can be seen from (11.88c) and (11.86).

It is a well-known fact that the turbulent fluxes exceed the corresponding molecular fluxes by several orders of magnitude. Therefore, it is customary to ignore the molecular fluxes since these are always added to the corresponding turbulent fluxes. Ignoring the molecular fluxes, however, is not a necessity. If the molecular fluxes are totally ignored the turbulent fluxes are given by

$$\boxed{\begin{aligned} \mathbf{T} &= \mathbf{i}_3 \cdot \mathbb{R} = -\mathbf{i}_3 \cdot \overline{\rho \mathbf{v}''\mathbf{v}''} = -\overline{\rho w''\mathbf{v}''} \\ H &= \mathbf{i}_3 \cdot \mathbf{J}_{\rm t}^{\theta} = \widetilde{\Pi}\overline{\rho c_{p,0}\theta w''} \\ W &= \mathbf{i}_3 \cdot \mathbf{J}_{\rm t}^{\Pi} = \overline{\rho c_{p,0} w''\theta \Pi'} \approx 0 \\ Q &= \mathbf{i}_3 \cdot \mathbf{J}_{\rm t}^{1} = \overline{\rho w'' q} \end{aligned}} \tag{13.27}$$

as follows from (11.36), (11.38), and (11.65).

In boundary-layer theory it is customary to introduce the scaling variables u_*, T_*, and q_* by means of the defining equations

$$\boxed{\tau = \overline{\rho} u_*^2, \qquad H = -\overline{\rho} c_{p,0} u_* T_*, \qquad Q = -\overline{\rho} u_* q_*} \tag{13.28}$$

The quantity u_* has the dimension of a velocity and is therefore known as the *frictional velocity*. The variable T_* has the dimension of temperature while q_* is dimensionless. If these scaling variables are known then τ, H, and Q may be determined since the density of the Prandtl layer is considered known also.

13.2.2 The Richardson number

Let us rewrite the complete budget equation for the turbulent kinetic energy (11.46) without using the specific Prandtl-layer conditions. Substituting from (11.70) for $\overline{\mathbf{v}'' \cdot \nabla p}$ and ignoring the term involving the pressure fluctuations, we find

$$\frac{\widehat{D}}{Dt}(\overline{\rho}\widehat{k}) + \nabla \cdot (\mathbf{k}_{\rm t} - \overline{\mathbf{v}'' \cdot \mathbb{J}}) = \nabla\widehat{\mathbf{v}} \cdot\cdot (\overline{\mathbb{J}} + \mathbb{R}) - \overline{\epsilon} - \mathbf{J}_{\rm t}^{\theta} \cdot \nabla \ln \widetilde{\Pi} \tag{13.29}$$

The right-hand side then represents the source for the turbulent kinetic energy, which can be written in the form

$$Q_{\widehat{k}} = \nabla\widehat{\mathbf{v}} \cdot\cdot (\overline{\mathbb{J}} + \mathbb{R}) \left(1 - \frac{\mathbf{J}_{\rm t}^{\theta} \cdot \nabla \ln \widetilde{\Pi}}{\nabla\widehat{\mathbf{v}} \cdot\cdot (\overline{\mathbb{J}} + \mathbb{R})} \right) - \overline{\epsilon} \tag{13.30}$$

Without proof we accept that the frictional term $\nabla\widehat{\mathbf{v}} \cdot\cdot (\overline{\mathbb{J}} + \mathbb{R})$ is positive definite so that it permanently produces kinetic energy. While the energy dissipation $\overline{\epsilon} \geq 0$ permanently transforms $\widehat{k}$ into internal energy, see Figure 13.2, the third term on the right-hand side of (13.29) can have either sign, thus producing or destroying turbulent kinetic energy. We will investigate the situation more closely. The fraction on the right-hand side of (13.30) is called the *flux Richardson number*:

$$\boxed{Ri_{\rm f} = \frac{\mathbf{J}_{\rm t}^{\theta} \cdot \nabla \ln \widetilde{\Pi}}{\nabla\widehat{\mathbf{v}} \cdot\cdot (\overline{\mathbb{J}} + \mathbb{R})} = \begin{cases} <1 & \text{production of } \widehat{k} \\ >1 & \text{destruction of } \widehat{k} \end{cases}} \tag{13.31}$$

Ignoring for the moment the contribution of $\overline{\epsilon}$ in (13.30), we see that turbulent kinetic energy is produced if $Ri_{\mathrm{f}} < 1$; otherwise dissipation takes place.

Let us now employ the Prandtl-layer assumptions. According to (13.12) and (13.14) the denominator of (13.31) reduces to

$$\nabla\widehat{\mathbf{v}}\cdot\cdot(\overline{\mathbb{J}} + \mathbb{R}) = \tau\,\frac{\partial\widehat{u}}{\partial z} \tag{13.32}$$

Observing that the exchange coefficient is a positive quantity, we easily recognize with the help of (13.12) that the denominator of the flux Richardson number cannot be negative. From (13.15) we obtain for the numerator of (13.31)

$$\mathbf{J}_{\mathrm{t}}^{\theta}\cdot\nabla\ln\widetilde{\Pi} = -\frac{gH}{c_{p,0}\overline{\overline{T}}} \tag{13.33}$$

Therefore, applying (13.26) and (13.28), the flux Richardson number may be expressed as

$$Ri_{\mathrm{f}} = -\frac{g}{\overline{\overline{T}}}\frac{H}{c_{p,0}\tau\,\dfrac{\partial\widehat{u}}{\partial z}} = \frac{g}{\overline{\overline{T}}}\frac{T_*}{u_*\,\dfrac{\partial\widehat{u}}{\partial z}} = \frac{g}{\overline{\overline{T}}}\frac{K^{\theta}}{K^{v}}\frac{\dfrac{\partial\widetilde{\theta}}{\partial z}}{\left(\dfrac{\partial\widehat{u}}{\partial z}\right)^2} \tag{13.34}$$

In the Prandtl layer $\overline{\overline{T}}$ may also be replaced by $\widetilde{\theta}$. It is customary to introduce the *turbulent Prandtl number* as the ratio of the two exchange coefficients:

$$\boxed{Pr = \frac{K^{v}}{K^{\theta}}} \tag{13.35}$$

so that the flux Richardson number can also be written as

$$\boxed{Ri_{\mathrm{f}} = \frac{Ri}{Pr} \quad \text{with} \quad Ri = \frac{g}{\overline{\overline{T}}}\frac{\dfrac{\partial\widetilde{\theta}}{\partial z}}{\left(\dfrac{\partial\widehat{u}}{\partial z}\right)^2}} \tag{13.36}$$

The variable Ri is known as the *gradient Richardson number* or simply the *Richardson number*, which was introduced in the previous chapter, see equation (12.77). Inspection of (13.34) clearly indicates that the numerator is a measure for the production or suppression of *buoyancy energy*. The denominator represents the part of the turbulence that is mechanically induced by shear forces of the basic air current.

For statically unstable air the gradient of the potential temperature is negative, so $Ri_{\mathrm{f}} < 0$. This causes an increase of turbulence. In a statically stable atmosphere the gradient of the potential temperature is positive so that $Ri_{\mathrm{f}} > 0$. If $\partial\widetilde{\theta}/\partial z$ is large enough this results in a sizable reduction of turbulent kinetic energy. If the flux Richardson number becomes larger than one, the first term on the right-hand side of (13.30) changes sign. The flux Richardson number $Ri_{\mathrm{f}} = 1$ is considered to be an upper limit of the so-called *critical flux Richardson number* $Ri_{\mathrm{f,c}}$. Some modern theoretical work on the basis of detailed observations indicates that the critical flux Richardson number should be close to 0.25. However, this value is still uncertain. For flux Richardson numbers larger than this value the transition from turbulent to laminar flow takes place. The flux Richardson number will be fundamental to the discussion of stability in the next few sections.

13.3 The Monin–Obukhov similarity theory of the neutral Prandtl layer

Before deriving various relationships for the non-neutral or diabatic Prandtl layer we will summarize some conditions and relations pertaining to neutral stratification. The relationships for the neutral Prandtl layer will then be generalized to accommodate arbitrary stratifications. In case that the Prandtl layer is characterized by static neutral stratification we may write

$$\frac{\partial\widetilde{\theta}}{\partial z} = 0, \qquad T_* = 0, \qquad J_{\mathrm{t}}^{\theta} = H = 0 \Longrightarrow Ri_{\mathrm{f}} = Ri = 0 \tag{13.37}$$

As stated in equations (13.12) and (13.24) the stress τ, the heat flux H, and the moisture flux Q do not change with height within the Prandtl layer. For reasons of symmetry we will also assume that the quantity E as defined in (13.13) will be constant with height so that we have

$$\frac{\partial E}{\partial z} = 0, \qquad \frac{\partial H}{\partial z} = 0, \qquad \frac{\partial Q}{\partial z} = 0, \qquad \frac{\partial \tau}{\partial z} = 0 \tag{13.38}$$

Using the conditions stated in (13.37) and (13.38), the budget equations for the turbulent kinetic and internal energy (13.25) reduce to

$$\tau\,\frac{\partial\widehat{u}}{\partial z} = \overline{\epsilon}, \qquad \frac{\partial W}{\partial z} = \overline{\epsilon} \Longrightarrow \frac{\partial W}{\partial z} = \tau\,\frac{\partial\widehat{u}}{\partial z} = \overline{\rho}u_*^2\,\frac{\partial\widehat{u}}{\partial z} \tag{13.39}$$

On multiplying this equation by the factor $z/(\overline{\rho}u_*^3)$ we obtain

$$\frac{z}{\overline{\rho}u_*^3}\frac{\partial W}{\partial z} = \frac{z\overline{\epsilon}}{\overline{\rho}u_*^3} = \frac{z}{u_*}\frac{\partial \widehat{u}}{\partial z} = z\frac{\partial}{\partial z}\left(\frac{\widehat{u}}{u_*}\right) \tag{13.40}$$

This equation will be the starting point for some of the analysis to follow.

The reader unfamiliar with the fundamental ideas of similarity theory is invited to consult Appendix A to this chapter. There the Buckingham Π theorem is introduced, where Π is a dimensionless number. This theory is a useful tool in many branches of science. The Monin–Obukhov (MO) similarity hypothesis (Monin and Obukhov, 1954) consists of two essential parts.

(i) In the neutral Prandtl layer there exists a unique relation among the height z, the frictional velocity u_*, and the vertical velocity gradient $\partial\widehat{u}/\partial z$:

$$F_1\left(z, u_*, \frac{\partial \widehat{u}}{\partial z}\right) = 0 \tag{13.41}$$

From dimensional analysis it can be shown that there exists a dimensionless number Π so that we have

$$\Pi = \frac{z}{u_*}\frac{\partial \widehat{u}}{\partial z} = \frac{1}{k} \quad \text{or} \quad \frac{kz}{u_*}\frac{\partial \widehat{u}}{\partial z} = 1 \tag{13.42}$$

The universal constant $k = 0.4$ is the *Von Karman constant*. Details are given in Appendix A, Example 1.

(ii) In analogy to (13.41) we assume that corresponding statements are true for the temperature distribution:

$$F_2\left(z, T_*, \frac{\partial \widetilde{\theta}}{\partial z}\right) = 0, \qquad \frac{kz}{T_*}\frac{\partial \widetilde{\theta}}{\partial z} = 1 \tag{13.43}$$

and for the moisture distribution:

$$F_3\left(z, q_*, \frac{\partial \widehat{q}}{\partial z}\right) = 0, \qquad \frac{kz}{q_*}\frac{\partial \widehat{q}}{\partial z} = 1 \tag{13.44}$$

Equation (13.43) must be viewed as a limit statement since both T_* and $\partial\widetilde{\theta}/\partial z$ are zero in the neutral Prandtl layer. Equations (13.43) and (13.44) will not be needed at present but will be used later.

A number of interesting and important conclusions may be drawn from the MO similarity theory.

13.3.1 The functions $\widehat{u}$, $\overline{\epsilon}$, and W

From (13.40) and (13.42) it follows immediately that

$$\frac{kz}{\overline{\rho}u_*^3}\frac{\partial W}{\partial z} = \frac{kz}{u_*}\frac{\partial \widehat{u}}{\partial z} = kz\frac{\partial}{\partial z}\left(\frac{\widehat{u}}{u_*}\right) = 1 \tag{13.45}$$

Integration of the last equation of (13.45) gives the well-known *logarithmic wind profile*:

$$\boxed{\widehat{u}(z) = \frac{u_*}{k}\ln\left(\frac{z}{z_0}\right), \qquad \widehat{u}(z = z_0) = 0, \qquad z \geq z_0} \tag{13.46}$$

At the roughness height z_0 the wind speed vanishes. There exist extensive tables in which values of the estimated roughness height are given. For example, for short grass z_0 is in the range 0.01–0.04 m, whereas for long grass $z_0 = 0.10$ m might be an acceptable value.

Similarly we find expressions for W:

$$W(z) - W(z_0) = \frac{\overline{\rho}u_*^3}{k}\ln\left(\frac{z}{z_0}\right), \qquad z \geq z_0 \tag{13.47}$$

so that the dissipation of energy $\overline{\epsilon}$ from (13.39) is given by

$$\overline{\epsilon}(z) = \frac{\overline{\rho}u_*^3}{kz}, \qquad z \geq z_0 \tag{13.48}$$

We need to point out that the Prandtl-layer theory does not permit the evaluation of $W(z_0)$.

13.3.2 The phenomenological coefficient K^v

We are now ready to obtain an expression for the exchange coefficient K^v for the neutral Prandtl layer. From (13.26) and (13.28) we obtain

$$K^v = \frac{u_*^2}{\partial\widehat{u}/\partial z}, \qquad z \geq z_0 \tag{13.49}$$

which is valid for the diabatic and for the neutral Prandtl layer. Utilizing (13.45) we immediately find for the neutral Prandtl layer

$$\boxed{K^v = kzu_*, \qquad z \geq z_0} \tag{13.50}$$

showing that, within the neutral Prandtl layer, the exchange coefficient increases linearly with height. This is the assumption we have made in Section 12.6.

13.3.3 The characteristic length or the mixing length

The ratio formed by the frictional velocity and the vertical gradient of the horizontal velocity has the dimension of a length. We use this ratio to define the *mixing length* l for neutral conditions:

$$\boxed{l = \frac{u_*}{\partial \widehat{u}/\partial z} \quad \Longrightarrow \quad l = kz} \tag{13.51}$$

The last expression follows immediately from (13.45). Later we will assume that the first expression is also valid in the diabatic Prandtl layer. The physical meaning of the mixing length will be briefly illuminated in Appendix B of this chapter. From (13.28) and (13.51) we find

$$\tau = \overline{\rho} u_*^2 = \overline{\rho} l^2 \left(\frac{\partial \widehat{u}}{\partial z} \right)^2 \tag{13.52}$$

Using the definition of the mixing length, equations (13.48) and (13.50) can be rewritten as

$$\overline{\epsilon} = \frac{\overline{\rho} u_*^3}{l}, \qquad u_* = \frac{K^v}{l} \quad \Longrightarrow \quad \overline{\epsilon} = \frac{\overline{\rho}\,(K^v)^3}{l^4} \tag{13.53}$$

showing the relation among the dissipation of energy, the exchange coefficient, and the mixing length. Finally, we can rewrite (13.53) to obtain

$$\boxed{(K^v)^2 = \frac{\overline{\epsilon}}{\overline{\rho}} \frac{l^4}{K^v}} \tag{13.54}$$

In (12.59b) we have derived the Heisenberg relation for the exchange coefficient from spectral analysis using the conditions of isotropic turbulence. This relation may be written as

$$(K^v)^2 = \overline{\epsilon}_{\mathrm{M}} \frac{b^3 l^4}{a K^v} = \overline{\epsilon}_{\mathrm{M}} \frac{(l')^4}{K^v} \tag{13.55}$$

For ease of comparison we have introduced the mixing length $(l')^2 = l^2 \sqrt{b^3/a}$ into this equation. Apart from a constant that includes the density, these two formulations obtained in entirely different ways are identical. Therefore, equation (13.54) is also known as *Heisenberg's relation*. This leads us to assume that (13.54) is valid also for non-neutral conditions.

Finally, we will derive another formula for the mixing length, which is attributed to Von Karman. We proceed as follows. We differentiate (13.45) with respect to z and recall that the frictional velocity is a constant. This yields

$$\frac{\partial^2 \widehat{u}}{\partial z^2} = -\frac{k}{l} \frac{\partial \widehat{u}}{\partial z} \tag{13.56}$$

From (13.45), (13.50), and (13.54) we obtain

$$\frac{\partial \widehat{u}}{\partial z} = \frac{K^v}{l^2} = \sqrt{\frac{\overline{\epsilon}}{\overline{\rho} K^v}} \tag{13.57}$$

The latter two equations lead directly to the *Von Karman relation* of the mixing length:

$$l = -k \frac{\dfrac{\partial \widehat{u}}{\partial z}}{\dfrac{\partial^2 \widehat{u}}{\partial z^2}} = -\frac{k}{\dfrac{\partial}{\partial z}\left[\ln\left(\dfrac{\partial \widehat{u}}{\partial z}\right)\right]} = -\frac{2k}{\dfrac{\partial}{\partial z}\left[\ln\left(\dfrac{\overline{\epsilon}}{\overline{\rho} K^v}\right)\right]} \tag{13.58}$$

We should realize that the two definitions of the mixing length (13.51) and (13.58) are not independent of each other. If we try to extrapolate the mixing-length formulas of the neutral Prandtl layer to the non-neutral or diabatic Prandtl layer we must select one of these formulations.

13.4 The Monin–Obukhov similarity theory of the diabatic Prandtl layer

We will now consider the diabatic or non-neutral Prandtl layer. The starting point of the analysis is the general Prandtl-layer energy budget. Addition of the last two equations of (13.25) and utilizing (13.51) and (13.52) results in

$$\frac{\partial E}{\partial z} + \frac{\partial W}{\partial z} = \frac{\partial E}{\partial z} + \overline{\epsilon} + \frac{\overline{\rho} g u_* T_*}{\overline{T}} = \overline{\rho} u_*^2 \frac{\partial \widehat{u}}{\partial z} \tag{13.59}$$

where we have expressed the heat flux according to (13.28). We consider not only the density to be independent of height but also the average temperature $\overline{T}$ whenever it appears in undifferentiated form. We multiply all terms of this equation by $kz/(\overline{\rho} u_*^3)$ and find the relations

$$\frac{kz}{\overline{\rho} u_*^3}\frac{\partial E}{\partial z} + \frac{kz}{\overline{\rho} u_*^3}\frac{\partial W}{\partial z} = \frac{kz}{\overline{\rho} u_*^3}\frac{\partial E}{\partial z} + \frac{kz\overline{\epsilon}}{\overline{\rho} u_*^3} + \frac{kzgT_*}{u_*^2 \overline{T}} = \frac{kz}{u_*}\frac{\partial \widehat{u}}{\partial z} \tag{13.60}$$

Each term of this equation now is dimensionless, which is most easily verified by inspecting the last term on the right-hand side. The last term on the left-hand side can be written as z/L_*, where L_* has the dimension of a length with

$$\boxed{L_* = \frac{u_*^2 \overline{T}}{gkT_*} = -\frac{\overline{\rho} c_{p,0} u_*^3 \overline{T}}{gkH}} \tag{13.61}$$

This length is known as the *Monin–Obukhov length* and is of special significance in the work to follow. Within the Prandtl layer L_* is a height-independent quantity. Inserting L_* into (13.60) gives

$$\frac{kz}{\overline{\rho}u_*^3}\frac{\partial E}{\partial z}+\frac{kz}{\overline{\rho}u_*^3}\frac{\partial W}{\partial z}=\frac{kz}{\overline{\rho}u_*^3}\frac{\partial E}{\partial z}+\frac{kz\overline{\epsilon}}{\overline{\rho}u_*^3}+\frac{z}{L_*}=\frac{kz}{u_*}\frac{\partial\widehat{u}}{\partial z} \tag{13.62}$$

We introduce the dimensionless vertical coordinate

$$\xi = z/L_* \tag{13.63}$$

Any function S depending solely on ξ and possibly on universal numbers is called a *Monin–Obukhov function*:

$$\boxed{S(\xi)=S\left(\frac{z}{L_*}\right)} \tag{13.64}$$

The basic idea of the similarity theory of the diabatic Prandtl layer is that the relationship (13.41) pertaining to the neutral atmosphere may be extended to the non-neutral atmosphere by including the variable L_* in the wind-profile analysis. In other words, it is assumed that there exists a unique relationship among the variables $z,\ u_*,\ \partial\widehat{u}/\partial z$, and L_*. The functional relation describing this is expressed by

$$F\left(z, u_*, \frac{\partial\widehat{u}}{\partial z}, L_*\right)=0 \tag{13.65}$$

Using dimensionless analysis, we show in Appendix A, Example 2, that the last expression of (13.62) is a MO function. Furthermore, the MO theory also assumes that not only the last term of (13.62) but also every other term occurring in this equation is a MO function so that we may write

$$\boxed{\begin{aligned} &\frac{kz}{u_*}\frac{\partial\widehat{u}}{\partial z}=S_u(\xi), &&\frac{kz}{\overline{\rho}u_*^3}\frac{\partial E}{\partial z}=S_E(\xi)\\ &\frac{kz}{\overline{\rho}u_*^3}\frac{\partial W}{\partial z}=S_W(\xi), &&\frac{kz\overline{\epsilon}}{\overline{\rho}u_*^3}=S_\epsilon(\xi)\end{aligned}} \tag{13.66}$$

By integrating the first of these equations we obtain the vertical wind profile of the diabatic Prandtl layer. This topic will be discussed later. Using these definitions equation (13.62) can now be rewritten in the form of two independent equations:

$$S_E(\xi)+S_W(\xi)=S_E(\xi)+S_\epsilon(\xi)+\xi=S_u(\xi) \tag{13.67}$$

In order to obtain information on the temperature profile and on the specific humidity profile we additionally assume relations anlogous to (13.66):

Table 13.1. *Sign conventions for various quantities as functions of the atmospheric stratification*

Stratification	$\partial\widetilde{\theta}/\partial z$	H	$T_*,\ \xi,\ Ri_f$	L_*
Stable	>0	<0	>0	>0
Neutral	0	0	0	$\pm\infty$
Unstable	<0	>0	<0	<0

$$\frac{kz}{T_*}\frac{\partial\widetilde{\theta}}{\partial z} = S_T(\xi), \qquad \frac{kz}{q_*}\frac{\partial\widehat{q}}{\partial z} = S_q(\xi) \tag{13.68}$$

We close this section by summarizing in Table 13.1 the sign conventions for various important quantities. It will be seen, for example, that, for stable and unstable stratifications, the heat flux H and the MO length L_* have opposite signs. For neutral conditions the heat flux is zero and L_* is $\pm\infty$. An important reference is Obukhov (1971), who reviews the basic concepts of the similarity theory.

13.4.1 The determination of the Monin–Obukhov functions

We recall that there exist only two independent equations (13.67) to determine the four MO functions $S_E(\xi),\ S_\epsilon(\xi),\ S_u(\xi)$, and $S_W(\xi)$. Therefore, it will be necessary to make additional assumptions. Since L_* approaches $\pm\infty$ for neutral conditions, the dimensionless height ξ approaches zero and the MO functions (with the exception of $S_E(\xi)$) must be equal to unity so that they reduce to the corresponding relations of the neutral Prandtl layer. In summary, we have for $\xi = 0$

$$\begin{aligned}
\frac{kz}{u_*}\frac{\partial\widehat{u}}{\partial z} &= 1 \implies S_u(\xi = 0) = 1\\
\frac{kz}{\overline{\rho}u_*^3}\frac{\partial W}{\partial z} &= 1 \implies S_W(\xi = 0) = 1\\
\frac{kz\overline{\epsilon}}{\overline{\rho}u_*^3} &= 1 \implies S_\epsilon(\xi = 0) = 1\\
\frac{kz}{\overline{\rho}u_*^3}\frac{\partial E}{\partial z} &= 0 \implies S_E(\xi = 0) = 0\\
\frac{kz}{T_*}\frac{\partial\widetilde{\theta}}{\partial z} &= 1 \implies S_T(\xi = 0) = 1\\
\frac{kz}{q_*}\frac{\partial\widehat{q}}{\partial z} &= 1 \implies S_q(\xi = 0) = 1
\end{aligned} \tag{13.69}$$

Next we will introduce the definition of the MO length L_* into the flux Richardson number (13.34). This results in

$$Ri_{\rm f} = \frac{\xi}{\dfrac{kz}{u_*}\dfrac{\partial \widehat{u}}{\partial z}} = \frac{\xi}{S_u(\xi)} \tag{13.70}$$

showing that the MO function $S_u(\xi)$ for the velocity may be expressed in terms of the dimensionless number $Ri_{\rm f}$.

On the basis of observational data various empirical relations have been proposed for the various MO functions. For $S_u(\xi)$ we use

$$S_u(\xi) = (1 - 2\alpha Ri_{\rm f})^{-1/4} \approx 1 + \frac{\alpha}{2} Ri_{\rm f} \approx 1 + \frac{\alpha}{2}\xi \quad \text{with} \quad \alpha = 7 \tag{13.71}$$

which is due to Ellison (1957). Utilizing only the first two terms of the expansion of the fourth root we obtain an expression attributed to Monin and Obukhov (1954). This approximation will now be used to generalize the logarithmic wind law (13.46). From (13.66) we find immediately

$$\frac{kz}{u_*}\frac{\partial \widehat{u}}{\partial z} = S_u(\xi) = 1 + \frac{\alpha}{2}\xi \tag{13.72}$$

Integration with respect to height gives

$$\boxed{\widehat{u} = \frac{u_*}{k}\left[\ln\left(\frac{z}{z_0}\right) + \frac{\alpha(z - z_0)}{2L_*}\right]} \tag{13.73}$$

which is known as the *log–linear wind profile*. Since the expansion of the MO function $S_u(\xi)$ was discontinued after the linear term, we may expect reasonable accuracy in the reproduction of the wind field only for conditions not too far removed from neutral stratification. Figure 13.3 shows qualitatively the vertical wind profile in the Prandtl layer for stable, neutral, and unstable stratification. For unstable stratification with $L_* < 0$ the wind speed for a given height is less than that for neutral conditions. For stable stratification with $L_* > 0$ the wind speed for a given height is larger than that for neutral conditions.

13.4.2 The KEYPS and the extended KEYPS formula

There exists a semi-empirical equation for the determination of S_u, which is known as the KEYPS formula. This curious name is the contraction of the names of the

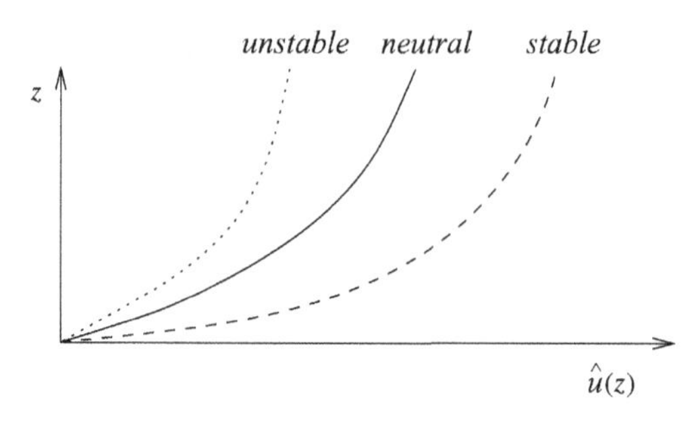

Fig. 13.3 Logarithmic and log–linear wind profiles for different stratifications.

five scientists Kondo, Ellison, Yamamoto, Panowski, and Sellers who derived this formula independently. Since this equation is used frequently for Prandtl-layer investigations it will be derived next. All that needs to be done is to take the fourth power of (13.71), yielding

$$\boxed{S_u^4 - 2\alpha\xi S_u^3 = 1} \tag{13.74}$$

Solutions to this equation give information on the MO function S_u. Some remarks on this formula may be helpful. There should be agreement between the KEYPS formula and the more general equation (13.67). On multiplying this equation by S_u^3 we obtain

$$S_u^4 - \xi S_u^3\left(1 + \frac{S_E}{\xi}\right) = S_\epsilon S_u^3 \tag{13.75}$$

Agreement with the KEYPS formula requires the validity of

$$S_\epsilon S_u^3 = 1, \qquad 1 + \frac{S_E}{\xi} = 2\alpha \tag{13.76}$$

as follows from comparison with (13.74). Replacing both MO functions in the first equation of (13.76) with the help of (13.66) yields

$$\frac{kz\overline{\epsilon}}{\overline{\rho}u_*^3} = \left(\frac{u_*}{kz\dfrac{\partial\widehat{u}}{\partial z}}\right)^3 = \left(\frac{K^v}{kzu_*}\right)^3 \tag{13.77}$$

where (13.49) has also been used to eliminate $\partial\widehat{u}/\partial z$. We now compare (13.77) with (13.54), which is considered a valid statement since it can also be derived from spectral theory without using Prandtl-layer assumptions. We find agreement only if the mixing length in the diabatic Prandtl layer is also given by $l = kz$. This result, however, is in conflict with equation (13.51) due to (13.66). Therefore, some criticism of the KEYPS equation or of Ellison's formula (13.71) is justified.

Finally, let us consider (13.67) showing the expected relation between $S_E(\xi)$ and the other MO functions. Measurements indicate that, at least in the unstable Prandtl layer, the MO function S_E does not linearly depend on ξ as required by (13.76).

In the following we will attempt to eliminate the defect in the KEYPS equation. The starting point of the analysis is (13.53). A relationship of this form was also derived from spectral analysis (see also (13.55)) for the special case of isotropic turbulence but independent of Prandtl-layer assumptions. Therefore, we have reason to believe that (13.53) can be generalized. On combining this equation with (13.49), which is valid for the diabatic as well as for the neutral Prandtl layer, we obtain

$$\frac{\overline{\epsilon}}{\overline{\rho}} = \frac{u_*^6}{l^4\left(\dfrac{\partial \widehat{u}}{\partial z}\right)^3} \Longrightarrow \frac{kz\overline{\epsilon}}{\overline{\rho}u_*^3}\left(\frac{l}{kz}\right)^4 = \left[\frac{kz}{u_*}\left(\frac{\partial \widehat{u}}{\partial z}\right)\right]^{-3} \tag{13.78}$$

In this equation the characteristic or mixing length l is still undetermined. In fact, all we know about the mixing length is that $l = kz$ in the neutral Prandtl layer. By introducing into (13.78) the MO relations (13.66) we obtain

$$S_\epsilon\left(\frac{l}{kz}\right)^4 = \frac{1}{S_u^3} \quad \text{or} \quad \left(\frac{l}{kz}\right)^4 = \frac{1}{S_\epsilon S_u^3} \tag{13.79}$$

Since $S_\epsilon(\xi)$ and $S_u(\xi)$ depend only on the dimensionless height, we consider the expression $l/(kz)$ as the definition of another MO function, that is

$$S_l(\xi) = \frac{l}{kz}, \qquad S_l(\xi = 0) = 1 \tag{13.80}$$

We also require that $S_l(\xi = 0) = 1$ since in the neutral case $l/(kz) = 1$. Thus, we have introduced an additional MO function that is related to S_ϵ and S_u. From (13.79) we find

$$S_\epsilon S_u^3 S_l^4 = 1 \tag{13.81}$$

which should also be valid in the diabatic Prandtl layer. This relation makes it possible to eliminate $S_\epsilon(\xi)$ from the budget equation (13.67). Substitution of $S_\epsilon(\xi)$ then gives

$$S_E + S_\epsilon + \xi = S_E + \frac{1}{S_l^4 S_u^3} + \xi = S_u \tag{13.82}$$

Rewriting the latter expression results in the extended KEYPS equation:

$$\boxed{S_u^4 - \xi S_u^3\left(1 + \frac{S_E}{\xi}\right) = \frac{1}{S_l^4}} \tag{13.83}$$

This equation is considered to be rigorously correct within the framework of the Prandtl-layer theory. Unless we are able to determine S_l as a function of the dimensionless height we have not gained anything. There exists an empirical relation for S_l due to Takeuchi and Yokoyama (1963), which reduces to $S_l(\xi = 0) = 1$ as it should. However, in some parts of the unstable Prandtl layer problems arise with that empirical formula. In the following we are not going to discuss this formula, but we will try to find an analytic relation between S_l and $S_u(\xi)$.

13.4.3 An analytic relation between the MO Functions S_l and S_u

In order to find an analytic relation between S_l and S_u we must first derive a relation for the mixing length that is valid also in the diabatic atmosphere. We may select either the Prandtl relation (13.51) or the Von Karman relation (13.58). These two equations are not independent in the neutral Prandtl layer. First we use the Prandtl relation. Dividing (13.51) by kz and utilizing (13.66) and (13.80) gives

$$\frac{l}{kz} = \frac{1}{\dfrac{kz}{u_*}\dfrac{\partial \widehat{u}}{\partial z}} = \frac{1}{S_u} \Longrightarrow l = \frac{kz}{S_u}, \qquad S_l = \frac{1}{S_u} \tag{13.84}$$

It is more difficult but also straightforward to apply the Von Karman relation (13.58). On rewriting the logarithm in (13.58) by means of (13.49), (13.66), and (13.81) as

$$\frac{\overline{\epsilon}}{\overline{\rho}K^v} = \frac{S_\epsilon S_u u_*^2}{(kz)^2} = \frac{u_*^2}{\left(kzS_l^2 S_u\right)^2} \tag{13.85}$$

we obtain from the Von Karman relation (13.58)

$$l = -\frac{2k}{\dfrac{\partial}{\partial z}\left[\ln\!\left(\dfrac{\overline{\epsilon}}{\overline{\rho}K^v}\right)\right]} = \frac{k}{\dfrac{\partial}{\partial z}\left[\ln\left(zS_l^2 S_u\right)\right]} \Longrightarrow$$

$$S_l = \frac{l}{kz} = \frac{1}{z\dfrac{\partial}{\partial z}\left[\ln\left(zS_l^2 S_u\right)\right]} = \frac{1}{\xi\dfrac{d}{d\xi}\left[\ln\left(\xi S_l^2 S_u\right)\right]} = \frac{1}{1 + \dfrac{2\xi}{S_l}\dfrac{dS_l}{d\xi} + \dfrac{\xi}{S_u}\dfrac{dS_u}{d\xi}} \tag{13.86}$$

since $\xi = z/L_*$ and $L_* =$ constant. The last equation may be rewritten as

$$2\xi\frac{dS_l}{d\xi} + S_l\left(1 + \frac{\xi}{S_u}\frac{dS_u}{d\xi}\right) = 1 \tag{13.87}$$

This ordinary linear differential equation can be solved without difficulty by standard methods. The solution is

$$S_l(\xi) = \frac{1}{2\sqrt{\xi S_u(\xi)}}\int_0^\xi \sqrt{\frac{S_u(\xi')}{\xi'}}\,d\xi' \tag{13.88}$$

We will now summarize how we find the various MO functions. Suppose that the function S_u is known. An empirical formula for S_u will be given soon. Then S_l can be found either from (13.84) or from (13.88). S_ϵ then follows from (13.81) and S_E and S_W from (13.67). The MO functions S_T and S_q (13.68) cannot be found in this manner, but must be obtained in some other way.

From numerous measurements various empirical formulas have been derived for the MO functions. Frequently the so-called *Dyer–Businger equations* are used to state S_u for stable, neutral, and unstable stratification. These are

$$\begin{aligned} S_u &= 1 + 5\xi & \xi &\geq 0, & &\text{stable stratification} \\ S_u &= (1 - 15\xi)^{-1/4} & \xi &\leq 0, & &\text{unstable stratification} \end{aligned} \tag{13.89}$$

The MO function S_T for the transport of sensible heat is usually given in the form

$$\begin{aligned} S_T &= (1 + 5\xi) & \xi &\geq 0, & &\text{stable stratification} \\ S_T &= (1 - 15\xi)^{-1/2} & \xi &\leq 0, & &\text{unstable stratification} \end{aligned} \tag{13.90}$$

Moreover, often it is assumed that $S_q = S_T$.

A search of the literature shows that not all authors use identical empirical equations for the MO functions but mostly something similar to them. Let us now turn to the defining equation (13.35) of the Prandtl number. Measurements show that, for neutral conditions, this number should be close to 1.35. Sometimes the right-hand sides of (13.90) are multiplied by the factor $1/1.35 = 0.74$. This, however, would violate the requirement $S_T(\xi = 0) = 1$.

13.5 Application of the Prandtl-layer theory in numerical prognostic models

A brief outline of how the Prandtl-layer theory can be applied to numerical weather prediction and mesoscale analysis will be given. The lowest surface of the numerical grid within the atmosphere is selected to coincide with the roughness height z_0, which is assumed to be known. A neighboring surface is fixed somewhere within the Prandtl layer, for example, at the height h. We assume that the numerical model is capable of calculating mean values of the horizontal velocity, the temperature (or potential temperature), the density, the pressure, and the specific humidity at these two surfaces. These mean variables are called the external parameters whereas L_* is known as an internal parameter. By the methods of the previous section we are in a position to calculate all required MO functions. Let us assume that all of these are at our disposal when needed. On integrating the relations (13.66) and (13.68) between the roughness height z_0 and h we find

$$\begin{aligned}
\widehat{u}(h) &= \frac{u_*}{k}\int_{z_0}^{h} S_u(\xi)\frac{dz}{z}, \qquad \widehat{u}(z_0) = 0 \\
\widetilde{\theta}(h) - \widetilde{\theta}(z_0) &= \frac{T_*}{k}\int_{z_0}^{h} S_T(\xi)\frac{dz}{z} \quad \text{with} \quad \xi = \frac{z}{L_*}
\end{aligned} \tag{13.91}$$

The left-hand sides are the external parameters which are considered known. The scaling parameters, the undifferentiated temperature, and the density are treated as constants in the height integrations within the Prandtl layer. With (13.61) and (13.91) we have three equations for the determination of u_*, T_*, and L_*. The solution of this system is best carried out by an iterative procedure. Momentarily let us assume that the MO functions S_u and S_T can be expressed as linear functions of ξ, which is the case for the stable Prandtl layer as expressed by (13.89) and (13.90). For the unstable case the expansion of the MO functions would have to be discontinued after the linear term. Therefore, we may write for the two MO functions the following two expressions:

$$S_u \approx 1 + \alpha_u \xi, \qquad S_T \approx 1 + \alpha_T \xi \tag{13.92}$$

The quantities α_u and α_T are the expansion coefficients.

By substituting (13.92) into (13.91) we may carry out the integration. The result is

$$\boxed{\begin{aligned}
\widehat{u}(h) &= \frac{u_*^{(n)}}{k}\left[\ln\left(\frac{h}{z_0}\right) + \frac{\alpha_u(h - z_0)}{L_*^{(n-1)}}\right] \\
\widetilde{\theta}(h) - \widetilde{\theta}(z_0) &= \frac{T_*^{(n)}}{k}\left[\ln\left(\frac{h}{z_0}\right) + \frac{\alpha_T(h - z_0)}{L_*^{(n-1)}}\right] \\
L_*^{(n-1)} &= \frac{\overline{T}}{gk}\left(\frac{u_*^2}{T_*}\right)^{n-1}, \qquad L_*^{(0)} = \infty
\end{aligned}} \tag{13.93}$$

The first equation of (13.93) was already stated earlier in (13.73) and describes the log–linear wind profile. Since the system (13.93) must be solved iteratively for u_* and T_*, we have added the iteration index as the superscript n. With the exception of u_* and T_* all remaining quantities are known. The heights z_0 and h are specified. The scaling height L_* is evaluated at the previous time step. To get the iteration started, we assume that we have neutral conditions so that $L_*^{(0)} = \infty$. Equations (13.93) can be easily solved for $u_*^{(n)}$ and $T_*^{(n)}$. Let us now assume that u_*, T_*, and L_* have been found by iterating sufficiently many times that u_* and T_* no longer change within a prescribed tolerance. This is sufficient to calculate the stress τ and the heat flux H from equation (13.28).

If the linear expansions of the MO functions S_u and S_T are considered insufficient, for unstable stratification the full expressions must be used. Equation (13.91) can still be solved by the same iterative procedure, but the integrals must be solved numerically. The remaining undetermined quantities will be discussed in the next section.

13.6 The fluxes, the dissipation of energy, and the exchange coefficients

The MO function S_W can be computed as described in Section 13.4.3 and is considered known. The height integration of (13.66) results in

$$W(h) - W(z_0) = \frac{\overline{\rho} u_*^3}{k} \int_{z_0}^{h} S_W \frac{dz}{z} \tag{13.94}$$

The definition of this heat flux is given by (13.24). Owing to the pressure fluctuation Π', W is expected to be very small and often may be neglected altogether. $W(z_0)$ is unknown and cannot be determined from Prandtl-layer theory. We assume that the turbulent part of $W(z_0)$ vanishes and approximate $W(z_0)$ by the molecular conduction of heat as

$$W(z_0) = -l_c \left(\frac{\partial \overline{T}}{\partial z} \right)_{z_0} \approx -l_c \frac{\overline{T}(h) - \overline{T}(z_0)}{h - z_0} \tag{13.95}$$

where l_c is the heat-conduction coefficient.

The quantity E appearing in the budget equation (13.25) for the turbulent kinetic energy cannot be calculated from the differential equation (13.66) since an integration constant for some point in the Prandtl layer is not available. The calculation of E, however, is not at all necessary since it does not appear in the prognostic system. It does, however, appear in the prognostic equation for the turbulent kinetic energy $\widehat{k}$, which is a subgrid quantity; see (11.46) and (13.13).

The dissipation of energy $\overline{\epsilon}$ given in (13.66) can be determined for all heights in the Prandtl layer since the MO function S_ϵ is considered known.

The flux of water vapor Q can be found from (13.28). q_* is obtained by solving the corresponding differential equation (13.68). Integration gives

$$\widehat{q}(h) - \widehat{q}(z_0) = \frac{q_*}{k} \int_{z_0}^{h} S_q \frac{dz}{z} \tag{13.96}$$

Since $q(h)$ and $q(z_0)$ are known external parameters and the MO function $S_q = S_T$ is known also, the height-constant scaling parameter q_* can be calculated and then be substituted into (13.28) to find the moisture flux.

Most mesoscale models resolve the Prandtl layer by including additional grid surfaces. The phenomenological coefficients, i.e. the exchange coefficients, may

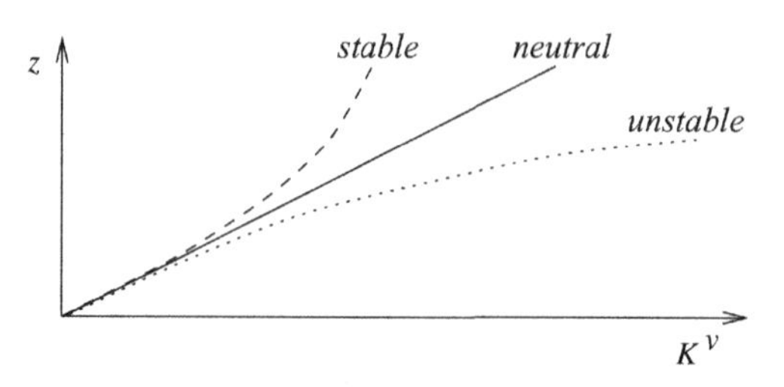

Fig. 13.4 Height profiles of the exchange coefficient.

be determined easily for these grid surfaces. There exist two formulas from which each exchange coefficient can be found. K^v is obtained from (13.49) and (13.66); K^θ and K^q follow from combining the relations listed in (13.26), (13.28), and (13.68). In summary we have

$$\boxed{K^v = \frac{u_*^2}{\dfrac{\partial \widehat{u}}{\partial z}} = \frac{kzu_*}{S_u(\xi)}, \qquad K^\theta = \frac{u_* T_*}{\dfrac{\partial \widetilde{\theta}}{\partial z}} = \frac{kzu_*}{S_T(\xi)}, \qquad K^q = \frac{u_* q_*}{\dfrac{\partial \widehat{q}}{\partial z}} = \frac{kzu_*}{S_q(\xi)}} \tag{13.97}$$

The vertical distribution of the exchange coefficient K^v is shown qualitatively in Figure 13.4. For a given height K^v is smaller for stable than it is for neutral and unstable stratification.

13.7 The interface condition at the earth's surface

The starting point in the derivation is the prognostic equation (11.49) for the enthalpy. Owing to the Prandtl-layer condition (13.5) the budget operator vanishes. We make the addtional assumption that only isobaric processes are admitted and that transitions between the turbulent kinetic energy and the internal energy do not take place. In view of Figure 13.2 and equation (12.63) we may summarize the situation by writing

$$\frac{\widehat{D}}{Dt}(\overline{\rho}\widehat{h}) = 0, \qquad \frac{\widehat{d}\,\overline{p}}{dt} = 0, \qquad \overline{\mathbf{v}'' \cdot \nabla p} + \overline{\epsilon} = \mathbf{J}_{\mathrm{t}}^{\theta} \cdot \nabla \ln \widetilde{\Pi} + \overline{\epsilon} = 0 \tag{13.98}$$

where $H = J_{\mathrm{t}}^{\theta}$. With these assumptions equation (11.49) reduces to the following divergence expression:

$$\nabla \cdot \left[\mathbf{J}_{\mathrm{s,t}}^{h} + \overline{\mathbf{J}}_{\mathrm{s}}^{h} + \left(\mathbf{J}_{\mathrm{t}}^{n} + \overline{\mathbf{J}}^{n}\right)\widetilde{h}_n + \overline{\mathbf{F}}_{\mathrm{R}}\right] = 0 \tag{13.99}$$

where the sensible enthalpy fluxes defined in (11.61) have been used. According to equation (9.6) we must replace the regular divergence by the surface divergence, which is repeated for convenience:

$$\nabla \cdot \boldsymbol{\Psi} \longrightarrow \mathbf{i}_3 \cdot (\boldsymbol{\Psi}_{\mathrm{a}} - \boldsymbol{\Psi}_{\mathrm{g}}) \tag{13.100}$$

The subscripts a and g stand for atmosphere and ground. By ignoring within the atmosphere the mean molecular sensible enthalpy flux $\overline{\mathbf{J}}_{\mathrm{s}}^{h}$ and replacing the turbulent sensible enthalpy flux $\mathbf{J}_{\mathrm{s,t}}^{h}$ by means of (11.69) we find

$$\mathbf{i}_3 \cdot \left(\mathbf{J}_{\mathrm{s,t}}^{h} + \overline{\mathbf{J}}_{\mathrm{s}}^{h}\right)_{\mathrm{a}} = \mathbf{i}_3 \cdot \left(\mathbf{J}_{\mathrm{t}}^{\theta} + \mathbf{J}_{\mathrm{t}}^{\Pi}\right) = J_{\mathrm{t}}^{\theta} \tag{13.101}$$

The turbulent Exner flux $\mathbf{J}_{\mathrm{t}}^{\Pi}$ vanishes since only isobaric processes are admitted. Assuming that fog does not form, we use the Prandtl-layer formulation (13.20) together with (13.24) to obtain for the enthalpy term

$$\mathbf{i}_3 \cdot \left[\left(\mathbf{J}_{\mathrm{t}}^{n} + \overline{\mathbf{J}}^{n}\right)\widetilde{h}_n\right]_{\mathrm{a}} = l_{21} Q \tag{13.102}$$

Turbulent and latent heat fluxes within the ground are ignored so that

$$\mathbf{i}_3 \cdot \left[\mathbf{J}_{\mathrm{s,t}}^{h} + \overline{\mathbf{J}}_{\mathrm{s}}^{h} + \left(\mathbf{J}_{\mathrm{t}}^{n} + \overline{\mathbf{J}}^{n}\right)\widetilde{h}_n\right]_{\mathrm{g}} = \mathbf{i}_3 \cdot \overline{\mathbf{J}}_{\mathrm{s,g}}^{h} = -\rho_{\mathrm{g}} c_{\mathrm{g}} K_{\mathrm{g}} \mathbf{i}_3 \cdot \nabla T_{\mathrm{g}} = -\rho_{\mathrm{g}} c_{\mathrm{g}} K_{\mathrm{g}} \frac{\partial T_{\mathrm{g}}}{\partial z} \tag{13.103}$$

The terms ρ_{g} and c_{g} are the density and specific heat of the ground while K_{g} is the thermal diffusivity coefficient. The radiative net flux is given by

$$\mathbf{i}_3 \cdot \overline{\mathbf{F}}_{\mathrm{R}} = \sigma T_{\mathrm{S}}^4 - F_{\mathrm{s}} - F_{\mathrm{l}} \tag{13.104}$$

where T_{S} in the Stefan–Boltzmann law refers to the surface temperature of the earth. F_{s} is the short-wave global net flux (accounting for the albedo of the ground) and F_{l} is the long-wave downward flux. The heat balance at the earth's surface is then given by

$$\sigma T_{\mathrm{S}}^4 - F_{\mathrm{s}} - F_{\mathrm{l}} + J_{\mathrm{t}}^{\theta} - l_{21} Q + \rho_{\mathrm{g}} c_{\mathrm{g}} K_{\mathrm{g}} \frac{\partial T_{\mathrm{g}}}{\partial z} = 0 \tag{13.105}$$

In this equation all energy fluxes directed from the surface into the atmosphere or into the ground are counted as positive.

A more complete formulation of the heat balance at the earth's surface and the moisture balance within the soil is given by Panhans (1976), for example. All modern investigations include the treatment of a vegetation–soil model; see for example Deardorff (1978), Pielke (1984), Sellers *et al.* (1986), and Siebert *et al.*

(1992). Owing to the complexity of such models, we refrain from discussing this subject.

We will give a simple example of a heat-transport problem to which the surface-balance equation (13.105) can be applied. We assume nocturnal conditions and ignore latent-heat effects. Often the wind profile in the surface layer may be expressed by the power law

$$\widehat{u} = \widehat{u}_1\left(\frac{z}{z_1}\right)^m \quad \text{with} \quad m = \begin{cases} 0.3 & \text{strong stable stratification} \\ 0.14 & \text{neutral stratification} \\ 0.05 & \text{strong unstable stratification} \end{cases} \tag{13.106}$$

where the stability exponents have been obtained from observations. z_1 is the anemometer height at which the wind velocity is measured. We wish to find the corresponding vertical profile of the exchange coefficient. Assuming constant values of τ and $\overline{\rho}$, we obtain from (13.26) and (13.106)

$$\frac{\tau}{\overline{\rho}} = K^{\upsilon}\frac{\partial \widehat{u}}{\partial z}, \qquad \frac{1}{\widehat{u}}\frac{\partial \widehat{u}}{\partial z} = \frac{m}{z} \implies K^{\upsilon} = az^{1-m}, \qquad a = \frac{\tau z_1^m}{\overline{\rho}\,\widehat{u}_1 m} = \text{constant} \tag{13.107}$$

The heat equation (11.60) will be strongly simplified by assuming that we have a dry-air atmosphere and by ignoring the molecular heat flux $\overline{\mathbf{J}}_{\mathrm{s}}^{h}$. Furthermore, let us disregard the transformation between the turbulent kinetic and the internal energy so that the heat equation reduces to the simple form

$$\overline{\rho}c_{p,0}\frac{\overline{T}}{\widetilde{\theta}}\frac{\widehat{d}\,\widetilde{\theta}}{dt} + \nabla\cdot\mathbf{J}_{\mathrm{s,t}}^{h} = 0 \tag{13.108}$$

In this simple case we assume that $\mathbf{J}_{\mathrm{s,t}}^{h}$ may be parameterized analogously to equation (13.26). Ignoring the advection of the potential temperature $\widetilde{\theta}$, we find with $\overline{T} \approx \widetilde{\theta}$

$$\frac{\partial \widetilde{\theta}}{\partial t} = \frac{\partial}{\partial z}\left(az^{1-m}\frac{\partial \widetilde{\theta}}{\partial z}\right) \tag{13.109}$$

For the soil a similar equation is valid:

$$\frac{\partial T_{\mathrm{g}}}{\partial t} = K_{\mathrm{g}}\frac{\partial^2 T_{\mathrm{g}}}{\partial z^2} \tag{13.110}$$

These two equations have been the starting point of many investigations of the calm nocturnal boundary layer. These two one-dimensional equations are coupled by means of the interface condition

$$-\rho_{\mathrm{g}}c_{\mathrm{g}}K_{\mathrm{g}}\frac{\partial T_{\mathrm{g}}}{\partial z} = -a\overline{\rho}c_{p,0}z^{1-m}\frac{\partial \widetilde{\theta}}{\partial z} + R_{\mathrm{N}}, \qquad z = 0,\ t > 0 \tag{13.111}$$

with $R_N = \sigma T_S^4 - F_l$, $a = 0.07$ and $m = \frac{1}{3}$. In this equation vertical gradients must be understood as limit statements. The radiative net flux at the ground is computed with the help of some radiative-transfer model fully accounting for the presence of absorbing and emitting gases. The reader will notice the internal inconsistency in the system (13.111) since the derivation of (13.109) assumed that we have a dry atmosphere that does not absorb and emit any infrared radiation. This imperfection must be accepted if we wish to tackle the problem analytically. The solution of this problem is not trivial by any means, but may be obtained with the help of Laplace transforms. Omitting details, which may be found in Zdunkowski and Kandelbinder (1997), here we just present the solution as

$$\widetilde{\theta}(z,t) = \widetilde{\theta}_0 - \frac{R_N(n)^{m/n}\exp\left(-\dfrac{z^n}{2an^2t}\right)}{\overline{\rho}c_{p,0}\Gamma\left(\dfrac{1}{n}\right)a^{1/(2n)}z^{1/2}} \sum_{l=0}^{\infty}(-1)^l\delta^{-(1+l)}t^{\alpha}W_{-\alpha,\beta}\left(\frac{z^n}{n^2at}\right)$$

$$\text{with} \quad \alpha = \frac{m+2+l(1-m)}{2n}, \qquad \beta = \frac{m}{2n}, \qquad n = m+1$$

$$\delta = \frac{\rho_g c_g\sqrt{K_g}\,\Gamma\left(\dfrac{m}{n}\right)}{\overline{\rho}c_{p,0}\Gamma\left(\dfrac{1}{n}\right)a^{1/n}n^{(1-m)/n}} \tag{13.112}$$

The potential temperature $\widetilde{\theta}_0$ is the constant-height initial potential temperature. The function $W_{\alpha,\beta}(z)$ is the Whitaker function, which can be computed directly or evaluated with the help of tables given in the *Handbook of Mathematical Functions* by Abramowitz and Segun (1968). The change of the potential temperature with time after sunset is shown in Figure 13.5. A numerical model (discussed in the original paper) is also used to show the influence of the radiative heating of the air, which is very small. This justifies to some extent the dry-air assumption leading to (13.109). Figure 13.5 shows that the cooling of the air is strongest near the earth's surface.

13.8 The Ekman layer – the classical approach

The major characteristics of the Ekman layer were listed in the introduction to this chapter. Many successful numerical studies, too numerous to list here, were performed in order to simulate the temperature and the wind fields of the entire boundary layer. These studies were based on the use of different types of closure assumptions to simulate the eddy exchange coefficient. Figure 13.6 shows a typical height distribution of the exchange coefficient K after O'Brien (1970). In the lower

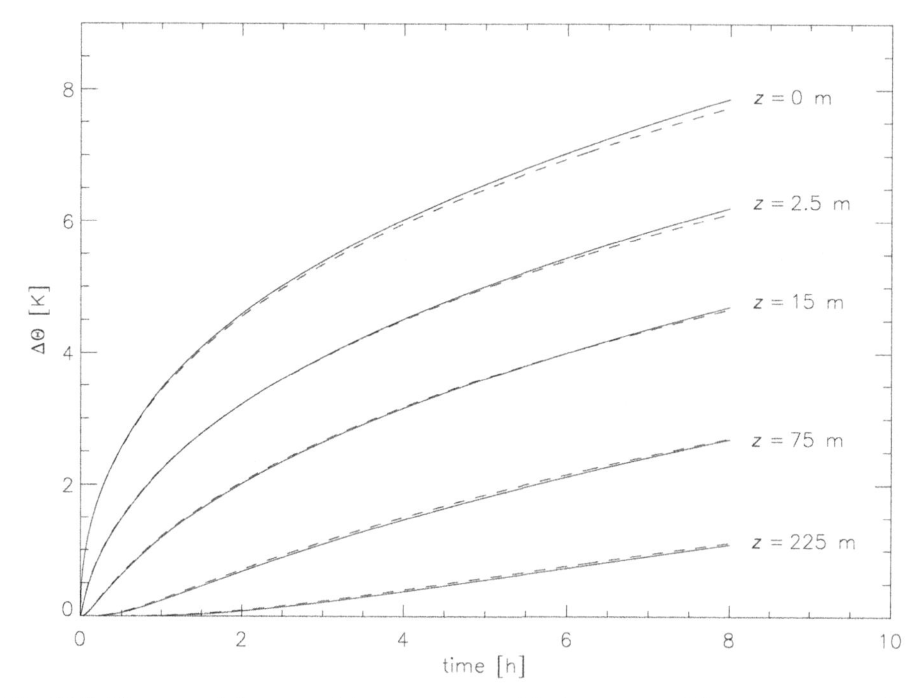

Fig. 13.5 Changes of the potential temperature with time counted from sunset at various heights with (dashed lines) and without (solid lines) radiative heating.

part of the boundary layer K increases more or less linearly with height until it reaches a maximum value $K_{\max}$ at h_1. Above h_2 the exchange coefficient decreases with height.

There is a region between the linear increase and the linear decrease where K is approximately constant with height. The linear approximation of K shown in Figure 13.6 will be used below to obtain an analytic solution of the wind field. First we will present Taylor's (1915) very simple analytic solution, which is based on the assumption that the exchange coefficient is constant with height throughout the entire boundary layer. The Taylor solution is already capable of capturing the gross features of the wind profile in the Ekman layer.

The starting point of the analysis is the averaged equation of motion (11.35d):

$$\frac{\widehat{D}}{Dt}(\overline{\rho}\,\widehat{\mathbf{v}}) + \nabla \cdot (p\mathbb{E} - \overline{\mathbb{J}} - \mathbb{R}) = -\overline{\rho}(\nabla\phi + 2\boldsymbol{\Omega} \times \widehat{\mathbf{v}}) \tag{13.113}$$

The theory considers only the mean horizontal motion, which is subjected to various constraints in order to obtain a simple solution of the vertical wind profile. These

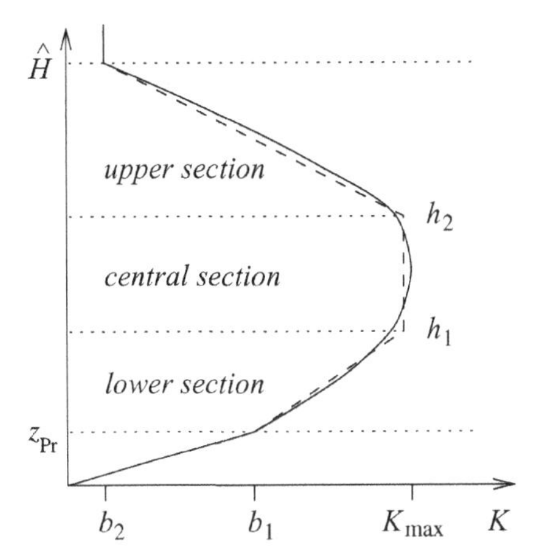

Fig. 13.6 Typical height variation of the exchange coefficient in the boundary layer (full line) and the linear approximation (dashed line). At the top of the boundary layer a residual value is expected.

constraints are

$$\frac{\widehat{d}\ \dots}{dt} = 0, \qquad \overline{\rho} = \text{constant}, \qquad K_{\rm v}^{v} = K = \text{constant}, \qquad \widehat{\mathbf{v}} = \widehat{\mathbf{v}}_{\rm h}(z), \qquad \widehat{w} = 0 \tag{13.114}$$

Thus the acceleration of the mean horizontal wind is zero while the density and the exchange coefficient are independent of spatial coordinates. Moreover, it is assumed that the geostrophic wind is independent of height and that the actual wind becomes geostrophic as z approaches infinity. These assumptions imply that the horizontal pressure gradient is constant with height also. Application of (11.92) results in

$$\nabla \cdot (\overline{\mathbb{J}} + \mathbb{R}) = \mathbf{i}_3 \cdot \frac{\partial}{\partial z}\left[\overline{\rho} K\left(\mathbf{i}_3 \frac{\partial \widehat{\mathbf{v}}_{\rm h}}{\partial z} + \frac{\partial \widehat{\mathbf{v}}_{\rm h}}{\partial z}\mathbf{i}_3\right)\right] = \overline{\rho} K \frac{\partial^2 \widehat{\mathbf{v}}_{\rm h}}{\partial z^2} \tag{13.115}$$

where $\widehat{\mathbf{v}}_{\rm h} = \widehat{u}\mathbf{i} + \widehat{v}\mathbf{j}$.

The equation of motion for the tangential plane (2.38) will now be modified by including the conditions (13.114) and by adding frictional effects according to (13.115). On replacing (u, v) by the average values $(\widehat{u}, \widehat{v})$ we find

$$\begin{aligned} -f\,\widehat{v} &= -\frac{1}{\overline{\rho}}\frac{\partial \overline{p}}{\partial x} + K\frac{\partial^2 \widehat{u}}{\partial z^2} \\ f\widehat{u} &= -\frac{1}{\overline{\rho}}\frac{\partial \overline{p}}{\partial y} + K\frac{\partial^2 \widehat{v}}{\partial z^2} \end{aligned} \tag{13.116}$$

As the next step we replace the components of the pressure-gradient force by the corresponding components of the geostrophic wind:

$$u_{\rm g} = -\frac{1}{\overline{\rho} f}\frac{\partial \overline{p}}{\partial y}, \qquad v_{\rm g} = \frac{1}{\overline{\rho} f}\frac{\partial \overline{p}}{\partial x},$$
$$\text{since} \quad \mathbf{v}_{\rm g} = \frac{1}{\overline{\rho} f}\mathbf{e}_r \times \nabla_{\rm h}\overline{p} \tag{13.117}$$

We choose our coordinate system in such a way that the pressure does not vary in the x-direction so that $v_{\rm g}$ vanishes. Substituting (13.117) into (13.116) gives the following coupled system of second-order linear differential equations:

$$K\frac{d^2\widehat{u}}{dz^2} = -f\widehat{v}, \qquad K\frac{d^2\widehat{v}}{dz^2} = f(\widehat{u} - u_{\rm g}) \tag{13.118}$$

Since z is the only independent variable, we have replaced the partial by the total derivative. These two equations can be combined to give a single equation by multiplying the second equation of (13.118) by $\sqrt{-1} = i$ and then adding the result to the first equation. This gives the following second-order differential equation:

$$\frac{d^2}{dz^2}(\widehat{u} + i\widehat{v} - u_{\rm g}) - \frac{if}{K}(\widehat{u} + i\widehat{v} - u_{\rm g}) = 0 \tag{13.119}$$

since the geostrophic wind was assumed to be height-independent. The general solution of (13.119) is easily found by standard methods and is given by

$$\begin{aligned}\widehat{u} + i\widehat{v} - u_{\rm g} &= C_1\exp\left(\sqrt{\frac{if}{K}}z\right) + C_2\exp\left(-\sqrt{\frac{if}{K}}z\right)\\ &= C_1\exp\left(\sqrt{\frac{f}{2K}}z\right)\exp\left(i\sqrt{\frac{f}{2K}}z\right)\\ &\quad + C_2\exp\left(-\sqrt{\frac{f}{2K}}z\right)\exp\left(-i\sqrt{\frac{f}{2K}}z\right)\end{aligned} \tag{13.120}$$

since $\sqrt{i} = (1+i)/\sqrt{2}$. Application of the boundary conditions at $z = 0$, $\widehat{u}+i\widehat{v} = 0$, and at $z \longrightarrow \infty$, $\widehat{u} + i\widehat{v} = u_{\rm g}$, gives the required integration constants $C_1 = 0$ and $C_2 = -u_{\rm g}$, and (13.120) results in

$$\widehat{u} + i\widehat{v} = u_{\rm g}\{1 - \exp(-Az)\,[\cos(Az) - i\sin(Az)]\} \quad \text{with} \quad A = \sqrt{\frac{f}{2K}} \tag{13.121}$$

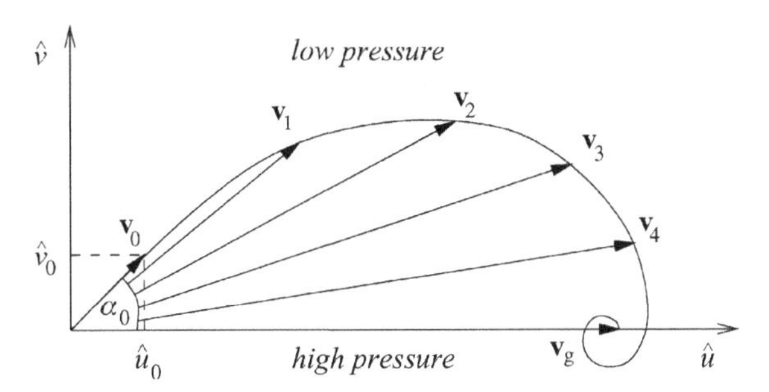

Fig. 13.7 The Ekman spiral, showing turning of the wind vector with height.

The solution may be separated by comparing the real and imaginary parts:

$$\boxed{\widehat{u} = u_g\left[1 - \exp(-Az)\cos(Az)\right], \qquad \widehat{v} = u_g \exp(-Az)\sin(Az)} \tag{13.122}$$

Taking $\widehat{v}$ as the ordinate and $\widehat{u}$ as the abscissa we obtain a nomogram, which is displayed in Figure 13.7. This nomogram is the celebrated *Ekman spiral* showing that the wind vector is turning clockwise in the northern hemisphere. For various heights the horizontal wind vector is shown, indicating that, with increasing height, the horizontal wind approaches the geostrophic wind. The largest *cross-isobar angle* α_0 is found at the earth's surface. α_0 can be found from the ratio of the wind-vector components. For $z = 0$ this gives an indeterminant form 0/0. By using (13.122) and applying L'Hospital's rule we find

$$\tan\alpha_0 = \lim_{z\to 0}\left(\frac{\widehat{v}}{\widehat{u}}\right) = \lim_{z\to 0}\left(\frac{u_g\exp(-Az)\sin(Az)}{u_g\left[1 - \exp(-Az)\cos(Az)\right]}\right) = 1 \implies \alpha_0 = 45^\circ \tag{13.123}$$

This value is too large and does not agree with observations. The height at which the actual wind coincides for the first time with the direction of the geostrophic wind is known as the *geostrophic wind height* z_g. There the wind speed exceeds the geostrophic wind by a few percent. The height of the planetary boundary layer is often defined by z_g. From the condition $\widehat{v}(z_g) = 0$, implying that $Az_g = \pi$, we find the height of the boundary layer as

$$z_g = \pi/A \tag{13.124}$$

Assuming that $K = 5\ \mathrm{m^2\ s^{-1}}$, which is a reasonable midlatitude value, and $f = 10^{-4}\ \mathrm{s^{-1}}$ we find that $z_g = 1000$ m. For different values of K this height may be twice as large or just half as large.

We will now investigate the influence of the turbulent viscosity, represented by the eddy exchange coefficient K, on the Ekman profile more closely. Applying the general definition (13.9) of the stress vector to equation (11.92), assuming that the exchange coefficient is constant with height, we find

$$\mathbf{T} = \mathbf{i}_3 \cdot (\mathbb{J} + \mathbb{R}) = \overline{\rho} K \frac{\partial \widehat{\mathbf{v}}_{\mathrm{h}}}{\partial z} \implies \frac{1}{\overline{\rho}} \frac{\partial \mathbf{T}}{\partial z} = K \frac{\partial^2 \widehat{\mathbf{v}}_{\mathrm{h}}}{\partial z^2} \tag{13.125}$$

The second equation follows immediately since the density was assumed to be constant with height. This expression is the frictional force per unit mass due to the eddy viscosity of the air. Equation (13.116) can be written in the following form:

$$-\overline{\rho} K \frac{\partial^2 \widehat{\mathbf{v}}_{\mathrm{h}}}{\partial z^2} = -\nabla_{\mathrm{h}} \overline{p} - \overline{\rho} f \mathbf{i}_3 \times \widehat{\mathbf{v}}_{\mathrm{h}} \tag{13.126}$$

To eliminate the pressure gradient we use the definition of the geostrophic wind (13.117). Taking the cross product $\mathbf{i}_3 \times \mathbf{v}_{\mathrm{g}}$ and applying the Grassmann rule, we obtain immediately

$$\mathbf{i}_3 \times \mathbf{v}_{\mathrm{g}} = \frac{1}{\overline{\rho} f} \mathbf{i}_3 \times (\mathbf{i}_3 \times \nabla_{\mathrm{h}} p) = -\frac{1}{\overline{\rho} f} \nabla_{\mathrm{h}} p \tag{13.127}$$

Substitution of this expression into (13.126) yields

$$K \frac{\partial^2 \widehat{\mathbf{v}}_{\mathrm{h}}}{\partial z^2} = f \mathbf{i}_3 \times (\widehat{\mathbf{v}}_{\mathrm{h}} - \mathbf{v}_{\mathrm{g}}) \tag{13.128}$$

which can be easily interpreted with the help of Figure 13.8. The vectorial difference $\Delta \widehat{\mathbf{v}}_{\mathrm{h}} = \widehat{\mathbf{v}}_{\mathrm{h}} - \mathbf{v}_{\mathrm{g}}$ between the horizontal and the geostrophic wind vectors is known as *the geostrophic wind deviation*. The Coriolis force is perpendicular to the actual wind vector and the pressure gradient force is perpendicular to the geostrophic wind vector. By adding the frictional force we obtain the balance of forces. The frictional force is perpendicular to the geostrophic deviation. The magnitude of the geostrophic deviation decreases with height and is zero at the top of the boundary layer since the frictional force vanishes at the level z_{g}.

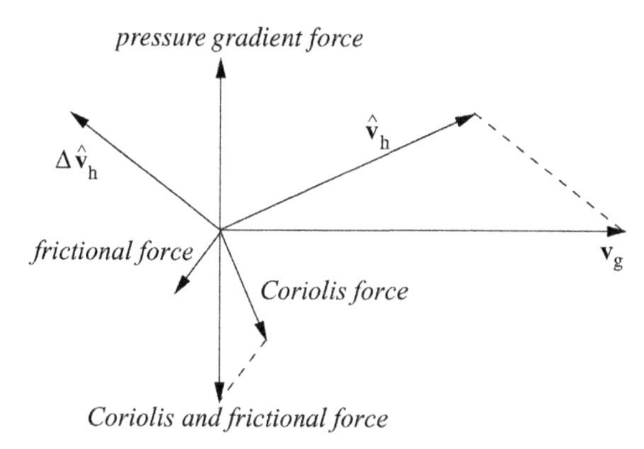

Fig. 13.8 The balance of forces for unaccelerated flow in the Ekman layer.

A few critical remarks on the solution (13.122) will be helpful. As stated above, the cross-isobar angle $\alpha = 45^\circ$ is too large to be in reasonable agreement with observations, for which α generally varies from 15° to 25°. The reason for this is that the exchange coefficient was assumed to be constant with height whereas in reality it varies rapidly with height in the lower 10–20 m, as we know from Prandtl-layer theory and from Figure 13.6. Therefore, the solution (13.122) should not be extended to the ground but only to some height above the ground that is often taken as the height of the anemometer level. We know from Prandtl-layer theory that, in the surface layer, the wind shear is along the wind vector itself. Using this as a lower boundary condition, and if $z = 0$ denotes the height of a well-exposed anemometer, we obtain a more realistic solution for the Ekman spiral. The cross-isobar angle then is not fixed at the unrealistically high value of 45° but is smaller. The solution to the problem is given, for example, in Petterssen (1956). We will not discuss this situation, but we present another solution that is based on the piecewise linear distribution of the exchange coefficient as shown in Figure 13.6. The price we have to pay for the more realistic approximation of the exchange coefficient is a more complicated analytic solution.

13.9 The composite Ekman layer

As stated above, there are many numerical solutions with which to model the atmospheric boundary layer. We will now show that it is also possible to obtain an analytic solution for the wind profile by subdividing the Ekman layer into different sections, in each of which sections K varies linearly with height. This treatment allows the exchange coefficient to vary with height in a reasonable manner. The right-hand side of equation (13.115) is now given by

$$\nabla \cdot (\overline{\mathbb{J}} + \mathbb{R}) = \overline{\rho} \frac{\partial}{\partial z}\left(K \frac{\partial \widehat{\mathbf{v}}_{\rm h}}{\partial z} \right) \tag{13.129}$$

Instead of (13.118) we have to solve the system

$$\frac{d\left(K \dfrac{d\widehat{u}}{dz} \right)}{dz} = -f\widehat{v}, \qquad \frac{d\left(K \dfrac{d\widehat{v}}{dz} \right)}{dz} = f(\widehat{u} - u_{\rm g}) \tag{13.130}$$

Proceeding as before, these differential equations will be combined to give a single equation:

$$K \frac{d^2}{dz^2}(\widehat{u} + i\widehat{v} - u_{\rm g}) + \frac{dK}{dz}\frac{d}{dz}(\widehat{u} + i\widehat{v} - u_{\rm g}) - if(\widehat{u} + i\widehat{v} - u_{\rm g}) = 0 \tag{13.131}$$

Introducing the abbreviation $g = \widehat{u} + i\widehat{v} - u_g$ we may also write

$$K \frac{d^2 g}{dz^2} + \frac{dK}{dz}\frac{dg}{dz} - ifg = 0 \tag{13.132}$$

Comparison with (13.119) shows that now an additional term appears.

In the following we will derive the solution for the case that the Ekman layer is subdivided into three parts as shown in Figure 13.6. In the central section of the vertical profile of the exchange coefficient the classical Ekman solution applies. For the upper section $z > h_2$ and the lower section $z_{Pr} < z < h_1$ the linear approximations of the exchange coefficient K are given by

$$\begin{aligned}
z_{Pr} \le z \le h_1: &\qquad K_1 = b_1 + a_1(z - z_{Pr}) > 0 \quad \text{with} \quad a_1 = \frac{K_{max} - b_1}{h_1 - z_{Pr}} \\
h_2 \le z \le \widehat{H}: &\qquad K_2 = b_2 + a_2(\widehat{H} - z) > 0 \quad \text{with} \quad a_2 = \frac{K_{max} - b_2}{\widehat{H} - h_2}
\end{aligned} \tag{13.133}$$

where we have admitted a residual value b_2 at the top of the boundary layer.

In order to transform (13.132) into a Bessel-type differential equation for which the solution is known, we have to introduce suitable transformation variables. Let us first direct our attention to the lower section, transforming the exchange coefficient K_1 as

$$\sqrt{K_1} = \sqrt{b_1 + a_1(z - z_{Pr})} = \xi > 0 \tag{13.134}$$

The transformations

$$\begin{aligned}
\frac{d\xi}{dz} &= \frac{a_1}{2\xi}, &\qquad \frac{d^2\xi}{dz^2} &= -\frac{a_1^2}{4\xi^3} \\
\frac{dg}{dz} &= \frac{dg}{d\xi}\frac{d\xi}{dz}, &\qquad \frac{d^2 g}{dz^2} &= \frac{d^2\xi}{dz^2}\frac{dg}{d\xi} + \left(\frac{d\xi}{dz}\right)^2 \frac{d^2 g}{d\xi^2}
\end{aligned} \tag{13.135}$$

follow from the differentiation rules, so (13.132) may be written as

$$\frac{d^2 g}{d\xi^2} + \frac{1}{\xi}\frac{dg}{d\xi} - \frac{4ifg}{a_1^2} = 0 \tag{13.136}$$

The solution to this equation is given in various textbooks on differential equations. We refer to Magnus and Oberhettinger (1948), where a very general form of Bessel's equation and its solution are given. These two equations are

$$\begin{aligned}
&\frac{d^2 u}{dz^2} + (2\alpha - 2\nu\beta + 1)\frac{1}{z}\frac{du}{dz} + \left(\beta^2\gamma^2 z^{2\beta-2} + \frac{\alpha(\alpha - 2\beta\nu)}{z^2}\right)u = 0 \\
&u = z^{\beta\nu-\alpha} Z_\nu\left(\gamma z^\beta\right)
\end{aligned} \tag{13.137}$$

The symbol Z stands for the cylinder function. In our situation we have $\alpha = 0,\ \beta = 1,\ \nu = 0$, and $\gamma^2 = -4if/a_1^2$, so the solution function is of zeroth order and is given by

$$g_l = Z_0\left(l_l i^{3/2}\xi\right) \quad \text{with} \quad l_l = \frac{2\sqrt{f}}{a_1} \tag{13.138}$$

where l_l is a constant applying to the lower section. The complete solution can be constructed from any pair of independent particular solutions. The form

$$g_l = c_l J_0\left(l_l i^{3/2}\xi\right) + d_l K_0\left(l_l i^{1/2}\xi\right) \tag{13.139}$$

where J_0 is a Bessel function of the first kind and K_0 a modified Bessel function of the second kind is suitable for our purposes. As before, the subscript l on the integration constants c and d refers to the lower section. Since the argument of the Bessel functions is complex, it is customary to introduce the form

$$\boxed{g_l = c_l[\mathrm{ber}_0(l_l\xi) + i\ \mathrm{bei}_0(l_l\xi)] + d_l[\mathrm{ker}_0(l_l\xi) + i\ \mathrm{kei}_0(l_l\xi)]} \tag{13.140}$$

where $\mathrm{ber}_0(x)$, $\mathrm{bei}_0(x)$, $\mathrm{ker}_0(x)$, and $\mathrm{kei}_0(x)$ are the Kelvin functions referring to the real and imaginary parts of the Bessel functions J_0 and K_0, respectively.[1] The function $\mathrm{ber}_0(x) + i\ \mathrm{bei}_0(x)$ is finite at the origin but becomes infinite as x becomes infinite. $\mathrm{ker}_0(x) + i\ \mathrm{kei}_0(x)$ is infinite at the origin but approaches zero as x becomes infinite. An excellent introductory discussion on Bessel functions is given in Wylie (1966).

We now turn to the transformation and the solution of (13.132) in the upper section. We apply the transformation

$$\sqrt{K_2} = \sqrt{b_2 + a_2(\widehat{H} - z)} = \eta > 0 \tag{13.141}$$

From the differentiations

$$\frac{d\eta}{dz} = -\frac{a_2}{2\eta}, \qquad \frac{d^2\eta}{dz^2} = -\frac{a_2^2}{4\eta^3} \tag{13.142}$$

there follows the differential equation

$$\frac{d^2 g}{d\eta^2} + \frac{1}{\eta}\frac{dg}{d\eta} - \frac{4ifg}{a_2^2} = 0 \tag{13.143}$$

From (13.137) with $\alpha = 0,\ \beta = 1,\ \nu = 0$, and $\gamma^2 = -4if/a_2^2$ we find the solution

$$g_u = Z_0\left(l_u i^{3/2}\eta\right) \quad \text{with} \quad l_u = \frac{2\sqrt{f}}{a_2} \tag{13.144}$$

[1] Many textbooks omit the suffix 0 from the functions $\mathrm{ber}_0(x)$, $\mathrm{bei}_0(x)$, $\mathrm{ker}_0(x)$, and $\mathrm{kei}_0(x)$.

which is a cylinder function of zeroth order. The subscript u on the constant l_u refers to the upper section. The general solution is given by

$$g_u = c_u J_0\big(l_u i^{3/2}\eta\big) + d_u\big[K_0\big(l_u i^{1/2}\eta\big)\big] \tag{13.145}$$

Analogously to (13.140) we may also write

$$\boxed{g_u = c_u[\mathrm{ber}_0(l_u\eta) + i\,\mathrm{bei}_0(l_u\eta)] + d_u[\mathrm{ker}_0(l_u\eta) + i\,\mathrm{kei}_0(l_u\eta)]} \tag{13.146}$$

Now we have three solutions for the composite boundary layer. For the lower section we have the solution (13.140). For the middle layer with a constant exchange coefficient K_{max} the solution (13.120) applies. It is restated in a convenient form as

$$\boxed{g_m = c_m \exp(Az)\,[\cos(Az) + i\,\sin(Az)] + d_m \exp(-Az)\,[\cos(Az) - i\,\sin(Az)]} \tag{13.147}$$

with $A = \sqrt{f/(2K_{max})}$. For the upper section we have the solution (13.146). The g_l, g_m, and g_u occurring in (13.140), (13.146), and (13.147) are complex numbers, so

$$u(z) = \Re[g(z) + u_g], \qquad v(z) = \Im[g(z)] \tag{13.148}$$

There are six constants in the composite solution to be evaluated, so six boundary statements must be at our disposal. These will be stated next:

(i) The log–linear wind profile (13.73) must hold within the Prandtl layer. We require that, at the top of the Prandtl layer z_{Pr}, the derivatives of the log–linear wind profile and of the solution for the lower section coincide.
(ii) At the top of the Ekman layer the wind becomes geostrophic.
(iii) The wind components at the interfaces h_1 and h_2 are continuous.
(iv) The derivatives of the wind components are continuous at these interfaces.

The continuity of the derivatives of wind components eliminates kinks in the wind profile. This is the same thing as requiring that the stresses are continuous at the interfaces. The composite-boundary-layer problem has the undesirable feature that the derivatives dK/dz are discontinuous at the interfaces.

The derivatives of the solution functions appearing in (13.140) and (13.146) are

$$\begin{aligned}
\frac{d\,\mathrm{ber}_0(x)}{dx} &= \frac{1}{\sqrt{2}}[\mathrm{ber}_1(x) + \mathrm{bei}_1(x)], & \frac{d\,\mathrm{bei}_0(x)}{dx} &= \frac{1}{\sqrt{2}}[-\mathrm{ber}_1(x) + \mathrm{bei}_1(x)] \\
\frac{d\,\mathrm{ker}_0(x)}{dx} &= \frac{1}{\sqrt{2}}[\mathrm{ker}_1(x) + \mathrm{kei}_1(x)], & \frac{d\,\mathrm{kei}_0(x)}{dx} &= \frac{1}{\sqrt{2}}[-\mathrm{ker}_1(x) + \mathrm{kei}_1(x)]
\end{aligned} \tag{13.149}$$

Application of these equations to the lower and upper sections leads to

$$x = l_1\xi = l_1\sqrt{b_1 + a_1(z - z_{\mathrm{Pr}})}, \qquad \frac{dx}{dz} = \frac{l_1 a_1}{2\xi}, \qquad \frac{d}{dz} = \frac{dx}{dz}\frac{d}{dx}$$

$$\frac{d}{dz}\begin{pmatrix} \mathrm{ber}_0(x) \\ \mathrm{bei}_0(x) \\ \mathrm{ker}_0(x) \\ \mathrm{kei}_0(x) \end{pmatrix}_l = \frac{l_1 a_1}{2\sqrt{2}\xi}\begin{pmatrix} \mathrm{ber}_1(x) + \mathrm{bei}_1(x) \\ -\mathrm{ber}_1(x) + \mathrm{bei}_1(x) \\ \mathrm{ker}_1(x) + \mathrm{kei}_1(x) \\ -\mathrm{ker}_1(x) + \mathrm{kei}_1(x) \end{pmatrix} \tag{13.150a}$$

and

$$x = l_{\mathrm{u}}\eta = l_{\mathrm{u}}\sqrt{b_2 + a_2(\widehat{H} - z)}, \qquad \frac{dx}{dz} = -\frac{l_{\mathrm{u}} a_2}{2\eta}, \qquad \frac{d}{dz} = \frac{dx}{dz}\frac{d}{dx}$$

$$\frac{d}{dz}\begin{pmatrix} \mathrm{ber}_0(x) \\ \mathrm{bei}_0(x) \\ \mathrm{ker}_0(x) \\ \mathrm{kei}_0(x) \end{pmatrix}_{\mathrm{u}} = -\frac{l_{\mathrm{u}} a_2}{2\sqrt{2}\eta}\begin{pmatrix} \mathrm{ber}_1(x) + \mathrm{bei}_1(x) \\ -\mathrm{ber}_1(x) + \mathrm{bei}_1(x) \\ \mathrm{ker}_1(x) + \mathrm{kei}_1(x) \\ -\mathrm{ker}_1(x) + \mathrm{kei}_1(x) \end{pmatrix} \tag{13.150b}$$

The functions $\mathrm{ber}_0(x)$, $\mathrm{bei}_0(x)$, $\mathrm{ker}_0(x)$, $\mathrm{kei}_0(x)$, $\mathrm{ber}_1(x)$, $\mathrm{bei}_1(x)$, $\mathrm{kei}_1(x)$, and $\mathrm{ker}_1(x)$ are tabulated, for example, in the *Handbook of Mathematical Functions* by Abramowitz and Segun (1968).

In order to evaluate the six integration constants it is best to introduce a compact notation. All parts appearing in (13.140), (13.146), and (13.147) are abbreviated as

$$\begin{aligned} &A_z = \mathrm{ber}_0[l_1\xi(z)], && B_z = \mathrm{bei}_0[l_1\xi(z)], && C_z = \mathrm{ker}_0[l_1\xi(z)] \\ &D_z = \mathrm{kei}_0[l_1\xi(z)], && E_z = \mathrm{ber}_0[l_{\mathrm{u}}\eta(z)], && F_z = \mathrm{bei}_0[l_{\mathrm{u}}\eta(z)] \\ &G_z = \mathrm{ker}_0[l_{\mathrm{u}}\eta(z)], && H_z = \mathrm{kei}_0[l_{\mathrm{u}}\eta(z)], && I_z = \exp(Az)\cos(Az) \\ &J_z = \exp(Az)\sin(Az), && K_z = \exp(-Az)\cos(Az), && L_z = -\exp(-Az)\sin(Az) \end{aligned} \tag{13.151}$$

By using this compact notation it is a simple matter to write down the six boundary statements:

At $z = z_{\mathrm{Pr}}$ – continuity of stresses:

$$\frac{u_*}{k}\left(\frac{1}{z_{\mathrm{Pr}}} + \frac{\alpha}{2L_*}\right) = c_1\frac{d}{dz}(A_z + iB_z)_{z_{\mathrm{Pr}}} + d_1\frac{d}{dz}(C_z + iD_z)_{z_{\mathrm{Pr}}} \tag{13.152}$$

At $z = h_1$ – continuity of velocities:

$$c_1(A_z + iB_z)_{h_1} + d_1(C_z + iD_z)_{h_1} = c_{\mathrm{m}}(I_z + iJ_z)_{h_1} + d_{\mathrm{m}}(K_z + iL_z)_{h_1} \tag{13.153}$$

At $z = h_2$ – continuity of velocities:

$$u_g + c_u(E_z + iF_z)_{h_2} + d_u(G_z + iH_z)_{h_2} = u_g + c_m(I_z + iJ_z)_{h_2} + d_m(K_z + iL_z)_{h_2} \tag{13.154}$$

At $z = \widehat{H}$ – continuity of velocities:

$$0 = c_u(E_z + iF_z)_{\widehat{H}} + d_u(G_z + iH_z)_{\widehat{H}} \tag{13.155}$$

At $z = h_1$ – continuity of stresses:

$$c_l \frac{d}{dz}(A_z + iB_z)_{h_1} + d_l \frac{d}{dz}(C_z + iD_z)_{h_1} = c_m \frac{d}{dz}(I_z + iJ_z)_{h_1} + d_m \frac{d}{dz}(K_z + iL_z)_{h_1} \tag{13.156}$$

At $z = h_2$ – continuity of stresses:

$$c_u \frac{d}{dz}(E_z + iF_z)_{h_2} + d_u \frac{d}{dz}(G_z + iH_z)_{h_2} = c_m \frac{d}{dz}(I_z + iJ_z)_{h_2} + d_m \frac{d}{dz}(K_z + iL_z)_{h_2} \tag{13.157}$$

These equations are sufficient to evaluate the real and imaginary parts of the constants. Once the constants for a given distribution of the exchange coefficient are known, we can use the various solutions to construct the wind profile.

In principle it is possible to extend the solution method by representing the vertical profiles of the exchange coefficients by more than three linear parts. For each additional part one obtains two more integration constants that have to be determined by formulating additional boundary conditions analogously to those stated above.

In order to investigate the quality of the solution for the composite Ekman layer, numerous case studies aiming to obtain an optimal choice of the subdivision of the exchange coefficient profile have been performed. Some examples will now be presented. Figure 13.9 depicts the wind profiles for the classical approach (curve 2) and three different solutions of the composite Ekman layer (curves 3–5) corresponding to an approximation of the exchange-coefficient profile by three, four, and five linear sections. Moreover, the results are compared with a numerical solution resulting from the O'Brien exchange coefficient (curve 1) which is treated here as the reference case. The upper left-hand panel shows the vertical distributions of the exchange coefficient. The vertical profiles of the horizontal wind components are depicted in the other two panels. The geostrophic wind was chosen as $(u_g, v_g) = (10, 0)$ m s^{-1}.

From all case studies the following conclusions are drawn.

(i) The composite solutions yield a distinct improvement over the classical Ekman solution. The three-section solution is already sufficient to obtain a substantial improvement. An additional refinement of the exchange-coefficient profile to four or five sublayers has no great effect.

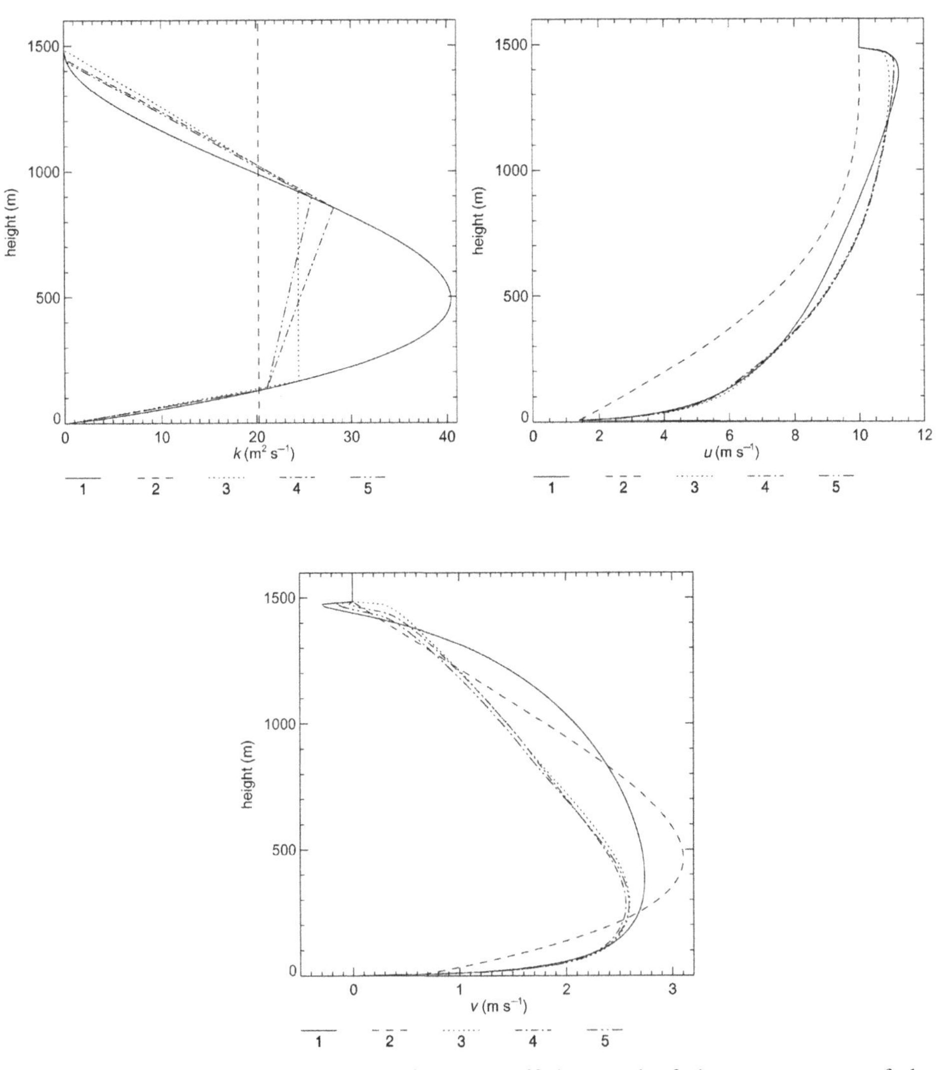

Fig. 13.9 Vertical profiles of the exchange coefficient and of the components of the horizontal wind. Curve 1: the profile according to O'Brien (1970); curve 2: the classical solution; curves 3, 4, and 5: composite solutions with three, four, and five linear sections.

(ii) Increasing the resolution of the composite profiles in the middle region of the boundary layer has only a minor influence on the results.

(iii) A rough estimate of the maximum value of the exchange coefficient is sufficient to construct the composite profiles since the results are barely affected by the particular choice of this value.

(iv) The undershooting of the v-component at the top of the boundary layer is obtained only if the resolution of the exchange-coefficient profile is fine enough there.

(v) The cross-isobar angle assumes reasonable values.

The analytic solution can also be used to check the accuracy of other approximate methods for obtaining the vertical wind profile in the boundary layer. The mathematical development of this section is based on our own unpublished lecture notes.

13.10 Ekman pumping

We conclude this chapter by studying in a very approximate manner the interaction of the atmospheric boundary layer with the free atmosphere; see Figure 13.1. From Figure 13.7 we recognize that the horizontal wind vector below the geostrophic-wind height has a component pointing from high to low pressure. This deviation is responsible for a mass transport perpendicular to the isobars, resulting in an equalization of the large-scale pressure field. It is well known from synoptic observations that the large-scale geostrophic wind is horizontally not uniform. The task ahead is to investigate the influence of the horizontal variation of the geostrophic wind on the Ekman layer. As before, to keep things simple, we assume that the geostrophic wind is directed to the east so that $v_g = 0$. The geostrophic-wind component u_g, however, varies in the northerly direction so that $u_g = u_g(y)$. Furthermore, we assume that the exchange coefficient K and the Coriolis parameter f do not vary in space. From the classical Ekman solution (13.122) we find the wind variation in the northerly direction to be given by

$$\frac{\partial \widehat{u}}{\partial y} = \frac{\partial u_g}{\partial y}\big[1 - \exp(-Az)\cos(Az)\big], \qquad \frac{\partial \widehat{v}}{\partial y} = \frac{\partial u_g}{\partial y}\exp(-Az)\sin(Az) \tag{13.158}$$

Since $\widehat{u}$ does not vary in the easterly direction the continuity equation for an incompressible fluid reduces to

$$\frac{\partial \widehat{v}}{\partial y} + \frac{\partial \widehat{w}}{\partial z} = 0 \tag{13.159}$$

Height integration gives the vertical wind component at the geostrophic-wind height:

$$\widehat{w}(z_g) = -\int_0^{z_g} \frac{\partial u_g}{\partial y}\exp(-Az)\sin(Az)\,dz \tag{13.160}$$

with $\widehat{w}(z = 0) = 0$. Hence, the variation of the wind component v in the northerly direction induces a vertical wind. Assuming barotropic conditions so that the geostrophic wind u_g is height-independent, we can carry out the integration without difficulty. The result is

$$\widehat{w}(z_g) = -\frac{\partial u_g}{\partial y}\,\frac{1+\exp(-\pi)}{2A} = \zeta_g\sqrt{K'}$$
$$\text{with}\quad \zeta_g = -\frac{\partial u_g}{\partial y},\qquad \sqrt{K'} = \frac{1+\exp(-\pi)}{2A} \approx \sqrt{\frac{K}{2f}} \tag{13.161}$$

where use of (13.121) was made. In this equation the *geostrophic vorticity* ζ_g has been introduced. The sign of $\widehat{w}(z_g)$ is controlled by the sign of the geostrophic vorticity. For cyclonic circulations $\zeta_g > 0$ so that $\widehat{w}(z_g) > 0$; for anticyclonic circulations $\zeta_g < 0$ so that $\widehat{w}(z_g) < 0$. For typical large-scale systems the geostrophic vorticity is about 10^{-5} s^{-1}. Assuming a value of $K = 5\text{–}10$ m^2 s^{-1} and a midlatitude Coriolis parameter $f = 10^{-4}$ s^{-1}, we find that the vertical velocity amounts to a few tenths of a centimeter per second. Certainly, such small vertical velocities cannot be measured. Nevertheless, these small values of $\widehat{w}$ are sufficient to influence significantly the life times of synoptic systems.

According to (13.161) the vertical motion induced by boundary-layer friction causes a large-scale vertical motion leading to the breakdown of high-pressure systems while low-pressure systems are filling up. Thus the large-scale pressure gradient becomes very small so that the geostrophic wind and the geostrophic vorticity cease to exist. It takes large-scale processes to create new pressure gradients.

Let us estimate how long it takes, for example, to fill up a cyclonic system. Instead of computing the change in pressure we may just as well compute the change with time of the geostrophic vorticity. We start with a simplified form of the barotropic vorticity equation (10.146):

$$\frac{d\eta_g}{dt} = \frac{d}{dt}(\zeta_g + f) = -\eta_g\,\nabla_h\cdot\widehat{\mathbf{v}}_h = \eta_g\,\frac{\partial\widehat{w}}{\partial z} \approx f\,\frac{\partial\widehat{w}}{\partial z} \tag{13.162}$$

since usually $\zeta_g \ll f$. Ignoring the spatial variation of the Coriolis parameter, we find

$$\frac{d\zeta_g}{dt} = f\,\frac{\partial\widehat{w}}{\partial z} \tag{13.163}$$

This expression will be integrated with respect to height from the top of the boundary layer z_g to the top of the atmosphere z_T, yielding

$$\int_{z_g}^{z_T}\frac{d\zeta_g}{dt}\,dz = f\left[\widehat{w}(z_T) - \widehat{w}(z_g)\right] \tag{13.164}$$

Assuming that the geostrophic wind is independent of height even above z_g, the geostrophic vorticity is height-independent also, so (13.164) can be integrated directly, yielding

$$(z_T - z_g)\,\frac{d\zeta_g}{dt} = f\left[\widehat{w}(z_T) - \widehat{w}(z_g)\right] \tag{13.165}$$

The vertical velocity vanishes at the top of the atmosphere and $z_T - z_g$ is about z_T. Therefore, we find

$$\frac{d\zeta_g}{dt} = -\frac{\zeta_g}{z_T}\sqrt{\frac{Kf}{2}} \tag{13.166}$$

where we have replaced $w(z_g)$ with the help of (13.161). This differential equation can be solved immediately to give

$$\boxed{\zeta_g(t) = \zeta_g(0)\exp\left(-\sqrt{\frac{Kf}{2z_T^2}}t\right)} \tag{13.167}$$

showing that the vorticity is decreasing exponentially in time. The reason for this is the existence of the ageostrophic wind component which resulted from the turbulent viscosity of the air. This behavior of the atmosphere is called *Ekman pumping*.

Let us estimate the time it takes for the vorticity to decrease to $1/e$ of its original value. Obviously this relaxation time t_e can be estimated from

$$t_e = \sqrt{\frac{2z_T^2}{Kf}} \tag{13.168}$$

Using $z_T = 10^4$ m, $K = 10$ m^2 s^{-1}, and $f = 10^{-4}$ s^{-1} we find that t_e amounts to about four days. It takes about nine days for the geostrophic vorticity to decay to 10% of its original value, which is in rough agreement with the life time of a low-pressure system.

Finally, let us recall that no frictional effects were included in the simplified vorticity equation (13.162). Therefore, the frictional effect observed within the Ekman layer on low- and high-pressure systems is indirect. On retracing the steps leading to (13.161) we find that a vertical circulation above the Ekman layer was induced due to the divergence of the ageostrophic wind component. By means of the divergence term on the right-hand side of the vorticity equation we finally obtained (13.166), showing that the vorticity is decreasing with time due to the turbulent viscosity of the air. This indirect frictional effect on the synoptic systems is known as the *spin down* and t_e is the *spin-down time*.

Etling (1996) continued this discussion by posing the question of whether the residual turbulent viscosity existing in the free atmosphere is sufficient to stop the circulation in low- and high-pressure systems. He drew the conclusion that the effect of the turbulent viscosity in the free atmosphere plays an unimportant role in the dynamics of synoptic systems. The 10% of the entire atmospheric mass contained in the atmospheric boundary layer is mainly responsible for the destruction of the atmospheric pressure systems.

13.11 Appendix A: Dimensional analysis

By necessity our discussion must be brief. There exist many excellent books on dimensional analysis. We refer to *Dimensionless Analysis and Theory of Models* by Langhaar (1967), where numerous practical examples are given.

13.11.1 The dimensional matrix

Let us consider a fluid problem involving the variables velocity V, length L, force F, density ρ, dynamic molecular viscosity μ, and acceleration due to gravity g. These variables can be expressed in terms of the fundamental variables mass M, length L, and time T as expressed by the following array of numbers, which is called the *dimensional matrix*:

$$\begin{array}{c|cccccc} & V & L & F & \rho & \mu & g \\ \hline M & 0 & 0 & 1 & 1 & 1 & 0 \\ L & 1 & 1 & 1 & -3 & -1 & 1 \\ T & -1 & 0 & -2 & 0 & -1 & -2 \end{array} \tag{13.169}$$

Consider, for example, the force F. The dimension of force is $\mathrm{M}^1\,\mathrm{L}^1\,\mathrm{T}^{-2}$, explaining the entries in the force column. The dimensions of the remaining columns can be written down analogously.

Any product Π of the variables $V, L, F, \ldots, g$ has the form

$$\Pi = V^{k_1} L^{k_2} F^{k_3} \rho^{k_4} \mu^{k_5} g^{k_6} \tag{13.170}$$

Whatever the values of the k's may be, the corresponding dimension of Π is given by

$$\begin{aligned} [\Pi] &= \left[L^1T^{-1}\right]^{k_1}\left[L^1\right]^{k_2}\left[M^1L^1T^{-2}\right]^{k_3}\left[M^1L^{-3}\right]^{k_4}\left[M^1L^{-1}T^{-1}\right]^{k_5}\left[L^1T^{-2}\right]^{k_6} \\ &= \left[M^{k_3+k_4+k_5}\right]\left[L^{k_1+k_2+k_3-3k_4-k_5+k_6}\right]\left[T^{-k_1-2k_3-k_5-2k_6}\right] \end{aligned} \tag{13.171}$$

If the product Π is required to be dimensionless we must demand that the exponents of the various basic variables add up to zero:

$$k_3+k_4+k_5=0, \quad k_1+k_2+k_3-3k_4-k_5+k_6=0, \quad -k_1-2k_3-k_5-2k_6=0 \tag{13.172}$$

Note that the coefficients multiplying the k_i, including the zeros, in each equation are a row of numbers in the dimensional matrix (13.169). Therefore, the equations for the exponents of a dimensionless product can be written down directly from the dimensional matrix. This set of three equations in six unknowns is underdetermined, possessing an infinite number of solutions. In this case we may arbitrarily assign values to three of the k's, say (k_1, k_2, k_3), and then solve the system of equations for the remaining k_i. The solution is given by

$$k_4 = \tfrac{1}{3}(k_1+2k_2+3k_3), \quad k_5 = \tfrac{1}{3}(-k_1-2k_2-6k_3), \quad k_6 = \tfrac{1}{3}(-k_1+k_2) \tag{13.173}$$

We choose (k_1, k_2, k_3) values such that fractions will be avoided. Three choices of numbers for (k_1, k_2, k_3) together with the resulting values for (k_4, k_5, k_6) are

$$\begin{aligned} &k_1 = 1, && k_2 = 1, && k_3 = 0 \Longrightarrow k_4 = 1, && k_5 = -1, && k_6 = 0 \\ &k_1 = -2, && k_2 = -2, && k_3 = 1 \Longrightarrow k_4 = -1, && k_5 = 0, && k_6 = 0 \\ &k_1 = 2, && k_2 = -1, && k_3 = 0 \Longrightarrow k_4 = 0, && k_5 = 0, && k_6 = -1 \end{aligned} \tag{13.174}$$

All six k_i are then substituted into (13.170). For each choice of (k_1, k_2, k_3) with the resulting (k_4, k_5, k_6) we obtain a dimensionless universal Π-number. These numbers are known as the *Reynolds number* Re, the *pressure number* P, and the *Froude number* Fr:

$$\Pi_1 = Re = \frac{VL\rho}{\mu}, \qquad \Pi_2 = P = \frac{F}{V^2L^2\rho} = \frac{p}{V^2\rho}, \qquad \Pi_3 = Fr = \frac{V^2}{Lg} \tag{13.175}$$

The procedure is quite arbitrary; any values might be chosen for (k_1, k_2, k_3). Suppose that we choose $k_1 = 10$, $k_2 = -5$, and $k_3 = 8$. Then we get $k_4 = 8$, $k_5 = -16$, and $k_6 = -5$. The resulting dimensionless number is given as

$$\Pi = V^{10}L^{-5}F^8\rho^8\mu^{-16}g^{-5} \tag{13.176}$$

This product looks very complicated but it does not really give new information since we can write this expression as the product

$$\Pi = P^8 Re^{16} Fr^5 \tag{13.177}$$

Regardless of the values we assign to (k_1, k_2, k_3), the resulting dimensionless product can be expressed as a product of powers of P, Re, and Fr. This fact and the condition that P, Re, and Fr are independent of each other characterize them as a complete set or group of dimensionless products. We may define a complete set as follows.

Definition: A set of dimensionless products of given variables is complete, if each product in the set is independent of the others and every other dimensionless product of the variables is the product of powers of dimensionless products in the set.

Application of linear algebra to the theory of dimensional analysis resulted in the following theorem.

Theorem: The number of dimensionless products in a complete set is equal to the total number n of variables minus the rank r of the dimensional matrix.

This theorem does not tell us the exact form of the dimensional products but it does tell us the number of universal products we should be looking for. The theorem is of great help. For the present case we have $n = 6$ variables. The rank of a matrix is defined as the order of the largest nonzero determinant that can be obtained from the elements of the matrix. In case of the dimensional matrix (13.169) the rank is $r = 3$, so the complete set consists of three independent dimensionless numbers. These are the products Re, P, and Fr which are considered to be universal numbers.

13.11.2 The Buckingham Π-theorem

Much of the theory of dimensional analysis is contained in this celebrated theorem which applies to dimensionally homogeneous equations. In simple words, an equation is dimensionally homogeneous if each term in the equation has the same dimension. Empirical equations are not necessarily dimensionally homogeneous.

Theorem: If an equation is dimensionally homogeneous, it can be reduced to a relationship among members of a complete set of dimensionless products.

If n variables are connected by an unknown dimensionally homogeneous equation Buckingham's Π-theorem allows us to conclude that the equation can be expressed in the form of a relationship among $n - r$ dimensionless products, where $n - r$ is the number of products in the complete set. It turns out that, in many cases, r is the number of fundamental dimensions in a problem. This was also the case above.

13.11.3 Examples from boundary-layer theory

Example 1: Observational data show that the wind profile in the neutral Prandtl layer is determined by z, u_*, and $d\widehat{u}/dz$. Our task is to find the functional relation among these three quantities. The dimensional matrix

$$\begin{array}{cccc} & z & u_* & d\widehat{u}/dz \\ L & 1 & 1 & 0 \\ T & 0 & -1 & -1 \end{array} \tag{13.178}$$

allows us to write down two linear equations from which the k_i may be determined:

$$k_1 + k_2 = 0, \qquad -k_2 - k_3 = 0 \tag{13.179}$$

The number of variables is $n = 3$ and the rank is $r = 2$. The rank and the number of fundamental variables coincides. With $n - r = 1$ we expect one dimensionless universal Π-number. Choosing $k_1 = 1$, we obtain $k_2 = -1$ and $k_3 = 1$. The universal number is taken as $1/k$, where k is the Von Karman constant. The result is the logarithmic wind profile

$$\Pi = \frac{z}{u_*}\frac{d\widehat{u}}{dz} = \frac{1}{k} \tag{13.180}$$

Example 2: For the non-neutral Prandtl layer an additonal variable is needed, which is the MO stability length L_*. Now the dimensional matrix must be found from the four variables $z,\ u_*,\ d\widehat{u}/dz$, and L_*:

$$\begin{array}{ccccc} & z & u_* & d\widehat{u}/dz & L_* \\ L & 1 & 1 & 0 & 1 \\ T & 0 & -1 & -1 & 0 \end{array} \tag{13.181}$$

Since the rank is $r = 2$, we expect two universal Π-numbers. We have two equations in four unknowns:

$$k_1 + k_2 + k_4 = 0, \qquad -k_2 - k_3 = 0 \tag{13.182}$$

We specify $(k_1,\ k_2)$ and obtain $(k_3,\ k_4)$:

$$\begin{array}{lllll} k_1 = 1, & k_2 = 0 \implies k_3 = 0, & k_4 = -1, & \Pi_1 = \dfrac{z}{L_*} \\ k_1 = 1, & k_2 = -1 \implies k_3 = 1, & k_4 = 0, & \Pi_2 = \dfrac{kz}{u_*}\dfrac{d\widehat{u}}{dz} \end{array} \tag{13.183}$$

For convenience we have included the constant k in Π_2. Formal application of the Buckingham Π-theorem gives

$$f_1(\Pi_1, \Pi_2) = 0 \quad \text{or} \quad \Pi_2 = S(\Pi_1) \implies \frac{kz}{u_*}\frac{d\widehat{u}}{dz} = S\left(\frac{z}{L_*}\right) \tag{13.184}$$

From boundary-layer theory we know that, for the present problem, the MO function $S(z/L_*)$ is a universal number.

13.12 Appendix B: The mixing length

The mixing length was introduced by equation (13.51) on purely dimensional grounds. Prandtl (1925) introduced this concept in analogy to the mean free path of the kinetic gas theory. For a given density of molecules (number of molecules per unit volume) there exists an average distance that a molecule may traverse before it collides with another molecule. This is the *mean free path*, which is about 10^{-7} m for standard atmospheric conditions. During each collision momentum will be exchanged. Moreover, molecular collisions cause the molecular viscosity or internal friction.

Prandtl introduced a mixing length l for turbulent motion, analogous to the free path on the molecular scale, by assuming that a portion of fluid or an eddy originally at a certain level in the fluid suddenly breaks away and then travels a certain distance. While the eddy is traveling it conserves most of its momentum until it mixes with the mean flow at some other level. Let us assume that we have an average horizontal velocity at level z. An eddy originating at this level carries the horizontal momentum of this level. After traversing the mixing length l' at the level $z + l'$ it will cause the turbulent fluctuation

$$\mathbf{v}_{\mathrm{h}}'' = \widehat{\mathbf{v}}_{\mathrm{h}}(z) - \widehat{\mathbf{v}}_{\mathrm{h}}(z + l') = -l' \frac{\partial \widehat{\mathbf{v}}_{\mathrm{h}}}{\partial z} \tag{13.185}$$

if the Taylor expansion is discontinued after the linear term. According to (13.9) and (11.36) the stress vector of the eddy is defined by

$$\mathbf{T} = \mathbf{i}_3 \cdot \mathbb{R} = -\overline{\rho w'' \mathbf{v}''} = -\overline{\rho}\,\overline{w'' \mathbf{v}_{\mathrm{h}}''} \tag{13.186}$$

We have ignored the molecular contribution and the density fluctuations and have retained only the horizontal part of the fluctuation vector. On substituting the fluctuation $\mathbf{v}_{\mathrm{h}}''$ into (13.186) we find that the stress vector of the eddy may be expressed in terms of the vertical gradient of the mean velocity:

$$\mathbf{T} = \overline{\rho}\,\overline{l' w''} \frac{\partial \widehat{\mathbf{v}}_{\mathrm{h}}}{\partial z} \tag{13.187}$$

For continuity of mass we require

$$\left|\mathbf{v}_{\mathrm{h}}''\right| = \left|w''\right| \tag{13.188}$$

For the upward and downward eddies we must have

$$w'' > 0, \qquad l' > 0 \qquad \text{or} \qquad w'' < 0, \qquad l' < 0 \tag{13.189}$$

so that the product of the velocity fluctuation w'' and the mixing length l' is always a positive quantity and the signs of the corresponding $w''$and l' are always identical. From (13.185) it follows that the fluctuation may be expressed as

$$w'' = l' \left| \frac{\partial \widehat{\mathbf{v}}_{\mathrm{h}}}{\partial z} \right| \tag{13.190}$$

Thus the stress vector of the eddy may be written as

$$\mathbf{T} = \overline{\rho l' l'} \left| \frac{\partial \widehat{\mathbf{v}}_{\rm h}}{\partial z} \right| \frac{\partial \widehat{\mathbf{v}}_{\rm h}}{\partial z} = \overline{\rho} l^2 \left| \frac{\partial \widehat{\mathbf{v}}_{\rm h}}{\partial z} \right| \frac{\partial \widehat{\mathbf{v}}_{\rm h}}{\partial z} \quad \text{with} \quad l^2 = \overline{l' l'} \tag{13.191}$$

where l is the mean mixing length. Taking the scalar product $\mathbf{i}_1 \cdot \mathbf{T}$ and assuming that the mean flow is along the x-axis only, we obtain

$$\mathbf{i}_1 \cdot \mathbf{T} = \tau = \overline{\rho} l^2 \left(\frac{\partial \widehat{u}}{\partial z} \right)^2 = A^v \frac{\partial \widehat{u}}{\partial z} \quad \text{with} \quad A^v = \overline{\rho} l^2 \frac{\partial \widehat{u}}{\partial z} = \overline{\rho} K^v \tag{13.192}$$

This expression is the definition (13.52). A^v is known as the *Austauschkoeffizient*. The similarity to the molecular situation becomes even more apparent on comparison with (1.12) defining the molecular stress tensor. On choosing the velocity vector $\mathbf{v}_{\rm a} = \widehat{u}(z)\mathbf{i}_1$ we find for the molecular stress the expression

$$\tau_{\rm mol} = \mathbf{i}_1 \cdot (\mathbf{i}_3 \cdot \mathbb{J}) = \mu \frac{\partial \widehat{u}}{\partial z} \tag{13.193}$$

which is identical in form with (13.192).

Our faith in Prandtl's formulation of l should not be unlimited since the stress vector depends only on the local vertical gradient, which is not always the case, as follows from the transilient-mixing formulation mentioned briefly in Chapter 11. This appendix follows Pichler (1997) to some extent. Similar treatments may be found in many other textbooks.

13.13 Problems

13.1: Apply the Prandtl layer conditions to equation (11.45) for the mean motion. Show that this results in the condition τ = constant.

13.2: In case of thermal turbulence for windless conditions (local free convection) the dimensional matrix may be constructed from the variables $\partial\widetilde{\theta}/\partial z$, $H/(\overline{\rho} c_{p,0})$, $g/\overline{T}$, and z. Find the Π-number.

13.3: Let us return to the previous chapter. In the inertial subrange ϵ can be expressed as $\epsilon(k) = f_1(k, \overline{\epsilon}_{\rm M})$.
(a) Use Buckingham's Π-theorem to find the Π-number and then set $\Pi = \kappa_{\rm K}$ to obain ϵ.
(b) Now include the dissipation range so that $\epsilon(k) = f_2(k, \overline{\epsilon}_{\rm M}, \nu)$. Find the Π-numbers.

13.4:
(a) Show that (13.88) satisfies the differential equation (13.87).
(b) Suppose that $\xi \to 0$. Is $S_1(0)$ defined and what is its value? Use equation (13.88) to prove your result.

13.5: Find S_1 from (13.84) and (13.88) for $L_* = 8$ m representing a stable atmosphere. Choose $z = 10$ m. Discuss your results by comparing them with the empirical value $S_1 = 0.872$.

13.6: It has been postulated that, within the Prandtl layer,

$$\widehat{u''w''} = -k^2 \frac{\left(\dfrac{\partial \widehat{u}}{\partial z}\right)^4}{\left(\dfrac{\partial^2 \widehat{u}}{\partial z^2}\right)^2}$$

Assume that $-\overline{\rho}\widehat{u''w''} = \tau$. Integrate this expression to find the logarithmic wind profile for neutral conditions.

13.7: Solve the Ekman-spiral problem with the conditions stated in the text. Instead of $\widehat{u}(z = 0) = 0$, $\widehat{v}(z = 0) = 0$ use the lower boundary condition

$$z \to 0: \qquad \widehat{u} + i\widehat{v} = C \frac{\partial}{\partial z}(\widehat{u} + i\widehat{v})$$

where C is a real constant. The cross-isobar angle α_0 at the ground may be specified. Find the height of the geostrophic wind level. Hint: Observe that the integration constants in equation (13.120) in general are complex numbers.

14
Wave motion in the atmosphere

14.1 The representation of waves

It is well known that the nonlinear system of the atmospheric equations includes a great number of very complex wave motions occurring on various spatial and time scales. In this chapter we wish to treat some simple types of wave motion that can be isolated from the linearized system of atmospheric equations. This system is obtained with the help of perturbation theory. Before introducing this theory we will briefly discuss the wave concept. For simplicity we are going to employ rectangular coordinates.

A periodic process occurring in time or in space is called an *oscillation*. If both time and space are involved we speak of *wave motion*. Let us consider the scalar wave equation

$$\nabla^2 U = \frac{1}{c^2}\frac{\partial^2 U}{\partial t^2} \tag{14.1a}$$

where U describes some atmospheric field or a component of a vector field and c the speed of propagation of the wave. For simplicity let us consider only the z-direction of propagation so that (14.1a) reduces to

$$\frac{\partial^2 U}{\partial z^2} = \frac{1}{c^2}\frac{\partial^2 U}{\partial t^2} \tag{14.1b}$$

A solution to this equation is given by

$$U(z, t) = U_0 \cos(kz - \omega t) \qquad \text{with} \qquad \omega/k = c \tag{14.2}$$

provided that the ratio of the constants ω and k is equal to the constant c. The particular solution (14.2) is known as a *plane harmonic wave* whose amplitude is U_0.

A graph of the function $U(z, t)$ is shown in Figure 14.1. For a given value of the spatial coordinate z the wave function $U(z, t)$ varies harmonically in time. The

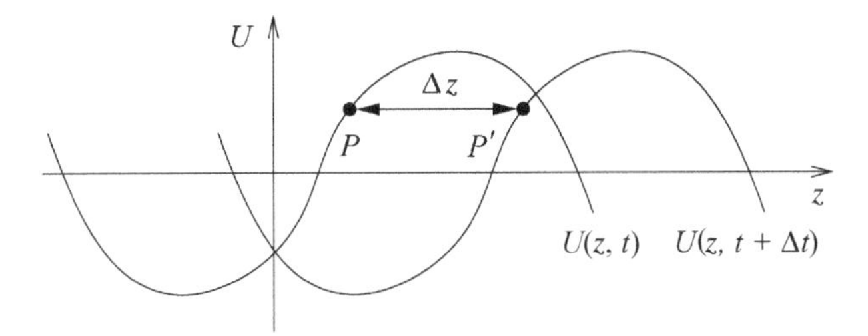

Fig. 14.1 A graph of U versus z at times t and $t + \Delta t$.

angular frequency of the variation with time is the constant ω. At a given instant of time the wave varies sinusoidally with z. Since the argument of the trigonometric function must be dimensionless, the constant k, known as the *wavenumber*, must have the dimension of an inverse length. Thus, k is the number of complete wave cycles in a distance of 2π units. From a study of the graph of $U(z,t)$ we see that, at a certain instant in time, the curve is a certain cosine function whereas at $t + \Delta t$ the entire curve is displaced by the distance $\Delta z = c\,\Delta t$ in the z-direction. Δz is the distance between any two points of equal phase, as shown in Figure 14.1. For this reason c is called the *phase speed.*

Let us now return to the three-dimensional scalar wave equation (14.1) which is satisfied by the three-dimensional plane harmonic wave function

$$U(x,y,z,t) = U_0 \cos(\mathbf{k}\cdot\mathbf{r} - \omega t) \quad \text{with} \quad \mathbf{k} = \mathbf{i}_1 k_x + \mathbf{i}_2 k_y + \mathbf{i}_3 k_z \tag{14.3}$$

Here $\mathbf{r}$ is the position vector and $\mathbf{k}$ the propagation vector or *wave vector.* Often the solution is written in the following equivalent form

$$U = U_0 \cos\omega\left(\frac{\mathbf{k}\cdot\mathbf{r}}{kc} - t\right) = U_0 \cos\omega\left(\frac{\mathbf{n}\cdot\mathbf{r}}{c} - t\right) \tag{14.4}$$

where $\mathbf{n} = \mathbf{k}/k$ is the unit normal.

In order to interpret equation (14.3) let us consider constant values of the argument of the cosine function

$$\mathbf{k}\cdot\mathbf{r} - \omega t = k_x x + k_y y + k_z z - \omega t = \text{constant} \tag{14.5}$$

which describes a set of planes called *surfaces of constant phase.* Consider the plane shown in Figure 14.2, where the vector $\mathbf{r}$ points to a general point on the plane while $\mathbf{k}$ is a vector that is perpendicular to the plane. The vector $\mathbf{u}$ in the plane is perpendicular to $\mathbf{k}$ so that their scalar product vanishes:

$$\mathbf{k}\cdot\mathbf{u} = \mathbf{k}\cdot(\mathbf{r} - \mathbf{k}) = 0 \tag{14.6a}$$

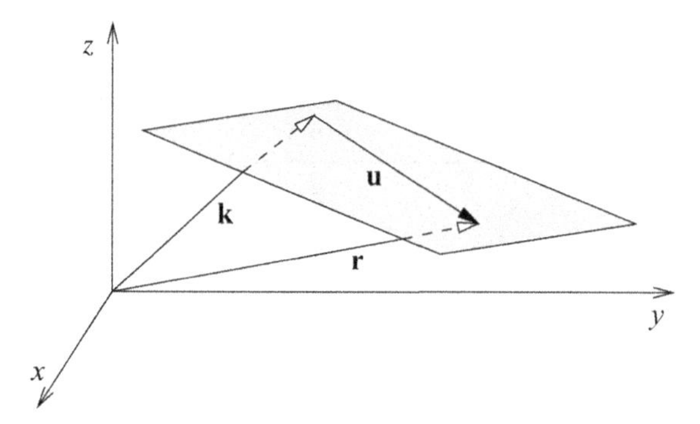

Fig. 14.2 The equation of the plane $\mathbf{k} \cdot \mathbf{r} = k^2$.

Since $\mathbf{k} \cdot \mathbf{k} = k_x^2 + k_y^2 + k_z^2 = k^2$ we obtain

$$\mathbf{k} \cdot \mathbf{r} = k_x x + k_y y + k_z z = k^2 \tag{14.6b}$$

The normal form of the equation of the plane is given by

$$\frac{k_x}{k} x + \frac{k_y}{k} y + \frac{k_z}{k} z = k \tag{14.6c}$$

Hence equation (14.5), indeed, is the equation representing plane surfaces of constant phase.

Let us return to the argument of the cosine function of equation (14.2), which is called the *phase of the harmonic wave*. There is no reason why the magnitude of the wave could not be anything one would like it to be at time $t = 0$ and at $z = 0$. This can be achieved by shifting the cosine function by introducing an initial phase ϵ so that the phase is given by

$$\varphi = kz - \omega t + \epsilon \tag{14.7}$$

Without loss of generality we will set $\epsilon = 0$. When we envision a harmonic wave sweeping by, we determine its speed by observing the motion of a point at which the magnitude of the disturbance remains constant. Thus, the speed of the wave is the speed at which the condition of constant phase travels, or

$$\left(\frac{dz}{dt}\right)_{\varphi=\text{constant}} = -\frac{\left(\dfrac{\partial \varphi}{\partial t}\right)_z}{\left(\dfrac{\partial \varphi}{\partial z}\right)_t} = \frac{\omega}{k} = \frac{2\pi \nu}{k} = \frac{L}{\tau} = c \quad \text{with} \quad L = \frac{2\pi}{k}, \quad \tau = \frac{1}{\nu} \tag{14.8}$$

Here L is the wavelength and τ the period of the wave.

In passing we would like to remark that the functions $\cos(kr - \omega t)$ have constant values on a sphere of radius r at a given time. As t increases the functions would represent spherically expanding waves except for the fact that they are not solutions of the wave equation. However, it is easy to verify that the function $U(r, t) = (1/r)\cos(kr - \omega t)$ is a solution of the wave equation

$$\frac{\partial^2(Ur)}{\partial r^2} = \frac{1}{c^2}\frac{\partial^2(Ur)}{\partial t^2} \tag{14.9}$$

14.2 The group velocity

In connection with the transport of energy by waves we need to discuss briefly the concept of the group velocity. When dealing with trigonometric functions it is often convenient to use the complex notation. Instead of (14.3) we introduce

$$U = U_0 \exp[i(\mathbf{k} \cdot \mathbf{r} - \omega t)] \tag{14.10}$$

It is understood that the real part is the actual physical quantity being represented. Now let us consider two harmonic waves that have slightly different angular frequencies $\omega + \Delta\omega$ and $\omega - \Delta\omega$. The corresponding wavenumbers will, in general, also differ. These shall be denoted by $k + \Delta k$ and $k - \Delta k$. Let us assume, in particular, that the two waves have the same amplitudes U_0 and are traveling in the same direction, which is taken to be the z-direction. Superposition of the two waves gives

$$U = U_0 \exp[i(k + \Delta k)z - i(\omega + \Delta\omega)t] + U_0 \exp[i(k - \Delta k)z - i(\omega - \Delta\omega)t] \tag{14.11a}$$

which can be rewritten as

$$U = U_0 \exp[i(kz - \omega t)]\,\{\exp[i(\Delta k\, z - \Delta\omega\, t)] + \exp[-i(\Delta k\, z - \Delta\omega\, t)]\} \tag{14.11b}$$

Using the Euler formula we obtain

$$U = 2U_0 \exp[i(kz - \omega\, t)]\cos(\Delta k\, z - \Delta\omega\, t) \tag{14.11c}$$

This expression can be regarded as a single wave described by $2U_0\, \exp[i(kz - \omega t)]$, which has a modulation envelope $\cos(\Delta k\, z - \Delta\omega\, t)$ as shown in Figure 14.3.

Generalization to three dimensions results in

$$U = 2U_0 \exp[i(\mathbf{k} \cdot \mathbf{r} - \omega\, t)]\cos(\Delta\mathbf{k} \cdot \mathbf{r} - \Delta\omega\, t) \tag{14.11d}$$

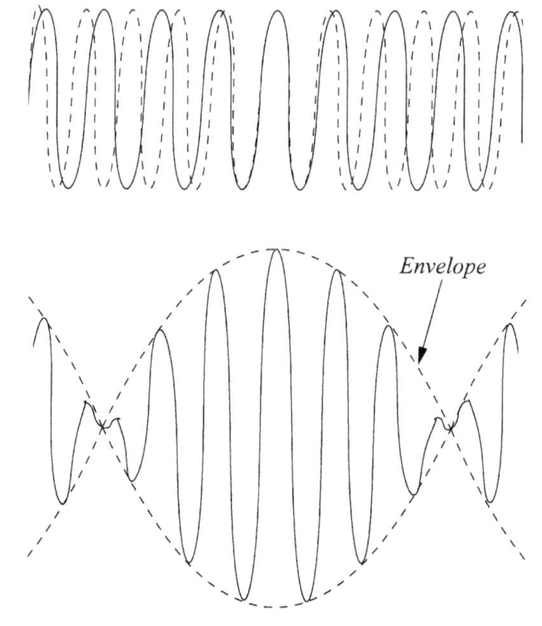

Fig. 14.3 The envelope of the combination of two harmonic waves.

where now $2U_0 \exp[i(\mathbf{k} \cdot \mathbf{r} - \omega t)]$ is the single wave and $\cos(\Delta\mathbf{k} \cdot \mathbf{r} - \Delta\omega t)$ the modulation envelope. Inspection of (14.11c) shows that the modulation amplitude does not travel with the phase velocity ω/k but at the rate $c_{\text{gr}} = \Delta\omega/\Delta k$, which is called the group velocity. In the limit c_{gr} is given by

$$\boxed{c_{\text{gr}} = \frac{d\omega}{dk}} \tag{14.12}$$

Utilizing (14.8) we find

$$c_{\text{gr}} = \frac{d\omega}{dk} = \frac{d}{dk}(kc) = c + k\,\frac{dc}{dk} = c - L\,\frac{dc}{dL} \tag{14.13}$$

showing that the group velocity differs from the phase velocity only if the phase velocity c explicitly depends on the wavelength. Rossby waves, these are large-scale synoptic waves, to be discussed later, exhibit this behavior. The dependency of the angular frequency on the wavenumber, that is $\omega = \omega(k)$, is known as the *dispersion relation*. Whenever $dc/dL > 0$ one speaks of *normal dispersion*; if $dc/dL < 0$ the dispersion is called *anomalous dispersion*. Very informative discussions on wave motion and group velocity can be found in many textbooks on optics and elsewhere. We refer to Fowles (1968).

The generalization of the group velocity to three dimensions is carried out by expanding the total derivative $d\omega$ as

$$\begin{aligned} d\omega &= \frac{\partial \omega}{\partial k_x} dk_x + \frac{\partial \omega}{\partial k_y} dk_y + \frac{\partial \omega}{\partial k_z} dk_z = \nabla_k \omega \cdot d\mathbf{k} \\ \text{with} \quad \nabla_k \omega &= \frac{\partial \omega}{\partial k_x} \mathbf{i}_1 + \frac{\partial \omega}{\partial k_y} \mathbf{i}_2 + \frac{\partial \omega}{\partial k_z} \mathbf{i}_3 \end{aligned} \tag{14.14}$$

Defining the components of the group velocity vector $\mathbf{c}_{\text{gr}}$ as

$$c_{\text{gr},x} = \frac{\partial \omega}{\partial k_x}, \qquad c_{\text{gr},y} = \frac{\partial \omega}{\partial k_y}, \qquad c_{\text{gr},z} = \frac{\partial \omega}{\partial k_z} \tag{14.15a}$$

we obtain

$$\mathbf{c}_{\text{gr}} = \nabla_k \omega \tag{14.15b}$$

14.3 Perturbation theory

In order to isolate some simple wave forms we need to linearize the basic atmospheric equations by means of the so-called *perturbation method*. The set of equations to be linearized consists of the three equations representing frictionless motion on the tangential plane,

$$\begin{aligned} \frac{du}{dt} &= \frac{\partial u}{\partial t} + u\frac{\partial u}{\partial x} + v\frac{\partial u}{\partial y} + w\frac{\partial u}{\partial z} = -\alpha\frac{\partial p}{\partial x} + fv \\ \frac{dv}{dt} &= \frac{\partial v}{\partial t} + u\frac{\partial v}{\partial x} + v\frac{\partial v}{\partial y} + w\frac{\partial v}{\partial z} = -\alpha\frac{\partial p}{\partial y} - fu \\ \frac{dw}{dt} &= \frac{\partial w}{\partial t} + u\frac{\partial w}{\partial x} + v\frac{\partial w}{\partial y} + w\frac{\partial w}{\partial z} = -\alpha\frac{\partial p}{\partial z} - g \end{aligned} \tag{14.16}$$

the continuity equation,

$$\alpha\left(\frac{\partial u}{\partial x} + \frac{\partial v}{\partial y} + \frac{\partial w}{\partial z}\right) - \left(\frac{\partial \alpha}{\partial t} + u\frac{\partial \alpha}{\partial x} + v\frac{\partial \alpha}{\partial y} + w\frac{\partial \alpha}{\partial z}\right) = 0 \tag{14.17}$$

and the first law of thermodynamics, which will be approximated by assuming that we are dealing with adiabatic processes:

$$de + p\,d\alpha = 0 \tag{14.18}$$

Here, $\alpha = 1/\rho$ is the specific volume. Since we are going to investigate the behavior of dry air only, we do not need to consider the equations for partial concentrations.

Treating the air as an ideal gas, we may substitute the ideal-gas law and the differential for the internal energy

$$p\alpha = R_0 T, \qquad de = c_{v,0}\, dT \tag{14.19}$$

into (14.18) to obtain, after some slight rearrangements,

$$p\kappa\left(\frac{\partial \alpha}{\partial t} + u\,\frac{\partial \alpha}{\partial x} + v\,\frac{\partial \alpha}{\partial y} + w\,\frac{\partial \alpha}{\partial z}\right) + \alpha\left(\frac{\partial p}{\partial t} + u\,\frac{\partial p}{\partial x} + v\,\frac{\partial p}{\partial y} + w\,\frac{\partial p}{\partial z}\right) = 0 \tag{14.20}$$

with $R_0 = c_{p,0} - c_{v,0}$ and $\kappa = c_{p,0}/c_{v,0}$.

We will now summarize the perturbation method.

(i) Any variable ψ is decomposed into a part representing the basic state ψ_0 and another part describing the disturbance ψ', which is also called the *perturbation*:

$$\psi = \psi_0 + \psi' \tag{14.21}$$

(ii) The basic state ψ_0 is considered known and must satisfy the original system of nonlinear equations.
(iii) The total motion, i.e. the basic field plus the disturbance, must satisfy the system of nonlinear equations.
(iv) The perturbations ψ' are assumed to be very small in comparison with the basic state ψ_0.
(v) Products of perturbations are ignored. This implies the linearization.

On applying (14.21) to the variables of motion and to the thermodynamic variables we find

$$(u, v, w, p, \alpha, T) = (u_0, v_0, w_0, p_0, \alpha_0, T_0) + (u', v', w', p', \alpha', T') \tag{14.22}$$

As an example we demonstrate the perturbation procedure by linearizing a simplified form of the advection equation:

$$\frac{\partial u}{\partial t} + u\,\frac{\partial u}{\partial x} = 0 \tag{14.23}$$

We split u according to (14.22) and obtain

$$\frac{\partial}{\partial t}(u_0 + u') + (u_0 + u')\,\frac{\partial}{\partial x}(u_0 + u') = 0 \tag{14.24a}$$

Owing to assumption (ii) we may write

$$\frac{\partial u_0}{\partial t} + u_0\,\frac{\partial u_0}{\partial x} = 0 \tag{14.24b}$$

By subtracting (14.24b) from (14.24a) we obtain the linearized equation

$$\frac{\partial u'}{\partial t} + u' \frac{\partial u_0}{\partial x} + u_0 \frac{\partial u'}{\partial x} = 0 \tag{14.24c}$$

from which, according to (v), the nonlinear term $u' \partial u'/\partial x$ has been omitted. In this very simple case the linearization procedure is very brief. For the more complex equations of the complete atmospheric system this procedure is very tedious, so a shortcut method would be welcome.

Such a method is the so-called *Bjerkness linearization procedure*, which results directly in the linearized equations. Suppose that we wish to linearize the product ab. This is done by varying the factor a to give $\delta a = a'$ and then multiplying a' by the basic state factor b_0. This is followed by obtaining another product term by multiplying the basic state a_0 by the variation $\delta b = b'$ of the factor b. In principle this is the product rule of differential calculus, which may be written as

$$\delta(ab) = (ab)' = (\delta a) b_0 + a_0\, \delta b = a' b_0 + a_0 b' \tag{14.25a}$$

Suppose that we wish to linearize the triple products abc occurring in (14.20). This is best done by combining two factors to give $d = bc$ and then using the rule (14.25a):

$$\delta(abc) = \delta(ad) = (\delta a) d_0 + a_0\, \delta d = a' b_0 c_0 + a_0 (b' c_0 + b_0 c') \tag{14.25b}$$

Application of the Bjerkness linearization rule results in the following system of atmospheric equations, which can be checked quickly and easily for accuracy. The equations of atmospheric motion are given by

$$\begin{aligned}
\frac{\partial u'}{\partial t} &= -u' \frac{\partial u_0}{\partial x} - v' \frac{\partial u_0}{\partial y} - w' \frac{\partial u_0}{\partial z} - u_0 \frac{\partial u'}{\partial x} - v_0 \frac{\partial u'}{\partial y} \\
&\quad - w_0 \frac{\partial u'}{\partial z} - \alpha' \frac{\partial p_0}{\partial x} - \alpha_0 \frac{\partial p'}{\partial x} + f v' \\
\frac{\partial v'}{\partial t} &= -u' \frac{\partial v_0}{\partial x} - v' \frac{\partial v_0}{\partial y} - w' \frac{\partial v_0}{\partial z} - u_0 \frac{\partial v'}{\partial x} - v_0 \frac{\partial v'}{\partial y} \\
&\quad - w_0 \frac{\partial v'}{\partial z} - \alpha' \frac{\partial p_0}{\partial y} - \alpha_0 \frac{\partial p'}{\partial y} - f u' \\
\frac{\partial w'}{\partial t} &= -u' \frac{\partial w_0}{\partial x} - v' \frac{\partial w_0}{\partial y} - w' \frac{\partial w_0}{\partial z} - u_0 \frac{\partial w'}{\partial x} - v_0 \frac{\partial w'}{\partial y} \\
&\quad - w_0 \frac{\partial w'}{\partial z} - \alpha' \frac{\partial p_0}{\partial z} - \alpha_0 \frac{\partial p'}{\partial z}
\end{aligned} \tag{14.26}$$

The adiabatic equation is

$$\begin{aligned}
&p'\kappa\left(\frac{\partial\alpha_0}{\partial t}+u_0\frac{\partial\alpha_0}{\partial x}+v_0\frac{\partial\alpha_0}{\partial y}+w_0\frac{\partial\alpha_0}{\partial z}\right)\\
&+\alpha'\left(\frac{\partial p_0}{\partial t}+u_0\frac{\partial p_0}{\partial x}+v_0\frac{\partial p_0}{\partial y}+w_0\frac{\partial p_0}{\partial z}\right)\\
&+p_0\kappa\left(\frac{\partial\alpha'}{\partial t}+u'\frac{\partial\alpha_0}{\partial x}+v'\frac{\partial\alpha_0}{\partial y}+w'\frac{\partial\alpha_0}{\partial z}+u_0\frac{\partial\alpha'}{\partial x}+v_0\frac{\partial\alpha'}{\partial y}+w_0\frac{\partial\alpha'}{\partial z}\right)\\
&+\alpha_0\left(\frac{\partial p'}{\partial t}+u'\frac{\partial p_0}{\partial x}+v'\frac{\partial p_0}{\partial y}+w'\frac{\partial p_0}{\partial z}+u_0\frac{\partial p'}{\partial x}+v_0\frac{\partial p'}{\partial y}+w_0\frac{\partial p'}{\partial z}\right)=0
\end{aligned}\tag{14.27}$$

which is easily identified by the appearance of the factor $\kappa = c_{p,0}/c_{v,0}$. The continuity equation is given by

$$\begin{aligned}
&\alpha'\left(\frac{\partial u_0}{\partial x}+\frac{\partial v_0}{\partial y}+\frac{\partial w_0}{\partial z}\right)+\alpha_0\left(\frac{\partial u'}{\partial x}+\frac{\partial v'}{\partial y}+\frac{\partial w'}{\partial z}\right)\\
&-\left(\frac{\partial\alpha'}{\partial t}+u'\frac{\partial\alpha_0}{\partial x}+v'\frac{\partial\alpha_0}{\partial y}+w'\frac{\partial\alpha_0}{\partial z}+u_0\frac{\partial\alpha'}{\partial x}+v_0\frac{\partial\alpha'}{\partial y}+w_0\frac{\partial\alpha'}{\partial z}\right)=0
\end{aligned}\tag{14.28}$$

Finally the linearized ideal-gas law is given by

$$p'\alpha_0 + p_0\alpha' = R_0T' \tag{14.29}$$

In order to isolate certain types of wave motion we are going to use trial solutions of the type

$$U = U_0\exp[i(k_x x + k_y y + k_z z - \omega t)] \tag{14.30}$$

In order to make the system of linearized equations (14.26)–(14.29) more manageable, it is customary to introduce some simplifications without disturbing the basic physics contained in the original set of linearized equations. These simplifications are the following.

(i) We restrict the propagation of plane waves to the (x, z)-plane: $v = 0$, $\partial/\partial y = 0$.
(ii) We assume that there is a constant basic current given by u_0 = constant so that $v_0 = w_0 = 0$ and hence $v' = 0$.
(iii) We assume that, in the basic state, the atmosphere is in hydrostatic balance: $\alpha_0\,\partial p_0/\partial z = -g$.
(iv) We assume that isothermal conditions pertain for the basic field so that $\partial\alpha_0/\partial z = g/p_0$.
(v) We assume that the rotation of the earth may be ignored by setting $f = 0$.
(vi) The basic thermodynamic variables are taken to be independent of x and t.

Application of these assumptions results in the following simplified set of linearized equations:

$$
\begin{aligned}
&\text{(a)} & \frac{\partial u'}{\partial t} + u_0 \frac{\partial u'}{\partial x} + \alpha_0 \frac{\partial p'}{\partial x} &= 0 \\
&\text{(b)} & \delta\left(\frac{\partial w'}{\partial t} + u_0 \frac{\partial w'}{\partial x}\right) + \alpha_0 \frac{\partial p'}{\partial z} - g\frac{\alpha'}{\alpha_0} &= 0 \\
&\text{(c)} & \alpha_0\left(\frac{\partial p'}{\partial t} + u_0 \frac{\partial p'}{\partial x}\right) - gw' + p_0\kappa\left(\frac{\partial \alpha'}{\partial t} + u_0 \frac{\partial \alpha'}{\partial x} + w'\frac{\partial \alpha_0}{\partial z}\right) &= 0 \\
&\text{(d)} & \alpha_0\left(\frac{\partial u'}{\partial x} + \frac{\partial w'}{\partial z}\right) - \left(\frac{\partial \alpha'}{\partial t} + u_0 \frac{\partial \alpha'}{\partial x} + w'\frac{\partial \alpha_0}{\partial z}\right) &= 0
\end{aligned}
\tag{14.31}
$$

Following Haltiner and Williams (1980), we have introduced the quantity δ into the equation for vertical motion. If $\delta = 1$ this equation remains unchanged; by setting $\delta = 0$ we ignore certain vertical-acceleration terms.

We shall now transform (14.31) by introducing the relative pressure $q = p'/p_0$ and the relative density $s = \rho'/\rho_0 = -\alpha'/\alpha_0$. The latter relation may be easily obtained by linearizing the equation $\rho\alpha = 1$. Using the abbreviation $d_1/dt = \partial/\partial t + u_0\, \partial/\partial x$ and matrix notation, we find without difficulty

$$
\begin{pmatrix}
\dfrac{d_1}{dt} & 0 & R_0T_0\dfrac{\partial}{\partial x} & 0 \\
0 & \delta\dfrac{d_1}{dt} & R_0T_0\dfrac{\partial}{\partial z} - g & g \\
0 & \dfrac{g(\kappa-1)}{R_0T_0} & \dfrac{d_1}{dt} & -\kappa\dfrac{d_1}{dt} \\
\dfrac{\partial}{\partial x} & \dfrac{\partial}{\partial z} - \dfrac{g}{R_0T_0} & 0 & \dfrac{d_1}{dt}
\end{pmatrix}
\begin{pmatrix} u' \\ w' \\ q \\ s \end{pmatrix} = 0
\tag{14.32}
$$

Inspection shows that all coefficients multiplying the variables (u', w', q, s) or the differential operators are constants, so the operator method may be applied to solve the homogeneous system of differential equations (14.32). We will demonstrate the procedure in the following sections.

14.4 Pure sound waves

The system (14.32) contains sound waves and gravity waves, including interactions between these two types of wave. In order to isolate pure sound waves we have to set $g = 0$. In the general case we set $\delta = 1$. Nonzero values of the unknown functions (u', w', q, s) of the homogeneous system can be obtained by setting the determinant of the four-by-four matrix in (14.32) equal to zero:

$$\begin{vmatrix} \dfrac{d_1}{dt} & 0 & R_0 T_0 \dfrac{\partial}{\partial x} & 0 \\ 0 & \dfrac{d_1}{dt} & R_0 T_0 \dfrac{\partial}{\partial z} & 0 \\ 0 & 0 & \dfrac{d_1}{dt} & -\kappa \dfrac{d_1}{dt} \\ \dfrac{\partial}{\partial x} & \dfrac{\partial}{\partial z} & 0 & \dfrac{d_1}{dt} \end{vmatrix} = 0 \tag{14.33}$$

The evaluation of the determinant gives the partial differential equations

$$\left(\frac{\partial}{\partial t} + u_0 \frac{\partial}{\partial x}\right)^2 \left[\left(\frac{\partial}{\partial t} + u_0 \frac{\partial}{\partial x}\right)^2 - \kappa R_0 T_0 \left(\frac{\partial^2}{\partial x^2} + \frac{\partial^2}{\partial z^2}\right)\right] \psi_j(x, z, t) = 0 \tag{14.34}$$

which applies to the four variables $(\psi_1, \psi_2, \psi_3, \psi_4) = (u', w', q, s)$. Since we are dealing with a constant-coefficient system we may assume the validity of wave solutions of the type

$$\psi_j = A_j \exp\left[i(k_x x + k_z z - \omega t)\right] \tag{14.35}$$

where the A_j are constant coefficients. The system (14.31) is linear so that any linear combination of solutions is a solution also. Therefore, it is sufficient for our purposes to consider a single harmonic. Substituting (14.35) into (14.34) and rearranging the result gives the fourth degree frequency equation

$$(k_x u_0 - \omega)^2[-(k_x u_0 - \omega)^2 + k^2 \kappa R_0 T_0] = 0 \quad \text{with} \quad k^2 = k_x^2 + k_z^2 \tag{14.36}$$

in agreement with Haltiner and Williams (1980), who assumed that $\alpha_0 =$ constant and $p_0 =$ constant.

We will now discuss the four roots of this equation. The quadratic first factor results in

$$\omega_{1,2} = 2\pi \nu_{1,2} = k_x u_0 = \frac{2\pi}{L_x} u_0 \tag{14.37a}$$

From this equation it follows that the phase speed

$$c_x = \frac{k_x}{k} u_0 \tag{14.37b}$$

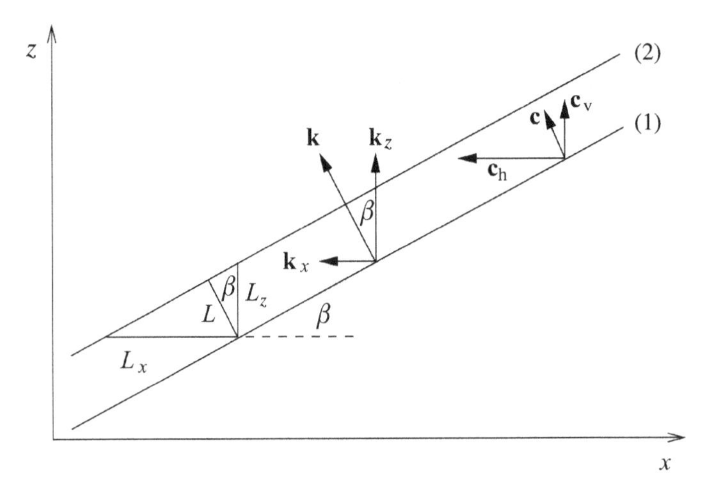

Fig. 14.4 Propagation of plane harmonic waves in the (x, z)-plane.

corresponds to simple advection. The third and fourth roots are given by

$$\omega_{3,4} = 2\pi\nu_{3,4} = k_x u_0 \pm k\sqrt{\kappa R_0 T_0} \tag{14.37c}$$

The corresponding phase velocities are

$$c_{3,4} = \nu_{3,4} L = \frac{\omega_{3,4} L}{2\pi} = \frac{k_x u_0}{k} \pm c_{\mathrm{L}} \quad \text{with} \quad c_{\mathrm{L}} = \sqrt{\kappa R_0 T_0} \tag{14.37d}$$

It is seen that $c_{3,4}$ consists of two parts including the basic current u_0 and the *Laplace speed of sound* c_{L} which is independent of direction and amounts to about 330 m s^{-1}.

With the help of Figure 14.4 we are going to investigate the propagation of plane harmonic waves in the (x, z)-plane. The wave vector $\mathbf{k}$ is normal to the *equiphase surfaces* (phase lines $k_x x + k_z z$ = constant in the (x, z)-plane). The wavelength L is the normal distance between two of these surfaces, (1) and (2); the distances L_x and L_z are the distances between these two surfaces in the x- and z-directions. The displacement of equiphase surfaces occurs with the phase speed c. The horizontal and vertical displacement speeds are denoted by c_{h} and c_{v}. From the trigonometric relations we find

$$\begin{aligned} \sin\beta &= \frac{L}{L_x} = \frac{c}{c_{\mathrm{h}}} = \frac{k_x}{k}, \qquad & \cos\beta &= \frac{L}{L_z} = \frac{c}{c_{\mathrm{v}}} = \frac{k_z}{k} \\ L &= \frac{L_x L_z}{\sqrt{L_x^2 + L_z^2}}, & c &= \frac{\omega}{k} = \frac{\omega}{\sqrt{k_x^2 + k_z^2}} \end{aligned} \tag{14.38}$$

If the phase speed c and β are given, we can calculate c_{h}. From Figure 14.4 it can be seen that the horizontal speed c_{h} increases with decreasing angle β.

14.5 Sound waves and gravity waves

Gravity waves arise from the differential effects of gravity on air parcels of different densities at the same level. It might be a better terminology to call them *buoyancy waves*. Let us reconsider the linearized system (14.32). For simplicity, without changing the essential physics, we set the basic flow speed $u_0 = 0$ so that d_1/dt degenerates to the partial derivative $\partial/\partial t$. The determinant of the four-by-four matrix of (14.32) now reduces to

$$\begin{vmatrix} \dfrac{\partial}{\partial t} & 0 & R_0 T_0 \dfrac{\partial}{\partial x} & 0 \\ 0 & \delta \dfrac{\partial}{\partial t} & R_0 T_0 \dfrac{\partial}{\partial z} - g & g \\ 0 & \dfrac{g(\kappa - 1)}{R_0 T_0} & \dfrac{\partial}{\partial t} & -\kappa \dfrac{\partial}{\partial t} \\ \dfrac{\partial}{\partial x} & \dfrac{\partial}{\partial z} - \dfrac{g}{R_0 T_0} & 0 & \dfrac{\partial}{\partial t} \end{vmatrix} = 0 \tag{14.39}$$

whose expansion is

$$\left[\delta \frac{\partial^4}{\partial t^4} - \kappa R_0 T_0 \left(\delta \frac{\partial^4}{\partial x^2\, \partial t^2} + \frac{\partial^4}{\partial z^2\, \partial t^2}\right) + \kappa g \frac{\partial^3}{\partial z\, \partial t^2} - g^2(\kappa - 1)\frac{\partial^2}{\partial x^2}\right]\psi_j = 0 \tag{14.40}$$

Here $(\psi_1, \psi_2, \psi_3, \psi_4) = (u', w', q, s)$. With the exception of one term all space derivatives in (14.40) are of second order. This asymmetry can be removed by substituting

$$\psi_j = \exp\left(\frac{gz}{2R_0 T_0}\right) F_j(x, z, t) \tag{14.41}$$

into (14.40) to obtain

$$\left[\delta \frac{\partial^4}{\partial t^4} - \kappa R_0 T_0 \left(\delta \frac{\partial^4}{\partial x^2\, \partial t^2} + \frac{\partial^4}{\partial z^2\, \partial t^2}\right) + \frac{\kappa g^2}{4R_0 T_0} \frac{\partial^2}{\partial t^2} - g^2(\kappa - 1)\frac{\partial^2}{\partial x^2}\right]F_j = 0 \tag{14.42}$$

where all space derivatives are now of second order. Substitution of the constant-coefficient harmonic-wave solution

$$F_j = A_j \exp[i(k_x x + k_z z - \omega t)] \tag{14.43}$$

into (14.42) results in the frequency equation

$$\delta\omega^4 - \omega^2\left(\kappa R_0 T_0 (\delta k_x^2 + k_z^2) + \frac{\kappa g^2}{4R_0 T_0}\right) + g^2(\kappa - 1)k_x^2 = 0 \tag{14.44}$$

Thus, the partial differential equation (14.42) was transformed into a fourth-degree algebraic equation for the circular frequency, permitting us to determine the phase velocities of the various waves.

For $\delta = 1$ representing the general case, equation (14.44) will be investigated from two points of view. In case (i) the appearance of gravity waves will be suppressed by setting $g = 0$; in case (ii) the gravity waves will be admitted.

(i) $g = 0$

This condition results in

$$\omega_{1,2} = 0, \qquad \omega_{3,4} = \pm k\sqrt{\kappa R_0 T_0} = \pm k c_{\mathrm{L}} \tag{14.45}$$

From (14.45) it follows that only two waves exist, namely the compressional waves. Thus, we have repeated the results shown in (14.37a) and (14.37c) for the special case that $u_0 = 0$.

(ii) $g \neq 0$

Now we obtain the solution

$$\omega_{1,2}^2 = \frac{1}{2}\left(\kappa R_0 T_0 k^2 + \frac{\kappa g^2}{4R_0 T_0}\right) \pm \sqrt{\frac{1}{4}\left(\kappa R_0 T_0 k^2 + \frac{\kappa g^2}{4R_0 T_0}\right)^2 - g^2(\kappa - 1)k_x^2} \tag{14.46}$$

for sound and gravity waves combined. The positive square root results in a solution for sound waves that are modified by gravitational effects. The negative square root may be viewed as a solution for gravity waves that are modified by sound effects.

We will conclude this section by considering some additional special cases of the frequency equation (14.44). These are the effect of incompressibility and the propagation of sound waves in the horizontal and vertical directions.

14.5.1 The effect of incompressibility

In textbooks on thermodynamics it is shown that, in an incompressible fluid where the coefficient of piezotropy γ_p^ρ approaches zero, the Laplace speed of sound c_{L} is infinitely large:

$$\gamma_p^\rho = \frac{1}{\kappa R_0 T_0} = \frac{1}{c_{\mathrm{L}}^2} = 0 \tag{14.47}$$

On setting $\delta = 1$ in (14.44) and multiplying this equation by the piezotropy coefficient, we find

$$\gamma_p^\rho \omega^4 - \omega^2\left(k^2 + \frac{g^2}{4R_0^2 T_0^2}\right) + \frac{g^2 k_x^2}{R_0 T_0} - g^2 k_x^2 \gamma_p^\rho = 0 \tag{14.48}$$

Permitting the piezotropy coefficient to approach zero to simulate incompressibility, we find the frequency equation for an incompressible medium to be

$$-\omega^2\left(R_0T_0k^2 + \frac{g^2}{4R_0T_0}\right) + g^2k_x^2 = 0 \tag{14.49}$$

Owing to the elimination of sound waves the frequency equation is now only of second degree and contains gravity waves only. The frequency of the gravity waves is given by

$$\omega_{1,2} = \pm\sqrt{\frac{g^2k_x^2}{R_0T_0k^2 + g^2/(4R_0T_0)}} \tag{14.50}$$

Now we wish to investigate the effect of static stability on gravity waves. For this purpose we replace the term $\kappa - 1$ in (14.44) by the vertical gradient of the potential temperature θ_0. Logarithmic differentiation of the basic-state potential-temperature equation, assuming that we have isothermal conditions of the basic state, and application of the hydrostatic equation results in

$$\kappa - 1 = \frac{\kappa R_0T_0}{g\theta_0}\frac{\partial\theta_0}{\partial z} \tag{14.51}$$

Substituting this expression into (14.44), setting $\delta = 1$, and observing that the scale height is given by $H = R_0T_0/g$, we find

$$\omega^4 - \omega^2\left(\kappa R_0T_0k^2 + \frac{\kappa g}{4H}\right) + k_x^2 g\frac{\kappa R_0T_0}{\theta_0}\frac{\partial\theta_0}{\partial z} = 0 \tag{14.52a}$$

This form of the frequency equation permits us to make a very useful approximation, which should be valid for a great number of situations. First we form the ratio $\kappa R_0T_0k_z^2/[\kappa g/(4H)]$ with $k^2 = k_x^2 + k_z^2$. If the largest vertical wavelength L_z does not significantly exceed the scale height H, as is often the case, this ratio is roughly 150. For $L_z < H$ the ratio is even larger, so that the term $\kappa g/(4H)$ in (14.52a) may be dropped in comparison with $\kappa R_0T_0k^2$ without a significant loss of accuracy. This yields the simplified form

$$\omega^4 - \omega^2\kappa R_0T_0k^2 + k_x^2 g\frac{\kappa R_0T_0}{\theta_0}\frac{\partial\theta_0}{\partial z} = 0 \tag{14.52b}$$

As before, we multiply this equation by the piezotropy coefficient. Permitting this coefficient to approach zero, we obtain the simplified frequency equation

$$\omega_{1,2} = \pm\frac{k_x}{k}\sqrt{\frac{g}{\theta_0}\frac{\partial\theta_0}{\partial z}}, \qquad u_0 = 0 \tag{14.53}$$

for the incompressible medium. If u_0 differs from zero, analogously to (14.37c) we must replace $\omega_{1,2}$ by

$$\boxed{\chi_{1,2} = \omega_{1,2} - u_0 k_x = \pm\frac{k_x}{k}\sqrt{\frac{g}{\theta_0}\frac{\partial\theta_0}{\partial z}}} \tag{14.54}$$

The term χ is known as the *intrinsic frequency*. In contrast to the circular frequency, the intrinsic frequency is measured by an observer drifting with the basic air current. The introduction of χ avoids the explicit appearance of the Doppler effect. The corresponding phase velocities are then given by

$$\begin{aligned} c_{1,2} &= \frac{\omega_{1,2}}{k} = \pm\frac{k_x}{k^2}\sqrt{\frac{g}{\theta_0}\frac{\partial\theta_0}{\partial z}}, \qquad u_0 = 0 \\ \text{or} \quad c_{1,2} &= \frac{\omega_{1,2}}{k} = \frac{k_x}{k}u_0 \pm \frac{k_x}{k^2}\sqrt{\frac{g}{\theta_0}\frac{\partial\theta_0}{\partial z}} \end{aligned} \tag{14.55}$$

One final remark on the approximation used in (14.52) might be helpful. Had we ignored the term $g^2/(4R_0T_0)$ in comparison with $R_0T_0k_z^2$ in (14.49), we would have obtained a slightly different approximation.

Inspection of (14.54) shows that, for stable stratification, $\partial\theta_0/\partial z > 0$, stable oscillations occur. An amplified disturbance occurs for superadiabatic conditions with $\partial\theta_0/\partial z < 0$ since $c_{1,2}$ is complex.

Finally, if the depth of the disturbance is large in comparison with the horizontal scale, that is $k_x^2 \gg k_z^2$, then k in (14.53) may be approximated by k_x and we obtain the *Brunt–Vaisala frequency*

$$\boxed{\omega_{\mathrm{Br}} = \sqrt{\frac{g}{\theta_0}\frac{\partial\theta_0}{\partial z}}} \tag{14.56}$$

which is usually discussed in connection with oscillations in vertical stability of isolated air parcels.

Next we wish to show that the system (14.32) contains two special cases describing sound waves propagating in the horizontal or in the vertical direction. For simplification we set the basic current $u_0 = 0$.

14.5.2 Horizontal sound waves

In the basic system (14.32) we set $g = 0$ and $w' = 0$ to eliminate gravity waves and vertical motion. This means that the remaining sound wave can propagate only in the x-direction so that the wave front is parallel to the (y, z)-plane. With these simplifications (14.32) reduces to the prognostic set

$$\frac{\partial u'}{\partial t} + R_0 T_0 \frac{\partial q}{\partial x} = 0, \qquad \frac{\partial q}{\partial t} - \kappa \frac{\partial s}{\partial t} = 0, \qquad \frac{\partial u'}{\partial x} + \frac{\partial s}{\partial t} = 0 \tag{14.57}$$

By combining the last two expressions we find

$$\frac{\partial u'}{\partial t} + R_0 T_0 \frac{\partial q}{\partial x} = 0, \qquad \frac{\partial u'}{\partial x} + \frac{1}{\kappa} \frac{\partial q}{\partial t} = 0 \tag{14.58}$$

from which we determine the perturbation velocity u' and the relative pressure disturbance q. By employing the operator method we obtain the wave equation for horizontally propagating sound waves:

$$\left(\frac{\partial^2}{\partial t^2} - \kappa R_0 T_0 \frac{\partial^2}{\partial x^2} \right) \psi_j = 0 \tag{14.59}$$

with $(\psi_1, \psi_2) = (u', q)$. The solution to (14.59), also known as *D'Alembert's solution*, can be written in the general form

$$\psi_j = \psi_j(x \pm c_{\mathrm{L}} t) \quad \text{with} \quad c_{\mathrm{L}}^2 = \kappa R_0 T_0 \tag{14.60}$$

where the disturbances propagate in the positive and negative x-directions with the speed of sound c_{L}.

14.5.3 Vertical sound waves

In the basic system (14.32) we set $\delta = 1,\ g = 0$, and $u' = 0$ to suppress gravity waves and the horizontal velocity. This results in the following system:

$$\frac{\partial w'}{\partial t} + R_0 T_0 \frac{\partial q}{\partial z} = 0, \qquad \frac{\partial q}{\partial t} - \kappa \frac{\partial s}{\partial t} = 0, \qquad \frac{\partial w'}{\partial z} + \frac{\partial s}{\partial t} = 0 \tag{14.61}$$

Eliminating again $\partial s / \partial t$ and applying the operator method to the remaining system, we obtain the wave equation

$$\left(\frac{\partial^2}{\partial t^2} - \kappa R_0 T_0 \frac{\partial^2}{\partial z^2} \right) \psi_j = 0 \tag{14.62}$$

with $(\psi_1, \psi_2) = (w', q)$. The solution is given by

$$\psi_j = \psi_j(z \pm c_{\mathrm{L}} t) \tag{14.63}$$

showing that the disturbances propagate in the positive and negative vertical directions with the speed of sound.

14.5.4 Hydrostatic filtering

In the basic set of the linearized equation (14.32) we now set $\delta = 0$. Thus, the term $(\partial/\partial t + u_0\,\partial/\partial x)w'$ disappears so that a prognostic equation is replaced by a diagnostic relation. This corresponds to *hydrostatic filtering*. Instead of (14.44) the frequency equation now reads

$$-\omega^2\left(R_0T_0k_z^2 + \frac{g^2}{4R_0T_0}\right) + g^2k_x^2\frac{\kappa - 1}{\kappa} = 0 \tag{14.64}$$

The comparison with the frequency equation (14.49) for an incompressible medium reveals far-reaching agreement, demonstrating that the hydrostatic approximation has eliminated to a large extent the effect of compressibility. This can also be seen by solving (14.64) for ω, yielding

$$\omega^2 = g^2\frac{\kappa - 1}{\kappa}\frac{k_x^2}{R_0T_0k_z^2 + g^2/(4R_0T_0)} \tag{14.65}$$

This expression shows that, for $g = 0$, not only gravity waves but also sound waves have completely been eliminated from the system since in this case $\omega = 0$. Comparison of (14.64) with (14.49) shows that the hydrostatic approximation has eliminated the quantity k_x^2 in comparison with k_z^2, meaning that $L_x \gg L_z$. From Figure 14.4 we recognize that, in the present situation, the angle of inclination of the wave front is very small, so the horizontal displacement velocity $c_{\rm h}$ is much larger than the phase speed c itself, whose direction is perpendicular to the wave front.

Finally, we wish to show from another point of view that hydrostatic filtering of the system (14.32) eliminates vertically propagating sound waves. This type of filtering is very important in numerical weather prediction. The reason for this is that the time step of integration that can be chosen is inversely proportional to the phase speed of the fastest waves contained in the system, which are the sound waves. More information on this topic will be given in the chapter on numerical procedures in weather prediction.

In order to actually show that the vertically propagating sound waves are filtered out of the system, we eliminate gravitational effects by setting $g = 0$ and set $u' = 0$ in (14.32). Thus, the remaining wave can propagate in the positive or negative z-direction only. With these assumptions (14.32) reduces to

$$R_0T_0\frac{\partial q}{\partial z} = 0, \qquad \frac{\partial q}{\partial t} - \kappa\frac{\partial s}{\partial t} = 0, \qquad \frac{\partial w'}{\partial z} + \frac{\partial s}{\partial t} = 0, \qquad u_0 = 0 \tag{14.66}$$

The first expression implies that the relative pressure disturbance is independent of height. Nevertheless, q might still be a function of time. If it turns out that q is a function of time so that $\partial q/\partial t \neq 0$, then the pressure disturbance occurs at all heights simultaneously. This means that the vertical phase speed of the pressure disturbance is infinitely large, thus leaving the atmosphere immediately.

To look at the problem from a slightly different point of view, we differentiate the second expression of (14.66) with respect to z and find

$$\frac{\partial}{\partial t}\frac{\partial q}{\partial z} = \kappa \frac{\partial}{\partial t}\frac{\partial s}{\partial z} = 0 \quad \text{since} \quad q \neq q(z) \tag{14.67}$$

From the third expression of (14.66) it now follows that

$$\frac{\partial^2 w'}{\partial z^2} = 0 \implies w' = az + b \tag{14.68}$$

The kinematic boundary condition states that $w'(z = 0) = 0$, so $b = 0$. To prevent w' becoming infinitely large with increasing z, we set $a = 0$ so that $w' = 0$. Therefore, according to (14.66), we have $\partial s/\partial t = 0$ and finally $\partial q/\partial t = 0$, so $q \neq q(t)$. Now q is independent of height and time so that there exists no (z, t)-relation for q, which would be required for a moving wave. We interpret this by saying that vertically propagating sound waves have been filtered out of the system. Horizontal sound waves have not been eliminated by hydrostatic filtering. They may result from a degeneration of horizontal gravity waves, as will be shown next.

14.5.5 Degeneration of horizontal gravity waves

We conclude this section by showing that horizontally propagating gravity waves may degenerate to sound waves. Setting $w' = 0$ in the system (14.32), we obtain with $u_0 = 0$ the prognostic system

$$\frac{\partial u'}{\partial t} + R_0 T_0 \frac{\partial q}{\partial x} = 0, \qquad \frac{\partial q}{\partial t} - \kappa \frac{\partial s}{\partial t} = 0, \qquad \frac{\partial u'}{\partial x} + \frac{\partial s}{\partial t} = 0 \tag{14.69}$$

consisting of the first equation of motion, the adiabatic equation, and the continuity equation. Furthermore, due to the hydrostatic filtering, the second row of (14.32) becomes a diagnostic relation:

$$\left(R_0 T_0 \frac{\partial}{\partial z} - g\right) q + gs = 0 \tag{14.70}$$

In order to apply the operator method, we form the required determinant:

$$\begin{vmatrix} \dfrac{\partial}{\partial t} & R_0 T_0 \dfrac{\partial}{\partial x} & 0 \\ 0 & \dfrac{\partial}{\partial t} & -\kappa \dfrac{\partial}{\partial t} \\ \dfrac{\partial}{\partial x} & 0 & \dfrac{\partial}{\partial t} \end{vmatrix} = 0 \tag{14.71}$$

Expansion of this determinant results in

$$\frac{\partial}{\partial t}\left(\frac{\partial^2}{\partial t^2} - c_L^2 \frac{\partial^2}{\partial x^2}\right)\psi_j = 0 \tag{14.72}$$

with $(\psi_1, \psi_2, \psi_3) = (u', q, s)$. Excluding the trivial solution in which the functions u', q, and s are constants, we obtain *D'Alembert's wave equation*

$$\boxed{\left(\frac{\partial^2}{\partial t^2} - c_L^2 \frac{\partial^2}{\partial x^2}\right)\psi_j = 0} \tag{14.73}$$

whose solution is given by (14.60). This equation shows that the disturbances are displaced in the positive or the negative x-direction.

Let us now specifically discuss the pressure disturbance $q = q(x \pm c_L t, z)$ in a little more detail by including the compatibility equation (14.70). Integration of the second expression of (14.69) with respect to time gives

$$\frac{q}{\kappa} = s + f(x, z, t_0) \tag{14.74}$$

where $f(x, z, t_0)$ is an arbitrary function. Initially, at time t_0, the relative pressure q and the relative density s are assumed to be zero so that f is zero also. Substitution of (14.74) into (14.70) yields

$$\frac{(\kappa - 1)g}{\kappa} q = R_0 T_0 \frac{\partial q}{\partial z} \tag{14.75}$$

which can be integrated to give

$$q = Q \exp\left(\frac{(\kappa - 1)gz}{\kappa R_0 T_0}\right) = Q \exp\left(\frac{(\kappa - 1)gz}{c_L^2}\right) \tag{14.76}$$

with $Q = q(x \pm c_L t, z = 0)$. Now we introduce the definition of the relative pressure disturbance $q = p'/p_0$ into (14.76). We also recall that the basic state pressure p_0 decays exponentially so that

$$p_0 = p_0(z = 0) \exp\left(-\frac{z}{H}\right) \quad \text{with} \quad H = \frac{R_0 T_0}{g} = \frac{c_L^2}{\kappa g} \tag{14.77}$$

Therefore, the pressure disturbance is given by

$$p' = p_0(z = 0) Q(x \pm c_L t, z = 0) \exp\left(-\frac{gz}{c_L^2}\right) \tag{14.78}$$

This equation describes the horizontal displacement of a pressure disturbance with a speed of displacement c_L. The disturbance decreases with height. At a height of 10 km the disturbance has decreased by about one third of its value at the surface of the earth. In this special case the pressure disturbance p' decreases exponentially with height in a horizontally displaced gravity wave that is moving with the speed of sound.

14.6 Lamb waves

By direct substitution it is easily shown that for $u_0 = 0$ the basic system (14.31) is satisfied by the following solution:

$$\begin{aligned}
u' &= S \exp\left(\frac{z}{H}\right) \exp\left(-\frac{z}{\kappa H}\right) \exp[ik_x(x - c_{\mathrm{L}}t)] \\
w' &= 0 \\
p' &= P \exp\left(-\frac{z}{\kappa H}\right) \exp[ik_x(x - c_{\mathrm{L}}t)] \\
\alpha' &= A \exp\left(\frac{2z}{H}\right) \exp\left(-\frac{z}{\kappa H}\right) \exp[ik_x(x - c_{\mathrm{L}}t)]
\end{aligned} \tag{14.79}$$

where the coefficients S, P, A are constants. This wave, known as the Lamb wave, has no vertical velocity and propagates horizontally with the speed of sound c_{L}. Waves propagating in the horizontal direction only are known as *trapped waves* or *evanescent waves*. As mentioned above, a wave moving with the speed of sound is important in numerical weather prediction since it places a severe restriction on the maximum time step. For details on Lamb waves see Haltiner and Williams (1980).

14.7 Lee waves

When air is forced to cross a mountain ridge under statically stable conditions, individual air parcels are displaced from their equilibrium level. In this case the air parcels begin to oscillate about their equilibrium positions. The consequence is the formation of a wave system in the lee region of the mountain. There exists a rather detailed theory for two-dimensional air flow crossing idealized mountains; see, for example, Queney (1948), Queney *et al.* (1960), and Smith (1979). The interested reader may consult Gossard and Hooke (1975), where this topic is treated in some detail. More realistic simulations are possible with the help of numerical models, which may be verified by means of satellite observations.

14.8 Propagation of energy

Wave motion is associated with the transport of physical quantities such as energy and momentum. In this section we will briefly derive an important equation for the transport of energy by gravity and acoustic waves. To keep the analysis simple but still sufficiently instructive, we will restrict our discussion to frictionless wave motion in the (x, z)-plane. Furthermore, we consider a scale of motion on which Coriolis effects are not important. We refer to Gossard and Hooke (1975) and Pichler (1997), where additional details may be found.

If p', u', w' are the perturbed field quantities due to wave motion, we must average the energy flux $\mathbf{F} = \mathbf{i}_1 F_x + \mathbf{i}_3 F_z$ (energy per unit area and time) over one complete cycle τ, yielding

$$\overline{\mathbf{F}} = \overline{p\mathbf{v}} = \frac{1}{\tau}\int_0^{\tau} p\mathbf{v}\,dt \tag{14.80}$$

In order to assume solutions of the type (14.35), we must have a constant-coefficient system of differential equations. To handle this particular problem, we transform the sytem (14.31) as shown next. First we set $\delta = 1$ and introduce the operator d_1/dt. This gives

$$\frac{d_1 u'}{dt} + \alpha_0 \frac{\partial p'}{\partial x} = 0 \tag{14.81a}$$

Next we apply the operator d/dt to (14.31b) and eliminate $d\alpha'/dt$ by means of (14.31c), yielding

$$\begin{aligned}\left(\frac{d_1^2}{dt^2} + \omega_{\mathrm{Br}}^2\right)w' + \alpha_0 \frac{d_1}{dt}\left(\frac{\partial}{\partial z} + \frac{g}{c_{\mathrm{L}}^2}\right)p' = 0 \\ \text{with} \quad \omega_{\mathrm{Br}}^2 = g\left(\frac{1}{\alpha_0}\frac{\partial \alpha_0}{\partial z} - \frac{g}{c_{\mathrm{L}}^2}\right)\end{aligned} \tag{14.81b}$$

Here we have written the Brunt–Vaisala frequency ω_{Br} as defined in (14.56) in a more suitable form. Finally, we combine (14.31c) and (14.31d) and obtain

$$\frac{d_1 p'}{dt} - \frac{g}{\alpha_0} w' + \frac{c_{\mathrm{L}}^2}{\alpha_0}\left(\frac{\partial u'}{\partial x} + \frac{\partial w'}{\partial z}\right) = 0 \tag{14.81c}$$

The wave solution (14.35) cannot be applied to (14.81) since α_0 depends on height. In order to obtain a constant-coefficient system, we transform the variables according to

$$u' = U\left(\frac{\alpha_0}{\alpha_{\mathrm{s}}}\right)^{1/2}, \qquad w' = W\left(\frac{\alpha_0}{\alpha_{\mathrm{s}}}\right)^{1/2}, \qquad p' = P\left(\frac{\alpha_0}{\alpha_{\mathrm{s}}}\right)^{1/2} \tag{14.82}$$

where α_{s} is a constant. Refer to Bretherton (1966). This transformation leads to the system

$$\begin{aligned}\frac{d_1 U}{dt} + \alpha_{\mathrm{s}} \frac{\partial P}{\partial x} = 0 \\ \left(\frac{d_1^2}{dt^2} + \omega_{\mathrm{Br}}^2\right)W + \alpha_{\mathrm{s}} \frac{d_1}{dt}\left(\frac{\partial}{\partial z} + E\right)P = 0 \\ \frac{\alpha_{\mathrm{s}}}{c_{\mathrm{L}}^2}\frac{d_1 P}{dt} + \frac{\partial U}{\partial x} + \left(\frac{\partial}{\partial z} - E\right)W = 0\end{aligned} \tag{14.83}$$

Here we have introduced the so-called *Eckart coefficient*

$$\boxed{E = -\frac{1}{2\alpha_0}\frac{\partial \alpha_0}{\partial z} + \frac{g}{c_{\rm L}^2}} \tag{14.84}$$

Assuming that we have a basic isothermal state, equation (14.83) is a constant-coefficient system so that the trial solution

$$\begin{pmatrix} U \\ W \\ P \end{pmatrix} = \begin{pmatrix} A_1 \\ A_2 \\ A_3 \end{pmatrix} \exp[i(k_x x + k_z z - \omega t)] \tag{14.85}$$

may be applied. The coefficients A_i are constant amplitudes. The resulting equations

$$\begin{aligned} &\text{(a)} \qquad \chi A_1 - \alpha_{\rm s} k_x A_3 = 0 \\ &\text{(b)} \quad (\omega_{\rm Br}^2 - \chi^2) A_2 + \alpha_{\rm s}\chi(k_z - iE) A_3 = 0 \\ &\text{(c)} \qquad k_x A_1 + (k_z + iE) A_2 - \tfrac{\alpha_{\rm s}}{c_{\rm L}^2}\chi A_3 = 0 \\ &\qquad\qquad \text{with} \quad \chi = \omega - k_x u_0 \end{aligned} \tag{14.86}$$

give the relations among the various A_i which are needed in order to find F_x and F_z.

Employing (14.80), we obtain

$$\overline{F_x} = \overline{p'u'} = A_1 A_3 \frac{\alpha_0}{\alpha_{\rm s}}\frac{1}{\tau}\int_0^\tau \cos^2(k_x x + k_z z - \omega t)\, dt \tag{14.87}$$

since only the real part of the wave is physically meaningful. By eliminating A_1 from this equation, we find that the energy flux $\overline{F_x}$ is proportional to the square of the pressure amplitude:

$$\boxed{\overline{F_x} = \frac{\alpha_{\rm s}}{2}\left(\frac{k_x}{\chi}\right) A_3^2} \tag{14.88}$$

In order to find the correlation $F_z = p'w'$ we proceed likewise. Inspection shows that equation (14.86b) is complex. In order to establish a relationship between p' and w' that is entirely real, we set the Eckart coefficient E equal to zero. Thus we obtain the following relation for the energy flux in the z-direction:

$$\overline{F_z} = \overline{p'w'} = A_2 A_3 \frac{1}{\tau}\int_0^\tau \cos^2(k_x x + k_z z - \omega t)\, dt \tag{14.89}$$

and, after eliminating A_2 by means of (14.86),

$$\boxed{\overline{F_z} = -\frac{\alpha_s \chi k_z A_3^2}{2(\omega_{Br}^2 - \chi^2)}} \tag{14.90}$$

As expected, the vertical energy transport $\overline{F_z}$ is proportional to the square of the pressure amplitude.

The next step is to establish the relation between the flow of energy and the group velocity. For the group velocity, relative to the basic current u_0, we replace ω in (14.15b) by $\chi = \omega - u_0 k_x$. This gives

$$\widetilde{\mathbf{c}}_{gr} = \mathbf{c}_{gr} - u_0 \mathbf{i}_1 = \nabla_k \chi = \frac{\partial \chi}{\partial k_x}\mathbf{i}_1 + \frac{\partial \chi}{\partial k_z}\mathbf{i}_3 \tag{14.91}$$

We will now find the relationship between the flow of energy and the group velocity for gravity waves. From (14.54) and (14.56) we obtain

$$\chi^2 = \frac{k_x^2}{k^2}\omega_{Br}^2 \implies \frac{\partial \chi}{\partial k_x} = \frac{\chi k_z^2}{k_x k^2}, \qquad \frac{\partial \chi}{\partial k_z} = -\frac{\chi k_z}{k^2} \tag{14.92}$$

Hence, for gravity waves we have $\omega_{Br}^2 > \chi^2$. Owing to this inequality, equation (14.90) admits a very interesting and useful interpretation. If $\chi > 0$ and the wave motion is in the upward direction ($k_z > 0$), then the flow of energy ($\overline{F_z} < 0$) is in the downward direction. If the wave motion is in the downward direction ($k_z < 0$), then the flow of energy is upward ($\overline{F_z} > 0$).

For acoustic waves the situation is different. In this case the directions of the wave motion and the energy flux are identical. For $k_z < 0$ we have $\overline{F_z} < 0$ and for $k_z > 0$ we find $\overline{F_z} > 0$.

We are now ready to state the relationship between the flow of energy and the group velocity. On substituting the derivative expressions listed in (14.92) into (14.91) we find immediately

$$\widetilde{\mathbf{c}}_{gr} = \frac{\chi k_z}{k^2}\left(\frac{k_z}{k_x}\mathbf{i}_1 - \mathbf{i}_3\right) \tag{14.93}$$

On combining the components $\overline{F_x}$ and $\overline{F_z}$ of the energy flux $\overline{\mathbf{F}}$, we find

$$\overline{\mathbf{F}} = \frac{\alpha_s A_3^2}{2}\left(\frac{k_x}{\chi}\mathbf{i}_1 - \frac{\chi k_z}{\omega_{Br}^2 - \chi^2}\mathbf{i}_3\right) \tag{14.94}$$

After eliminating the square of the Brunt–Vaisala frequency in (14.94) with the help of (14.92), we obtain after some slight rearrangements the equation

$$\overline{\mathbf{F}} = \frac{k_x^2 \alpha_s A_3^2}{2k_z \chi}\left(\frac{k_z}{k_x}\mathbf{i}_1 - \mathbf{i}_3\right) \tag{14.95}$$

Comparison of this equation with (14.93) shows that the energy flux and the group velocity have the same direction so that (14.93) and (14.95) may be combined to give

$$\overline{\mathbf{F}} = \frac{\alpha_s A_3^2}{2}\left(\frac{k_x}{k_z}\right)^2\left(\frac{k}{\chi}\right)^2 \widetilde{\mathbf{c}}_{gr} \tag{14.96}$$

We see that the energy flux and the group velocity are unidirectional. This is true not only for gravity waves but also for wave motion in general.

14.9 External gravity waves

In this section the atmosphere is modeled by a one-layer, homogeneous, and incompressible fluid. Surface waves due to gravitational effects closely approximate ocean waves. Since these waves occur at the outer boundary of the fluid they are called external gravity waves. For simplicity we again restrict the motion to the (x, z)-plane so that the waves can propagate along the x-axis only. Since the fluid is homogeneous and incompressible, we set $\alpha' = 0$ so that $\alpha = \alpha_0 =$ constant. In this particular situation the first law of thermodynamics does not apply, so (14.31c) must be ignored. Thus, the equation system (14.31) reduces to a system of three equations, given by

$$\begin{aligned}
&\text{(a)} \qquad \frac{\partial u'}{\partial t} + u_0 \frac{\partial u'}{\partial x} + \alpha_0 \frac{\partial p'}{\partial x} = 0 \\
&\text{(b)} \qquad \delta\left(\frac{\partial w'}{\partial t} + u_0 \frac{\partial w'}{\partial x}\right) + \alpha_0 \frac{\partial p'}{\partial z} = 0 \\
&\text{(c)} \qquad \frac{\partial u'}{\partial x} + \frac{\partial w'}{\partial z} = 0
\end{aligned} \tag{14.97}$$

Again we assume that the perturbed flow may be represented by harmonic waves but now we admit that the amplitudes A_j are height-dependent:

$$\begin{pmatrix} u' \\ w' \\ p' \end{pmatrix} = \begin{pmatrix} A_1(z) \\ A_2(z) \\ \rho_0 A_3(z) \end{pmatrix} \exp[ik_x(x - ct)] \tag{14.98}$$

The density ρ_0, multiplying the amplitude A_3, is irrelevant insofar as the solution is concerned and has been introduced for mathematical convenience.

Substituting (14.98) into (14.97) gives

$$\begin{pmatrix} u_0 - c & 0 & 1 \\ 0 & ik_x\delta(u_0 - c) & \dfrac{d}{dz} \\ ik_x & \dfrac{d}{dz} & 0 \end{pmatrix} \begin{pmatrix} A_1(z) \\ A_2(z) \\ A_3(z) \end{pmatrix} = 0 \tag{14.99}$$

If ρ_0 had not been included, then the specific volume of the basic state would have appeared explicitly. Equation (14.99) is a coupled set that can be solved most easily by means of the operator method. As in the previous cases, we set the determinant of the coefficient matrix equal to zero,

$$\begin{vmatrix} u_0 - c & 0 & 1 \\ 0 & ik_x\delta(u_0 - c) & \dfrac{d}{dz} \\ ik_x & \dfrac{d}{dz} & 0 \end{vmatrix} = 0 \tag{14.100}$$

and then evaluate the determinant. The resulting operator

$$\left(\frac{d^2}{dz^2} - k_x^2\delta\right)A_j(z) = 0 \tag{14.101}$$

applies to all A_j ($j = 1, 2, 3$). For the general case $\delta = 1$, the eigenvalues (characteristic values) are $\pm k_x$ so that the solution is of the form

$$A_j(z) = C_1 \exp(k_x z) + C_2 \exp(-k_x z) \tag{14.102}$$

Now we must evaluate the integration constants C_1 and C_2 by applying proper boundary conditions. We begin with the case A_2. From (14.98) and (14.102) the vertical velocity perturbation is given by

$$w'(z, t) = \left[C_1 \exp(k_x z) + C_2 \exp(-k_x z)\right] \exp[ik_x(x - ct)] \tag{14.103}$$

By applying the kinematic boundary condition (9.43a) and canceling out the form factor of the harmonic wave $\exp(ik_x(x - ct)$ which cannot be zero, we obtain

$$w'(0) = 0 = C_1 + C_2 \implies C_1 = -C_2 = A \tag{14.104}$$

so

$$A_2 = A[\exp(k_x z) - \exp(-k_x z)] \tag{14.105}$$

The amplitude A_1 is found from the last equation of (14.99), yielding

$$A_1 = -\frac{1}{ik_x}\frac{dA_2}{dz} = iA[\exp(k_x z) + \exp(-k_x z)] \tag{14.106}$$

From the first equation of (14.99) we obtain the amplitude A_3:

$$A_3 = -(u_0 - c)A_1 = -iA(u_0 - c)[\exp(k_x z) + \exp(-k_x z)] \tag{14.107}$$

The next step is to eliminate the integration constant A. This is done by application of the dynamic boundary condition (9.44c). For a frictionless fluid and a free material surface, the dynamic boundary condition is given by

$$p(H) = 0, \qquad \left(\frac{dp}{dt}\right)_H = 0 \tag{14.108a}$$

where H is the height of the medium, which varies with the distance x and with time. Since H in general differs very little from the average height $\overline{H}$ of the medium, we may use the approximate boundary condition

$$p(\overline{H}) = p_0(\overline{H}) + p'(\overline{H}) = 0, \qquad \left(\frac{dp}{dt}\right)_{\overline{H}} = \left(\frac{\partial p}{\partial t} + u\,\frac{\partial p}{\partial x} + w\,\frac{\partial p}{\partial z}\right)_{\overline{H}} = 0 \tag{14.108b}$$

The required perturbation pressure p' is found by linearizing the previous equation to give

$$\left(\frac{\partial p'}{\partial t} + u_0\,\frac{\partial p'}{\partial x} + w'\,\frac{\partial p_0}{\partial z}\right)_{\overline{H}} = 0 \tag{14.109}$$

If the pressure p_0 of the basic field were permitted to vary with x, an acceleration of the flow in the x-direction would have to take place, in contrast to the assumption that u_0 is constant. For this reason the pressure gradient of the basic field in the x-direction must be set equal to zero, that is $\partial p_0/\partial x = 0$.

After substituting for p' according to (14.98), using (14.107), we find

$$-k_x c(u_0 - c) + k_x u_0(u_0 - c) - g\tanh(k_x\overline{H}) = 0 \tag{14.110}$$

since the constant A cancels out. The phase speed of the surface gravity waves propagating in the x-direction is then given by

$$c = u_0 \pm \sqrt{\frac{g}{k_x}\tanh(k_x\overline{H})} \tag{14.111}$$

Two special cases of external gravity waves forming at the free surface are of particular interest.

14.9.1 Case I

Suppose that the lateral extent of the disturbance L_x is very large in comparison with the depth $\overline{H}$ of the homogeneous layer, i.e. $L_x \gg \overline{H}$. In this case the argument of the hyperbolic tangent approaches zero. From the Taylor expansion of tanh we then find

$$\lim_{k_x\overline{H}\to 0} \tanh(k_x\overline{H}) = k_x\overline{H} \tag{14.112}$$

so the phase speed of the so-called *shallow-water waves* or *long waves* is given by

$$c = u_0 \pm \sqrt{g\overline{H}} \quad \text{if} \quad \overline{H} \ll L_x \tag{14.113}$$

This formula, first given by Lagrange, is correct to within one percent if $\overline{H} < 0.024L_x$. For the homogeneous atmosphere we may replace the argument $g\overline{H}$ by R_0T_0 so that the phase speed is given by

$$c = u_0 \pm \sqrt{R_0T_0} \tag{14.114}$$

In the absence of the basic current u_0 the long waves propagate with the *Newton speed of sound* $c_{\mathrm{N}} = \sqrt{R_0T_0}$, which is only slightly less than the Laplace speed of sound c_{L}.

14.9.2 Case II

Suppose that the depth of the fluid medium is large in comparison with the horizontal extent of the disturbance, i.e. $L_x \ll \overline{H}$. In this case the hyperbolic tangent approaches the value 1, so the phase speed of the so-called *deep-water wave* is given by

$$c = u_0 \pm \sqrt{\frac{gL_x}{2\pi}} \quad \text{if} \quad \overline{H} \gg L_x \tag{14.115}$$

This formula may be used with an error of about one percent if $\overline{H} > 0.4L_x$. For the linear theory it is possible to determine the trajectories (orbits) of the fluid particles for progressive waves; see LeMéhaute (1976). For the shallow-water waves the fluid-particle trajectories are elongated ellipses whereas for the deep-water waves they are nearly circular.

Case I dealing with $L_x \gg \overline{H}$ is equivalent to the hydrostatic approximation. To confirm this, we set $\delta = 0$ in (14.101) and obtain the simple second-order differential equation

$$\frac{d^2A_2}{dz^2} = 0 \tag{14.116}$$

which can be integrated to give the linear profile

$$A_2(z) = D_1z + D_2 \tag{14.117}$$

so the vertical velocity perturbation is given by

$$w'(z) = (D_1z + D_2)\exp[ik_x(x - ct)] \tag{14.118}$$

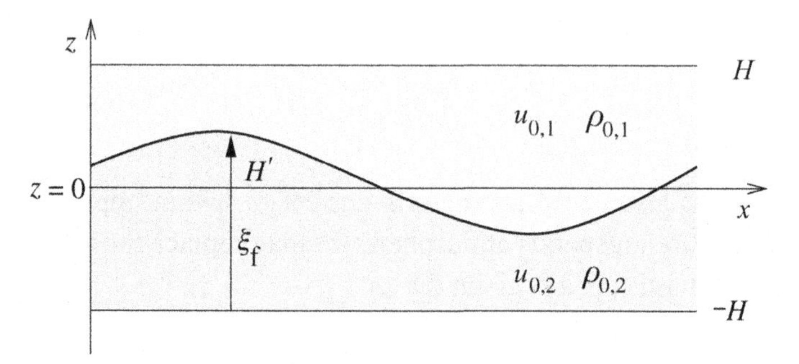

Fig. 14.5 Formation of an internal gravity wave on a zeroth-order discontinuity surface of density and velocity.

Proceeding as above, we find $D_2 = 0$. Putting $D_1 = D$, we obtain

$$\begin{aligned}
&\text{(a)} \quad A_2 = Dz \\
&\text{(b)} \quad A_1 = -\frac{1}{ik_x}\frac{dA_2}{dz} = \frac{iD}{k_x} \\
&\text{(c)} \quad A_3 = -(u_0 - c)A_1 = -(u_0 - c)\frac{iD}{k_x}
\end{aligned} \tag{14.119}$$

wherein (14.119b), and (14.119c) follow from (14.99). Knowing A_3, we also know p' from (14.98). On substituting p' into the linearized dynamic boundary condition (14.109) the remaining constant D cancels out and we obtain, as expected, equation (14.113).

14.10 Internal gravity waves

We are going to deepen our understanding of gravity waves by discussing *internal gravity waves*, which may develop on a zeroth-order discontinuity surface both for density and for velocity. Waves of this type are known as *Helmholtz waves*. The theory presented here has also been worked out for less restrictive conditions, under which the density may vary in a simple manner; see, for example, Gossard and Hooke (1975). The situation is depicted in Figure 14.5 stating the terminology which is used.

Again we are dealing with the solution to equation (14.101), which must be obtained for each of the two layers $k = 1, 2$:

$$\left(\frac{d^2}{dz^2} - k_x^2\right)A_{j,k}(z) = 0, \qquad j = 1, 2, 3, \quad k = 1, 2 \tag{14.120}$$

The required kinematic boundary conditions for the two rigid boundaries at $z = H$ and $z = -H$ are

$$z = H: \quad w_1' = 0, \qquad z = -H: \quad w_2' = 0 \tag{14.121}$$

while the vertical velocity at the discontinuity surface is given by equation (9.41). Setting the generalized height $\xi_f = H + H'(x, t)$, see Figure 14.5, we find

$$w_k' = \frac{d_k H'}{dt} = \frac{\partial \xi_f}{\partial t} + u_k \frac{\partial \xi_f}{\partial x}, \qquad k = 1, 2 \tag{14.122}$$

We assume that a wave will form at the boundary and be in phase with the remaining disturbances. Hence, H' may be written as

$$H' = A \exp[ik_x(x - ct)] \tag{14.123}$$

Linearization of (14.122) yields

$$w_k' = \frac{\partial H'}{\partial t} + u_{0,k} \frac{\partial H'}{\partial x} = -(c - u_{0,k}) \frac{\partial H'}{\partial x}, \qquad k = 1, 2 \tag{14.124}$$

For brevity we also introduce the operator Q_k:

$$Q_k = \frac{\partial}{\partial t} + u_{0,k} \frac{\partial}{\partial x} = -(c - u_{0,k}) \frac{\partial}{\partial x} \tag{14.125}$$

By forming the ratio of the perturbations of the vertical velocity for the two layers we find

$$\frac{w_1'}{w_2'} = \frac{Q_1 H'}{Q_2 H'} = \frac{c - u_{0,1}}{c - u_{0,2}} \tag{14.126}$$

indicating that this ratio involves the phase speed and the two constant horizontal velocities of the basic unperturbed fluid currents. This ratio will be needed shortly.

Next we apply the dynamic boundary condition at the interface:

$$p_1 - p_2 = 0, \qquad \frac{d_k}{dt}(p_1 - p_2) = 0 \quad \text{with} \quad \frac{d_k}{dt} = \frac{\partial}{\partial t} + u_k \frac{\partial}{\partial x} + w_k \frac{\partial}{\partial z} \tag{14.127}$$

Application of the Bjerkness linearization rule gives

$$\begin{aligned} &\frac{\partial p_1'}{\partial t} + u_{0,k} \frac{\partial p_1'}{\partial x} + w_k' \frac{\partial p_{0,1}}{\partial z} - \left(\frac{\partial p_2'}{\partial t} + u_{0,k} \frac{\partial p_2'}{\partial x} + w_k' \frac{\partial p_{0,2}}{\partial z} \right) = 0 \\ &\text{or} \quad Q_k(p_1' - p_2') - g(\rho_{0,1} - \rho_{0,2}) w_k' = 0 \end{aligned} \tag{14.128}$$

where the hydrostatic equation has been used to obtain the second relation.

We are now ready to determine the perturbations of the vertical velocities in each layer. The solution of (14.120) is analogous to (14.103), so the vertical perturbations are given by

$$\begin{aligned} w_1'(z) &= A_{2,1}(z)\exp[ik_x(x-ct)] \\ &= [C_1\exp(k_x z) + C_2\exp(-k_x z)]\exp[ik_x(x-ct)] \\ w_2'(z) &= A_{2,2}(z)\exp[ik_x(x-ct)] \\ &= [D_1\exp(k_x z) + D_2\exp(-k_x z)]\exp[ik_x(x-ct)] \end{aligned} \tag{14.129}$$

The constants C_1, C_2, D_1, and D_2 will now be determined from the boundary conditions. Application of the kinematic boundary condition (14.121) to the first equation of (14.129) requires that $A_{2,1}(H) = 0$, from which

$$C_1\exp(k_x H) = -C_2\exp(-k_x H) = C/2 \tag{14.130}$$

follows immediately. Therefore, the amplitude $A_{2,1}(z)$ is given by

$$A_{2,1}(z) = \frac{C}{2}\{\exp[k_x(z-H)] - \exp[-k_x(z-H)]\} = C\sinh[k_x(z-H)] \tag{14.131}$$

Expanding the hyperbolic sine function and contracting the constants gives for the disturbance w_1'

$$\begin{aligned} &w_1'(z) = A[\sinh(k_x z) - \tanh(k_x H)\cosh(k_x z)]\exp[ik_x(x-ct)] \\ &\text{with} \quad A = C\cosh(k_x H) = \text{constant} \end{aligned} \tag{14.132}$$

Application of the kinematic boundary condition (14.121) for the lower layer results in

$$D_1\exp(-k_x H) = -D_2\exp(k_x H) = D/2 \tag{14.133}$$

and

$$A_{2,2}(z) = \frac{D}{2}\{\exp[k_x(z+H)] - \exp[-k_x(z+H)]\} = D\sinh[k_x(z+H)] \tag{14.134}$$

Hence, the perturbation velocity w_2' may be written as

$$\begin{aligned} &w_2'(z) = B[\sinh(k_x z) + \tanh(k_x H)\cosh(k_x z)]\exp[ik_x(x-ct)] \\ &\text{with} \quad B = D\cosh(k_x H) = \text{constant} \end{aligned} \tag{14.135}$$

In order to find a relation between the integration constants A and B, we form the ratio (14.126) and obtain

$$\frac{w_1'(z=0)}{w_2'(z=0)} = -\frac{A}{B} = \frac{c-u_{0,1}}{c-u_{0,2}} \implies (c-u_{0,2})A + (c-u_{0,1})B = 0 \tag{14.136}$$

For the determination of the phase speed c of the wave we need another relation between the constants A and B. To obtain this relation we apply the linearized dynamic boundary condition (14.128) to the layer $k = 1$. From (14.125) we have

$$\begin{aligned} Q_1(p_1' - p_2') &= -(c - u_{0,1}) \frac{\partial}{\partial x}(p_1' - p_2') \\ &= -(c - u_{0,1})\left[-\rho_{0,1}\left(\frac{\partial}{\partial t} + u_{0,1}\frac{\partial}{\partial x}\right)u_1' + \rho_{0,2}\left(\frac{\partial}{\partial t} + u_{0,2}\frac{\partial}{\partial x}\right)u_2'\right] \\ &= -(c - u_{0,1})\left[\rho_{0,1}(c - u_{0,1})\frac{\partial u_1'}{\partial x} - \rho_{0,2}(c - u_{0,2})\frac{\partial u_2'}{\partial x}\right] \end{aligned} \tag{14.137}$$

The pressure gradient for each layer has been eliminated with the help of the original perturbation equation (14.97a). Using the continuity equation (14.97c) together with (14.128) in (14.137), we find

$$-(c - u_{0,1})\left[-\rho_{0,1}(c - u_{0,1})\frac{\partial w_1'}{\partial z} + \rho_{0,2}(c - u_{0,2})\frac{\partial w_2'}{\partial z}\right] = g(\rho_{0,1} - \rho_{0,2})w_1' \tag{14.138}$$

Had we used $k = 2$ in (14.128), then, owing to (14.126) we would have obtained the same result.

The vertical partial derivatives of the perturbation velocities w_1' and w_2' at the boundary $z = 0$ follow from (14.132) and (14.135):

$$\left(\frac{\partial w_1'}{\partial z}\right)_{z=0} = k_x A \exp[ik_x(x - ct)], \qquad \left(\frac{\partial w_2'}{\partial z}\right)_{z=0} = k_x B \exp[ik_x(x - ct)] \tag{14.139}$$

By substituting these expressions into (14.138) we obtain the second relation between A and B:

$$\left[(c - u_{0,1})^2\rho_{0,1}k_x + g(\rho_{0,1} - \rho_{0,2})\tanh(k_x H)\right]A - (c - u_{0,1})(c - u_{0,2})\rho_{0,2}k_x B = 0 \tag{14.140}$$

Together with equation (14.136) we now have two homogeneous equations for the unknowns A and B. Arranged in matrix form we may write

$$\begin{pmatrix} (c - u_{0,1})^2\rho_{0,1}k_x + g(\rho_{0,1} - \rho_{0,2})\tanh(k_x H) & -(c - u_{0,1})(c - u_{0,2})\rho_{0,2}k_x \\ c - u_{0,2} & c - u_{0,1} \end{pmatrix} \times \begin{pmatrix} A \\ B \end{pmatrix} = 0 \tag{14.141}$$

Nonzero values of the integration constants A and B are possible only if the determinant of the matrix of this expression vanishes. Evaluating the determinant yields the two roots of the phase speed of the wave at the interface between the two fluid layers:

$$c_{1,2} = \frac{\rho_{0,1}u_{0,1} + \rho_{0,2}u_{0,2}}{\rho_{0,1} + \rho_{0,2}} \pm \sqrt{\frac{g}{k_x}\frac{\rho_{0,2} - \rho_{0,1}}{\rho_{0,1} + \rho_{0,2}}\tanh(k_x H) - \frac{\rho_{0,1}\rho_{0,2}(u_{0,1} - u_{0,2})^2}{(\rho_{0,1} + \rho_{0,2})^2}} \tag{14.142}$$

The first term is known as the convective term while the square root is known as the dynamic term. Owing to the complexity of this equation we will consider special cases that are more easily interpreted than the original equation.

First of all, it will be noticed that (14.142) reduces to the one-layer solution (14.111) on setting $\rho_{0,1} = 0$. For shallow-water waves, $L_x \gg H$, we find

$$c_{1,2} = \frac{\rho_{0,1}u_{0,1} + \rho_{0,2}u_{0,2}}{\rho_{0,1} + \rho_{0,2}} \pm \sqrt{gH\frac{\rho_{0,2} - \rho_{0,1}}{\rho_{0,1} + \rho_{0,2}} - \frac{\rho_{0,1}\rho_{0,2}(u_{0,1} - u_{0,2})^2}{(\rho_{0,1} + \rho_{0,2})^2}} \tag{14.143}$$

which reduces to (14.113) if $\rho_{0,1} = 0$.

For deep-water waves, $L_x \ll H$, we find

$$c_{1,2} = \frac{\rho_{0,1}u_{0,1} + \rho_{0,2}u_{0,2}}{\rho_{0,1} + \rho_{0,2}} \pm \sqrt{\frac{g}{k_x}\frac{\rho_{0,2} - \rho_{0,1}}{\rho_{0,1} + \rho_{0,2}} - \frac{\rho_{0,1}\rho_{0,2}(u_{0,1} - u_{0,2})^2}{(\rho_{0,1} + \rho_{0,2})^2}} \tag{14.144}$$

which reduces to (14.115) for $\rho_{0,1} = 0$.

Inspection of (14.142) shows that the expression under the square root might be negative, giving a complex value for the phase speed of the wave. Whenever this happens the wave is said to be unstable. The conditions for stable and unstable wave motion are given by

$$\begin{aligned} \text{Stable waves:}\ (u_{0,1} - u_{0,2})^2 &\le \frac{gL_x}{2\pi}\frac{\rho_{0,2}^2 - \rho_{0,1}^2}{\rho_{0,1}\rho_{0,2}}\tanh\left(\frac{2\pi H}{L_x}\right) \\ \text{Unstable waves:}\ (u_{0,1} - u_{0,2})^2 &> \frac{gL_x}{2\pi}\frac{\rho_{0,2}^2 - \rho_{0,1}^2}{\rho_{0,1}\rho_{0,2}}\tanh\left(\frac{2\pi H}{L_x}\right) \end{aligned} \tag{14.145}$$

Unstable Helmholtz waves are known as *Kelvin–Helmholtz waves*.

We conclude this section by considering three special cases that follow from inspection of (14.142). If the velocities of the basic currents in each layer are equal but the densities differ, that is $u_{0,1} = u_{0,2}$, $\rho_{0,1} \neq \rho_{0,2}$, then the wave is unstable only if the density of the upper layer exceeds the density of the lower layer. The result is obvious since overturning would take place immediately. If the densities in each layer are the same but the basic air currents differ, i.e. $u_{0,1} \neq u_{0,2}$, $\rho_{0,1} = \rho_{0,2}$, the wave is always unstable because the dynamic term is always imaginary. These waves are called *shearing waves* since the wave velocity depends only on the wind shear.

Finally, we consider the stationary wave by setting the phase speed c equal to zero. Furthermore, we assume that we are dealing with deep-water waves, $L_x \ll H$. In this special case the wavelength L_x is expressed by

$$L_x = \frac{2\pi}{g} \frac{\rho_{0,1} u_{0,1}^2 + \rho_{0,2} u_{0,2}^2}{\rho_{0,2} - \rho_{0,1}} \tag{14.146}$$

Using the ideal-gas law and recalling that, at the interface, the dynamic boundary condition must hold ($p_1 = p_2 = p$), we find the wavelength of the so-called *billow clouds* to be

$$L_x = \frac{2\pi}{g} \frac{u_{0,1}^2 T_2 + u_{0,2}^2 T_1}{T_1 - T_2} \tag{14.147}$$

These clouds are observed to form at the boundary of an inversion. Assuming that the clouds move with the mean velocity $(u_{0,1} + u_{0,2})/2$ of the lower and the upper layer, the wave velocity vanishes in a coordinate system moving with the velocity of the cloud system. Condensation and cloud formation take place when the air is ascending while the sky is clear where the wave motion causes descent of air. Comparison with observations shows that this simple theory overestimates the wavelengths of the billow clouds. As early as 1931 Haurwitz improved the above theory by permitting a variation in density of both air masses. This resulted in a better agreement between theory and observations. If the difference in temperature at the top of the inversion is about 4 K and the corresponding change in the wind velocity 5.5 m s^{-1}, the wavelength of the billow cloud is about 600 m. In this case the above theory agrees reasonably well with observations.

14.11 Nonlinear waves in the atmosphere

There are other types of wave motion that cannot be handled using the methods described so far. Instead of linear equations, nonlinear equations must be solved, with their accompanying complexities. In this section we merely wish to make a few comments on *solitary waves*, which are described by the so-called *Korteweg–de Vries equation*. A quite informative mathematical introduction to this topic is to be found, for example, in Keener (1988). Here we follow the introduction given by Panchev (1985), who treats some aspects of meteorological applications in connection with solitary waves.

A solitary wave is a localized perturbation propagating in a dispersive medium without change of shape. If solitary waves collide without changing shape they

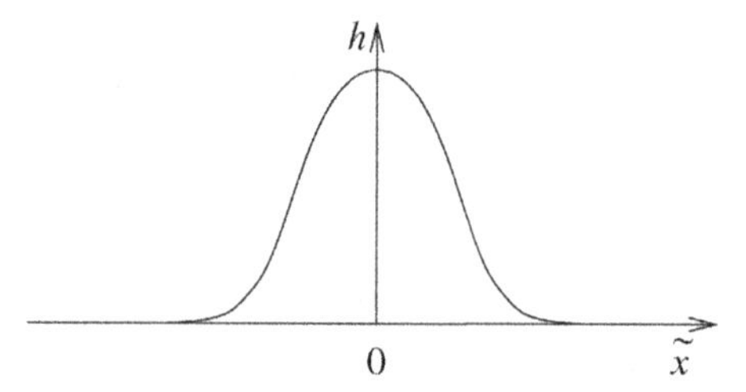

Fig. 14.6 The shape of the solitary wave.

are called *solitons*. The Korteweg–de Vries equation, in the shallow-water approximation, describes the evolution of long surface gravity waves with small amplitudes. In the one-dimensional case this equation may be written as

$$\frac{\partial h}{\partial t} + (C_0 + C_1 h)\frac{\partial h}{\partial x} + C_2\frac{\partial^3 h}{\partial x^3} = 0 \tag{14.148}$$

Here C_0, C_1, and C_2 are constants and $h(x, t)$ is the deviation from the free surface. In the linear approximation ($C_1 = 0$) there exists the solution

$$h(x, t) = A\exp[i(k_x x - \omega t)] \quad \text{with} \quad \omega = C_0 k_x - C_2 k_x^3 \tag{14.149}$$

where A is the amplitude of the disturbance. The dispersion relation shows that the angular frequency ω depends on the wavenumber k_x only, not on the amplitude. In the general case equation (14.148) also has a periodic solution that can be expressed in terms of elliptic cosine functions (*Jacobi's elliptical function*). It is quite remarkable that, in addition to the periodic solution, equation (14.148) has a particular solution that can be expressed in terms of the hyperbolic secant function

$$h(x, t) = A\ \mathrm{sech}^2\left[\widetilde{k_x}(x - ct)\right] \tag{14.150}$$

with

$$\widetilde{k_x} = \sqrt{\frac{AC_1}{12C_2}}$$

and

$$c = C_0 + \frac{AC_1}{3}$$

The angular frequency depends not only on the wavenumber, as in the linear case, but also on the amplitude. This is the most important property of dispersive nonlinear waves.

Figure 14.6 shows the shape of the function $h(\tilde{x})$, that is the real part of $h(\tilde{x})$, where $\tilde{x} = k_x(x - ct)$. In the extremely far wings h approaches the value zero; at

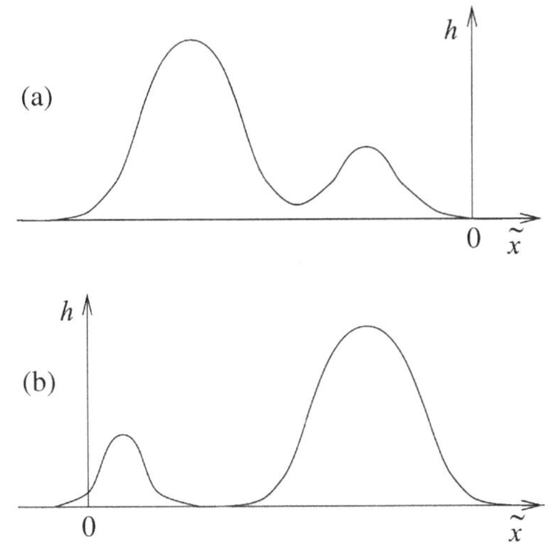

Fig. 14.7 Interaction of two solitons.

the center of the disturbance $h = A$. A disturbance, i.e. the solitary wave, moves in the positive x-direction with a speed c that is proportional to the amplitude A. Linear equations do not behave in this way. From mathematical theory we know that the superposition of solutions of linear systems gives another solution. Owing to this fact, a neutral (not decaying or amplifying) linear wave can overtake another neutral linear wave without the shape of either wave changing. In contrast to this, the superposition of solutions of nonlinear equations does not give a new solution. Nevertheless, many solitary waves exhibit aspects of linear behavior. When one solitary wave overtakes another, they interact nonlinearly. After separating they retain their original shapes as shown in Figure 14.7. The soliton with the larger amplitude overtakes a soliton of smaller amplitude as shown in part (b) of Figure 14.7. The occurrence of nonlinearity follows from the observation that, after the interaction, the solitons are not located at those coordinates where they would have been found had no interaction taken place.

Finally, we briefly describe a meteorological application. From a careful analysis of the surface pressure field, sometime in 1951 in Kansas, USA, a pressure disturbance of 3.4 hPa was observed to propagate with a speed of about 21 m s^{-1} up to a distance of 800 km. Abdullah (1955) explains this phenomenon as an internal gravitational solitary wave resulting from an impulsive motion of a quasi-stationary cold front in the thermal inversion layer. Further observations of this type have been presented by Christie *et al.* (1978).

14.12 Problems

14.1: Sketch the profile of the wave $U = U_0 \sin(kx - \omega t + \epsilon)$ with $\epsilon = \pi$.

14.2: A standing or stationary wave can be described as the product $U = U_1(x)U_2(t)$. Show that the superposition of two waves of the same amplitude and the same period moving in opposite directions can be written in this way. Assume that we are dealing with simple sine waves and give the explicit form of U.

14.3: Let α_0 = constant, p_0 = constant, and u_0 = constant. Eliminate α' in (14.31) and verify that pure sound waves move according to (14.37d). Assume that $\delta = 1$.

14.4: Let $\delta = 1$, $u_0 = 0$, and $g = 0$ in the system (14.31). Introduce the transformation (14.82). Instead of α_0 = constant assume that the relative change in height $(1/\alpha_0)\,\partial\alpha_0/\partial z$ is constant. Does this result in the phase speed of a pure sound wave?

14.5: Show that the system of Lamb waves (14.79) satisfies the basic linearized system (14.31). For simplification set $u_0 = 0$.

14.6: Modify equation (14.142) for the case in which both fluids are infinitely deep.

14.7: In shallow water the phase speed is given by (14.113) if the earth's rotation is ignored. Estimate the effect of the earth's rotation. In this case $c = \sqrt{g\overline{H} + f^2/k_x^2}$ if $u_0 = 0$.

14.8: Find an approximate wavelength of a lee wave forming behind a symmetric mountain in a stable atmosphere. This wavelength is fairly well approximated by a standing wave produced by vertical oscillations in a homogeneous current at a height twice the height of the mountain. Assume that $u_0 = 10$ m s^{-1}, the observed lapse rate is -10 K km^{-1}, and T_0(3000 m) is 240 K.

14.9: Show that $\omega_{\rm Br}^2 = (g/\alpha_0)\,\partial\alpha_0/\partial z - g^2/c_{\rm L}^2$.

15

The barotropic model

Barotropic and baroclinic atmospheric processes manifest themselves in the numerous facets of large-scale weather phenomena. Typical examples are the formation and propagation of synoptic waves having wavelengths of several thousand kilometers and the characteristic life cycles of high- and low-pressure systems. The barotropic and baroclinic physics provides the physical basis of numerical weather prediction. In this chapter we will consider various aspects of barotropic models. It is realized that the prediction of the daily weather by means of barotropic models is no longer practiced by the national weather services. Nevertheless, by discussing the mathematical theory of the barotropic physics we can learn very well how physical variables are interconnected and how much care must be taken to construct even a very simple prediction model. The first numerical barotropic weather-prediction model was introduced by the renowned meteorologist C. G. Rossby and by the famous mathematician John von Neumann. Baroclinic models will be described in some detail in later chapters.

Barotropic models are short-range-prediction models that include only the reversible part of atmospheric physics. The consequence is that the atmosphere is treated as a one-component gas consisting of dry air. The irreversible physics such as non-adiabatic heating and cloud formation is not taken into account.

15.1 The basic assumptions of the barotropic model

The name of the model is derived from the assumption that the atmosphere is in a barotropic state throughout the prediction period. The condition of barotropy by itself, however, is not sufficient to construct a barotropic prediction model; additional assumptions are mandatory. The model described here rests on three basic assumptions. These are

(i) the validity of the condition of barotropy for the entire prediction period:

$$\rho = \rho(p) \quad \text{or} \quad \nabla\rho \times \nabla p = 0 \quad \text{for} \quad t \geq t_0 \tag{15.1}$$

(this is known as the *condition of autobarotropy*)

(ii) the validity of the hydrostatic equation at all times

$$\frac{\partial p}{\partial z} = -g\rho \quad \text{for} \quad t \geq t_0 \tag{15.2}$$

(iii) the horizontal wind is independent of height

$$\frac{\partial \mathbf{v}_\mathrm{h}}{\partial z} = 0 \quad \text{for} \quad t \geq t_0 \tag{15.3}$$

The assumption of hydrostatic equilibrium is also an integral part of most baroclinic large-scale-weather-prediction models. It can be shown that conditions (i) and (ii) imply that the horizontal pressure-gradient force is independent of height. This is equivalent to stating that the geostrophic wind is height-independent or that the thermal wind vanishes. In contrast to this, the strength of the thermal wind is a measure of atmospheric developments in baroclinic models. Moreover, the hydrostatic assumption implies the removal of the rapidly moving vertical sound waves. This is important since the filtering of the rapidly moving "noise" waves permits a larger time step in the numerical integration of the model equations. Insofar as condition (iii) is concerned, it may be replaced by a weaker statement. Indeed, it is sufficient to require that the horizontal wind is height-independent at the beginning of the prediction period since (i) and (ii) automatically guarantee that the horizontal wind remains height-independent at all later times. We will prove this statement at the end of Section 15.2.1.

There also exists the so-called *equivalent barotropic model*, in which the actual horizontal wind changes with height in a prescribed manner. We are not going to discuss this type of model since it requires various empirical modifications that are not entirely consistent with the remaining physics of the model.

A few remarks on the assumptions involved in the model may be helpful. Assumption (i) does not permit the formation of isobaric temperature gradients. All development processes are thermally inactive. Thus we ignore the first and second laws of thermodynamics. Only one thermodynamic variable, either pressure or density, needs to be controlled.

Assumption (ii), as we have demonstrated in the previous chapter, eliminates vertically propagating sound waves. If the wavelength of the disturbance becomes comparable to the depth of the homogeneous atmosphere, the hydrostatic assumption becomes unrealistic.

Assumption (iii) together with assumptions (i) and (ii) reduces the model atmosphere to a spatially two-dimensional system. In contrast to this, all baroclinic models are three-dimensional, thus simulating atmospheric processes more realistically. Hence, the barotropic model applies to one pressure level only, which is usually taken as the 500-hPa surface. There the model physics applies best. Owing to the special assumptions of the model the barotropic model suppresses the synoptically relevant interactions of height-dependent velocity divergences and temperature varations which are responsible for the transformation of potential energy into kinetic energy on a significant scale. Development processes such as frontogenesis and occlusions are suppressed by the barotropic model. At most, transformations between the kinetic energy of the mean flow and the disturbances are taking place.

Finally, it should be realized that there are numerous variants of the barotropic model. We are going to discuss unfiltered and filtered barotropic models, in which the gravitational surface waves are eliminated by means of diagnostic relations. From the historical point of view it is interesting to remark that the first barotropic model using actual meteorological data was successfully applied by Charney and Eliassen (1949). They used a filtered linearized version of the model.

15.2 The unfiltered barotropic prediction model

15.2.1 The general barotropic model

We will now consider the barotropic model without the elimination of the rapidly moving surface gravity waves. The elimination of these fast waves is called *noise filtering*. Let us consider a large enough section of the atmosphere so that the hydrostatic equation applies. For simplicity, we represent the field of motion on a tangential plane fixed to the earth where the motion is described in terms of Cartesian coordinates. The surface of the earth is the rigid lower boundary $z_s(x, y)$ while the upper boundary is assumed to be a free surface, $H(x, y, t)$. The pressure at the earth's surface will be denoted by $p_s(x, y, t)$; the hydrostatic pressure at the free surface vanishes as depicted in Figure 15.1.

At an arbitrary reference level z the hydrostatic pressure is given by

$$p(x, y, z, t) = g \int_z^H \rho \, dz' \tag{15.4a}$$

so that the surface pressure may be written as

$$p_s(x, y, t) = p(x, y, z, t) + g \int_{z_s}^z \rho \, dz' \tag{15.4b}$$

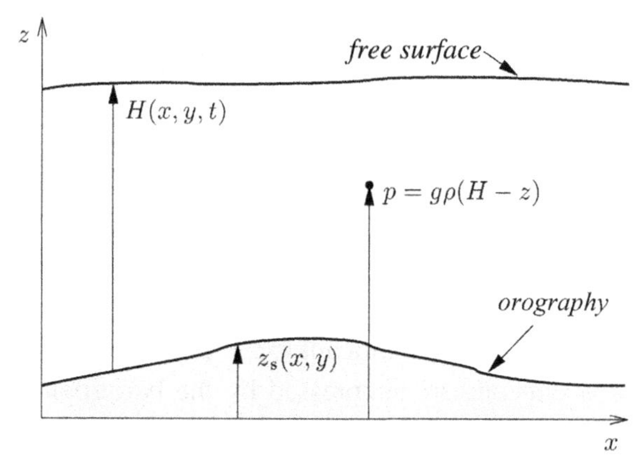

Fig. 15.1 A vertical cross-section of the atmosphere in the (x, z)-plane.

Since the horizontal pressure gradient force is height-independent we may compute this gradient everywhere, including at the earth's surface. We must keep in mind, however, that the occurrence of orography will modify the surface pressure gradient. In order to handle this situation we apply the Leibniz rule to (15.4b). We first carry out the partial differentiations with respect to the horizontal coordinates. Remembering that z_s also depends on x and y, we obtain

$$\begin{aligned}\frac{\partial}{\partial x}[p_s(x,y,t)] &= \frac{\partial}{\partial x}[p(x,y,z,t)] + g\int_{z_s}^{z}\frac{\partial \rho}{\partial x}\,dz' - g\rho_s\frac{\partial}{\partial x}[z_s(x,y)] \\ \frac{\partial}{\partial y}[p_s(x,y,t)] &= \frac{\partial}{\partial y}[p(x,y,z,t)] + g\int_{z_s}^{z}\frac{\partial \rho}{\partial y}\,dz' - g\rho_s\frac{\partial}{\partial y}[z_s(x,y)]\end{aligned} \tag{15.5a}$$

with $\rho_s = \rho(p(z_s), t)$. Combining these equations leads to

$$\nabla_h p(x,y,z,t) = \nabla_h p_s - g\int_{z_s}^{z}\nabla_h \rho\,dz' + g\rho_s\,\nabla_h z_s \tag{15.5b}$$

On letting z approach z_s the integral will vanish. Thus, in the limit $z \longrightarrow z_s$ we find

$$\boxed{\left[\frac{1}{\rho}\nabla_h p\right]_{z=z_s} = \frac{1}{\rho_s}\nabla_h p_s + \nabla_h \phi_s} \tag{15.6}$$

Here we have introduced the geopotential $\phi_s(x, y)$ of the earth's surface, which is a time-independent quantity and may be computed with the help of a suitable map, i.e. $\phi_s(x, y) = g z_s(x, y)$.

Since in the barotropic model the horizontal pressure gradient is independent of height, we may introduce (15.6) into (2.29) to obtain the equation of motion for frictionless flow on the tangential plane as

$$\boxed{\left(\frac{\partial \mathbf{v}_{\rm h}}{\partial t}\bigg|_{\mathbf{q}_i}\right)_{\rm h} + \mathbf{v}_{\rm h}\cdot\nabla_{\rm h}\mathbf{v}_{\rm h} + f\mathbf{i}_3\times\mathbf{v}_{\rm h} + \frac{1}{\rho_{\rm s}}\nabla_{\rm h}p_{\rm s} + \nabla_{\rm h}\phi_{\rm s} = 0} \tag{15.7}$$

Recall that the vertical line attached to the local time derivative simply means that the basis vectors are not to be differentiated with respect to time. Their time dependency is already contained in the Coriolis term. As usual, the sum of the local acceleration term and the advection term can be combined to give the horizontal acceleration. To complete the prognostic system we must involve the continuity equation

$$\frac{\partial \rho}{\partial t} + \nabla_{\rm h}\cdot(\rho\mathbf{v}_{\rm h}) + \frac{\partial}{\partial z}(\rho w) = 0 \tag{15.8}$$

In order to apply the kinematic boundary conditions, this equation will be integrated from the lower to the upper boundary of the atmosphere, yielding

$$\int_{z_{\rm s}}^{H}\frac{\partial \rho}{\partial t}\,dz + \int_{z_{\rm s}}^{H}\nabla_{\rm h}\cdot(\rho\mathbf{v}_{\rm h})\,dz + (\rho w)\bigg|_{z_{\rm s}}^{H} = 0 \tag{15.9}$$

Reference to equations (9.43b) and (9.43c) shows that these boundary conditions may be written as

$$w(H) = \frac{\partial H}{\partial t} + \mathbf{v}_{\rm h}\cdot\nabla_{\rm h}H, \qquad w(z_{\rm s}) = \mathbf{v}_{\rm h}\cdot\nabla_{\rm h}z_{\rm s} \tag{15.10}$$

The first term of (15.9) will now be given a more suitable form with the help of the Leibniz rule

$$\frac{\partial}{\partial t}\int_{z_{\rm s}}^{H}\rho\,dz = \int_{z_{\rm s}}^{H}\frac{\partial \rho}{\partial t}\,dz + \rho_H\frac{\partial H}{\partial t} \tag{15.11}$$

with $\rho_H = \rho(p(H), t)$. The second term in (15.9) will also be rewritten with the help of the Leibniz rule. The differentiation with respect to the coordinate x yields

$$\frac{\partial}{\partial x}\int_{z_{\rm s}}^{H}\rho u\,dz = \int_{z_{\rm s}}^{H}\frac{\partial}{\partial x}(\rho u)\,dz + (\rho u)\bigg|_{H}\frac{\partial H}{\partial x} - (\rho u)\bigg|_{z_{\rm s}}\frac{\partial z_{\rm s}}{\partial x} \tag{15.12}$$

The differentiation with respect to y is accomplished by simply replacing x by the coordinate y. Therefore, the second term of (15.12) can be written as

$$\begin{aligned}\int_{z_{\rm s}}^{H}\nabla_{\rm h}\cdot(\rho\mathbf{v}_{\rm h})\,dz &= \nabla_{\rm h}\cdot\int_{z_{\rm s}}^{H}(\rho\mathbf{v}_{\rm h})\,dz - (\rho u)\bigg|_{H}\frac{\partial H}{\partial x}\\ &\quad + (\rho u)\bigg|_{z_{\rm s}}\frac{\partial z_{\rm s}}{\partial x} - (\rho v)\bigg|_{H}\frac{\partial H}{\partial y} + (\rho v)\bigg|_{z_{\rm s}}\frac{\partial z_{\rm s}}{\partial y}\end{aligned} \tag{15.13}$$

Substituting (15.10), (15.11), and (15.13) into (15.9) gives

$$\begin{aligned}
&\frac{\partial}{\partial t}\int_{z_s}^{H}\rho\,dz-\rho_H\frac{\partial H}{\partial t}+\nabla_h\cdot\int_{z_s}^{H}\rho\mathbf{v}_h\,dz-\rho_H\mathbf{v}_h\cdot\nabla_h H\\
&\quad+\rho_s\mathbf{v}_h\cdot\nabla_h z_s+\rho_H\left(\frac{\partial H}{\partial t}+\mathbf{v}_h\cdot\nabla_h H\right)-\rho_s\mathbf{v}_h\cdot\nabla_h z_s=0
\end{aligned}\tag{15.14}$$

in which several terms cancel out. Finally we obtain

$$\frac{\partial}{\partial t}\int_{z_s}^{H}\rho\,dz+\nabla_h\cdot\int_{z_s}^{H}\rho\mathbf{v}_h\,dz=0\tag{15.15}$$

On multiplying both sides of (15.15) by the acceleration due to gravity g and recalling that the horizontal velocity is height-independent, we find in view of (15.4a) the surface-pressure tendency equation

$$\boxed{\frac{\partial p_s}{\partial t}+\nabla_h\cdot(\mathbf{v}_h p_s)=0}\tag{15.16}$$

For the model variables $\mathbf{v}_h(x, y, t)$ and $p_s(x, y, t)$ we have with equations (15.7) and (15.16) the required prognostic equations for the barotropic model. With the help of the condition of barotropy $\rho = \rho(p_s)$ the barotropic model may be integrated in principle if proper initial conditions are provided.

It is of great advantage for the numerical integration of the barotropic model equations that the model variables are independent of height. For practical applications, however, a suitable connection between the model variables $\mathbf{v}_h(x, y, t)$ and $p_s(x, y, t)$ for the behavior of the real atmosphere must be established. For the surface pressure this connection follows immediately. Consequently, the predicted horizontal wind should correspond to the surface wind. However, it can hardly be expected that the model produces a reliable replication of the observed surface wind. Experience shows that the vertical average of the observed wind or the wind at some intermediate level is a better representation of the large-scale motion than is the surface wind. Therefore, it may be expected that the two-dimensional barotropic model will produce more realistic pressure and wind fields at the 500-hPa pressure level which is located near the so-called "*level of nondivergence*" at which the observed divergence of the wind velocity is negligibly small. Since the two-dimensional barotropic model is incapable of producing meteorologically significant amounts of divergence, best results should be expected if the model is applied to the 500-hPa level.

Before closing this section, we will now prove the statement that the horizontal wind remains height-independent if it is height-independent initially at time t_0.

For this purpose let us formally rewrite equation (15.7) as $\partial \mathbf{v}_h/\partial t = \mathbf{F}(x, y, t)$, where the right-hand side is independent of height. Let us now formally carry out the integration over time. After the first time step Δt the right-hand side is height-independent again:

$$\mathbf{v}_h(x, y, t_0 + \Delta t) = \mathbf{v}_h(x, y, t_0) + \Delta t\, \mathbf{F}(x, y, t_0) \tag{15.17}$$

On repeating this procedure for the following time steps the right-hand side remains height-independent. Hence, it may be concluded for the barotropic model that the horizontal wind remains height-independent if it is height-independent initially.

15.2.2 The barotropic model of the homogeneous atmosphere

In place of the surface pressure $p_s(x, y, t)$ it is of advantage to introduce the height of the free surface $H(x, y, t)$ as a model variable. In order to construct a finite-height model it is necessary to specify the barotropy function $\rho(p)$ in a suitable manner. The condition of barotropy must not necessarily be invertible to guarantee barotropy. The simplest condition of barotropy is given by assuming that we have a homogeneous model atmosphere specified by

$$\rho = \rho_s = \text{constant}, \qquad \nabla \rho = 0, \qquad \frac{d\rho}{dt} = 0 \tag{15.18}$$

For a ground temperature of 273 K the pressure scale height or the height of the homogeneous atmosphere is about 8000 m. Instead of the height of the free surface itself we may just as well use the corresponding geopotential so that the upper and lower boundaries of the model atmosphere are specified by

$$\phi(x, y, t) = gH(x, y, t), \qquad \phi_s(x, y) = gz_s(x, y) \tag{15.19}$$

Owing to the constant-density assumption the surface pressure is given by

$$p_s = g\rho(H - z_s) = \rho(\phi - \phi_s) \tag{15.20}$$

Introducing this expression into (15.7) leads to the equation of horizontal motion assuming the form

$$\boxed{\left(\left.\frac{\partial \mathbf{v}_h}{\partial t}\right|_{q_i}\right)_h + \mathbf{v}_h \cdot \nabla_h \mathbf{v}_h + f\mathbf{i}_3 \times \mathbf{v}_h + \nabla_h \phi = 0} \tag{15.21}$$

Owing to (15.18) the continuity equation (15.8) reduces to the statement that the three-dimensional velocity divergence vanishes. Height integration of the reduced continuity equation between the lower and upper boundaries gives

$$\int_{z_s}^{H} \frac{\partial w}{\partial z}\, dz = w(H) - w(z_s) = -\nabla_h \cdot \mathbf{v}_h \int_{z_s}^{H} dz = -(H - z_s)\, \nabla_h \cdot \mathbf{v}_h \tag{15.22}$$

since the horizontal velocity is height-independent. By introducing the kinematic boundary conditions (15.10) into (15.22) and recalling that z_s is time-independent we obtain

$$\frac{\partial}{\partial t}(H - z_s) = -\nabla_h \cdot [\mathbf{v}_h(H - z_s)] \tag{15.23a}$$

Multiplying this equation by the acceleration of gravity results in the corresponding expressions for the geopotentials:

$$\boxed{\frac{\partial}{\partial t}(\phi - \phi_s) = -\nabla_h \cdot [\mathbf{v}_h(\phi - \phi_s)]} \tag{15.23b}$$

By multiplying both sides of (15.23a) by the constant model density, we obtain

$$\frac{\partial \rho_A}{\partial t} = -\nabla_h \cdot (\mathbf{v}_h \rho_A) \quad \text{or} \quad \frac{d\rho_A}{dt} + \rho_A \nabla_h \cdot \mathbf{v}_h = 0 \tag{15.23c}$$

where $\rho_A = \rho(H - z_s)$ is the density per unit area. This form of the continuity equation for the two-dimensional space closely resembles the normal continuity equation (15.8) for the three-dimensional space, where the density represents mass per unit volume.

Let us consider the prognostic set (15.21) and (15.23b), where the density does not appear at all as a parameter. At first glance there seems to be no connection between the model variables $\phi(x, y, t)$ and the horizontal velocity $\mathbf{v}_h(x, y, t)$ on the one hand and the variables of the real atmosphere on the other. It is postulated and verified by experience that the height-independent horizontal wind vector of the model atmosphere is a reasonable approximation to the height-averaged wind field of the real atmosphere. This wind is assumed to apply to the so-called *equivalent barotropic level*, which for practical reasons is taken as the 500-hPa pressure level. As mentioned previously, this surface is located near the level of nondivergence which approximately divides the mass of the atmosphere into two equal parts. This results in the following correspondence of variables:

$$\mathbf{v}_h = \mathbf{v}(p = 500\,\text{hPa}), \qquad \phi = \phi(p = 500\,\text{hPa}) \tag{15.24}$$

Various empirical reductions of the geopotential of the earth's surface may be necessary in order to prevent the orography modifying the predicted fields too strongly. These modifications are based on experience and are of entirely empirical character.

The model consisting of equations (15.21) and (15.23b) is best described by a cylinder filled with water having a free surface that is rotating about its vertical axis of symmetry. The assumption of autobarotropy is satisfied reasonably well by pure water and deviations from the hydrostatic pressure are small as long as the flow velocities are not too large. Height-independent horizontal velocities, however, cannot be realized in general. In this rotating cylinder characteristic flow patterns evolve with embedded vortices. In addition to this, one observes expanding ring-type waves of short periods, which are superimposed on the larger-scale characteristic flow patterns.

In close correspondence, the model equations of the homogeneous atmosphere describe relevant characteristic synoptic flow patterns and the displacement of ridges and troughs. Superimposed on the large-scale features are high-speed external gravity waves of short periods. These waves, which we have described as "*meteorological noise*," have little direct influence on the processes associated with the formation of the synoptic waves. As mentioned before, due to the existence of the rapidly moving external gravity waves, the time integration of the model necessitates the use of very small time steps. Moreover, the amplitudes of the noise waves may be large enough to falsify meteorologically relevant results. In a later section we will briefly discuss how to filter out the noise waves without significantly modifying the synoptic waves.

15.2.3 The energy budget of the homogeneous model and the available potential energy

First of all we define the kinetic energy

$$K_{\rm A} = \int_{z_{\rm s}}^{H} \rho \frac{\mathbf{v}_{\rm h}^2}{2}\, dz = \rho(H - z_{\rm s})\frac{\mathbf{v}_{\rm h}^2}{2} = \rho_{\rm A}\frac{\mathbf{v}_{\rm h}^2}{2} \tag{15.25}$$

and the potential energy

$$P_{\rm A} = \int_{z_{\rm s}}^{H} g\rho z\, dz = g\rho\frac{H^2 - z_{\rm s}^2}{2} = \rho_{\rm A}\frac{\phi + \phi_{\rm s}}{2} \tag{15.26}$$

contained in a vertical column of unit cross-sectional area. These definitions are then used to set up the budget equations. It will be shown that only limited amounts of potential energy can be transformed into kinetic energy, otherwise the principle of conservation of mass would be violated. The amount of potential energy that can be transformed into kinetic energy is known as the *available potential energy*.

First we rewrite (15.21) as

$$\left(\left.\frac{d\mathbf{v}_{\rm h}}{dt}\right|_{\mathbf{q}_i}\right)_{\rm h} + f\mathbf{i}_3 \times \mathbf{v}_{\rm h} = -\nabla_{\rm h}\phi \tag{15.27a}$$

and, in component form, as

$$\begin{aligned} \frac{\partial u}{\partial t} + u\frac{\partial u}{\partial x} + v\frac{\partial u}{\partial y} - fv &= -\frac{\partial \phi}{\partial x} \\ \frac{\partial v}{\partial t} + u\frac{\partial v}{\partial x} + v\frac{\partial v}{\partial y} + fu &= -\frac{\partial \phi}{\partial y} \end{aligned} \tag{15.27b}$$

Scalar multiplication of (15.27a) by $\rho_A \mathbf{v}_h$, using simplified notation, gives

$$\rho_A \frac{d}{dt}\left(\frac{\mathbf{v}_h^2}{2}\right) = -\rho_A \mathbf{v}_h \cdot \nabla_h \phi \tag{15.28}$$

As expected, the Coriolis term vanishes since the Coriolis force cannot perform any work. The left-hand side of (15.28) may be rewritten in budget form. With the help of the two-dimensional continuity equation (15.23c) we find immediately

$$\rho_A \frac{d}{dt}\left(\frac{\mathbf{v}_h^2}{2}\right) = \frac{\partial}{\partial t}\left(\rho_A \frac{\mathbf{v}_h^2}{2}\right) + \nabla_h \cdot \left(\rho_A \mathbf{v}_h \frac{\mathbf{v}_h^2}{2}\right) \tag{15.29}$$

Expressing the geopotential in terms of $\rho_A = \rho(H - z_s)$, equation (15.19) yields

$$\phi = \phi_s + g\rho_A/\rho \tag{15.30}$$

After a slight rearrangement of terms we obtain from (15.28)

$$\frac{\partial}{\partial t}\left(\rho_A \frac{\mathbf{v}_h^2}{2}\right) + \nabla_h \cdot \left(\rho_A \mathbf{v}_h \frac{\mathbf{v}_h^2}{2} + \mathbf{v}_h \frac{g\rho_A^2}{2\rho}\right) = \frac{g\rho_A^2}{2\rho} \nabla_h \cdot \mathbf{v}_h - \rho_A \mathbf{v}_h \cdot \nabla_h \phi_s \tag{15.31}$$

which has the desired budget form. Denoting the flux of the kinetic energy by

$$\mathbf{F}_{K_A} = \mathbf{v}_h K_A + \mathbf{v}_h \frac{g\rho_A^2}{2\rho} \tag{15.32}$$

and the corresponding source by

$$Q_{K_A} = \frac{g\rho_A^2}{2\rho} \nabla_h \cdot \mathbf{v}_h - \rho_A \mathbf{v}_h \cdot \nabla_h \phi_s \tag{15.33}$$

we may write the budget equation for the kinetic energy in the usual form

$$\boxed{\frac{\partial K_A}{\partial t} + \nabla_h \cdot \mathbf{F}_{K_A} = Q_{K_A}} \tag{15.34}$$

Next we derive the budget equation for the potential energy by first multiplying (15.23c) by the factor $(\phi + \phi_s)/2$. Recalling that ϕ_s is independent of time, we obtain immediately

$$\frac{\partial P_A}{\partial t} - \frac{\rho_A}{2} \frac{\partial \phi}{\partial t} = -\nabla_h \cdot (\mathbf{v}_h P_A) + \frac{\rho_A}{2} \mathbf{v}_h \cdot \nabla_h(\phi + \phi_s) \tag{15.35}$$

Replacing $\partial\phi/\partial t$ in this expression by means of (15.23b) yields

$$\begin{aligned} \frac{\partial P_A}{\partial t} + \nabla_h \cdot (\mathbf{v}_h P_A) &= -\frac{\rho_A}{2} \nabla_h \cdot [\mathbf{v}_h(\phi - \phi_s)] + \frac{\rho_A}{2} \mathbf{v}_h \cdot \nabla_h(\phi + \phi_s) \\ &= -\frac{g\rho_A^2}{2\rho} \nabla_h \cdot \mathbf{v}_h + \rho_A \mathbf{v}_h \cdot \nabla_h \phi_s = Q_{P_A} \end{aligned} \tag{15.36}$$

where (15.30) was used to obtain the second equation. In (15.36) the source term Q_{P_A} of the potential energy of the column has been introduced. Comparison of Q_{P_A} with Q_{K_A} as given by (15.33) shows that $Q_{P_A} = -Q_{K_A}$. This yields the desired budget equation

$$\boxed{\frac{\partial P_A}{\partial t} + \nabla_h \cdot (\mathbf{v}_h P_A) = -Q_{K_A}} \tag{15.37}$$

If $Q_{K_A} > 0$ then kinetic energy is produced from potential energy and vice versa.

We will now show that only a limited amount of the potential energy P_A contained in a column of air can be transformed into kinetic energy due to the constraint of conservation of mass. Let us decompose the height of the free surface H into its mean value $\overline{H}$ and the fluctuation H':

$$H = \overline{H} + H' \tag{15.38}$$

According to equation (15.26) the potential energy is given by

$$P_A = \frac{g\rho}{2}\left(\overline{H}^2 + 2\overline{H}H' + H'^2 - z_s^2\right) \tag{15.39}$$

Let us now consider a materially closed region of cross-section S. On averaging the potential energy we find

$$\overline{P}_A = \frac{g\rho}{2S}\int_S \left(\overline{H}^2 + 2\overline{H}H' + H'^2 - z_s^2\right) dS' \tag{15.40}$$

This equation may be simplified since the average over the fluctuation must vanish:

$$\frac{1}{S}\int_S H'\, dS' = \frac{1}{S}\int_S H\ dS' - \frac{1}{S}\int_S \overline{H}\ dS' = \overline{H} - \overline{H} = 0 \tag{15.41}$$

Therefore, the average value of the potential energy within the domain defined by the upper and lower boundaries as well as by the side walls is given by

$$\overline{P}_A = \frac{g\rho}{2}\left(\overline{H}^2 + \overline{H'^2} - z_s^2\right) \quad \text{with} \quad \overline{H'^2} = \frac{1}{S}\int_S H'^2\, dS' \tag{15.42}$$

Since the quantity $\overline{H}^2 - z_s^2$ is fixed, it defines the minimum amount of potential energy contained within the region:

$$\overline{P}_{\min} = \frac{g\rho}{2}\left(\overline{H}^2 - z_s^2\right) \tag{15.43}$$

The potential energy which is free or available for transformation into kinetic energy leads to the definition of the *available potential energy*,

$$\boxed{P_{av} = \overline{P}_A - \overline{P}_{\min} = \tfrac{1}{2} g\rho \overline{H'^2}} \tag{15.44}$$

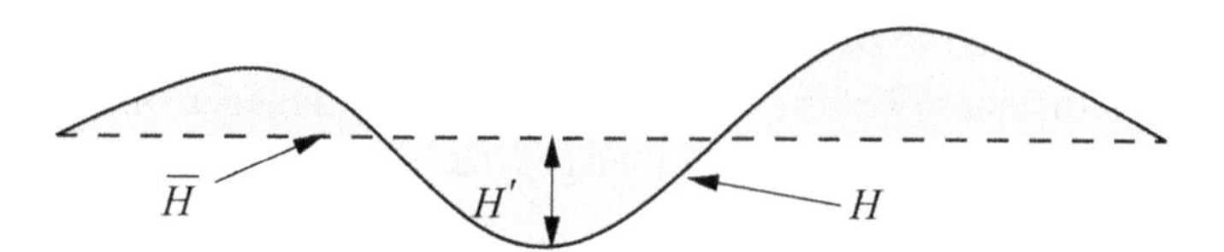

Fig. 15.2 Available potential energy.

which is proportional to the mean value of the square of the fluctuation; see Figure 15.2.

It will be recognized that the transformation is solely due to the appearance of short-periodic noise processes. This transformation has nothing to do with baroclinic developments, whereby the kinetic energy is gained from potential energy due to a rearrangement of cold and warm air in the same volume.

To complete the discussion, we must show that the potential energy cannot become smaller than $\overline{P}_{\text{min}}$ if the principle of conservation of mass is not violated. The mass contained within the volume is defined by

$$M = \int_S \rho(H - z_s)\, dS' = \rho S(\overline{H} - \overline{z_s}) \tag{15.45}$$

From

$$\frac{\partial M}{\partial t} = \rho S \frac{\partial \overline{H}}{\partial t} = 0 \quad \text{or} \quad \frac{\partial \overline{H}}{\partial t} = 0 \tag{15.46}$$

it follows that conservation of mass is equivalent to the statement that the average height of the model region does not change with time. Therefore, $\overline{P}_{\text{min}}$ cannot change with time either:

$$\frac{\partial \overline{P}_{\text{min}}}{\partial t} = g\rho \overline{H} \frac{\partial \overline{H}}{\partial t} = \frac{g\overline{H}}{S} \frac{\partial M}{\partial t} = 0 \tag{15.47}$$

We will close this section by remarking that the available potential energy of baroclinic processes has been investigated by E. Lorenz (1955) and van Miegham (1973). For adiabatic processes invariance of the potential temperature must be observed.

15.2.4 Properties of the homogeneous model

In order to survey all types of wave motions admitted by the barotropic model, we apply the linearization method described in the previous chapter to the basic model equations (15.21) and (15.23b). For simplicity we ignore the occurrence of orographic effects by setting $\phi_s = 0$. Furthermore, we assume that we have a stationary basic state (denoted by the overbar) that is independent of the coordinates (x, t) and we hold the Coriolis parameter constant, i.e. $f \approx f_0$:

$$\overline{u} = -\frac{1}{f_0}\frac{\partial \overline{\phi}}{\partial y}, \qquad \overline{\phi}(y) = g\overline{H}(y) \tag{15.48}$$

Now we are able to obtain analytic solutions that are easy to interpret. Superimposed on the basic current are the periodic disturbances $u'(x, t)$, $v'(x, t)$ and $\phi'(x, t)$. This results in the basic set of equations describing the motion and the mass continuity:

$$\begin{aligned}\frac{\partial u'}{\partial t} + \overline{u}\frac{\partial u'}{\partial x} + \frac{\partial \phi'}{\partial x} - f_0 v' &= 0\\ \frac{\partial v'}{\partial t} + \overline{u}\frac{\partial v'}{\partial x} + f_0 u' &= 0\\ \frac{\partial \phi'}{\partial t} + \overline{u}\frac{\partial \phi'}{\partial x} + \overline{\phi}_0\frac{\partial u'}{\partial x} - \overline{u} f_0 v' &= 0\end{aligned} \tag{15.49}$$

Here the geopotential of the mean height $\overline{\phi}(y)$ has been approximated by an average value $\overline{\phi}_0$ = constant. This is not quite consistent mathematically but it makes it possible to take a solution of the form

$$\begin{pmatrix} u' \\ v' \\ \phi' \end{pmatrix} = \begin{pmatrix} u^* \\ v^* \\ \phi^* \end{pmatrix} \exp[ik_x(x - ct)] \tag{15.50}$$

where the amplitudes are constants. Substitution of (15.50) into (15.49) results in the cubic equation

$$(c - \overline{u})^3 - \left(\overline{\phi}_0 + \frac{f_0^2}{k_x^2}\right)(c - \overline{u}) - \frac{f_0^2}{k_x^2}\overline{u} = 0 \tag{15.51}$$

from which the unknown phase velocities can be determined. This cubic equation in the variable $c - \overline{u}$ is in the reduced form of the general cubic equation. The coefficients in the reduced form (15.51) satisfy a condition that guarantees the existence of three real solutions. One of the three waves, whose phase speed, say c_1, is of the order of the mean wind speed $\overline{u}$, is of meteorological relevance. Therefore, we may apply the inequality

$$|c_1 - \overline{u}|^2 \ll \overline{\phi} + f_0^2/k_x^2 \tag{15.52}$$

to obtain an approximate expression for c_1. Owing to (15.52) the first term in (15.51) may be neglected in comparison with the second term and we obtain the approximate solution for c_1

$$c_1 \approx \overline{u}\frac{\overline{\phi}}{\overline{\phi} + f_0^2/k_x^2} \tag{15.53}$$

Inspection shows that shorter waves ($L_x < 3000$ km) move practically with the mean wind speed $\overline{u}$ whereas longer waves move with a lesser speed. The remaining two waves $c_{2,3}$ are external or surface gravity waves moving approximately with the Newtonian speed of sound so that the inequality

$$(c_{2,3} - \overline{u})^3 \gg \overline{u} f_0^2/k_x^2 \tag{15.54}$$

is satisfied. Therefore, in obtaining $c_{2,3}$, the last term in (15.51) may be neglected. The resulting phase speeds are approximately given by

$$\boxed{c_{2,3} = \overline{u} \pm \sqrt{\overline{\phi} + f_0^2/k_x^2}} \tag{15.55}$$

demonstrating that the gravity waves are moving in the positive and negative x-directions. Comparison with equation (14.113) shows that the speed of the gravitational surface waves is slightly modified due to the rotation of the earth which resulted in the term f_0^2/k_x^2 under the square root in (15.55).

Finally, by neglecting gravitational effects, i.e. putting $g = 0$, and by setting $\overline{u} = 0$ we obtain the so-called *inertial waves*. In this case (15.49) reduces to two equations for the components of the horizontal wind field:

$$\frac{\partial u'}{\partial t} - f_0 v' = 0, \qquad \frac{\partial v'}{\partial t} + f_0 u' = 0 \tag{15.56}$$

Taking the solution in the form (15.50), we find the frequency equation

$$\begin{vmatrix} -ik_x c_\mathrm{I} & -f_0 \\ f_0 & -ik_x c_\mathrm{I} \end{vmatrix} = -k_x^2 c_\mathrm{I}^2 + f_0^2 = 0 \tag{15.57}$$

The phase speed for pure inertial waves is then

$$\boxed{c_\mathrm{I} = \pm f_0/k_x} \tag{15.58}$$

For midlatitudes and a wavelength of 3000 km the phase speed c_I is about 50 m s^{-1}. In a later chapter we will discuss the inertial waves in more detail.

15.2.5 Adaptation of the initial data

In the previous section we have shown that the barotropic model equations admit two different types of processes. These are noise processes and the so-called meteorologically relevant *Rossby processes*. On weather maps drawn entirely on the basis of observational data, noise processes are totally absent. Therefore, it is desirable to eliminate noise processes that might be produced by numerical procedures of weather prediction.

A well-proven method for eliminating noise processes from the solutions of the barotropic model equations (15.21) and (15.23) is to require that, at time $t = 0$, the divergence of the wind vector and its time derivative vanish:

$$D_{\mathrm{h},0} = (\nabla_\mathrm{h} \cdot \mathbf{v}_\mathrm{h})_{t=0} = 0, \qquad \left(\frac{\partial D_\mathrm{h}}{\partial t}\right)_{t=0} = 0 \tag{15.59}$$

These filter conditions do not influence the prognostic system. What happens is that the application of (15.59) restricts the free specification of the initial wind velocity $\mathbf{v}_\mathrm{h}(t = 0)$ and the geopotential $\phi(t = 0)$. To assure that the condition (15.59) truly holds we introduce the stream function ψ, see equation (7.21), by means of

$$\mathbf{v}_\mathrm{h}(t = 0) = \mathbf{i}_3 \times \nabla_\mathrm{h}\psi \quad \text{or} \quad u(t = 0) = -\frac{\partial \psi}{\partial y}, \qquad v(t = 0) = \frac{\partial \psi}{\partial x} \tag{15.60}$$

Simple substitution verifies that D_h is zero. The wind defined by this equation is tangential to the lines ψ_i = constant. The geostrophic wind possesses the same property, but in this case the geopotential assumes the role of the stream function.

In order to apply (15.59) we must derive an expression wherein the divergence appears explicitly. This is done by scalar multiplication of the equation of motion (15.21) by the two-dimensional nabla operator. The result is

$$\frac{\partial D_\mathrm{h}}{\partial t} + \nabla_\mathrm{h}\mathbf{v}_\mathrm{h} \cdot\cdot \nabla_\mathrm{h}\mathbf{v}_\mathrm{h} + \mathbf{v}_\mathrm{h} \cdot \nabla_\mathrm{h} D_\mathrm{h} + \nabla_\mathrm{h} \cdot (f\mathbf{i}_3 \times \mathbf{v}_\mathrm{h}) = -\nabla_\mathrm{h} \cdot \nabla_\mathrm{h}\phi = -\nabla_\mathrm{h}^2\phi \tag{15.61}$$

where now the divergence and its time derivative appear explicitly so that the condition (15.59) can be enforced. According to

$$\nabla_\mathrm{h}\mathbf{v}_\mathrm{h} \cdot\cdot \nabla_\mathrm{h}\mathbf{v}_\mathrm{h} = D_\mathrm{h}^2 - 2J(u, v) \quad \text{with} \quad J(u, v) = \begin{vmatrix} \dfrac{\partial u}{\partial x} & \dfrac{\partial v}{\partial x} \\ \dfrac{\partial u}{\partial y} & \dfrac{\partial v}{\partial y} \end{vmatrix} \tag{15.62}$$

the double scalar product in (15.61) can be replaced, resulting in

$$\frac{\partial D_\mathrm{h}}{\partial t} + D_\mathrm{h}^2 - 2J(u, v) + \mathbf{v}_\mathrm{h} \cdot \nabla_\mathrm{h} D_\mathrm{h} + \nabla_\mathrm{h} \cdot (f\mathbf{i}_3 \times \mathbf{v}_\mathrm{h}) = -\nabla_\mathrm{h}^2\phi \tag{15.63}$$

Here we have also introduced the Jacobian J for brevity. This is the *barotropic divergence equation*. For $t = 0$, by applying the conditions (15.59) we obtain the so-called *balance equation*

$$-2J(u, v)\Big|_{t=0} + \nabla_{\rm h} \cdot (f\mathbf{i}_3 \times \mathbf{v}_{\rm h})\Big|_{t=0} = -\nabla_{\rm h}^2\phi\Big|_{t=0} \tag{15.64a}$$

Without the nonlinear term $2J(u, v)|_{t=0}$ the balance equation reduces to the geostrophic wind relation, as may be easily verified.

In order to insure that the divergence of the horizontal wind vector is zero, we have to introduce the stream function (15.60) into (15.64a) . This results in

$$-2J\left(\frac{\partial\psi}{\partial x}, \frac{\partial\psi}{\partial y}\right) - \nabla_{\rm h} \cdot (f\,\nabla_{\rm h}\psi) = -\nabla_{\rm h}^2\phi\Big|_{t=0} \tag{15.64b}$$

or, by expanding the Jacobian, we find

$$2\left[-\frac{\partial^2\psi}{\partial x^2}\frac{\partial^2\psi}{\partial y^2} + \left(\frac{\partial^2\psi}{\partial x\,\partial y}\right)^2\right] - f\,\nabla_{\rm h}^2\psi - \nabla_{\rm h} f \cdot \nabla_{\rm h}\psi = -\nabla_{\rm h}^2\phi\Big|_{t=0} \tag{15.64c}$$

This is a partial differential equation of the Monge–Ampère type.

We summarize: Before starting the integration over time of the barotropic prediction equations (15.21) and (15.23), it is necessary to solve the balance equation (15.64c). From a given geopotential field $\phi(t = 0)$ the stream function ψ must be calculated, from which the initial velocities $u(t = 0)$ and $v(t = 0)$ can be obtained by using (15.60). Now the wind field is free from divergence, and it is so adjusted to the geopotential field $\phi(t = 0)$ that the time derivative of the divergence, as required by (15.59), also vanishes at time $t = 0$. This mathematical process suppresses the undesirable noise processes.

We will not discuss the numerical evaluation of the prediction equations and of the balance equation but leave this to textbooks on numerical weather prediction.

15.3 The filtered barotropic model

By assuming that the conditions listed in (15.59) are valid at all times, not only at $t = 0$, we obtain the so-called *filtered barotropic model* wherein the high-speed gravity waves are completely eliminated. The model condition is now given by

$$D_{\rm h} = \nabla_{\rm h} \cdot \mathbf{v}_{\rm h} = 0 \quad \text{for} \quad t \geq 0 \tag{15.65}$$

which replaces the continuity equation (15.23). Equation (15.65) and the equation of motion

$$\left(\left.\frac{\partial \mathbf{v}_{\mathrm{h}}}{\partial t}\right|_{\mathbf{q}_i}\right)_{\mathrm{h}} + \mathbf{v}_{\mathrm{h}} \cdot \nabla_{\mathrm{h}} \mathbf{v}_{\mathrm{h}} + f \mathbf{i}_3 \times \mathbf{v}_{\mathrm{h}} = -\nabla_{\mathrm{h}} \phi \tag{15.66}$$

constitute the basic predictive system. The physical model described by this mathematical system consists of a rotating homogeneous incompressible fluid embedded between two plane-parallel plates. Owing to the incompressibility of the fluid, vertical motion cannot take place, so the horizontal divergence D_{h} vanishes. Thus we are dealing with frictionless purely horizontal motion. The upper surface is no longer a free surface where the pressure vanishes. Instead, the pressure at the top of the fluid layer must differ from zero and be changing, for example, in the y-direction so that a basic current in the x-direction may form. At the top of the fluid the geopotential can be interpreted as $\overline{p}/\overline{\rho}$ plus an irrelevant constant. This simple physical model would require that some pistons are part of the upper plate so that the pressure may be varied horizontally.

The system (15.65) and (15.66) is unsuitable for the numerical integration. Instead, we should use the balance equations (15.64a) and (15.64b), extending their validity to all times $t \geq 0$, i.e.

$$-2J(u, v) + \nabla_{\mathrm{h}} \cdot (f \mathbf{i}_3 \times \mathbf{v}_{\mathrm{h}}) = -\nabla_{\mathrm{h}}^2 \phi, \qquad t \geq 0 \tag{15.67a}$$

or utilize the stream function

$$\boxed{2J\left(\frac{\partial \psi}{\partial x}, \frac{\partial \psi}{\partial y}\right) + \nabla_{\mathrm{h}} \cdot (f\, \nabla_{\mathrm{h}} \psi) = \nabla_{\mathrm{h}}^2 \phi, \qquad t \geq 0} \tag{15.67b}$$

together with the equation of motion (15.66) as the prognostic system.

Nevertheless, due to computer limitations, numerical errors resulting in small amounts of divergence may occur. To avoid this type of numerical problem it is customary to integrate the barotropic vorticity equation

$$\frac{\partial \zeta}{\partial t} + \mathbf{v}_{\mathrm{h}} \cdot \nabla_{\mathrm{h}} \eta = 0 \quad \text{with} \quad \eta = \zeta + f \tag{15.68}$$

which is repeated from (10.147), instead of the equation of motion. By introducing the stream function (15.60) into the definition of the relative vorticity, we obtain

$$\zeta = \frac{\partial v}{\partial x} - \frac{\partial u}{\partial y} = \frac{\partial^2 \psi}{\partial x^2} + \frac{\partial^2 \psi}{\partial y^2} = \nabla_{\mathrm{h}}^2 \psi \tag{15.69}$$

Substituting this equation into (15.68) gives the *divergence-free vorticity equation*

$$\nabla_{\mathrm{h}}^2 \frac{\partial \psi}{\partial t} + \frac{\partial \psi}{\partial x} \frac{\partial}{\partial y} (\nabla_{\mathrm{h}}^2 \psi + f) - \frac{\partial \psi}{\partial y} \frac{\partial}{\partial x} (\nabla_{\mathrm{h}}^2 \psi + f) = 0 \tag{15.70}$$

which may be rewritten as

$$\boxed{\nabla_{\mathrm{h}}^2 \frac{\partial \psi}{\partial t} + J\left(\psi, \nabla_{\mathrm{h}}^2 \psi + f\right) = 0} \tag{15.71}$$

This equation and the balance equation (15.67b) constitute the prognostic system of the filtered model. The entire prognostic process is taken care of by the vorticity equation (15.71) with the prediction variable ψ. Since only ψ appears, one must solve a Poisson equation for each time step. Equation (15.67b) is used only if information about the pressure field which is represented by the geopotential field is desired.

Because the filtered model does not permit the appearance of noise processes there is no exchange of potential and kinetic energy. Therefore, in a materially closed region the total kinetic energy must be conserved. Which energetic changes are still possible in the filtered model? Apparently only those processes which redistribute kinetic energy between waves and vortices of differing dimensions are permitted. Charney *et al.* (1950) used (15.71) to carry out successfully numerical forecasts.

Summarizing, we may say that the unfiltered and the filtered barotropic models are different in principle. Therefore, it is quite surprising that these two models produce nearly the same predictions. This justifies the filtering of noise waves.

15.4 Barotropic instability

15.4.1 Shearing instability as a limiting case of barotropic instability

Again we consider the previous case in which the constant-density barotropic fluid is embedded between two plane-parallel plates so that the horizontal divergence $D_{\mathrm{h}} = 0$. We must imagine that the pressure variation is imposed at the top of the model. We assume that the basic current is directed parallel to the positive x-axis and that the region of flow along the y-axis extends to infinity in the positive and negative directions as shown in Figure 15.3. Between $y = -a$ and $y = a$ the model assumes that we have a linear wind shear, thus dividing the flow region into three subregions. Since the basic current is in the x-direction, in each of the three regions the mean geopotential $\overline{\phi}_j$, $j = 1, 2, 3$, cannot vary in the x-direction, otherwise there would be a mean flow in the y-direction. Again the Coriolis parameter is held constant within the entire region. It should be expected that a disturbance of the flow field will occur in the shear region. For the three regions $j = 1, 2, 3$ the

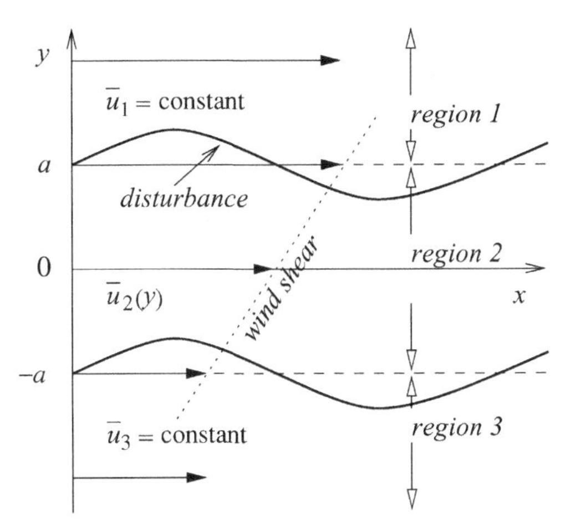

Fig. 15.3 Formation of a wave in the shearing zone.

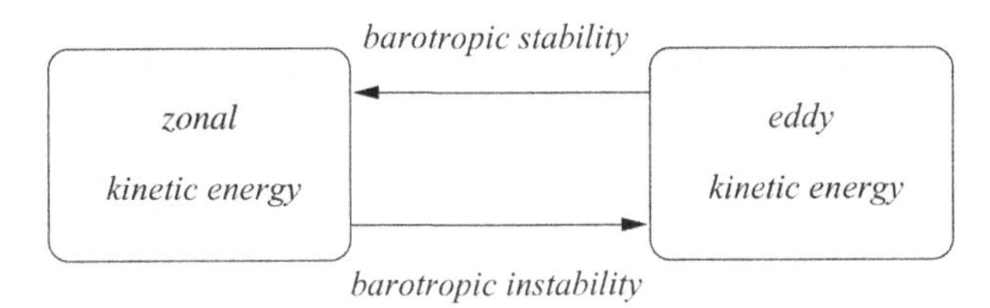

Fig. 15.4 Barotropic stability and instability.

horizontal flow field is given by

$$\overline{u}_j = u_{j,0} + u_{j,1} y \qquad u_{j,0} = -\frac{1}{f_0}\frac{\partial \overline{\phi}_j}{\partial y} = \text{constant}_j \qquad u_{1,1} = u_{3,1} = 0$$
$$u_{2,1} = \frac{\overline{u}_2(a) - \overline{u}_2(-a)}{2a} \qquad \overline{u}_2(a) = \overline{u}_1 \qquad \overline{u}_2(-a) = \overline{u}_3 \tag{15.72}$$

Since the fluid is bounded by the lower and the upper plate there is no available potential energy. The only transformation of energy which is possible is an exchange of kinetic energy between the vortices, called kinetic energy of the eddies K', and the kinetic energy of the zonal current $\overline{K}$. If the exchange is from K' to $\overline{K}$, one speaks of *barotropic stability*. *Barotropic instability* occurs if the flow of kinetic energy is in the opposite direction, that is from the zonal current to the eddies. The situation is displayed schematically in Figure 15.4.

In order to obtain an analytic solution to the problem we must linearize the horizontal equations of motion (15.27b). For ease of notation we introduce the operator Q_j for the three regions:

$$Q_j = \frac{\partial}{\partial t} + \overline{u}_j \frac{\partial}{\partial x}, \qquad j = 1, 2, 3 \tag{15.73a}$$

From (15.72) we find

$$\frac{\partial Q_1}{\partial y} = Q_1 \frac{\partial}{\partial y}, \qquad \frac{\partial Q_2}{\partial y} = Q_2 \frac{\partial}{\partial y} + u_{2,1} \frac{\partial}{\partial x}, \qquad \frac{\partial Q_3}{\partial y} = Q_3 \frac{\partial}{\partial y} \tag{15.73b}$$

Keeping in mind that the basic current is along the x-direction, the individual derivatives are linearized as

$$\begin{aligned} \left(\frac{du_j}{dt}\right)' &= \frac{\partial u_j'}{\partial t} + \overline{u}_j \frac{\partial u_j'}{\partial x} + v_j' \frac{\partial \overline{u}_j}{\partial y} = Q_j u_j' + v_j' \frac{\partial \overline{u}_j}{\partial y} \\ \left(\frac{dv_j}{dt}\right)' &= \frac{\partial v_j'}{\partial t} + \overline{u}_j \frac{\partial v_j'}{\partial x} = Q_j v_j', \qquad j = 1, 2, 3 \end{aligned} \tag{15.74}$$

For the three regions we now have to deal with the following set of equations:

Region 1

$$Q_1 u_1' = -\frac{\partial \phi_1'}{\partial x} + f_0 v_1', \qquad Q_1 v_1' = -\frac{\partial \phi_1'}{\partial y} - f_0 u_1', \qquad \frac{\partial u_1'}{\partial x} + \frac{\partial v_1'}{\partial y} = 0 \tag{15.75a}$$

Region 2

$$Q_2 u_2' + u_{2,1} v_2' = -\frac{\partial \phi_2'}{\partial x} + f_0 v_2', \qquad Q_2 v_2' = -\frac{\partial \phi_2'}{\partial y} - f_0 u_2', \qquad \frac{\partial u_2'}{\partial x} + \frac{\partial v_2'}{\partial y} = 0 \tag{15.75b}$$

Region 3

$$Q_3 u_3' = -\frac{\partial \phi_3'}{\partial x} + f_0 v_3', \qquad Q_3' v_3' = -\frac{\partial \phi_3'}{\partial y} - f_0 u_3', \qquad \frac{\partial u_3'}{\partial x} + \frac{\partial v_3'}{\partial y} = 0 \tag{15.75c}$$

In each region the condition of vanishing divergence has been added. In this set there are six unknown variables (u_j', v_j', $j = 1, 2, 3$), requiring six conditions in order to solve the problem. These conditions are the behaviors of the solution at $y = \pm\infty$, two kinematic boundary-surface conditions at $y = \pm a$, and two dynamic boundary-surface conditions at $y = \pm a$.

Instead of using the equations of motion directly, it is easier to find the solution to the problem (15.75) by employing the barotropic vorticity equation. This can be done directly by linearizing (15.68) and keeping in mind that the mean geopotential cannot vary in the x-direction. In each region we are going to obtain the linearized vorticity equation by taking the partial derivative with respect to x of the second equation of motion and subtracting the partial derivative with respect to y of the first equation. For the second region this will now be demonstrated. From (15.73b) and (15.75b) we find

$$\begin{aligned} Q_2 \frac{\partial u_2'}{\partial y} + u_{2,1} \frac{\partial u_2'}{\partial x} + u_{2,1} \frac{\partial v_2'}{\partial y} &= -\frac{\partial^2 \phi_2'}{\partial x\, \partial y} + f_0 \frac{\partial v_2'}{\partial y} \\ Q_2 \frac{\partial v_2'}{\partial x} &= -\frac{\partial^2 \phi_2'}{\partial x\, \partial y} - f_0 \frac{\partial u_2'}{\partial x} \end{aligned} \tag{15.76}$$

Subtraction results in

$$Q_2\left(\frac{\partial v_2'}{\partial x} - \frac{\partial u_2'}{\partial y}\right) - u_{2,1}\left(\frac{\partial u_2'}{\partial x} + \frac{\partial v_2'}{\partial y}\right) = -f_0\left(\frac{\partial u_2'}{\partial x} + \frac{\partial v_2'}{\partial y}\right) \tag{15.77}$$

Since the divergence must vanish, we find

$$Q_2\left(\frac{\partial v_2'}{\partial x} - \frac{\partial u_2'}{\partial y}\right) = 0 \tag{15.78}$$

Hence, the Q_2 term is operating on the vorticity of this region. With the help of (15.75a) and (15.75c) it may easily be verified that one obtains analogous expressions for regions 1 and 3 so that in general we have

$$Q_j\left(\frac{\partial v_j'}{\partial x} - \frac{\partial u_j'}{\partial y}\right) = 0, \qquad j = 1, 2, 3 \tag{15.79}$$

This equation can also be written in the form $d\eta_j'/dt = 0$ so that vorticity is conserved. The physically interesting solution in connection with the problem at hand is to set the vorticity itself equal to zero:

$$\frac{\partial v_j'}{\partial x} - \frac{\partial u_j'}{\partial y} = 0, \qquad j = 1, 2, 3 \tag{15.80}$$

On differentiating this equation with respect to x and the divergence condition with respect to y, we find upon elimination of the mixed partial derivatives the Laplace equation

$$\frac{\partial^2 v_j'}{\partial x^2} + \frac{\partial^2 v_j'}{\partial y^2} = 0, \qquad j = 1, 2, 3 \tag{15.81}$$

which must hold in each region j. We assume the validity of a solution of the form

$$v_j' = V_j(y) \exp[ik_x(x - ct)], \qquad j = 1, 2, 3 \tag{15.82}$$

If the phase speed c happens to be real, then the solution corresponds to a stable wave. If c is complex then the amplitude will grow and barotropic instability will occur. We will now state the solution conditions.

(1) Lateral boundary conditions at infinity:

$$\boxed{y = \infty: \quad v_1' = 0, \qquad y = -\infty: \quad v_3' = 0} \tag{15.83}$$

(2) Kinematic boundary conditions at $y = \pm a$

The kinematic boundary condition (9.41) also applies to the horizontal situation. Therefore, we may write for region 1 and 2 at $y = a$ the conditions for the generalized y-coordinate ξ_a:

$$v_1'(a) = \frac{\partial \xi_a}{\partial t} + u_{1,0}\frac{\partial \xi_a}{\partial x}, \qquad v_2'(a) = \frac{\partial \xi_a}{\partial t} + u_{2,0}\frac{\partial \xi_a}{\partial x} \tag{15.84a}$$

Analogously we find the kinematic boundary condition at $y = -a$:

$$v_2'(-a) = \frac{\partial \xi_{-a}}{\partial t} + u_{2,0}\frac{\partial \xi_{-a}}{\partial x}, \qquad v_3'(-a) = \frac{\partial \xi_{-a}}{\partial t} + u_{3,0}\frac{\partial \xi_{-a}}{\partial x} \tag{15.84b}$$

Since $u_{1,0}=u_{2,0}$ at $y=a$ and $u_{2,0}=u_{3,0}$ at $y=-a$, we find the kinematic boundary conditions at $y = \pm a$:

$$\boxed{v_1'(a) = v_2'(a), \qquad v_2'(-a) = v_3'(-a)} \tag{15.85}$$

(3) Dynamic boundary conditions at $y = \pm a$

The dynamic boundary condition, see Section 9.5, requires that the pressure must be continuous across the boundary in order to prevent infinitely large forces occurring. In the present situation the pressure is represented by the geopotential. Therefore, we may write the dynamic boundary conditions in the form

$$y = a: \quad \frac{d}{dt}(\phi_1 - \phi_2) = 0, \qquad y = -a: \quad \frac{d}{dt}(\phi_2 - \phi_3) = 0 \tag{15.86}$$

or, expanded and linearized for each side, as

$$\left(\frac{d\phi_j}{dt}\right)' = \frac{\partial \phi_j'}{\partial t} + \overline{u}_j\frac{\partial \phi_j'}{\partial x} + v_j'\frac{\partial \overline{\phi}_j}{\partial y}, \qquad j = 1, 2, 3 \tag{15.87}$$

This form follows upon recalling that the averaged geopotential does not vary in the x-direction. By taking the partial derivative with respect to x of this expression, for each j we find

$$\frac{\partial}{\partial x}\left(\frac{d\phi_j}{dt}\right)' = \frac{\partial}{\partial x}\left(\frac{\partial}{\partial t} + \overline{u}_j\frac{\partial}{\partial x}\right)\phi_j' + \frac{\partial}{\partial x}\left(v_j'\frac{\partial \overline{\phi}_j}{\partial y}\right) = Q_j\frac{\partial \phi_j'}{\partial x} + \frac{\partial \overline{\phi}_j}{\partial y}\frac{\partial v_j'}{\partial x} \tag{15.88}$$

Inspection of Figure 15.4 shows that, at $y = \pm a$, the basic currents coincide, that is $\overline{u}_1(a) = \overline{u}_2(a)$ and $\overline{u}_2(-a) = \overline{u}_3(-a)$, so that $Q_1(a) = Q_2(a)$ and $Q_2(-a) = Q_3(-a)$. Thus, the partial derivative with respect to x of (15.86) is given by

$$
\begin{aligned}
y = a: &\qquad Q_1\left(\frac{\partial \phi'_1}{\partial x} - \frac{\partial \phi'_2}{\partial x}\right) + \left(\frac{\partial \overline{\phi}_1}{\partial y} - \frac{\partial \overline{\phi}_2}{\partial y}\right)\frac{\partial v'_1}{\partial x} = 0 \\
y = -a: &\qquad Q_3\left(\frac{\partial \phi'_2}{\partial x} - \frac{\partial \phi'_3}{\partial x}\right) + \left(\frac{\partial \overline{\phi}_2}{\partial y} - \frac{\partial \overline{\phi}_3}{\partial y}\right)\frac{\partial v'_3}{\partial x} = 0
\end{aligned}
\tag{15.89}
$$

where we have used the kinematic boundary condition (15.85). Owing to the facts that $u_{1,0}(a) = u_{2,0}(a)$ and $u_{2,0}(-a) = u_{3,0}(-a)$, it follows from (15.72) that

$$
\begin{aligned}
y = a: &\qquad \frac{\partial \overline{\phi}_1}{\partial y} - \frac{\partial \overline{\phi}_2}{\partial y} = -f_0 u_{1,0} + f_0 u_{2,0} = 0 \\
y = -a: &\qquad \frac{\partial \overline{\phi}_2}{\partial y} - \frac{\partial \overline{\phi}_3}{\partial y} = -f_0 u_{2,0} + f_0 u_{3,0} = 0
\end{aligned}
\tag{15.90}
$$

Therefore, in each equation of (15.89) the second term vanishes.

As in the case of (15.79), at $y = \pm a$ we consider the differential equations

$$
y = a: \qquad \frac{\partial \phi'_1}{\partial x} - \frac{\partial \phi'_2}{\partial x} = 0, \qquad y = -a: \qquad \frac{\partial \phi'_2}{\partial x} - \frac{\partial \phi'_3}{\partial x} = 0
\tag{15.91}
$$

On replacing $\partial \phi'_j/\partial x$ in these expressions by means of (15.75a) and (15.75b) and utilizing the kinematic boundary conditions (15.85) we obtain

$$
\begin{aligned}
y = a: &\qquad -Q_1 u'_1 + Q_2 u'_2 + u_{2,1} v'_2 = 0 \\
y = -a: &\qquad -Q_2 u'_2 + Q_3 u'_3 - u_{2,1} v'_2 = 0
\end{aligned}
\tag{15.92}
$$

Since the Laplace equation (15.81), which needs to be solved, is expressed in the variable v'_j, we must eliminate the variables u'_j in terms of the variables v'_j of the filtering condition stated for each region. This task is accomplished by partial differentiation of (15.92) with respect to x and then replacing $\partial u'_j/\partial x$ by $-\partial v'_j/\partial y$ since $D_\text{h} = 0$. This results in the dynamic boundary conditions

$$
\boxed{
\begin{aligned}
&\text{(a)} \quad y = a: &\qquad Q_1 \frac{\partial v'_1}{\partial y} - Q_2 \frac{\partial v'_2}{\partial y} + u_{2,1}\frac{\partial v'_2}{\partial x} = 0 \\
&\text{(b)} \quad y = -a: &\qquad Q_2 \frac{\partial v'_2}{\partial y} - Q_3 \frac{\partial v'_3}{\partial y} - u_{2,1}\frac{\partial v'_2}{\partial x} = 0
\end{aligned}
}
\tag{15.93}
$$

We are now ready to determine the phase velocity of the disturbance. Substitution of (15.82) into (15.81) gives the ordinary linear differential equation

$$\frac{d^2 V_j}{dy^2} - k_x^2 V_j = 0, \qquad j = 1, 2, 3 \tag{15.94}$$

which can easily be solved by standard methods. The characteristic values of this equation are k_x and $-k_x$, so the solution for each of the three regions can be written down immediately as

$$V_j = A_j \exp(-k_x y) + B_j \exp(k_x y), \qquad j = 1, 2, 3 \tag{15.95}$$

The constants B_1 and A_3 must vanish in order to keep the solution bounded for $y \longrightarrow \pm\infty$. Using the trial solution (15.82) for (15.81), we obtain the perturbation velocities

$$\begin{aligned} v_1' &= A_1 \exp(-k_x y) \exp[ik_x(x - ct)] \\ v_2' &= [A_2 \exp(-k_x y) + B_2 \exp(k_x y)] \exp[ik_x(x - ct)] \\ v_3' &= B_3 \exp(k_x y) \exp[ik_x(x - ct)] \end{aligned} \tag{15.96}$$

which still contain the unknown constants A_1, A_2, B_2, B_3. Owing to the kinematic boundary surface conditions (15.85) we find the following relations among the remaining constants:

$$\begin{aligned} y = a: \quad v_1' = v_2' &\Longrightarrow A_1 \exp(-k_x a) = A_2 \exp(-k_x a) + B_2 \exp(k_x a) \\ y = -a: \quad v_2' = v_3' &\Longrightarrow B_3 \exp(-k_x a) = A_2 \exp(k_x a) + B_2 \exp(-k_x a) \end{aligned} \tag{15.97}$$

So far we have used the conditions at infinity and the kinematic boundary conditions.

Now we are going to involve equations (15.93), which resulted from the dynamic boundary conditions. We will show how to evaluate (15.93a). For $y = a$ we need to substitute the expressions

$$\begin{aligned} Q_1 \left.\frac{\partial v_1'}{\partial y}\right|_{y=a} &= ik_x^2(c - \overline{u}_1) A_1 \exp(-k_x a) \exp[ik_x(x - ct)] \\ Q_2 \left.\frac{\partial v_2'}{\partial y}\right|_{y=a} &= ik_x^2(c - \overline{u}_1)[A_2 \exp(-k_x a) - B_2 \exp(k_x a)] \exp[ik_x(x - ct)] \end{aligned} \tag{15.98}$$

into (15.93a). The constant $A_1 \exp(-k_x a)$ can be replaced by means of (15.97). After some easy manipulations we obtain the upper equation of the matrix system

$$\begin{pmatrix} u_{2,1} \exp(-k_x a) & [2k_x(c - \overline{u}_1) + u_{2,1}] \exp(k_x a) \\ [2k_x(c - \overline{u}_3) - u_{2,1}] \exp(k_x a) & -u_{2,1} \exp(-k_x a) \end{pmatrix} \begin{pmatrix} A_2 \\ B_2 \end{pmatrix} = 0 \tag{15.99}$$

In complete analogy the lower equation of this system has been obtained from (15.93b), wherein now the constant $B_3 \exp(-k_x a)$ has been replaced with the help of (15.97).

Nonzero values of A_2 and B_2 are possible only if the determinant of the homogeneous equations (15.99) vanishes. On solving the determinant of this two-by-two matrix for the phase velocity we obtain

$$\boxed{c_{1,2} = \overline{u} \pm \sqrt{\frac{(\Delta\overline{u})^2}{4}\left(1 - \frac{1}{ak_x} + \frac{1 - \exp(-4k_x a)}{4a^2k_x^2}\right)}} \tag{15.100}$$

where the abbreviations

$$\overline{u} = \overline{u}_2(y = 0) = \frac{\overline{u}_1 + \overline{u}_3}{2}, \qquad \Delta\overline{u} = 2au_{2,1} = \overline{u}_1 - \overline{u}_3 \tag{15.101}$$

have been introduced.

Equation (15.100) will now be discussed. If the expression within the large parentheses is less than zero then the phase velocity will be a complex number so that barotropic instability occurs. Numerical evaluation of this expression shows that

$$1 - \frac{1}{ak_x} + \frac{1 - \exp(-4k_x a)}{4a^2k_x^2} < 0 \quad \text{for} \quad k_x a = \frac{2a\pi}{L_x} < 0.64 \tag{15.102}$$

Since $2a$ is the width of the shear zone, we see that the ratio of this width to the wavelength L_x determines whether the wave is barotropically stable or not. Broad zones of wind shear with short wavelength of the disturbance are barotropically stable since they do not satisfy the inequality occurring in (15.102). Laterally small zones of wind shear with a large wavelength are barotropically unstable.

Assume that the shear zone shrinks to zero. In this case a velocity jump occurs at $y = 0$, where the basic current abruptly changes from $\overline{u}_1$ to $\overline{u}_3$. Letting $a \to 0$, we find upon using L'Hospital's rule or by expanding the exponential at least to the quadratic term that

$$\boxed{c_{1,2} = \overline{u} \pm \sqrt{-\frac{(\Delta\overline{u})^2}{4}}} \tag{15.103}$$

Since the phase velocity is always complex this is the unstable solution yielding the shearing waves. It is noteworthy that the same result has been obtained for gravity waves at discontinuities by setting $u_{0,1} \neq u_{0,2}$ and $\rho_{0,1} = \rho_{0,2}$ in (14.142).

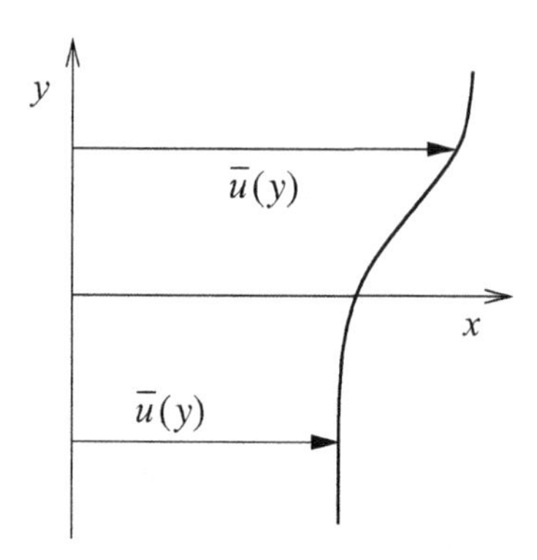

Fig. 15.5 A continuous velocity profile $\overline{u}(y)$.

15.4.2 Conditions for the occurrence of barotropic instability

Let us now consider a continuous velocity profile of the basic current; see Figure 15.5. For the continuous velocity profile it is possible to derive a necessary condition for the occurrence of barotropic instability. The basic model assumptions are

$$\nabla_{\rm h} \cdot \mathbf{v}_{\rm h} = 0, \qquad \overline{u} = \overline{u}(y), \qquad f = f_0 + \beta y, \qquad \beta = \frac{\partial f}{\partial y} = \text{constant} \quad (15.104)$$

Again the barotropic fluid is embedded between two plane-parallel plates so that no potential energy can be transformed into kinetic energy. The decisive change in the model assumptions (15.72), apart from the velocity profile, is that now the Coriolis parameter is permitted to change with y. The parameter β is known as the *Rossby parameter*.

Starting with the vorticity equation (15.71) and linearizing the Coriolis parameter after Rossby, we obtain

$$\frac{\partial}{\partial t}\left(\nabla_{\rm h}^2 \psi\right) + J\left(\psi, \nabla_{\rm h}^2 \psi\right) + \beta \frac{\partial \psi}{\partial x} = 0 \qquad (15.105)$$

We assume that the stream function varies according to

$$\psi = \overline{\psi}(y) + \psi'(x, y, t) \qquad (15.106)$$

so that the velocity components of the basic current can be expressed by

$$\overline{u} = -\frac{\partial \overline{\psi}}{\partial y}, \qquad \overline{v} = \frac{\partial \overline{\psi}}{\partial x} = 0 \qquad (15.107)$$

In order to obtain an analytic solution to the problem we must linearize (15.105). This results in the following equation:

$$\frac{\partial}{\partial t}(\nabla_{\mathrm{h}}^2\psi') + \overline{u}\,\frac{\partial}{\partial x}(\nabla_{\mathrm{h}}^2\psi') - \frac{d^2\overline{u}}{dy^2}\frac{\partial\psi'}{\partial x} + \beta\,\frac{\partial\psi'}{\partial x} = 0 \tag{15.108}$$

which can be solved by assuming the solution

$$\psi' = F(y)\exp[ik_x(x-ct)] \tag{15.109}$$

We now introduce the absolute vorticity of the basic current

$$\overline{\eta} = -\frac{d\overline{u}}{dy} + f \tag{15.110}$$

Differentiation of this equation with respect to y leads to the introduction of the Rossby parameter β:

$$\frac{d\overline{\eta}}{dy} = -\frac{d^2\overline{u}}{dy^2} + \beta \tag{15.111}$$

Hence, we speak of a representation in the *β-plane*. Substituting (15.109) and (15.111) into (15.108) gives the desired differential equation for the amplitude $F(y)$ of the disturbance:

$$\boxed{\frac{d^2F}{dy^2} - \left(k_x^2 + \frac{1}{c-\overline{u}}\frac{d\overline{\eta}}{dy}\right)F = 0} \tag{15.112}$$

This is the famous *Orr–Sommerfeld equation* for frictionless flow, which cannot be solved for arbitrary values of $\overline{u}(y)$.

To make the solution of (15.112) more tractable we assume that we have a channel flow that is bounded by rigid walls at $y = \pm a$. We may think of the walls of the channel as latitude circles. At the rigid walls the normal components of the velocity must vanish:

$$v(y=\pm a) = \left.\frac{\partial\psi}{\partial x}\right|_{y=\pm a} = \left.\frac{\partial\psi'}{\partial x}\right|_{y=\pm a} = 0 \quad \text{since} \quad \frac{\partial\overline{\psi}}{\partial x} = 0 \tag{15.113}$$

so the amplitude of the disturbance must vanish also, that is $F(y = \pm a) = 0$. In order to have a very general solution we permit not only the phase velocity c but also the amplitude F to be complex:

$$c = c_{\mathrm{r}} + ic_{\mathrm{i}}, \qquad F = F_{\mathrm{r}} + iF_{\mathrm{i}} \tag{15.114}$$

Substitution of (15.114) into (15.112) gives

$$\frac{d^2F_r}{dy^2}+i\,\frac{d^2F_i}{dy^2}-\left(k_x^2+\frac{c_r-ic_i-\overline{u}}{\lambda^2}\frac{d\overline{\eta}}{dy}\right)(F_r+iF_i)=0 \tag{15.115}$$

with $\lambda^2=(c_r-\overline{u})^2+c_i^2$. By separating the real and the imaginary parts we find

$$\begin{aligned}
&\text{(a)}\quad \frac{d^2F_r}{dy^2}-\left(k_x^2+\frac{c_r-\overline{u}}{\lambda^2}\frac{d\overline{\eta}}{dy}\right)F_r-\frac{c_i}{\lambda^2}\frac{d\overline{\eta}}{dy}F_i=0\\
&\text{(b)}\quad \frac{d^2F_i}{dy^2}-\left(k_x^2+\frac{c_r-\overline{u}}{\lambda^2}\frac{d\overline{\eta}}{dy}\right)F_i+\frac{c_i}{\lambda^2}\frac{d\overline{\eta}}{dy}F_r=0
\end{aligned} \tag{15.116}$$

The next step is to multiply (15.116a) by F_i and (15.116b) by $-F_r$ and then add the resulting equations together. Application of the identity

$$F_i\frac{d^2F_r}{dy^2}-F_r\frac{d^2F_i}{dy^2}=\frac{d}{dy}\left(F_i\frac{dF_r}{dy}-F_r\frac{dF_i}{dy}\right) \tag{15.117}$$

and then integrating over the width of the channel yields

$$\int_{-a}^{a}\frac{d}{dy}\left(F_i\frac{dF_r}{dy}-F_r\frac{dF_i}{dy}\right)dy=c_i\int_{-a}^{a}\frac{F_i^2+F_r^2}{\lambda^2}\frac{d\overline{\eta}}{dy}\,dy \tag{15.118}$$

Since $F(y=\pm a)=0$, the left-hand side of (15.118) must vanish. Thus, the remaining part

$$\boxed{c_i\int_{-a}^{a}\left(\frac{F_i^2+F_r^2}{\lambda^2}\right)\frac{d\overline{\eta}}{dy}\,dy=0} \tag{15.119}$$

determines whether the flow is barotropically stable or unstable. If $c_i=0$ then the flow is always stable. If, however, $c_i\neq 0$ then unstable solutions exist only if the integral is zero. This type of stability treatment originated with Rayleigh. For comparison let us recall the result of the previous section, where we considered a situation with a linear wind shear and a constant Coriolis parameter. In this case $d\overline{\eta}/dy=0$ and the instability was determined solely by c_i.

Let us now discuss the stability integral (15.119). Since the expression in parentheses is positive definite, it is the product $c_i\,d\overline{\eta}/dy$ that determines the stability. The derivative of the absolute vorticity was given by (15.111), permitting three possibilities:

$$\frac{d\overline{\eta}}{dy}=-\frac{d^2\overline{u}}{dy^2}+\beta \gtreqless 0 \tag{15.120}$$

(i) The flow field is always stable if $d\overline{\eta}/dy$ does not change sign in the region of integration so that the integral cannot vanish. If $d\overline{\eta}/dy>0$ or $d\overline{\eta}/dy<0$ between $-a\leq y\leq a$

then c_i must be zero so that (15.119) remains valid. This condition is sufficient for barotropic stability since (15.109) results in stable oscillations.

(ii) If c_i differs from zero then $d\overline{\eta}/dy$ must have at least one zero in order to satisfy (15.119). This condition is necessary but not sufficient for barotropic instability to occur. Thus the necessary condition for barotropic instability is that $\beta - d^2\overline{u}/dy^2 = 0$ somewhere in the range of integration. Suppose that $d\overline{\eta}/dy$ has a zero to make the integrand zero. The flow field is not necessarily unstable since c_i could be zero also so that (15.119) vanishes under all circumstances. According to the results of the previous section, we may speculate that a sufficient condition for instability would result from the requirement that the wavelength of the disturbance is large in comparison with the lateral extent of the shear zone.

(iii) Suppose that c_i differs from zero and that $d\overline{\eta}/dy = 0$ everywhere. If the Coriolis parameter is assumed to be fixed, then $d\overline{\eta}/dy = -d^2\overline{u}/dy^2 = 0$. In this case we have either a vorticity maximum or a vorticity minimum. Therefore, the wind profile must have an inflexion point. This is a particular example of *inflexion-point instability*. This type of stability investigation goes back to Lord Rayleigh (1880) for a nonrotating system. Kuo (1951) has extended the theory by including the β-term.

15.5 The mechanism of barotropic development

Let us return to the divergence-free vorticity equation (15.105). This nonlinear equation is satisfied by the so-called *Neamtan solution* (Neamtan, 1946)

$$\psi = \overline{\psi} + \psi' \quad \text{with} \quad \begin{aligned} \overline{\psi} &= \text{constant} - u_0 y + A\cos(\overline{k}y) \\ \psi' &= B\cos[k_x(x - ct)]\cos(k_y y) \end{aligned} \tag{15.121}$$

The term $\overline{\psi}$ represents the stream function of the basic current, $\overline{k} = 2\pi/\overline{L}$ is the wavenumber, and $\overline{L}$ the wavelength of the basic current. The term ψ' is the stream function of the disturbance which is superimposed on the basic flow.

We will now investigate the Neamtan solution. Assuming that

$$\overline{k} = \sqrt{k_x^2 + k_y^2} \tag{15.122}$$

the disturbance moves with the constant phase velocity

$$c = u_0 - \beta/\overline{k}^2 \tag{15.123}$$

which is known as the *phase velocity of the Rossby wave*. The validity of (15.123) will be shown later. Note also that, in this particular situation, the wave moves without change of shape since c is independent of y. This situation is known as the indifferent case.

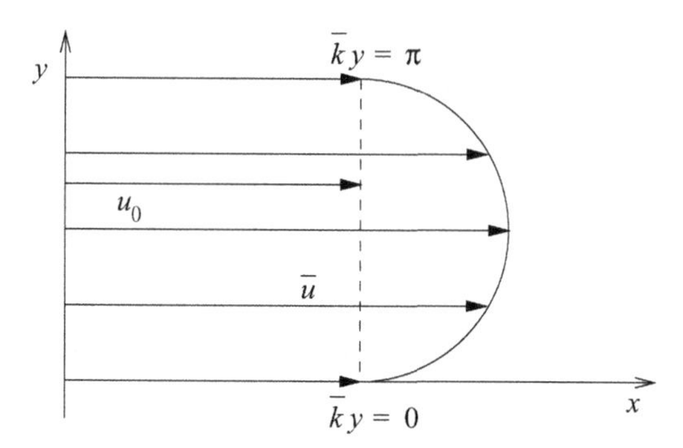

Fig. 15.6 A schematic view of the basic flow field $\overline{u}(y)$ for the Neamtan solution.

The basic current itself is obtained by differentiating the basic current part of the stream function with respect to y; see Figure 15.6.

$$\overline{u} = -\frac{\partial \overline{\psi}}{\partial y} = u_0 + \overline{k} A \sin(\overline{k} y) \tag{15.124}$$

Since the indifference of the flow field results from the equality (15.122), we may expect that the formation of barotropic stability or instability in some way depends on the inequality

$$\overline{k} \neq \sqrt{k_x^2 + k_y^2} \tag{15.125}$$

For this general case it is impossible to determine the phase velocity c.

Let us reconsider the vorticity equation (15.68) in the form

$$\frac{\partial \zeta}{\partial t} + u\frac{\partial \zeta}{\partial x} + v\frac{\partial \zeta}{\partial y} + \beta v = 0 \tag{15.126}$$

We split the vorticity and the velocity components:

$$\zeta = \overline{\zeta} + \zeta', \qquad u = \overline{u} + u', \qquad v = v' \quad \text{since} \quad \overline{v} = \frac{\partial \overline{\psi}}{\partial x} = 0 \tag{15.127}$$

We substitute these into (15.126) but retain the nonlinear terms and find

$$\frac{\partial \zeta'}{\partial t} + \frac{\partial \overline{\zeta}}{\partial t} = -\left[\overline{u}\frac{\partial \zeta'}{\partial x} + \left(u'\frac{\partial \zeta'}{\partial x} + v'\frac{\partial \zeta'}{\partial y}\right) + v'\frac{\partial \overline{\zeta}}{\partial y} + \beta v'\right] \tag{15.128}$$

Note that $\partial\overline{\zeta}/\partial x = 0$ since $\overline{\psi} = \overline{\psi}(y)$. By averaging this equation and recalling that the mean over the fluctuation vanishes, i.e.

$$\overline{\frac{\partial \zeta'}{\partial t}} = 0, \qquad \overline{\frac{\partial \zeta'}{\partial x}} = 0, \qquad \overline{v'} = 0 \tag{15.129}$$

we find

$$\frac{\partial \overline{\zeta}}{\partial t} = -\left(\overline{u' \frac{\partial \zeta'}{\partial x} + v' \frac{\partial \zeta'}{\partial y}}\right) = -\overline{\mathbf{v}'_{\mathrm{h}} \cdot \nabla_{\mathrm{h}} \zeta'} \tag{15.130}$$

Thus, the change of $\overline{\zeta}$ with time results from the advection of the perturbation vorticity due to the perturbation velocity.

Let us now consider the Neamtan formula at the initial time $t = 0$. For clarity we have collected the basic relationships needed to evaluate (15.130) in

$$\begin{aligned} &u' = -\frac{\partial \psi'}{\partial y}, \qquad v' = \frac{\partial \psi'}{\partial x}, \qquad \zeta' = \nabla_{\mathrm{h}}^2 \psi' = -(k_x^2 + k_y^2)\psi' \\ &\frac{\partial \zeta'}{\partial x} = -(k_x^2 + k_y^2)v', \quad \frac{\partial \zeta'}{\partial y} = (k_x^2 + k_y^2)u', \quad \frac{\partial \overline{\zeta}}{\partial y} = \overline{k}^3 A \sin(\overline{k} y) \end{aligned} \tag{15.131}$$

The reason why these expressions are valid at $t = 0$ is that in general the phase velocity c depends on y. Performing the required differentiations and substituting the resulting expressions into (15.130) results in

$$\left(\frac{\partial \overline{\zeta}}{\partial t}\right)_{t=0} = 0 \tag{15.132}$$

We conclude that, at $t = 0$, the nonlinearity of the vorticity equation (15.126) is removed due to the Neamtan solution. Again using (15.131), we find at $t = 0$ from (15.128) for the change with time of the perturbation vorticity that

$$\left(\frac{\partial \zeta'}{\partial t}\right)_{t=0} = -\left(\overline{u} - \frac{1}{k_x^2 + k_y^2}\frac{\partial \overline{\zeta}}{\partial y} - \frac{\beta}{k_x^2 + k_y^2}\right)\frac{\partial \zeta'}{\partial x} \tag{15.133}$$

This formula has the form of an advection equation, so the expression within the large parentheses is the momentary phase velocity $c(y)_{t=0}$ of the vorticity in the x-direction. By substituting $\overline{u}$ according to (15.124) and replacing the partial derivative of the mean vorticity, we obtain the final form for the phase speed at $t = 0$:

$$\boxed{c(y)_{t=0} = \left(u_0 - \frac{\beta}{k_x^2 + k_y^2}\right) + \overline{k} A\left(1 - \frac{\overline{k}^2}{k_x^2 + k_y^2}\right)\sin(\overline{k} y)} \tag{15.134}$$

There are three cases that need to be discussed.

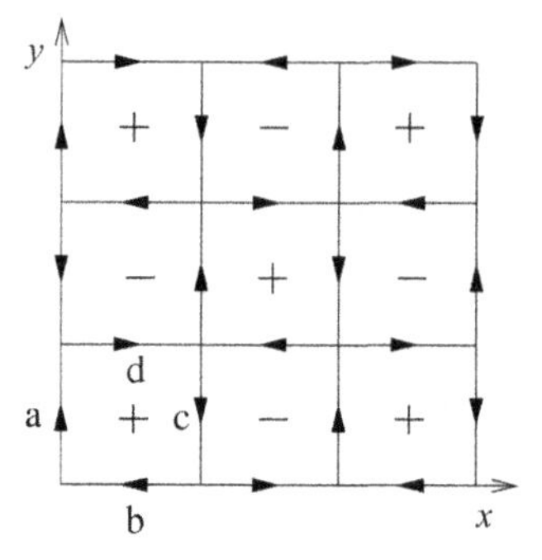

Fig. 15.7 Positive and negative areas of the perturbation field $\cos(k_x x)\cos(k_y y)$ at $t = 0$ with $k_x x = k_y y$. $u' = 0$ at positions a and c; $v' = 0$ at positions b and d.

$$\text{(i)} \qquad \overline{k}^2 = k_x^2 + k_y^2$$

The disturbance moves with a constant velocity given by (15.123), which is independent of y. The phase velocity and the disturbance do not adapt to the profile of the basic current. In this case of indifference the disturbance moves without change of shape.

Before we discuss the second case let us obtain a picture of the disturbance ψ' of (15.121) for the special case that $k_x x = k_y y$; see Figure 15.7. Positive and negative areas correspond to regions of high and low pressure, thus defining the sense of the circulation. On the boundaries of each region one of the two velocity components, u' or v', is zero so that the product $u'v'$ is zero. Thus, the zonal mean of the perturbation product is zero also. However, the product $u'v'$, which is usually nonzero, can be interpreted as the meridional transport of the mean zonal momentum along a latitudinal circle. The direction and the magnitude of the transport strongly depend on the structure of the disturbance. As shown in the appendix to this chapter, the kinetic energy of the flow may be decomposed into a zonal part $\overline{K}$ plus a perturbation part K'. There it is demonstrated that the change of $\overline{K}$ with time is given by

$$\frac{\partial \overline{K}}{\partial t} = -\frac{1}{WL}\int_0^L \int_0^W \overline{u}\,\frac{\partial}{\partial y}(\overline{u'v'})\,dy\,dx \tag{15.135}$$

where W and L represent the width and the length of the channel. Since in the present case $\overline{u'v'}$ is zero, there is no meridional transport of zonal kinetic energy. Hence, $\overline{K}$ does not change with time.

$$\text{(ii)} \qquad \overline{k}^2 < k_x^2 + k_y^2$$

This means that the wavelengths L_x and L_y are small in comparison with the wavelength of the basic current. According to (15.134) the phase velocity and thus the shape of the perturbation field adjust to the shape of the basic flow profile

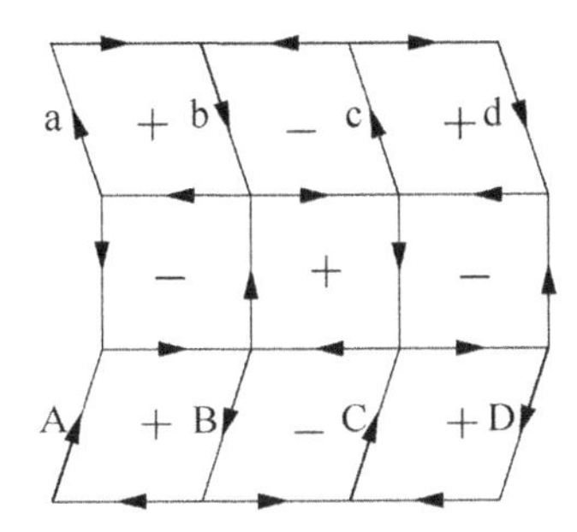

Fig. 15.8 Meridional transport of momentum, $\partial \overline{u'v'}/\partial y < 0$.

(see Figure 15.6) since the factor multiplying $\sin(\overline{k}y)$ remains positive. The largest contribution occurs at $\overline{k}y = \pi/2$. The situation is displayed schematically in Figure 15.8. To illustrate, let us consider the flow at points (a, c) and (b, d). In both cases $u'v' < 0$, so the zonal mean of $u'v'$ is also less than zero. The central part of Figure 15.8 repeats the situation of case I, in which the meridional transport of momentum vanishes. Consider the points (A, C) and (B, D) in the lower section of Figure 15.8. In both cases we have $u'v' > 0$, so the zonal mean of $u'v'$ is also larger than zero. In summary, the situation is characterized by $\partial \overline{u'v'}/\partial y < 0$. Thus, for the configuration of the perturbed part of the flow field shown in Figure 15.8 the zonal kinetic energy increases with time due to the meridional transport of momentum. This leads to the formation of a jet stream. Referring to Figure 15.4, this situation expresses barotropic stability.

From energy balances it follows that the real (baroclinic) atmosphere in the majority of cases is barotropically stable. Friction would cause the western flow to slow down if kinetic energy of the perturbations were not transferred to the kinetic energy of the basic current. To state it differently, barotropic stability is an essential requirement for the maintenance of the westward wind drift. All barotropic models exhibit the tendency of zonalization.

$$\text{(iii)} \qquad \overline{k}^2 > k_x^2 + k_y^2$$

In this case L_x and L_y are not small in comparison with the wavelength of the basic current. Now the term multiplying $\sin(\overline{k}y)$ in (15.134) is negative. The contributions of the basic current and the perturbations to the phase velocity occur with opposite signs, in contrast to the previous case. Therefore, the disturbance has the opposite shape, as shown in Figure 15.9. According to (15.135) this results in a depletion of the zonal kinetic energy and in an accumulation of K'. According to Figure 15.4 this corresponds to barotropic instability. This situation is relatively rare and occurs only in connection with large wavelengths of the disturbances. It is known that this type of process contributes to the formation and maintenance of blocking highs.

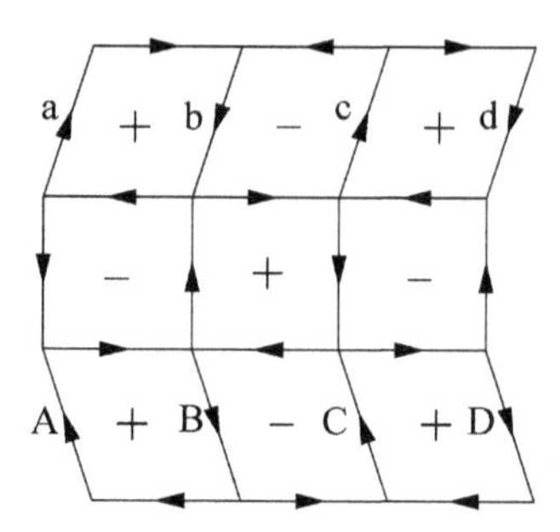

Fig. 15.9 Meridional transport of momentum, $\partial\overline{u'v'}/\partial y > 0$.

15.6 Appendix

In this appendix we are going to briefly derive a formula for the transformation of zonal to perturbation kinetic energy and vice versa for the filtered model. Since for the moment we are interested in midlatitude flow only, we limit the flow to a channel of width W and length L. We may think of this restricted atmospheric flow as being bounded by two latitude circles. Moreover, the northern and southern boundaries of the channel are treated as rigid walls. We also assume that the Coriolis parameter varies according to (15.104). The average kinetic energy of the horizontal flow within the channel is then given by

$$K = \frac{1}{WL}\int_0^L\int_0^W \frac{\mathbf{v}_\mathrm{h}^2}{2}\,dy\,dx = \frac{1}{WL}\int_0^L\int_0^W \frac{(\nabla_\mathrm{h}\psi)^2}{2}\,dy\,dx \tag{15.136}$$

since the divergence-free wind is given by $\mathbf{v}_\mathrm{h} = \mathbf{i}_3 \times \nabla_\mathrm{h}\psi$. In view of

$$\nabla_\mathrm{h}\psi \cdot \nabla_\mathrm{h}\frac{\partial\psi}{\partial t} = \nabla_\mathrm{h}\cdot\left(\psi\,\nabla_\mathrm{h}\frac{\partial\psi}{\partial t}\right) - \psi\,\nabla_\mathrm{h}^2\frac{\partial\psi}{\partial t} \tag{15.137}$$

the change with time of the kinetic energy is given by

$$\frac{\partial K}{\partial t} = -\frac{1}{WL}\int_0^L\int_0^W \psi\,\frac{\partial\zeta}{\partial t}\,dy\,dx \tag{15.138}$$

Application of the two-dimensional Gaussian divergence theorem (M6.34) shows that the divergence term of (15.137) does not contribute to the change in energy. This results from the assumption that we have rigid walls and cyclic boundary conditions in the x-direction. Using (15.68) and recalling that the horizontal divergence of the velocity field is zero in this equation, we find

$$\frac{\partial K}{\partial t} = \frac{1}{WL}\int_0^L\int_0^W \nabla_\mathrm{h}\cdot(\mathbf{v}_\mathrm{h}\psi\eta)\,dy\,dx \tag{15.139}$$

since $\mathbf{v}_h \cdot \nabla_h \psi = 0$. By again applying the divergence theorem and the adopted boundary conditions we find that the change with time of the kinetic energy vanishes completely:

$$\frac{\partial K}{\partial t} = 0 \tag{15.140}$$

Next we split the horizontal velocity by introducing the zonal average $\overline{\mathbf{v}}_h$ and the perturbation velocity $\mathbf{v}_h'$:

$$\mathbf{v}_h = \overline{\mathbf{v}}_h + \mathbf{v}_h', \qquad \overline{\mathbf{v}}_h = \frac{1}{L}\int_0^L \mathbf{v}_h \, dx, \qquad \overline{\mathbf{v}_h'} = 0 \tag{15.141}$$

Analogously, we split the kinetic energy into the zonal and perturbation parts:

$$K = \overline{K} + K' = \frac{1}{WL}\int_0^L \int_0^W \frac{(\overline{\mathbf{v}}_h)^2 + (\mathbf{v}_h')^2}{2} \, dy \, dx \tag{15.142}$$

In view of (15.138) we may write for the change of $\overline{K}$ with time

$$\frac{\partial \overline{K}}{\partial t} = -\frac{1}{WL}\int_0^L \int_0^W \overline{\psi}\,\frac{\partial \overline{\zeta}}{\partial t} \, dy \, dx = \frac{1}{WL}\int_0^L \int_0^W \overline{\psi}\;\overline{\mathbf{v}_h' \cdot \nabla_h \zeta'} \, dy \, dx \tag{15.143}$$

since the zonal average of $\partial \overline{\zeta}/\partial t$ has the form (15.130). On rewriting this expression we find

$$\begin{aligned}
\frac{\partial \overline{K}}{\partial t} &= \frac{1}{WL}\int_0^L \int_0^W \nabla_h \cdot (\overline{\psi}\;\overline{\mathbf{v}_h' \zeta'}) \, dy \, dx - \frac{1}{WL}\int_0^L \int_0^W \overline{\mathbf{v}_h' \zeta'} \cdot \nabla_h \overline{\psi} \, dy \, dx \\
&= -\frac{1}{WL}\int_0^L \int_0^W \overline{v'\zeta'}\,\frac{\partial \overline{\psi}}{\partial y} \, dy \, dx
\end{aligned} \tag{15.144}$$

where the first integral on the right-hand side must vanish due to the divergence theorem of Gauss and due to the boundary conditions.

Equation (15.144) can be rewritten giving a formula that consists of three integrals:

$$\begin{aligned}
\frac{\partial \overline{K}}{\partial t} = &-\frac{1}{WL}\int_0^L \int_0^W \overline{u}\,\frac{\partial}{\partial y}(\overline{u'v'}) \, dy \, dx + \frac{1}{WL}\int_0^L \int_0^W \overline{u}\;\overline{v'\,\frac{\partial v'}{\partial x}} \, dy \, dx \\
&+ \frac{1}{WL}\int_0^L \int_0^W \overline{u}\;\overline{u'\,\frac{\partial v'}{\partial y}} \, dy \, dx
\end{aligned} \tag{15.145}$$

Only the first of these differs from zero, so

$$\frac{\partial \overline{K}}{\partial t} = -\frac{1}{WL}\int_0^L \int_0^W \overline{u}\,\frac{\partial}{\partial y}(\overline{u'v'}) \, dy \, dx \tag{15.146}$$

This is the desired equation for the change with time of the zonal kinetic energy. The other two integrals in (15.145) are zero, as follows immediately from partial integration and by application of the boundary conditions. Finally, due to (15.140), we find

$$\frac{\partial \overline{K}}{\partial t} = -\frac{\partial K'}{\partial t} \tag{15.147}$$

showing the only possible way for transformation of kinetic energy to occur.

Several parts of this chapter are based on the synopsis *The Barotropic Model* reported in *Promet* (1973), which is a publication of the German Weather Service. It refers to articles by Reiser (1973), Edelmann (1973), Edelmann and Reiser (1973), and Tiedtke (1973).

15.7 Problems

15.1: Show that the conditions (15.1) and (15.2) guarantee that the horizontal pressure gradient in the barotropic model atmosphere is independent of height.

Hint: Start with the relation $\nabla \times [(1/\rho)\nabla p] = 0$. Prove it.

15.2: Show that any two of the three relations

$$\nabla\rho \times \nabla p = 0, \qquad \frac{\partial}{\partial z}\left(\frac{1}{\rho}\nabla_{\rm h} p\right) = 0, \qquad \frac{1}{\rho}\frac{\partial p}{\partial z} = f(z)$$

imply the correctness of the third one.

Hint: First prove that $\nabla_{\rm h}\rho \times \nabla_{\rm h} p = 0$.

15.3: Verify equation (15.62).

15.4: Verify equation (15.103).

15.5: An upper-level midlatitude jet profile may be approximated by

$$\overline{u} = \overline{u}_0 \operatorname{sech}^2(y/y_0)$$

Plot the wind profile $\overline{u}/\overline{u}_0$ (x-axis) versus y/y_0.

(a) Suppose that $\beta = 0$. Is the necessary condition for barotropic instability satisfied?

(b) Suppose that $\beta \neq 0$. Find the necessary condition for barotropic instability.

16

Rossby waves

The daily weather maps for any extended part of the globe in middle and high latitudes show well-defined dynamic systems, which normally move from west to east. The speed of displacement of these systems differs from the mean wind speed of the air current. Moreover, the structure of these systems varies considerably with height. Near the earth's surface the motion systems exhibit much complexity, their predominant features being closed cyclonic and anticyclonic patterns of irregular shape. In contrast to this, in the middle and upper troposphere as well as in the lower stratosphere (200–700 hPa), the dynamic systems usually consist of relatively simple smooth wave-shaped patterns. Typically one finds about four or five waves around the hemisphere. Their thermal structure is characterized by cold troughs and warm ridges. The dynamics of these relatively long waves was first studied by Rossby (1939) and in some generalization by Haurwitz (1940). Therefore, these waves are called *Rossby–Haurwitz waves* or more simply *Rossby waves.*

16.1 One- and two-dimensional Rossby waves

Rossby employed the filtered barotropic model which we have studied in some detail in the previous chapter. This model assumes that there is horizontal frictionless flow between plane-parallel plates. The medium is thought to be an incompressible homogeneous fluid of constant density so that, due to the upper boundary condition, sound waves and gravity waves cannot form. Nevertheless, horizontal pressure gradients are possible.

The starting point of the analysis is the divergence-free barotropic vorticity equation (15.70), which is restated here for convenience:

$$\nabla_{\rm h}^2 \frac{\partial \psi}{\partial t} + \frac{\partial \psi}{\partial x}\frac{\partial}{\partial y}(\nabla_{\rm h}^2 \psi + f) - \frac{\partial \psi}{\partial y}\frac{\partial}{\partial x}(\nabla_{\rm h}^2 \psi + f) = 0$$
$$\text{with} \quad f = f_0 + \beta y, \qquad \beta = \frac{2\Omega \cos\varphi}{a} = \text{constant} \tag{16.1}$$

where a is the mean radius of the earth.

With Rossby we shall assume that the Coriolis parameter varies linearly in the y-direction; see also equation (15.104). In order to obtain an analytic solution for the speed of displacement of the long waves we will linearize the vorticity equation. We assume the validity of a linearization of the form

$$\psi = \overline{\psi}(y) + \psi'(x, y, t), \qquad \overline{\psi} = \text{constant} - \overline{u}y, \qquad \overline{u} = -\frac{\partial \overline{\psi}}{\partial y} = \text{constant} \tag{16.2}$$

where $\overline{u}$ is the mean wind speed of the air current. The corresponding linear variant of the vorticity equation in the so-called *β-plane* is

$$\left(\frac{\partial}{\partial t} + \overline{u}\,\frac{\partial}{\partial x}\right)\nabla_{\mathrm{h}}^2 \psi' + \beta\,\frac{\partial \psi'}{\partial x} = 0 \tag{16.3}$$

Substituting the wave-type disturbance

$$\psi'(x, y, t) = A \exp[i(k_x x + k_y y - \omega t)] \tag{16.4}$$

into (16.3) gives the characteristic or frequency equation

$$\omega = k_x \overline{u} - \frac{\beta k_x}{k_x^2 + k_y^2} \tag{16.5}$$

Division of the circular frequency by the wavenumber k_x or by k_y yields the components of the phase velocity $\mathbf{c}$ in the x- and y-directions:

$$\boxed{\begin{aligned} c_x &= \frac{\omega}{k_x} = \overline{u} - \frac{\beta}{k_x^2 + k_y^2} \\ c_y &= \frac{\omega}{k_y} = \frac{k_x}{k_y}\overline{u} - \frac{\beta k_x / k_y}{k_x^2 + k_y^2} = \frac{k_x}{k_y} c_x \end{aligned}} \tag{16.6}$$

From the mathematical point of view it is interesting to remark that the Rossby waves satisfy not only the linearized equation (16.3) but also the complete nonlinearized equation (16.1). By combining terms (16.1) may be written as

$$\nabla_{\mathrm{h}}^2 \frac{\partial \psi}{\partial t} + J(\psi, \nabla_{\mathrm{h}}^2 \psi) + \beta\,\frac{\partial \psi}{\partial x} = 0 \tag{16.7}$$

The Jacobian is given by

$$\begin{aligned} J(\psi, \nabla_{\mathrm{h}}^2 \psi) &= \frac{\partial \psi'}{\partial x}\,\frac{\partial}{\partial y}(\nabla_h^2 \psi') - \frac{\partial \psi'}{\partial y}\,\frac{\partial}{\partial x}(\nabla_h^2 \psi') - \frac{\partial \overline{\psi}}{\partial y}\,\frac{\partial}{\partial x}(\nabla_h^2 \psi') \\ &= -ik_x \overline{u}(k_x^2 + k_y^2)\psi' \end{aligned} \tag{16.8}$$

Substitution of (16.8) into (16.7) together with (16.2) and (16.4) again yields the frequency equation (16.5).

First let us discuss the original Rossby formula by assuming that $k_y = 0$ in (16.4). In this case and with $\overline{u} = 0$ we speak of *pure Rossby waves*. We then find for the more general case with $\overline{u} \neq 0$

$$\boxed{c_x = \overline{u} - \frac{\beta}{k_x^2} = \overline{u} - \beta \frac{L_x^2}{4\pi^2}} \tag{16.9}$$

In the midlatitudes, for a wide range of values of L_x, this formula is in reasonable agreement with observations.

The simplest situation occurs if we assume that the wind speed $\overline{u}$ vanishes. In this case the Rossby waves propagate from east to west as implied by the minus sign in (16.9). Usually the zonal flow is large enough with $\overline{u} > 0$ that the Rossby waves propagate to the east. However, relative to the basic zonal current, the waves still move to the west. Furthermore, we recognize that the phase velocity of the Rossby wave equals the mean wind speed $\overline{u}$ if the Coriolis parameter is not permitted to vary with latitude. For high geographical latitudes and very short waves this is in rough agreement with reality. For lower geographical latitudes and increasing wavelength L_x the wave lags behind the zonal air current.

There exists a certain wavelength $L_{x,\mathrm{stat}}$

$$L_{x,\mathrm{stat}} = 2\pi\sqrt{\frac{\overline{u}}{\beta}} = 2\pi\sqrt{\frac{a\overline{u}}{2\Omega\cos\varphi}} \tag{16.10}$$

at which the wave becomes stationary. For a latitude $\varphi = 45^\circ$ and $\overline{u} = 10\ \mathrm{m\ s^{-1}}$ the stationary wavelength is about 5000 km and for $\overline{u} = 20\ \mathrm{m\ s^{-1}}$ it amounts to 7000 km.

Introducing (16.10) into (16.9), the wave speed may be expressed as

$$c_x = \frac{\beta}{4\pi^2}\left(L_{x,\mathrm{stat}}^2 - L_x^2\right) \tag{16.11}$$

From this equation it follows that waves of length $L_x > L_{x,\mathrm{stat}}$ are retrogressive, i.e. they are moving from east to west. Let N represent the number of waves around a latitude circle so that

$$NL_x = 2\pi a\cos\varphi \tag{16.12}$$

The so-called *velocity deficit* of the retrogressive wave $c_x - \overline{u}$ may be found, for example, from the equation

$$c_x - \overline{u} = -\frac{2\Omega a\cos^3\varphi}{N^2} \tag{16.13}$$

If, for a given latitude, the wave speed is zero, we may calculate the so-called *stationary wavenumber* $N_{\rm stat}$ so that the velocity deficit may also be determined with the help of

$$c_x = 2\Omega a \cos^3\varphi \left(\frac{1}{N_{\rm stat}^2} - \frac{1}{N^2}\right) \tag{16.14}$$

The velocity deficit may assume considerable values. For example, for a latitude $\varphi = 45^\circ$ and $N = 3$ the deficit amounts to -36.5 m s^{-1}; for $N = 6$ and for the same latitude the deficit is only -9.1 m s^{-1}. The largest deficits are found at low latitudes and for very long waves. For $\varphi = 30^\circ$ and $N = 3$ the deficit is -67 m s^{-1}.

Compared with observations, particularly for the very long waves, the velocity deficit is clearly too high. This very unrealistic behavior of the Rossby displacement formula may be traced back to the simplicity of the model which assumed the existence of a fixed upper plate. This resulted in a vanishing horizontal divergence of the velocity field. If the rigid lid is replaced by a free surface, then the Rossby formula (16.9) assumes a modified form, as will be shown next. For convenience we restate the two equations needed for the analysis. These are the barotropic vorticity equation

$$\frac{d\eta}{dt} + \eta\,\nabla_{\rm h}\cdot\mathbf{v}_{\rm h} = 0 \tag{16.15}$$

repeated from (10.146), and the time-change equation for a free surface H

$$\frac{dH}{dt} + H\,\nabla_{\rm h}\cdot\mathbf{v}_{\rm h} = 0 \tag{16.16}$$

repeated from (15.23a). Since we wish to find a simple analytic solution to the problem we assume that the ground is flat. By eliminating the divergence of the horizontal wind field between these two equations we find

$$\frac{\partial\eta}{\partial t} + u\,\frac{\partial\eta}{\partial x} + v\,\frac{\partial\eta}{\partial y} - \frac{\eta}{H}\left(\frac{\partial H}{\partial t} + u\,\frac{\partial H}{\partial x} + v\,\frac{\partial H}{\partial y}\right) = 0 \tag{16.17a}$$

which can also be written as the conservation statement

$$\boxed{\frac{d}{dt}\left(\frac{\eta}{H}\right) = 0} \tag{16.17b}$$

This is the potential vorticity equation which was derived earlier; see Section 10.5.8. It simply states that the potential vorticity of each individual air parcel is conserved in a divergent barotropic flow. Hence, the number of minima and maxima of the absolute vorticity cannot change, so a true cyclogenesis cannot be predicted by the theory of barotropic flow. It should be emphasized, however, that in many regions for long time periods the atmosphere acts nearly as a barotropic medium. Many developments that appear to be new simply result from a redistribution of already existing extremals of the absolute vorticity.

Equations (16.17a) and (16.17b) were obtained from the equations for the barotropic vorticity and the free surface without any additional assumptions, so gravity waves are not eliminated. A reasonable filter for this type of noise is the geostrophic approximation of the wind field. By introducing the geostrophic wind components

$$u_{\rm g} = -\frac{g}{f_0}\frac{\partial H}{\partial y}, \qquad v_{\rm g} = \frac{g}{f_0}\frac{\partial H}{\partial x} \tag{16.18}$$

we find the following expressions for the absolute geostrophic vorticity $\eta_{\rm g}$:

$$\eta_{\rm g} = \zeta_{\rm g} + f \quad \text{with} \quad \zeta_{\rm g} = \frac{\partial v_{\rm g}}{\partial x} - \frac{\partial u_{\rm g}}{\partial y} = \frac{g}{f_0}\nabla_{\rm h}^2 H \tag{16.19}$$

It can be seen that the Coriolis parameter f has been replaced by the average value f_0 in the expression for $\zeta_{\rm g}$ but not in the expression for $\eta_{\rm g}$. This treatment ensures that the β-effect will still be accounted for in the model and that at the same time analytic solutions may be obtained.

By substituting (16.18) and (16.19) into (16.17a) we obtain the barotropic model equation

$$\begin{aligned}\nabla_{\rm h}^2\frac{\partial H}{\partial t} &- \frac{\partial H}{\partial y}\frac{\partial}{\partial x}\left(\frac{g}{f_0}\nabla_{\rm h}^2 H + f\right) + \frac{\partial H}{\partial x}\frac{\partial}{\partial y}\left(\frac{g}{f_0}\nabla_{\rm h}^2 H + f\right)\\ &- \frac{f_0}{gH}\left(\frac{g}{f_0}\nabla_{\rm h}^2 H + f\right)\frac{\partial H}{\partial t} = 0\end{aligned} \tag{16.20}$$

This partial differential equation for the tendency $\partial H/\partial t$ is of the Helmholtz type and can be solved, for example, by a numerical procedure known as the *relaxation method*.

In order to obtain an analytic solution for the displacement of a wave-type disturbance, we must linearize equation (16.20). We assume that a simple disturbance (perturbation) is embedded in a zonal current. For this purpose we decompose the H-field into a part $\overline{H}(y)$ plus a perturbation $H'(x, t)$:

$$H(x, y, t) = \overline{H}(y) + H'(x, t) \tag{16.21}$$

In analogy to the geostrophic wind component $u_{\rm g}$ we may write for the zonal current

$$-\frac{g}{f_0}\frac{\partial \overline{H}}{\partial y} = \overline{u} = \text{constant} \tag{16.22a}$$

Integration yields

$$\overline{H}(y) = H_0 - \overline{u} f_0 y/g \tag{16.22b}$$

where H_0 is a constant.

Substituting (16.21) into (16.20) gives the linearized equation

$$\left(\frac{\partial^2}{\partial x^2} - \frac{f_0^2}{gH_0}\right)\frac{\partial H'}{\partial t} + \beta\,\frac{\partial H'}{\partial x} + \overline{u}\,\frac{\partial^3 H'}{\partial x^3} = 0 \tag{16.23}$$

Wherever H appears in undifferentiated form we have replaced it by the mean height H_0 of the free surface. Moreover, we have approximated $ff_0 \approx f_0^2$.

Assuming that the disturbance is a wave propagating in the x-direction,

$$H' = A\exp[ik_x(x - ct)] \tag{16.24}$$

we find that the phase velocity of a Rossby wave, modified by gravitational effects, is given by

$$\boxed{c_x = \frac{\overline{u} - \beta\dfrac{L_x^2}{4\pi^2}}{1 + \dfrac{f_0^2 L_x^2}{gH_0 4\pi^2}}} \tag{16.25}$$

Sometimes this wave is also called a *mixed Rossby wave.*

We now compare (16.25) with (16.9), which applies to a fixed upper boundary. We recognize that the stationary wavelength has not changed. For all nonstationary waves the magnitude of the phase velocity c_x is smaller than that predicted by (16.9). The velocity deficit with respect to the basic current increases somewhat for waves propagating from west to east. However, the speed of the extremely fast retrogressive waves is considerably less, resulting in better agreement with observations. For the above example, assuming a latitude $\varphi = 30°$, $N = 3$, and $H_0 = 8000$ m, the deficit is now -56.5 m s^{-1} instead of -67 m s^{-1} as computed from (16.13). Had we chosen $H_0 = 2000$ m, the velocity deficit would have been reduced further, to a value of only -40 m s^{-1}.

16.2 Three-dimensional Rossby waves

The starting point of the analysis is the baroclinic vorticity equation (10.77) in the pressure coordinate system, which is rewritten as

$$\frac{\partial \zeta}{\partial t} + \mathbf{v}_{\mathrm{h}}\cdot\nabla_{\mathrm{h}}\eta + \omega\,\frac{\partial \zeta}{\partial p} = \left(\mathbf{e}_{\mathrm{r}} \times \frac{\partial \mathbf{v}_{\mathrm{h}}}{\partial p}\right)\cdot\nabla_{\mathrm{h}}\omega + \eta\,\frac{\partial \omega}{\partial p} \tag{16.26}$$

For simplicity we have omitted the suffix p since confusion is unlikely to occur. With sufficient accuracy we will now replace the velocity vector by the geostrophic wind vector and the vorticity by the geostrophic vorticity

$$\mathbf{v}_{\mathrm{h}} \longrightarrow \mathbf{v}_{\mathrm{g}}, \qquad \zeta \longrightarrow \zeta_{\mathrm{g}} = \frac{1}{f_0}\,\nabla_{\mathrm{h}}^2\phi, \qquad \eta \longrightarrow \eta_{\mathrm{g}} = \zeta_{\mathrm{g}} + f \tag{16.27}$$

In view of this approximation and a consequential treatment of the various terms appearing in the vorticity equation, we obtain the quasi-geostrophic approximation of the vorticity equation in the form

$$\boxed{\left(\frac{\partial}{\partial t} + \mathbf{v}_{\mathrm{g}} \cdot \nabla_{\mathrm{h}}\right)(\zeta_{\mathrm{g}} + f) = f_0 \frac{\partial \omega}{\partial p}} \tag{16.28}$$

As before, on the left-hand side the Coriolis parameter f is permitted to vary with latitude to include the β-effect. The steps leading to (16.28) will be given later when we discuss the quasi-geostrophic theory in some detail.

In order to include in a rough approximation the vertical structure of the atmosphere, we must involve the first law of thermodynamics. The approximate adiabatic form for dry air is given by

$$\rho c_{p,0} \frac{dT}{dt} = \frac{dp}{dt} = \omega \tag{16.29}$$

It is useful to introduce the definition of the *static stability* σ_0 into the previous equation:

$$\sigma_0 = -\frac{1}{\rho} \frac{\partial \ln \theta}{\partial p} = \frac{R_0}{p}\left(\frac{R_0 T}{c_{p,0} p} - \frac{\partial T}{\partial p}\right) \tag{16.30}$$

After a few easy steps we find

$$\left(\frac{\partial}{\partial t} + \mathbf{v}_{\mathrm{g}} \cdot \nabla_{\mathrm{h}}\right) T - \frac{p}{R_0} \sigma_0 \omega = 0 \tag{16.31}$$

By means of the hydrostatic equation we replace the temperature T in terms of the geopotential ϕ:

$$T = -\frac{p}{R_0} \frac{\partial \phi}{\partial p} \tag{16.32}$$

so that the first law of thermodynamics can be written as

$$\boxed{\left(\frac{\partial}{\partial t} + \mathbf{v}_{\mathrm{g}} \cdot \nabla_{\mathrm{h}}\right) \frac{\partial \phi}{\partial p} + \sigma_0 \omega = 0} \tag{16.33}$$

Recall that we are using the p system, so the operators $\partial/\partial t$ and ∇_{h} are applied at constant pressure.

In order to obtain an analytic solution to our problem we must linearize equations (16.28) and (16.33). With $\mathbf{v}_{\mathrm{g}} = \overline{u}_{\mathrm{g}}\mathbf{i} + \overline{v}_{\mathrm{g}}\mathbf{j}$, $\overline{u}_{\mathrm{g}} = \overline{u}$ we have

$$\mathbf{v}_{\mathrm{g}} \cdot \nabla_{\mathrm{h}} \zeta_{\mathrm{g}} \Longrightarrow \overline{u} \frac{\partial}{\partial x}\left(\frac{1}{f_0} \nabla_{\mathrm{h}}^2 \phi'\right), \qquad \mathbf{v}_{\mathrm{g}} \cdot \nabla_{\mathrm{h}} f \Longrightarrow \frac{\beta}{f_0} \frac{\partial \phi'}{\partial x} \tag{16.34}$$

With this linearization the vorticity equation obtains the form

$$\left(\frac{\partial}{\partial t}+\overline{u}\,\frac{\partial}{\partial x}\right)\nabla_{\rm h}^2\phi'+\beta\,\frac{\partial\phi'}{\partial x}=f_0^2\,\frac{\partial\omega'}{\partial p} \tag{16.35}$$

while the first law of thermodynamics may be written in linearized form as

$$\left(\frac{\partial}{\partial t}+\overline{u}\,\frac{\partial}{\partial x}\right)\frac{\partial\phi'}{\partial p}+\sigma_0\omega'=0 \tag{16.36}$$

By eliminating the generalized vertical velocity, equations (16.35) and (16.36) may be combined to give the single equation

$$\left(\frac{\partial}{\partial t}+\overline{u}\,\frac{\partial}{\partial x}\right)\left(\nabla_{\rm h}^2\phi'+\frac{f_0^2}{\overline{\sigma}_0}\,\frac{\partial^2\phi'}{\partial p^2}\right)+\beta\,\frac{\partial\phi'}{\partial x}=0 \tag{16.37}$$

In order to obtain this equation we have replaced the stability parameter σ_0 by an average value $\overline{\sigma}_0$ for the vertical layer under consideration. Assuming as a solution of (16.37) the three-dimensional wave in the form

$$\phi'(x,y,p,t)=A\exp[i(k_x x+k_y y+k_p p-\omega t)] \tag{16.38}$$

we obtain the desired dispersion relation for the intrinsic frequency χ:

$$\chi=\omega-\overline{u}k_x=-\frac{\beta k_x}{k_x^2+k_y^2+\left(f_0^2/\overline{\sigma}_0\right)k_p^2} \tag{16.39}$$

The component of the phase velocity $\mathbf{c}$ of the three-dimensional wave along the x-axis is given by

$$\boxed{c_x=\frac{\omega}{k_x}=\overline{u}-\frac{\beta}{k_x^2+k_y^2+\left(f_0^2/\overline{\sigma}_0\right)k_p^2}} \tag{16.40a}$$

Compared with the two-dimensional phase velocity of (16.5), the value of c_x is modified due to the presence of the vertical wavenumber k_p which has the dimension hPa^{-1}. Along the y-axis, in analogy to the two-dimensional case (16.6), we find

$$\boxed{c_y=\frac{\omega}{k_y}=\left(\overline{u}-\frac{\beta}{k_x^2+k_y^2+\left(f_0^2/\overline{\sigma}_0\right)k_p^2}\right)\frac{k_x}{k_y}=c_x\frac{k_x}{k_y}} \tag{16.40b}$$

The phase speed c_p along the vertical pressure axis,

$$\boxed{c_p=\frac{\omega}{k_p}=\left(\overline{u}-\frac{\beta}{k_x^2+k_y^2+\left(f_0^2/\overline{\sigma}_0\right)k_p^2}\right)\frac{k_x}{k_p}=c_x\frac{k_x}{k_p}} \tag{16.40c}$$

has the units $\mathrm{hPa\ s}^{-1}$.

Let us now return to equation (16.40a) to obtain the condition for the standing wave. Setting c_x equal to zero, we find

$$\text{Standing wave:} \qquad k_{p,s}^2 = \left(\frac{\beta}{\overline{u}} - \left(k_x^2 + k_y^2\right)\right)\frac{\overline{\sigma}_0}{f_0^2} \tag{16.41}$$

showing how the vertical and the horizontal wavenumbers must be related for standing waves to form. Compare this result with (16.10). Inspection of (16.38) shows that vertical propagation of Rossby waves is possible only if k_p is a real quantity, i.e. $k_p^2 > 0$. In order to state this condition for the more general case we solve (16.40a) and obtain

$$k_p^2 = \left(\frac{\beta}{\overline{u} - c_x} - \left(k_x^2 + k_y^2\right)\right)\frac{\overline{\sigma}_0}{f_0^2} \tag{16.42}$$

For the large-scale motion which is being considered here, the vertical stability is larger than zero. From (16.42) we recognize that the condition $k_p^2 > 0$ holds whenever the inequality

$$\frac{\beta}{\overline{u} - c_x} > k_x^2 + k_y^2 \tag{16.43}$$

is satisfied. This requires that, for a large positive value of the velocity deficit $\overline{u} - c_x$, vertically propagating Rossby waves cannot occur as might be the case for strong westerly flow. For easterly flows with a negative value of the velocity deficit, the vertical propagation of Rossby waves is not permitted by (16.42). These conclusions are supported by observations.

Finally we observe that (16.40a) reduces to (16.6) if $k_p = 0$.

16.3 Normal-mode considerations

Additional information about vertical Rossby waves can be obtained by simplifying the basic equations (16.35) and (16.36). Following Wiin-Nielsen (1975) we set the basic flow velocity $\overline{u} = 0$ and ignore any y-dependency of the pertinent variables. However, we permit the amplitudes of the geopotential and the generalized vertical velocity to depend on pressure. The system of equations to be solved is then given by

$$\begin{aligned} &\text{(a)} \qquad \frac{\partial}{\partial t}\frac{\partial^2 \phi'}{\partial x^2} + \beta\frac{\partial \phi'}{\partial x} = f_0^2\frac{\partial \omega'}{\partial p} \\ &\text{(b)} \qquad \frac{\partial}{\partial t}\frac{\partial \phi'}{\partial p} + \sigma_0\omega' = 0 \end{aligned} \tag{16.44}$$

Now we consider perturbations of the form

$$\begin{pmatrix} \phi' \\ \omega' \end{pmatrix} = \begin{pmatrix} A(p) \\ B(p) \end{pmatrix} \exp[ik_x(x - ct)] \tag{16.45}$$

which are referred to as *normal-mode solutions*. By substituting (16.45) into (16.44) we obtain the first-order differential-equation system

$$\left(k_x^3 c + k_x \beta\right) iA = f_0^2 \frac{\partial B}{\partial p}, \qquad ik_x c \frac{\partial A}{\partial p} = \sigma_0 B \tag{16.46}$$

Since A and B depend only on the single variable p we may replace the partial by the total derivative. By obvious steps we combine this system to give a single second-order differential equation:

$$\frac{d^2 B}{dp^2} - \frac{c + \beta/k_x^2}{c(f_0/k_x)^2} \sigma_0 B = 0 \tag{16.47}$$

We now introduce the abbreviations

$$c_{\mathrm{R}} = \frac{\beta}{k_x^2}, \qquad c_{\mathrm{I}} = \frac{f_0}{k_x}, \qquad a^2 = \sigma_0 p_0^2, \qquad p_{\mathrm{r}} = \frac{p}{p_0} \tag{16.48}$$

into (16.47) and obtain

$$\frac{d^2 B}{dp_{\mathrm{r}}^2} + l^2(p_{\mathrm{r}}) B = 0 \quad \text{with} \quad l^2(p_{\mathrm{r}}) = -\frac{c + c_{\mathrm{R}}}{c c_{\mathrm{I}}^2} a^2(p_{\mathrm{r}}) \tag{16.49}$$

Note that, according to (16.30) and (16.48), $a^2 = a^2(p_{\mathrm{r}})$ so that we also have $l^2 = l^2(p_{\mathrm{r}})$. We recognize that $-c_{\mathrm{R}}$ and c_{I} represent the phase velocities of the pure Rossby waves and of the inertial waves. The amplitude B of the vertical velocity is now defined with respect to the relative pressure p_{r}. Obviously the solution of (16.49) depends on the behavior of $l^2(p_{\mathrm{r}})$.

A simple wave-type solution can be obtained by requiring that $l(p_{\mathrm{r}})$ is a constant and that $l^2 > 0$. In this case the characteristic values of (16.49) are imaginary, thus resulting in the solution

$$B(p_{\mathrm{r}}) = C_1 \cos(l p_{\mathrm{r}}) + C_2 \sin(l p_{\mathrm{r}}) \tag{16.50}$$

Imposing the boundary conditions

$$B(p_{\mathrm{r}} = 0) = 0, \quad B(p_{\mathrm{r}} = 1) = 0 \implies C_1 = 0, \quad l = l_n = n\pi, \quad n = 0, 1, \ldots \tag{16.51}$$

which require that the generalized vertical velocities vanish at the top and at the base of the atmosphere, means that we must select $l = l_n = n\pi$ to make the sine function vanish.

With $l_n = n\pi$, $n = 0, 1, \ldots$ from (16.49) we find a whole spectrum of phase velocities:

$$\boxed{c_n = -\frac{a^2 c_{\mathrm{R}}}{a^2 + n^2 \pi^2 c_{\mathrm{I}}^2}} \tag{16.52}$$

In case of $n = 0$ the phase velocity $c_0 = -c_R = -\beta/k_x^2$. This is the zeroth normal mode of the pure Rossby wave. This solution is consistent with the requirement that pure Rossby waves form in the absence of vertical velocities. For $n = 1, 2, \ldots$ higher-order vertical modes are obtained. The corresponding pure waves also propagate from east to west since $c_n < 0$, but they move more slowly than in case of the zeroth normal mode. The amplitudes of the generalized velocities for normal modes in this case differ from zero as shown in

$$B(p_r) = C_2 \sin(l_n p_r) \neq 0 \tag{16.53}$$

For large values of n equation (16.52) can be approximated by

$$c_n \approx -\frac{a^2 c_R}{n^2\pi^2 c_I^2} = -\frac{a^2\beta}{n^2\pi^2 f_0^2} \tag{16.54}$$

showing that c_n no longer depends on the wavelength of the wave. This particular situation is known as the *ultra-long-wave approximation.*

We will now show that the approximation leading to (16.54) is equivalent to ignoring the time-dependent term in the vorticity equation (16.44a). This assumption implies an approximate balance between the β-effect and the divergence of the flow. In fact, the expression $\partial\omega'/\partial p$ represents vertical stretching, which is coupled with the occurrence of divergence. This also follows directly from the continuity equation in pressure coordinates which will be formally derived in a later chapter.

The mathematical development is the same as that above leading to the form (16.49) but now $l^2(p_r)$ has a different meaning so that we formally replace $B(p_r)$ by $B_1(p_r)$. Instead of (16.49) we now obtain

$$\frac{d^2 B_1}{dp_r^2} + l_1^2 B_1 = 0 \quad \text{with} \quad l_1^2 = -\frac{a^2\beta}{c f_0^2} \tag{16.55}$$

Assuming that l_1 is a constant and that $l_1^2 > 0$, we again obtain a wave solution, which is given by

$$B_1(p_r) = D_1 \cos(l p_r) + D_2 \sin(l p_r) \tag{16.56}$$

By imposing the boundary conditions that the vertical velocity vanishes at the top of the atmosphere and at the ground where $p = p_0$, we immediately find the conditions

$$B(p_r = 0) = 0, \quad B(p_r = 1) = 0 \implies D_1 = 0, \quad l = n\pi, \quad n = 0, 1, \ldots \tag{16.57}$$

so that c_n is given by

$$c_n = -\frac{a^2\beta}{n^2\pi^2 f_0^2} \tag{16.58}$$

This verifies the above statement that ignoring the term a^2 in the denominator of (16.52) is equivalent to ignoring the time-dependency in the vorticity equation (16.44a).

16.4 Energy transport by Rossby waves

It is well known that the horizontal as well as the vertical transport of energy by Rossby waves plays an important role in the maintenance of the general circulation of the troposphere and the lower sections of the stratosphere. We have previously shown that the direction of the group velocity coincides with the direction of the transport of energy.

We begin by repeating the fundamental equation (14.91):

$$\widetilde{c}_{g,x} = \frac{\partial \chi}{\partial k_x} = \frac{\partial \omega}{\partial k_x} - \overline{u} = c_{g,x} - \overline{u} \tag{16.59}$$

giving the relation between the group velocity (indicated by a tilde), relative to the basic current, and the intrinsic frequency χ. Using the basic frequency equations (16.39), we obtain immediately the components of the group velocity for the three directions. The specialization to lower dimensions is obvious. Note that, in case of the vertical direction, we are dealing with a generalized velocity. From

$$\boxed{c_{g,x} = \frac{\partial \omega}{\partial k_x} = \overline{u} + \beta \frac{k_x^2 - k_y^2 - (f_0^2/\overline{\sigma}_0)k_p^2}{\left[k_x^2 + k_y^2 + (f_0^2/\overline{\sigma}_0)k_p^2\right]^2}} \tag{16.60}$$

we see that, for pure Rossby waves ($k_y = k_p = 0$) and $\overline{u} = 0$, the energy flux is always downstream so that $c_{g,x} > 0$. Comparison with (16.40a) shows that, for $\overline{u} = 0$, the phase and group velocities have opposite directions. In the general case, however, $c_{g,x}$ may be either positive or negative so that energy may flow in both directions.

From

$$\boxed{c_{g,y} = \frac{\partial \omega}{\partial k_y} = \frac{2\beta k_x k_y}{\left[k_x^2 + k_y^2 + (f_0^2/\overline{\sigma}_0)k_p^2\right]^2}} \tag{16.61}$$

we recognize that meridional transport of energy is possible only if k_y differs from zero.

The basic requirement for meridional transport of energy to occur is that the axis of troughs (ridges) must be inclined with respect to the south–north direction. For a southward transport ($k_y < 0$) the axis must assume a direction from the south-west to the north-east. In case of a northward transport of energy ($k_y > 0$) the axis must be directed from the south-east to the north-west. Comparison with equation (16.40b) shows that, for pure Rossby waves, the phase velocity c_y and the group velocity $c_{g,y}$ are opposite in direction.

Let us now consider the vertical transport of energy. From

$$\boxed{c_{g,p} = \frac{\partial \omega}{\partial k_p} = \frac{2\beta k_x \left(f_0^2/\overline{\sigma}_0\right) k_p}{\left[k_x^2 + k_y^2 + \left(f_0^2/\overline{\sigma}_0\right) k_p^2\right]^2}} \tag{16.62}$$

it follows that, for a positive vertical wavenumber ($k_p > 0$), the transport of energy is upward. If the vertical wavenumber is negative ($k_p < 0$), then the transport is downward. For $\overline{u} = 0$ we recognize immediately that c_p and $c_{g,p}$ have opposite signs (see equation (16.40c)).

16.5 The influence of friction on the stationary Rossby wave

There is much observational evidence that Rossby waves may be amplified by external forcing due to large-scale thermal inhomogeneities resulting, for example, from the distribution of continents and oceans. Inspection of weather maps shows that their dimensions are of the same order as the lengths of stationary Rossby waves. Not only does forcing play an important role in the dynamics of the Rossby waves, but also dissipating factors are of importance. As in any physical problem, dissipation is usually very difficult to treat mathematically if realistic situations are assumed. In order to assess the influence of friction between the atmosphere and the earth's surface, we will accept the simple *Guldberg and Mohn scheme*, which assumes that the frictional force is proportional to the wind velocity. Therefore, we must include the frictional effect in the vorticity equation. We proceed by including on the right-hand sides of the horizontal components of the equation of motion (15.27b) the terms $-ru$ and $-rv$, where r is a frictional factor. We then derive the divergence-free vorticity equation analogously to (15.20), but now the term $r\zeta$ will be added. With Panchev (1985) we assume that we are dealing with stationary conditions; we also ignore any y-dependence of the variables. This leads to

$$u \frac{\partial \zeta}{\partial x} + \beta v + r\zeta = 0 \tag{16.63a}$$

In order to obtain an analytic solution, we linearize this equation with $\overline{u} =$ constant, $\overline{v} = 0$ and obtain

$$\overline{u} \frac{d\zeta'}{dx} + \beta v' + r\zeta' = 0 \tag{16.63b}$$

Since $\zeta' = dv'/dx$ we obtain an ordinary second-order differential equation for v':

$$\overline{u} \frac{d^2 v'}{dx^2} + \beta v' + r \frac{dv'}{dx} = 0 \tag{16.64}$$

Assuming that, at $x = 0$, the velocity $v' = 0$, and $(dv'/dx)_0 = \zeta_0$, which is the initial vorticity, the solution to the frictional problem is given by

$$v' = \frac{\zeta_0}{k'_{\text{stat}}} \exp\!\left(-\frac{r}{2\overline{u}}x\right) \sin(k'_{\text{stat}}x) \quad \text{with} \quad k'_{\text{stat}} = \sqrt{\frac{\beta}{\overline{u}} - \frac{r^2}{4\overline{u}^2}} \tag{16.65}$$

Comparison of $L'_{\text{stat}} = 2\pi/k'_{\text{stat}}$ with (16.10) shows that the stationary wavenumber has been modified due to the existence of surface friction. Accepting Panchev's value $r = 10^{-6}\ \text{s}^{-1}$, the amplitude of the wave decreases to about half its value for $\overline{u} = 10\ \text{m s}^{-1}$ if $x = 2L'_{\text{stat}}$ for a midlatitude situation. We may conclude that the frictional effect surely counteracts quite efficiently any forcing that would produce amplification of the Rossby wave.

16.6 Barotropic equatorial waves

In this section we will briefly consider Rossby waves forming at the equator of the earth. The starting points of the analysis are the shallow-water equations (15.23b) and (15.27b), which will be linearized and simplified. Assuming that the basic current is zero means that we must also set the inclination of the mean height of the fluid equal to zero. For a flat ground the basic equations are

$$\frac{\partial u'}{\partial t} - \beta y v' = -\frac{\partial \phi'}{\partial x}, \qquad \frac{\partial v'}{\partial t} + \beta y u' = -\frac{\partial \phi'}{\partial y}, \qquad \frac{\partial \phi'}{\partial t} + \overline{\phi}\left(\frac{\partial u'}{\partial x} + \frac{\partial v'}{\partial y}\right) = 0 \tag{16.66}$$

The Coriolis parameter is approximated according to (16.1) with $f_0 = 0$. We wish to obtain a solution of the form

$$\begin{pmatrix} u' \\ v' \\ \phi' \end{pmatrix} = \begin{pmatrix} \widehat{u}(y) \\ \widehat{v}(y) \\ \widehat{\phi}(y) \end{pmatrix} \exp[i(k_x x - \omega t)] \tag{16.67}$$

Substitution of this equation into (16.66) results in the coupled system

$$\begin{gathered} -i\omega\widehat{u} - \beta y\widehat{v} = -ik_x\widehat{\phi}, \qquad -i\omega\widehat{v} + \beta y\widehat{u} = -\frac{d\widehat{\phi}}{dy}, \\ -i\omega\widehat{\phi} + \overline{\phi}\left(ik_x\widehat{u} + \frac{d\widehat{v}}{dy}\right) = 0 \end{gathered} \tag{16.68}$$

consisting of one purely algebraic equation and two differential equations. At this point it is of advantage to introduce the dimensionless form of (16.68). The only dimensional parameters in this set of equations are β and $\overline{\phi}$, which will be combined to introduce scales of time and length:

$$\tau = \frac{1}{\beta^{1/2}\overline{\phi}^{1/4}}, \qquad l = \frac{\overline{\phi}^{1/4}}{\beta^{1/2}} \tag{16.69}$$

These scales are then used to define the dimensionless variables

$$\begin{aligned}
&\widetilde{u} = \widehat{u}\frac{\tau}{l} = \frac{\widehat{u}}{\overline{\phi}^{1/2}}, \qquad \widetilde{v} = \widehat{v}\frac{\tau}{l} = \frac{\widehat{v}}{\overline{\phi}^{1/2}}, \qquad \widetilde{y} = \frac{y}{l} = y\frac{\beta^{1/2}}{\overline{\phi}^{1/4}} \\
&\widetilde{k}_x = k_x l = k_x\frac{\overline{\phi}^{1/4}}{\beta^{1/2}}, \qquad \widetilde{\phi} = \widehat{\phi}\frac{\tau^2}{l^2} = \frac{\widehat{\phi}}{\overline{\phi}}, \qquad \widetilde{\omega} = \omega\tau = \frac{\omega}{\beta^{1/2}\overline{\phi}^{1/4}} \\
&\frac{d\widetilde{\phi}}{d\widetilde{y}} = \frac{1}{\overline{\phi}^{3/4}\beta^{1/2}}\frac{d\widehat{\phi}}{dy}, \qquad \frac{d\widetilde{v}}{d\widetilde{y}} = \frac{d\left(\widehat{v}/\overline{\phi}^{1/2}\right)}{d\left(y\beta^{1/2}/\overline{\phi}^{1/4}\right)} = \frac{1}{\beta^{1/2}\overline{\phi}^{1/4}}\frac{d\widehat{v}}{dy}
\end{aligned} \tag{16.70}$$

By introducing these variables into (16.68), we find the equivalent dimensionless set

$$\begin{aligned}
&-i\widetilde{\omega}\widetilde{u} - \widetilde{y}\widetilde{v} + i\widetilde{k}_x\widetilde{\phi} = 0, \qquad -i\widetilde{\omega}\widetilde{v} + \widetilde{y}\widetilde{u} + \frac{d\widetilde{\phi}}{d\widetilde{y}} = 0, \\
&-i\widetilde{\omega}\widetilde{\phi} + i\widetilde{k}_x\widetilde{u} + \frac{d\widetilde{v}}{d\widetilde{y}} = 0
\end{aligned} \tag{16.71}$$

Eliminating $\widetilde{u}$ and $\widetilde{\phi}$ by obvious steps yields

$$\boxed{\frac{d^2\widetilde{v}}{d\widetilde{y}^2} + \left(\widetilde{\omega}^2 - \widetilde{k}_x^2 - \frac{\widetilde{k}_x}{\widetilde{\omega}} - \widetilde{y}^2\right)\widetilde{v} = 0} \tag{16.72}$$

This is *Schrödinger's equation* for the simple harmonic oscillator. Application of the natural conditions $\widetilde{v}(\pm\infty) = 0$ gives the general solution

$$\widetilde{v}(\widetilde{y}) = A_n H_n(\widetilde{y}) \exp\left(-\frac{\widetilde{y}^2}{2}\right) \tag{16.73}$$

to (16.72) in terms of the Hermite polynomials $H_n(\widetilde{y})$ upon assuming the validity of the dispersion relation

$$\widetilde{\omega}^2 - \widetilde{k}_x^2 - \widetilde{k}_x/\widetilde{\omega} = 2n + 1, \qquad n = 0, 1, \ldots \tag{16.74}$$

The lowest-order Hermite polynomials are $H_0 = 1$, $H_1 = 2\widetilde{y}$, and $H_2 = 4\widetilde{y}^2 - 2$.

It will be observed that (16.74) is a cubic equation in the reduced form. Conditions under which this equation provides three real and unequal roots are known. In the present case the three real roots correspond to different types of waves. From the type of the basic equations (16.66) we should expect two inertial gravity waves and one Rossby wave. The solutions can be separated by considering two approximate cases.

Let us consider the conditions stated for the first approximate case

$$\widetilde{\omega}^2 - \widetilde{k}_x^2 \gg \widetilde{k}_x/\widetilde{\omega} \tag{16.75}$$

so that the dimensionless frequency equation (16.74) reduces to

$$\widetilde{\omega}_n = \pm\sqrt{2n + 1 + \widetilde{k}_x^2} \tag{16.76}$$

The lowest possible frequency is found by setting $n = 0$. On returning to the dimensional form by means of (16.70) we obtain

$$\omega_n = \pm\sqrt{(2n+1)\beta\overline{\phi}^{1/2} + \overline{\phi}k_x^2} \tag{16.77}$$

The phase speeds of the two waves are given by

$$\boxed{c_n = \frac{\omega_n}{k_x} = \pm\sqrt{(2n+1)\frac{\beta\overline{\phi}}{k_x^2} + \overline{\phi}}} \tag{16.78}$$

Recalling that at the equator the Coriolis parameter is very small, for $\overline{u} = 0$ this equation is very similar to (15.55). Clearly, (16.78) describes a pair of eastward- and westward-propagating inertial gravity waves.

The conditions for the second approximate case are

$$\widetilde{\omega}^2 \ll \widetilde{k}_x/\widetilde{\omega} + \widetilde{k}_x^2 \tag{16.79}$$

so that (16.74) reduces to the frequency equation

$$\widetilde{\omega}_n = -\frac{\widetilde{k}_x}{\widetilde{k}_x^2 + 2n + 1} \tag{16.80}$$

The dimensional form is given by

$$\omega_n = -\frac{\beta k_x}{k_x^2 + (2n+1)k_{\text{scale}}^2} \quad \text{with} \quad k_{\text{scale}} = \frac{1}{l} = \frac{\beta^{1/2}}{\overline{\phi}^{1/4}} \tag{16.81}$$

Here the wavenumber scale k_{scale} has been introduced. Assuming that $\overline{\phi} = 10^5$ m^2 s^{-2}, the corresponding wavelength scale l is about 4000 km. By dividing ω_n by k_x we obtain the phase speed

$$\boxed{c_n = -\frac{\beta}{k_x^2 + (2n+1)k_{\text{scale}}^2}} \tag{16.82}$$

We observe that, unlike the inertial gravity waves, this type of wave is unidirectional and propagating in the westward direction only, as is implied by the negative sign. Therefore, we are dealing with a Rossby-type wave since all Rossby waves move from east to west if the basic current $\overline{u} = 0$, as follows from (16.9). This section originates from the work of Matsuno (1966).

16.7 The principle of geostrophic adjustment

In the previous chapter we considered the properties of the homogeneous atmosphere by discussing the solution of the system (15.49). We found that the frequency equation has three real roots. One of these represents the meteorologically interesting wave moving with a speed c_1, see (15.53), which is of the order of the wind velocity. This is not an actual Rossby wave since the Coriolis parameter f was not permitted to vary with latitude. Nevertheless, loosely speaking, we sometimes call the slow-moving waves displaced with c_1 also Rossby waves. Had we permitted f to vary with the coordinate y, $\overline{u}$ would have had to be replaced by $\overline{u} - \beta/k_x$. The remaining two roots resulted in the phase velocities $c_{2,3}$ of the external gravity waves, see (15.55), moving at approximately the Newtonian speed of sound.

The solution of the predictive meteorological equations requires a complete set of initial data. These observational data are usually measured independently with a certain observational error. In contrast, the meteorological variables are connected by a system of prognostic and diagnostic equations that must be satisfied by the meteorological variables at all times, including the initial time. Owing to observational errors, the initital data will introduce a perturbation, which cannot be completely avoided. Observational evidence shows that the large-scale atmospheric motion is quasi-geostrophic and quasi-static, implying a balance among the Coriolis force, the pressure-gradient force, and the gravitational force. If this balance is disturbed in some region by frontogenesis or by some other phenomenon, fast wave motion is generated and perturbation energy is exported to other regions of the atmosphere. After a period of adjustment the quasi-balance is restored. In fact, this process of adjustment operates continually to maintain a state of approximate geostrophic balance.

Let us reconsider the linearized shallow-water equations assuming that we have a resting basic state. From (15.23b) with $\phi_s = 0$ and from (15.27b) we obtain directly

$$\frac{\partial u'}{\partial t} - fv' = -\frac{\partial \phi'}{\partial x}, \qquad \frac{\partial v'}{\partial t} + fu' = -\frac{\partial \phi'}{\partial y}, \qquad \frac{\partial \phi'}{\partial t} + \overline{\phi}\left(\frac{\partial u'}{\partial x} + \frac{\partial v'}{\partial y}\right) = 0 \tag{16.83}$$

We could have found this equation also from (15.66) by replacing β by f. In order to simplify the notation, the primes on u, v, and ϕ will be omitted henceforth. Note well that the gradient of the mean geopotential must be zero for a resting mean state to exist, that is $\overline{\phi}$ = constant. In order to simplify the system we set $\partial/\partial x = 0$, thus obtaining a problem in one-dimensional space:

$$\begin{aligned} &\text{(a)} \quad \frac{\partial u}{\partial t} - fv = 0 \\ &\text{(b)} \quad \frac{\partial v}{\partial t} + fu = -\frac{\partial \phi}{\partial y} \\ &\text{(c)} \quad \frac{\partial \phi}{\partial t} + \overline{\phi}\frac{\partial v}{\partial y} = 0 \end{aligned} \tag{16.84}$$

It turns out that the result we are going to obtain is quite general and does not depend on the dimension of the adjustment process. For the treatment of the higher-dimensional adjustment processes see Panchev (1985).

The initial conditions will be specified by

$$\begin{aligned} & v(y, t = 0) = \overline{v} = 0, \qquad \nabla\phi(y, t = 0) = 0 \\ |y| \le a: \quad & u(y, t = 0) = \overline{u}, \qquad |y| > a: \quad \overline{u} = 0 \end{aligned} \tag{16.85}$$

We need to point out that the initial fields of the velocity and the geopotential are not balanced since, on the line segment $|y| \le a$, the basic state velocity $\overline{u} \ne 0$ while the gradient of the geopotential ϕ is assumed to vanish. This inconsistency stimulates a disturbance that will propagate away from the region of imbalance.

We proceed by eliminating u and ϕ in (16.84) and obtain the partial differential equation

$$\frac{\partial^2 v}{\partial t^2} + f^2 v - \overline{\phi}\frac{\partial^2 v}{\partial y^2} = 0 \tag{16.86}$$

If we seek a solution of the form

$$v = V \exp[i(k_y y - \omega t)], \qquad V = \text{constant} \tag{16.87}$$

we find the frequency equation

$$\boxed{\omega = \pm\sqrt{f^2 + \overline{\phi}k_y^2}} \tag{16.88}$$

This dispersion relation shows that inertial gravity waves, which will eventually propagate out of the imbalance region, are generated. After a period of adjustment the geostrophic balance will be restored.

In order to study the adjustment process itself, it is not sufficient to consider the frequency equation; we must actually solve an initial-value problem. We will now solve equation (16.86) subject to the initial conditions

$$t = 0, \quad |y| \le a: \quad v(y, 0) = 0, \qquad \left(\frac{\partial v}{\partial t}\right)_{t=0} = -\overline{u}f \tag{16.89}$$

noting that the second condition is formulated with the help of (16.84) and (16.85). The method of solution we choose is based on operational calculus. By introducing the pair of Fourier transforms

$$\begin{aligned} &\text{(a)} \quad \widetilde{v}(k_y, t) = \int_{-\infty}^{\infty} v(y, t) \exp(-ik_y y)\, dy \\ &\text{(b)} \quad v(y, t) = \frac{1}{2\pi} \int_{-\infty}^{\infty} \widetilde{v}(k_y, t) \exp(ik_y y)\, dk_y \end{aligned} \tag{16.90}$$

we transform the partial differential equation (16.86) into an ordinary second-order differential equation. We could also apply the Laplace-transform method to find the solution to this equation. Multiplying (16.86) by $\exp(-ik_y y)$ and integrating over y from $-\infty$ to $+\infty$ as required by (16.90a), we obtain

$$\int_{-\infty}^{\infty} \frac{\partial^2 v}{\partial y^2} \exp(-ik_y y)\, dy = \frac{f^2}{\overline{\phi}} \int_{-\infty}^{\infty} v \exp(-ik_y y)\, dy + \frac{1}{\overline{\phi}} \frac{\partial^2}{\partial t^2} \int_{-\infty}^{\infty} v \exp(-ik_y y)\, dy \tag{16.91}$$

We have extracted the partial derivative with respect to time from under the integral sign since the integration is over y and the limits of the integral are independent of t. In the appendix to this chapter it is briefly shown how to take the Fourier transform of a derivative and of a constant. With reference to the appendix, see equation (16.117), we immediately find the differential equation

$$\frac{1}{\overline{\phi}} \frac{d^2 \widetilde{v}}{dt^2} + \frac{f^2}{\overline{\phi}} \widetilde{v} = (ik_y)^2 \widetilde{v} \tag{16.92}$$

in the transformed plane, which is now an ordinary differential equation. Using (16.88), we may write equation (16.92) more succinctly as

$$\frac{d^2 \widetilde{v}}{dt^2} + \omega^2 \widetilde{v} = 0 \tag{16.93}$$

The solution to this equation can be written down immediately:

$$\widetilde{v}(k_y, t) = A(k_y) \cos(\omega t) + B(k_y) \sin(\omega t) \tag{16.94}$$

Since the differentiation variable is the time t, the integration constants, in general, should depend on k_y.

From the boundary condition (16.89) it follows that

$$\widetilde{v}(k_y, 0) = 0 \implies A(k_y) = 0 \tag{16.95a}$$

From (16.89) we obtain

$$\frac{\partial}{\partial t}[v(y, 0)] = -\overline{u} f = \text{constant} \tag{16.95b}$$

so that from (16.120) derived in the appendix we find the integration constant $B(k_y)$:

$$\frac{\partial}{\partial t}\int_{-\infty}^{\infty} v(y,0)\exp(-ik_y y)\,dy = \frac{\partial}{\partial t}[\widetilde{v}(k_y,0)] = -\frac{2\overline{u}f}{k_y}\sin(k_y a)$$
$$\Longrightarrow\ B(k_y) = -\frac{2\overline{u}f}{k_y\omega}\sin(k_y a) \tag{16.96}$$

Hence, $\widetilde{v}(k_y,t)$ is known. The Fourier integral (16.90b) then provides the solution to the differential equation (16.93), which is given by

$$v(y,t) = -\frac{\overline{u}f}{\pi}\int_{-\infty}^{\infty}\frac{\sin(k_y a)\sin(\omega t)}{k_y\omega}\exp(ik_y y)\,dk_y \tag{16.97}$$

By splitting the integral into two parts ranging from $-\infty$ to 0 and from 0 to $+\infty$ we get the final form of the solution:

$$\boxed{v(y,t) = -\frac{2\overline{u}f}{\pi}\int_{0}^{\infty}\frac{\sin(k_y a)\sin(\omega t)}{k_y\omega}\cos(k_y y)\,dk_y} \tag{16.98}$$

The imagninary part of (16.97) vanishes, as can readily be verified. If desired, $u(y,t)$ and $\phi(y,t)$ can easily be found from (16.84a) and (16.48c).

We now direct our attention to the behavior of $v(y,t)$ as $t\to\infty$. For convenience we consider the central point $y=0$, and then change the integration variable from k_y to ω using the positive root (16.88). After a few obvious steps we find

$$v(0,t) = -\frac{2\overline{u}f}{\overline{\phi}}\int_{f}^{\infty}\frac{\omega\sin\left(a\sqrt{\gamma}\right)}{\gamma}\frac{\sin(\omega t)}{\pi\omega}\,d\omega = -\frac{2\overline{u}f}{\overline{\phi}}\int_{f}^{\infty} g(\omega)\frac{\sin(\omega t)}{\pi\omega}\,d\omega \tag{16.99}$$

where the abbreviations $\gamma = (\omega^2 - f^2)/\overline{\phi}$ and $g(\omega) = \omega\sin\left(a\sqrt{\gamma}\right)/\gamma$ have been introduced. The integrand is quite suitable for use of the Dirac delta function in the form

$$\delta(\omega) = \lim_{t\to\infty}\frac{\sin(\omega t)}{\pi\omega} \tag{16.100}$$

In the limit $t\to\infty$ we may therefore write for (16.99)

$$v(0,t\to\infty) = -\frac{2\overline{u}f}{\overline{\phi}}\int_{f}^{\infty} g(\omega)\delta(\omega-0)\,d\omega = 0 \tag{16.101}$$

where the integral has been evaluated by means of (M6.80). We obtain the identical result for the negative sign of the root in (16.88).

Let us now consider the tendencies of u, ϕ, and v at the central point $y = 0$ as the time t approaches infinity. The result is

$$\text{(a)}\ \left.\frac{\partial u}{\partial t}\right|_{t\to\infty} = 0, \qquad \text{(b)}\ \left.\frac{\partial \phi}{\partial t}\right|_{t\to\infty} = 0, \qquad \text{(c)}\ \left.\frac{\partial v}{\partial t}\right|_{t\to\infty} = 0 \tag{16.102}$$

For the stationary values $u_\infty = u(y, t \to \infty)$, $v_\infty = v(y, t \to \infty)$, and $\phi_\infty = \phi(y, t \to \infty)$, characterized by vanishing partial derivatives with respect to time, we find

$$\boxed{u_\infty = -\frac{1}{f}\frac{\partial \phi_\infty}{\partial y}, \qquad v_\infty = 0, \qquad \phi_\infty = \phi_\infty(y)} \tag{16.103}$$

Owing to the assumption that only the space variable y is considered, ϕ_∞ must be independent of x. If this were not the case v_∞ would differ from zero. The important point to observe is that, in case of stationarity, we have geostrophic balance.

In order to evaluate u_∞, we must derive the profile function for ϕ_∞. We proceed by observing that the system (16.84) has an invariant. This follows simply by combining parts (16.84a) and (16.84c) and by treating the Coriolis parameter as a constant. The result is

$$\frac{\partial}{\partial t}\left(\frac{\partial u}{\partial y} + f\frac{\phi}{\overline{\phi}}\right) = 0 \tag{16.104}$$

from which it follows that

$$\frac{\partial u}{\partial y} + f\frac{\phi}{\overline{\phi}} = \text{constant} \tag{16.105}$$

The similarity to the potential vorticity defined by equation (10.149) is apparent. Application of the conservation theorem (16.105) for the cases $t \to \infty$ and $t = 0$ gives

$$\frac{du_\infty}{dy} + f\frac{\phi_\infty}{\overline{\phi}} = \frac{d\overline{u}}{dy} + f\frac{\phi(y, 0)}{\overline{\phi}} \tag{16.106}$$

With Rossby we introduce the characteristic length $L_{\rm R}$,

$$L_{\rm R} = \overline{\phi}^{1/2}/f \tag{16.107}$$

now known as the *Rossby deformation radius*, and we obtain the second-order ordinary differential equation

$$\frac{d^2\phi_\infty}{dy^2} - \frac{\phi_\infty}{L_{\rm R}^2} = -f\frac{d\overline{u}}{dy} - \frac{\phi(y, 0)}{L_{\rm R}^2} \tag{16.108}$$

This differential equation can be solved by specifying proper conditions involving $\overline{u}$ and the geopotential ϕ. On applying the conditions

$$\begin{aligned} \phi(y,0) &= 0 \qquad |y| \le \infty \\ \phi_\infty(y) &= 0 \qquad |y| \to \infty \\ \overline{u}(y) &= \begin{cases} 0 & |y| \ge L \\ U\big[1+\cos(\pi y/L)\big] & |y| < L \end{cases} \end{aligned} \tag{16.109}$$

where U and L are fixed values of velocity and length, equation (16.108) transforms to give

$$\frac{d^2\phi_\infty}{dy^2} - \frac{\phi_\infty}{L_{\rm R}^2} = \begin{cases} 0 & |y| \ge L \\ (\pi U f/L)\sin(\pi y/L) & |y| < L \end{cases} \tag{16.110}$$

The eigenvalues (characteristic values) of the homogeneous part of (16.110) are given by $\pm L_{\rm R}$, so the general solution of the homogeneous part of this differential equation can be written down immediately. The particular solution $\widehat{\phi}_\infty$ may be found using the method of undetermined coefficients by choosing a solution of the type of the right-hand side of (16.110)

$$\widehat{\phi}_\infty = A\sin\left(\frac{\pi y}{L}\right) \Longrightarrow A = -\frac{\pi L U f}{\pi^2 + L^2/L_{\rm R}^2} \tag{16.111}$$

wherein the constant A has been obtained by substituting $\widehat{\phi}_\infty$ into (16.110). For the various segments on the y-axis we obtain the complete solution as

$$\boxed{\phi_\infty(y) = \begin{cases} B\exp(-y/L_{\rm R}) & y \ge L \\ C\exp(y/L_{\rm R}) & y \le -L \\ D\exp(y/L_{\rm R}) + E\exp(-y/L_{\rm R}) + \widehat{\phi}_\infty & |y| < L \end{cases}} \tag{16.112}$$

The various integration constants can be found from the requirement that ϕ_∞ and $\partial\phi_\infty/\partial y$ are continuous at $y = \pm L$. They are given by

$$\begin{aligned} B &= -C = \pi A\frac{L_{\rm R}}{L}\sinh\left(\frac{L}{L_{\rm R}}\right) \\ D &= -E = \frac{\pi A}{2}\frac{L_{\rm R}}{L}\exp\left(-\frac{L}{L_{\rm R}}\right) \end{aligned} \tag{16.113}$$

Inspection of the integration constants shows that the ratio $L/L_{\rm R}$ plays a dominant role. Therefore, we are going to consider two asymptotic cases. For the stationary velocity field we obtain

$$\boxed{u_\infty = -\frac{1}{f}\frac{\partial\phi_\infty}{\partial y} = \begin{cases} 0 & L \gg L_{\rm R} \\ \overline{u}(y) & L \ll L_{\rm R} \end{cases}} \tag{16.114}$$

Hence, for $t \rightarrow \infty$ we obtain for the limited line segment $L \ll L_{\rm R}$ that the initial and final velocity fields nearly coincide. For $L \gg L_{\rm R}$ the final velocity field is mainly dominated by the mass field which is represented by the stationary geopotential. The verification of the latter equation is left as an exercise.

Several sections of this chapter quite closely follow Panchev (1985) and also Pichler (1997) in slightly modified form.

16.8 Appendix

In this appendix we briefly derive the Fourier transform of a derivative and of the constant A. The Fourier transform $\widetilde{v}_1(y,t)$ of the partial derivative $\partial v(y,t)/\partial y$ is given by

$$
\begin{aligned}
\widetilde{v}_1(k_y,t) &= \int_{-\infty}^{\infty} \frac{\partial v(y,t)}{\partial y} \exp(-ik_y y)\,dy \\
&= v(y,t)\exp(-iky)\Big|_{-\infty}^{\infty} + ik_y \int_{-\infty}^{\infty} v(y,t)\exp(-ik_y y)\,dy
\end{aligned}
\tag{16.115}
$$

Assuming that $v(y,t)$ vanishes at $y = \pm\infty$,

$$
\lim_{y\rightarrow\pm\infty} v(y,t) = 0 \tag{16.116}
$$

the transform of the partial derivative of $v(y,t)$ is given by the second term on the right-hand side of (16.116). Generalizing, we obtain for the nth derivative of the function $v(y,t)$

$$
\widetilde{v}_n(k_y,t) = (ik_y)^n \widetilde{v}(k_y,t) \tag{16.117}
$$

provided that all the integrated parts vanish at $y = \pm\infty$.

If the function $v(y,t)$ is specified as

$$
v(y) = \begin{cases} A & |y| \leq a \\ 0 & |y| > a \end{cases} \tag{16.118}
$$

the Fourier transform is given by

$$
\begin{aligned}
\widetilde{v}(k_y) &= A\int_{-a}^{a} \exp(-ik_y y)\,dy = \frac{A[\exp(ik_y a) - \exp(-ik_y a)]}{ik_y} \\
&= \frac{2A\sin(k_y a)}{k_y}, \qquad k_y \neq 0
\end{aligned}
\tag{16.119}
$$

16.9 Problems

16.1: Assume that the stream function

$$\psi(x, y, t) = -u_0 y + A \sin(ay + \varepsilon) + B \sin[k_x(x - ct - \mu y)]$$

satisfies the nonlinear divergence-free vorticity equation. Which conditions must be satisified? The basic current is expressed by the first two terms on the right-hand side of the above equation. Find an expression for the velocity of the zonal basic current.

16.2: Show that the static stability can be written as

$$\sigma_0 = \frac{\partial^2 \phi}{\partial p^2} + \frac{1}{\kappa p} \frac{\partial \phi}{\partial p} \quad \text{with} \quad \kappa = \frac{c_p}{c_v}$$

16.3: Derive equation (16.55).

16.4: Prove that equation (16.71) is equivalent to equation (16.72).

16.5: Verify equation (16.73) for $n = 2$.

16.6: Verify equation (16.102). Show that (16.102b) is independent of Y. To verify (16.102c), make a proper transformation.

16.7: Solve the differential equation (16.110) subject to the conditions (16.109) to verify the solution (16.112).

16.8: Verify equation (16.114) for the section $|y| < L$ defined by equation (16.112).

17

Inertial and dynamic stability

In Sections 17.1–17.3 we consider inertial frictionless horizontal motion of an air parcel in a basic zonal geostrophic flow field. Dynamic-stability criteria for atmospheric motions are derived in Sections 17.4–17.7. The criteria for inertial and dynamic stability will be derived by means of the so-called *air-parcel-dynamic method*. This method considers an isolated air parcel that is subjected to a virtual displacement that leaves the state variables of the basic field undisturbed. The air pressure acting on the parcel is assumed to be identical with the air pressure of the unperturbed surroundings.

17.1 Inertial motion in a horizontally homogeneous pressure field

The simplest case is frictionless horizontal inertial motion in a homogeneous pressure field. In this case the basic current is absent. The horizontal components of the equation of motion on the tangential plane, assuming that we have a constant Coriolis parameter f_0, are given by

$$\begin{aligned} \frac{du}{dt} - f_0 v &= 0 \\ f_0 u + \frac{dv}{dt} &= 0 \end{aligned} \quad \text{or} \quad \begin{pmatrix} \dfrac{d}{dt} & -f_0 \\ f_0 & \dfrac{d}{dt} \end{pmatrix} \begin{pmatrix} u \\ v \end{pmatrix} = 0 \tag{17.1}$$

We shall find the solution of this system by application of the operator method, converting the system (17.1) into a second-order differential equation:

$$\left(\frac{d^2}{dt^2} + f_0^2 \right) \begin{pmatrix} u \\ v \end{pmatrix} = 0 \tag{17.2}$$

Because of the positive sign in front of f_0^2 we have the well-known vibrational differential equation with the general solution

$$\begin{aligned} u &= A\cos(f_0 t) + B\sin(f_0 t) \\ v &= C\cos(f_0 t) + D\sin(f_0 t) \end{aligned} \tag{17.3}$$

The frequency of the vibrational motion is often called the *inertial frequency*:

$$\omega = 2\pi/\tau = f_0 \tag{17.4}$$

τ is the period of the vibration. To evaluate the four constants and to find the trajectory we choose as initial conditions

$$t = 0: \qquad x = x_0, \quad u = 0; \qquad y = y_0 = 0, \quad v = v_0 \tag{17.5}$$

Using (17.5) we find from (17.3) and the differential equations of the flow

$$A = D = 0, \qquad B = v_0, \qquad C = v_0 \tag{17.6}$$

The solution to (17.2) is then given by

$$u = v_0 \sin(f_0 t), \qquad v = v_0 \cos(f_0 t) \tag{17.7}$$

Integration over time yields

$$x - x_0 = \frac{v_0}{f_0}[1 - \cos(f_0 t)], \qquad y = \frac{v_0}{f_0}\sin(f_0 t) \tag{17.8}$$

Elimination of the time dependency by squaring (17.8) results in the trajectory of the inertial motion:

$$\left[x - \left(x_0 + \frac{v_0}{f_0}\right)\right]^2 + y^2 = \frac{v_0^2}{f_0^2} \tag{17.9}$$

which is depicted in Figure 17.1. This figure represents circular anticyclonic motion of inertial frequency f_0. This type of motion is known as the *circle of inertia*. By necessity the motion is stable since the air parcel returns to its initial position.

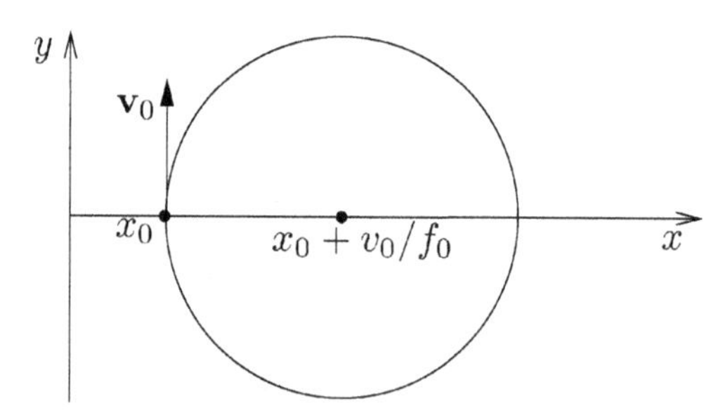

Fig. 17.1 The trajectory of an air parcel with vanishing pressure gradient, inertial motion.

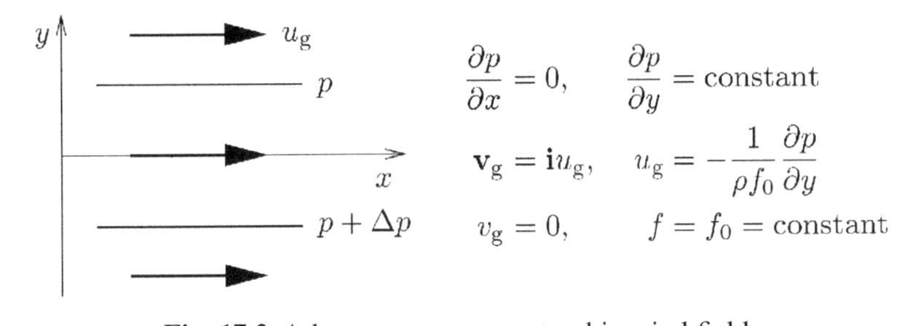

Fig. 17.2 A homogeneous geostrophic wind field.

17.2 Inertial motion in a homogeneous geostrophic wind field

In contrast to the previous section, we assume the existence of a constant geostrophic air current. An isolated air parcel embedded in it is subjected to a virtual displacement. Owing to the parcel-dynamic condition, the pressure existing within the moving air parcel is identical to that of the unperturbed surroundings. The situation is depicted in Figure 17.2.

The equation of motion of the isolated air parcel, in constrast to (17.1), is now given by

$$\frac{d}{dt}(u_g - u) + f_0 v = 0, \qquad -f_0(u_g - u) + \frac{dv}{dt} = 0 \tag{17.10}$$

since the geostrophic wind is assumed to be constant. Application of the operator method gives

$$\left(\frac{d^2}{dt^2} + f_0^2\right)\begin{pmatrix} u_g - u \\ v \end{pmatrix} = 0 \tag{17.11}$$

Again we find a differential equation of pure vibration of inertial frequency $\omega = f_0$ whose general solution is given by

$$u_g - u = A\cos(f_0 t) + B\sin(f_0 t), \qquad v = C\cos(f_0 t) + D\sin(f_0 t) \tag{17.12}$$

We apply the initial conditions

$$t = 0: \qquad x = x_0, \quad u = u_g; \qquad y = y_0 = 0, \quad v = v_0 \tag{17.13}$$

to obtain the integration constants and the trajectory. From (17.12) and the differential equation of the flow field we find

$$A = D = 0, \qquad B = -v_0, \qquad C = v_0 \tag{17.14}$$

yielding

$$u = u_g + v_0\sin(f_0 t), \qquad v = v_0\cos(f_0 t) \tag{17.15}$$

Integration over time leads to

$$x - x_0 = u_g t + \frac{v_0}{f_0}[1 - \cos(f_0 t)], \qquad y = \frac{v_0}{f_0}\sin(f_0 t) \tag{17.16}$$

This parametric representation of the trajectory of the isolated air parcel can be transformed onto a moving (ξ, y)-coordinate system in which the ξ-axis is displaced in the positive x-direction with the geostrophic velocity u_g. The new coordinate ξ is then given by

$$\xi = x - u_g t \quad \text{with} \quad \xi(t = 0) = \xi_0 = x_0 \tag{17.17}$$

In the moving system equation (17.16) reads

$$\xi = \xi_0 + \frac{v_0}{f_0}[1 - \cos(f_0 t)], \qquad y = \frac{v_0}{f_0}\sin(f_0 t) \tag{17.18}$$

Elimination of the time dependency results in the circular trajectory

$$\left[\xi - \left(\xi_0 + \frac{v_0}{f_0}\right)\right]^2 + y^2 = \frac{v_0^2}{f_0^2} \tag{17.19}$$

in the moving coordinate frame. Backward transformation to the resting coordinate system yields a cycloid as the trajectory for the inertial motion of the isolated air parcel as a superposition of the circular motion and the uniform basic air current. Again we have a stable inertial motion since the motion repeats itself.

17.3 Inertial motion in a geostrophic shear wind field

17.3.1 Stability considerations

We now consider the inertial motion of an isolated parcel of air in a basic geostrophic flow of horizontal shear $u_g = u_g(y)$ as shown in Figure 17.3.

The horizontal equations of motion for the particle velocity (u, v) assuming the parcel-dynamic condition are given by

$$\frac{du}{dt} - f_0 v = 0, \qquad \frac{dv}{dt} + f_0(u - u_g) = 0 \tag{17.20}$$

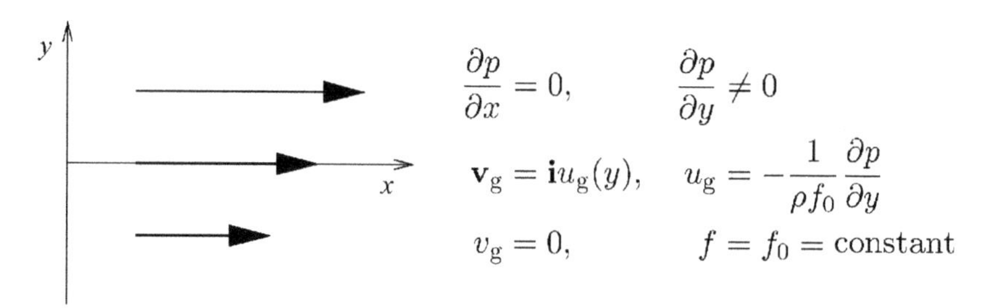

Fig. 17.3 A geostrophic shear wind field.

We may introduce the invariant quantity I into the u-component equation by means of

$$\frac{du}{dt} - f_0 v = \frac{du}{dt} - f_0 \frac{dy}{dt} = \frac{d}{dt}(u - f_0 y) = \frac{dI}{dt} = 0 \quad \text{with} \quad I = u - f_0 y \tag{17.21}$$

showing that, for the perturbed motion of the air parcel, the quantity I is individually conserved. Correspondingly, for the basic current we define the quantity

$$\bar{I} = \bar{u} - f_0 y \quad \text{or} \quad I_g = u_g - f_0 y \tag{17.22}$$

The isolated air parcel is displaced latitudinally from its original position somewhere on the x-axis at $x = x_0$ and $y = y_0 = 0$ to position y. If it does not return to its original position we observe the case of *inertial instability*. We conclude from (17.21) that the displaced particle retains its invariant quantity I of the original position at $y = 0$, which is $I_g(0)$, or

$$I(y) = I(0) = I_g(0) \tag{17.23}$$

In the undisturbed surroundings at position y, however, we have

$$I_g(y) \approx I_g(0) + \frac{dI_g}{dy}(0)y \tag{17.24}$$

From the above equations it follows immediately that

$$u(y) - u_g(y) = I(y) - I_g(y) = -\frac{dI_g}{dy}(0)y \tag{17.25}$$

This expression will be introduced into the equation of motion (17.20). Before doing so we will introduce the definition of the *absolute geostrophic vorticity* η_g. With $v_g = 0$ we have

$$\eta_g = \zeta_g + f_0 = -\frac{\partial u_g}{\partial y} + f_0 \tag{17.26}$$

From (17.22) we find

$$-\frac{dI_g}{dy}(0) = -\frac{\partial u_g}{\partial y}(0) + f_0 = \eta_g(0) \tag{17.27}$$

Owing to the assumption about the geostrophic wind field the relative vorticity is a pure shear vorticity. Equation (17.25) can now be written as

$$u(y) - u_g(y) = \eta_g(0)y \tag{17.28}$$

On introducing this expression into the v-component of the equation of motion (17.20) we obtain a differential equation for the perturbed motion in the y-direction:

$$\frac{dv}{dt} + f_0(u - u_g) = \frac{d^2 y}{dt^2} + f_0 \eta_g(0) y = 0 \tag{17.29}$$

whose well-known solution properties are

$$\eta_g(0) \begin{cases} >0 & \text{vibration solution} & \text{stable inertial motion} \\ =0 & \text{transition} & \text{the indifferent case} \\ <0 & \text{exponential solution} & \text{unstable inertial motion} \end{cases} \tag{17.30}$$

To make the physical interpretation easier, this stability criterion will be rewritten with the help of the relative shear vorticity (17.27) as

$$\boxed{\frac{\partial u_g}{\partial y}(0) \begin{cases} <f_0 & \text{stable inertial motion} \\ =f_0 & \text{the indifferent case} \\ >f_0 & \text{unstable inertial motion} \end{cases}} \tag{17.31}$$

This criterion states that inertial instability occurs only if the anticyclonic wind shear of the basic geostrophic current exceeds a critical value. This effect is often observed at the southern edge of the jet stream.

17.3.2 Determination of the trajectories

Since $u_g = u_g(y)$ we have from the Taylor expansion (discontinued after the linear term)

$$u_g(y) = u_g(0) + \frac{\partial u_g}{\partial y}(0) y, \qquad \frac{du_g}{dt} = v \frac{\partial u_g}{\partial y}(0) = -v\zeta_g(0) \tag{17.32}$$

so that the u-component of the equation of motion (17.20) can be written as

$$\frac{d}{dt}(u - u_g) - (\zeta_g(0) + f_0)v = 0, \quad \text{where} \quad \zeta_g(0) + f_0 = \eta_g(0) \tag{17.33}$$

By applying the operator method to the set consisting of (17.33) and the v-component of (17.20) we obtain at once

$$\left(\frac{d^2}{dt^2} + \omega^2\right)\begin{pmatrix} u - u_g \\ v \end{pmatrix} = 0 \quad \text{with} \quad \omega^2 = f_0 \eta_g(0) \tag{17.34}$$

The general solution is given by

$$u - u_g = A\cos(\omega t) + B\sin(\omega t), \qquad v = C\cos(\omega t) + D\sin(\omega t) \tag{17.35}$$

where ω refers to the *natural frequency* of the system. The four constants and the trajectory are evaluated with the help of the following initial conditions:

$$\begin{aligned} t = 0: \qquad & x = x_0, \quad u = u_g(y = 0); \quad y = y_0 = 0, \quad v = v_0 \implies \\ & A = D = 0, \qquad B = \frac{v_0 \eta_g(0)}{\omega} = \frac{v_0 \omega}{f_0}, \qquad C = v_0 \end{aligned} \tag{17.36}$$

where (17.20) and (17.35) have been utilized. In order to obtain the trajectory we use (17.36) in (17.35) and integrate the resulting equation over time. After introducing the moving coordinate system, as before, by means of

$$\xi = x - u_g(0)t, \qquad \xi_0 = x(t = 0) = x_0 \tag{17.37}$$

we find that the equation of the trajectory is given by

$$\begin{aligned} & \left[\xi - \left(\xi_0 + \frac{v_0}{\eta_g(0)}\right)\right]^2 + y^2 \frac{f_0}{\eta_g(0)} = \frac{v_0^2}{\eta_g(0)^2} \\ \text{since} \qquad & u_g(y) \approx u_g(0) + \frac{\partial u_g}{\partial y}(0) y = u_g(0) + \frac{\partial u_g}{\partial y}(0) \frac{v_0}{\omega} \sin(\omega t) \end{aligned} \tag{17.38}$$

For the trajectory of the air parcel in the moving system we distinguish three cases:

$$\eta_g(0) \begin{cases} > 0 & \text{ellipse} & \text{stable case} \\ = f_0 > 0 & \text{circle} & \text{stable case} \\ < 0 & \text{hyperbola} & \text{unstable case} \end{cases} \tag{17.39}$$

For the cases in which the displaced air parcels eventually return to their original positions (circle, ellipse) we have the stable situation. In case of the hyperbola the air parcel continues to be displaced from the original position, so we have the unstable situation. For $\eta_g(0) = f_0$ (17.38) reduces to (17.19). The special situation $\eta_g(0) = 0$ will be discussed in Problem 2.

17.4 Derivation of the stability criteria in the geostrophic wind field

Hydrostatic stability is a special stability case of atmospheric motion characterizing the vertical displacement of an air parcel. In contrast to this, inertial stability concerns a purely horizontal displacement in case of an indifferent hydrostatic equilibrium. Both effects taken together lead to consideration of the so-called *dynamic stability*. The stability of the geostrophic wind field was first investigated by Kleinschmidt (1941a and b) and was thoroughly discussed by Van Mieghem (1951).

The fundamental assumption is the existence of a basic geostrophic shear wind field,

$$\overline{u}(y) = u_{\rm g}(y) \neq 0, \qquad \overline{v} = \overline{w} = 0 \tag{17.40}$$

On this basic field of motion with $\partial p/\partial x = 0$ we superimpose an adiabatic perturbation, which is assumed to obey the system of equations

$$\begin{aligned} &\text{(a)} \quad \frac{dI}{dt} = 0 \\ &\text{(b)} \quad \frac{dv}{dt} = -\frac{1}{\rho}\frac{\partial p}{\partial y} - f_0 u \\ &\text{(c)} \quad \frac{dw}{dt} = -\frac{1}{\rho}\frac{\partial p}{\partial z} - g \end{aligned} \tag{17.41}$$

with

$$I = u - f_0(y - y_0), \qquad \overline{I} = I_{\rm g} = u_{\rm g} - f_0(y - y_0) \tag{17.42}$$

where the overbar refers to the mean flow. It is customary to introduce the Exner function Π as a variable instead of the pressure p by means of

$$\begin{aligned} &\text{(a)} \quad \Pi = c_{p,0}\left(\frac{p}{p_0}\right)^{k_0}, \qquad k_0 = R_0/c_{p,0} \\ &\text{(b)} \quad \frac{1}{\rho} = \theta R_0 p_0^{-k_0} p^{k_0-1} \\ &\text{(c)} \quad \frac{1}{\rho}\frac{\partial p}{\partial s} = c_{p,0}\theta \frac{\partial}{\partial s}\left(\frac{p}{p_0}\right)^{k_0} = \theta\frac{\partial \Pi}{\partial s}, \qquad s = x, y, z \end{aligned} \tag{17.43}$$

Using (17.42) and (17.43c), equations (17.41b) and (17.41c) then read

$$\begin{aligned} \frac{dv}{dt} &= -f_0 I - f_0^2(y - y_0) - \theta\frac{\partial \Pi}{\partial y} \\ \frac{dw}{dt} &= -g - \theta\frac{\partial \Pi}{\partial z} \end{aligned} \tag{17.44}$$

and represent the motion in the (y, z)-plane while equation (17.41a) remains unaltered.

For ease of mathematical manipulation the two equations for the perturbed motion in the (y, z)-plane will be symmetrized in order to represent the motion by a vector. For this purpose we define two new functions by means of

$$\begin{aligned} &L = f_0(y - y_0), && \frac{\partial L}{\partial y} = f_0, && \frac{\partial L}{\partial z} = 0 \\ &H = \frac{f_0^2}{2}(y - y_0)^2 + g(z - z_0), && \frac{\partial H}{\partial y} = f_0^2(y - y_0), && \frac{\partial H}{\partial z} = g \end{aligned} \tag{17.45}$$

Using the definitions of the new functions, the equation of motion reads

$$\begin{aligned}\frac{dI}{dt} &= 0\\ \frac{dv}{dt} &= -I\,\frac{\partial L}{\partial y} - \frac{\partial H}{\partial y} - \theta\,\frac{\partial \Pi}{\partial y}\\ \frac{dw}{dt} &= -I\,\frac{\partial L}{\partial z} - \frac{\partial H}{\partial z} - \theta\,\frac{\partial \Pi}{\partial z}\end{aligned} \tag{17.46}$$

Combination of the (y, z)-components of the perturbed motion gives

$$\frac{d\mathbf{v}_{\rm v}}{dt} = -I\,\nabla_{\rm v} L - \nabla_{\rm v} H - \theta\,\nabla_{\rm v}\Pi$$
$$\text{with}\quad \mathbf{v}_{\rm v} = \mathbf{j}v + \mathbf{k}w, \qquad \nabla_{\rm v} = \mathbf{j}\,\frac{\partial}{\partial y} + \mathbf{k}\,\frac{\partial}{\partial z} \tag{17.47}$$

The suffix v denotes the vertical (y, z)-plane. Introduction of the arbitrary displacement s in the (y, z)-plane transforms (17.47) into

$$\frac{d^2 s}{dt^2} = -I\,\frac{\partial L}{\partial s} - \frac{\partial H}{\partial s} - \theta\,\frac{\partial \Pi}{\partial s} \tag{17.48}$$

Assuming that we have parcel-dynamic conditions, not only $\partial L/\partial s$ and $\partial H/\partial s$ but also $\partial \Pi/\partial s$ of the perturbed field are equivalent to the corresponding values of the basic field. Therefore, due to the equilibrium conditions of the undisturbed surroundings of the displaced particle in the (y, z)-plane, we have

$$0 = -\overline{I}\,\frac{\partial L}{\partial s} - \frac{\partial H}{\partial s} - \overline{\theta}\,\frac{\partial \Pi}{\partial s} \tag{17.49}$$

where $\overline{I}$ and $\overline{\theta}$ refer to the basic field variables. Subtraction of (17.49) from (17.48) eliminates $\partial H/\partial s$, resulting in

$$\frac{d^2 s}{dt^2} = -(I - \overline{I})\,\frac{\partial L}{\partial s} - (\theta - \overline{\theta})\,\frac{\partial \Pi}{\partial s} \tag{17.50}$$

We now assume that the perturbed motion takes place from $y_0 = 0$ with invariant I and θ. At the original position the perturbed and the basic field quantities are identical, so, due to the invariance of I and θ, we have at position s

$$I(s) = \overline{I}(0), \qquad \theta(s) = \overline{\theta}(0) \tag{17.51}$$

In contrast to this, we find for the basic field at position s

$$\overline{I}(s) \approx \overline{I}(0) + \frac{\partial \overline{I}}{\partial s}(0)s, \qquad \overline{\theta}(s) \approx \overline{\theta}(0) + \frac{\partial \overline{\theta}}{\partial s}(0)s \tag{17.52}$$

where we have discontinued the Taylor expansion after the linear term. All consequences derived from the following equations are then valid only to within this approximation. Using this information, equation (17.50) transforms into

$$\frac{d^2 s}{dt^2} - \left(\frac{\partial \bar{I}}{\partial s}(0) \frac{\partial L}{\partial s} + \frac{\partial \overline{\theta}}{\partial s}(0) \frac{\partial \Pi}{\partial s} \right) s = 0$$

$$\text{with} \quad \frac{\partial \Pi}{\partial s} = \frac{\partial \overline{\Pi}}{\partial s} = \frac{\partial}{\partial s} \left(\overline{\Pi}(0) + \frac{\partial \overline{\Pi}}{\partial s}(0) s \right) = \frac{\partial \overline{\Pi}}{\partial s}(0) = \text{constant} \tag{17.53}$$

describing the perturbed motion in the (y, z)-plane. This equation has the well-known solution properties

$$\frac{\partial \bar{I}}{\partial s}(0) \frac{\partial L}{\partial s} + \frac{\partial \overline{\theta}}{\partial s}(0) \frac{\partial \Pi}{\partial s} \begin{cases} <0 & \text{vibration solution} & \text{dynamic stability} \\ =0 & \text{transition} & \text{the indifferent case} \\ >0 & \text{exponential solution} & \text{dynamic instability} \end{cases} \tag{17.54}$$

We will briefly discuss two examples representing two important special cases.

Example 1: Static stability The displacement occurs in the vertical direction: $s = z - z_0$. Since $\partial L / \partial z = 0$, dynamic stability is obtained if

$$\frac{\partial \overline{\theta}}{\partial z}(0) \frac{\partial \Pi}{\partial z} < 0 \tag{17.55}$$

Since $\partial \Pi / \partial z < 0$ we must have $\partial \overline{\theta} / \partial z\,(0) > 0$, which is the well-known *criterion of hydrostatic stability*.

Example 2: Inertial stability The displacement occurs in the horizontal direction y along an isobaric surface, so $(\partial \Pi / \partial y)_\Pi = 0$. Now (17.54) yields dynamic stability if

$$\frac{\partial \bar{I}}{\partial y}(0) f_0 < 0 \quad \text{or} \quad \left(\frac{\partial u_{\text{g}}}{\partial y}(0) - f_0 \right) f_0 < 0 \tag{17.56}$$

Since $f_0 > 0$ we have again the inertial-stability criterion (17.31):

$$\frac{\partial u_{\text{g}}}{\partial y}(0) < f_0 \tag{17.57}$$

17.5 Sectorial stability and instability

It is somewhat easier to treat the problem of sectorial stability and instability by eliminating the Exner function in (17.54). We proceed by eliminating $\partial \Pi / \partial s$ by

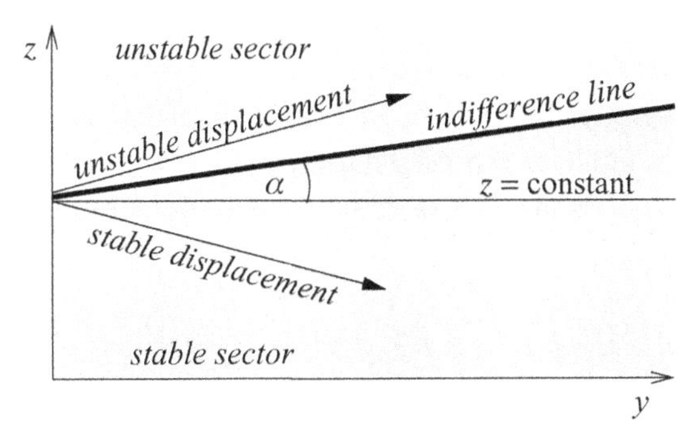

Fig. 17.4 An indifference line separating stable and unstable sectors.

obvious steps from (17.48) and (17.49) and obtain

$$\frac{1}{\theta}\frac{d^2 s}{dt^2} = -\left(\frac{I}{\theta} - \frac{\bar{I}}{\bar{\bar{\theta}}}\right)\frac{\partial L}{\partial s} - \left(\frac{1}{\theta} - \frac{1}{\bar{\bar{\theta}}}\right)\frac{\partial H}{\partial s} \tag{17.58}$$

Steps analogous to (17.51) and (17.52) with a slight additional approximation result in the differential equation of the displacement in the form

$$\frac{d^2 s}{dt^2} - \theta\left[\frac{\partial}{\partial s}\left(\frac{\bar{I}}{\bar{\bar{\theta}}}\right)(0)\,\frac{\partial L}{\partial s} + \frac{\partial}{\partial s}\left(\frac{1}{\bar{\bar{\theta}}}\right)(0)\,\frac{\partial H}{\partial s}\right]s = 0 \tag{17.59}$$

whose solution properties are given by

$$\left[\frac{\partial}{\partial s}\left(\frac{\bar{I}}{\bar{\bar{\theta}}}\right)(0)\,\frac{\partial L}{\partial s} + \frac{\partial}{\partial s}\left(\frac{1}{\bar{\bar{\theta}}}\right)(0)\,\frac{\partial H}{\partial s}\right]\begin{cases} <0 & \text{dynamic stability} \\ =0 & \text{the indifferent case} \\ >0 & \text{dynamic instability} \end{cases}$$
$$\text{with}\quad \frac{\partial}{\partial s} = \cos\alpha\,\frac{\partial}{\partial y} + \sin\alpha\,\frac{\partial}{\partial z} \tag{17.60}$$

We have included in (17.60) a well-known transformation of the partial derivatives relating the direction s to directions y and z by means of an angle α, which will be defined now. It stands to reason that the regions of stability and instability in the (y, z)-plane will be separated by a line of indifference; see Figure 17.4. We wish to find the direction of this *indifference line*.

In this general display the regions of stability and instability could have been interchanged. Application of the transformation listed in (17.60) together with (17.45) to the condition of indifference at the point $(y_0 = 0, z)$ yields

$$\frac{f_0}{g}\left[\frac{\partial}{\partial y}\left(\frac{\bar{I}}{\bar{\bar{\theta}}}\right)(0) + \tan\alpha\,\frac{\partial}{\partial z}\left(\frac{\bar{I}}{\bar{\bar{\theta}}}\right)(0)\right] + \tan\alpha\,\frac{\partial}{\partial y}\left(\frac{1}{\bar{\bar{\theta}}}\right)(0)$$
$$+ \tan^2\alpha\,\frac{\partial}{\partial z}\left(\frac{1}{\bar{\bar{\theta}}}\right)(0) = 0 \tag{17.61}$$

This relation is a quadratic equation for $\tan\alpha$, which defines at $(y_0 = 0, z)$ the inclination of the indifference line with respect to the y-axis. Therefore, there exist two directions for the indifferent behavior of the displaced air parcel.

As the next step we find an expression for the inclination of an isentropic surface with respect to the y-axis at $y_0 = 0$. This is easily done by using the differential expression

$$d\left(\frac{1}{\overline{\overline{\theta}}}\right) = \frac{\partial}{\partial z}\left(\frac{1}{\overline{\overline{\theta}}}\right) dz + \frac{\partial}{\partial y}\left(\frac{1}{\overline{\overline{\theta}}}\right) dy = 0 \tag{17.62}$$

from which it follows that

$$\left(\frac{dz}{dy}\right)_{\overline{\theta}=\text{constant}} = -\frac{\dfrac{\partial}{\partial y}\left(\dfrac{1}{\overline{\overline{\theta}}}\right)}{\dfrac{\partial}{\partial z}\left(\dfrac{1}{\overline{\overline{\theta}}}\right)} = \tan\alpha_{\overline{\theta}} = \text{constant} \tag{17.63}$$

On dividing (17.61) by $[\partial(1/\overline{\theta})/\partial z](0)$ we easily find

$$\tan^2\alpha + \tan\alpha\left[-\tan\alpha_{\overline{\theta}} + \frac{f_0}{g}\frac{\dfrac{\partial}{\partial z}\left(\dfrac{\overline{I}}{\overline{\overline{\theta}}}\right)(0)}{\dfrac{\partial}{\partial z}\left(\dfrac{1}{\overline{\overline{\theta}}}\right)(0)}\right] + \frac{f_0}{g}\frac{\dfrac{\partial}{\partial y}\left(\dfrac{\overline{I}}{\overline{\overline{\theta}}}\right)(0)}{\dfrac{\partial}{\partial z}\left(\dfrac{1}{\overline{\overline{\theta}}}\right)(0)} = 0 \tag{17.64}$$

where use of equation (17.63) has been made. To finish our analysis we first return to the equilibrium condition (17.49) which will be applied to the special directions y and z, as is shown next:

$$\begin{aligned}
&\text{(a)}\quad s = y\colon\quad 0 = -\overline{I}\,\frac{\partial L}{\partial y} - \frac{\partial H}{\partial y} - \overline{\theta}\,\frac{\partial \Pi}{\partial y} = -\overline{I} f_0 - f_0^2(y - y_0) - \overline{\theta}\,\frac{\partial \Pi}{\partial y}\\
&\text{(b)}\quad s = z\colon\quad 0 = -\overline{I}\,\frac{\partial L}{\partial z} - \frac{\partial H}{\partial z} - \overline{\theta}\,\frac{\partial \Pi}{\partial z} = -g - \overline{\theta}\,\frac{\partial \Pi}{\partial z}
\end{aligned} \tag{17.65}$$

On dividing both equations by $\overline{\theta}$ and differentiating (17.65a) with respect to z and (17.65b) with respect to y we obtain at $y_0 = 0$ the expressions

$$\begin{aligned}
&\text{(a)}\quad 0 = -f_0\,\frac{\partial}{\partial z}\left(\frac{\overline{I}}{\overline{\overline{\theta}}}\right)(0) - \frac{\partial^2 \Pi}{\partial y\,\partial z}(0)\\
&\text{(b)}\quad 0 = -g\,\frac{\partial}{\partial y}\left(\frac{1}{\overline{\overline{\theta}}}\right)(0) - \frac{\partial^2 \Pi}{\partial y\,\partial z}(0)
\end{aligned} \tag{17.66}$$

We eliminate the Exner function by subtraction and find

$$\frac{\partial}{\partial z}\left(\frac{\overline{I}}{\overline{\overline{\theta}}}\right)(0) = \frac{g}{f_0}\frac{\partial}{\partial y}\left(\frac{1}{\overline{\overline{\theta}}}\right)(0) \tag{17.67}$$

By using this expression in the third term from the left in (17.64) we find with the help of (17.63)

$$\frac{f_0}{g}\frac{\dfrac{\partial}{\partial z}\left(\dfrac{\overline{I}}{\overline{\overline{\theta}}}\right)^{(0)}}{\dfrac{\partial}{\partial z}\left(\dfrac{1}{\overline{\overline{\theta}}}\right)^{(0)}} = \frac{\dfrac{\partial}{\partial y}\left(\dfrac{1}{\overline{\overline{\theta}}}\right)^{(0)}}{\dfrac{\partial}{\partial z}\left(\dfrac{1}{\overline{\overline{\theta}}}\right)^{(0)}} = -\tan\alpha_{\overline{\theta}} \tag{17.68}$$

and therefore

$$\tan^2\alpha - 2\tan\alpha\tan\alpha_{\overline{\theta}} + \frac{f_0}{g}\frac{\dfrac{\partial}{\partial y}\left(\dfrac{\overline{I}}{\overline{\overline{\theta}}}\right)^{(0)}}{\dfrac{\partial}{\partial z}\left(\dfrac{1}{\overline{\overline{\theta}}}\right)^{(0)}} = 0 \tag{17.69}$$

The solution of this quadratic equation is given by

$$\tan\alpha_{1,2} = \tan\alpha_{\overline{\theta}} \pm \sqrt{\tan^2\alpha_{\overline{\theta}} - \frac{f_0}{g}\frac{\dfrac{\partial}{\partial y}\left(\dfrac{\overline{I}}{\overline{\overline{\theta}}}\right)^{(0)}}{\dfrac{\partial}{\partial z}\left(\dfrac{1}{\overline{\overline{\theta}}}\right)^{(0)}}} \tag{17.70}$$

The existence of two indifference lines in the (y, z)-plane requires that the discriminant be greater than zero. If this is the case then the two indifference lines are arranged as shown in Figure 17.5.

In order to investigate the indifference behavior, the discriminant in (17.70) will be reformulated. This is best accomplished by introducing the potential temperature as the vertical coordinate. In (M4.51) it was shown that, for an arbitrary function

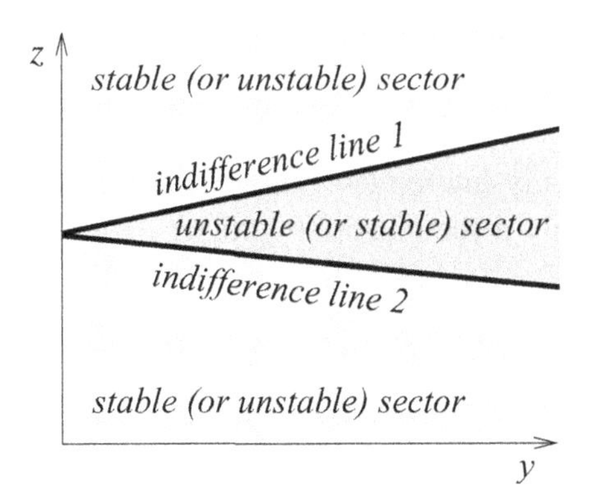

Fig. 17.5 The arrangement of two indifference lines.

ψ, the following transformation relation is valid:

$$\left(\frac{\partial \psi}{\partial y}\right)_z = \left(\frac{\partial \psi}{\partial y}\right)_\theta - \frac{\partial \psi}{\partial z}\left(\frac{\partial z}{\partial y}\right)_\theta, \qquad \left(\frac{\partial z}{\partial y}\right)_\theta = \tan\alpha_{\overline{\theta}} \tag{17.71}$$

Application of this formula at $y_0 = 0$ gives

$$\left[\frac{\partial}{\partial y}\left(\frac{\overline{I}}{\overline{\overline{\theta}}}\right)(0)\right]_z = \left[\frac{\partial}{\partial y}\left(\frac{\overline{I}}{\overline{\overline{\theta}}}\right)(0)\right]_\theta - \frac{\partial}{\partial z}\left(\frac{\overline{I}}{\overline{\overline{\theta}}}\right)(0)\tan\alpha_{\overline{\theta}} \tag{17.72}$$

Utilizing (17.72) together with (17.67) yields the second part of the discriminant as

$$\begin{aligned}\frac{f_0}{g}\frac{\left[\dfrac{\partial}{\partial y}\left(\dfrac{\overline{I}}{\overline{\overline{\theta}}}\right)(0)\right]_z}{\dfrac{\partial}{\partial z}\left(\dfrac{1}{\overline{\overline{\theta}}}\right)(0)} &= \frac{f_0}{g}\frac{\left[\dfrac{\partial}{\partial y}\left(\dfrac{\overline{I}}{\overline{\overline{\theta}}}\right)(0)\right]_\theta}{\dfrac{\partial}{\partial z}\left(\dfrac{1}{\overline{\overline{\theta}}}\right)(0)} - \frac{\left[\dfrac{\partial}{\partial y}\left(\dfrac{1}{\overline{\overline{\theta}}}\right)(0)\right]_z}{\dfrac{\partial}{\partial z}\left(\dfrac{1}{\overline{\overline{\theta}}}\right)(0)}\tan\alpha_{\overline{\theta}} \\ &= -\frac{f_0\overline{\theta}(0)}{g}\frac{\left(\dfrac{\partial \overline{I}}{\partial y}\right)_\theta(0)}{\dfrac{\partial\overline{\theta}}{\partial z}(0)} + \tan^2\alpha_{\overline{\theta}} = \frac{f_0\overline{\theta}(0)\overline{\eta}_\theta(0)}{g\dfrac{\partial\overline{\theta}}{\partial z}(0)} + \tan^2\alpha_{\overline{\theta}}\end{aligned} \tag{17.73}$$

In this expression the vorticity $\overline{\eta}_\theta(0)$ has been introduced according to (17.42):

$$\left(\frac{\partial \overline{I}}{\partial y}\right)_\theta(0) = \left(\frac{\partial u_{\mathrm{g}}}{\partial y}\right)_\theta(0) - f_0 = -\overline{\eta}_\theta(0) \tag{17.74}$$

Instead of (17.70) we finally obtain

$$\tan\alpha_{1,2} = \tan\alpha_{\overline{\theta}} \pm \sqrt{-\frac{f_0\overline{\theta}(0)\overline{\eta}_\theta(0)}{g\dfrac{\partial\overline{\theta}}{\partial z}(0)}} \tag{17.75}$$

which can be more easily interpreted. This equation reveals that the indifferent behavior is possible only if the potential vorticity expression of the basic field at $y_0 = 0$ is restricted by

$$\frac{\overline{\eta}_\theta(0)}{\dfrac{\partial\overline{\theta}}{\partial z}(0)} < 0 \tag{17.76}$$

This condition is the requirement for the existence of *partial* or *sectorial stability* or instability. Hence there always exist two displacement directions s_1 and s_2 defining the indifference lines separating stable and unstable regions for displacements in the (y, z)-plane. If, in contrast to (17.76),

$$\frac{\overline{\eta}_\theta(0)}{\dfrac{\partial\overline{\theta}}{\partial z}(0)} > 0 \tag{17.77}$$

then no indifference lines exist. Therefore, we have either total stability or total instability. If this is the case at the point being considered, i.e. at $y_0 = 0$ of the basic field, then, independently of the selected direction s of the virtual displacement in the (y, z)-plane, we have total dynamic stability for all directions if

$$\frac{\partial\overline{\theta}}{\partial z}(0) > 0 \quad \text{and} \quad \overline{\eta}_\theta(0) > 0 \tag{17.78}$$

and total dynamic instability for all directions if

$$\frac{\partial\overline{\theta}}{\partial z}(0) < 0 \quad \text{and} \quad \overline{\eta}_\theta(0) < 0 \tag{17.79}$$

On the other hand, the validity of (17.76) results in the formation of four sectors with alternating regions of stability and instability.

17.6 Sectorial stability for normal atmospheric conditions

In order to investigate the sectorial stability for normal atmospheric conditions we assume that we have hydrostatic stability, which is characterized by $[\partial\overline{\theta}(0)/\partial z] > 0$. According to (17.76) we then must have $\overline{\eta}_\theta(0) < 0$. We now wish to find the stable and unstable sectors. Obviously stable and unstable sectors are separated by lines of indifference. Now we consider the displacement of an air parcel along a line $\overline{\theta}$ = constant and consider the stability equation (17.60). In this particular situation the second term of (17.60) vanishes and we obtain the relation

$$\frac{1}{\overline{\theta}}\left(\frac{\partial\overline{I}}{\partial s}\right)_{\overline{\theta}}(0)\,\frac{\partial L}{\partial s}\begin{cases} <0 & \text{dynamic stability} \\ =0 & \text{the indifferent case} \\ >0 & \text{dynamic instability}\end{cases} \tag{17.80}$$

Now we expand $\partial/\partial s$ as shown in (17.60) by using (17.45) and find

$$\frac{f_0\cos^2\alpha_{\overline{\theta}}}{\overline{\theta}}\left(\frac{\partial\overline{I}}{\partial y}\right)_{\overline{\theta}}(0)\begin{cases} <0 & \text{dynamic stability} \\ =0 & \text{the indifferent case} \\ >0 & \text{dynamic instability}\end{cases} \tag{17.81}$$

Note that, according to (17.42), $\overline{I}$ is independent of the vertical coordinate z. Using (17.74) we may rewrite this formula and find the very useful expression

$$-\frac{f_0\cos^2\alpha_{\overline{\theta}}}{\overline{\theta}}\overline{\eta}_{\overline{\theta}}(0)\begin{cases} <0 & \text{dynamic stability} \\ =0 & \text{the indifferent case} \\ >0 & \text{dynamic instability}\end{cases} \tag{17.82}$$

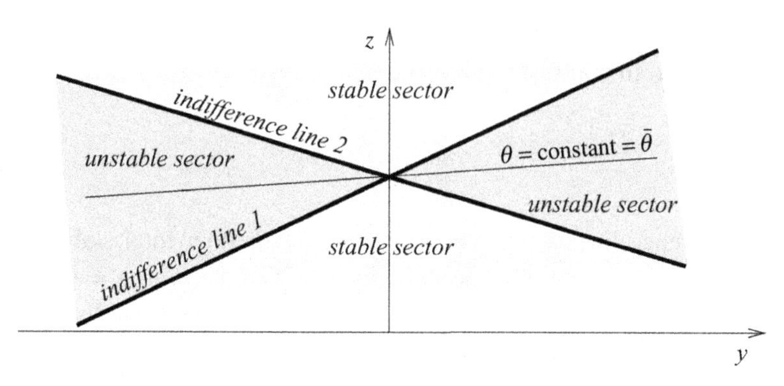

Fig. 17.6 The arrangement of stable and unstable sectors for the normal atmosphere.

Since $\overline{\eta}_\theta(0) < 0$ the lower unequality applies, so the displacement along the line $\overline{\theta} = $ constant is unstable. Thus, the line $\overline{\theta} = $ constant is located in the unstable sector as shown in Figure 17.6.

It is of some interest to determine the conditions which satisfy the requirement $\eta_\theta(0) < 0$. We have previously given an expression relating the absolute vorticities in the p and θ systems, see Section (10.4.5). This expression has the form

$$\eta_\theta = \eta_p - \frac{g}{f_0 T} \frac{(\nabla_{\mathrm{h},p} T)^2}{\gamma_\mathrm{a} - \gamma_\mathrm{g}} \tag{17.83}$$

We recognize that the horizontal temperature gradient is mainly responsible for satisfying the requirement $\overline{\eta}_\theta(0) < 0$. We will now estimate the limiting value for the occurrence of sectorial stability. Assuming the validity of

$$\overline{\eta}_p = \overline{\zeta}_p + f_0 \approx f_0 \tag{17.84}$$

we find for midlatitude situations, setting (17.83) equal to zero, that the limiting value for the occurrence of sectorial stability is

$$|\nabla_{\mathrm{h},p} T| \approx 2K/100 \text{ km} \tag{17.85}$$

Larger values of $|\nabla_{\mathrm{h},p} T|$ are common in frontal zones.

The reader interested in a more comprehensive discussion on dynamic stability is referred to the detailed discussions presented by Van Mieghem (1951) in the *Compendium of Meteorology* and in the textbook *Dynamic Meteorology and Weather Forecasting* by Godske *et al.* (1957).

17.7 Sectorial stability and instability with permanent adaptation

In Section 17.4 we derived the equation for the perturbed motion (17.50) of an air parcel. All results obtained so far followed from the invariance of I and the potential temperature θ. Ertel (1943) approached the stability problem in a different way by giving up the requirement that I is invariant. Instead, he assumed that the quantity I of the perturbed motion permanently assumed the value $\overline{I}$ of the basic field at each position. The truth probably lies between the assumptions of invariance and the permanent adaptation. Ertel's assumption simplifies equation (17.50), which now reads

$$\frac{d^2 s}{dt^2} = -(\theta - \overline{\theta})\frac{\partial \Pi}{\partial s} \tag{17.86}$$

Using (17.51) and (17.52), the differential equation for the perturbed motion is given by

$$\frac{d^2 s}{dt^2} - \frac{\partial \overline{\theta}}{\partial s}(0)\frac{\partial \Pi}{\partial s} s = 0 \tag{17.87}$$

The solution properties of this equation are given by

$$\frac{\partial \overline{\theta}}{\partial s}(0)\frac{\partial \Pi}{\partial s}\begin{cases} <0 & \text{dynamic stability} \\ =0 & \text{the indifferent case} \\ >0 & \text{dynamic instability} \end{cases} \tag{17.88}$$

We consider two examples.

(i) Vertical displacement, $s = z - z_0$. Because $\partial \Pi/\partial z < 0$ we must have $\partial \overline{\theta}/\partial z > 0$ for dynamic stability.
(ii) We consider two arbitrary directions s_1 and s_2 of displacement in the (y, z)-plane as shown in Figure 17.7. Indifference lines are isentropes and lines of constant Exner functions. Displacement s_1 results in dynamic instability whereas displacement s_2 represents dynamic stability. The figure refers to average conditions in the tropospheric west-wind region.

In case of permanent adaptation of the quantity I to average field conditions the unstable sector is located between isentropes and lines of $\Pi =$ constant. For a useful physical interpretation of Ertel's theory we are compelled to borrow a result from the general circulation theory. Let us designate rising motion by the vertical velocity $\omega' < 0$ in the p system. In case of southerly flow we expect heating $\theta' > 0$. For the production of kinetic energy we must have

$$\overline{\theta' \omega'} < 0 \tag{17.89}$$

where the primed quantities refer to the deviation of properly defined mean values and the overbar refers to the average of the correlation. Equation (17.89) states that rising motion must be accompanied by heating. This requirement is satisfied by displacement s_1 so that dynamic instability is coupled with the production of

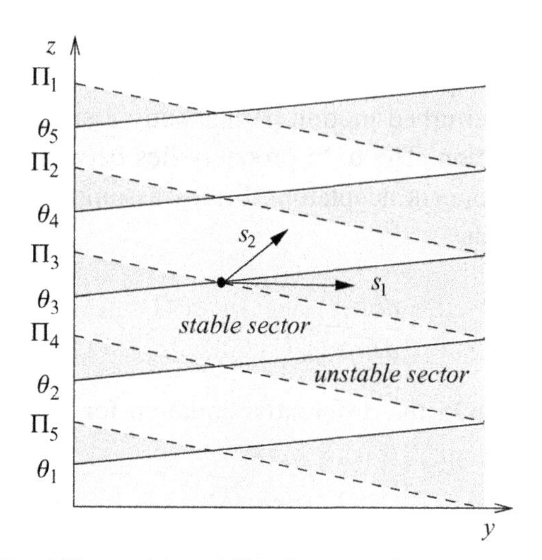

Fig. 17.7 Sectors of stability and instability in case of permanent adaptation. Unstable displacement s_1: $\partial\overline{\theta}/\partial s < 0, \partial\Pi/\partial s < 0$; stable displacement s_2: $\partial\overline{\theta}/\partial s > 0, \partial\Pi/\partial s < 0$.

kinetic energy. The situation differs in case of the stable displacement s_2. Here the advective temperature change due to southerly flow is overcompensated by a very strong rising air current ($\omega' < 0$) causing a cooling effect ($\theta' < 0$). Therefore, a stable dynamic displacement s_2 due to

$$\overline{\theta'\omega'} > 0 \tag{17.90}$$

is connected with a reduction of the kinetic energy of the flow. For further details on the theory of stability see Ertel *et al.* (1941).

17.8 Problems

17.1: Perform in detail all steps between equations (17.35) and (17.38).

17.2: Discuss the special case $\eta_g(0) = 0$, see equation (17.34), and find the trajectory. Express your answer in the moving coordinate system.

17.3: Show in detail the steps involved in going from (17.60) to (17.63).

17.4: Derive equation (17.81).

18

The equation of motion in general coordinate systems

18.1 Introduction

The numerical investigation of specific meteorological problems requires the selection of a suitable coordinate system. In many cases the best choice is quite obvious. Attempts to use the same coordinate system for entirely different geometries usually introduce additional mathematical complexities, which should be avoided. For example, it is immediately apparent that the rectangular Cartesian system is not well suited for the treatment of problems with spherical symmetry. The inspection of the metric fundamental quantities g_{ij} or g^{ij} and their derivatives helps to decide which coordinate system is best suited for the solution of a particular problem. The study of the motion over irregular terrain may require a terrain-following coordinate system. However, it is not clear from the beginning whether the motion is best described in terms of covariant or contravariant measure numbers. We will discuss this situation later.

From the thermo-hydrodynamic system of equations consisting of the dynamic equations, the continuity equation, the heat equation, and the equation of state we will direct our attention mostly toward the equation of motion using covariant and contravariant measure numbers. We will also briefly derive the continuity equation in general coordinates. In addition we will derive the equation of motion using physical measure numbers and assuming that the curvilinear coordinate lines are orthogonal.

The starting point of the analysis is the equation of motion in the absolute coordinate system. The description of the motion in a rotating and time-dependent coordinate system, in general, requires knowledge of the rotational velocity of a point in the atmosphere as well as the deformation velocity of the material surface on which the point is located. Finally, we must compute the velocity of an air parcel relative to this point. For further details review Chapter 1.

In order to obtain the general form of the equation of motion, it is very convenient to use the method of covariant differentiation. The reader who is not familiar with this method may find the necessary mathematical background in Chapter M5.

18.2 The covariant equation of motion in general coordinate systems

Before proceeding with the derivation of the equation of motion in covariant coordinates, it will be best to present some useful expressions that will be needed later. We begin the discussion by writing down the individual derivative of the rotational part of $\mathbf{v}_P$ in the absolute and the relative coordinate systems. In contrast to the invariant individual derivative which is independent of any particular coordinate system, the local derivative does depend on the coordinate system. The local time derivative $\partial \mathbf{v}_\Omega/\partial t$ in the absolute coordinate system is zero, as was shown in (1.45). Therefore, it follows from

$$\frac{d\mathbf{v}_\Omega}{dt} = \left(\frac{\partial \mathbf{v}_\Omega}{\partial t}\right)_{x^i} + \mathbf{v}_\mathrm{A}\cdot\nabla\mathbf{v}_\Omega = (\mathbf{v}_P + \mathbf{v})\cdot\nabla\mathbf{v}_\Omega = \left(\frac{\partial \mathbf{v}_\Omega}{\partial t}\right)_{q^i} + \mathbf{v}\cdot\nabla\mathbf{v}_\Omega \tag{18.1}$$

that the local time derivative of the rotational part of the motion in the general q^i system is given by

$$\left(\frac{\partial \mathbf{v}_\Omega}{\partial t}\right)_{q^i} = \frac{\partial \mathbf{v}_\Omega}{\partial t} = \mathbf{v}_P\cdot\nabla\mathbf{v}_\Omega \tag{18.2}$$

and in terms of measure numbers by

$$\frac{\partial}{\partial t}\left(\underset{\Omega}{W_m}\mathbf{q}^m\right) = W^n\mathbf{q}_n\cdot\mathbf{q}^r\,\nabla_r\left(\underset{\Omega}{W_m}\mathbf{q}^m\right) \quad \text{with} \quad W^i = \underset{\Omega}{W^i} + \underset{\mathrm{D}}{W^i} \tag{18.3}$$

On applying the rules of covariant differentiation followed by scalar multiplication by $\mathbf{q}_k$, we obtain

$$\frac{\partial}{\partial t}\underset{\Omega}{W_k} = W^n\,\nabla_n\underset{\Omega}{W_k} \tag{18.4}$$

Using this relation together with (M5.17) and splitting W^i, we immediately find the expression

$$\begin{aligned}
\frac{\partial}{\partial t}\underset{\Omega}{W_k} &= \underset{\Omega}{W_n}\,\nabla_k W^n + W^n\,\nabla_n\underset{\Omega}{W_k} \\
&= \underset{\Omega}{W_n}\,\nabla_k\underset{\Omega}{W^n} + \underset{\Omega}{W_n}\,\nabla_k\underset{\mathrm{D}}{W^n} + \underset{\Omega}{W^n}\,\nabla_n\underset{\Omega}{W_k} + \underset{\mathrm{D}}{W^n}\,\nabla_n\underset{\Omega}{W_k} \\
&= \underset{\Omega}{W_n}\,\nabla_k\underset{\mathrm{D}}{W^n} + \underset{\mathrm{D}}{W^n}\,\nabla_n\underset{\Omega}{W_k} + \underset{\Omega}{W^n}\left(\nabla_n\underset{\Omega}{W_k} + \nabla_k\underset{\Omega}{W_n}\right)
\end{aligned} \tag{18.5}$$

which is not yet in the form most useful to our analysis. Since $\nabla \mathbf{v}_\Omega$ is an antisymmetric dyadic we recognize at once that

$$\nabla \mathbf{v}_\Omega + \widetilde{\nabla \mathbf{v}_\Omega} = \mathbf{q}^m \mathbf{q}^n (\nabla_m \underset{\Omega}{W_n} + \nabla_n \underset{\Omega}{W_m}) = 0 \implies \nabla_i \underset{\Omega}{W_j} + \nabla_j \underset{\Omega}{W_i} = 0 \tag{18.6}$$

so that the expression in parentheses of (18.5) vanishes. Additionally, with the help of (M5.4) and (M5.6), we find that the remaining part of equation (18.5) may be formulated as

$$\frac{\partial}{\partial t} \underset{\Omega}{W_k} = \underset{\Omega}{W_n} \nabla_k \underset{D}{W^n} + \underset{D}{W^n} \nabla_n \underset{\Omega}{W_k} = \underset{\Omega}{W_n}(\nabla_k \underset{D}{W^n} + \underset{D}{W^r} \Gamma^n_{rk}) + \underset{D}{W^n}(\nabla_n \underset{\Omega}{W_k} - \underset{\Omega}{W_r} \Gamma^r_{nk}) \tag{18.7}$$

The two parts involving the Christoffel symbols add up to zero:

$$\underset{\Omega}{W_n} \underset{D}{W^r} \Gamma^n_{rk} - \underset{D}{W^n} \underset{\Omega}{W_r} \Gamma^r_{nk} = 0 \tag{18.8}$$

which is most easily recognized by interchanging the summation symbols n and r in the second term. We have finally found the important relation

$$\boxed{\frac{\partial}{\partial t} \underset{\Omega}{W_k} = \underset{\Omega}{W_n} \nabla_k \underset{D}{W^n} + \underset{D}{W^n} \nabla_n \underset{\Omega}{W_k}} \tag{18.9}$$

which is very useful to our work. In fact, it should be observed that the local time derivative of $\underset{\Omega}{W_k}$ does not vanish if the deformation velocity of the coordinate surfaces differs from zero. Otherwise, in the absence of deformation, $\partial \underset{\Omega}{W_k}/\partial t = 0$.

We begin the derivation of the equation of motion in covariant measure numbers by writing down the equation of absolute motion, which is repeated for convenience from (1.11):

$$\frac{d\mathbf{v}_\mathrm{A}}{\mathrm{dt}} = -\frac{1}{\rho} \nabla p - \nabla \phi_\mathrm{a} + \frac{1}{\rho} \nabla \cdot \mathbb{J} \tag{18.10}$$

with $\mathbf{v}_\mathrm{A} = \mathbf{q}^n (v_n + W_n)$. The covariant individual time derivative follows immediately:

$$\mathbf{q}^n \frac{dv_{\mathrm{A},n}}{dt} = \mathbf{q}^n \left(\frac{dv_n}{dt} + \frac{dW_n}{dt} \right) = -\frac{1}{\rho} \mathbf{q}^n \frac{\partial p}{\partial q^n} - \mathbf{q}^n \frac{\partial \phi_\mathrm{a}}{\partial q^n} + \frac{1}{\rho} \nabla \cdot \mathbb{J} \tag{18.11}$$

Scalar multiplication of this equation by the covariant basis vector $\mathbf{q}_k$ gives

$$\frac{dv_{\mathrm{A},k}}{dt} = \frac{dv_k}{dt} + \frac{dW_k}{dt} = -\frac{1}{\rho} \frac{\partial p}{\partial q^k} - \frac{\partial \phi_\mathrm{a}}{\partial q^k} + \frac{1}{\rho} \mathbf{q}_k \cdot \nabla \cdot \mathbb{J} \tag{18.12}$$

With the help of (M5.40) we may write

$$\frac{dv_k}{dt} = \frac{\partial v_k}{\partial t} + v^n \nabla_n v_k - v^n v_m \Gamma^m_{kn} - v_n \nabla_k W^n \tag{18.13}$$

The term involving the Christoffel symbol can be rewritten in a more convenient manner simply by employing the rules of raising and lowering the appropriate indices. This yields

$$v^n v_m \Gamma^m_{kn} = v^n g_{mr} v^r \Gamma^m_{kn} = v^n v^r \Gamma_{knr} = v^n v^m \Gamma_{knm} = \frac{1}{2} v^n v^m \frac{\partial g_{nm}}{\partial q^k} \tag{18.14}$$

where use of (M3.40) has been made. By combining the first and the second term on the right-hand side of (18.13) to give the individual acceleration, we may finally write for the covariant derivative

$$\frac{dv_k}{dt} = \frac{dv_k}{dt} - \frac{1}{2} v^n v^m \frac{\partial g_{nm}}{\partial q^k} - v_n \nabla_k W^n \tag{18.15}$$

The second term on the right-hand side is known as the *metric acceleration* since it involves the metric fundamental quantities. This fictitious acceleration does not result from the interaction of an air particle with other bodies but rather stems from the particular choice of the coordinate system which is used to describe the motion of the particle.

Next we must find suitable expressions for the individual time derivatives of the measure numbers W_k measuring the acceleration of a point P in the q^i-coordinate system that is moving with the velocity $\mathbf{v}_P$ relative to the absolute system. In the previous sections we have already set up the relationships which are needed. With the help of (M5.39) we obtain immediately

$$\frac{dW_k}{dt} = \frac{d}{dt} \underset{\Omega}{W_k} + \frac{d}{dt} \underset{\mathrm{D}}{W_k} = \frac{\partial}{\partial t} \underset{\Omega}{W_k} + \frac{\partial}{\partial t} \underset{\mathrm{D}}{W_k} + v^n \nabla_n \underset{\Omega}{W_k} + v^n \nabla_n \underset{\mathrm{D}}{W_k} \tag{18.16}$$

On replacing the covariant time derivatives by means of (M5.37) we find

$$\begin{aligned}\frac{dW_k}{dt} =& \frac{\partial}{\partial t} \underset{\Omega}{W_k} - \underset{\mathrm{D}}{W_n} \nabla_k \underset{\Omega}{W^n} - \underset{\Omega}{W_n} \nabla_k \underset{\mathrm{D}}{W^n} \\ &+ \frac{\partial}{\partial t} \underset{\mathrm{D}}{W_k} - \underset{\mathrm{D}}{W_n} \nabla_k \underset{\Omega}{W^n} - \underset{\mathrm{D}}{W_n} \nabla_k \underset{\mathrm{D}}{W^n} + v^n \nabla_n \underset{\Omega}{W_k} + v^n \nabla_n \underset{\mathrm{D}}{W_k}\end{aligned} \tag{18.17}$$

By utilizing (M5.49) the last two terms in (18.17) can be rewritten as

$$\begin{aligned} v^n(\nabla_n \underset{\Omega}{W_k} + \nabla_n \underset{\mathrm{D}}{W_k}) &= 2v^n(\underset{\Omega}{\omega_{nk}} + \underset{\mathrm{D}}{\omega_{nk}}) + v^n(\nabla_k \underset{\Omega}{W_n} + \nabla_k \underset{\mathrm{D}}{W_n}) \\ &= 2v^n(\underset{\Omega}{\omega_{nk}} + \underset{\mathrm{D}}{\omega_{nk}}) + v_n(\nabla_k \underset{\Omega}{W^n} + \nabla_k \underset{\mathrm{D}}{W^n}) \end{aligned} \tag{18.18}$$

where the final step resulted from raising and lowering of indices, and the fact that the metric fundamental quantities may be treated as constants in covariant differentiation.

Again by raising and lowering indices and by renaming them whenever needed, we find without any difficulty

$$\begin{aligned} &\underset{\Omega}{W_n} \nabla_k \underset{\Omega}{W^n} = \nabla_k(\underset{\Omega}{W_n} \underset{\Omega}{W^n}) - \underset{\Omega}{W^n} \nabla_k \underset{\Omega}{W_n} = \nabla_k(\underset{\Omega}{W_n} \underset{\Omega}{W^n}) - \underset{\Omega}{W_n} \nabla_k \underset{\Omega}{W^n} \\ \Longrightarrow\ &\underset{\Omega}{W_n} \nabla_k \underset{\Omega}{W^n} = \nabla_k\left(\frac{\underset{\Omega}{W_n} \underset{\Omega}{W^n}}{2}\right) = \nabla_k\left(\frac{\mathbf{v}_\Omega^2}{2}\right) = \nabla_k\left(\frac{\mathbf{v}_\Omega^2}{2}\right) \end{aligned} \tag{18.19}$$

In this expression we have used the facts that $\mathbf{v}_\Omega^2/2$ is a scalar field function and that, for any scalar field function, the ordinary and covariant differential operators are identical since no basis vector needs to be extracted. By applying identical operations we find

$$\underset{D}{W}_n \nabla_k \underset{D}{W}^n = \nabla_k\left(\frac{\mathbf{v}_D^2}{2}\right) = \nabla_k\left(\frac{\mathbf{v}_D^2}{2}\right) \tag{18.20}$$

We may now combine the following three terms occurring in (18.17) as

$$\begin{aligned}\frac{\partial}{\partial t}\underset{\Omega}{W}_k - \underset{\Omega}{W}_n \nabla_k \underset{D}{W}^n - \underset{D}{W}_n \nabla_k \underset{\Omega}{W}^n &= \underset{D}{W}^n \nabla_n \underset{\Omega}{W}_k + \underset{\Omega}{W}_n \nabla_k \underset{D}{W}^n - \underset{\Omega}{W}_n \nabla_k \underset{D}{W}^n - \underset{D}{W}_n \nabla_k \underset{\Omega}{W}^n \\ &= \underset{D}{W}^n \nabla_n \underset{\Omega}{W}_k - \underset{D}{W}_n \nabla_k \underset{\Omega}{W}^n = \underset{D}{W}^n \nabla_n \underset{\Omega}{W}_k - \underset{D}{W}^n \nabla_k \underset{\Omega}{W}_n = 2\underset{D}{W}^n \underset{\Omega}{\omega}_{nk}\end{aligned} \tag{18.21}$$

Here we have replaced the local time derivative of $\underset{\Omega}{W}_k$ with the help of (18.7) and have used the defining relation (M5.48) for $\underset{\Omega}{\omega}_{ij}$. On substituting (18.19), (18.20), and (18.21) into (18.17) we obtain for the individual covariant time derivative of W_k

$$\frac{dW_k}{dt} = 2\underset{D}{W}^n \underset{\Omega}{\omega}_{nk} - \nabla_k\left(\frac{\mathbf{v}_\Omega^2}{2} + \frac{\mathbf{v}_D^2}{2}\right) + 2v^n(\underset{\Omega}{\omega}_{nk} + \underset{D}{\omega}_{nk}) + v_n \nabla_k W^n + \frac{\partial}{\partial t}\underset{D}{W}_k \tag{18.22}$$

Expressions (18.15) and (18.22) are now used in (18.11), yielding the covariant form of the equation of relative motion with repect to a general time-dependent coordinate system:

$$\boxed{\begin{aligned}\frac{dv_k}{dt} - \frac{v^n v^m}{2}\frac{\partial g_{mn}}{\partial q^k} + 2v^n(\underset{\Omega}{\omega}_{nk} + \underset{D}{\omega}_{nk}) &= -\frac{1}{\rho}\frac{\partial p}{\partial q^k} - \frac{\partial \phi}{\partial q^k} + \frac{1}{\rho}\mathbf{q}_k\cdot\nabla\cdot\mathbb{J} \\ &\quad + \frac{\partial}{\partial q^k}\left(\frac{\mathbf{v}_D^2}{2}\right) - \frac{\partial}{\partial t}\underset{D}{W}_k - 2\underset{D}{W}^n \underset{\Omega}{\omega}_{nk}\end{aligned}} \tag{18.23}$$

where the geopotential ϕ has been introduced according to

$$\phi = \phi_a - \frac{\mathbf{v}_\Omega^2}{2} = \phi_a + \phi_z, \qquad \phi_z = -\frac{\mathbf{v}_\Omega^2}{2} \tag{18.24}$$

Closer inspection of this equation shows that several terms also contain contravariant coordinates. Since the leading term, i.e. the individual time derivative dv_k/dt, contains the covariant velocity v_k, the entire equation is called the *covariant form of the equation of motion.*

Finally we consider an important simplification by requiring that the coordinate lines are rigid so that the deformation velocity $\mathbf{v}_D = 0$. In this case (18.23) reduces to

$$\boxed{\frac{dv_k}{dt} - \frac{v^n v^m}{2}\frac{\partial g_{mn}}{\partial q^k} + 2v^n \underset{\Omega}{\omega}_{nk} = -\frac{1}{\rho}\frac{\partial p}{\partial q^k} - \frac{\partial \phi}{\partial q^k} + \frac{1}{\rho}\mathbf{q}_k\cdot\nabla\cdot\mathbb{J}} \tag{18.25}$$

In some cases the covariant equation may be less convenient for practical applications than its contravariant counterpart. In the following section we are going to derive the contravariant form of the equation of relative motion. Which of these two equations is more easily applied can be decided by inspecting the metric fundamental quantities g_{ij} or g^{ij} and their derivatives.

18.3 The contravariant equation of motion in general coordinate systems

The derivation of the contravariant representation of relative motion follows the procedure of the previous section. We begin the derivation by stating the individual change with time of the absolute velocity in terms of the contravariant measure number v_{A}^k

$$\frac{d}{dt}\left(v_{\mathrm{A}}^n \mathbf{q}_n\right) = \mathbf{q}_n \frac{d v_{\mathrm{A}}^n}{dt} \tag{18.26}$$

The components of the gradient operator appearing as part of the individual time derivative are

$$\nabla = \mathbf{q}_m \nabla^m = \mathbf{q}_m \frac{\partial}{\partial q_m} = g^{mn} \mathbf{q}_m \frac{\partial}{\partial q^n} \tag{18.27}$$

The equation of absolute motion (18.10) in terms of the contravariant measure number v_{A}^k is given analogously to (18.12) by

$$\frac{d v_{\mathrm{A}}^k}{dt} = \frac{d v^k}{dt} + \frac{d W^k}{dt} = -\frac{g^{nk}}{\rho}\frac{\partial p}{\partial q^n} - g^{nk}\frac{\partial \phi_{\mathrm{a}}}{\partial q^n} + \frac{1}{\rho}\mathbf{q}^k \cdot \nabla \cdot \mathbb{J} \tag{18.28}$$

The next step is to derive an expression for the individual covariant change with time of the contravariant relative velocity v^k. Application of (18.11), (18.16), and (M5.4) results in

$$\begin{aligned}
\frac{d v^k}{dt} &= \frac{\partial v^k}{\partial t} + v^n \nabla_n v^k = \frac{\partial}{\partial t} v^k + v^n \nabla_n W^k + v^n \nabla_n v^k \\
&= \frac{\partial v^k}{\partial t} + v^n \nabla_n W^k + v^n \nabla_n v^k + v^n v^m \Gamma_{nm}^k \\
&= \frac{d v^k}{dt} + g^{km} v^n \nabla_n W_m + v^n v^m \Gamma_{nm}^k
\end{aligned} \tag{18.29}$$

The covariant derivative of the contravariant measure numbers W^k will not be derived anew but will be found more briefly by raising the covariant index. Utilizing (18.22) we find

$$\begin{aligned}
\frac{d W^k}{dt} = g^{km}\frac{d W_m}{dt} &= g^{km}\left[2 \underset{\mathrm{D}}{W^n} \underset{\Omega}{\omega_{nm}} - \nabla_m\left(\frac{\mathbf{v}_{\Omega}^2}{2} + \frac{\mathbf{v}_{\mathrm{D}}^2}{2}\right)\right] \\
&\quad + g^{km}\left(2 v^n (\underset{\Omega}{\omega_{nm}} + \underset{\mathrm{D}}{\omega_{nm}}) + v_n \nabla_m W^n + \frac{\partial}{\partial t} \underset{\mathrm{D}}{W_m}\right)
\end{aligned} \tag{18.30}$$

By application of the addition theorem of the velocities in the form $v_{\mathrm{A}}^k = v^k + W^k$, we obtain from (18.29) and (18.30)

$$\frac{dv_{\rm A}^k}{dt} = \frac{dv^k}{dt} + g^{km} v^n \nabla_n \underset{\rm D}{W}_m + v^n v^m \Gamma_{nm}^k + g^{km} \left[2 \underset{\rm D}{W}^n \underset{\Omega}{\omega}_{nm} - \nabla_m \left(\frac{\mathbf{v}_\Omega^2}{2} + \frac{\mathbf{v}_{\rm D}^2}{2} \right) \right]$$
$$+ g^{km} \left(2v^n (\underset{\Omega}{\omega}_{nm} + \underset{\rm D}{\omega}_{nm}) + v_n \nabla_m \underset{\rm D}{W}^n + \frac{\partial}{\partial t} \underset{\rm D}{W}_m \right) \tag{18.31}$$

Utilizing (M5.49) and (18.6), three of the terms occurring on the right-hand side of this expression may be combined according to

$$v^n \nabla_n \underset{\rm D}{W}_m + v_n \nabla_m \underset{\rm D}{W}^n + 2v^n \underset{\rm D}{\omega}_{nm} = v^n \left[\nabla_n (\underset{\Omega}{W}_m + \underset{\rm D}{W}_m) + \nabla_m (\underset{\Omega}{W}_n + \underset{\rm D}{W}_n) + 2\underset{\rm D}{\omega}_{nm} \right]$$
$$= v^n \left(\nabla_n \underset{\rm D}{W}_m + \nabla_m \underset{\rm D}{W}_n + 2\underset{\rm D}{\omega}_{nm} \right) = 2v^n \nabla_n \underset{\rm D}{W}_m \tag{18.32}$$

Substitution of (18.31) together with (18.32) into (18.28) yields the final form of the *contravariant equation of motion* for a general time-dependent coordinate system:

$$\boxed{\begin{aligned} \frac{dv^k}{dt} + v^m v^n \Gamma_{mn}^k + 2v^n (\nabla_n \underset{\rm D}{W}^k + \underset{\Omega}{\omega}_n{}^k) &= -g^{km} \left(\frac{1}{\rho} \frac{\partial p}{\partial q^m} + \frac{\partial \phi}{\partial q^m} \right) + \frac{1}{\rho} \mathbf{q}^k \cdot \nabla \cdot \mathbb{J} \\ &+ g^{km} \left[\frac{\partial}{\partial q^m} \left(\frac{\mathbf{v}_{\rm D}^2}{2} \right) - \frac{\partial}{\partial t} \underset{\rm D}{W}_m - 2 \underset{\rm D}{W}^n \underset{\Omega}{\omega}_{nm} \right] \end{aligned}} \tag{18.33}$$

For the special case of a rigid system the deformation velocity vanishes, so the above equation reduces to

$$\boxed{\frac{dv^k}{dt} + v^m v^n \Gamma_{mn}^k + 2v^n \underset{\Omega}{\omega}_n{}^k = -g^{km} \left(\frac{1}{\rho} \frac{\partial p}{\partial q^m} + \frac{\partial \phi}{\partial q^m} \right) + \frac{1}{\rho} \mathbf{q}^k \cdot \nabla \cdot \mathbb{J}} \tag{18.34}$$

Summarizing, we may state that with equations (18.23) and (18.33) we have obtained the components of the equation of motion in covariant and contravariant representations in a very general form. We may apply these forms to any time-dependent coordinate system. No restriction has been made regarding the velocity $\mathbf{v}_P$. If we wish to write down the equation of motion in a particular coordinate system, all we must do is procure information about the metric fundamental quantities g_{ij} or g^{ij} and the velocity $\mathbf{v}_P$. If this information is available, we are in a position to write down at once the components of the equation of motion either in the covariant or in the contravariant form. Thus, the laborious method of obtaining the component representation of the equation of motion for each coordinate system separately can be entirely avoided. Which system is simpler to use for the numerical solution depends on the metric fundamental quantities and on the final form of the equation. In the next chapters we will gain some practice in applying the various coordinate systems.

18.4 The equation of motion in orthogonal coordinate systems

In this section we are going to assume that the q^i-coordinate lines form an orthogonal system so that $g_{ij} = 0,\ i \neq j$, and $g_{ii}g^{ii} = 1$. Whenever orthogonal systems are used it is customary and useful to introduce the physical measure numbers since there is no difference between covariant and contravariant physical measure numbers. Furthermore, unit vectors will be employed instead of basis vectors. We will transform the covariant form of the equation of motion (18.23) into physical measure numbers.

The way we proceed is to multiply (18.23) by the factor $1/\sqrt{g_{kk}}$. For clarity we transform each term of (18.23) separately. All mathematical steps are summarized in

$$
\begin{aligned}
\frac{1}{\sqrt{g_{kk}}}\frac{dv_k}{dt} &= \frac{d}{dt}\left(\frac{v_k}{\sqrt{g_{kk}}}\right) - v_k\frac{d}{dt}\left(\frac{1}{\sqrt{g_{kk}}}\right) = \frac{d}{dt}\left(\frac{v_k}{\sqrt{g_{kk}}}\right) + \frac{v_k}{g_{kk}}\frac{d\sqrt{g_{kk}}}{dt} \\
&= \frac{d\overset{*}{v}{}^k}{dt} + \overset{*}{v}{}^k\left(\frac{1}{\sqrt{g_{kk}}}\frac{\partial\sqrt{g_{kk}}}{\partial t} + \frac{\overset{*}{v}{}^n}{\sqrt{g_{kk}}}\frac{\partial\sqrt{g_{kk}}}{\partial\overset{*}{q}{}^n}\right) \\
-\frac{1}{2}\frac{v^n v^n}{\sqrt{g_{kk}}}\frac{\partial g_{nn}}{\partial q^k} &= -\frac{v^n v^n}{\sqrt{g_{kk}}}\frac{g_{nn}}{\sqrt{g_{nn}}}\frac{\partial\sqrt{g_{nn}}}{\partial q^k} = -\frac{\overset{*}{v}{}^n\overset{*}{v}{}^n}{\sqrt{g_{nn}}}\frac{\partial\sqrt{g_{nn}}}{\partial\overset{*}{q}{}^k} \\
\frac{2v^n}{\sqrt{g_{kk}}}\left(\underset{\Omega}{\omega}_{nk} + \underset{D}{\omega}_{nk}\right) &= \frac{2v^n}{\sqrt{g_{kk}}}\left(\underset{\Omega}{\omega}_{nk} + \underset{D}{\omega}_{nk}\right)\frac{\sqrt{g_{nn}}}{\sqrt{g_{nn}}} = 2\overset{*}{v}{}^n\left(\underset{\Omega}{\overset{*}{\omega}}_{nk} + \underset{D}{\overset{*}{\omega}}_{nk}\right) \\
\frac{1}{\sqrt{g_{kk}}}\left[-\frac{1}{\rho}\frac{\partial p}{\partial q^k} - \frac{\partial\phi}{\partial q^k} + \frac{\partial}{\partial q^k}\left(\frac{\mathbf{v}_D^2}{2}\right)\right] &= -\frac{1}{\rho}\frac{\partial p}{\partial\overset{*}{q}{}^k} - \frac{\partial\phi}{\partial\overset{*}{q}{}^k} + \frac{\partial}{\partial\overset{*}{q}{}^k}\left(\frac{\mathbf{v}_D^2}{2}\right) \\
-\frac{1}{\sqrt{g_{kk}}}\frac{\partial}{\partial t}\underset{D}{W}_k &= -\frac{\partial}{\partial t}\left(\frac{\underset{D}{W}_k}{\sqrt{g_{kk}}}\right) + \underset{D}{W}_k\frac{\partial}{\partial t}\left(\frac{1}{\sqrt{g_{kk}}}\right) = -\frac{\partial}{\partial t}\underset{D}{\overset{*}{W}}_k - \frac{\underset{D}{\overset{*}{W}}{}^k}{\sqrt{g_{kk}}}\frac{\partial\sqrt{g_{kk}}}{\partial t} \\
-\frac{2\underset{D}{W}{}^n\underset{\Omega}{\omega}_{nk}}{\sqrt{g_{kk}}} &= -\frac{2\underset{D}{W}{}^n\underset{\Omega}{\omega}_{nk}}{\sqrt{g_{kk}}}\frac{\sqrt{g_{nn}}}{\sqrt{g_{nn}}} = -2\underset{D}{\overset{*}{W}}{}^n\underset{\Omega}{\overset{*}{\omega}}_{nk}
\end{aligned}
\tag{18.35}
$$

Recall that k is not a summation index. Substituting the various terms of (18.35) into (18.23) gives the desired equation for orthogonal coordinate systems:

$$\boxed{\begin{aligned}&\frac{d\overset{*}{v}^{k}}{dt}+\overset{*}{v}^{n}\left(\frac{\overset{*}{v}^{k}}{\sqrt{g_{kk}}}\frac{\partial\sqrt{g_{kk}}}{\partial\overset{*}{q}^{n}}-\frac{\overset{*}{v}^{n}}{\sqrt{g_{nn}}}\frac{\partial\sqrt{g_{nn}}}{\partial\overset{*}{q}^{k}}\right)+2\overset{*}{v}^{n}(\underset{\Omega}{\overset{*}{\omega}}_{nk}+\underset{\mathrm{D}}{\overset{*}{\omega}}_{nk})\\&=-\frac{1}{\rho}\frac{\partial p}{\partial\overset{*}{q}^{k}}-\frac{\partial\phi}{\partial\overset{*}{q}^{k}}+\frac{1}{\rho}\mathbf{e}_{k}\cdot\nabla\cdot\mathbb{J}\\&+\frac{\partial}{\partial\overset{*}{q}^{k}}\left(\frac{\mathbf{v}_{\mathrm{D}}^{2}}{2}\right)-\frac{\partial}{\partial t}\underset{\mathrm{D}}{\overset{*}{W}}_{k}-(\overset{*}{v}^{k}+\underset{\mathrm{D}}{\overset{*}{W}}^{k})\frac{1}{\sqrt{g_{kk}}}\frac{\partial\sqrt{g_{kk}}}{\partial t}-2\underset{\mathrm{D}}{\overset{*}{W}}^{n}\underset{\Omega}{\overset{*}{\omega}}_{nk}\end{aligned}} \tag{18.36}$$

If the relative system is moving like a rigid body then the terms involving $\mathbf{v}_{\mathrm{D}}$ and $\partial\sqrt{g_{kk}}/\partial t$ vanish and we obtain

$$\boxed{\begin{aligned}&\frac{d\overset{*}{v}^{k}}{dt}+\overset{*}{v}^{n}\left(\frac{\overset{*}{v}^{k}}{\sqrt{g_{kk}}}\frac{\partial\sqrt{g_{kk}}}{\partial\overset{*}{q}^{n}}-\frac{\overset{*}{v}^{n}}{\sqrt{g_{nn}}}\frac{\partial\sqrt{g_{nn}}}{\partial\overset{*}{q}^{k}}\right)+2\overset{*}{v}^{n}\underset{\Omega}{\overset{*}{\omega}}_{nk}\\&=-\frac{1}{\rho}\frac{\partial p}{\partial\overset{*}{q}^{k}}-\frac{\partial\phi}{\partial\overset{*}{q}^{k}}+\frac{1}{\rho}\mathbf{e}_{k}\cdot\nabla\cdot\mathbb{J}\end{aligned}} \tag{18.37}$$

It will be observed that the expression in parentheses vanishes if $n = k$. This simplifies equations (18.36) and (18.37) if they are written down separately for $k = 1, 2, 3$.

It might be of interest to expand the frictional term as

$$\begin{aligned}\mathbf{e}_{k}\cdot\nabla\cdot\mathbb{J}&=\frac{\mathbf{e}_{k}}{\sqrt{g}}\cdot\frac{\partial}{\partial q^{n}}(\sqrt{g}J^{nm}\mathbf{q}_{m})=\frac{\mathbf{e}_{k}}{\sqrt{g}}\cdot\frac{\partial}{\partial q^{n}}\left(\sqrt{g}\frac{\overset{*}{J}^{nm}}{\sqrt{g_{nn}}\sqrt{g_{mm}}}\mathbf{q}_{m}\right)\\&=\mathbf{e}_{k}\cdot\left[\frac{1}{\sqrt{g}}\nabla_{n}\left(\frac{\sqrt{g}}{\sqrt{g_{nn}}}\overset{*}{J}^{nm}\right)\mathbf{e}_{m}+\frac{\overset{*}{J}^{nm}}{\sqrt{g_{nn}}}\nabla_{n}\mathbf{e}_{m}\right]\end{aligned} \tag{18.38}$$

by using the general divergence formula (M3.56). The last term involves the spatial derivative of the unit vector $\mathbf{e}_j$ which, according to (M4.45), can be written as

$$\nabla_{j}\mathbf{e}_{i}=\mathbf{e}_{j}\,\overset{*}{\nabla}_{i}\sqrt{g_{jj}}-\delta^{i}_{j}\mathbf{e}_{m}\,\overset{*}{\nabla}_{m}\sqrt{g_{ii}} \tag{18.39}$$

Using this equation in (18.38) yields

$$\mathbf{e}_{k}\cdot\nabla\cdot\mathbb{J}=\mathbf{e}_{k}\cdot\mathbf{e}_{m}\left[\frac{\sqrt{g_{nn}}}{\sqrt{g}}\overset{*}{\nabla}_{n}\left(\frac{\sqrt{g}}{\sqrt{g_{nn}}}\overset{*}{J}^{nm}\right)+\frac{\overset{*}{J}^{mn}}{\sqrt{g_{mm}}}\overset{*}{\nabla}_{n}\sqrt{g_{mm}}-\frac{\overset{*}{J}^{nn}}{\sqrt{g_{nn}}}\overset{*}{\nabla}_{m}\sqrt{g_{nn}}\right] \tag{18.40}$$

so that the final form of the frictional term is given by

$$\mathbf{e}_k\cdot\nabla\cdot\mathbb{J} = \frac{\sqrt{g_{nn}}}{\sqrt{g}}\overset{*}{\nabla}_n\left(\frac{\sqrt{g}}{\sqrt{g_{nn}}}\overset{*}{J}{}^{nk}\right) + \frac{\overset{*}{J}{}^{kn}}{\sqrt{g_{kk}}}\overset{*}{\nabla}_n\sqrt{g_{kk}} - \frac{\overset{*}{J}{}^{nn}}{\sqrt{g_{nn}}}\overset{*}{\nabla}_k\sqrt{g_{nn}} \tag{18.41}$$

Hence, if the form of the frictional dyadic and the metric tensor g_{ij} are known we can evaluate equation (18.41). However, for practical purposes one will often be satisfied with simple parameterizations. If the measure numbers of the frictional dyadic are given in Cartesian coordinates, we can transform these by well-known rules to the general q^i system.

Finally, we will briefly show that the term $(1/\sqrt{g_{kk}})\,\partial\sqrt{g_{kk}}/\partial t$ contained in the general form (18.36) of the equation of motion disappears if $\mathbf{v}_{\mathrm{D}} = 0$. Starting with

$$\frac{\partial \mathbf{q}_i}{\partial t} = \nabla_i \mathbf{v}_P = \mathbf{q}_i\cdot\nabla\mathbf{v}_P \tag{18.42}$$

and utilizing (M5.44), we can write for the local change with time of the metric tensor g_{ij}

$$\begin{aligned}\frac{\partial g_{ij}}{\partial t} &= \frac{\partial}{\partial t}(\mathbf{q}_i\cdot\mathbf{q}_j) = \mathbf{q}_i\cdot(\mathbf{q}_j\cdot\nabla\mathbf{v}_P) + \mathbf{q}_j\cdot(\mathbf{q}_i\cdot\nabla\mathbf{v}_P) = \mathbf{q}_i\cdot\mathbf{v}_P\overset{\curvearrowleft}{\nabla}\cdot\mathbf{q}_j + (\mathbf{q}_i\cdot\nabla\mathbf{v}_P)\cdot\mathbf{q}_j \\ &= \mathbf{q}_i\cdot(\nabla\mathbf{v}_P + \mathbf{v}_P\overset{\curvearrowleft}{\nabla})\cdot\mathbf{q}_j = \mathbf{q}_i\cdot 2\mathbb{D}_P\cdot\mathbf{q}_j = 2d_{ij}\end{aligned} \tag{18.43}$$

Hence, the desired term of (18.36) can be written as

$$\frac{1}{\sqrt{g_{kk}}}\frac{\partial\sqrt{g_{kk}}}{\partial t} = \frac{d_{kk}}{g_{kk}} \tag{18.44}$$

showing that it is equal to zero if the deformational part $\mathbb{D}_P$ of the dyadic $\nabla\mathbf{v}_P$ vanishes due to (18.6).

The steps leading to the continuity equation in general coordinates are very simple. From (M6.67) we have

$$\frac{D\rho}{Dt} = \frac{d\rho}{dt} + \rho\,\nabla\cdot\mathbf{v}_A = \frac{d\rho}{dt} + \rho\,\nabla\cdot\mathbf{v} + \rho\,\nabla\cdot\mathbf{v}_P = 0 \tag{18.45}$$

By expanding the individual derivative in the relative system we find

$$\frac{\partial\rho}{\partial t} + \mathbf{v}\cdot\nabla\rho + \rho\,\nabla\cdot\mathbf{v} + \rho\,\nabla\cdot\mathbf{v}_P = 0 \tag{18.46}$$

On combining the second and third terms to give a divergence expression for the quantity $\rho\mathbf{v}$ and applying the general equation (M3.56) for the divergence, we find the desired form of the continuity equation:

$$\boxed{\frac{\partial\rho}{\partial t} + \frac{1}{\sqrt{g}}\frac{\partial}{\partial q^n}(\sqrt{g}\,\dot{q}^n\rho) + \frac{\rho}{\sqrt{g}}\frac{\partial\sqrt{g}}{\partial t} = 0} \tag{18.47}$$

For a rigid system the last term of (18.47) vanishes, resulting in the usual form

$$\boxed{\frac{\partial \rho}{\partial t} + \frac{1}{\sqrt{g}}\frac{\partial}{\partial q^n}(\sqrt{g}\dot{q}^n \rho) = 0} \tag{18.48}$$

which can be found in many textbooks.

18.5 Lagrange's equation of motion

Another method for the representation of the covariant form of the equation of motion is due to Lagrange. We will derive Lagrange's equation of motion in the form best suited for meteorological applications. The entire method is based on knowledge of the absolute kinetic energy permitting the determination of the metric fundamental quantitites and the velocity $\mathbf{v}_P$.

The starting point of the analysis is the well-known relation

$$(d\mathbf{r})^2 = g_{mn}\, dq^m\, dq^n \tag{18.49}$$

The kinetic energy per unit mass can be written as

$$K_{\rm A} = \frac{1}{2}\left(\frac{d\mathbf{r}}{dt}\right)^2 = \frac{\mathbf{v}_A^2}{2} = \frac{\dot{x}^n \dot{x}^n}{2} \tag{18.50}$$

in the absolute Cartesian coordinate system and as

$$K_{\rm A} = \frac{g_{mn}}{2}\dot{q}_{\rm A}^m \dot{q}_{\rm A}^n = \frac{1}{2}(g_{mn}\dot{q}^m + \underset{\Omega}{W_n} + \underset{\rm D}{W_n})(\dot{q}^n + \underset{\Omega}{W^n} + \underset{\rm D}{W^n}) \tag{18.51}$$

in the general q^i system. This expression may also be formulated as

$$\boxed{K_{\rm A} = \frac{g_{mn}}{2}\dot{q}^m \dot{q}^n + \dot{q}^n(\underset{\Omega}{W_n} + \underset{\rm D}{W_n}) + \frac{1}{2}(\mathbf{v}_\Omega^2 + \mathbf{v}_{\rm D}^2) + \underset{\Omega}{W_m}\underset{\rm D}{W^m}} \tag{18.52}$$

since $\mathbf{v}_\Omega^2 = \underset{\Omega}{W_n}\underset{\Omega}{W^n}$ and $\mathbf{v}_{\rm D}^2 = \underset{\rm D}{W_n}\underset{\rm D}{W^n}$. Inspection of (18.52) shows that the absolute kinetic energy is a quadratic inhomogeneous form in the variables $\dot{q}^i$. The coefficients of the quadratic term are the metric fundamental quantities g_{ij}; the coefficients of the linear term are the covariant measure numbers $\underset{\Omega}{W_k}$ and $\underset{\rm D}{W_k}$ of the velocity $\mathbf{v}_P$. The kinetic energy of the relative system appears as the separate term $(g_{mn}/2)\dot{q}^m\dot{q}^n$.

The Lagrange equation will be derived by transforming the individual terms of the absolute equation of motion

$$\frac{d\dot{x}^i}{dt} = -\frac{1}{\rho}\frac{\partial p}{\partial x^i} - \frac{\partial \phi_{\rm a}}{\partial x^i} + \frac{1}{\rho}\mathbf{i}_i \cdot \nabla \cdot \mathbb{J} \tag{18.53}$$

expressed in the rectangular Cartesian system to the q^i system. Now we apply the *method of contraction* by multiplying the latter equation by $\partial x^i/\partial q^k$ and then adding the result from $i = 1$ to $i = 3$:

$$\sum_{i=1}^{3} \frac{\partial x^i}{\partial q^k}\frac{d\dot{x}^i}{dt} = -\frac{1}{\rho}\sum_{i=1}^{3} \frac{\partial x^i}{\partial q^k}\frac{\partial p}{\partial x^i} - \sum_{i=1}^{3} \frac{\partial x^i}{\partial q^k}\frac{\partial \phi_{\mathrm{a}}}{\partial x^i} + \frac{1}{\rho}\sum_{i=1}^{3} \frac{\partial x^i}{\partial q^k}\mathbf{i}_i \cdot \nabla \cdot \mathbb{J} \tag{18.54}$$

The same expression can also be written more briefly by applying the Einstein summation rule:

$$\begin{aligned} \frac{\partial x^n}{\partial q^k}\frac{d\dot{x}^n}{dt} &= -\frac{1}{\rho}\frac{\partial x^n}{\partial q^k}\frac{\partial p}{\partial x^n} - \frac{\partial x^n}{\partial q^k}\frac{\partial \phi_{\mathrm{a}}}{\partial x^n} + \frac{1}{\rho}\frac{\partial x^n}{\partial q^k}\mathbf{i}_n \cdot \nabla \cdot \mathbb{J} \\ &= -\frac{1}{\rho}\frac{\partial p}{\partial q^k} - \frac{\partial \phi_{\mathrm{a}}}{\partial q^k} + \frac{1}{\rho}\mathbf{q}_k \cdot \nabla \cdot \mathbb{J} \end{aligned} \tag{18.55}$$

Let us rewrite the left-hand side as

$$\begin{aligned} \frac{\partial x^n}{\partial q^k}\frac{d\dot{x}^n}{dt} &= \frac{d}{dt}\left(\frac{\partial x^n}{\partial q^k}\dot{x}^n\right) - \dot{x}^n \frac{d}{dt}\left(\frac{\partial x^n}{\partial q^k}\right) \\ &= \frac{d}{dt}\left(\frac{\partial x^n}{\partial q^k}\dot{x}^n\right) - \dot{x}^n\left(\frac{\partial^2 x^n}{\partial t\,\partial q^k} + \dot{q}^m \frac{\partial^2 x^n}{\partial q^m\,\partial q^k}\right) \end{aligned} \tag{18.56}$$

We will now attempt to introduce the kinetic energy K_{A}, which is a function of $\dot{q}^k$ and q^k. We need to recall that the metric tensor g_{ij} and the measure numbers W_k and W^k are functions of q^k only. Implicitly the contravariant velocities $\dot{q}^k$ also depend on q^k and on the time t. In the Lagrange treatment the kinetic energy will be considered as a function of the independent variables $\dot{q}^k$ and q^k so that $K_{\mathrm{A}} = K_{\mathrm{A}}(\dot{q}^k, q^k, t)$. For clarity we also introduce the special partial derivative operator

$$\left.\frac{\partial}{\partial q^k}\right|_{\dot{q}^i} = \left(\frac{\partial}{\partial q^k}\right)_{\dot{q}^1,\dot{q}^2,\dot{q}^3} \tag{18.57}$$

indicating that the differentiation $\partial/\partial q^k$ is performed for constant-velocity components $\dot{q}^1, \dot{q}^2, \dot{q}^3$. Using this convention, the last term in (18.56) can be rewritten as

$$\dot{x}^n \left.\frac{\partial}{\partial q^k}\right|_{\dot{q}^i}\left(\frac{\partial x^n}{\partial t} + \dot{q}^m \frac{\partial x^n}{\partial q^m}\right) = \dot{x}^n \left.\frac{\partial \dot{x}^n}{\partial q^k}\right|_{\dot{q}^i} = \left.\frac{\partial}{\partial q^k}\right|_{\dot{q}^i}\left(\frac{\dot{x}^n\dot{x}^n}{2}\right) = \left.\frac{\partial K_{\mathrm{A}}}{\partial q^k}\right|_{\dot{q}^i} \tag{18.58}$$

The next step in the derivation is made easy by applying the transformation rule

$$\underset{x}{A}^k = \frac{\partial x^k}{\partial q^m}\underset{q}{A}^m \tag{18.59}$$

which is repeated from (M4.15). On identifying A^k as the measure number $\dot{x}^k$ of the absolute velocity $\mathbf{v}_A$ we obtain

$$\dot{x}^k = \frac{\partial x^k}{\partial q^m}(\dot{q}^m + W^m) \tag{18.60}$$

Next we differentiate $\dot{x}^k$ with respect to $\dot{q}^i$:

$$\frac{\partial \dot{x}^k}{\partial \dot{q}^i} = \frac{\partial x^k}{\partial q^m}\delta_i^m = \frac{\partial x^k}{\partial q^i} \tag{18.61}$$

since $\partial x^k/\partial q^m$ and W^m surely do not depend on $\dot{q}^i$. Substituting this expression together with (18.58) into (18.56) yields

$$\frac{\partial x^n}{\partial q^k}\frac{d\dot{x}^n}{dt} = \frac{d}{dt}\left(\frac{\partial \dot{x}^n}{\partial \dot{q}^k}\dot{x}^n\right) - \left.\frac{\partial K_\text{A}}{\partial q^k}\right|_{\dot{q}^i} = \frac{d}{dt}\left(\frac{\partial K_\text{A}}{\partial \dot{q}^k}\right) - \left.\frac{\partial K_\text{A}}{\partial q^k}\right|_{\dot{q}^i} \tag{18.62}$$

so (18.55) may be written as

$$\frac{d}{dt}\left(\frac{\partial K_\text{A}}{\partial \dot{q}^k}\right) - \left.\frac{\partial K_\text{A}}{\partial q^k}\right|_{\dot{q}^i} = -\frac{1}{\rho}\frac{\partial p}{\partial q^k} - \frac{\partial \phi_\text{a}}{\partial q^k} + \frac{1}{\rho}\mathbf{q}_k\cdot\nabla\cdot\mathbb{J} \tag{18.63}$$

Since the potential ϕ_a of the gravitational attraction is assumed to be a function of the position q^i only, we may rewrite equation (18.63) by introducing the *Lagrangian function* L, defined by

$$\boxed{L = K_\text{A} - \phi_\text{a} = \frac{1}{2}g_{mn}\dot{q}^m\dot{q}^n + \dot{q}^n(\underset{\Omega}{W_n} + \underset{\text{D}}{W_n}) + \frac{1}{2}\mathbf{v}_\text{D}^2 + \underset{\Omega}{W_m}\underset{\text{D}}{W^m} - \phi} \tag{18.64}$$

As usual, we have combined the potential of the gravitational attraction ϕ_a with the centrifugal potential $\phi_z = \mathbf{v}_\Omega^2/2$ to give the geopotential $\phi = \phi_\text{a} - \phi_z$ representing the effective gravitational force. Thus, we obtain

$$\boxed{\frac{d}{dt}\left(\frac{\partial L}{\partial \dot{q}^k}\right) - \left.\frac{\partial L}{\partial q^k}\right|_{\dot{q}^i} = -\frac{1}{\rho}\frac{\partial p}{\partial q^k} + \frac{1}{\rho}\mathbf{q}_k\cdot\nabla\cdot\mathbb{J}} \tag{18.65}$$

which is known as *Lagrange's equation of motion in general coordinates* or *Langrange's equation of motion of the second kind.* We must observe that the derivatives $\partial L/\partial \dot{q}^k$ and $\partial L/\partial q^k|_{\dot{q}^i}$ are again functions of $\dot{q}^i$ and q^i.

Finally, we expand the individual time operator $d/dt = \partial/\partial t + \dot{q}^n\,\partial/\partial q^n$, which requires differentiations with respect to q^i. Moreover, the velocity $\dot{q}^i$ must no longer be considered a constant. On rewriting (18.65) in detail we find

$$\frac{\partial}{\partial t}\left(\frac{\partial L}{\partial \dot{q}^k}\right) + \dot{q}^n\frac{\partial}{\partial q^n}\left(\frac{\partial L}{\partial \dot{q}^k}\right) - \left.\frac{\partial L}{\partial q^k}\right|_{\dot{q}^i} = -\frac{1}{\rho}\frac{\partial p}{\partial q^k} + \frac{1}{\rho}\mathbf{q}_k\cdot\nabla\cdot\mathbb{J} \tag{18.66}$$

The required derivatives

$$\frac{\partial L}{\partial \dot{q}^k} = g_{mk}\dot{q}^m + \underset{\Omega}{W_k} + \underset{\mathrm{D}}{W_k} = v_k + \underset{\Omega}{W_k} + \underset{\mathrm{D}}{W_k} \tag{18.67}$$

and

$$\begin{aligned}\left.\frac{\partial L}{\partial q^k}\right|_{\dot{q}^i} &= \frac{\dot{q}^m \dot{q}^n}{2}\frac{\partial g_{mn}}{\partial q^k} + \dot{q}^n \frac{\partial}{\partial q^k}\left(\underset{\Omega}{W_n} + \underset{\mathrm{D}}{W_n}\right) + \frac{\partial}{\partial q^k}\left(\frac{\mathbf{v}_{\mathrm{D}}^2}{2}\right) \\ &\quad + \underset{\Omega}{W_n}\frac{\partial}{\partial q^k}\underset{\mathrm{D}}{W^n} + \underset{\mathrm{D}}{W^n}\frac{\partial}{\partial q^k}\underset{\Omega}{W_n} - \frac{\partial \phi}{\partial q^k}\end{aligned} \tag{18.68}$$

follow directly from (18.37). It will be observed that the metric fundamental quantity g_{ij} occurs twice since $g_{ij} = g_{ji}$, so the factor $\frac{1}{2}$ no longer appears in (18.67).

We will now prove that Lagrange's equation (18.65) is a special form of the covariant equation of motion (18.23). We proceed by first taking the individual time derivative of (18.67) to obtain

$$\frac{d}{dt}\left(\frac{\partial L}{\partial \dot{q}^k}\right) = \frac{dv_k}{dt} + \frac{\partial}{\partial t}\underset{\Omega}{W_k} + \dot{q}^n \frac{\partial}{\partial q^n}\underset{\Omega}{W_k} + \frac{\partial}{\partial t}\underset{\mathrm{D}}{W_k} + \dot{q}^n \frac{\partial}{\partial q^n}\underset{\mathrm{D}}{W_k} \tag{18.69}$$

With (18.68) and (18.69) the left-hand side of (18.65) can be written as

$$\begin{aligned}\frac{d}{dt}\left(\frac{\partial L}{\partial \dot{q}^k}\right) - \left.\frac{\partial L}{\partial q^k}\right|_{\dot{q}^i} &= \frac{dv_k}{dt} - \frac{\dot{q}^m \dot{q}^n}{2}\frac{\partial g_{mn}}{\partial q^k} \\ &\quad + \dot{q}^n\left[\frac{\partial}{\partial q^n}\left(\underset{\Omega}{W_k} + \underset{\mathrm{D}}{W_k}\right) - \frac{\partial}{\partial q^k}\left(\underset{\Omega}{W_n} + \underset{\mathrm{D}}{W_n}\right)\right] + \frac{\partial \phi}{\partial q^k} + \frac{\partial}{\partial t}\underset{\mathrm{D}}{W_k} \\ &\quad - \frac{\partial}{\partial q^k}\left(\frac{\mathbf{v}_{\mathrm{D}}^2}{2}\right) + \frac{\partial}{\partial t}\underset{\Omega}{W_k} - \underset{\Omega}{W_n}\frac{\partial}{\partial q^k}\underset{\mathrm{D}}{W^n} - \underset{\mathrm{D}}{W^n}\frac{\partial}{\partial q^k}\underset{\Omega}{W_n}\end{aligned} \tag{18.70}$$

According to (M5.47) the term in brackets is given by $2(\underset{\Omega}{\omega_{nk}} + \underset{\mathrm{D}}{\omega_{nk}})$. Furthermore, the last three terms of (18.70) may be reformulated by means of (18.25), (18.27), and (18.9), yielding $2\underset{\mathrm{D}}{W^n}\underset{\Omega}{\omega_{nk}}$. By substituting the resulting equation into (18.65) we finally obtain

$$\begin{aligned}\frac{dv_k}{dt} - \frac{\dot{q}^m \dot{q}^n}{2}\frac{\partial g_{mn}}{\partial q^k} + 2\dot{q}^n(\underset{\Omega}{\omega_{nk}} + \underset{\mathrm{D}}{\omega_{nk}}) &= -\frac{1}{\rho}\frac{\partial p}{\partial q^k} - \frac{\partial \phi}{\partial q^k} + \frac{1}{\rho}\mathbf{q}_k \cdot \nabla \cdot \mathbb{J} \\ &\quad + \frac{\partial}{\partial q^k}\left(\frac{\mathbf{v}_{\mathrm{D}}^2}{2}\right) - \frac{\partial}{\partial t}\underset{\mathrm{D}}{W_k} - 2\underset{\mathrm{D}}{W^n}\underset{\Omega}{\omega_{nk}}\end{aligned} \tag{18.71}$$

Comparison with the general covariant form of the equation of motion (18.23) shows that these two equations are completely identical. The Langrange method is very powerful, as we shall see from various applications to be described later.

18.6 Hamilton's equation of motion

We conclude this chapter by briefly discussing Hamilton's equation of motion. Let us consider Lagrange's equation in the form for absolute motion. In this case (18.65) can be written as

$$\frac{d}{dt}\left(\frac{\partial L}{\partial \dot{q}^k_{\mathrm{A}}}\right) - \left.\frac{\partial L}{\partial q^k}\right|_{\dot{q}^i_{\mathrm{A}}} = -\frac{1}{\rho}\frac{\partial p}{\partial q^k} + \frac{1}{\rho}\mathbf{q}_k \cdot \nabla \cdot \mathbb{J} \tag{18.72}$$

Now the components $\dot{q}^i_{\mathrm{A}}$, $i = 1, 2, 3$, are held constant when the differentiation with respect to q^k is carried out. Thus, the Lagrangian function is given by

$$L\left(\dot{q}^k_{\mathrm{A}}, q^k\right) = K_{\mathrm{A}}\left(\dot{q}^k_{\mathrm{A}}, q^k\right) - \phi_{\mathrm{a}}(q^k) = \frac{g_{mn}}{2}\dot{q}^m_{\mathrm{A}}\dot{q}^n_{\mathrm{A}} - \phi_{\mathrm{a}}(q^k) \tag{18.73}$$

from which it follows that

$$\frac{\partial L}{\partial \dot{q}^k_{\mathrm{A}}} = \frac{1}{2}\left(g_{mn}\delta^m_k \dot{q}^n_{\mathrm{A}} + g_{mn}\delta^n_k \dot{q}^m_{\mathrm{A}}\right) = g_{kn}\dot{q}^n_{\mathrm{A}} = \dot{q}_{k,\mathrm{A}} \tag{18.74}$$

Let us now introduce the *Hamiltonian function* H, which is defined by

$$H(\dot{q}_{k,\mathrm{A}}, q^k) = K_{\mathrm{A}}(\dot{q}_{k,\mathrm{A}}, q^k) + \phi_{\mathrm{a}}(q^k) = \frac{g^{mn}}{2}\dot{q}_{m,\mathrm{A}}\dot{q}_{n,\mathrm{A}} + \phi_{\mathrm{a}}(q^k) \tag{18.75}$$

To facilitate the operations we have expressed the kinetic energy K_{A} in terms of covariant velocity components. We now differentiate H with respect to the covariant velocity component $\dot{q}_{k,\mathrm{A}}$ and obtain

$$\frac{\partial H}{\partial \dot{q}_{k,\mathrm{A}}} = \frac{1}{2}\left(g^{mn}\delta^k_m \dot{q}_{n,\mathrm{A}} + g^{mn}\delta^k_n \dot{q}_{m,\mathrm{A}}\right) = \frac{1}{2}\left(g^{kn}\dot{q}_{n,\mathrm{A}} + g^{km}\dot{q}_{m,\mathrm{A}}\right) = \dot{q}^k_{\mathrm{A}} \tag{18.76}$$

This results in the contravariant velocity component $\dot{q}^k_{\mathrm{A}}$ which appears in the kinetic energy used in the Lagrange function L in (18.73).

Next we find from (18.73) the derivative of L with respect to q^k, where the contravariant velocity component $\dot{q}^i_{\mathrm{A}}$ is held constant. This yields

$$\left.\frac{\partial L}{\partial q^k}\right|_{\dot{q}^i_{\mathrm{A}}} = \frac{\dot{q}^m_{\mathrm{A}}\dot{q}^n_{\mathrm{A}}}{2}\frac{\partial g_{mn}}{\partial q^k} - \frac{\partial \phi_{\mathrm{a}}}{\partial q^k} \tag{18.77}$$

Holding the covariant velocity constant, we find from (18.75)

$$\left.\frac{\partial H}{\partial q^k}\right|_{\dot{q}_{i,\mathrm{A}}} = \frac{\dot{q}_{m,\mathrm{A}}\dot{q}_{n,\mathrm{A}}}{2}\frac{\partial g^{mn}}{\partial q^k} + \frac{\partial \phi_{\mathrm{a}}}{\partial q^k} \tag{18.78}$$

So far we have not found any important relation between L and H. In order to discover these, we reformulate the first term on the right-hand side of (18.78) with the help of well-known relations, giving

$$\dot{q}_{m,\mathrm{A}}\dot{q}_{n,\mathrm{A}}\frac{\partial g^{mn}}{\partial q^k} = \dot{q}^r_{\mathrm{A}} g_{mr}\dot{q}_{s,\mathrm{A}}\frac{\partial g^{ms}}{\partial q^k} = -\dot{q}^r_{\mathrm{A}}\dot{q}_{s,\mathrm{A}} g^{ms}\frac{\partial g_{mr}}{\partial q^k} = -\dot{q}^r_{\mathrm{A}}\dot{q}^m_{\mathrm{A}}\frac{\partial g_{mr}}{\partial q^k}$$
$$\text{since}\quad g_{mr}g^{ms} = \delta^s_r \Longrightarrow \frac{\partial g_{mr}}{\partial q^k} g^{ms} + g_{mr}\frac{\partial g^{ms}}{\partial q^k} = 0 \tag{18.79}$$

Therefore, equation (18.78) can be rewritten in the form

$$\left.\frac{\partial H}{\partial q^k}\right|_{\dot{q}_{i,\mathrm{A}}} = -\frac{\dot{q}^m_{\mathrm{A}}\dot{q}^n_{\mathrm{A}}}{2}\frac{\partial g_{mn}}{\partial q^k} + \frac{\partial \phi_{\mathrm{a}}}{\partial q^k} \tag{18.80}$$

Comparison of this equation with (18.77) gives the desired relation

$$\left.\frac{\partial H}{\partial q^k}\right|_{\dot{q}_{i,\mathrm{A}}} = -\left.\frac{\partial L}{\partial q^k}\right|_{\dot{q}^i_{\mathrm{A}}} \tag{18.81}$$

Substituting (18.74) into (18.72) gives Lagrange's form of the equation of motion:

$$\boxed{\frac{d\dot{q}_{k,\mathrm{A}}}{dt} = \left.\frac{\partial L}{\partial q^k}\right|_{\dot{q}^i_{\mathrm{A}}} - \frac{1}{\rho}\frac{\partial p}{\partial q^k} + \frac{1}{\rho}\mathbf{q}_k\cdot\nabla\cdot\mathbb{J}} \tag{18.82}$$

Finally, with the help of (18.81), we obtain Hamilton's form of the equation of motion:

$$\boxed{\frac{d\dot{q}_{k,\mathrm{A}}}{dt} = -\left.\frac{\partial H}{\partial q^k}\right|_{\dot{q}_{i,\mathrm{A}}} - \frac{1}{\rho}\frac{\partial p}{\partial q^k} + \frac{1}{\rho}\mathbf{q}_k\cdot\nabla\cdot\mathbb{J}} \tag{18.83}$$

We now look at the same problem from a different point of view. Using the basic definitions of L and H stated in (18.73) and (18.75), we find the relationship between H and L. Since the g_{ij} depend on q^k and $\dot{q}_{\mathrm{A},k} = g_{kn}\dot{q}^n_{\mathrm{A}}$ we may also write

$$H = 2K_{\mathrm{A}} - L = g_{mn}\dot{q}^m_{\mathrm{A}}\dot{q}^n_{\mathrm{A}} - L\left(\dot{q}^k_{\mathrm{A}}, q^k\right) = \dot{q}_{n,\mathrm{A}}\dot{q}^n_{\mathrm{A}} - L\left(\dot{q}^k_{\mathrm{A}}, q^k\right) \tag{18.84}$$

Let us consider H as a function of $(\dot{q}_{\mathrm{A},k}, q^k)$. Then the variation of H results in

$$\delta H = \frac{\partial H}{\partial \dot{q}_{n,\mathrm{A}}}\delta\dot{q}_{n,\mathrm{A}} + \left.\frac{\partial H}{\partial q^n}\right|_{\dot{q}_{i,\mathrm{A}}}\delta q^n = \dot{q}_{n,\mathrm{A}}\,\delta\dot{q}^n_{\mathrm{A}} + \dot{q}^n_{\mathrm{A}}\,\delta\dot{q}_{n,\mathrm{A}} - \frac{\partial L}{\partial \dot{q}^n_{\mathrm{A}}}\delta\dot{q}^n_{\mathrm{A}} - \left.\frac{\partial L}{\partial q^n}\right|_{\dot{q}^i_{\mathrm{A}}}\delta q^n \tag{18.85}$$

Here the first and the third term on the right-hand side cancel out according to (18.74). Inspection of the latter equation shows that

$$\boxed{\frac{\partial H}{\partial \dot{q}_{k,\mathrm{A}}} = \dot{q}^k_{\mathrm{A}},\qquad \left.\frac{\partial H}{\partial q^k}\right|_{\dot{q}_{i,\mathrm{A}}} = -\left.\frac{\partial L}{\partial q^k}\right|_{\dot{q}^i_{\mathrm{A}}}} \tag{18.86}$$

thus verifying the relations (18.76) and (18.81).

Equations (18.76) and (18.83),

$$\boxed{\frac{\partial H}{\partial \dot{q}_{k,\mathrm{A}}} = \dot{q}^k_{\mathrm{A}}, \qquad \left.\frac{\partial H}{\partial q^k}\right|_{\dot{q}_{i,\mathrm{A}}} = -\frac{d\dot{q}_{k,\mathrm{A}}}{dt} - \frac{1}{\rho}\frac{\partial p}{\partial q^k} + \frac{1}{\rho}\mathbf{q}_k \cdot \nabla \cdot \mathbb{J}} \tag{18.87}$$

are known as *Hamilton's canonical equations of motion*. They consist of $2n$ first-order differential equations whereas Lagrange's equation consists of n second-order differential equations. In our case $n = 3$. The velocity component $\dot{q}_{k,\mathrm{A}}$ may be considered as the momentum for unit mass. Hamilton's canonical equations are very general and hold for the case that the potential-energy function also depends on $\dot{q}^k$ (the tidal problem) and for systems in which L explicitly depends on time, but in these cases the total energy is no longer necessarily H.

In our furture work we will have many opportunities to work with Lagrange's equation of motion, but we refrain from using the Hamilton formulation, which is an extremely important tool in quantum mechanics. In our work the Hamilton formulation has no decisive advantage over the Lagrange method. A very simple but illuminating example will be given next in order to obtain Hamilton's equations of motion for the one-dimensional harmonic oscillator. The equation of the harmonic oscillator is of great importance in many branches of physics and in physical meteorology, but it will usually be obtained by elementary considerations.

Recalling that there is no difference between covariant and contravariant coordinates in the Cartesian system, H assumes a particularly simple form. Since we are not dealing with unit mass, we write for the kinetic energy K_{A} and the potential energy V

$$K_{\mathrm{A}} = \frac{m\dot{x}^2}{2}, \qquad V = \frac{kx^2}{2} \tag{18.88}$$

where m is the mass of the oscillating particle and k is *Hooke's constant*. Now the Hamiltonian function is given by

$$H = \frac{m\dot{x}^2}{2} + \frac{kx^2}{2} \tag{18.89}$$

resulting in Hamilton's equations of motion as

$$\frac{\partial H}{\partial (m\dot{x})} = \dot{x}, \qquad \left.\frac{\partial H}{\partial x}\right|_{\dot{x}^i} = -m\frac{d\dot{x}}{dt} = kx \tag{18.90}$$

The first expression is simply an identity while the second expression is the well-known *equation of the harmonic oscillator*:

$$m\,\frac{d\dot{x}}{dt} + kx = 0 \tag{18.91}$$

18.7 Appendix

The material of this appendix is not essential for understanding the following chapters and may therefore be omitted. However, it helps us to grasp the consequences of the metric simplification which will be introduced soon.

Let us form the *Riemann–Christoffel tensor* or the *covariant curvature tensor.* First we form the covariant expression $\boldsymbol{\nabla}_i(\boldsymbol{\nabla}_j A_k) - \boldsymbol{\nabla}_j(\boldsymbol{\nabla}_i A_k)$, where A_k is any covariant tensor of rank 1. For the first term, according to Section M5.6, we may write

$$
\begin{aligned}
\boldsymbol{\nabla}_i(\boldsymbol{\nabla}_j A_k) &= \nabla_i(\boldsymbol{\nabla}_j A_k) - (\boldsymbol{\nabla}_m A_k)\Gamma_{ij}^m - (\boldsymbol{\nabla}_j A_m)\Gamma_{ik}^m \\
&= \nabla_i(\nabla_j A_k - A_n \Gamma_{jk}^n) - \Gamma_{ij}^m(\nabla_m A_k - A_n \Gamma_{mk}^n) - \Gamma_{ik}^m(\nabla_j A_m - A_n \Gamma_{jm}^n) \\
&= \underline{\nabla_i(\nabla_j A_k)} - \underline{\Gamma_{jk}^n \nabla_i A_n} - A_n \nabla_i \Gamma_{jk}^n - \underline{\Gamma_{ij}^m \nabla_m A_k} \\
&\quad + \underline{A_n \Gamma_{ij}^m \Gamma_{mk}^n} - \underline{\Gamma_{ik}^m \nabla_j A_m} + A_n \Gamma_{ik}^m \Gamma_{jm}^n
\end{aligned}
\tag{18.92}
$$

The second term is found by interchanging the free indices i and j. We obtain

$$
\begin{aligned}
\boldsymbol{\nabla}_j(\boldsymbol{\nabla}_i A_k) = {} & \underline{\nabla_j(\nabla_i A_k)} - \underline{\Gamma_{ik}^n \nabla_j A_n} - A_n \nabla_j \Gamma_{ik}^n - \underline{\Gamma_{ji}^m \nabla_m A_k} \\
& + \underline{A_n \Gamma_{ji}^m \Gamma_{mk}^n} - \underline{\Gamma_{jk}^m \nabla_i A_m} + A_n \Gamma_{jk}^m \Gamma_{im}^n
\end{aligned}
\tag{18.93}
$$

Next we subtract (18.93) from (18.92). The underlined terms cancel out since $\Gamma_{ij}^k = \Gamma_{ji}^k$ and due to the fact that $\nabla_i \nabla_j = \nabla_j \nabla_i$. Thus, we obtain

$$
\boldsymbol{\nabla}_i(\boldsymbol{\nabla}_j A_k) - \boldsymbol{\nabla}_j(\boldsymbol{\nabla}_i A_k) = A_n(\Gamma_{ik}^m \Gamma_{jm}^n - \nabla_i \Gamma_{jk}^n - \Gamma_{jk}^m \Gamma_{im}^n + \nabla_j \Gamma_{ik}^n) = A_n R_{kij}^n \tag{18.94}
$$

The expression

$$
\boxed{R_{kij}^l = \Gamma_{ik}^m \Gamma_{jm}^l - \nabla_i \Gamma_{jk}^l - \Gamma_{jk}^m \Gamma_{im}^l + \nabla_j \Gamma_{ik}^l} \tag{18.95}
$$

is known as the Riemann–Christoffel tensor or the *curvature tensor*.

In the Cartesian coordinate system we obviously have $R_{kij}^l = 0$. Thus, in the Euclidean space the interchange rule

$$
\boldsymbol{\nabla}_i \boldsymbol{\nabla}_j = \boldsymbol{\nabla}_j \boldsymbol{\nabla}_i \tag{18.96}
$$

is valid. In atmospheric dynamics we often use simplified metric forms resulting from the assumption that the radius of the earth–atmosphere system is equal to the mean radius of the earth itself. By assuming that the atmosphere has zero vertical extent we are leaving the Euclidean metric. Therefore, there exists no relation between the q^i coordinates of the chosen system and the Cartesian coordinates, so $q^i \neq q^i(x^1, x^2, x^3, t)$. In this case the curvature tensor does not vanish and the interchange rule (18.96) concerning the covariant derivatives is no longer valid.

18.8 Problems

18.1: Perform in detail the steps involved in going from (18.65) to (18.68).

18.2: Show the steps involved in going from (18.78) to (18.79).

18.3: Express the Euler expansion d/dt with the help of covariant, contravariant, and physical measure numbers of the velocity in the spherical coordinate system with $q^1 = \lambda$, $q^2 = \varphi$, and $q^3 = r$.

18.4: Express the continuity equation for the rigid spherical coordinate system by employing

(a) physical measure numbers of the velocity, and
(b) contravariant measure numbers of the velocity.

18.5: Consider the spherical coordinate system shown in Figure 1.2. For rigid rotation ($\mathbf{v}_{\mathrm{D}} = 0$) the kinetic energy is given by

$$K_{\mathrm{A}} = \frac{1}{2}\left[r^2 \cos(\dot{\lambda} + \Omega)^2 + r^2\dot{\varphi}^2 + \dot{r}^2\right]$$

(a) From the metric fundamental quantities (see Section M4.2.1) find $\underset{\Omega}{W}_i$, $\underset{\Omega}{W}^j$, Γ_{ij}^k with $i, j, k = 1, 2, 3$.

(b) Find the elements of the matrices $(\underset{\Omega}{\omega}_{ij})$ and $(\underset{\Omega}{\overset{*}{\omega}}_{ij})$.

19

The geographical coordinate system

In this chapter we will present the equation of motion and the continuity equation in component form for the *geographical coordinate system.* This is a spherical coordinate system that is rotating with constant angular velocity Ω about the polar axis $\varphi = \pi/2$. The horizontal coordinate lines to be used are the geographical length λ and the geographical latitude φ. The vertical coordinate is taken as the distance r from the center of the earth to the point under consideration. The horizontal coordinate lines are curved but orthogonal to each other. The deformation of coordinate surfaces will not be taken into account. A graphical representation of the geographical system is shown in Figures 1.2 and 1.4 of Chapter 1.

19.1 The equation of motion

All we need do is employ the metric tensor as well as the velocity $\mathbf{v}_P$ of the geographical system in connection with the general formulations of the equations of motion which we have presented in the previous chapter. The covariant forms of the metric fundamental quantities have been derived previously; see (1.73). Since we are dealing with an orthogonal coordinate system, we obtain the contravariant form from the basic relation $g_{ii} = 1/g^{ii}$ so that

$$\begin{aligned} &g_{11} = r^2\cos^2\varphi, \qquad g_{22} = r^2, \qquad g_{33} = 1, \qquad g_{ij} = 0, \quad i \neq j \\ &g^{11} = \frac{1}{r^2\cos^2\varphi}, \qquad g^{22} = \frac{1}{r^2}, \qquad g^{33} = 1, \qquad g^{ij} = 0, \quad i \neq j \end{aligned} \tag{19.1}$$

Of fundamental importance is the absolute kinetic energy $K_{\rm A}$ of the system:

$$\boxed{K_{\rm A} = \frac{1}{2}\left(\frac{d\mathbf{r}}{dt}\right)^2 = \frac{\mathbf{v}_{\rm A}^2}{2} = \frac{1}{2}\left(r^2\cos^2\varphi\,\dot{\lambda}_{\rm A}^2 + r^2\dot{\varphi}_{\rm A}^2 + \dot{r}_{\rm A}^2\right)} \tag{19.2}$$

where $d\mathbf{r}$ is taken from (1.72). The components of the contravariant velocity are given by

$$\dot{q}_{\mathrm{A}}^{1} = \dot{\lambda}_{\mathrm{A}} = \dot{\lambda} + \Omega, \qquad \dot{q}_{\mathrm{A}}^{2} = \dot{\varphi}_{\mathrm{A}} = \dot{\varphi}, \qquad \dot{q}_{\mathrm{A}}^{3} = \dot{r}_{\mathrm{A}} = \dot{r} \tag{19.3}$$

The angular velocity Ω of the rotating system has been added to $\dot{\lambda}$ since we are considering the motion in the absolute reference frame. The covariant and contravariant components of the velocity $\mathbf{v}_P = \mathbf{v}_\Omega$ are then given by

$$\begin{aligned} &\underset{\Omega}{W}^1 = \Omega, && \underset{\Omega}{W}^2 = 0, && \underset{\Omega}{W}^3 = 0 \\ &\underset{\Omega}{W}_1 = \Omega r^2 \cos^2\varphi, && \underset{\Omega}{W}_2 = 0, && \underset{\Omega}{W}_3 = 0 \end{aligned} \tag{19.4}$$

There is another way to obtain the metric fundamental quantities as well as the components $\underset{\Omega}{W}_i$. According to (M4.38), the kinetic energy K_{A} of the system provides all the information which is needed. First we split K_{A} into the kinetic energy K of relative motion and the kinetic energy K_P of the rotating system, $K_{\mathrm{A}} = K + K_P$, with

$$\begin{aligned} K_{\mathrm{A}} &= \frac{\mathbf{v}_{\mathrm{A}}^2}{2} = \frac{(\mathbf{v} + \mathbf{v}_P)^2}{2} = \frac{\mathbf{v}^2}{2} + \mathbf{v}\cdot\mathbf{v}_P + \frac{\mathbf{v}_P^2}{2} \\ K &= \frac{\mathbf{v}^2}{2} = \frac{1}{2}(r^2\cos^2\varphi\,\dot{\lambda}^2 + r^2\dot{\varphi}^2 + \dot{r}^2) \\ K_P &= \mathbf{v}\cdot\mathbf{v}_P + \frac{\mathbf{v}_P^2}{2} = \Omega r^2\cos^2\varphi\,\dot{\lambda} + \frac{\Omega^2}{2} r^2\cos^2\varphi \end{aligned} \tag{19.5}$$

From these expressions we then obtain

$$\begin{aligned} &g_{11} = \frac{\partial^2 K}{\partial\dot{\lambda}^2} = r^2\cos^2\varphi, && g_{22} = \frac{\partial^2 K}{\partial\dot{\varphi}^2} = r^2, && g_{33} = \frac{\partial^2 K}{\partial\dot{r}^2} = 1 \\ &\underset{\Omega}{W}_1 = \frac{\partial K_P}{\partial\dot{\lambda}} = \Omega r^2\cos^2\varphi, && \underset{\Omega}{W}_2 = \frac{\partial K_P}{\partial\dot{\varphi}} = 0, && \underset{\Omega}{W}_3 = \frac{\partial K_P}{\partial\dot{r}} = 0 \end{aligned} \tag{19.6}$$

The covariant form of the equation of motion for rigid rotation is given by (18.25). In order to evaluate this equation, we need to calculate the components of the tensor $\underset{\Omega}{\omega}_{ij}$ as defined in (M5.48). Inspection shows that this tensor is antisymmetric or skew-symmetric, as it is often called, so that

$$\underset{\Omega}{\omega}_{ij} = \frac{1}{2}\left(\frac{\partial}{\partial q^i}\underset{\Omega}{W}_j - \frac{\partial}{\partial q^j}\underset{\Omega}{W}_i\right) = -\underset{\Omega}{\omega}_{ji} \tag{19.7}$$

For consistency with our previous notation we put $(q^1, q^2, q^3) = (\lambda, \varphi, r)$. For ease of reference we summarize the necessary components of the rotational tensor in

$$
\begin{array}{lll}
\underset{\Omega}{\omega}_{11} = 0, & \underset{\Omega}{\omega}_{12} = \Omega r^2 \cos\varphi \sin\varphi, & \underset{\Omega}{\omega}_{13} = -\Omega r \cos^2\varphi \\
\underset{\Omega}{\omega}_{21} = -\underset{\Omega}{\omega}_{12}, & \underset{\Omega}{\omega}_{22} = 0, & \underset{\Omega}{\omega}_{23} = 0 \\
\underset{\Omega}{\omega}_{31} = -\underset{\Omega}{\omega}_{13}, & \underset{\Omega}{\omega}_{32} = 0, & \underset{\Omega}{\omega}_{33} = 0 \\
\\
\underset{\Omega}{\omega}_i{}^k = g^{km}\underset{\Omega}{\omega}_{im}, & \underset{\Omega}{\overset{*}{\omega}}_{ij} = \sqrt{g^{ii}}\sqrt{g^{jj}}\underset{\Omega}{\omega}_{ij} &
\end{array}
\tag{19.8}
$$

In order to write down the contravariant form of the equation of motion, we need to specify the Christoffel symbols of the second kind whose definition is repeated here:

$$
\Gamma^k_{ij} = \frac{g^{kn}}{2}\left(\frac{\partial g_{in}}{\partial q^j} + \frac{\partial g_{jn}}{\partial q^i} - \frac{\partial g_{ij}}{\partial q^n}\right) = \Gamma^k_{ji}
\tag{19.9}
$$

Even though the Christoffel symbols have the outward appearance of a tensor, they do not obey the transformation laws for tensors. Nevertheless, it is still possible to raise and lower the indices. In the most general case we would have 27 nonzero Christoffel symbols. For orthogonal systems the easy but tedious computational work is drastically reduced. For the geographical system the Christoffel symbols are

$$
\begin{array}{lll}
\Gamma^1_{11} = 0, & \Gamma^1_{12} = -\tan\varphi, & \Gamma^1_{13} = 1/r \\
\Gamma^1_{21} = \Gamma^1_{12}, & \Gamma^1_{22} = 0, & \Gamma^1_{23} = 0 \\
\Gamma^1_{31} = \Gamma^1_{13}, & \Gamma^1_{32} = 0, & \Gamma^1_{33} = 0 \\
\\
\Gamma^2_{11} = \cos\varphi\sin\varphi, & \Gamma^2_{12} = 0, & \Gamma^2_{13} = 0 \\
\Gamma^2_{21} = 0, & \Gamma^2_{22} = 0, & \Gamma^2_{23} = 1/r \\
\Gamma^2_{31} = 0, & \Gamma^2_{32} = \Gamma^2_{23}, & \Gamma^2_{33} = 0 \\
\\
\Gamma^3_{11} = -r\cos^2\varphi & \Gamma^3_{12} = 0, & \Gamma^3_{13} = 0 \\
\Gamma^3_{21} = 0, & \Gamma^3_{22} = -r, & \Gamma^3_{23} = 0 \\
\Gamma^3_{31} = 0, & \Gamma^3_{32} = 0, & \Gamma^3_{33} = 0
\end{array}
\tag{19.10}
$$

Using the above information it is a simple task to write down the covariant form of the equation of motion:

$$\begin{aligned}
\frac{dv_1}{dt} + \underline{2\Omega r \cos^2\varphi\, \dot{r}} - 2\Omega r^2 \cos\varphi \sin\varphi\, \dot{\varphi} &= -\frac{1}{\rho}\frac{\partial p}{\partial \lambda} - \frac{\partial \phi}{\partial \lambda} + \frac{1}{\rho}\mathbf{q}_\lambda \cdot \nabla \cdot \mathbb{J} \\
\frac{dv_2}{dt} + r^2 \cos\varphi \sin\varphi\, \dot{\lambda}^2 + 2\Omega r^2 \cos\varphi \sin\varphi\, \dot{\lambda} &= -\frac{1}{\rho}\frac{\partial p}{\partial \varphi} - \frac{\partial \phi}{\partial \varphi} + \frac{1}{\rho}\mathbf{q}_\varphi \cdot \nabla \cdot \mathbb{J} \\
\frac{dv_3}{dt} - \underline{r \cos^2\varphi\, \dot{\lambda}^2} - \underline{r\dot{\varphi}^2} - \underline{2\Omega r \cos^2\varphi\, \dot{\lambda}} &= -\frac{1}{\rho}\frac{\partial p}{\partial r} - \frac{\partial \phi}{\partial r} + \frac{1}{\rho}\mathbf{q}_r \cdot \nabla \cdot \mathbb{J} \\
\frac{d}{dt} = \frac{\partial}{\partial t} + \frac{v_1}{r^2 \cos^2\varphi}\frac{\partial}{\partial \lambda} &+ \frac{v_2}{r^2}\frac{\partial}{\partial \varphi} + v_3 \frac{\partial}{\partial r}
\end{aligned} \tag{19.11}$$

and the contravariant form:

$$\begin{aligned}
\frac{d\dot{\lambda}}{dt} + \underline{\frac{2}{r}(\dot{\lambda} + \Omega)\dot{r}} - 2(\dot{\lambda} + \Omega)\tan\varphi\, \dot{\varphi} &= -\frac{1}{r^2\cos^2\varphi}\left(\frac{1}{\rho}\frac{\partial p}{\partial \lambda} + \frac{\partial \phi}{\partial \lambda}\right) + \frac{1}{\rho}\mathbf{q}^\lambda \cdot \nabla \cdot \mathbb{J} \\
\frac{d\dot{\varphi}}{dt} + \cos\varphi \sin\varphi\, (\dot{\lambda} + 2\Omega)\dot{\lambda} + \underline{\frac{2\dot{\varphi}\dot{r}}{r}} &= -\frac{1}{r^2}\left(\frac{1}{\rho}\frac{\partial p}{\partial \varphi} + \frac{\partial \phi}{\partial \varphi}\right) + \frac{1}{\rho}\mathbf{q}^\varphi \cdot \nabla \cdot \mathbb{J} \\
\frac{d\dot{r}}{dt} - \underline{r\cos^2\varphi\, (\dot{\lambda} + 2\Omega)\dot{\lambda}} - \underline{r\dot{\varphi}^2} &= -\left(\frac{1}{\rho}\frac{\partial p}{\partial r} + \frac{\partial \phi}{\partial r}\right) + \frac{1}{\rho}\mathbf{q}^r \cdot \nabla \cdot \mathbb{J} \\
\frac{d}{dt} = \frac{\partial}{\partial t} + \dot{\lambda}\frac{\partial}{\partial \lambda} &+ \dot{\varphi}\frac{\partial}{\partial \varphi} + \dot{r}\frac{\partial}{\partial r}
\end{aligned} \tag{19.12}$$

just by following the general equations (18.25) and (18.34). The reader may decide for himself which form he likes best and which set of equations seems easier to use. It should be noted that the underlined terms are usually omitted in practical applications. The justification for omitting these terms will be given a little later.

In equation (19.11) the leading terms are written as dv_i/dt while the remaining velocity variables appearing in this equation are $(\dot{\lambda}, \dot{\varphi}, \dot{r})$. If it is desired to use the contravariant components of the velocity $(\dot{\lambda}, \dot{\varphi}, \dot{r})$ everywhere in this equation, we may replace the v_i by using the definitions

$$v_1 = g_{1n}v^n = r^2\cos^2\varphi\, \dot{\lambda}, \qquad v_2 = g_{2n}v^n = r^2\dot{\varphi}, \qquad v_3 = g_{3n}v^n = \dot{r} \tag{19.13}$$

We will now derive the three components of the equation of motion in physical coordinates. Since deformation of the coordinate surfaces is ignored ($\mathbf{v}_D = 0$) the equation of motion in the form (18.37) may be applied. For the chosen geographical coordinate system we have

$$
\begin{aligned}
&\overset{*}{v}^k = \sqrt{g_{kk}}\dot{q}^k \quad \Longrightarrow \overset{*}{v}^1 = r\cos\varphi\,\dot{\lambda}, \qquad \overset{*}{v}^2 = r\dot{\varphi}, \qquad \overset{*}{v}^3 = \dot{r} \\
&\frac{\partial}{\partial \overset{*}{q}^k} = \frac{1}{\sqrt{g_{kk}}}\frac{\partial}{\partial q^k} \Longrightarrow \frac{\partial}{\partial \overset{*}{q}^1} = \frac{1}{r\cos\varphi}\frac{\partial}{\partial\lambda}, \qquad \frac{\partial}{\partial \overset{*}{q}^2} = \frac{1}{r}\frac{\partial}{\partial\varphi}, \qquad \frac{\partial}{\partial \overset{*}{q}^3} = \frac{\partial}{\partial r}
\end{aligned}
\tag{19.14}
$$

Recall that, in this particular system, there is no difference between covariant and contravariant physical measure numbers. Using (19.8) and the information (19.6) on the metric tensor g_{ij}, it is a simple matter to obtain the component equations in physical measure numbers by successively setting $k = 1, 2, 3$ in (18.37). Denoting the velocity components by $\overset{*}{v}^1 = u$, $\overset{*}{v}^2 = v$, and $\overset{*}{v}^3 = w$, we obtain

$$
\boxed{
\begin{aligned}
\frac{du}{dt} + \underline{\frac{uw}{r}} - \frac{uv}{r}\tan\varphi - fv + \underline{lw} &= -\frac{1}{r\cos\varphi}\left(\frac{1}{\rho}\frac{\partial p}{\partial\lambda} + \frac{\partial\phi}{\partial\lambda}\right) + \frac{1}{\rho}\mathbf{e}_\lambda\cdot\nabla\cdot\mathbb{J} \\
\frac{dv}{dt} + \underline{\frac{uw}{r}} + \frac{u^2}{r}\tan\varphi + fu &= -\frac{1}{r}\left(\frac{1}{\rho}\frac{\partial p}{\partial\varphi} + \frac{\partial\phi}{\partial\varphi}\right) + \frac{1}{\rho}\mathbf{e}_\varphi\cdot\nabla\cdot\mathbb{J} \\
\frac{dw}{dt} - \underline{\frac{(u^2+v^2)}{r}} - \underline{lu} &= -\left(\frac{1}{\rho}\frac{\partial p}{\partial r} + \frac{\partial\phi}{\partial r}\right) + \frac{1}{\rho}\mathbf{e}_r\cdot\nabla\cdot\mathbb{J} \\
\frac{d}{dt} = \frac{\partial}{\partial t} + \frac{u}{r\cos\varphi}\frac{\partial}{\partial\lambda} &+ \frac{v}{r}\frac{\partial}{\partial\varphi} + w\frac{\partial}{\partial r}
\end{aligned}
}
\tag{19.15}
$$

which is the most frequently used form of the equation of motion. For brevity we have also introduced the Coriolis parameters $f = 2\Omega\sin\varphi$ and $l = 2\Omega\cos\varphi$; see Figure 1.4 and equation (1.81). It should be noted that the underlined terms are usually omitted.

Certainly, it would have been possible to obtain the equation of motion in spherical coordinates from the vector equation directly, as we have already demonstrated in Section 1.6 for physical measure numbers. However, by using the general forms of the equation of motion derived in Chapter 18, we avoid entirely any problems arising from the differentiation of the basis or unit vectors with respect to time and space.

19.2 Application of Lagrange's equation of motion

Now we wish to demonstrate how to obtain the components of the covariant form of the equation of motion by using the powerful Lagrange formalism. At this point it will be sufficient to demonstrate the method for $k = 1$. For this situation equation (18.65) can be written as

$$\frac{d}{dt}\left(\frac{\partial L}{\partial \dot{\lambda}}\right) - \left.\frac{\partial L}{\partial \lambda}\right|_{\dot{q}^i} = -\frac{1}{\rho}\frac{\partial p}{\partial \lambda} + \frac{1}{\rho}\mathbf{q}_\lambda \cdot \nabla \cdot \mathbb{J} \tag{19.16}$$

where the Lagrangian is given by

$$L = K_{\mathrm{A}} - \phi_{\mathrm{a}} = \tfrac{1}{2}(r^2 \cos^2\varphi\, \dot{\lambda}^2 + r^2\dot{\varphi}^2 + \dot{r}^2 + 2\Omega r^2 \cos^2\varphi\, \dot{\lambda} + r^2 \cos^2\varphi\, \Omega^2) - \phi_{\mathrm{a}} \tag{19.17}$$

From this equation we easily find

$$\begin{aligned} \frac{\partial L}{\partial \dot{\lambda}} &= r^2 \cos^2\varphi\, \dot{\lambda} + \Omega r^2 \cos^2\varphi = v_1 + \Omega r^2 \cos^2\varphi \\ \left.\frac{\partial L}{\partial \lambda}\right|_{\dot{q}^i} &= -\frac{\partial \phi_{\mathrm{a}}}{\partial \lambda} = -\frac{\partial \phi}{\partial \lambda} \quad \text{with} \quad \phi = \phi_{\mathrm{a}} - \phi_z = \phi_{\mathrm{a}} - \frac{\Omega^2}{2} r^2 \cos^2\varphi \end{aligned} \tag{19.18}$$

which are then substituted into (19.16). With very little effort we find the equation of motion for the covariant measure number v_1:

$$\frac{dv_1}{dt} + 2\Omega r \cos^2\varphi\, \dot{r} - 2\Omega r^2 \cos\varphi \sin\varphi\, \dot{\varphi} = -\frac{1}{\rho}\frac{\partial p}{\partial \lambda} - \frac{\partial \phi}{\partial \lambda} + \frac{1}{\rho}\mathbf{q}_\lambda \cdot \nabla \cdot \mathbb{J} \tag{19.19}$$

which is, of course, identical with the first equation in (19.11).

In order to obtain the corresponding equation in physical measure numbers, we transform (19.19). However, there may be situations in which it is preferable to use (18.37) or even (18.36) when the deformation velocity $\mathbf{v}_{\mathrm{D}}$ cannot be ignored. To transform (19.19) into physical velocity variables we use the relations

$$v_1 = r \cos\varphi\, u, \qquad \dot{\varphi} = v/r, \qquad \dot{r} = w \tag{19.20}$$

Substitution of (19.20) into (19.19) gives

$$\frac{d}{dt}(r \cos\varphi\, u) + lr \cos\varphi\, w - lr \sin\varphi\, v = -\frac{1}{\rho}\frac{\partial p}{\partial \lambda} - \frac{\partial \phi}{\partial \lambda} + \frac{1}{\rho}\mathbf{q}_\lambda \cdot \nabla \cdot \mathbb{J} \tag{19.21}$$

Simple differentiation of the first term finally results in the first equation of (19.15). Similarly, we may use the Lagrange method to verify the remainder of (19.15).

19.3 The first metric simplification

Now we wish to explain the meaning of the underlined terms of the previous equations. Observing that the vertical extent of the atmosphere is very small in comparison with the radius a of the earth, it seems reasonable to introduce the so-called *first metric simplification*. By this we mean that, in coordinate systems having r or z as a vertical coordinate, we appear to be justified in replacing the variable radius r by the fixed radius $r = a$ whenever r appears in undifferentiated form, i.e.

$$g_{11} = a^2 \cos^2 \varphi, \quad g_{22} = a^2, \quad g_{33} = 1, \quad g_{ii} = 1/g^{ii}, \quad g_{ij} = 0 \text{ for } i \neq j \tag{19.22}$$

This simply means that the components of the metric tensor g_{ij} and g^{ij} become independent of height. From (19.22) it follows that

$$(d\mathbf{r})^2 = a^2 \cos^2 \varphi \, (d\lambda)^2 + a^2 (d\varphi)^2 + (dr)^2 \tag{19.23}$$

Equations (19.22) and (19.23) describe a metric system in which various spherical surfaces in the atmosphere, actually having different radii of curvature r, are forced to assume the same radius of curvature a. This system cannot be visualized in three-dimensional Euclidean space. We have already made some remarks on the effect of this approximation in the appendix to Chapter 18.

The first metric simplification also leads to a change in the definition of the absolute kinetic energy K_{A}. Instead of (19.2) we now write

$$K_{\text{A}} = \tfrac{1}{2}[a^2 \cos^2 \varphi \, (\dot{\lambda}^2 + 2\dot{\lambda}\Omega) + a^2 \dot{\varphi}^2 + \dot{r}^2] + \tfrac{1}{2} a^2 \cos^2 \varphi \, \Omega^2 \tag{19.24}$$

and for the functional determinant

$$\sqrt{g} = \sqrt{|g_{ij}|} = \sqrt{\left| \frac{\partial^2 K_{\text{A}}}{\partial \dot{q}^i \, \partial \dot{q}^j} \right|} = a^2 \cos \varphi \tag{19.25}$$

Specifically, if we were to repeat the derivations leading to equations (19.10), (19.11), and (19.13), but using the g_{ij} as stated in (19.22), the underlined terms would not appear at all.

Comparison of equation (19.15), omitting the underlined terms, with equation (2.26) shows that they are identical. Thus, the use of the first metric simplification gives the same result as the scale analysis leading to (2.26). It should be noted, however, that, in (2.26b), we have also ignored the latitudinal dependency of the geopotential due to the assumption (1.83b).

19.4 The coordinate simplification

In this section we wish to briefly review and elaborate the discussion on the geopotential presented in Section 1.6. From the physical point of view it would be desirable to replace the spherical coordinate system which we have used to describe the atmospheric motion by another orthogonal coordinate system. This new coordinate system should be adapted to the force fields of the rotating earth in such a way that the vertical coordinate q^3 should follow the force lines resulting from the gravitational attraction of the earth and from the centrifugal force.

In this system surfaces q^3 = constant would coincide with surfaces of constant geopotential $\phi = \phi_{\mathrm{a}} - \frac{1}{2}(\Omega^2 r^2 \cos^2\varphi)$, which may be well approximated by a rotational ellipsoid of very small eccentricity. Such a surface ϕ = constant would be the earth's surface if it were not rigid, but instead had a freely movable surface mass such as water, and if it were subjected only to the gravitational pull of the earth and to the centrifugal force. The analytic consequence of this special idealized surface would cause the horizontal derivatives of the geopotential to vanish.

Such an ideal coordinate system to describe the atmospheric motion would be a spheroidal coordinate system, which is characterized by a rather complicated metric tensor. The description of the metric fundamental quantities would then require the specification of the geocentric latitude, see Figure 1.3, and the eccentricity of the earth. Moreover, hyperbolic functions would arise in the description of the g_{ii} instead of the trigonometric functions appearing in (19.1). However, the difference between the geographical and the geocentric latitude is very small and it would be very impractical to introduce an elliptic coordinate system to describe the flow instead of the simple spherical coordinate system we have used so far. For this reason we will continue to use the spherical coordinate system, but we retain the advantage of the elliptic system by assuming that, in the relevant section of the atmosphere, surfaces of ϕ = constant coincide with surfaces of r = constant. Thus, the horizontal derivatives vanish:

$$\frac{\partial \phi}{\partial \lambda} = 0, \qquad \frac{\partial \phi}{\partial \varphi} = 0 \Longrightarrow \phi = \phi(r) \tag{19.26}$$

Furthermore, we may assume that, in the region relevant to atmospheric weather systems, the geopotential is a linear function of $q^3 = r$:

$$\phi = gr + \text{constant} \tag{19.27}$$

From this it follows that

$$\frac{\partial \phi}{\partial r} = g \tag{19.28}$$

where we take $g = 9.81\ \mathrm{m\ s^{-1}}$ as a sufficiently representative value. We will call the approximations (19.26)–(19.28) the *coordinate simplification*.

19.5 The continuity equation

In the absence of deformation of the coordinate surfaces, $\mathbf{v}_{\mathrm{D}} = 0$, the functional determinant for the rigid rotation of the spherical coordinate system is given by $\sqrt{g} = r^2 \cos\varphi$ and the continuity equation in the form (18.48) is valid. Thus, we immediately obtain

$$\frac{\partial \rho}{\partial t} + \frac{1}{r^2 \cos\varphi}\left(\frac{\partial}{\partial \lambda}(\rho r^2 \cos\varphi\, \dot{\lambda}) + \frac{\partial}{\partial \varphi}(\rho r^2 \cos\varphi\, \dot{\varphi}) + \frac{\partial}{\partial r}(\rho r^2 \cos\varphi\, \dot{r})\right) = 0 \tag{19.29}$$

On combining the various terms we obtain the continuity equation with contravariant velocity components in the form

$$\begin{gathered}\frac{1}{\rho r^2 \cos\varphi}\frac{d}{dt}(\rho r^2 \cos\varphi) + \frac{\partial \dot{\lambda}}{\partial \lambda} + \frac{\partial \dot{\varphi}}{\partial \varphi} + \frac{\partial \dot{r}}{\partial r} = 0 \\ \text{with} \quad \frac{d}{dt} = \frac{\partial}{\partial t} + \dot{\lambda}\frac{\partial}{\partial \lambda} + \dot{\varphi}\frac{\partial}{\partial \varphi} + \dot{r}\frac{\partial}{\partial r}\end{gathered} \tag{19.30}$$

By introducing the relations (19.14) into (19.29), we find the continuity equation expressed in physical velocity components:

$$\begin{gathered}\frac{1}{\rho}\frac{d\rho}{dt} + \frac{1}{r\cos\varphi}\left(\frac{\partial u}{\partial \lambda} + \frac{\partial}{\partial \varphi}(v\cos\varphi)\right) + \frac{\partial w}{\partial r} + \frac{2w}{r} = 0 \\ \text{with} \quad \frac{d}{dt} = \frac{\partial}{\partial t} + \frac{u}{r\cos\varphi}\frac{\partial}{\partial \lambda} + \frac{v}{r}\frac{\partial}{\partial \varphi} + w\frac{\partial}{\partial r}\end{gathered} \tag{19.31}$$

Suppose that we repeat the derivation of the continuity equation leading to (19.30) but now using the first metric simplification. This simply means that we replace r^2 by a^2. In this case the functional determinant (19.25) must be used, so equation (19.30) simplifies to the form

$$\boxed{\begin{gathered}\frac{1}{\rho \cos\varphi}\frac{d}{dt}(\rho\cos\varphi) + \frac{\partial \dot{\lambda}}{\partial \lambda} + \frac{\partial \dot{\varphi}}{\partial \varphi} + \frac{\partial \dot{r}}{\partial r} = 0 \\ \text{with} \quad \frac{d}{dt} = \frac{\partial}{\partial t} + \dot{\lambda}\frac{\partial}{\partial \lambda} + \dot{\varphi}\frac{\partial}{\partial \varphi} + \dot{r}\frac{\partial}{\partial r}\end{gathered}} \tag{19.32}$$

which is used in many practical applications. Finally, utilizing the first metric simplification, equation (19.31) can be written as

$$\boxed{\begin{gathered}\frac{1}{\rho}\frac{d\rho}{dt} + \frac{1}{a\cos\varphi}\left(\frac{\partial u}{\partial \lambda} + \frac{\partial}{\partial \varphi}(v\cos\varphi)\right) + \frac{\partial w}{\partial r} = 0 \\ \text{with} \quad \frac{d}{dt} = \frac{\partial}{\partial t} + \frac{u}{a\cos\varphi}\frac{\partial}{\partial \lambda} + \frac{v}{a}\frac{\partial}{\partial \varphi} + w\frac{\partial}{\partial r}\end{gathered}} \tag{19.33}$$

It can be seen that the term $2w/r$ in (19.31) has vanished in (19.33). This form is also used for many practical applications.

19.6 Problems

19.1: Use equation (18.52) and the information given in Section 19.1 to verify the validity of the second equation of (19.11). Then use Lagrange's equation of motion to show that the second equation of (19.11) is correct. By a direct transformation obtain the second equation of (19.12) from the second equation of (19.11).

19.2: Use the first metric simplification in conjunction with Lagrange's equation of motion to show that the underlined term in the first equation of (19.11) drops out. Transform this equation to physical measure numbers of the velocity to show that the two underlined terms in the first equation of (19.15) drop out.

19.3: Use the first metric simplification to show that the term $2w/r$ in equation (19.31) vanishes.

19.4: The generalized coordinates of the elliptic coordinate system may be denoted by $q^i = \lambda', \varphi', r'$. The equations for transformation between the Cartesian and the elliptic coordinates are then given by

$$
\begin{aligned}
x^1 &= \epsilon \cosh r' \cos\varphi' \cos(\lambda' + \Omega), & 0 \le \lambda' \le 2\pi \\
x^2 &= \epsilon \cosh r' \cos\varphi' \sin(\lambda' + \Omega), & -\pi/2 \le \varphi' \le \pi/2 \\
x^3 &= \epsilon \sinh r' \sin\varphi', & 0 \le r' \le \infty
\end{aligned}
$$

where ϵ is the excentricity.

(a) Show that the kinetic energy of the elliptic coordinate system is given by

$$
K_A = \frac{\epsilon^2}{2}[(\cosh^2 r' - \cos^2\varphi')(\dot{\varphi}'^2 + \dot{r}'^2) + \cosh^2 r' \cos^2\varphi' (\dot{\lambda}' + \Omega)^2]
$$

(b) Find g_{11} from the definition (M4.8).

(c) Find the fundamental quantities g_{ij}.

20
The stereographic coordinate system

For the analysis and depiction of meteorological data it is useful and customary to map the surface of the earth onto a plane. Therefore, it is advisable for purposes of numerical weather prediction to formulate and evaluate the atmospheric equations in *stereographic coordinates*. Such a map projection should represent the spherical surface as accurately as possible, but obviously some features will be lost. It is extremely important to preserve the angle between intersecting curves such as the right angle between latitude circles and meridians. Maps possessing this desirable and valuable property are called *conformal*. If distances were preserved by mapping the sphere onto the projection plane, the map would be called *isometric*. Mapping from the sphere to the stereographic plane is conformal but not isometric. In order to remove this deficiency to a tolerable level, a *scale factor* m, also called the *image scale*, will be introduced.

20.1 The stereographic projection

We will now describe the sphere-to-plane mapping by introducing a projection plane that is parallel to the equatorial plane. On the projection plane we may construct a Cartesian (x, y, z)-coordinate system with a square rectangular grid. The stereographic Cartesian coordinate system differs from the regular Cartesian coordinates since the metric fundamental quantities are not $g_{ij} = \delta_{ij}$. In fact, they still contain the Gaussian curvature $1/r$ of the spherical coordinate system, showing that in reality we have not left the sphere.

Conformal mapping means that the angle between two lines intersecting at a certain point on the sphere is reproduced in magnitude and sense by the angle between the corresponding curves on the plane. Infinitesimally small triangles around such points on the sphere are mapped onto similar infinitesimally small triangles on the plane. Moreover, a small circle remains a circle and an ellipse

Fig. 20.1 The infinitesimal surroundings of a point P_r on the sphere and of the point P_E on the projection plane.

transforms into an ellipse such that the ratio of the lengths of the major and minor axes is preserved. This is demonstrated in Figure 20.1 for the infinitesimally small area of elliptical shape. The point P_r is somewhere on the sphere, and P_E is located on the projection plane symbolized by the subscript E. Since the ratio of a line element on the sphere ds_r to the line element mapped onto the projection plane ds_E is independent of the orientation of the line element we have

$$\frac{(ds_r)_1}{(ds_E)_1} = \frac{(ds_r)_2}{(ds_E)_2} \tag{20.1}$$

We will use this information to define the scale factor.

Let us consider the plane E_0 defined by the angle of latitude φ_0 shown in Figure 20.2. The conformal stereographic map results from projecting an arbitrary spherical reference surface within the atmosphere, defined by the radius r, onto the plane E. This plane is parallel to E_0 and is located a distance $r - r_0$ above E_0, where r_0 is the mean radius of the earth. For a central projection the line element ds_r, defined by the latitudinal angle φ, now has a length ds_{r_0} on the surface of the earth. Obviously, the latitude and the longitude (φ, λ) do not change. The infinitesimal line element ds_{r_0} is first projected stereographically onto the plane E_0, yielding ds_{E_0}, and then onto the plane E without any distortion, resulting in ds_E. From Figure 20.2 we see that the projection source is the south pole. Furthermore, we have $ds_{r_0} = -r_0\, d\varphi$ and $ds_{E_0} = dR$.

We will now define the *scale* or *image factors* for the earth's surface $m_0 = m(r_0)$ and the spherical reference surface $m = m(r)$ by means of

$$m_0 = \frac{ds_{E_0}}{ds_{r_0}} = -\frac{dR}{r_0\, d\varphi}, \qquad m = \frac{ds_E}{ds_r} = -\frac{dR}{r\, d\varphi} \Longrightarrow m = m_0 \frac{r_0}{r} \tag{20.2}$$

As will be seen, the scale factors are independent of the longitude λ.

In order to establish the dependency of m on the latitude φ, we consider the rectangular triangle formed by the sides $l_0 + r_0$, R, and the projection ray extending from the south pole at angle γ measured from the polar axis; see Figure 20.2. From the figure we immediately find

$$\begin{aligned} &\text{(a)} \qquad \gamma = \frac{1}{2}\left(\frac{\pi}{2} - \varphi\right) = \frac{\chi}{2}, \qquad c = r_0 + l_0 = r_0(1 + \sin\varphi_0) \\ &\text{(b)} \quad \tan\gamma = \frac{R}{r_0 + l_0} = \frac{R}{c} = \frac{\sin\chi}{1 + \cos\chi} = \frac{\cos\varphi}{1 + \sin\varphi} \end{aligned} \tag{20.3}$$

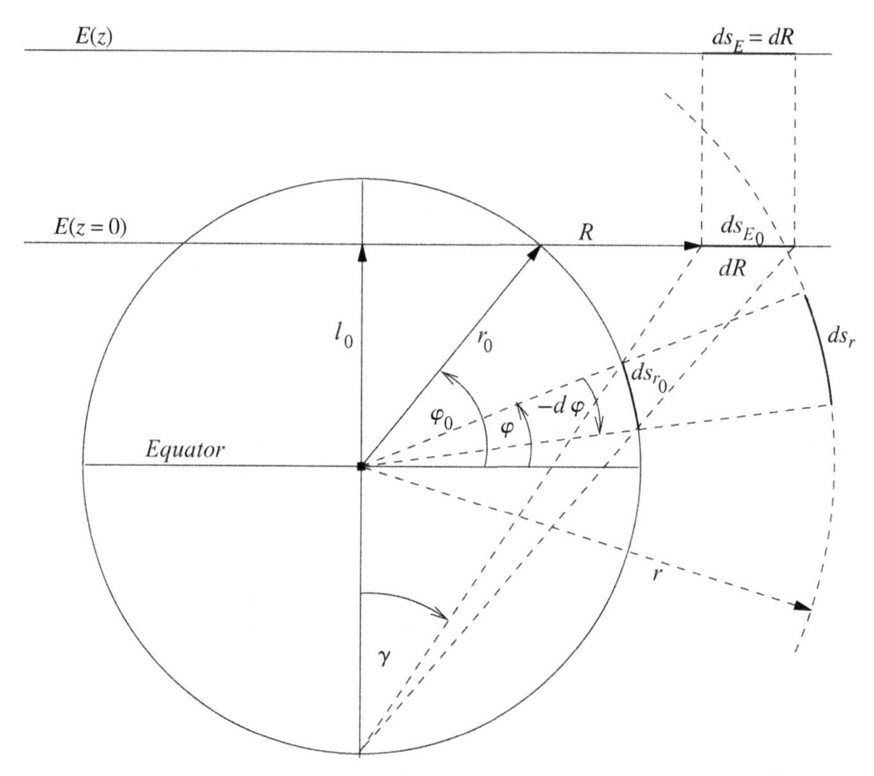

Fig. 20.2 Stereographic projection of a line element ds_r on the longitudinal circle with radius r onto the projection plane $E(z)$ with height $z = r - r_0$. The line element on E is ds_E.

where the angle χ is the co-latitude $\pi/2 - \varphi$. Equation (20.3b) makes use of a well-known trigonometric identity. From (20.2) and (20.3b) upon differentiation we obtain

$$m_0(\varphi) = \frac{c}{r_0} \frac{1}{1 + \sin \varphi} \tag{20.4}$$

showing that m_0 depends on the latitude only. By eliminating the constant c/r_0 we find the desired form

$$m_0 = \frac{1 + \sin \varphi_0}{1 + \sin \varphi} \tag{20.5}$$

where φ_0 is a fixed but arbitrary latitude. The Weather Services of Germany and the USA use $\varphi_0 = 60^\circ$ and $\varphi_0 = 90^\circ$, respectively.

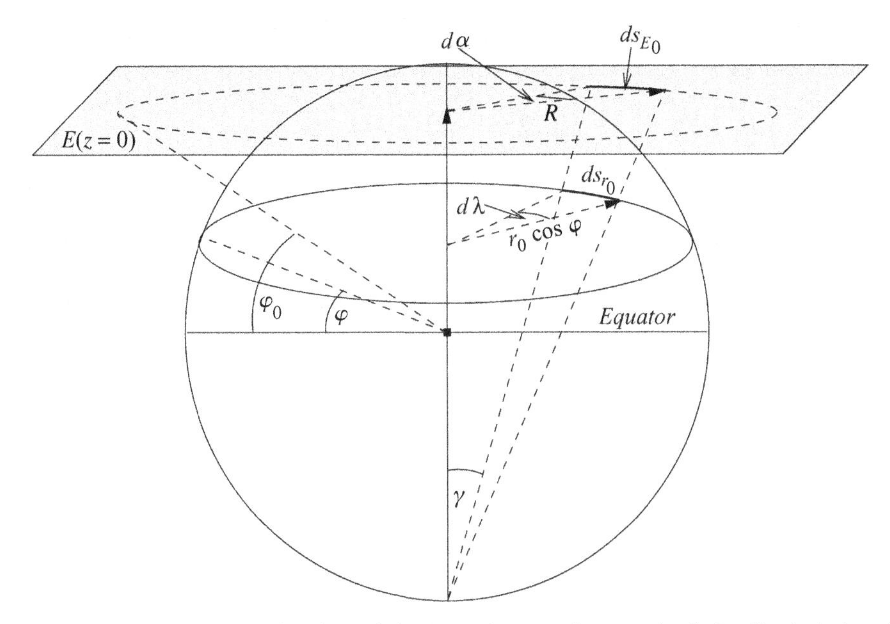

Fig. 20.3 Stereographic projection of the line element ds_{r_0} on the latitudinal circle with radius $r_0 \cos\varphi$ onto the projection plane $E(z = 0)$. The line element on $E(z = 0)$ is ds_{E_0}.

It will be useful to derive additional forms of the scale factor m. On replacing the term $1/(1 + \sin\varphi)$ in (20.4) by means of (20.3b) we find

$$m_0 = \frac{R}{r_0 \cos\varphi} \tag{20.6}$$

Substitution of (20.4) and (20.6) into the expression $\sin\varphi = \sqrt{1 - \cos^2\varphi}$ gives

$$m_0 = \frac{c^2 + R^2}{2cr_0} \tag{20.7}$$

Finally, we project stereographically a line element of a latitudinal circle onto the plane $E(z = 0) = E_0$ as shown in Figure 20.3. Using the definition of m_0 given in (20.6) we may write

$$m_0 = \frac{ds_{E_0}}{ds_{r_0}} = \frac{R\,d\alpha}{r_0 \cos\varphi\, d\lambda} = \frac{R}{r_0 \cos\varphi} \quad \text{since} \quad d\alpha = d\lambda \tag{20.8}$$

Here we have used the cylindrical coordinates R and α. Summarizing, we may write for the scale factor m

$$\boxed{m = \frac{ds_E}{ds_r} = -\frac{dR}{r\,d\varphi} = \frac{R}{r\cos\varphi} = \frac{c}{r}\frac{1}{1 + \sin\varphi}} \tag{20.9}$$

Inspection of (20.9) shows that m depends not only on the latitude φ but also on the radial distance r. The explicit dependence of m on r follows from (20.9):

$$\frac{\partial m}{\partial r} = \frac{\partial m}{\partial z} = -\frac{m}{r} \tag{20.10}$$

If we use the metric simplification $r = r_0 = a$ then $m = m_0$. Later we will make use of this approximation since it will simplify our work without significant loss of accuracy.

In order to introduce the Coriolis parameter $f = 2\Omega \sin\varphi$ into the equation of motion in stereographic coordinates, we need to derive an auxiliary relation. First we write the trigonometric function $\sin\varphi$ with the help of equations (20.2) and (20.6) in the form

$$\sin\varphi = -\frac{1}{2\cos\varphi}\frac{d\cos^2\varphi}{d\varphi} = \frac{m_0^2}{2R}\frac{d}{dR}\left(\frac{R^2}{m_0^2}\right) \tag{20.11}$$

from which it follows immediately that

$$\sin\varphi = 1 + \frac{m_0^2 R}{2}\frac{d}{dR}\left(\frac{1}{m_0^2}\right) \tag{20.12}$$

Thus, we have expressed the sine function in terms of the cylindrical coordinate R.

For the horizontal position vector in cylindrical coordinates $\mathbf{R} = \mathbf{q}_R R$ we may write

$$\mathbf{R}\cdot\nabla = \mathbf{q}_R R\cdot\left(\mathbf{q}^R \frac{\partial}{\partial R} + \mathbf{q}^\alpha \frac{\partial}{\partial \alpha} + \mathbf{q}^z \frac{\partial}{\partial z}\right) = R\frac{\partial}{\partial R} \tag{20.13}$$

Thus, we may express the sine function in (20.12) as

$$\sin\varphi = 1 + \frac{m_0^2}{2}\mathbf{R}\cdot\nabla\left(\frac{1}{m_0^2}\right) \tag{20.14}$$

where m_0 depends on R only, see (20.7), so that the total derivative in (20.12) may be replaced by the partial derivative.

20.2 Metric forms in stereographic coordinates

Our first task is to derive an expression for the functional determinant in the stereographic system by relating the various derivatives appearing in the geographical system with coordinates (λ, φ, r) and the stereographic system with cylindrical coordinates (R, α, z).

Let us consider the unspecified function ψ which will be expressed in both systems. Expressing ψ as a function of the longitude, i.e. $\psi(\lambda) = \psi\,[\alpha(\lambda)]$, we obtain

$$\left(\frac{\partial\psi}{\partial\lambda}\right)_{\varphi,r} = \left(\frac{\partial\psi}{\partial\alpha}\right)_{R,z}\frac{d\alpha}{d\lambda} = \left(\frac{\partial\psi}{\partial\alpha}\right)_{R,z} \Longrightarrow \frac{1}{r\cos\varphi}\left(\frac{\partial\psi}{\partial\lambda}\right)_{\varphi,r} = \frac{m}{R}\left(\frac{\partial\psi}{\partial\alpha}\right)_{R,z} \tag{20.15}$$

where the geographical part $r\cos\varphi$ of (20.9) is appended to the geographical system and m/R to the stereographic system. We proceed similarly with the remaining partial derivatives. Since the longitude λ and the angle α correspond to each other, as do r and z, we may write

$$\left(\frac{\partial\psi}{\partial\varphi}\right)_{\lambda,r} = \left(\frac{\partial\psi}{\partial R}\right)_{\alpha,z}\frac{dR}{d\varphi} \Longrightarrow -\frac{1}{r}\left(\frac{\partial\psi}{\partial\varphi}\right)_{\lambda,r} = m\left(\frac{\partial\psi}{\partial R}\right)_{\alpha,z} \tag{20.16}$$

where again use of (20.9) has been made. Finally

$$\left(\frac{\partial\psi}{\partial r}\right)_{\lambda,\varphi} = \left(\frac{\partial\psi}{\partial z}\right)_{R,\alpha} \tag{20.17}$$

is the obvious statement that $dr = dz$.

In order to find the functional determinant for the stereographic system we use the well-known transformation formula (M4.21) which leads to

$$\sqrt{g}\,\big|_{R,\alpha,z} = \sqrt{g}\,\big|_{\lambda\varphi,r}\left|\frac{\partial(\lambda,\varphi,r)}{\partial(R,\alpha,z)}\right| = r^2\cos\varphi\begin{vmatrix} 0 & 1 & 0 \\ -\dfrac{\cos\varphi}{R} & 0 & 0 \\ 0 & 0 & 1 \end{vmatrix} \tag{20.18}$$

The various elements of the determinant can then be found by identifying in ψ (20.15)–(20.17) with λ, φ, and r successively. For the case that $\psi = \lambda$ this will now be demonstrated:

$$\begin{aligned} m\left(\frac{\partial\lambda}{\partial R}\right)_{\alpha,z} &= -\frac{1}{r}\left(\frac{\partial\lambda}{\partial\varphi}\right)_{\lambda,r} = 0 \\ \left(\frac{\partial\lambda}{\partial\alpha}\right)_{R,z} &= \frac{R}{mr\cos\varphi} = 1 \\ \left(\frac{\partial\lambda}{\partial z}\right)_{R,\alpha} &= \left(\frac{\partial\lambda}{\partial r}\right)_{\lambda,\varphi} = 0 \end{aligned} \tag{20.19}$$

Expanding the determinant and using (20.9), we get the desired result

$$\sqrt{g}\Big|_{R,\alpha,z} = R/m^2 \tag{20.20}$$

The next task is to find the metric fundamental quantities g_{ij} for the stereographic system in cylindrical coordinates. From (20.9) we find the basic relation for the surfaces $r =$ constant or $z =$ constant

$$(ds)^2_{\text{spherical}} = \frac{1}{m^2}(ds)^2_{\text{stereographic}} \tag{20.21}$$

In general we may write

$$(d\mathbf{r})^2 = (ds)^2_{\text{spherical}} + (dr)^2 = \frac{1}{m^2}(ds)^2_{\text{stereographic}} + (dz)^2 \tag{20.22}$$

We apply this formula to the stereographic cylindrical coordinates and find

$$(d\mathbf{r})^2 = g_{mn}\, dq^m\, dq^n = \frac{1}{m^2}\big[(dR)^2 + R^2(d\alpha)^2\big] + (dz)^2 \tag{20.23}$$

since the arclength element $d\mathbf{s}_{\text{stereographic}} = dR\,\mathbf{e}_R + R\,d\alpha\,\mathbf{e}_\alpha$. Hence, we are dealing with an orthogonal system. The covariant and contravariant metric fundamental quantities g_{ij} and g^{ij} are easily found as

$$\begin{aligned} &g_{11} = 1/m^2, \quad g_{22} = R^2/m^2, \quad g_{33} = 1, \quad g_{ij} = 0 \quad \text{for} \quad i \neq j \\ &g^{11} = m^2, \quad g^{22} = m^2/R^2, \quad g^{33} = 1, \quad g^{ij} = 0 \quad \text{for} \quad i \neq j \end{aligned} \tag{20.24}$$

since $g_{ii}g^{ii} = 1$. We proceed analogously with the Cartesian system. From (20.23) we find with $x = R\cos\alpha$ and $y = R\sin\alpha$

$$(d\mathbf{r})^2 = g_{mn}\, dq^m\, dq^n = \frac{1}{m^2}\big[(dx)^2 + (dy)^2\big] + (dz)^2 \tag{20.25}$$

so that

$$\begin{aligned} &g_{11} = 1/m^2, \quad g_{22} = 1/m^2, \quad g_{33} = 1, \quad g_{ij} = 0 \quad \text{for} \quad i \neq j \\ &g^{11} = m^2, \quad g^{22} = m^2, \quad g^{33} = 1, \quad g^{ij} = 0 \quad \text{for} \quad i \neq j \end{aligned} \tag{20.26}$$

Equation (20.14) can also be used to state the sine function in terms of the Cartesian coordinates (x, y) by expressing the horizontal position vector $\mathbf{R}$ as

$$\mathbf{R} = x\mathbf{q}_x + y\mathbf{q}_y \tag{20.27}$$

Therefore, we find

$$\mathbf{R}\cdot\nabla = (x\mathbf{q}_x + y\mathbf{q}_y)\cdot\left(\mathbf{q}^x\frac{\partial}{\partial x} + \mathbf{q}^y\frac{\partial}{\partial y} + \mathbf{q}^z\frac{\partial}{\partial z}\right) = x\frac{\partial}{\partial x} + y\frac{\partial}{\partial y} \tag{20.28}$$

so that

$$\sin\varphi = 1 + \frac{m_0^2}{2}\left[x\frac{\partial}{\partial x}\left(\frac{1}{m_0^2}\right) + y\frac{\partial}{\partial y}\left(\frac{1}{m_0^2}\right)\right] \tag{20.29}$$

20.3 The absolute kinetic energy in stereographic coordinates

There are several ways to derive an expression for the absolute kinetic energy in the stereographic system. One possibility is to transform the absolute kinetic energy of the geographical system,

$$K_{\rm A} = \frac{1}{2}\left[r^2 \cos^2\varphi\,(\dot\lambda + \Omega)^2 + r^2\dot\varphi^2 + \dot r^2\right] \tag{20.30}$$

to the stereographic system. From equation (20.9) and the transformation relations $d\lambda = d\alpha,\ dr = dz$ one obtains

$$\dot\lambda = \dot\alpha, \qquad \dot\varphi = -\frac{\dot R}{mr}, \qquad \dot r = \dot z, \qquad \cos\varphi = \frac{R}{mr} \tag{20.31}$$

Substituting (20.31) into (20.30) results in the absolute kinetic energy of the stereographic system in cylindrical coordinates:

$$\begin{gathered} K_{\rm A} = K + K_P = \frac{1}{2}\left(\frac{\dot R^2 + R^2(\dot\alpha^2 + 2\,\dot\alpha\Omega)}{m^2} + \dot z^2\right) - \phi_z \\ \text{with} \quad K = \frac{1}{2}\left(\frac{\dot R^2 + R^2\,\dot\alpha^2}{m^2} + \dot z^2\right), \qquad K_P = \frac{R^2\,\dot\alpha\Omega}{m^2} - \phi_z, \qquad \phi_z = -\frac{R^2\Omega^2}{2m^2} \end{gathered} \tag{20.32}$$

where ϕ_z represents the centrifugal potential.

Some useful relations between the Cartesian and the cylindrical coordinates are summarized in

$$\begin{gathered} x = x^1 = R\cos\alpha, \qquad y = x^2 = R\sin\alpha, \qquad z = x^3 \\ R = \sqrt{x^2 + y^2}, \qquad \dot R = \frac{x\dot x + y\dot y}{\sqrt{x^2 + y^2}} \\ \frac{d\tan\alpha}{dt} = \frac{\dot\alpha}{\cos^2\alpha} = \frac{d}{dt}\left(\frac{y}{x}\right), \qquad \dot\alpha^2 = \frac{\dot x^2 + \dot y^2}{x^2 + y^2} - \frac{(x\dot x + y\dot y)^2}{(x^2 + y^2)^2} \end{gathered} \tag{20.33}$$

Substituting these expressions into (20.32) gives the absolute kinetic energy of the stereographic system in Cartesian coordinates:

$$\begin{gathered} K_{\rm A} = K + K_P = \frac{1}{2}\left(\frac{\dot x^2 + \dot y^2 + 2\Omega(\dot y x - \dot x y)}{m^2} + \dot z^2\right) - \phi_z \\ \text{with} \quad K = \frac{1}{2}\left(\frac{\dot x^2 + \dot y^2}{m^2} + \dot z^2\right), \qquad K_P = \frac{\Omega(\dot y x - \dot x y)}{m^2} - \phi_z, \\ \phi_z = -\frac{(x^2 + y^2)\Omega^2}{2m^2} \end{gathered} \tag{20.34}$$

The covariant measure numbers of the velocities $\mathbf{v}_P$ are obtained by differentiating K_P with respect to the contravariant measure numbers of the relative velocity.

For the cylindrical coordinates from (20.32) we have

$$(\underset{\Omega}{W_1}, \underset{\Omega}{W_2}, \underset{\Omega}{W_3}) = \left(0, \frac{R^2\Omega}{m^2}, 0\right), \qquad (\underset{\Omega}{W^1}, \underset{\Omega}{W^2}, \underset{\Omega}{W^3}) = (0, \Omega, 0) \tag{20.35}$$

In this expression the contravariant measure numbers are obtained by raising the indices of the covariant measure numbers. From the relation $g_{ii} = \partial^2 K/(\partial \dot{q}^i \partial \dot{q}^i)$ the validity of (20.24) may be easily checked.

For the kinetic energy in Cartesian coordinates as given by (20.34) the corresponding result is

$$(\underset{\Omega}{W_1}, \underset{\Omega}{W_2}, \underset{\Omega}{W_3}) = \left(-\frac{\Omega y}{m^2}, \frac{\Omega x}{m^2}, 0\right), \qquad (\underset{\Omega}{W^1}, \underset{\Omega}{W^2}, \underset{\Omega}{W^3}) = (-\Omega y, \Omega x, 0) \tag{20.36}$$

Again (20.26) may be obtained from $g_{ii} = \partial^2 K/(\partial \dot{q}^i \partial \dot{q}^i)$. Finally, in stereographic cylindrical coordinates the measure numbers of the relative velocity are

$$(v^1, v^2, v^3) = (\dot{R}, \dot{\alpha}, \dot{z}), \qquad (v_1, v_2, v_3) = \left(\frac{\dot{R}}{m^2}, \frac{R^2 \dot{\alpha}}{m^2}, \dot{z}\right) \tag{20.37}$$

whereas in Cartesian coordinates they are given by

$$(v^1, v^2, v^3) = (\dot{x}, \dot{y}, \dot{z}), \qquad (v_1, v_2, v_3) = \left(\frac{\dot{x}}{m^2}, \frac{\dot{y}}{m^2}, \dot{z}\right) \tag{20.38}$$

On utilizing these results in cylindrical coordinates the components of the absolute velocity are given by

$$\left(v_A^1, v_A^2, v_A^3\right) = (\dot{R}, \dot{\alpha} + \Omega, \dot{z}), \qquad (v_{A,1}, v_{A,2}, v_{A,3}) = \left(\frac{\dot{R}}{m^2}, \frac{R^2(\dot{\alpha} + \Omega)}{m^2}, \dot{z}\right) \tag{20.39}$$

In the Cartesian coordinate system we obtain analogously

$$\begin{aligned} \left(\dot{x}_A^1, \dot{x}_A^2, \dot{x}_A^3\right) &= (\dot{x} - \Omega y, \dot{y} + \Omega x, \dot{z}) \\ (\dot{x}_{A,1}, \dot{x}_{A,2}, \dot{x}_{A,3}) &= \left(\frac{\dot{x} - \Omega y}{m^2}, \frac{\dot{y} + \Omega x}{m^2}, \dot{z}\right) \end{aligned} \tag{20.40}$$

20.4 The equation of motion in the stereographic Cartesian coordinates

Before we evaluate the general equations in the covariant, the contravariant, and the physical coordinate systems we need to state the components of the rotational tensor and the Christoffel symbols of the second kind. In order to simplify the pertinent equations we make use of the first metric simplification by setting $r = a$

so that $m = m_0$. Substituting (20.26) into (19.9) yields for the Christoffel symbols

$$
\begin{aligned}
&\Gamma^1_{11} = -\frac{1}{2m_0^2}\frac{\partial m_0^2}{\partial x}, && \Gamma^1_{12} = -\frac{1}{2m_0^2}\frac{\partial m_0^2}{\partial y}, && \Gamma^1_{13} = 0 \\
&\Gamma^1_{21} = \Gamma^1_{12}, && \Gamma^1_{22} = \frac{1}{2m_0^2}\frac{\partial m_0^2}{\partial x}, && \Gamma^1_{23} = 0 \\
&\Gamma^1_{31} = 0, && \Gamma^1_{32} = 0, && \Gamma^1_{33} = 0 \\
&\Gamma^2_{11} = \frac{1}{2m_0^2}\frac{\partial m_0^2}{\partial y}, && \Gamma^2_{12} = -\frac{1}{2m_0^2}\frac{\partial m_0^2}{\partial x}, && \Gamma^2_{13} = 0 \\
&\Gamma^2_{21} = \Gamma^2_{12}, && \Gamma^2_{22} = -\frac{1}{2m_0^2}\frac{\partial m_0^2}{\partial y}, && \Gamma^2_{23} = 0 \\
&\Gamma^2_{31} = 0, && \Gamma^2_{32} = 0, && \Gamma^2_{33} = 0 \\
&&& \Gamma^3_{ij} = 0 \quad \text{for} \quad i, j = 1, 2, 3
\end{aligned}
\tag{20.41}
$$

From equations (M5.48), (20.29), and (20.36) we find

$$
\begin{aligned}
&\underset{\Omega}{\omega}_{11} = 0, && \underset{\Omega}{\omega}_{12} = \frac{f}{2m_0^2}, && \underset{\Omega}{\omega}_{13} = 0 \\
&\underset{\Omega}{\omega}_{21} = -\underset{\Omega}{\omega}_{12}, && \underset{\Omega}{\omega}_{22} = 0, && \underset{\Omega}{\omega}_{23} = 0 \\
&\underset{\Omega}{\omega}_{31} = 0, && \underset{\Omega}{\omega}_{32} = 0, && \underset{\Omega}{\omega}_{33} = 0 \\
&\underset{\Omega}{\omega}_{i}{}^{k} = g^{km}\underset{\Omega}{\omega}_{im}, && && \underset{\Omega}{\overset{*}{\omega}}_{ij} = \sqrt{g^{ii}}\sqrt{g^{jj}}\underset{\Omega}{\omega}_{ij}
\end{aligned}
\tag{20.42}
$$

The computational labor leading to these equations is not excessive since we are dealing with orthogonal coordinate systems.

Utilizing (20.26), (20.38), and (20.42) it is easy to evaluate the covariant equations of motion for rigid rotation (18.25):

$$
\boxed{
\begin{aligned}
&\frac{dv_1}{dt} + \frac{v_1^2 + v_2^2}{2}\frac{\partial m_0^2}{\partial x} - f v_2 = -\frac{1}{\rho}\frac{\partial p}{\partial x} - \frac{\partial \phi}{\partial x} + \frac{1}{\rho}\mathbf{q}_1 \cdot \nabla \cdot \mathbb{J} \\
&\frac{dv_2}{dt} + \frac{v_1^2 + v_2^2}{2}\frac{\partial m_0^2}{\partial y} + f v_1 = -\frac{1}{\rho}\frac{\partial p}{\partial y} - \frac{\partial \phi}{\partial y} + \frac{1}{\rho}\mathbf{q}_2 \cdot \nabla \cdot \mathbb{J} \\
&\frac{dv_3}{dt} = -\frac{1}{\rho}\frac{\partial p}{\partial z} - \frac{\partial \phi}{\partial z} + \frac{1}{\rho}\mathbf{q}_3 \cdot \nabla \cdot \mathbb{J} \\
&\text{with} \quad \frac{d}{dt} = \frac{\partial}{\partial t} + m_0^2\left(v_1\frac{\partial}{\partial x} + v_2\frac{\partial}{\partial y}\right) + v_3\frac{\partial}{\partial z}
\end{aligned}
}
\tag{20.43}
$$

To get some exercise with Lagrange's formulation, we verify the first equation of (20.43). From the Lagrangian function L,

$$L = K_{\mathrm{A}} - \phi_a = \frac{1}{2}\left(\frac{\dot{x}^2 + \dot{y}^2 + 2\Omega\,(x\dot{y} - \dot{x}y)}{m_0^2} + \dot{z}\right) - \phi$$
$$\text{with} \quad \phi = \phi_{\mathrm{a}} - \frac{(x^2 + y^2)\Omega^2}{m_0^2} \tag{20.44a}$$

we find with little effort

$$\begin{aligned}
\frac{\partial L}{\partial \dot{x}} &= \frac{\dot{x} - \Omega y}{m_0^2} \\
\frac{d}{dt}\left(\frac{\partial L}{\partial \dot{x}}\right) &= \frac{d}{dt}\left(\frac{\dot{x}}{m_0^2}\right) - \Omega\,\frac{d}{dt}\left(\frac{y}{m_0^2}\right) \\
&= \frac{dv_1}{dt} - \frac{\Omega\,\dot{y}}{m_0^2} - \Omega y\left[\dot{x}\frac{\partial}{\partial x}\left(\frac{1}{m_0^2}\right) + \dot{y}\frac{\partial}{\partial y}\left(\frac{1}{m_0^2}\right)\right] \\
\left.\frac{\partial}{\partial x}\right|_{\dot{x},\dot{y}} L &= \frac{\Omega\,\dot{y}}{m_0^2} + \frac{1}{2}[\dot{x}^2 + \dot{y}^2 + 2\Omega(x\dot{y} - \dot{x}y)]\,\frac{\partial}{\partial x}\left(\frac{1}{m_0^2}\right) - \frac{\partial \phi}{\partial x}
\end{aligned} \tag{20.44b}$$

Substitution of (20.44b) into the Lagrange equation of motion (18.65) yields

$$\begin{aligned}
\frac{d}{dt}\left(\frac{\partial L}{\partial \dot{x}}\right) - \left.\frac{\partial}{\partial x}\right|_{\dot{x},\dot{y}} L &= \frac{dv_1}{dt} - \frac{2\Omega}{m_0^2}\,\dot{y}\left\{1 + \frac{m_0^2}{2}\left[x\,\frac{\partial}{\partial x}\left(\frac{1}{m_0^2}\right) + y\,\frac{\partial}{\partial y}\left(\frac{1}{m_0^2}\right)\right]\right\} \\
\frac{\dot{x}^2 + \dot{y}^2}{2m_0^4}\,\frac{\partial m_0^2}{\partial x} + \frac{\partial \phi}{\partial x} &= -\frac{1}{\rho}\frac{\partial p}{\partial x} + \frac{1}{\rho}\mathbf{q}_1 \cdot \nabla \cdot \mathbb{J}
\end{aligned} \tag{20.45a}$$

Utilizing (20.29) and (20.38), we have

$$\begin{aligned}
\frac{2\Omega}{m_0^2}\,\dot{y}\left\{1 + \frac{m_0^2}{2}\left[x\,\frac{\partial}{\partial x}\left(\frac{1}{m_0^2}\right) + y\,\frac{\partial}{\partial y}\left(\frac{1}{m_0^2}\right)\right]\right\} &= v_2 2\Omega \sin\varphi = f v_2 \\
\frac{\dot{x}^2 + \dot{y}^2}{2m_0^4}\,\frac{\partial\, m_0^2}{\partial x} &= \frac{v_1^2 + v_2^2}{2}\,\frac{\partial m_0^2}{\partial x}
\end{aligned} \tag{20.45b}$$

Equation (20.45a) agrees with the first equation of (20.43).

The contravariant system is obtained by evaluating (18.34) with the help of (20.41) and (20.42):

$$\boxed{\begin{aligned}
&\frac{d\dot{x}}{dt} + \frac{\dot{y}^2 - \dot{x}^2}{2m_0^2}\frac{\partial m_0^2}{\partial x} - \frac{\dot{x}\dot{y}}{m_0^2}\frac{\partial m_0^2}{\partial y} - f\dot{y} = -\frac{m_0^2}{\rho}\frac{\partial p}{\partial x} - m_0^2\frac{\partial \phi}{\partial x} + \frac{1}{\rho}\mathbf{q}^1\cdot\nabla\cdot\mathbb{J} \\
&\frac{d\dot{y}}{dt} + \frac{\dot{x}^2 - \dot{y}^2}{2m_0^2}\frac{\partial m_0^2}{\partial y} - \frac{\dot{x}\dot{y}}{m_0^2}\frac{\partial m_0^2}{\partial x} + f\dot{x} = -\frac{m_0^2}{\rho}\frac{\partial p}{\partial y} - m_0^2\frac{\partial \phi}{\partial y} + \frac{1}{\rho}\mathbf{q}^2\cdot\nabla\cdot\mathbb{J} \\
&\frac{d\dot{z}}{dt} = -\frac{1}{\rho}\frac{\partial p}{\partial z} - \frac{\partial \phi}{\partial z} + \frac{1}{\rho}\mathbf{q}^3\cdot\nabla\cdot\mathbb{J} \\
&\text{with}\quad \frac{d}{dt} = \frac{\partial}{\partial t} + \dot{x}\frac{\partial}{\partial x} + \dot{y}\frac{\partial}{\partial y} + \dot{z}\frac{\partial}{\partial z}
\end{aligned}} \tag{20.46}$$

The covariant system and the contravariant system are equivalent.

Whenever we are dealing with orthogonal coordinate systems, physical velocity components offer some advantages. Owing to the rigidity of motion the basic equation (18.37) is easy to evaluate. The required relations are summarized as

$$\begin{aligned}
&\overset{*}{v}{}^1 = u = \dot{x}/m_0, \qquad \overset{*}{v}{}^2 = v = \dot{y}/m_0, \qquad \overset{*}{v}{}^3 = w = \dot{z} \\
&\frac{\partial}{\partial \overset{*}{q}{}^1} = m_0\frac{\partial}{\partial x}, \qquad \frac{\partial}{\partial \overset{*}{q}{}^2} = m_0\frac{\partial}{\partial y}, \qquad \frac{\partial}{\partial \overset{*}{q}{}^3} = \frac{\partial}{\partial z}
\end{aligned} \tag{20.47}$$

Hence, the equations of motion in stereographic Cartesian coordinates with physical measure numbers are given by

$$\boxed{\begin{aligned}
&\frac{du}{dt} + v\left(v\frac{\partial m_0}{\partial x} - u\frac{\partial m_0}{\partial y}\right) - fv = -\frac{m_0}{\rho}\frac{\partial p}{\partial x} - m_0\frac{\partial \phi}{\partial x} + \frac{1}{\rho}\mathbf{e}_1\cdot\nabla\cdot\mathbb{J} \\
&\frac{dv}{dt} - u\left(v\frac{\partial m_0}{\partial x} - u\frac{\partial m_0}{\partial y}\right) + fu = -\frac{m_0}{\rho}\frac{\partial p}{\partial y} - m_0\frac{\partial \phi}{\partial y} + \frac{1}{\rho}\mathbf{e}_2\cdot\nabla\cdot\mathbb{J} \\
&\frac{dw}{dt} = -\frac{1}{\rho}\frac{\partial p}{\partial z} - \frac{\partial \phi}{\partial z} + \frac{1}{\rho}\mathbf{e}_3\cdot\nabla\cdot\mathbb{J} \\
&\text{with}\quad \frac{d}{dt} = \frac{\partial}{\partial t} + m_0\left(u\frac{\partial}{\partial x} + v\frac{\partial}{\partial y}\right) + w\frac{\partial}{\partial z}
\end{aligned}} \tag{20.48}$$

If the coordinate approximation is applied, i.e. $\phi = \phi(z)$, then the partial derivatives of ϕ with respect to x and y vanish in (20.43), (20.46), and (20.48).

We will now reexamine the geographical system. At the north pole this system is undefined due to the appearance of $\tan\varphi$ in equations (2.26), (19.12), and

(19.15). In contrast to this, the $\tan\varphi$ term does not appear in the stereographic coordinate system. It should be observed that this term does not appear in the covariant geographical system (19.11) either. This is certainly an advantage of the covariant geographical system over the contravariant formulation.

Suppose that we wish to evaluate the system (19.15) numerically by using finite-difference methods. The longitudinal term $\partial/\partial \overset{*}{q}{}^1 = (1/r\cos\varphi)\,\partial/\partial\lambda$ appearing in the expansion of d/dt is latitude-dependent, so the numerical grid becomes latitude-dependent also. This is an undesirable property. In contrast to this, the Cartesian grid of the stereographic system is distorted only very slightly, which is favorable for the numerical evaluation of the prognostic equations.

We now wish to establish the relation between the Cartesian coordinates of the stereographic system and the coordinates of the geographical system. From (20.3), (20.5), and (20.33) we have

$$x = R\cos\lambda, \qquad y = R\sin\lambda, \qquad R = \frac{1+\sin\varphi_0}{1+\sin\varphi}a\cos\varphi = m_0 a\cos\varphi \tag{20.49a}$$

In order to estimate the distances of the Cartesian grid in terms of the variables of the geographical system, we expand dx and dy as

$$dx = \left(\frac{\partial x}{\partial\varphi}\right)_\lambda d\varphi + \left(\frac{\partial x}{\partial\lambda}\right)_\varphi d\lambda, \qquad dy = \left(\frac{\partial y}{\partial\varphi}\right)_\lambda d\varphi + \left(\frac{\partial y}{\partial\lambda}\right)_\varphi d\lambda \tag{20.49b}$$

Evaluating the partial derivatives with the help of (20.49a) yields

$$\begin{pmatrix} dx \\ dy \end{pmatrix} = m_0 \begin{pmatrix} -\sin\lambda & -\cos\lambda \\ \cos\lambda & -\sin\lambda \end{pmatrix} \begin{pmatrix} d\overset{*}{\lambda} \\ d\overset{*}{\varphi} \end{pmatrix} \tag{20.49c}$$

with $d\overset{*}{\lambda} = a\cos\varphi\, d\lambda$ and $d\overset{*}{\varphi} = a\, d\varphi$. It should be observed that the matrix stated in (20.49c) is orthogonal, so the inversion of the system to find $(d\overset{*}{\lambda}, d\overset{*}{\varphi})$ is easily accomplished. For additional details see Haltiner and Williams (1980).

20.5 The equation of motion in stereographic cylindrical coordinates

As we have seen, it is possible to introduce not only Cartesian coordinates but also cylindrical coordinates in the stereographic plane. We will proceed as in Section 20.4. Again we use the approximation $m = m_0$. The Christoffel symbols in cylindrical coordinates are given by

$$
\begin{aligned}
&\Gamma^1_{11} = \frac{m_0^2}{2}\frac{\partial}{\partial R}\left(\frac{1}{m_0^2}\right), && \Gamma^1_{12} = 0, && \Gamma^1_{13} = 0 \\
&\Gamma^1_{21} = 0, && \Gamma^1_{22} = -\frac{m_0^2}{2}\frac{\partial}{\partial R}\left(\frac{R^2}{m_0^2}\right), && \Gamma^1_{23} = 0 \\
&\Gamma^1_{31} = 0, && \Gamma^1_{32} = 0, && \Gamma^1_{33} = 0 \\
&\Gamma^2_{11} = 0, && \Gamma^2_{12} = \frac{m_0^2}{2R^2}\frac{\partial}{\partial R}\left(\frac{R^2}{m_0^2}\right), && \Gamma^2_{13} = 0 \\
&\Gamma^2_{21} = \Gamma^2_{12}, && \Gamma^2_{22} = 0, && \Gamma^2_{23} = 0 \\
&\Gamma^2_{31} = 0, && \Gamma^2_{32} = 0, && \Gamma^2_{33} = 0 \\
&&& \Gamma^3_{ij} = 0 \quad \text{for} \quad i, j = 1, 2, 3
\end{aligned}
\tag{20.50}
$$

Comparison with (20.41) shows that more Christoffel symbols are zero than in the Cartesian system. From (M5.48) and (19.7) we find

$$
\begin{aligned}
&\underset{\Omega}{\omega}_{11} = 0, && \underset{\Omega}{\omega}_{12} = \frac{Rf}{2m_0^2}, && \underset{\Omega}{\omega}_{13} = 0 \\
&\underset{\Omega}{\omega}_{21} = -\underset{\Omega}{\omega}_{12}, && \underset{\Omega}{\omega}_{22} = 0, && \underset{\Omega}{\omega}_{23} = 0 \\
&\underset{\Omega}{\omega}_{31} = 0, && \underset{\Omega}{\omega}_{32} = 0, && \underset{\Omega}{\omega}_{33} = 0 \\
&\underset{\Omega}{\omega}_i{}^k = g^{km}\underset{\Omega}{\omega}_{im}, &&& \underset{\Omega}{\overset{*}{\omega}}_{ij} = \sqrt{g^{ii}}\sqrt{g^{jj}}\underset{\Omega}{\omega}_{ij}
\end{aligned}
\tag{20.51}
$$

The rotational dyadic exhibits the same simplicity as (20.42).

Substitution of (20.24), (20.38), and (20.51) into (18.25) results in the covariant equations of motion

$$
\boxed{
\begin{aligned}
\frac{dv_1}{dt} + \frac{R^2v_1^2 + v_2^2}{2R^2}\frac{\partial m_0^2}{\partial R} - \frac{v_2^2 m_0^2}{R^3} - \frac{fv_2}{R} &= -\frac{1}{\rho}\frac{\partial p}{\partial R} - \frac{\partial \phi}{\partial R} + \frac{1}{\rho}\mathbf{q}_R\cdot\nabla\cdot\mathbb{J} \\
\frac{dv_2}{dt} + fv_1R &= -\frac{1}{\rho}\frac{\partial p}{\partial \alpha} - \frac{\partial \phi}{\partial \alpha} + \frac{1}{\rho}\mathbf{q}_\alpha\cdot\nabla\cdot\mathbb{J} \\
\frac{dv_3}{dt} &= -\frac{1}{\rho}\frac{\partial p}{\partial z} - \frac{\partial \phi}{\partial z} + \frac{1}{\rho}\mathbf{q}_z\cdot\nabla\cdot\mathbb{J} \\
\text{with} \quad \frac{d}{dt} &= \frac{\partial}{\partial t} + m_0^2\left(v_1\frac{\partial}{\partial R} + \frac{v_2}{R^2}\frac{\partial}{\partial \alpha}\right) + v_3\frac{\partial}{\partial z}
\end{aligned}
}
\tag{20.52}
$$

On substituting (20.50) and (20.51) into (18.34), the contravariant counterparts are analogously given as

$$
\boxed{
\begin{aligned}
\frac{d\dot{R}}{dt} - \frac{\dot{R}^2 - \dot{\alpha}^2 R^2}{2m_0^2}\frac{\partial m_0^2}{\partial R} - \dot{\alpha}^2 R - f\dot{\alpha}R &= -\frac{m_0^2}{\rho}\frac{\partial p}{\partial R} - m_0^2\frac{\partial \phi}{\partial R} + \frac{1}{\rho}\mathbf{q}^R\cdot\nabla\cdot\mathbb{J} \\
\frac{d\dot{\alpha}}{dt} - \frac{\dot{\alpha}\dot{R}}{m_0^2}\frac{\partial m_0^2}{\partial R} + \frac{2\dot{\alpha}\dot{R}}{R} + \frac{f\dot{R}}{R} &= -\frac{m_0^2}{R^2\rho}\frac{\partial p}{\partial \alpha} - \frac{m_0^2}{R^2}\frac{\partial \phi}{\partial \alpha} + \frac{1}{\rho}\mathbf{q}^\alpha\cdot\nabla\cdot\mathbb{J} \\
\frac{d\dot{z}}{dt} &= -\frac{1}{\rho}\frac{\partial p}{\partial z} - \frac{\partial \phi}{\partial z} + \frac{1}{\rho}\mathbf{q}^z\cdot\nabla\cdot\mathbb{J} \\
\text{with}\quad \frac{d}{dt} &= \frac{\partial}{\partial t} + \dot{R}\frac{\partial}{\partial R} + \dot{\alpha}\frac{\partial}{\partial \alpha} + \dot{z}\frac{\partial}{\partial z}
\end{aligned}
}
\tag{20.53}
$$

In order to obtain the equations of motion in physical measure numbers we evaluate (18.37). Utilizing (20.24), (20.51), and

$$
\begin{aligned}
&\overset{*}{v}{}^1 = u = \dot{R}/m_0, &&\overset{*}{v}{}^2 = v = R\dot{\alpha}/m_0, &&\overset{*}{v}{}^3 = w = \dot{z} \\
&\frac{\partial}{\partial \overset{*}{q}{}^1} = m_0\frac{\partial}{\partial R}, &&\frac{\partial}{\partial \overset{*}{q}{}^2} = \frac{m_0}{R}\frac{\partial}{\partial \alpha}, &&\frac{\partial}{\partial \overset{*}{q}{}^3} = \frac{\partial}{\partial z}
\end{aligned}
\tag{20.54}
$$

we find the desired equations of motion

$$
\boxed{
\begin{aligned}
\frac{du}{dt} + v^2\frac{\partial m_0}{\partial R} - v^2\frac{m_0}{R} - fv &= -\frac{m_0}{\rho}\frac{\partial p}{\partial R} - m_0\frac{\partial \phi}{\partial R} + \frac{1}{\rho}\mathbf{e}_1\cdot\nabla\cdot\mathbb{J} \\
\frac{dv}{dt} - uv\frac{\partial m_0}{\partial R} + uv\frac{m_0}{R} + fu &= -\frac{m_0}{R\rho}\frac{\partial p}{\partial \alpha} - \frac{m_0}{R}\frac{\partial \phi}{\partial \alpha} + \frac{1}{\rho}\mathbf{e}_2\cdot\nabla\cdot\mathbb{J} \\
\frac{dw}{dt} &= -\frac{1}{\rho}\frac{\partial p}{\partial z} - \frac{\partial \phi}{\partial z} + \frac{1}{\rho}\mathbf{e}_3\cdot\nabla\cdot\mathbb{J} \\
\text{with}\quad \frac{d}{dt} &= \frac{\partial}{\partial t} + m_0\left(u\frac{\partial}{\partial R} + \frac{v}{R}\frac{\partial}{\partial \alpha}\right) + w\frac{\partial}{\partial z}
\end{aligned}
}
\tag{20.55}
$$

The degree of complexity of this set of equations is about the same as that of (20.48). If the coordinate approximation is applied, i.e. $\phi = \phi(z)$, then the partial derivatives of ϕ with respect to R and α vanish in (20.52), (20.53), and (20.55).

20.6 The continuity equation

In order to solve any prognostic system describing the atmospheric motion we also need to solve the equation of continuity in addition to the heat equation and the ideal-gas law. In this section we will briefly derive the various versions of the continuity equation. For the system describing rigid rotation the last term in (18.47) vanishes. This equation can be easily rewritten in the form

$$\frac{d\rho}{dt} + \frac{\rho}{\sqrt{g}} \frac{\partial}{\partial q^n}(\sqrt{g}\dot{q}^n) = 0 \tag{20.56}$$

For the various cases the continuity equation can now be obtained very easily.

Covariant stereographic Cartesian coordinates

$$\frac{1}{\rho}\frac{d\rho}{dt} + m_0^2\left(\frac{\partial v_1}{\partial x} + \frac{\partial v_2}{\partial y}\right) + \frac{\partial v_3}{\partial z} = 0 \tag{20.57}$$

Contravariant stereographic Cartesian coordinates

$$\frac{1}{\rho}\frac{d\rho}{dt} + m_0^2\left[\frac{\partial}{\partial x}\left(\frac{\dot{x}}{m_0^2}\right) + \frac{\partial}{\partial y}\left(\frac{\dot{y}}{m_0^2}\right)\right] + \frac{\partial \dot{z}}{\partial z} = 0 \tag{20.58}$$

Physical stereographic Cartesian coordinates

$$\frac{1}{\rho}\frac{d\rho}{dt} + m_0\left(\frac{\partial u}{\partial x} + \frac{\partial v}{\partial y}\right) - \left(u\frac{\partial m_0}{\partial x} + v\frac{\partial m_0}{\partial y}\right) + \frac{\partial w}{\partial z} = 0 \tag{20.59}$$

Covariant stereographic cylindrical coordinates

$$\frac{1}{\rho}\frac{d\rho}{dt} + \frac{m_0^2}{R}\left[\frac{\partial}{\partial R}(Rv_1) + \frac{\partial}{\partial \alpha}\left(\frac{v_2}{R}\right)\right] + \frac{\partial v_3}{\partial z} = 0 \tag{20.60}$$

Contravariant stereographic cylindrical coordinates

$$\frac{1}{\rho}\frac{d\rho}{dt} + \frac{m_0^2}{R}\left[\frac{\partial}{\partial R}\left(\frac{R\dot{R}}{m_0^2}\right) + \frac{\partial}{\partial \alpha}\left(\frac{R\dot{\alpha}}{m_0^2}\right)\right] + \frac{\partial \dot{z}}{\partial z} = 0 \tag{20.61}$$

Physical stereographic cylindrical coordinates

$$\frac{1}{\rho}\frac{d\rho}{dt} + m_0\left(\frac{\partial u}{\partial R} + \frac{1}{R}\frac{\partial v}{\partial \alpha}\right) + \frac{\partial w}{\partial z} + \frac{m_0 u}{R} - u\frac{\partial m_0}{\partial R} = 0 \tag{20.62}$$

In a later chapter we will integrate in principle various versions of the atmospheric equations and then recognize more fully the importance of the continuity equation.

20.7 The equation of motion on the tangential plane

In Section 2.5 we have introduced equations decribing the motion on the tangential plane. This was done by fixing a rectangular coordinate system at the point of observation on the surface of the earth. The z-axis is pointing to the local zenith so that the x- and y- axes describe a tangential plane. The x-axis is pointing to the east and the y-axis to the north, so the motion is described with respect to a right-handed system. Going directly from the geographical coordinate system (19.15) to the equation of motion on the tangential plane is not possible even if we delete the underlined terms. Even on taking the radius of the earth as infinitely large to approximate a plane, we still have a problem at the north pole. On the other hand, if we set $m_0 = 1$, which corresponds to $\varphi = \varphi_0$ in (20.5), we obtain the equations of motion with reference to the tangential plane directly. For the Cartesian system we find from (20.48)

$$\boxed{\begin{aligned} \frac{du}{dt} - fv &= -\frac{1}{\rho}\frac{\partial p}{\partial x} + \frac{1}{\rho}\mathbf{e}_1\cdot\nabla\cdot\mathbb{J} \\ \frac{dv}{dt} + fu &= -\frac{1}{\rho}\frac{\partial p}{\partial y} + \frac{1}{\rho}\mathbf{e}_2\cdot\nabla\cdot\mathbb{J} \\ \frac{dw}{dt} &= -\frac{1}{\rho}\frac{\partial p}{\partial z} - \frac{\partial \phi}{\partial z} + \frac{1}{\rho}\mathbf{e}_3\cdot\nabla\cdot\mathbb{J} \end{aligned}} \tag{20.63}$$

The corresponding equations of the cylindrical system follow from (20.55) as

$$\boxed{\begin{aligned} \frac{du}{dt} - \frac{v^2}{R} - fv &= -\frac{1}{\rho}\frac{\partial p}{\partial R} + \frac{1}{\rho}\mathbf{e}_1\cdot\nabla\cdot\mathbb{J} \\ \frac{dv}{dt} + \frac{uv}{R} + fu &= -\frac{1}{R\rho}\frac{\partial p}{\partial \alpha} + \frac{1}{\rho}\mathbf{e}_2\cdot\nabla\cdot\mathbb{J} \\ \frac{dw}{dt} &= -\frac{1}{\rho}\frac{\partial p}{\partial z} - \frac{\partial \phi}{\partial z} + \frac{1}{\rho}\mathbf{e}_3\cdot\nabla\cdot\mathbb{J} \end{aligned}} \tag{20.64}$$

In (20.63) and (20.64) use of the coordinate approximation $\phi = \phi(z)$ has been made.

Geostrophic motion is a very simple but also a very useful approximation to real flow, describing unaccelerated frictionless motion parallel to the isobars. By neglecting in (20.63) frictional effects and setting the horizontal accelerations $du/dt = 0$ and $dv/dt = 0$ we obtain the well-known geostrophic wind relations

$$-fv_{\rm g} = -\frac{1}{\rho}\frac{\partial p}{\partial x}, \qquad fu_{\rm g} = -\frac{1}{\rho}\frac{\partial p}{\partial y} \tag{20.65}$$

It is also possible to give an analytic solution to horizontal frictionless motion on a horizontal plane if the advection term is ignored in the expansion of d/dt. In this case (20.63) reduces to

$$\frac{\partial u}{\partial t} - fv = -fv_{\mathrm{g}}, \qquad \frac{\partial v}{\partial t} + fu = fu_{\mathrm{g}} \tag{20.66}$$

By employing the somewhat complicated method of two-dimensional Laplace transforms, see Voelker and Doetsch (1950), Chapter 5, we find the solution to the coupled system

$$\begin{aligned} u(z,t) &= u(z,0)\cos(ft) + v(z,0)\sin(ft) \\ &\quad - f\int_0^t \cos(ft')\,v_{\mathrm{g}}(z,t-t')\,dt' + f\int_0^t \sin(ft')\,u_{\mathrm{g}}(z,t-t')\,dt' \\ v(z,t) &= v(z,0)\cos(ft) - u(z,0)\sin(ft) \\ &\quad + f\int_0^t \cos(ft')\,u_{\mathrm{g}}(z,t-t')\,dt' + f\int_0^t \sin(ft')\,v_{\mathrm{g}}(z,t-t')\,dt' \end{aligned} \tag{20.67}$$

where the geostrophic wind components may be dependent on height and time. For negative arguments u_{g} and v_{g} are set equal to zero. The solution (20.67) may be verified by substituting (20.67) into (20.66).

20.8 The equation of motion in Lagrangian enumeration coordinates

We will now return to Section 3.5 to express the equation of motion in terms of the Lagrangian enumeration coordinates. We assume that we have frictionless flow and apply the Cartesian coordinate system of the stereographic projection. The transformation will be carried out with the help of Lagrange's form of the equation of motion.

We recall the transformation rule (M4.15) for covariant measure numbers of vectors for the transformation from the Cartesian x^i system to a general a^i system. In the present context the a^i are the Lagrangian enumeration coordinates. This rule is given by

$$\underset{a\,k}{A} = \underset{x\,n}{A}\,\frac{\partial x^n}{\partial a^k} \tag{20.68}$$

Application of this formula to the velocity results in

$$\underset{a\,k}{V} = \underset{x\,n}{V}\,\frac{\partial x^n}{\partial a^k} \tag{20.69}$$

where the Cartesian components (20.38) are given by

$$\underset{x\,k}{V} = \left(\frac{\dot{x}^1}{m^2}, \frac{\dot{x}^2}{m^2}, \dot{x}^3\right) \tag{20.70}$$

Using the stringent metric simplification $m = 1$ gives

$$\underset{a}{V}_k = \dot{x}^n \frac{\partial x^n}{\partial a^k} \tag{20.71}$$

It is also possible to use $m = m(x^1, x^2)$ but the solution is more complicated and less readily interpreted.

The contravariant basis vectors in the Cartesian and the Lagrangian system will be arranged as

$$(\mathbf{i}^i) = \begin{pmatrix} \mathbf{i}^1 \\ \mathbf{i}^2 \\ \mathbf{i}^3 \end{pmatrix}, \qquad (\mathbf{a}^i) = \begin{pmatrix} \mathbf{a}^1 \\ \mathbf{a}^2 \\ \mathbf{a}^3 \end{pmatrix} \tag{20.72}$$

According to (M4.9) the transformation rule is given by

$$(\mathbf{i}^i) = \left(\frac{\partial x^i}{\partial a^j}\right)(\mathbf{a}^i) = (T^i_{\ j})(\mathbf{a}^i) = \left(\frac{\partial x^i}{\partial a^n}\,\mathbf{a}^n\right) \tag{20.73}$$

Here i and j denote the row and the column of the transformation matrix, respectively. For the column of the basis vectors in the Lagrangian system we obtain the following relation:

$$(T^i_{\ j})^{-1} = \left(\frac{\partial x^i}{\partial a^j}\right)^{-1} = \left(\frac{\partial a^i}{\partial x^j}\right), \qquad (\mathbf{a}^i) = \left(\frac{\partial x^i}{\partial a^j}\right)^{-1}(\mathbf{i}^i) \tag{20.74}$$

Using the rules of matrix inversion, we find for the contravariant basis vector in the Lagrangian system, see also (M3.14),

$$\mathbf{a}^i = \frac{M^i_{\ n}\mathbf{i}^n}{\left|\dfrac{\partial x^i}{\partial a^j}\right|} = \frac{M^i_{\ n}\mathbf{i}^n}{\sqrt{\underset{a}{g}}} \tag{20.75}$$

The $M^{\ j}_{i}$ are the elements of the adjoint matrix $(M^i_{\ j})$ of the transformation matrix $(T^i_{\ j})$.

In order to apply the Lagrangian form of the equation of motion, we need to find an expression for the derivative of an arbitrary field function ψ with respect to the enumeration coordinate a^k. Recalling that $\dot{a}^k = 0$, we may write

$$\boxed{\begin{aligned} \frac{\partial \dot{\psi}}{\partial \dot{a}^k} &= \lim_{\dot{a}^k \to 0}\left\{\frac{\partial}{\partial \dot{a}^k}\left(\frac{d\psi}{dt}\right)\right\} \\ &= \lim_{\dot{a}^k \to 0}\left\{\frac{\partial}{\partial \dot{a}^k}\left[\left(\frac{\partial \psi}{\partial t}\right)_{a^k} + \dot{a}^n \frac{\partial \psi}{\partial a^n}\right]\right\} \\ &= \delta^n_k \frac{\partial \psi}{\partial a^n} = \frac{\partial \psi}{\partial a^k} \end{aligned}} \tag{20.76}$$

which is a very useful expression.

According to (18.64) the Lagrangian function $L = K_{\mathrm{A}} - \phi_{\mathrm{a}}$ for the Cartesian coordinates of the stereographic projection is given by

$$L = \frac{1}{2m^2}[(\dot{x}^1)^2 + (\dot{x}^2)^2 + 2\Omega(\dot{x}^2 x^1 - x^2\dot{x}^1)] + \frac{(\dot{x}^3)^2}{2} + \frac{\Omega^2}{2m^2}\left[(x^1)^2 + (x^2)^2\right] - \phi_{\mathrm{a}} \tag{20.77}$$

We now make use of the metric simplification $m = 1$, which requires that Ω must be replaced by $\Omega \sin\varphi = f/2$ since we leave the stereographic plane and go to the tangential plane. Therefore, the metrically simplified form reads

$$\boxed{L = \tfrac{1}{2}[(\dot{x}^1)^2 + (\dot{x}^2)^2 + (\dot{x}^3)^2 + f(\dot{x}^2 x^1 - x^2\dot{x}^1)] - \phi} \tag{20.78}$$

In the basic system of Cartesian coordinates of the stereographic projection the Coriolis force may be written as

$$\begin{aligned} -2\overset{+}{\boldsymbol{\Omega}} \times \mathbf{v} &= -f\mathbf{a}_3 \times (\mathbf{a}_1\dot{x}^1 + \mathbf{a}_2\dot{x}^2 + \mathbf{a}_3\dot{x}^3) \\ &= \frac{f}{m^2}\left(\mathbf{a}^1\dot{x}^2 - \mathbf{a}^2\dot{x}^1\right) = \underset{x}{C}_1\mathbf{a}^1 + \underset{x}{C}_2\mathbf{a}^2 + \underset{x}{C}_3\mathbf{a}^3 = \underset{x}{C}_n\mathbf{a}^n \end{aligned} \tag{20.79}$$

The + above $\boldsymbol{\Omega}$ was placed there to show that, in general, the direction of the rotation differs from the rotational direction in the geographical coordinates. Whereas $\overset{+}{\boldsymbol{\Omega}}$ has a fixed direction and cannot be decomposed, the rotational vector $\boldsymbol{\Omega}$ can be split in two components. On setting $m = 1$ we find for the covariant measure numbers

$$\underset{x}{C}_1 = f\dot{x}^2, \qquad \underset{x}{C}_2 = -f\dot{x}^1, \qquad \underset{x}{C}_3 = 0 \tag{20.80}$$

Finally, according to (20.68), the transformation of these measure numbers to the Lagrangian enumeration coordinates yields

$$\underset{a}{C}_k = \underset{x}{C}_n \frac{\partial x^n}{\partial a^k} \tag{20.81}$$

Owing to the special properties ($\dot{a}^i = 0$) of the Lagrangian enumeration coordinates we must write the Lagrangian equation of motion in the form

$$\lim_{\dot{a}^k \to 0}\left\{\frac{d}{dt}\frac{\partial L}{\partial \dot{a}^k} - \left.\frac{\partial L}{\partial a^k}\right|_{\dot{a}^k}\right\} = -\frac{1}{\rho}\frac{\partial p}{\partial a^k} \tag{20.82}$$

The substitution of (20.78) into (20.82) is somewhat complex, therefore we shall proceed stepwise. Recalling (3.58) and (20.76), we may first write

$$\begin{aligned}
\lim_{\dot{a}^k\to 0}\left\{\frac{\partial L}{\partial \dot{a}^k}\right\} &= \lim_{\dot{a}^k\to 0}\left\{\dot{x}^1\frac{\partial \dot{x}^1}{\partial \dot{a}^k}+\dot{x}^2\frac{\partial \dot{x}^2}{\partial \dot{a}^k}+\dot{x}^3\frac{\partial \dot{x}^3}{\partial \dot{a}^k}+\frac{f}{2}\left(x^1\frac{\partial \dot{x}^2}{\partial \dot{a}^k}-x^2\frac{\partial \dot{x}^1}{\partial \dot{a}^k}\right)\right\}\\
&= \dot{x}^1\frac{\partial x^1}{\partial a^k}+\dot{x}^2\frac{\partial x^2}{\partial a^k}+\dot{x}^3\frac{\partial x^3}{\partial a^k}+\frac{f}{2}\left(x^1\frac{\partial x^2}{\partial a^k}-x^2\frac{\partial x^1}{\partial a^k}\right)\\
&= \left(\frac{\partial x^1}{\partial t}\right)_{a^k}\frac{\partial x^1}{\partial a^k}+\left(\frac{\partial x^2}{\partial t}\right)_{a^k}\frac{\partial x^2}{\partial a^k}+\left(\frac{\partial x^3}{\partial t}\right)_{a^k}\frac{\partial x^3}{\partial a^k}\\
&\quad+\frac{f}{2}\left(x^1\frac{\partial x^2}{\partial a^k}-x^2\frac{\partial x^1}{\partial a^k}\right)=A
\end{aligned} \tag{20.83}$$

Observing (3.58) once again, we obtain

$$\begin{aligned}
\lim_{\dot{a}^k\to 0}\left\{\frac{d}{dt}\left(\frac{\partial L}{\partial \dot{a}^k}\right)\right\} = \left(\frac{\partial A}{\partial t}\right)_{a^k} &= \left(\frac{\partial^2 x^1}{\partial t^2}\right)_{a^k}\frac{\partial x^1}{\partial a^k}+\left(\frac{\partial^2 x^2}{\partial t^2}\right)_{a^k}\frac{\partial x^2}{\partial a^k}\\
&\quad+\left(\frac{\partial^2 x^3}{\partial t^2}\right)_{a^k}\frac{\partial x^3}{\partial a^k}+\left(\frac{\partial x^1}{\partial t}\right)_{a^k}\frac{\partial^2 x^1}{\partial t\,\partial a^k}+\left(\frac{\partial x^2}{\partial t}\right)_{a^k}\frac{\partial^2 x^2}{\partial t\,\partial a^k}\\
&\quad+\left(\frac{\partial x^3}{\partial t}\right)_{a^k}\frac{\partial^2 x^3}{\partial t\,\partial a^k}+\frac{f}{2}\left[\left(\frac{\partial x^1}{\partial t}\right)_{a^k}\frac{\partial x^2}{\partial a^k}+x^1\frac{\partial^2 x^2}{\partial t\,\partial a^k}\right.\\
&\quad\left.-\left(\frac{\partial x^2}{\partial t}\right)_{a^k}\frac{\partial x^1}{\partial a^k}-x^2\frac{\partial^2 x^1}{\partial t\,\partial a^k}\right]
\end{aligned} \tag{20.84}$$

For the second term on the left-hand side of (20.82) we find from (20.78) the following expression:

$$\begin{aligned}
\lim_{\dot{a}^k\to 0}\left\{-\left.\frac{\partial L}{\partial a^k}\right|_{\dot{a}^k}\right\} &= -\dot{x}^1\frac{\partial \dot{x}^1}{\partial a^k}-\dot{x}^2\frac{\partial \dot{x}^2}{\partial a^k}-\dot{x}^3\frac{\partial \dot{x}^3}{\partial a^k}\\
&\quad-\frac{f}{2}\left(x^1\frac{\partial \dot{x}^2}{\partial a^k}+\dot{x}^2\frac{\partial x^1}{\partial a^k}-x^2\frac{\partial \dot{x}^1}{\partial a^k}-\dot{x}^1\frac{\partial x^2}{\partial a^k}\right)+\frac{\partial \phi}{\partial a^k}\\
&= -\left(\frac{\partial x^1}{\partial t}\right)_{a^k}\frac{\partial^2 x^1}{\partial t\,\partial a^k}-\left(\frac{\partial x^2}{\partial t}\right)_{a^k}\frac{\partial^2 x^2}{\partial t\,\partial a^k}-\left(\frac{\partial x^3}{\partial t}\right)_{a^k}\frac{\partial^2 x^3}{\partial t\,\partial a^k}\\
&\quad-\frac{f}{2}\left[x^1\frac{\partial^2 x^2}{\partial t\,\partial a^k}+\left(\frac{\partial x^2}{\partial t}\right)_{a^k}\frac{\partial x^1}{\partial a^k}-x^2\frac{\partial^2 x^1}{\partial t\,\partial a^k}\right.\\
&\quad\left.-\left(\frac{\partial x^1}{\partial t}\right)_{a^k}\frac{\partial x^2}{\partial a^k}\right]+\frac{\partial \phi}{\partial a^k}
\end{aligned} \tag{20.85}$$

Substitution of (20.84) and (20.85) into (20.82) results in a number of terms canceling out. The equation of motion in terms of the Lagrangian enumeration coordinates now reads

$$\left(\frac{\partial^2 x^1}{\partial t^2}\right)_{a^k}\frac{\partial x^1}{\partial a^k}+\left(\frac{\partial^2 x^2}{\partial t^2}\right)_{a^k}\frac{\partial x^2}{\partial a^k}+\left(\frac{\partial^2 x^3}{\partial t^2}\right)_{a^k}\frac{\partial x^3}{\partial a^k}$$
$$=-\frac{1}{\rho}\frac{\partial p}{\partial a^k}-\frac{\partial \phi}{\partial a^k}+f\left[\left(\frac{\partial x^2}{\partial t}\right)_{a^k}\frac{\partial x^1}{\partial a^k}-\left(\frac{\partial x^1}{\partial t}\right)_{a^k}\frac{\partial x^2}{\partial a^k}\right] \tag{20.86}$$
$$\text{with}\quad \left(\frac{\partial x^1}{\partial t}\right)_{a^k}=\dot{x}^1,\qquad \left(\frac{\partial x^2}{\partial t}\right)_{a^k}=\dot{x}^2$$

The last term on the right-hand side of this equation represents the Coriolis effect. With the help of (20.80) and (20.81) the Coriolis part of the equation of motion can then be written as

$$f\left(\dot{x}^2\frac{\partial x^1}{\partial a^k}-\dot{x}^1\frac{\partial x^2}{\partial a^k}\right)=\underset{x}{C}_1\frac{\partial x^1}{\partial a^k}+\underset{x}{C}_2\frac{\partial x^2}{\partial a^k}=\underset{a}{C}_k \tag{20.87}$$

so that (20.86) assumes the form

$$\boxed{\left(\frac{\partial^2 x^n}{\partial t^2}\right)_{a^k}\frac{\partial x^n}{\partial a^k}=-\frac{1}{\rho}\frac{\partial p}{\partial a^k}-\frac{\partial \phi}{\partial a^k}+\underset{a}{C}_k} \tag{20.88}$$

This equation can often be found in the literature, but usually without the Coriolis effect. Together with the continuity equation we now have a system of four scalar equations from which to determine the variables u, v, w, and ρ.

It is possible to write (20.88) in still another form by applying a Legendre transformation and by using (3.58) again. For the left-hand side of (20.88) we find

$$\left(\frac{\partial^2 x^n}{\partial t^2}\right)_{a^k}\frac{\partial x^n}{\partial a^k}=\left.\frac{\partial}{\partial t}\right|_{a^k}\left(\frac{\partial x^n}{\partial t}\frac{\partial x^n}{\partial a^k}\right)-\left[\frac{\partial}{\partial a^k}\left(\frac{\partial x^n}{\partial t}\right)_{a^k}\right]\left(\frac{\partial x^n}{\partial t}\right)_{a^k}$$
$$=\left.\frac{\partial}{\partial t}\right|_{a^k}\left(\dot{x}^n\frac{\partial x^n}{\partial a^k}\right)-\frac{1}{2}\frac{\partial}{\partial a^k}\left[\left(\frac{\partial x^n}{\partial t}\right)_{a^k}\left(\frac{\partial x^n}{\partial t}\right)_{a^k}\right] \tag{20.89}$$
$$=\left.\frac{\partial}{\partial t}\right|_{a^k}\left(\dot{x}^n\frac{\partial x^n}{\partial a^k}\right)-\frac{\partial}{\partial a^k}\left(\frac{\mathbf{v}^2}{2}\right)$$
$$\text{with}\quad \mathbf{v}^2=\underset{x}{g}_{nn}\,\dot{x}^n\dot{x}^n=\dot{x}^n\dot{x}^n\qquad (m=1\quad\text{or}\quad \underset{x}{g}_{ii}=1)$$

Substituting (20.89) into (20.88) results in the following version of the equation of motion in terms of the Lagrangian enumeration coordinates:

$$\left.\frac{\partial}{\partial t}\right|_{a^k}\left(\dot{x}^n\frac{\partial x^n}{\partial a^k}\right)=-\frac{1}{\rho}\frac{\partial p}{\partial a^k}+\frac{\partial}{\partial a^k}\left(\frac{\mathbf{v}^2}{2}-\phi\right)+\underset{a}{C}_k \tag{20.90}$$

By using the transformation equation (20.81), we find in place of (20.90) the very convenient form

$$\boxed{\frac{\partial}{\partial t}\bigg|_{a^k}\left(\underset{a}{v}_k\right) = -\frac{1}{\rho}\frac{\partial p}{\partial a^k} + \frac{\partial}{\partial a^k}\left(\frac{\mathbf{v}^2}{2} - \phi\right) + \underset{a}{C}_k} \tag{20.91}$$

Sometimes the pressure term is replaced by the thermodynamic relation

$$-\frac{1}{\rho}\,\delta p = T\,\delta s - \delta h \tag{20.92}$$

where s and h denote the specific entropy and the specific enthalpy. With $\delta = \partial/\partial a^k$ equation (20.91) is given the final form

$$\boxed{\frac{\partial}{\partial t}\bigg|_{a^k}\left(\underset{a}{v}_k\right) = \frac{\partial}{\partial a^k}\left(\frac{\mathbf{v}^2}{2} - \phi - h\right) + T\,\frac{\partial s}{\partial a^k} + \underset{a}{C}_k} \tag{20.93}$$

This equation, the continuity equation, the first law of thermodynamics, and the ideal-gas law give a complete system from which to determine the nonturbulent flow of unsaturated air.

20.9 Problems

20.1:
(a) Use Lagrange's equation of motion to verify the second equation of (20.43).
(b) Transform the first equation of (20.43) to obtain the first equation of (20.46).
(c) Transform the first equation of (20.46) to obtain the first equation of (20.48).

20.2: Verify that equation (20.67) is a solution by substituting $u(z, t)$ and $v(z, t)$ into the first equation of (20.66).

20.3: On the Mercator and Bond map the surface of the earth is represented by a Cartesian system in such a way that latitude circles are parallel to the x-axis and longitude circles are parallel to the y-axis. The representation is due to the differential transformation relations $dx = r\,d\lambda$ and $dy = r\,d\varphi/\cos\varphi$.
(a) Show that the representation is conformal. Find the map factor m.
(b) Starting with the rigidly rotating geographical coordinate system, find the following quantities: the metric fundamental form $(d\mathbf{r})^2$, kinetic energy K_{A}, metric fundamental quantities g_{ij}, g^{ij}, $\sqrt{g}$, and covariant and physical measure numbers of the relative velocity $\mathbf{v}$.
(c) Replace z by the pressure coordinate p and find $\underset{p}{\sqrt{g}}$. Assume the validity of the hydrostatic equation.

21

Orography-following coordinate systems

21.1 The metric of the η system

Suppose that we wish to model the air flow over a limited region of the earth's surface so that the effect of the earth's curvature on the flow may be ignored. In this case we may set the map factor $m_0 = 1$ so that the air flow refers to the tangential plane. The flow, however, might be strongly influenced by orographic effects. The question which now arises quite naturally is that of how the effects of the lower boundary on the flow should be formulated. It is always possible to state the lower boundary condition in the presence of orography by using the orthogonal Cartesian system but the formulation might be quite unwieldy. A far superior method for handling orography is to replace the Cartesian vertical coordinate z by a new vertical coordinate η, which is formulated in such a way that the surface of the earth coincides with a surface of the new vertical coordinate. We call this coordinate system the *η system*. The relation between the coordinate z and the new coordinate η is defined by the following transformation:

$$\eta = \frac{z - H}{H - h(x, y)} = \eta(x, y, z) \implies z(x, y, \eta) = \eta[H - h(x, y)] + H \qquad (21.1)$$

In this formula H represents a fixed upper boundary of the model region while $h(x, y)$ describes the orography. The new coordinate system is not orthogonal, so the equation of motion assumes a form that is more complicated than before. We admit only rigid rotation so that the coordinate surfaces of the η system do not deform. By introducing the η-coordinate the mathematical complexity is overcompensated by the effectiveness in the numerical evaluation of the flow model.

In order to visualize the transformation procedure we consider the idealized situation depicted in Figure 21.1. The covariant basis vectors $\mathbf{q}_x$ and $\mathbf{q}_\eta$ are tangential to the x- and η-coordinate lines while the contravariant basis vectors $\mathbf{q}^x$ and $\mathbf{q}^\eta$ point in the direction of the gradients ∇q^i, see (M3.27), since $\mathbf{q}^i = \nabla q^i$. We would

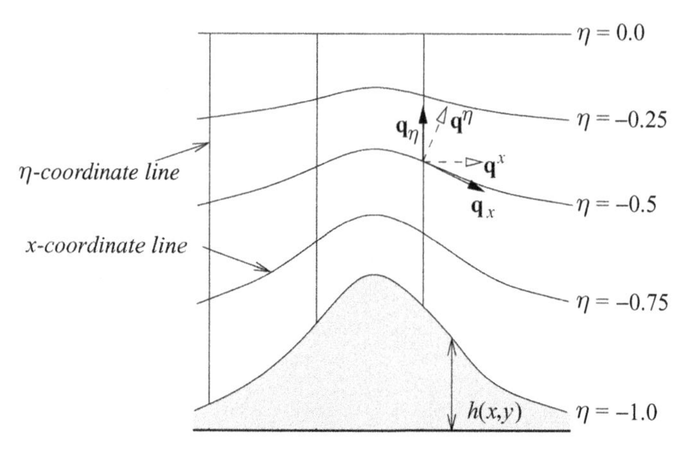

Fig. 21.1 A schematic representation of orography.

like to point out that we are dealing with a right-handed coordinate system. Had we taken the numerator of (21.1) as $H - z$ then the surface of the earth would be defined by $\eta = 1$, describing a left-handed coordinate system. As will be seen soon, in this case the functional determinant $\sqrt{g}$ assumes negative values.

The elements of the metric tensor or the metric fundamental quantities g_{ij} can be found by substituting

$$dz = \left(\frac{\partial z}{\partial x}\right)_{y,\eta} dx + \left(\frac{\partial z}{\partial y}\right)_{x,\eta} dy + \left(\frac{\partial z}{\partial \eta}\right)_{x,y} d\eta \tag{21.2}$$

into the metric fundamental form of the Cartesian system,

$$(d\mathbf{r})^2 = (dx)^2 + (dy)^2 + (dz)^2 \tag{21.3}$$

This yields

$$\begin{aligned}(d\mathbf{r})^2 = &\left[1 + \left(\frac{\partial z}{\partial x}\right)^2_{y,\eta}\right](dx)^2 + \left[1 + \left(\frac{\partial z}{\partial y}\right)^2_{x,\eta}\right](dy)^2 + \left(\frac{\partial z}{\partial \eta}\right)^2_{x,y}(d\eta)^2 \\ &+ 2\left(\frac{\partial z}{\partial x}\right)_{y,\eta}\left(\frac{\partial z}{\partial y}\right)_{x,\eta} dx\,dy + 2\left(\frac{\partial z}{\partial x}\right)_{y,\eta}\left(\frac{\partial z}{\partial \eta}\right)_{x,y} dx\,d\eta \\ &+ 2\left(\frac{\partial z}{\partial y}\right)_{x,\eta}\left(\frac{\partial z}{\partial \eta}\right)_{x,y} dy\,d\eta\end{aligned} \tag{21.4}$$

The required partial derivatives are obtained from the transformation (21.1) as

$$\left(\frac{\partial z}{\partial x}\right)_{y,\eta} = -\eta\frac{\partial h}{\partial x}, \qquad \left(\frac{\partial z}{\partial y}\right)_{x,\eta} = -\eta\frac{\partial h}{\partial y}, \qquad \left(\frac{\partial z}{\partial \eta}\right)_{x,y} = H - h \tag{21.5}$$

Comparison of (21.4) with the general relation $(d\mathbf{r})^2 = g_{nm}\, dq^n\, dq^m$ gives all components of the metric tensor:

$$
\begin{aligned}
g_{11} &= 1 + \eta^2\left(\frac{\partial h}{\partial x}\right)^2, & g_{12} &= \eta^2 \frac{\partial h}{\partial x}\frac{\partial h}{\partial y}, & g_{13} &= -\eta \frac{\partial h}{\partial x}(H - h) \\
g_{21} &= g_{12}, & g_{22} &= 1 + \eta^2\left(\frac{\partial h}{\partial y}\right)^2, & g_{23} &= -\eta \frac{\partial h}{\partial y}(H - h) \\
g_{31} &= g_{13}, & g_{32} &= g_{23}, & g_{33} &= (H - h)^2
\end{aligned}
\tag{21.6}
$$

In the work to follow we will also need the contravariant components g^{ij}. A convenient way to find the g^{ij} is to evaluate the transformation rule (M4.19):

$$
\underset{q}{g}^{ij} = \frac{\partial q^i}{\partial a^m}\frac{\partial q^j}{\partial a^n}\underset{a}{g}^{mn}, \qquad q^i = (x, y, \eta), \qquad a^i = (x, y, z), \qquad \underset{a}{g}^{ij} = \delta^{ij}
\tag{21.7}
$$

An example of (21.7) is given next in order to find the components $\underset{\eta}{g}^{23}$ of the metric tensor:

$$
\underset{\eta}{g}^{23} = \left(\frac{\partial y}{\partial x}\right)_{y,z}\left(\frac{\partial \eta}{\partial x}\right)_{y,z} + \left(\frac{\partial y}{\partial y}\right)_{x,z}\left(\frac{\partial \eta}{\partial y}\right)_{x,z} + \left(\frac{\partial y}{\partial z}\right)_{x,y}\left(\frac{\partial \eta}{\partial z}\right)_{x,y} = \frac{\eta}{H - h}\frac{\partial h}{\partial y}
$$

$$
\left(\frac{\partial \eta}{\partial x}\right)_{y,z} = \frac{\eta}{H - h}\frac{\partial h}{\partial x}, \qquad \left(\frac{\partial \eta}{\partial y}\right)_{x,z} = \frac{\eta}{H - h}\frac{\partial h}{\partial y}, \qquad \left(\frac{\partial \eta}{\partial z}\right)_{x,y} = \frac{1}{H - h}
\tag{21.8}
$$

The complete contravariant metric tensor g^{ij} is listed below:

$$
\begin{aligned}
g^{11} &= 1, & g^{12} &= 0, & g^{13} &= \frac{\eta}{H - h}\frac{\partial h}{\partial x} \\
g^{21} &= g^{12}, & g^{22} &= 1, & g^{23} &= \frac{\eta}{H - h}\frac{\partial h}{\partial y} \\
g^{31} &= g^{13}, & g^{32} &= g^{23}, & g^{33} &= \frac{1}{(H - h)^2}\left[1 + \eta^2\left(\frac{\partial h}{\partial x}\right)^2 + \eta^2\left(\frac{\partial h}{\partial y}\right)^2\right]
\end{aligned}
\tag{21.9}
$$

In order to obtain the continuity equation of the η system we need to find the functional determinant. This quantity is obtained from the general transformation rule (M4.21):

$$
\sqrt{\underset{q}{g}} = \sqrt{\underset{a}{g}}\left|\frac{\partial(a^1, a^2, a^3)}{\partial(q^1, q^2, q^3)}\right| \Longrightarrow \sqrt{\underset{\eta}{g}} = \sqrt{\underset{z}{g}}\left(\frac{\partial z}{\partial \eta}\right)_{x,y} = H - h, \qquad \sqrt{\underset{z}{g}} = 1
\tag{21.10}
$$

This expression confirms our previous statement that the functional determinant becomes negative if we choose $H - z$ instead of $z - H$ in the transformation relation (21.1) for going between η and z.

21.2 The equation of motion in the η system

In order to proceed efficiently, we will first obtain various parts of the equation of motion. Then it will be easy to write down the final covariant and contravariant forms in the η system. The relations between the contravariant and covariant velocity components are

$$\dot{q}^1 = v^1 = \dot{x}, \qquad \dot{q}^2 = v^2 = \dot{y}, \qquad \dot{q}^3 = v^3 = \dot{\eta}, \qquad v_i = g_{in} v^n \tag{21.11}$$

and can be used whenever needed. Next we wish to obtain the gradient of the geopotential

$$\phi = gz = g[\eta(H - h) + H] \tag{21.12}$$

yielding

$$\left(\frac{\partial \phi}{\partial x}\right)_{y,\eta} = -g\eta \frac{\partial h}{\partial x}, \quad \left(\frac{\partial \phi}{\partial y}\right)_{x,\eta} = -g\eta \frac{\partial h}{\partial y}, \quad \left(\frac{\partial \phi}{\partial \eta}\right)_{x,y} = g(H - h) \tag{21.13}$$

Here the usual symbol g has been utilized for the gravitational acceleration since confusion with the functional determinant $\sqrt{g}$ is unlikely.

Next, we wish to obtain the components of the rotational dyadic by employing equation (M5.48). The required components $\underset{\Omega}{W_i}$ of the rotational velocity $\mathbf{v}_\Omega$ are found with the help of (20.36). We are going to retain m_0 in the calculations and set $m_0 = 1$ at the end of the analysis. Application of the general transformation rule (M4.15) to the components of $\mathbf{v}_\Omega$ yields

$$\underset{q}{W}_i = \left(\frac{\partial a^n}{\partial q^i}\right)_{q^k} \underset{a}{W}_n \tag{21.14}$$

with $q^i = x, y, \eta$ and $a^i = x, y, z$. Both systems refer to the stereographic projection. From (20.36) we find

$$\underset{a}{W}_1 = \underset{\Omega}{W_1} = -\frac{\Omega y}{m_0^2}, \qquad \underset{a}{W}_2 = \underset{\Omega}{W_2} = \frac{\Omega x}{m_0^2}, \qquad \underset{a}{W}_3 = \underset{\Omega}{W_3} = 0 \tag{21.15}$$

Employing the transformation rule (21.14) yields

$$\underset{\eta}{W}_1 = -\frac{\Omega y}{m_0^2}, \qquad \underset{\eta}{W}_2 = \frac{\Omega x}{m_0^2}, \qquad \underset{\eta}{W}_3 = 0 \tag{21.16}$$

For example, we find for the component $\underset{\Omega}{\omega}_{12}$

$$
\begin{aligned}
\underset{\Omega}{\omega}_{12} &= \frac{1}{2}\left[\frac{\partial}{\partial x}\left(\frac{\Omega x}{m_0^2}\right) + \frac{\partial}{\partial y}\left(\frac{\Omega y}{m_0^2}\right)\right] \\
&= \frac{\Omega}{m_0^2}\left\{1 + \frac{m_0^2}{2}\left[x\,\frac{\partial}{\partial x}\left(\frac{1}{m_0^2}\right) + y\,\frac{\partial}{\partial y}\left(\frac{1}{m_0^2}\right)\right]\right\} \\
&= \frac{\Omega}{m_0^2}\sin\varphi = \frac{f}{2m_0^2}
\end{aligned}
\tag{21.17}
$$

The last step follows with the help of (20.29) so that the Coriolis parameter $f = 2\Omega\sin\varphi$ can be utilized.

The complete rotational tensor is given by

$$
\begin{aligned}
&\underset{\Omega}{\omega}_{11} = 0, && \underset{\Omega}{\omega}_{12} = \frac{f}{2m_0^2}, && \underset{\Omega}{\omega}_{13} = 0 \\
&\underset{\Omega}{\omega}_{21} = -\underset{\Omega}{\omega}_{12}, && \underset{\Omega}{\omega}_{22} = 0, && \underset{\Omega}{\omega}_{23} = 0 \\
&\underset{\Omega}{\omega}_{31} = 0, && \underset{\Omega}{\omega}_{32} = 0, && \underset{\Omega}{\omega}_{33} = 0
\end{aligned}
\tag{21.18}
$$

which agrees with (20.42). For the evaluation of the contravariant form of the equation of motion, given by (18.34), we need the relation

$$
\underset{\Omega}{\omega}_i{}^k = g^{km}\underset{\Omega}{\omega}_{im}
\tag{21.19}
$$

Finally, we calculate the Christoffel symbols according to equation (21.9). The evaluation is an easy but admittedly very tedious task. The result is

$$
\begin{aligned}
&\Gamma_{11}^{3} = -\frac{\eta}{H-h}\frac{\partial^2 h}{\partial x^2}, && \Gamma_{12}^{3} = -\frac{\eta}{H-h}\frac{\partial^2 h}{\partial x\,\partial y}, && \Gamma_{13}^{3} = -\frac{1}{H-h}\frac{\partial h}{\partial x} \\
&\Gamma_{21}^{3} = \Gamma_{12}^{3}, && \Gamma_{22}^{3} = -\frac{\eta}{H-h}\frac{\partial^2 h}{\partial y^2}, && \Gamma_{23}^{3} = -\frac{1}{H-h}\frac{\partial h}{\partial y} \\
&\Gamma_{31}^{3} = \Gamma_{13}^{3}, && \Gamma_{32}^{3} = \Gamma_{23}^{3}, && \Gamma_{33}^{3} = 0 \\
& && \Gamma_{1i}^{1} = 0, \quad \Gamma_{2i}^{2} = 0, && i = 1, 2, 3
\end{aligned}
\tag{21.20}
$$

Now we have all the parts needed in order to evaluate the general covariant form of the equations of motion (18.25) and general contravariant form (18.34). The covariant equations of motion of the η system are given by

$$
\begin{aligned}
&\frac{dv_1}{dt} - \eta^2\left[\frac{\dot{x}^2}{2}\frac{\partial}{\partial x}\left(\frac{\partial h}{\partial x}\right)^2 + \frac{\dot{y}^2}{2}\frac{\partial}{\partial x}\left(\frac{\partial h}{\partial y}\right)^2 + \dot{x}\dot{y}\frac{\partial}{\partial x}\left(\frac{\partial h}{\partial x}\frac{\partial h}{\partial y}\right)\right] \\
&\quad - \frac{\dot{\eta}^2}{2}\frac{\partial}{\partial x}(H-h)^2 + \dot{\eta}\eta\left[\dot{x}\frac{\partial}{\partial x}\left((H-h)\frac{\partial h}{\partial x}\right)\right. \\
&\quad \left. + \dot{y}\frac{\partial}{\partial x}\left((H-h)\frac{\partial h}{\partial y}\right)\right] - f\dot{y} = -\frac{1}{\rho}\frac{\partial p}{\partial x} + g\eta\frac{\partial h}{\partial x} + \frac{1}{\rho}\mathbf{q}_1\cdot\nabla\cdot\mathbb{J} \\
&\frac{dv_2}{dt} - \eta^2\left[\frac{\dot{x}^2}{2}\frac{\partial}{\partial y}\left(\frac{\partial h}{\partial x}\right)^2 + \frac{\dot{y}^2}{2}\frac{\partial}{\partial y}\left(\frac{\partial h}{\partial y}\right)^2 + \dot{x}\dot{y}\frac{\partial}{\partial y}\left(\frac{\partial h}{\partial x}\frac{\partial h}{\partial y}\right)\right] \\
&\quad - \frac{\dot{\eta}^2}{2}\frac{\partial}{\partial y}(H-h)^2 + \dot{\eta}\eta\left[\dot{x}\frac{\partial}{\partial y}\left((H-h)\frac{\partial h}{\partial x}\right)\right. \\
&\quad \left. + \dot{y}\frac{\partial}{\partial y}\left((H-h)\frac{\partial h}{\partial y}\right)\right] + f\dot{x} = -\frac{1}{\rho}\frac{\partial p}{\partial y} + g\eta\frac{\partial h}{\partial y} + \frac{1}{\rho}\mathbf{q}_2\cdot\nabla\cdot\mathbb{J} \\
&\frac{dv_3}{dt} - \eta\left[\dot{x}^2\left(\frac{\partial h}{\partial x}\right)^2 + \dot{y}^2\left(\frac{\partial h}{\partial y}\right)^2 + 2\dot{x}\dot{y}\frac{\partial h}{\partial x}\frac{\partial h}{\partial y}\right] \\
&\quad + \dot{\eta}\left(\dot{x}\frac{\partial h}{\partial x} + \dot{y}\frac{\partial h}{\partial y}\right)(H-h) = -\frac{1}{\rho}\frac{\partial p}{\partial \eta} - g(H-h) + \frac{1}{\rho}\mathbf{q}_3\cdot\nabla\cdot\mathbb{J} \\
&\text{with} \quad \frac{d}{dt} = \frac{\partial}{\partial t} + v_n g^{mn}\frac{\partial}{\partial q^m}
\end{aligned}
\tag{21.21}
$$

and those in the contravariant form are given by

$$
\begin{aligned}
&\frac{d\dot{x}}{dt} - f\dot{y} = -\frac{1}{\rho}\frac{\partial p}{\partial x} - \frac{1}{\rho}\left(\frac{\eta}{H-h}\frac{\partial h}{\partial x}\right)\frac{\partial p}{\partial \eta} + \frac{1}{\rho}\mathbf{q}^1\cdot\nabla\cdot\mathbb{J} \\
&\frac{d\dot{y}}{dt} + f\dot{x} = -\frac{1}{\rho}\frac{\partial p}{\partial y} - \frac{1}{\rho}\left(\frac{\eta}{H-h}\frac{\partial h}{\partial y}\right)\frac{\partial p}{\partial \eta} + \frac{1}{\rho}\mathbf{q}^2\cdot\nabla\cdot\mathbb{J} \\
&\frac{d\dot{\eta}}{dt} - \frac{\eta}{H-h}\left(\dot{x}^2\frac{\partial^2 h}{\partial x^2} + \dot{y}^2\frac{\partial^2 h}{\partial y^2} + 2\dot{x}\dot{y}\frac{\partial^2 h}{\partial x\,\partial y}\right) \\
&\qquad - \frac{2\dot{\eta}}{H-h}\left(\dot{x}\frac{\partial h}{\partial x} + \dot{y}\frac{\partial h}{\partial y}\right) + \frac{f\eta}{H-h}\left(\dot{x}\frac{\partial h}{\partial y} - \dot{y}\frac{\partial h}{\partial x}\right) \\
&\quad = -\frac{\eta}{H-h}\left(\frac{\partial h}{\partial x}\frac{\partial p}{\partial x} + \frac{\partial h}{\partial y}\frac{\partial p}{\partial y}\right) - \frac{g}{H-h} + \frac{1}{\rho}\mathbf{q}^3\cdot\nabla\cdot\mathbb{J} \\
&\qquad - \frac{1}{(H-h)^2}\left[1 + \eta^2\left(\frac{\partial h}{\partial x}\right)^2 + \eta^2\left(\frac{\partial h}{\partial y}\right)^2\right]\frac{\partial p}{\partial \eta} \\
&\text{with} \quad \frac{d}{dt} = \frac{\partial}{\partial t} + \dot{x}\frac{\partial}{\partial x} + \dot{y}\frac{\partial}{\partial y} + \dot{\eta}\frac{\partial}{\partial \eta}
\end{aligned}
\tag{21.22}
$$

It should be understood that $\dot{\eta}$ is the vertical velocity of a point mass. Comparison of (21.21) with (21.22) shows that the contravariant form has a less complicated structure than the covariant form and, therefore, is easier to apply. It should also be noted that the gradient of the geopotential does not appear in the horizontal components of the contravariant system. The covariant system, however, does contain contributions from the geopotential.

21.3 The continuity equation in the η system

The general form of the continuity equation (18.48) for rigid rotation is easily adapted to the η system since the functional determinant has already been given by (21.10). The following result is obtained:

$$\frac{d\rho}{dt}+\frac{\rho}{H-h}\left(\frac{\partial}{\partial x}[(H-h)\dot{x}]+\frac{\partial}{\partial y}[(H-h)\dot{y}]+\frac{\partial}{\partial \eta}[(H-h)\dot{\eta}]\right)=0 \quad (21.23)$$

If covariant velocity components are desired in the continuity equation, we need to substitute equation (21.11) into (21.23). This gives a very lengthy equation, which will not be presented here.

In this chapter we have demonstrated that it is very easy to obtain the equation of motion for a special coordinate system due to our knowledge of the general equations for the covariant and contravariant forms given by (18.23), (18.25), (18.33) and (18.34). It appears that Sommerville and Gal-Chen (1974) were the first to introduce the orography following-coordinates giving the exact form of the equation of motion. In this context we also recommend the important paper by Gal-Chen and Sommerville (1975). Many other papers have followed. Very often various approximations were used, such as forced orthogonalization. This can be done quite simply by setting the off-diagonal elements in the metric tensors g_{ij} and g^{ij} in (21.6) and (21.9) equal to zero.

21.4 Problems

21.1: Rewrite the first equation of (21.21) by using covariant velocity components. To keep the analysis simple, simplify the metric tensor by a forced orthogonalization.

22

The stereographic system with a generalized vertical coordinate

In the previous chapter we introduced the vertical coordinate η to handle orographic effects in mesoscale models. In the synoptic-scale models we are going to replace the height coordinate z which extends to infinity by a generalized vertical coordinate ξ. The introduction of ξ is motivated by the fact that we cannot integrate the predictive equations using z as a vertical coordinate to infinitely large heights. Replacing z by the atmospheric pressure p, for example, results in a finite range of the vertical coordinate. We will see that another advantage of the (x, y, p)-coordinate system is that the continuity equation is time-independent. There are other specific coordinate systems that we are going to discuss. Therefore, it seems of advantage to first set up the atmospheric equations in terms of the unspecified generalized vertical coordinate ξ. Later we will specify ξ as desired. We wish to point out that the introduction of the generalized coordinate is of advantage only if the hydrostatic equation is a part of the atmospheric system.

We will briefly state the consequences of the transformation from the stereographic (x, y, z)-coordinate system to the stereographic (x, y, ξ)-coordinate system, which henceforth will be called the *ξ system.*

(i) The hydrostatic approximation is not restricted to the hydrostatic equation itself but enters implicitly into the horizontal equations of motion and the continuity equation.

(ii) The boundary conditions at the earth's surface can be formulated rather easily. A surface ξ = constant may be arranged in such a way that it coincides with the orographic surface of the earth.

(iii) The infinite height range in the z-coordinate is usually replaced by a finite height range in the ξ-coordinate. Large-scale models normally employ covariant and physical velocity measure numbers to formulate the equation of motion. For this reason and to prevent the following chapters from becoming too lengthy, we will omit any discussion of contravariant forms. Nevertheless, we have given a sufficient background for any interested reader to formulate these by using the proper transformation rules. Furthermore, we will neglect friction in the large-scale flow.

22.1 The ξ transformation and resulting equations

As the original system we consider the orthogonal stereographic Cartesian system. The uniqueness of the transformation between the (x, y, z)-system and the (x, y, ξ)-system is guaranteed by the requirement that ξ is a monotonically increasing or decreasing function of z. Now the function $z = z(x, y, \xi, t)$ represents the height of the time-dependent ξ-surface. Although the original system is orthogonal, the effect of the coordinate transformation is that the new system is not orthogonal. In order to retain the convenience of the orthogonal systems we are forced to make a simplifying assumption regarding the metric tensor. We assume that the height of the ξ-surface is independent of the horizontal coordinates (x, y). We call this assumption the *second metric simplification* since it is used only to find a simplified form of the metric tensor. Hence, we have

$$z = z(\xi, t) \implies \dot{z} = \left(\frac{\partial z}{\partial t}\right)_\xi + \left(\frac{\partial z}{\partial \xi}\right)_t \dot{\xi} \tag{22.1}$$

We obtain the metric tensor by replacing $\dot{z}^2$ in the original system (20.34) with the help of (22.1). The absolute kinetic energy is given by $K_A = K + K_P$ with

$$\begin{aligned} K &= \frac{1}{2}\left[\frac{\dot{x}^2 + \dot{y}^2}{m_0^2} + \left(\frac{\partial z}{\partial \xi}\right)_t^2 \dot{\xi}^2\right] \\ K_P &= \frac{\Omega(x\dot{y} - \dot{x}y)}{m_0^2} + \frac{(x^2 + y^2)\Omega^2}{2m_0^2} + \frac{1}{2}\left(\frac{\partial z}{\partial t}\right)_\xi^2 + \left(\frac{\partial z}{\partial t}\right)_\xi \left(\frac{\partial z}{\partial \xi}\right)_t \dot{\xi} \end{aligned} \tag{22.2}$$

from which we obtain the covariant form of the metric tensor according to

$$g_{ij} = \frac{\partial^2 K_A}{\partial \dot{q}^i\, \partial \dot{q}^j} \implies g_{11} = \frac{1}{m_0^2}, \quad g_{22} = \frac{1}{m_0^2}, \quad g_{33} = \left(\frac{\partial z}{\partial \xi}\right)_t^2, \quad g_{ij} = 0, \quad i \neq j \tag{22.3}$$

Since all off-diagonal elements of g_{ij} are zero, we are dealing with an orthogonal system as desired. What we have done, in fact, is implemented a *forced orthogonalization* of the metric tensor. Had we permitted the height z of the ξ-surface to vary with the horizontal coordinates (x, y), every element g_{ij} would have been nonzero.

As the next step we are going to obtain the components of the velocity vector $\mathbf{v}_P$ which can be split into the two velocity vectors $\mathbf{v}_\Omega$ and $\mathbf{v}_D$ describing the rotation of the earth and the deformation of the material surfaces on which the point P is located. The components W_i of $\mathbf{v}_P$ can be found from

$$W_i = \frac{\partial K_P}{\partial \dot{q}^i} \implies W_1 = -\frac{\Omega y}{m_0^2}, \quad W_2 = \frac{\Omega x}{m_0^2}, \quad W_3 = \left(\frac{\partial z}{\partial t}\right)_\xi \left(\frac{\partial z}{\partial \xi}\right)_t \tag{22.4}$$

Utilizing this information, the components of $\mathbf{v}_P$ are given by

$$
\begin{aligned}
&(\underset{\Omega}{W_1}, \underset{\Omega}{W_2}, \underset{\Omega}{W_3}) = \left(-\frac{\Omega y}{m_0^2}, \frac{\Omega x}{m_0^2}, 0\right), &&(\underset{\Omega}{W^1}, \underset{\Omega}{W^2}, \underset{\Omega}{W^3}) = (-\Omega y, \Omega x, 0) \\
&(\underset{D}{W_1}, \underset{D}{W_2}, \underset{D}{W_3}) = \left(0, 0, \left(\frac{\partial z}{\partial t}\right)_\xi \left(\frac{\partial z}{\partial \xi}\right)_t\right), &&(\underset{D}{W^1}, \underset{D}{W^2}, \underset{D}{W^3}) = \left(0, 0, -\left(\frac{\partial \xi}{\partial t}\right)_z\right)
\end{aligned}
\tag{22.5}
$$

From the contravariant measure numbers of the relative velocity $(v^1, v^2, v^3) = (\dot{x}, \dot{y}, \dot{\xi})$ we obtain the covariant and physical measure numbers with the help of (22.3) as

$$
(v_1, v_2, v_3) = \left(\frac{\dot{x}}{m_0^2}, \frac{\dot{y}}{m_0^2}, \left(\frac{\partial z}{\partial \xi}\right)_t^2 \dot{\xi}\right), \quad (u, v, w) = \left(\frac{\dot{x}}{m_0}, \frac{\dot{y}}{m_0}, \left(\frac{\partial z}{\partial \xi}\right)_t \dot{\xi}\right) \tag{22.6}
$$

Since the ξ system is deformational, in contrast to the Cartesian stereographic system, we have to evaluate the general forms (18.23) (covariant) and (18.36) (physical) of the equations of motion. From (22.5) it can be seen that only the third component of the deformational velocity differs from zero. Furthermore, due to the second metric simplification $\underset{D}{W_3}$ is independent of (x, y). As a consequence of this we have, see (M5.48),

$$
\begin{aligned}
&\underset{D}{\omega_{ij}} = 0, \; i, j = 1, 2, 3, \quad \frac{\partial}{\partial x}\left(\frac{\mathbf{v}_D^2}{2}\right) = 0, \\
&\frac{\partial}{\partial y}\left(\frac{\mathbf{v}_D^2}{2}\right) = 0, \quad \frac{\partial}{\partial t}\underset{D}{W_1} = 0, \quad \frac{\partial}{\partial t}\underset{D}{W_2} = 0
\end{aligned}
\tag{22.7}
$$

The rotational tensor $\underset{\Omega}{\omega_{ij}}$ is identical with the corresponding tensor of the stereographic system and is given by (20.42). Comparison of (22.3) with (20.26) shows that g_{11} and g_{22} are also identical in these two systems since we are using $m = m_0$. With this information the evaluation of (18.50) shows that the horizontal components of the equations of motion in the ξ system are the same as those in the Cartesian stereographic system. However, it should be observed that, in the ξ system, all partial derivatives with respect to x, y, or t are taken at constant values of ξ, in contrast to the stereographic coordinate system, in which the independent variable z is held constant in these operations.

For the vertical component of the equation of motion we obtain from (18.23)

$$
\frac{dv_3}{dt} - \frac{\dot{\xi}^2}{2}\frac{\partial}{\partial \xi}\left(\frac{\partial z}{\partial \xi}\right)_t^2 + \frac{\partial}{\partial t}\underset{D}{W_3} - \frac{1}{2}\frac{\partial}{\partial \xi}\left[\left(\frac{\partial z}{\partial t}\right)_\xi^2 \left(\frac{\partial z}{\partial \xi}\right)_t^2\right] = -\frac{1}{\rho}\frac{\partial p}{\partial \xi} - \frac{\partial \phi}{\partial \xi} \tag{22.8}
$$

Had we permitted the height z of the ξ-surface to vary with (x, y) then this equation as well as the horizontal components would have assumed a very complicated form. Even in the simplified form the left-hand side of (22.8) is difficult to evaluate.

In Section 2.3 we discussed the hydrostatic approximation and showed by means of scale analysis that, for large-scale frictionless flow, the vertical equation of motion may be realistically approximated by the hydrostatic equation. However, for mesoscale motion the use of the hydrostatic equation is not permitted. We may find the hydrostatic equation in terms of the generalized coordinate ξ from

$$\frac{\partial p}{\partial z} = \frac{\partial p}{\partial \xi}\frac{\partial \xi}{\partial z} = -g\rho, \qquad \frac{\partial \phi}{\partial \xi} = \frac{\partial \phi}{\partial z}\frac{\partial z}{\partial \xi} = g\frac{\partial z}{\partial \xi} \implies \frac{\partial p}{\partial \xi} = -\rho\frac{\partial \phi}{\partial \xi} \tag{22.9}$$

The same equation is obtained by setting the left-hand side of (22.8) equal to zero. Assuming that we have hydrostatic conditions and frictionless flow, the covariant form of the equations of motion in the ξ system may be written as

$$\boxed{\begin{aligned} \frac{dv_1}{dt} + \frac{v_1^2 + v_2^2}{2}\frac{\partial m_0^2}{\partial x} - f v_2 &= -\frac{1}{\rho}\frac{\partial p}{\partial x} - \frac{\partial \phi}{\partial x} \\ \frac{dv_2}{dt} + \frac{v_1^2 + v_2^2}{2}\frac{\partial m_0^2}{\partial y} + f v_1 &= -\frac{1}{\rho}\frac{\partial p}{\partial y} - \frac{\partial \phi}{\partial y} \\ 0 &= -\frac{1}{\rho}\frac{\partial p}{\partial \xi} - \frac{\partial \phi}{\partial \xi} \\ \text{with} \quad \frac{d}{dt} = \frac{\partial}{\partial t} + m_0^2\left(v_1\frac{\partial}{\partial x} + v_2\frac{\partial}{\partial y}\right) &+ \dot{\xi}\frac{\partial}{\partial \xi} \end{aligned}} \tag{22.10}$$

After substituting the identities $u = m_0 v_1$ and $v = m_0 v_2$ into (22.10), by obvious steps we obtain the equations of motion in physical velocity components:

$$\boxed{\begin{aligned} \frac{du}{dt} + v\left(v\frac{\partial m_0}{\partial x} - u\frac{\partial m_0}{\partial y}\right) - fv &= -\frac{m_0}{\rho}\frac{\partial p}{\partial x} - m_0\frac{\partial \phi}{\partial x} \\ \frac{dv}{dt} - u\left(v\frac{\partial m_0}{\partial x} - u\frac{\partial m_0}{\partial y}\right) + fu &= -\frac{m_0}{\rho}\frac{\partial p}{\partial y} - m_0\frac{\partial \phi}{\partial y} \\ 0 &= -\frac{1}{\rho}\frac{\partial p}{\partial \xi} - \frac{\partial \phi}{\partial \xi} \\ \text{with} \quad \frac{d}{dt} = \frac{\partial}{\partial t} + m_0\left(u\frac{\partial}{\partial x} + v\frac{\partial}{\partial y}\right) &+ \dot{\xi}\frac{\partial}{\partial \xi} \end{aligned}} \tag{22.11}$$

As in the covariant case, the horizontal equations of motion agree with the corresponding equations (20.48) of the stereographic (x, y, z) system. Recall, however, that all partial derivatives with respect to x, y, or t in (22.11) are calculated at constant values of ξ.

Any realistic flow problem requires the involvement of the continuity equation, which will be derived next. The necessary prerequisite is knowledge of the functional determinant whose elements were given in (22.3) so that

$$\sqrt{g} = \sqrt{|g_{ij}|} = \frac{1}{m_0^2}\left(\frac{\partial z}{\partial \xi}\right)_t = \frac{1}{m_0^2 g}\left(\frac{\partial \phi}{\partial \xi}\right)_t = -\frac{1}{m_0^2 g \rho}\left(\frac{\partial p}{\partial \xi}\right)_t \tag{22.12}$$

follows immediately. Since the density ρ is time-dependent, the functional determinant $\sqrt{g}$ is time-dependent also, so the last term in (18.47) does not vanish, that is the ξ system is deformative. Rewriting (18.47) as

$$\frac{1}{\rho\sqrt{g}}\frac{d}{dt}(\rho\sqrt{g}) + \frac{\partial \dot{q}^n}{\partial q^n} = 0 \tag{22.13}$$

yields the contravariant form of the continuity equation:

$$\frac{m_0^2}{\left(\frac{\partial p}{\partial \xi}\right)_t}\frac{d}{dt}\left[\frac{1}{m_0^2}\left(\frac{\partial p}{\partial \xi}\right)_t\right] + \frac{\partial \dot{x}}{\partial x} + \frac{\partial \dot{y}}{\partial y} + \frac{\partial \dot{\xi}}{\partial \xi} = 0 \tag{22.14}$$

With the help of (22.6) we obtain the covariant form

$$\boxed{\frac{m_0^2}{\left(\frac{\partial p}{\partial \xi}\right)_t}\frac{d}{dt}\left[\frac{1}{m_0^2}\left(\frac{\partial p}{\partial \xi}\right)_t\right] + \frac{\partial}{\partial x}(m_0^2 v_1) + \frac{\partial}{\partial y}(m_0^2 v_2) + \frac{\partial \dot{\xi}}{\partial \xi} = 0} \tag{22.15}$$

and the form with physical velocity components

$$\boxed{\frac{m_0^2}{\left(\frac{\partial p}{\partial \xi}\right)_t}\frac{d}{dt}\left[\frac{1}{m_0^2}\left(\frac{\partial p}{\partial \xi}\right)_t\right] + \frac{\partial}{\partial x}(m_0 u) + \frac{\partial}{\partial y}(m_0 v) + \frac{\partial \dot{\xi}}{\partial \xi} = 0} \tag{22.16}$$

For convenience the contravariant velocity component $\dot{\xi}$ has been retained in the latter two equations.

In order to solve atmospheric flow problems, it is sometimes of advantage to combine the equation of motion and the continuity equation to give the so-called *stream-momentum form of the equation of motion*. This form is obtained by multiplying the covariant horizontal equations of motion by $\partial p/\partial \xi$ and the continuity equation by v_k, $k = 1, 2$. Adding the resulting equations yields

$$
\begin{aligned}
&\frac{\partial}{\partial t}\left(\frac{\partial p}{\partial \xi} v_1\right) + m_0^2\left[\frac{\partial}{\partial x}\left(\frac{\partial p}{\partial \xi} v_1^2\right) + \frac{\partial}{\partial y}\left(\frac{\partial p}{\partial \xi} v_1 v_2\right)\right] + \frac{\partial}{\partial \xi}\left(\frac{\partial p}{\partial \xi} v_1 \dot{\xi}\right) \\
&\qquad + \frac{\partial p}{\partial \xi}\frac{v_1^2 + v_2^2}{2}\frac{\partial m_0^2}{\partial x} - f\frac{\partial p}{\partial \xi} v_2 = -\frac{1}{\rho}\frac{\partial p}{\partial \xi}\frac{\partial p}{\partial x} - \frac{\partial p}{\partial \xi}\frac{\partial \phi}{\partial x} \\
&\frac{\partial}{\partial t}\left(\frac{\partial p}{\partial \xi} v_2\right) + m_0^2\left[\frac{\partial}{\partial x}\left(\frac{\partial p}{\partial \xi} v_2 v_1\right) + \frac{\partial}{\partial y}\left(\frac{\partial p}{\partial \xi} v_2^2\right)\right] + \frac{\partial}{\partial \xi}\left(\frac{\partial p}{\partial \xi} v_2 \dot{\xi}\right) \\
&\qquad + \frac{\partial p}{\partial \xi}\frac{v_1^2 + v_2^2}{2}\frac{\partial m_0^2}{\partial y} + f\frac{\partial p}{\partial \xi} v_1 = -\frac{1}{\rho}\frac{\partial p}{\partial \xi}\frac{\partial p}{\partial y} - \frac{\partial p}{\partial \xi}\frac{\partial \phi}{\partial y}
\end{aligned}
\tag{22.17}
$$

22.2 The pressure system

For any practical applications we need to specify ξ in order to obtain various prediction systems. In this section we choose p as the generalized vertical coordinate and thus obtain the so-called pressure system or simply the p *system*. In the following we will see that the atmospheric equations assume a simpler form in the p system than they do in any other ξ system. Therefore, the p system is often used for analytic and numerical studies. The (x, y, z)-coordinate system is a right-handed system. For the (x, y, ξ)-coordinate system to be right-handed also, the generalized coordinate ξ should also increase with height so that $\xi = -p$. It is customary, however, to set $\xi = p$ so that we may view this system as left-handed since z and p increase in opposite directions. It can be shown that the resulting equations of the p system are identical irrespective of whether we choose $\xi = -p$ or $\xi = p$. Therefore, we stick to the conventional notation and choose $\xi = p$. It should be observed that the p system as used here was subjected to a forced orthogonalization.

Since the pressure cannot vary along any constant-pressure surface, the covariant form of the equations of motion (22.10) simplifies to

$$
\boxed{
\begin{aligned}
&\frac{dv_1}{dt} + \frac{v_1^2 + v_2^2}{2}\frac{\partial m_0^2}{\partial x} - f v_2 = -\frac{\partial \phi}{\partial x} \\
&\frac{dv_2}{dt} + \frac{v_1^2 + v_2^2}{2}\frac{\partial m_0^2}{\partial y} + f v_1 = -\frac{\partial \phi}{\partial y} \\
&\frac{\partial \phi}{\partial p} = -\frac{1}{\rho} \\
&\text{with}\quad \frac{d}{dt} = \frac{\partial}{\partial t} + m_0^2\left(v_1\frac{\partial}{\partial x} + v_2\frac{\partial}{\partial y}\right) + \omega\frac{\partial}{\partial p}
\end{aligned}
}
\tag{22.18}
$$

where $\omega = dp/dt$. The system (22.11) with physical velocity components assumes the form

$$\boxed{\begin{aligned} \frac{du}{dt} + v\left(v\frac{\partial m_0}{\partial x} - u\frac{\partial m_0}{\partial y}\right) - fv &= -m_0\frac{\partial \phi}{\partial x} \\ \frac{dv}{dt} - u\left(v\frac{\partial m_0}{\partial x} - u\frac{\partial m_0}{\partial y}\right) + fu &= -m_0\frac{\partial \phi}{\partial y} \\ \frac{\partial \phi}{\partial p} &= -\frac{1}{\rho} \\ \text{with}\quad \frac{d}{dt} &= \frac{\partial}{\partial t} + m_0\left(u\frac{\partial}{\partial x} + v\frac{\partial}{\partial y}\right) + \omega\frac{\partial}{\partial p} \end{aligned}} \tag{22.19}$$

No further comments are necessary. Likewise the continuity equations (22.15) and (22.16) reduce to

$$m_0^2\left(\frac{\partial v_1}{\partial x} + \frac{\partial v_2}{\partial y}\right) + \frac{\partial \omega}{\partial p} = 0 \tag{22.20}$$

and

$$m_0\left(\frac{\partial u}{\partial x} + \frac{\partial v}{\partial y}\right) + \frac{\partial \omega}{\partial p} - \left(u\frac{\partial m_0}{\partial x} + v\frac{\partial m_0}{\partial y}\right) = 0 \tag{22.21}$$

From (22.6) we see that, for the special case $m_0 = 1$, the contravariant velocities $(\dot{x}, \dot{y})$, the covariant velocities (v_1, v_2), and the physical velocity components (u, v) are identical. In this case the equations of motion and the continuity equation reduce to

$$\boxed{\begin{aligned} \frac{du}{dt} - fv &= -\frac{\partial \phi}{\partial x} \\ \frac{dv}{dt} + fu &= -\frac{\partial \phi}{\partial y} \\ \frac{\partial \phi}{\partial p} &= -\frac{1}{\rho} \end{aligned}} \tag{22.22}$$

and

$$\boxed{\frac{\partial u}{\partial x} + \frac{\partial v}{\partial y} + \frac{\partial \omega}{\partial p} = 0} \tag{22.23}$$

describing the flow on the tangential plane.

The introduction of $\xi = p$ causes the continuity equation to assume the form of a diagnostic relation in which the time derivative does not appear explicitly. This form of the continuity equation implies that the air behaves like an incompressible fluid in the pressure system, as will be shown next. For simplicity let us assume

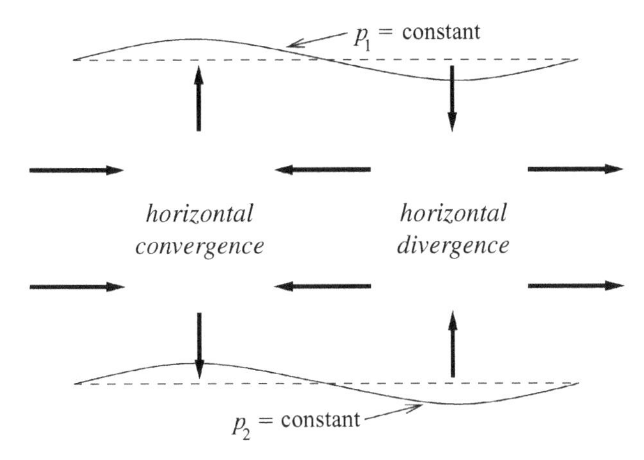

Fig. 22.1 An example of mass flow and corresponding vertical displacement of isobars. The initial pressure distribution is given by the dashed lines.

that $m_0 = 1$ so that the motion takes place on the tangential plane. Let us consider the mass M contained between the two constant-pressure surfaces $p_2 > p_1$. The mass contained in a volume defined by an area S and p_1 and p_2 can easily be found with the help of the hydrostatic equation and is given by

$$\begin{aligned} M &= \int_{z_1}^{z_2} \int_S \rho \, dx \, dy \, dz = -\frac{1}{g} \int_{z_1}^{z_2} \int_S \frac{\partial p}{\partial z} \, dx \, dy \, dz \\ &= \frac{1}{g} \int_{p_1}^{p_2} \int_S dx \, dy \, dp = \frac{S(p_2 - p_1)}{g} \end{aligned} \tag{22.24}$$

Since the pressure surfaces p_1 and p_2 are constant, M is also constant, so $dM/dt = 0$. Therefore, in a hydrostatic system using pressure as the vertical coordinate, horizontal convergence must be compensated by vertical divergence or vice versa. Figure 22.1 shows the vertical displacement of the isobars $p_1 =$ constant and $p_2 =$ constant resulting from horizontal convergence and divergence.

22.3 The solution scheme using the pressure system

In this section we are going to solve in principle the predictive system consisting of the equation of motion, the continuity equation, and the heat equation. In order to reduce the problem to the simplest form possible, we assume that we have dry adiabatic flow so that the heat equation reduces to a conservation equation for the potential temperature θ:

$$\frac{d\theta}{dt} = 0 \tag{22.25}$$

The equation of state is always a part of the total atmospheric forecast system. For dry air we have $p = \rho R_0 T$.

Before we explain the numerical solution procedure we must set up the proper boundary conditions. The lower boundary is the earth's surface where the surface pressure p_s is observed. Since p_s varies with the horizontal coordinates (x, y) and with time t, we may think of the lower boundary as an oscillating pressure surface. The upper boundary at $p = 0$ may be thought of as a rigid plane surface.

According to Section 9.4.3, the boundary conditions may be written in the following form:

$$
\begin{aligned}
&\text{Lower boundary:} \quad && p = p_s(x, y, t), \quad \frac{dp_s}{dt} = \omega_s \\
& && = \frac{\partial p_s}{\partial t} + m_0^2\left(v_{1,s}\frac{\partial p_s}{\partial x} + v_{2,s}\frac{\partial p_s}{\partial y}\right) \\
&\text{Upper boundary:} \quad && p = 0, \quad \omega(p = 0) = 0
\end{aligned}
\tag{22.26}
$$

It should be observed that, in the p system, the pressure surface $p = p_s(x, y, t)$ coincides with the time-independent geopotential $\phi = \phi_s(x, y)$ of the earth's surface. Hence, a variation in time of p_s is equivalent to an apparent vertical displacement of ϕ_s; see Figure 22.2. Owing to the vibrating lower boundary surface p_s external gravity waves are produced. This type of meteorological noise is particularly undesirable from the numerical point of view.

We wish to demonstrate the principle of the integration over time of the atmospheric system consisting of the covariant form of the equation of motion (22.18), the continuity equation (22.20), the heat equation (22.25) for dry adiabatic processes, and the equation of state. If we prefer to work with the physical velocity components we may proceed analogously.

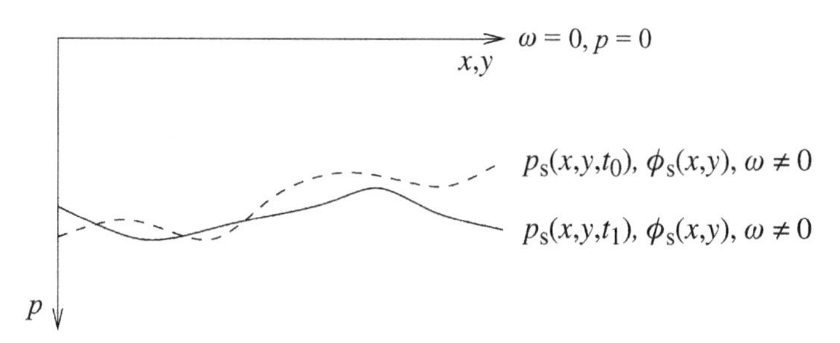

Fig. 22.2 The time dependency of the lower boundary $p = p_s(x, y, t)$ of the p system.

22.3.1 Initial data

At the starting time $t_0 = 0$ of the numerical integration we prescribe without any restriction initial values of the prognostic variables p_s and (v_1, v_2, θ) for $0 < p < p_s$.

22.3.2 The diagnostic part

The hydrostatic equation and the continuity equation do not contain time derivatives. Thus, they may be viewed as diagnostic relations or *compatibility conditions*. With the help of the hydrostatic equation in the form

$$\frac{\partial \phi}{\partial p} = -\frac{1}{\rho} = -\frac{R_0 T}{p} = -\frac{R_0}{p}\left(\frac{p}{p_0}\right)^{k_0}\theta = -\frac{R_0}{p_0^{k_0}}\frac{\theta}{p^{1-k_0}}, \qquad k_0 = \frac{R_0}{c_{p,0}} \tag{22.27}$$

we find a suitable expression for the geopotential, given by

$$\phi(x, y, p) = \phi_s(x, y) + \frac{R_0}{p_0^{k_0}} \int_p^{p_s} \frac{\theta}{p^{1-k_0}}\, dp \tag{22.28}$$

The suffix 0 refers to dry air. The continuity equation (22.20) permits the calculation of the generalized vertical velocity, yielding

$$\omega(x, y, p) = -m_0^2 \int_0^p \left(\frac{\partial v_1}{\partial x} + \frac{\partial v_2}{\partial y}\right) dp \tag{22.29}$$

In general, the integrals in (22.28) and (22.29) must be evaluated numerically.

22.3.3 The prognostic part

Let us assume that all required values of the variables ϕ and ω have been generated with the help of the initial data and the compatibility relations (22.28) and (22.29). The local tendencies of v_1, v_2, and θ can be written down from (22.18) and (22.25) by expanding the individual time derivative for each of these quantities. The tendency of p_s follows from (22.26). The complete set of prognostic equations is summarized as

$$
\begin{aligned}
\frac{\partial v_1}{\partial t} &= -m_0^2\left(v_1\frac{\partial v_1}{\partial x} + v_2\frac{\partial v_1}{\partial y}\right) - \omega\frac{\partial v_1}{\partial p} - \frac{v_1^2+v_2^2}{2}\frac{\partial m_0^2}{\partial x} + f v_2 - \frac{\partial \phi}{\partial x}\\
\frac{\partial v_2}{\partial t} &= -m_0^2\left(v_1\frac{\partial v_2}{\partial x} + v_2\frac{\partial v_2}{\partial y}\right) - \omega\frac{\partial v_2}{\partial p} - \frac{v_1^2+v_2^2}{2}\frac{\partial m_0^2}{\partial y} - f v_1 - \frac{\partial \phi}{\partial y}\\
\frac{\partial \theta}{\partial t} &= -m_0^2\left(v_1\frac{\partial \theta}{\partial x} + v_2\frac{\partial \theta}{\partial y}\right) - \omega\frac{\partial \theta}{\partial p}\\
\frac{\partial p_s}{\partial t} &= -m_0^2\left(v_{1,\mathrm{s}}\frac{\partial p_\mathrm{s}}{\partial x} + v_{2,\mathrm{s}}\frac{\partial p_\mathrm{s}}{\partial y}\right) + \omega_\mathrm{s}
\end{aligned}
\tag{22.30}
$$

The numerical integration is carried out by replacing the local time derivatives by means of suitable finite-difference quotients. After the first time step Δt, the prognostic variables $(v_1, v_2, \theta, p_\mathrm{s})$ are available at $t_1 = t_0 + \Delta t$. They now serve as initial values for the next time step as well as for the new evaluation of the diagnostic equations. This iteration must be continued until the entire prognostic period has been covered.

The integration which we have described in principle is the integration scheme which was used by Richardson (1922). However, he did not get any useful results because he employed time steps that were too large. The maximum allowable time step Δt is specified by the so-called Courant–Friedrichs–Lewy criterion, which will be explained later. When Richardson performed the numerical integrations this criterion was still unknown.

22.4 The solution to a simplified prediction problem

In this section we are going to discuss the solution of the atmospheric system by employing simplified boundary conditions. The method we are going to describe has been applied successfully by Hinkelmann (1959) and his research group to solve for the first time the atmospheric equations in their primitive (original) form. The historical aspect by itself would not be sufficient to justify the discussion. What is more important to us is to show that the simplifying assumptions often cannot be restricted to a certain part of the system. Consequences resulting from the simplifying assumptions must be carefully analyzed since they may influence the entire system.

In the previous section we did not place any restriction on the lower boundary surface $p_\mathrm{s}(x, y, t)$. This oscillating lower boundary pressure surface generates external gravity waves whose appearance may obscure the Rossby physics. Moreover, due to the high phase speed of the gravity waves, see Chapters 15 and 16, the integration scheme requires very short time steps. In order to suppress these external gravity waves we assume that the lower pressure surface is fixed, say at $p_0 = 1000$ hPa so that the generalized velocity ω now vanishes not only at the upper but also at the lower boundary. An immediate consequence of this assumption is that the horizontal velocities v_1 and v_2 at time $t_0 = 0$ cannot be prescribed without any restrictions as before. By evaluating (22.29), which must be valid for all times including $t_0 = 0$, we find the following unavoidable restriction on the velocity components:

$$\int_0^{p_0} \left(\frac{\partial v_1}{\partial x} + \frac{\partial v_2}{\partial y} \right) dp = 0 \tag{22.31}$$

so that

$$\frac{\partial}{\partial t}\left[\int_0^{p_0} \left(\frac{\partial v_1}{\partial x} + \frac{\partial v_2}{\partial y} \right) dp \right] = 0 \tag{22.32}$$

Equation (22.32) provides a new boundary condition for the geopotential ϕ since (22.28) is no longer valid due to the modified upper limit of the integral. In order to enforce the condition (22.32) we must employ the equation of motion. We use the stream-momentum formulation of the equation of motion in the p system, which is easily obtained from (22.17) as

$$\begin{aligned}
\frac{\partial v_1}{\partial t} &= -m_0^2\left(\frac{\partial}{\partial x}(v_1^2) + \frac{\partial}{\partial y}(v_1 v_2) \right) - \frac{\partial}{\partial p}(v_1 \omega) - \frac{v_1^2 + v_2^2}{2} \frac{\partial m_0^2}{\partial x} + f v_2 - \frac{\partial \phi}{\partial x} \\
\frac{\partial v_2}{\partial t} &= -m_0^2\left(\frac{\partial}{\partial x}(v_2 v_1) + \frac{\partial}{\partial y}(v_2^2) \right) - \frac{\partial}{\partial p}(v_2 \omega) - \frac{v_1^2 + v_2^2}{2} \frac{\partial m_0^2}{\partial y} - f v_1 - \frac{\partial \phi}{\partial y}
\end{aligned} \tag{22.33}$$

To simplify the notation, we introduce the following abbreviations:

$$\begin{aligned}
F_1 &= -m_0^2\left(\frac{\partial}{\partial x}(v_1^2) + \frac{\partial}{\partial y}(v_1 v_2) \right) - \frac{v_1^2 + v_2^2}{2} \frac{\partial m_0^2}{\partial x} + f v_2 \\
F_2 &= -m_0^2\left(\frac{\partial}{\partial x}(v_2 v_1) + \frac{\partial}{\partial y}(v_2^2) \right) - \frac{v_1^2 + v_2^2}{2} \frac{\partial m_0^2}{\partial y} - f v_1
\end{aligned} \tag{22.34}$$

Substituting (22.34) into (22.33) yields

$$\begin{aligned}
\frac{\partial v_1}{\partial t} &= F_1 - \frac{\partial}{\partial p}(v_1 \omega) - \frac{\partial \phi}{\partial x} \\
\frac{\partial v_2}{\partial t} &= F_2 - \frac{\partial}{\partial p}(v_2 \omega) - \frac{\partial \phi}{\partial y}
\end{aligned} \tag{22.35}$$

so (22.32) can be written as

$$\begin{aligned}
\frac{\partial}{\partial t}\left[\int_0^{p_0} \left(\frac{\partial v_1}{\partial x} + \frac{\partial v_2}{\partial y} \right) dp \right] &= \int_0^{p_0} \left(\frac{\partial F_1}{\partial x} + \frac{\partial F_2}{\partial y} \right) dp - \left[\frac{\partial}{\partial x}(v_1 \omega) + \frac{\partial}{\partial y}(v_2 \omega) \right]_0^{p_0} \\
&\quad - \nabla_h^2 \int_0^{p_0} \phi \, dp = 0
\end{aligned} \tag{22.36}$$

The second term on the right-hand side is zero due to the vanishing of generalized velocities ω at the lower and upper boundaries of the model.

Next we introduce the vertical mean of the geopotential,

$$\overline{\phi} = \frac{1}{p_0} \int_0^{p_0} \phi \, dp \tag{22.37}$$

into (22.36) and obtain

$$\nabla_{\mathrm{h}}^2 \overline{\phi} = \frac{1}{p_0} \int_0^{p_0} \left(\frac{\partial F_1}{\partial x} + \frac{\partial F_2}{\partial y} \right) dp \tag{22.38}$$

This Poisson-type equation is a balance equation for $\overline{\phi}$. Let us assume that proper boundary conditions have been stated for $\overline{\phi}$ and that (22.38) has been solved so that $\overline{\phi}(x, y)$ is a known quantity.

We will now integrate the hydrostatic equation (22.27) to find the geopotential $\phi(x, y, p)$. Instead of (22.28) we now obtain

$$\phi(x, y, p) = \phi(x, y, p_0) + \frac{R_0}{p_0^{k_0}} \int_p^{p_0} \frac{\theta}{p^{1-k_0}} \, dp \tag{22.39}$$

where $\phi(x, y, p_0)$ is still an unknown quantity for which we must find a suitable expression. Integration of (22.37) by parts results in

$$\overline{\phi}(x, y) = \phi(x, y, p_0) - \frac{1}{p_0} \int_0^{p_0} p \frac{\partial \phi}{\partial p} \, dp \tag{22.40}$$

By elimination of $\phi(x, y, p_0)$ between the latter two equations and also using (22.27), we find

$$\phi(x, y, p) = \overline{\phi}(x, y) + \frac{R_0}{p_0^{k_0}} \left(\int_p^{p_0} \frac{\theta}{p^{1-k_0}} \, dp - \int_0^{p_0} \frac{p^{k_0} \theta}{p_0} \, dp \right) \tag{22.41}$$

which can be evaluated to give the geopotental at all required coordinates.

The prediction now follows the integration cycle of the previous section. The tendencies of the quantities (v_1, v_2, θ) are taken from (22.30). The calculation of the geopotential makes use of (22.41) thus guaranteeing that the boundary condition (22.31) is satisfied.

22.5 The solution scheme with a normalized pressure coordinate

In this section we introduce the so-called σ *system*, which is defined by

$$\xi = p/p_{\mathrm{s}} = \sigma \tag{22.42}$$

This very useful coordinate transformation was introduced by Phillips (1957). Since the vertical coordinate σ also decreases with height, the σ system is also left-handed. The integration region is limited by $\sigma = 0$ and $\sigma = 1$. Thus, the mountainous orographic surface of the earth $\phi = \phi_s(x, y)$ coincides with the surface of the constant vertical coordinate $\sigma = 1$ where $p = p_s$. At the lower and at the upper boundary the generalized velocity $\dot{\sigma} = 0$. This statement results from the application of the kinematic boundary condition requiring that the normal component of the velocity must be zero at the boundaries. It is this simple form of the lower boundary condition that provides the advantage of the σ system over the p system. It should be clearly understood that $\dot{\sigma}$ is not zero because the derivative of a constant is zero. If this were the reason for $\dot{\sigma}$ vanishing at the boundaries, the generalized vertical velocity would be zero everywhere. This is, of course, not the case.

According to (22.10) the pressure gradient and the geopotential gradient must be evaluated along constant σ-surfaces. Therefore, the horizontal pressure gradients may be written as

$$\frac{1}{\rho}\frac{\partial p}{\partial x} = \frac{\sigma}{\rho}\frac{\partial p_s}{\partial x}, \qquad \frac{1}{\rho}\frac{\partial p}{\partial y} = \frac{\sigma}{\rho}\frac{\partial p_s}{\partial y} \tag{22.43}$$

The hydrostatic equation is given by

$$\frac{\partial \phi}{\partial \sigma} = \frac{\partial \phi}{\partial p}\frac{\partial p}{\partial \sigma} = -\frac{p_s}{\rho} = -\frac{R_0 T}{\sigma} \tag{22.44}$$

Therefore, the covariant equation of motion (22.10) assumes the form

$$\boxed{\begin{aligned}
&\frac{dv_1}{dt} + \frac{v_1^2 + v_2^2}{2}\frac{\partial m_0^2}{\partial x} - f v_2 = -\frac{\sigma}{\rho}\frac{\partial p_s}{\partial x} - \frac{\partial \phi}{\partial x} \\
&\frac{dv_2}{dt} + \frac{v_1^2 + v_2^2}{2}\frac{\partial m_0^2}{\partial y} + f v_1 = -\frac{\sigma}{\rho}\frac{\partial p_s}{\partial y} - \frac{\partial \phi}{\partial y} \\
&\qquad \frac{\partial \phi}{\partial \sigma} = -\frac{p_s}{\rho} \\
&\text{with} \quad \frac{d}{dt} = \frac{\partial}{\partial t} + m_0^2\left(v_1\frac{\partial}{\partial x} + v_2\frac{\partial}{\partial y}\right) + \dot{\sigma}\frac{\partial}{\partial \sigma}
\end{aligned}} \tag{22.45}$$

Note that the horizontal gradients of the geopotential do not vanish since they have to be evaluated along surfaces with σ = constant. The continuity equation is obtained from (22.15) simply by replacing the generalized coordinate ξ by σ. After expanding the individual derivative and rearranging terms we find

$$\boxed{\frac{\partial p_s}{\partial t} + m_0^2\left(\frac{\partial}{\partial x}(v_1 p_s) + \frac{\partial}{\partial y}(v_2 p_s)\right) + \frac{\partial}{\partial \sigma}(\dot{\sigma} p_s) = 0} \tag{22.46}$$

Owing to the validity of the hydrostatic equation for all times, the atmospheric fields ϕ and the unreduced surface pressure p_s uniquely determine the atmospheric mass field. We will now discuss the principle of the integration scheme.

22.5.1 Initial data

The geopotential of the earth's surface $\phi_s = (x, y)$ and the map factor $m_0(x, y)$ are considered known. Moreover, the Coriolis parameter f appears in the prognostic system. Since the latitude φ does not enter explicitly, we must express the $\sin\varphi$ function by means of (20.29) so that the Coriolis parameter may be written as $f = f(x, y)$. The three quantities ϕ_s, m_0, and f do not depend on time. For the initial time $t_0 = 0$, without any restriction, we prescribe the prognostic variables p_s at $\sigma = 1$ and (v_1, v_2, θ) for $0 \le \sigma \le 1$.

22.5.2 The diagnostic part

By using the ideal-gas law and the potential-temperature formula, the hydrostatic equation of the σ system can be rewritten as

$$\frac{\partial \phi}{\partial \sigma} = -\frac{R_0 T}{\sigma} = -R_0 \left(\frac{p_s}{p_0}\right)^{k_0} \sigma^{k_0 - 1}\theta \tag{22.47}$$

Integration between the limits σ and 1 gives the geopotential

$$\phi(x, y, \sigma) = \phi_s(x, y) + R_0 \left(\frac{p_s}{p_0}\right)^{k_0} \int_\sigma^1 (\sigma')^{k_0 - 1}\theta \, d\sigma' \tag{22.48}$$

Our next task is to find a suitable expression for the generalized vertical velocity $\dot{\sigma}$. Integrating the continuity equation (22.46) between the limits $\sigma = 0$ and $\sigma = 1$ gives

$$\frac{\partial p_s}{\partial t} + m_0^2 \frac{\partial}{\partial x}\left(\int_0^1 v_1 p_s \, d\sigma\right) + m_0^2 \frac{\partial}{\partial y}\left(\int_0^1 v_2 p_s \, d\sigma\right) = 0 \tag{22.49}$$

since $\dot{\sigma}$ vanishes at the boundaries of the model. We now eliminate the surface pressure tendency $\partial p_s/\partial t$ between (22.46) and (22.49) and then integrate the resulting equation between $\sigma = 0$ and σ. Recognizing that equation (22.49) is independent of σ, we find

$$\dot{\sigma} = -\frac{m_0^2}{p_s} \int_0^\sigma \left(\frac{\partial}{\partial x}(v_1 p_s) + \frac{\partial}{\partial y}(v_2 p_s)\right) d\sigma + \frac{m_0^2 \sigma}{p_s} \int_0^1 \left(\frac{\partial}{\partial x}(v_1 p_s) + \frac{\partial}{\partial y}(v_2 p_s)\right) d\sigma \tag{22.50}$$

Since (v_1, v_2, θ, p_s) are known at t_0, we may proceed with the numerical integration of (22.48) and of (22.50). Inspection of this equation shows that the condition $\dot{\sigma} = 0$ is satisfied at the lower and upper boundaries.

22.5.3 The prognostic part

Using (22.49) and eliminating the density of the air, the prognostic set is given by

$$
\begin{aligned}
\frac{\partial v_1}{\partial t} &= -m_0^2\left(v_1\frac{\partial v_1}{\partial x}+v_2\frac{\partial v_1}{\partial y}\right)-\dot{\sigma}\frac{\partial v_1}{\partial \sigma}-\frac{v_1^2+v_2^2}{2}\frac{\partial m_0^2}{\partial x}+fv_2\\
&\quad-\frac{R_0}{p_s}\left(\frac{p_s}{p_0}\right)^{k_0}\sigma^{k_0}\theta\frac{\partial p_s}{\partial x}-\frac{\partial \phi}{\partial x}\\
\frac{\partial v_2}{\partial t} &= -m_0^2\left(v_1\frac{\partial v_2}{\partial x}+v_2\frac{\partial v_2}{\partial y}\right)-\dot{\sigma}\frac{\partial v_2}{\partial \sigma}-\frac{v_1^2+v_2^2}{2}\frac{\partial m_0^2}{\partial y}-fv_1\\
&\quad-\frac{R_0}{p_s}\left(\frac{p_s}{p_0}\right)^{k_0}\sigma^{k_0}\theta\frac{\partial p_s}{\partial y}-\frac{\partial \phi}{\partial y}\\
\frac{\partial \theta}{\partial t} &= -m_0^2\left(v_1\frac{\partial \theta}{\partial x}+v_2\frac{\partial \theta}{\partial y}\right)-\dot{\sigma}\frac{\partial \theta}{\partial \sigma}\\
\frac{\partial p_s}{\partial t} &= -m_0^2\int_0^1\left(\frac{\partial}{\partial x}(v_1p_s)+\frac{\partial}{\partial y}(v_2p_s)\right)d\sigma
\end{aligned}
\tag{22.51}
$$

The partial derivatives may now be replaced by suitable finite-difference formulas and the iteration of this system may be started. Knowing (v_1, v_2, θ, p_s) after the first iteration step, new values of the diagnostic quantities ϕ and $\dot{\sigma}$ can be determined by means of (22.48) and (22.50). The iterations may be carried out in the same manner as we have described for the p system.

Finally, we wish to remark that, so far, all tendencies appeared explicitly. By this we mean that they are not embedded in differential operators. A consequence is that an increasing resolution of the grid results only in a linear growth of the numerical work. This is not the case in some other numerical schemes. The integration principle we have described in this section was used for a considerable period of time by the Weather Services of the USA and Germany.

22.6 The solution scheme with potential temperature as vertical coordinate

By selecting the potential temperature as the generalized vertical coordinate, often called the *isentropic vertical coordinate*, we require that the potential temperature is a monotonically increasing function. This requirement reflects the observed large-scale conditions. Thus, we may write

$$
\xi = \theta, \qquad \dot{\xi} = \dot{\theta} \tag{22.52}
$$

If the atmosphere were adiabatically stratified in a certain height interval, there could be no unique attachment of a potential temperature to height coordinates.

Using the potential-temperature formula, we find that the components of the pressure gradients along isentropic surfaces are given by

$$-\frac{1}{\rho}\frac{\partial p}{\partial x} = -c_{p,0}\frac{\partial T}{\partial x}, \qquad -\frac{1}{\rho}\frac{\partial p}{\partial y} = -c_{p,0}\frac{\partial T}{\partial y} \tag{22.53}$$

By introducing the so-called *Montgomery potential*,

$$M = c_{p,0}T + \phi \tag{22.54}$$

we may write for the right-hand sides of the horizontal equations of motion

$$-\frac{1}{\rho}\frac{\partial p}{\partial x} - \frac{\partial \phi}{\partial x} = -\frac{\partial M}{\partial x}, \qquad -\frac{1}{\rho}\frac{\partial p}{\partial y} - \frac{\partial \phi}{\partial x} = -\frac{\partial M}{\partial y} \tag{22.55}$$

According to (22.9) the hydrostatic equation for the θ system is given by

$$-\frac{1}{\rho}\frac{\partial p}{\partial \theta} - \frac{\partial \phi}{\partial \theta} = c_{p,0}\frac{T}{\theta} - c_{p,0}\frac{\partial T}{\partial \theta} - \frac{\partial \phi}{\partial \theta} = 0 \tag{22.56}$$

Introduction of the Montgomery potential leads to the final form of the hydrostatic equation:

$$\frac{\partial M}{\partial \theta} = c_{p,0}\frac{T}{\theta} \tag{22.57}$$

An attractive feature of the θ system is that, for dry adiabatic motion, the potential temperature of each air parcel is conserved, so $d\theta/dt = \dot{\theta} = 0$. The predictive set (22.10) can then be written as

$$\boxed{\begin{aligned}
\frac{\partial v_1}{\partial t} &= -m_0^2\left(v_1\frac{\partial v_1}{\partial x} + v_2\frac{\partial v_1}{\partial y}\right) - \frac{v_1^2 + v_2^2}{2}\frac{\partial m_0^2}{\partial x} + f v_2 - \frac{\partial M}{\partial x} \\
\frac{\partial v_2}{\partial t} &= -m_0^2\left(v_1\frac{\partial v_2}{\partial x} + v_2\frac{\partial v_2}{\partial y}\right) - \frac{v_1^2 + v_2^2}{2}\frac{\partial m_0^2}{\partial y} - f v_1 - \frac{\partial M}{\partial y} \\
\frac{\partial}{\partial t}\left(\frac{\partial p}{\partial \theta}\right) &= -m_0^2\left[\frac{\partial}{\partial x}\left(v_1\frac{\partial p}{\partial \theta}\right) + \frac{\partial}{\partial y}\left(v_2\frac{\partial p}{\partial \theta}\right)\right]
\end{aligned}} \tag{22.58}$$

The last expression is the continuity equation of the θ system.

The lower boundary condition on θ may be written as

$$\frac{d\theta_{\rm s}}{dt} = \frac{\partial \theta_{\rm s}}{\partial t} + m_0^2\left(v_{1,{\rm s}}\frac{\partial \theta_{\rm s}}{\partial x} + v_{2,{\rm s}}\frac{\partial \theta_{\rm s}}{\partial y}\right) = 0 \tag{22.59}$$

With increasing height the potential temperature increases to infinity while the pressure and density decrease to zero. By utilizing (22.56) the upper boundary condition may be written as

$$\lim_{\substack{\theta\to\infty \\ \rho\to 0}} \frac{\partial p}{\partial \theta} = c_{p,0} \lim_{\substack{\theta\to\infty \\ \rho\to 0}} \left(\rho\frac{\partial T}{\partial \theta} - \frac{\rho T}{\theta}\right) = 0 \tag{22.60}$$

where $\partial T/\partial \theta$ is finite.

In order to find the pressure tendency we integrate the continuity equation, i.e. the last equation of (22.58), between θ and $\theta \to \infty$ and obtain with the help of the Leibniz rule

$$\frac{\partial p}{\partial t} = m_0^2 \left(\frac{\partial}{\partial x} \int_\theta^\infty v_1 \frac{\partial p}{\partial \theta'} \, d\theta' + \frac{\partial}{\partial y} \int_\theta^\infty v_2 \frac{\partial p}{\partial \theta'} \, d\theta' \right) \tag{22.61}$$

For adiabatic flow the θ system has the decisive advantage that the vertical velocity vanishes everywhere so that we are dealing with two-dimensional motion. As soon as nonadiabatic effects must be taken into account this advantage is lost, so the θ system is not used for routine numerical investigations. Therefore, we will omit a detailed discussion of the integration cycle.

Finally, we would like to remark that other generalized vertical coordinates might be used, such as the density of the air. The theory for this system has been worked out, but it has not yet found many important applications.

22.7 Problems

22.1: By a direct transformation obtain the first equation of (22.11) from the first equation of (22.10). Compare your result with equation (20.48).

22.2: Show that equations (22.20) and (22.21) follow from (22.15) and (22.16).

22.3: Balloons drifting on surfaces of constant density could be used to collect global atmospheric data. To utilize these data, we need a predictive system using ρ as a vertical coordinate.
(a) Introduce the Montgomery potential for the ρ system $N = R_0 T + \phi$ to find the horizontal equation of motion in terms of the covariant velocity components. Obtain the hydrostatic equation. State the individual derivative d/dt for this system.
(b) Find the continuity equation.
(c) Formulate the upper and lower boundary conditions for the ρ system.
(d) Find an expression for the pressure tendency $\partial p/\partial t$.

22.4: Consider an idealized atmosphere consisting of an incompressible frictionless fluid with a free surface z_H. Assuming that $\mathbf{v}_\text{h}$ is independent of height, find a prognostic equation for the free surface.

22.5: Consider the equation of motion for a frictionless fluid in Cartesian coordinates ($m_0 = 1$) in simplified form, i.e.

$$\left.\frac{d\mathbf{v}_\text{h}}{dt}\right|_{\mathbf{q}_i} = -\frac{1}{\rho} \nabla_{\text{h},\xi} p - f\mathbf{k} \times \mathbf{v}_\text{h} - \nabla_{\text{h},\xi} \phi$$

where $\nabla_{\mathrm{h},\xi}$ is the horizontal nabla operator in the arbitrary ξ system. Show that, for the σ system, we may write the equation of motion in the form

$$\begin{aligned}\frac{\partial}{\partial t}(\Pi\mathbf{v}_{\mathrm{h}})\bigg|_{\mathbf{q}_i} &+ \nabla_{\mathrm{h},\sigma}\cdot(\Pi\mathbf{v}_{\mathrm{h}}\mathbf{v}_{\mathrm{h}}) + \frac{\partial}{\partial\sigma}(\Pi\mathbf{v}_{\mathrm{h}}\dot{\sigma})\\ &= -\nabla_{\mathrm{h},\sigma}(\Pi\phi) + \frac{\partial}{\partial\sigma}(\phi\sigma)\,\nabla_{\mathrm{h},\sigma}\Pi - \Pi f\mathbf{k}\times\mathbf{v}_{\mathrm{h}}\end{aligned}$$

where $\sigma = p/p_{\mathrm{s}}$ and $\Pi = p_{\mathrm{s}}/p_0$.

22.6: With the help of the Leibniz rule, verify equation (22.61).

22.7: Verify equation (22.46).

23

A quasi-geostrophic baroclinic model

23.1 Introduction

So far we have treated the so-called primitive equations of baroclinic systems which, in addition to typical meteorological effects, automatically include horizontally propagating sound waves as well as external and internal gravity waves. These waves produce high-frequency oscillations in the numerical solutions of the baroclinic systems, which are of no interest to the meteorologist. Thus, the tendencies of the various field variables are representative only of small time intervals of the order of minutes while the predicted weather tendencies should be representative of much longer time intervals.

In order to obtain meteorologically significant tendencies we are going to eliminate the meteorological noise from the primitive equations b y modifying the predictive system so that a longer time step in the numerical solution becomes possible. We recall that the vertically propagating sound waves are no longer a part of the solution since they are removed by the hydrostatic approximation. The noise filtering is accomplished by a diagnostic coupling of the horizontal wind field and the mass field while in reality at a given time these fields are independent of each other. The simplest coupling of the wind and mass field is the geostrophic wind relation. The mathematical systems resulting from the artificial inclusion of filter conditions are called *quasi-geostrophic systems* or, more generally, *filtered systems*. For such systems at the initial time $t_0 = 0$ only one variable, usually the geopotential, is specified without any restriction. The remaining dependent variables, for a given time, result from the employment of compatibility conditions with the geopotential, which is known as *diagnostic coupling*.

For reasons of expediency we select the p system. Since we are interested only in a qualitative discussion of the large-scale motion rather than in actual weather forecasts, we simply describe the flow on the tangential plane by setting the map factor $m_0 = 1$. This eliminates a number of terms and simplifies our work. The theory which will be presented in the following sections is indispensable for comprehending in some depth large-scale atmospheric motion and some of the problems associated with weather prediction. The actual weather prediction is carried out with the help of the primitive equations in some modification using adjusted and compatible initial fields. In contrast to this, some drastic modifications are required in order to obtain the quasi-geostrophic system. However, the advantage of the quasi-geostrophic system is that the resulting mathematical system is simple enough to promote the physical interpretation of the motion field. Moreover, we learn that approximations have to be applied with great care, otherwise inconsistencies may result, such as increases of the potential and kinetic energy at the same time.

We will now introduce and discuss in some detail the quasi-gestrophic theory, which is based on suitable modifications of the first law of thermodynamics and of the baroclinic vorticity equation. It will also be shown that the ageostrophic approximation of the wind field due to Philipps (1939) is a very useful tool for verifying some of the results of this theory. Numerous authors have contributed to the development of the quasi-geostrophic theory. The first group of important contributions includes the pioneering work of Charney (1947), Eady (1949), and Phillips (1956).

23.2 The first law of thermodynamics in various forms

We begin our work by writing the heat equation in the form

$$\rho c_p \frac{dT}{dt} - \frac{dp}{dt} = \rho \frac{đq}{dt} \tag{23.1}$$

where the term $\rho đq/dt$ includes all heat sources such as radiation, heat conduction, turbulent heat transport, phase changes of the water substance, and friction if the equations to be considered represent the mean atmospheric motion. This equation may also be expressed in terms of the potential temperature as

$$\frac{d \ln \theta}{dt} = \frac{1}{c_p T} \frac{đq}{dt} \tag{23.2}$$

It is customary to introduce the *static stability* σ_0 which, according to (16.30), is given by

$$\sigma_0 = -\frac{1}{\rho} \frac{\partial \ln \theta}{\partial p} = \frac{R_0}{p}\left(\frac{R_0 T}{c_p p} - \frac{\partial T}{\partial p}\right) = \frac{R_0^2 T}{g p^2}(\Gamma - \gamma) \tag{23.3}$$

Here Γ and γ are, respectively, the dry adiabatic and the actually observed geometric lapse rates. Expansion of (23.1) gives the form

$$\frac{\partial T}{\partial t} + \mathbf{v}_{\mathrm{h}} \cdot \nabla_{\mathrm{h}} T + \omega\left(\frac{\partial T}{\partial p} - \frac{R_0 T}{c_p p}\right) = \frac{1}{c_p} \frac{đq}{dt} \tag{23.4}$$

from which it follows that

$$\frac{\partial T}{\partial t} + \mathbf{v}_h \cdot \nabla_h T - \frac{p\sigma_0\omega}{R_0} = \frac{1}{c_p}\frac{\text{đ}q}{dt} \tag{23.5}$$

due to the introduction of the static stability by means of (23.3). Expanding (23.2) results in

$$\frac{\partial \ln\theta}{\partial t} + \mathbf{v}_h \cdot \nabla_h \ln\theta + \omega\frac{\partial \ln\theta}{\partial p} = \frac{1}{c_p T}\frac{\text{đ}q}{dt} \tag{23.6}$$

or, using the static stability, we find

$$\frac{\partial \ln\theta}{\partial t} + \mathbf{v}_h \cdot \nabla_h \ln\theta - p\sigma_0\omega = \frac{1}{c_p T}\frac{\text{đ}q}{dt} \tag{23.7}$$

Finally, we will write the heat equation in a form involving the geopotential ϕ. From the hydrostatic equation we first find

$$\frac{\partial\phi}{\partial p} = -\frac{1}{\rho} = -\frac{R_0 T}{p} \implies T = -\frac{p}{R_0}\frac{\partial\phi}{\partial p} \tag{23.8}$$

Remembering that all equations refer to the p system, we obtain from (23.5) after a few easy steps the expression

$$\frac{\partial}{\partial t}\left(\frac{\partial\phi}{\partial p}\right) + \mathbf{v}_h \cdot \nabla_h\left(\frac{\partial\phi}{\partial p}\right) + \sigma_0\omega = -\frac{R_0}{c_p p}\frac{\text{đ}q}{dt} \quad \text{with} \quad \sigma_0 = \frac{\partial^2\phi}{\partial p^2} + \frac{c_v}{c_p p}\frac{\partial\phi}{\partial p} \tag{23.9}$$

The latter form of σ_0 follows from substituting (23.8) into (23.3) with $R_0 = c_p - c_v$.

The physical interpretation of the heat equation is quite simple. The first term of equation (23.5) describes the temperature tendency, the second term expresses the change in temperature due to horizontal advection, and the third term describes convective processes or changes in temperature due to anisobaric vertical motion. The term $\text{đ}q/dt$ has already been explained above. Corresponding explanations apply to equations (23.7) and (23.9).

23.4 The vorticity and the divergence equation

In order to eliminate the meteorological noise, it is convenient to employ the baroclinic vorticity equation instead of the horizontal equation of motion itself. The vorticity equation is a scalar equation and, therefore, cannot be equivalent to the horizontal equation of motion, which is a vector equation. To express complete equivalence to the equation of horizontal motion we must also add the baroclinic divergence equation. A comparison of the filtered system with the exact system is facilitated by applying the unfiltered vorticity and divergence equations instead of employing the equivalent equation of horizontal motion itself.

To begin with, we repeat the baroclinic vorticity equation (10.77) for the p system, which, in its modified form, is one of the basic equations of the quasi-geostrophic theory. Suppressing for convenience the suffix p and using the continuity equation (22.23), we find

$$\boxed{\underset{\text{I}}{\frac{\partial \eta}{\partial t}} + \underset{\text{II}}{\mathbf{v}_{\text{h}} \cdot \nabla_{\text{h}} \eta} + \underset{\text{III}}{\eta \nabla_{\text{h}} \cdot \mathbf{v}_{\text{h}}} + \underset{\text{IV}}{\omega \frac{\partial \eta}{\partial p}} + \underset{\text{V}}{\mathbf{k} \cdot \left(\nabla_{\text{h}} \omega \times \frac{\partial \mathbf{v}_{\text{h}}}{\partial p} \right)} = 0} \tag{23.10}$$

where $\eta = \mathbf{k} \cdot \nabla_{\text{h}} \times \mathbf{v}_{\text{h}} + f$ is the absolute vorticity. This is the baroclinic vorticity equation in the unfiltered form. Terms I, II, III, and IV can be easily interpreted but term V is more difficult to comprehend. Term I stands for the local change of either the absolute or the relative vorticity, term II represents the horizontal advection of the absolute vorticity, term III describes the divergence or convergence of the horizontal velocity, and convective processes are expressed by term IV. The *twisting term* V, also called the tipping or the tilting term, implies that, due to rotational motion, the vertical component of the vector $\nabla \times \mathbf{v}$, i.e. the relative vorticity $\zeta = \mathbf{k} \cdot \nabla \times \mathbf{v} = \mathbf{k} \cdot \nabla_{\text{h}} \times \mathbf{v}_{\text{h}}$, increases at the expense of the horizontal component of the vector $\nabla \times \mathbf{v}$ which is represented by the vertical shear $\mathbf{k} \times \partial \mathbf{v}/\partial p$.

The expansion of the twisting term results in

$$\mathbf{k} \cdot \left(\nabla_{\text{h}} \omega \times \frac{\partial \mathbf{v}_{\text{h}}}{\partial p} \right) = \frac{\partial \omega}{\partial x} \frac{\partial v}{\partial p} - \frac{\partial \omega}{\partial y} \frac{\partial u}{\partial p} \tag{23.11}$$

In order to better understand the meteorological significance of the twisting term, we will consider the simplified physical situation shown in the upper part of Figure 23.1. At the point A the upward transport of low v values from lower layers will produce a local decrease of v with time, whereas at point B high v values of upper layers are transported downward, yielding a local increase of v with time. From (23.10) and (23.11) we see that, in this particular situation, the local change with time of the vorticity resulting from the twisting term is positive, as indicated in the lower part of the figure.

The baroclinic divergence equation, which is not yet at our disposal, will now be derived. For the barotropic system the divergence equation (15.63) was obtained from the horizontal equation of motion (15.21). The baroclinic form of the horizontal equation of motion in the p system formally differs from (15.21) only by virtue of the vertical advection term $\omega \partial/\partial p$ which appears in the expansion of the individual time derivative. This fact permits a shortcut in the derivation. We simply take the divergence of this term,

$$\nabla_{\text{h}} \cdot \left(\omega \frac{\partial \mathbf{v}_{\text{h}}}{\partial p} \right) = \omega \frac{\partial D}{\partial p} + \nabla_{\text{h}} \omega \cdot \frac{\partial \mathbf{v}_{\text{h}}}{\partial p} \quad \text{with} \quad D = \nabla_{\text{h}} \cdot \mathbf{v}_{\text{h}} \tag{23.12}$$

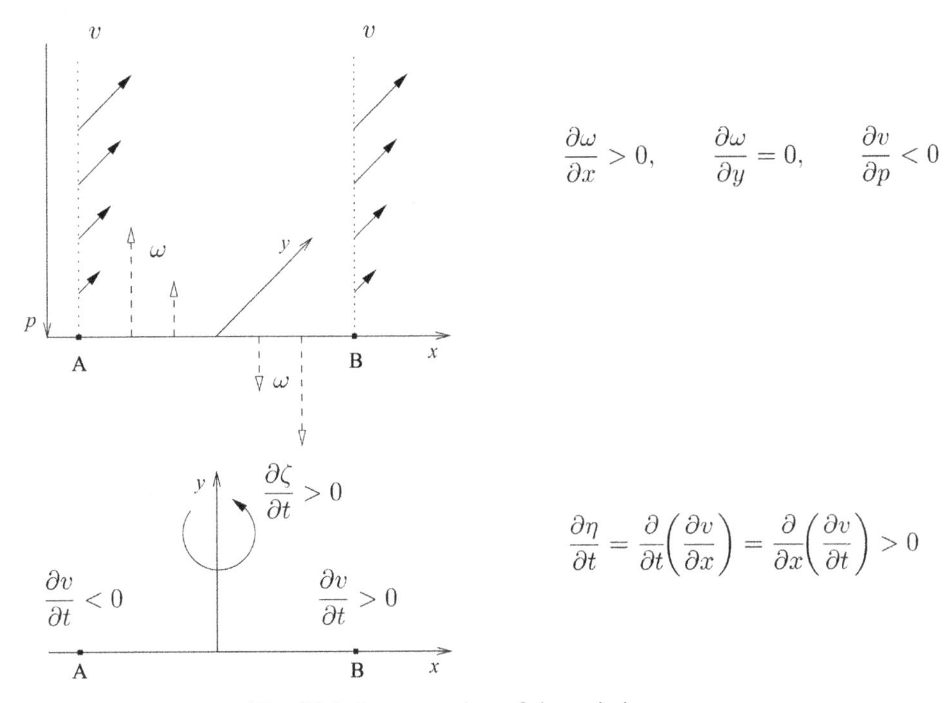

Fig. 23.1 Interpretation of the twisting term.

and then add the result to equation (15.63). This gives the baroclinic divergence equation

$$\boxed{\begin{aligned}&\frac{\partial D}{\partial t} + \mathbf{v}_{\rm h} \cdot \nabla_{\rm h} D + D^2 + \omega \frac{\partial D}{\partial p} + \nabla_{\rm h}\omega \cdot \frac{\partial \mathbf{v}_{\rm h}}{\partial p} + 2J(v, u) \\ &\quad - \mathbf{k} \cdot \left[\nabla_{\rm h} \times (f\mathbf{v}_{\rm h})\right] = -\nabla_{\rm h}^2 \phi\end{aligned}} \tag{23.13}$$

which will be used later. Once again, we consider equations (23.10) and (23.13) to be equivalent to the horizontal equation of motion.

23.5 The first and second filter conditions

As in Section 22.4 we introduce an artificial boundary condition by replacing the lower boundary by a rigid surface $p = p_0 = 1000$ hPa. This results in the *first filter condition* (22.31), which is not repeated here. Since we are studying the flow on the tangential plane, i.e. $m_0 = 1$, we conclude that the covariant velocities v_1 and v_2 are identical with the physical velocity components u and v. As discussed previously, the lower rigid boundary eliminates external gravity waves, which would form if the lower boundary were free to oscillate. The first filter condition states that, at the initial time $t_0 = 0$, the horizontal velocity components (u, v) cannot be prescribed without any restriction, but rather must obey the condition (22.31). We will now introduce the *second filter condition* to eliminate horizontal sound waves, internal gravity waves, and possibly inertial waves characterized by the frequency f/k. We proceed as follows.

With the help of the continuity equation of the p system (22.23) we replace the horizontal divergence $\nabla_{\rm h} \cdot \mathbf{v}_{\rm h}$ in term III of the unfiltered vorticity equation (23.10) by $-\partial\omega/\partial p$. In the remaining terms the wind is assumed to be nondivergent so that $\mathbf{v}_{\rm h}$ may be expressed by a stream function:

$$\mathbf{v}_{\rm h} = \mathbf{k} \times \nabla_{\rm h}\psi, \qquad u = -\frac{\partial \psi}{\partial y}, \qquad v = \frac{\partial \psi}{\partial x} \tag{23.14}$$

Since the divergence term is exempted from the filtering process we speak of *selective filtering*. On specifying the stream function as

$$\psi = \phi/f_0 \tag{23.15}$$

we see that the horizontal wind is very similar to the geostrophic wind. The only difference is that the Coriolis parameter f occurring in the definition of $\mathbf{v}_{\rm g}$ has been replaced by the constant f_0 representing an area average of the Coriolis parameter.

We have shown in (6.12) that the horizontal wind may be decomposed as

$$\mathbf{v}_{\rm h} = \mathbf{k} \times \nabla_{\rm h}\psi + \nabla_{\rm h}\chi \tag{23.16}$$

These two components represent the rotational part and the divergence part of the wind field. We might proceed by introducing equation (23.16) into the heat equation and the vorticity equation and then set $\psi = \phi/f_0$. This unselective filtering results in a less drastic simplification of the physics but in more complicated equations. For simplicity we will restrict ourselves to a model based on selective filtering.

In order to preserve important integral properties (averages expressed by integrals) of the filtered system, it is necessary to introduce additional modifications and simplifications into the predictive equations. We will now discuss in detail the filtering of the heat equation and the vorticity equation. The introduction of the stream function (23.15) into (23.14) results in a nondivergent wind, which we will call the geostrophic wind. The vorticity resulting from the geostrophic wind will be called the *geostrophic vorticity*:

$$\mathbf{v}_{\rm g} = \frac{1}{f_0}\mathbf{k} \times \nabla_{\rm h}\phi, \qquad \nabla_{\rm h} \cdot \mathbf{v}_{\rm g} = 0, \qquad \zeta_{\rm g} = \mathbf{k} \cdot \nabla_{\rm h} \times \mathbf{v}_{\rm g} = \frac{1}{f_0}\nabla_{\rm h}^2\phi \tag{23.17}$$

Moreover, we require that the advection is replaced by the geostrophic advection. If B represents a general field function then the geostrophic advection is given by

$$\mathbf{v}_g \cdot \nabla_h B = -\frac{1}{f_0}\frac{\partial \phi}{\partial y}\frac{\partial B}{\partial x} + \frac{1}{f_0}\frac{\partial \phi}{\partial x}\frac{\partial B}{\partial y} = \frac{1}{f_0} J(\phi, B) = J(\psi, B) \tag{23.18}$$

where the Jacobian has been introduced for brevity.

23.6 The geostrophic approximation of the heat equation

For simplicity we assume that we have adiabatic conditions so that the right-hand side of (23.9) vanishes. Geostrophic filtering requires the replacement of the actual horizontal wind by the geostrophic wind. Thus, the filtered heat equation is given by

$$\boxed{\frac{\partial}{\partial t}\left(\frac{\partial \phi}{\partial p}\right) + \frac{1}{f_0} J\left(\phi, \frac{\partial \phi}{\partial p}\right) = -\overline{\sigma}_0 \omega} \tag{23.19}$$

In this equation we have replaced the static stability $\sigma_0(x, y, p, t)$ by its horizontal average $\overline{\sigma}_0(p, t)$. This has been done in order to preserve some integral properties of the predictive system. Integral properties of the system are characteristic features that are obtained if we average the model equations over a certain atmospheric region.

A detailed but qualitative justification for the introduction of $\overline{\sigma}_0$ into (23.19) will now be given. To facilitate the discussion we employ the unfiltered heat equation in the temperature form (23.5), assuming that we have adiabatic conditions for consistency. Thus, we obtain

$$\frac{\partial T}{\partial t} + \nabla_h \cdot (\mathbf{v}_h T) = T\, \nabla_h \cdot \mathbf{v}_h + \frac{p}{R_0}\sigma_0 \omega \tag{23.20a}$$

The filtered heat equation is found by introducing the divergence-free geostrophic wind (23.17) into the previous equation:

$$\frac{\partial T}{\partial t} + \nabla_h \cdot (\mathbf{v}_g T) = \frac{p}{R_0}\sigma_0 \omega \tag{23.20b}$$

By means of the integral operator

$$\overline{B} = \frac{1}{A}\int_A B\, dx\, dy \tag{23.21}$$

we introduce the average of the arbitrary field function B over the horizontal area A which is a part of a pressure surface. The very small inclination of A relative to the (x, y)-plane has been ignored. Thus, the average horizontal divergence is defined by

$$\overline{\nabla_h \cdot \mathbf{v}_h} = \frac{1}{A}\int_A \nabla_h \cdot \mathbf{v}_h \, dx\, dy = \frac{1}{A}\oint_{\Gamma_A} \mathbf{v}_h \cdot d\mathbf{s}, \qquad d\mathbf{s} = d\mathbf{r} \times \mathbf{k} \tag{23.22a}$$

where we have used the two-dimensional divergence theorem (M6.34). Here $d\mathbf{s}$ is a vector line element perpendicular to Γ_A. Assuming that we have vanishing normal components of the horizontal and the geostrophic wind vector along the boundary of A, the average divergence is zero. If $d\mathbf{s}$ is normalized then $\mathbf{v}_h \cdot d\mathbf{s}$ is the normal component of $\mathbf{v}_h$ on Γ_A. As an example, we choose A as a rectangle with sides parallel to the coordinate axis. Then we obtain

$$\overline{\nabla_h \cdot \mathbf{v}_h} = \frac{1}{A}\oint_{\Gamma_A} (u\, dy - v\, dx) = 0, \qquad \overline{\nabla_h \cdot B\mathbf{v}_h} = 0, \qquad \overline{\nabla_h \cdot B\mathbf{v}_g} = 0 \tag{23.22b}$$

This type of situation occurs if A comprises a region with a periodic continuation of the wind field. Had we taken the average over a pressure surface enclosing the earth, we would have obtained the same result since a spherical surface has no boundary.

We will now apply the averaging formula (23.21) to the unfiltered and filtered forms of the heat equation, (23.20a) and (23.20b). Since the divergence terms vanish, we find for the unfiltered system

$$\frac{\partial \overline{T}}{\partial t} = \overline{TD} + \frac{p}{R_0}\overline{\sigma_0 \omega} \tag{23.23a}$$

and for the filtered system

$$\frac{\partial \overline{T}}{\partial t} = \frac{p}{R_0}\overline{\sigma_0 \omega} \tag{23.23b}$$

The two systems differ by the term $\overline{TD} = \overline{T\nabla_h \cdot \mathbf{v}_h}$. As the next step we will qualitatively estimate the effect produced by the term $\overline{TD}$ by considering the correlations between T and $D = \nabla_h \cdot \mathbf{v}_h$ for a development situation characterized by intensifying cyclones and anticyclones.

We proceed by dividing the atmosphere into an upper and a lower section. The separating pressure surface p_{lnd} is placed at the level of nondivergence, which is usually located between the 500- and 600-hPa pressure surfaces. Hence, we have $D(p_{\text{lnd}}) = 0$. Above or below p_{lnd} we assume that the sign of D is uniform in each subregion. In the real atmosphere more than one pressure level p_{lnd} may occur.

Figure 23.2 depicts the qualitative behaviors of the variables (D, T, ω, σ_0) in the subregions of rising and sinking air within the developing region A. At any

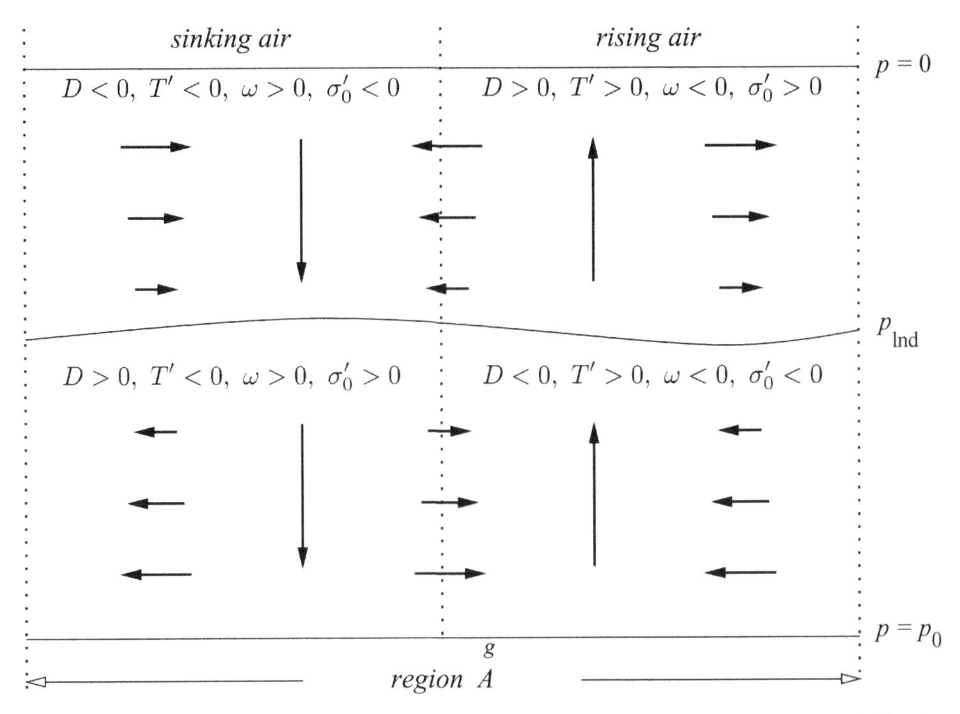

Fig. 23.2 Behaviors of variables in the subregions of rising and sinking air within the developing region A.

height the horizontal average values of the variables within A are $(\overline{D}, \overline{T}, \overline{\omega}, \overline{\sigma}_0)$ while the deviations from the average values are denoted by $(D', T', \omega', \sigma_0')$. For any pair of these variables the correlation product is $\overline{\psi\chi} = \overline{\psi}\,\overline{\chi} + \overline{\psi'\chi'}$ with $\psi = \overline{\psi} + \psi'$, $\chi = \overline{\chi} + \chi'$, and $\psi, \chi = T, D, \omega, \sigma_0$. Since $\overline{D} = 0$ and $\overline{\omega} = 0$ we have $D' = D$ and $\omega' = \omega$. The signs of D and ω in the subregions of rising and sinking air follow immediately from the assumed directions of flow indicated in the figure. In the subregion of sinking air motion we expect colder temperatures than the average temperature of the entire region A so that there $T < \overline{T}$ or $T' < 0$. In the subregion of rising air the opposite situation is observed, that is $T' > 0$. Note that T' is a temperature difference. Since $\overline{D} = 0$ we obtain $\overline{TD} = \overline{T'D'} = \overline{T'D}$. Analogously, due to $\overline{\omega} = 0$ we have $\overline{T\omega} = \overline{T'\omega'} = \overline{T'\omega}$ and $\overline{\sigma_0\omega} = \overline{\sigma_0'\omega'} = \overline{\sigma_0'\omega}$.

Let us consider the upper section of the atmosphere. In the upper subregion of sinking air we see that $D < 0$ and $T' < 0$ so that in this subregion the correlation product $\overline{TD}$ is positive. In the upper subregion of rising air we have $D > 0$ and $T' > 0$ so that there the correlation is also greater than zero. This means that, throughout the upper section of the atmosphere, the correlation of temperature and divergence is $\overline{TD} > 0$. We proceed similarly with the lower section and with the remaining variables.

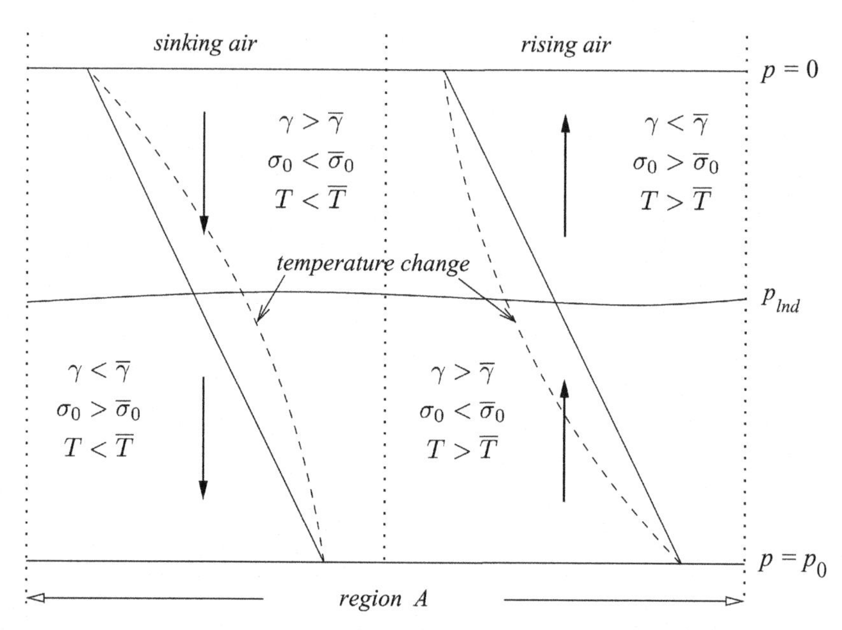

Fig. 23.3 Changes of the atmospheric stability σ_0 in the subregions of rising and sinking air within the developing region A. Full curves: initial temperature profiles; dashed curves: modified temperature profiles after vertical motion.

To study the correlation between the static stability σ_0 and the vertical velocity ω in the subregions of rising and sinking air, we refer to Figure 23.3. Assuming that $T < \overline{T}$ in the left section of the developing region and $T > \overline{T}$ in the right section we obtain sinking and rising air motion as discussed for Figure 23.2. At the upper and lower boundaries of the atmosphere ω vanishes, so the temperature remains constant there. Obviously, sinking air motion results in dry adiabatic heating whereas rising air motion causes dry adiabatic cooling. The maximum change in temperature is expected at the level of nondivergence where the vertical velocity is largest. According to (23.3) $\sigma_0 \propto \Gamma - \gamma$, so the relation between σ_0 and $\overline{\sigma}_0$ shown in Figure 23.3 is easily verified. This yields the σ_0' distribution shown in Figure 23.2 and therefore the correlations $\overline{\sigma_0\omega} < 0$ in the upper section and $\overline{\sigma_0\omega} > 0$ in the lower section of the atmosphere. In summary we expect the following correlations:

$$\boxed{\begin{array}{llll} p < p_{\rm lnd}: & \overline{TD} > 0, & \overline{\sigma_0\omega} < 0, & \overline{T\omega} < 0 \\ p > p_{\rm lnd}: & \overline{TD} < 0, & \overline{\sigma_0\omega} > 0, & \overline{T\omega} < 0 \end{array}} \tag{23.24}$$

Before continuing our discussion of the correlations it will be helpful to make the following observations.

(i) Assuming the validity of the hydrostatic equation, it is shown in textbooks on thermodynamics that the internal and the potential energy in an air column always change in the same sense. Thus, for a qualitative discussion we do not need to consider the internal energy separately.

(ii) In case of an atmospheric development the average kinetic energy in region A must increase. This occurs at the expense of the potential and the internal energy. Owing to the rising warm air and the sinking cold air, the center of mass of the system must decrease in height. At the end of this section we will verify mathematically that the negative correlation $\overline{T\omega} < 0$ is consistent with an increase of the average kinetic energy.

From the correlations (23.24) we conclude that, in the developing stage of region A in the upper and in the lower section, the two terms on the right-hand side of (23.23a) have opposite effects. The term $\overline{TD}$ results in heating of the upper section ($p < p_{\rm lnd}$) and cooling of the lower section of the atmosphere. Owing to the thermal rearrangement, we conclude from equations (23.23a) and (23.24) that

$$\left|\overline{TD}\right| > \frac{p}{R_0}\left|\overline{\sigma_0\omega}\right| \tag{23.25}$$

Owing to the geostrophic approximation, in the averaged filtered equation (23.23b) the term $\overline{TD}$ has disappeared. Since $(p/R)\overline{\sigma_0\omega}$ has an effect opposite to that of $(p/R)\overline{\sigma_0\omega} + \overline{TD}$, it then follows from (23.23a) and (23.23b) that the changes in temperature in the filtered and unfiltered systems take place in opposite directions.

In order to improve the energy balance of the filtered system, it seems logical to omit not only $\overline{TD}$ in order to preserve integral properties, but also the less effective term $(p/R)\overline{\sigma_0\omega}$ in (23.23b). This term drops out on replacing the static stability $\sigma_0(x, y, p, t)$ by an average value $\overline{\sigma}_0(p, t)$, so

$$\overline{\overline{\sigma}_0\omega} = \overline{\sigma}_0\overline{\omega} = 0 \tag{23.26}$$

This is the reason why $\overline{\sigma}_0$ was introduced into equation (23.19). Owing to this treatment the energy balance will certainly not be corrected but now the unphysical simultaneous increase of potential and kinetic energy is no longer possible.

We now want to show that the average kinetic energy in the developing region does indeed increase with time as required by the correlation product (23.24). We first obtain a prognostic equation for the kinetic energy of horizontal motion $K_{\rm h} = \mathbf{v}_{\rm h}^2/2$. In vector form the horizontal equation of motion is given by

$$\frac{\partial \mathbf{v}_{\rm h}}{\partial t} + \mathbf{v}_{\rm h}\cdot\nabla_{\rm h}\mathbf{v}_{\rm h} + \omega\frac{\partial \mathbf{v}_{\rm h}}{\partial p} + f\mathbf{k}\times\mathbf{v}_{\rm h} = -\nabla_{\rm h}\phi \tag{23.27}$$

Scalar multiplication of this equation by the horizontal velocity $\mathbf{v}_h$ results in

$$\frac{\partial K_h}{\partial t} + \mathbf{v}_h \cdot \nabla_h K_h + \omega \frac{\partial K_h}{\partial p} = -\mathbf{v}_h \cdot \nabla_h \phi \tag{23.28}$$

Owing to the continuity equation (22.23) we find the equivalent form

$$\frac{\partial K_h}{\partial t} + \nabla_h \cdot (\mathbf{v}_h K_h) + \frac{\partial}{\partial p}(\omega K_h) = -\nabla_h \cdot (\mathbf{v}_h \phi) + \phi \, \nabla_h \cdot \mathbf{v}_h \tag{23.29}$$

Since we are dealing with a developing region that is assumed to extend from the bottom to the top of the model atmosphere, we are motivated to introduce a volume average by means of

$$\overline{B}^V = \frac{1}{A p_0} \int_0^{p_0} \int_A B \, dx \, dy \, dp \tag{23.30}$$

Here B, as before, represents an arbitrary field function. The normal velocity vanishes at the boundaries of V since we are dealing with a closed system. Application of (23.30) to (23.29) immediately yields

$$\frac{\partial \overline{K_h}^V}{\partial t} = \overline{\phi \, \nabla_h \cdot \mathbf{v}_h}^V \tag{23.31}$$

Here we have used (23.22b) and the lower and upper boundary conditions with $\omega = 0$ at $p = 0$ and $p = p_0$. With the help of the continuity equation and with (23.8), the right-hand side of (23.31) can easily be rewritten as

$$\overline{\phi \, \nabla_h \cdot \mathbf{v}_h}^V = -\overline{\frac{\partial}{\partial p}(\omega \phi)}^V + \overline{\omega \frac{\partial \phi}{\partial p}}^V = -R_0 \overline{\frac{T\omega}{p}}^V \tag{23.32}$$

so the volume average of the kinetic energy may be stated in the form

$$\frac{\partial \overline{K_h}^V}{\partial t} = -R_0 \overline{\frac{T\omega}{p}}^V \tag{23.33}$$

Hence, if we require an increase in the kinetic energy $\overline{K}_h^V$, the vertical velocity and the temperature T must be negatively correlated everywhere in the atmosphere. Reference to (23.24) shows that this requirement is precisely satisfied.

23.7 The geostrophic approximation of the vorticity equation

In this section we derive a simplified form of the vorticity equation, which will be used in the quasi-geostrophic system. To begin with, we rewrite the baroclinic vorticity equation (23.10) by replacing the divergence term with the help of the continuity equation,

$$\frac{\partial \zeta}{\partial t} + \mathbf{v}_{\mathrm{h}} \cdot \nabla_{\mathrm{h}} \eta + \omega \frac{\partial \zeta}{\partial p} + \mathbf{k} \cdot \left(\nabla_{\mathrm{h}} \omega \times \frac{\partial \mathbf{v}_{\mathrm{h}}}{\partial p} \right) = \eta \frac{\partial \omega}{\partial p} \tag{23.34}$$

which may also be written in the equivalent form

$$\frac{\partial \zeta}{\partial t} + \nabla_{\mathrm{h}} \cdot \left(\eta \mathbf{v}_{\mathrm{h}} + \omega \frac{\partial \mathbf{v}_{\mathrm{h}}}{\partial p} \times \mathbf{k} \right) = 0 \tag{23.35}$$

Thus, the tendency of the vorticity can be expressed as the divergence of two vectors. We wish to retain this form in the quasi-geostrophic theory. In analogy to (23.34) the selective geostrophic approximation of the vorticity equation can be written as

$$\frac{\partial \zeta_{\mathrm{g}}}{\partial t} + \mathbf{v}_{\mathrm{g}} \cdot \nabla_{\mathrm{h}} \eta_{\mathrm{g}} + \omega \frac{\partial \zeta_{\mathrm{g}}}{\partial p} + \mathbf{k} \cdot \left(\nabla_{\mathrm{h}} \omega \times \frac{\partial \mathbf{v}_{\mathrm{g}}}{\partial p} \right) = \eta_{\mathrm{g}} \frac{\partial \omega}{\partial p} \tag{23.36}$$

It should be observed that $\partial \omega / \partial p$ appearing on the right-hand side of this equation is equivalent to $-\nabla \cdot \mathbf{v}_{\mathrm{h}}$, so the actual horizontal wind velocity is retained in this term while everywhere else it is replaced by the geostrophic wind. This is the reason why we speak of a selective geostrophic approximation.

We will now discuss the selective or quasi-geostrophic approximation of the vorticity equation. According to (23.22b) there exists the following integral property for the unfiltered system:

$$\overline{\nabla_{\mathrm{h}} \cdot (\eta \mathbf{v}_{\mathrm{h}})} = \overline{\mathbf{v}_{\mathrm{h}} \cdot \nabla_{\mathrm{h}} \eta} + \overline{\eta \nabla_{\mathrm{h}} \cdot \mathbf{v}_{\mathrm{h}}} = 0 \tag{23.37}$$

We wish to retain this integral property even after the introduction of the geostrophic approximation

$$\overline{\mathbf{v}_{\mathrm{h}} \cdot \nabla_{\mathrm{h}} \eta} \;\rightarrow\; \overline{\mathbf{v}_{\mathrm{g}} \cdot \nabla_{\mathrm{h}} \eta_{\mathrm{g}}}, \qquad \overline{\eta \nabla_{\mathrm{h}} \cdot \mathbf{v}_{\mathrm{h}}} \;\rightarrow\; -\overline{\eta_{\mathrm{g}} \frac{\partial \omega}{\partial p}} \tag{23.38}$$

While the first expression of (23.38) is already zero since $\overline{\mathbf{v}_{\mathrm{g}} \cdot \nabla_{\mathrm{h}} \eta_{\mathrm{g}}} = \overline{\nabla_{\mathrm{h}} \cdot (\mathbf{v}_{\mathrm{g}} \eta_{\mathrm{g}})} = 0$, the second expression does not vanish. However, this expression will also be zero if we replace η_{g} by a constant value. To preserve the integral property (23.37) in the filtered system, we neglect ζ_{g} in the expression for the absolute geostrophic vorticity $\eta_{\mathrm{g}} = \zeta_{\mathrm{g}} + f$ and replace f by the constant Coriolis parameter f_0. Under typical midlatitude atmospheric conditions $\zeta_g < f$. Now we obtain for the second expression of (23.38)

$$\overline{\eta \, \nabla_h \cdot \mathbf{v}_h} \;\rightarrow\; -\overline{\eta_g \frac{\partial \omega}{\partial p}} = -\overline{(\zeta_g + f)\frac{\partial \omega}{\partial p}} \approx -\overline{f_0 \frac{\partial \omega}{\partial p}} = f_0 \overline{\nabla_h \cdot \mathbf{v}_h} = 0 \qquad (23.39)$$

Hence, we also retain the integral property (23.37) in the filtered system.

Owing to the approximation (23.39) the term $\zeta_g \, \partial\omega/\partial p$ will be ignored on the right-hand side of (23.36). Let us now investigate the importance of the term $\omega \, \partial\zeta_g/\partial p$. Introducing the vertical average over the model atmosphere according to

$$\overline{B}^p = \frac{1}{p_0}\int_0^{p_0} B \, dp \qquad (23.40)$$

we obtain

$$\overline{\frac{\partial}{\partial p}(\omega\zeta_g)}^p = \frac{1}{p_0}\int_0^{p_0} \frac{\partial}{\partial p}(\omega\zeta_g)\, dp = 0 \qquad (23.41)$$

since the vertical velocity vanishes at the boundaries of the atmosphere. As a consequence of this we have

$$\left| \overline{\omega \frac{\partial \zeta_g}{\partial p}}^p \right| = \left| \overline{\zeta_g \frac{\partial \omega}{\partial p}}^p \right| \qquad (23.42)$$

Hence, on average, the term $\omega \, \partial\zeta_g/\partial p$ is of the same order of magnitude as the term $\zeta_g \, \partial\omega/\partial p$. Since the term $\zeta_g \, \partial\omega/\partial p$ has been neglected in (23.36), it seems logical to ignore the term $\omega \, \partial\zeta_g/\partial p$ on the left-hand side of this equation as well.

Finally we observe that the term $\omega \, \partial\zeta_g/\partial p$ plus the twisting term can be combined to give a divergence expression:

$$\omega \frac{\partial \zeta_g}{\partial p} + \mathbf{k} \cdot \left(\nabla_h \omega \times \frac{\partial \mathbf{v}_g}{\partial p} \right) = \nabla_h \cdot \left(\omega \frac{\partial \mathbf{v}_g}{\partial p} \times \mathbf{k} \right) \qquad (23.43)$$

Obviously, without the omitted term we cannot obtain a divergence expression for the tendency of the geostrophic vorticity. Thus, it stands to reason that we should omit the twisting term in (23.36) as well. This simplification is also justified because, in large-scale flow fields, $\nabla_h \omega$ is usually small in comparison with the remaining terms of the vorticity equation.

On introducing these simplifications into (23.36) we obtain the quasi-geostrophic approximation of the vorticity equation as

$$\boxed{\frac{\partial \zeta_g}{\partial t} + \mathbf{v}_g \cdot \nabla_h \eta_g = f_0 \frac{\partial \omega}{\partial p}} \qquad (23.44)$$

which, in analogy to (23.35), can be written in the desired divergence form

$$\frac{\partial \zeta_g}{\partial t} + \nabla_h \cdot (\eta_g \mathbf{v}_g + f_0 \mathbf{v}_h) = 0 \tag{23.45}$$

Utilizing (23.17) and (23.18), we may also write (23.44) as

$$\frac{\partial}{\partial t}(\nabla_h^2 \phi) + \frac{1}{f_0} J(\phi, \nabla_h^2 \phi) + \beta \frac{\partial \phi}{\partial x} = f_0^2 \frac{\partial \omega}{\partial p} \tag{23.46}$$

In order to give a simple physical interpretion of the vorticity equation, we integrate (23.44) over the depth of the atmosphere from $p = 0$ to $p = p_0$. Observing that the generalized vertical velocity vanishes at the lower and the upper boundaries of the model, we obtain the following tendency equation:

$$\frac{\partial \overline{\zeta_g}^p}{\partial t} = -\overline{\mathbf{v}_g \cdot \nabla_h(\zeta_g + f)}^p \tag{23.47}$$

This equation shows that, within the validity of the quasi-geostrophic theory, the change with time of the vertical average of the geostrophic vorticity in the p system depends solely on the horizontal advection of the absolute vorticity. A more detailed physical interpretation will be given later.

It should be realized that the derivation of the divergence form of the vorticity equation in the geostrophic approximation results in a severe reduction of physical significance in comparison with the original equation. We will show at the end of this chapter that the quasi-geostrophic approximation of the vorticity equation and other results of the quasi-geostrophic theory can be easily derived with the help of the ageostrophic wind approximation of Philipps (1939).

The heat equation (23.19) and the geostrophic vorticity equation (23.44) form the basis of the entire quasi-geostrophic theory. Both equations result from a drastic simplification of the original equations. In particular, the geostrophic approximations of the horizontal wind in the advection terms $\mathbf{v}_h \cdot \nabla_h T$ and $\mathbf{v}_h \cdot \nabla_h \eta$ make it impossible to realistically predict such processes as occlusion and frontogenesis and to form precise energy balances. However, it is possible to improve the theory.

23.8 The ω equation

The heat equation, the vorticity equation, and the divergence equation require knowledge of the vertical velocity ω. The integration of the continuity equation in the p system appears to give the required information:

$$\omega = -\int_0^p \nabla_h \cdot \mathbf{v}_h \, dp \tag{23.48}$$

However, routine measurements of the horizontal wind field are not sufficiently accurate to determine the divergence $\nabla_h \cdot \mathbf{v}_h$, so this equation is unsuitable for finding ω. Thus, we have to look for a more powerful method to determine ω.

We proceed by eliminating the tendency $\partial\phi/\partial t$ from the vorticity equation (23.46) and from the heat equation in the form (23.19). This is accomplished by the following mathematical operations. (1) We differentiate the vorticity equation with respect to p. (2) We apply the horizontal Laplacian to the heat equation. (3) We subtract one of the resulting equations from the other and find

$$\boxed{\begin{aligned}\overline{\sigma}_0 \nabla_h^2 \omega + f_0^2 \frac{\partial^2 \omega}{\partial p^2} &= \frac{\partial}{\partial p}\left[J\left(\phi, \frac{\nabla_h^2 \phi}{f_0}\right)\right] + \beta \frac{\partial^2 \phi}{\partial x \, \partial p} - \frac{1}{f_0} \nabla_h^2 \left[J\left(\phi, \frac{\partial \phi}{\partial p}\right)\right] \\ &= f_0 \frac{\partial}{\partial p}(\mathbf{v}_g \cdot \nabla_h \eta_g) - \nabla_h^2\left(\mathbf{v}_g \cdot \nabla_h \frac{\partial \phi}{\partial p}\right)\end{aligned}} \tag{23.49}$$

This equation is known as the *ω equation*. For large-scale motion the stability function $\overline{\sigma}_0 > 0$, so we are dealing with a partial differential equation of the elliptic type. Thus, we are confronted with a boundary-value problem permitting us to find ω if the geopotential field $\phi(x, y, p, t = \text{constant})$ is known at a fixed time. Textbooks on numerical analysis discuss the numerical procedures to be used. In earlier days this equation closed a gap in routine weather observations, which even today do not report the vertical velocity field. The mass field, represented by the geopotential, is measured relatively accurately. It is a simple matter to find the geostrophic wind from the geopotential field, but it requires quite a bit of numerical work to find the generalized vertical velocity from the ω equation. Sometimes the ω equation is called the *geostrophic relation for the vertical wind*.

It should be observed that the so-called *τ equation* is equivalent to the ω equation. By eliminating ω from the vorticity and the heat equation we find a second-order partial differential equation for the tendency $\partial\phi/\partial t$, which for simplicity is designated τ. Since we do not gain anything new, we omit a discussion of this boundary-value problem.

The principle of the numerical solution for the quasi-geostrophic system will now be summarized.

(i) At the initial time $t_0 = 0$, the geopotential $\phi(x, y, p, t_0 = 0)$ is assumed to be given so that $\mathbf{v}_g$ and ω can be determined by solving (23.17) and (23.49).

(ii) With ω known initially, a new ϕ-field at $t_1 = t_0 + \Delta t$ can be determined from the vorticity equation (23.46). The whole system can now be iterated.
(iii) If the temperature field is required at any time t, we have to solve the hydrostatic equation (23.8).
(iv) If it is sufficient to find the actual wind field in some approximation, we make use of the divergence equation (23.13). By assuming that the divergence D is zero, we obtain a balance equation of the Monge–Ampère type, which can be solved numerically. The problem may be further simplified by assuming that barotropic conditions apply for each layer so that $\partial \mathbf{v}_\mathrm{h}/\partial p$ vanishes. Since the divergence was set equal to zero, we may express the wind by means of the stream function. Thus, we have to solve the following equation:

$$2J\left(\frac{\partial \psi}{\partial x}, \frac{\partial \psi}{\partial y}\right) + \nabla_\mathrm{h} \cdot (f\,\nabla_\mathrm{h}\psi) = \nabla_\mathrm{h}^2 \phi \quad \text{with} \quad \nabla_\mathrm{h} \cdot (f\,\nabla_\mathrm{h}\psi) = -\beta u + f\zeta \tag{23.50}$$

We have shown that the quasi-geostrophic theory of filtering has reduced the prognostic system to a one-field system. The entire prognostic process is reduced to the determination of the geopotential field, which must be known at the initial time $t_0 = 0$. This fact verifies that all noise waves (horizontal sound waves, gravitational waves, inertial waves) have been eliminated, since they can appear only if more than one variable can be specified without restriction at the initial time $t_0 = 0$. In the next chapter we will actually show how to solve the quasi-geostrophic system for the simple case of a four-layer model. It should be realized that, with the availability of modern computers, the quasi-geostrophic theory is no longer used for actual weather forecasting. Nevertheless, the theory is of great value for the interpretation of the atmospheric mechanism.

In order to improve our understanding of the quasi-geostrophic theory we will extract the physical content of the ω equation. In a rough approximation we assume that the vertical velocity is given by a harmonic function of the form

$$\omega = A \sin\left(\frac{\pi p}{p_0}\right) \cos(k_x x) \cos(k_y y) \tag{23.51}$$

satisfying the boundary conditions $\omega(p = 0) = \omega(p = p_0) = 0$. Substituting (23.51) into (23.49) yields

$$-l^2 \omega = f_0 \frac{\partial}{\partial p}(\mathbf{v}_\mathrm{g} \cdot \nabla_\mathrm{h} \eta_\mathrm{g}) - \nabla_\mathrm{h}^2 \left(\mathbf{v}_\mathrm{g} \cdot \nabla_\mathrm{h} \frac{\partial \phi}{\partial p}\right) \tag{23.52}$$

where the correlation factor l^2 is given by

$$l^2 = f_0^2 \left(\frac{\pi}{p_0}\right)^2 + \overline{\sigma}_0 \left(k_x^2 + k_y^2\right) \tag{23.53}$$

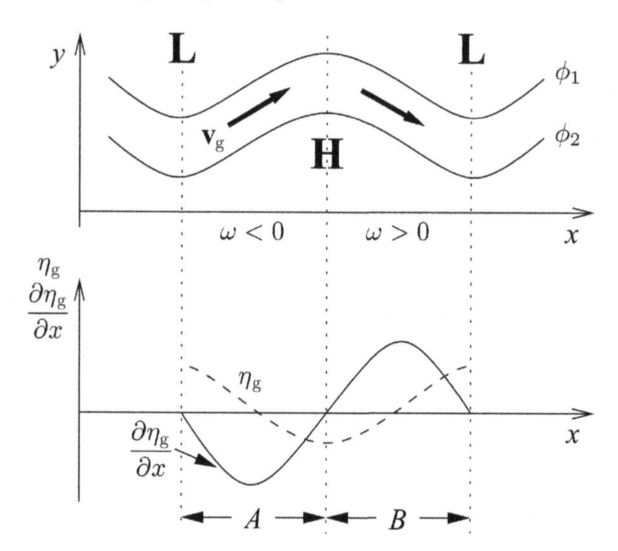

Fig. 23.4 A schematic intepretation of the term $\partial(\mathbf{v}_g \cdot \nabla_h \eta_g)/\partial p$ of the ω equation.

Since $\overline{\sigma}_0 > 0$ in large-scale motion, the quantity l^2 is positive. From (23.52) we see that the left-hand side of (23.49) is negatively correlated to ω.

We will now interpret this equation. Since in the quasi-geostrophic theory we have reduced the solution of the prognostic system to knowledge of the geopotential field, for the interpretation of the ω equation it is sufficient to prescribe the ϕ-field. In order to understand the effect of the first term on the right-hand side of (23.52), let us consider an idealized sinusoidal pattern of ϕ-contour lines as shown in the upper part of Figure 23.4. Since $\eta_g = (1/f_0)\nabla_h^2\phi + f$ the corresponding η_g curve has maximum and minimum values in the low- and high-pressure regions, respectively, as indicated by the dashed curve in the lower part of Figure 23.4. For simplicity the lower part of Figure 23.4 depicts the evolution of the curves along the x-axis only. From the η_g curve we immediately obtain the $\nabla_h \eta_g$ curve which is also shown in Figure 23.4. We conclude that, in region A of Figure 23.4, the term $\mathbf{v}_g \cdot \nabla_h \eta_g < 0$. For typical atmospheric situations $|\mathbf{v}_g|$ increases with height so that $\partial(\mathbf{v}_g \cdot \nabla_h \eta_g)/\partial z < 0$ and, therefore, $\partial(\mathbf{v}_g \cdot \nabla_h \eta_g)/\partial p > 0$. Hence, in region A the first term on the right-hand side of (23.52) induces rising air motion, that is $\omega < 0$. In region B we observe the opposite situation. Here $\mathbf{v}_g \cdot \nabla_h \eta_g > 0$ so that $\partial(\mathbf{v}_g \cdot \nabla_h \eta_g)/\partial p < 0$ and $\omega > 0$, thus indicating sinking air motion.

Next we consider the effect of the second term on the right-hand side of the ω equation. An assumed vertical distribution of the geopotential field is shown in the upper part of Figure 23.5. Since $\partial\phi/\partial p = -R_0 T/p$ we obtain regions of cold and warm air as indicated in the figure. This yields cold- and warm-air advection in regions A and B, respectively. From the given ϕ distribution we obtain curve a

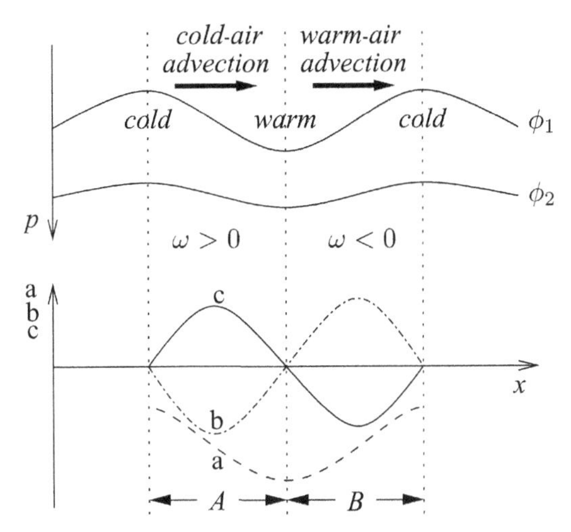

Fig. 23.5 A schematic intepretation of the term $\nabla_h^2(\mathbf{v}_g \cdot \nabla_h(\partial\phi/\partial p))$ of the ω equation. The three curves in the lower part of the figure describe the pattern of the following terms: a, $\partial\phi/\partial p$; b, $\mathbf{v}_g \cdot \nabla_h(\partial\phi/\partial p)$; and c, $\nabla_h^2(\mathbf{v}_g \cdot \nabla_h(\partial\phi/\partial p))$.

denoting the pattern of $\partial\phi/\partial p$. Curve b describes $\mathbf{v}_g \cdot \nabla_h(\partial\phi/\partial p)$ and finally the pattern of the term $\nabla_h^2[\mathbf{v}_g \cdot \nabla_h(\partial\phi/\partial p)]$ is indicated by curve c. According to (23.52) we conclude that sinking air motion is observed in regions of cold-air advection and rising air motion in regions of warm-air advection.

23.9 The Philipps approximation of the ageostrophic component of the horizontal wind

The actual weather is determined in large measure by the deviation of the actual horizontal wind from the geostrophic wind. We call this part of the horizontal wind the *ageostrophic wind component*, $\mathbf{v}_{ag} = \mathbf{v}_h - \mathbf{v}_g$. The ageostrophic wind component can be elegantly discussed with the help of an approximation introduced by Philipps (1939). Moreover, this approximation can be used to obtain most easily the geostrophic approximation of the vorticity equation.

We begin the analysis by introducing the geostrophic wind into the horizontal equation of motion (22.22), yielding

$$\frac{du}{dt} = f(v - v_g), \qquad \frac{dv}{dt} = -f(u - u_g) \quad \text{with} \quad u_g = -\frac{1}{f}\frac{\partial\phi}{\partial y}, \qquad v_g = \frac{1}{f}\frac{\partial\phi}{\partial x} \tag{23.54}$$

On multiplying the prognostic equation for v by $i = \sqrt{-1}$ and adding this to the prognostic equation for u, we find

$$\frac{dq}{dt} + ifq = ifq_g, \qquad \text{with} \quad q = u + iv, \quad q_g = u_g + iv_g \tag{23.55}$$

In this expression we approximate the Coriolis parameter f by the constant value f_0 and obtain

$$q = q_{g,0} - \frac{1}{if_0}\frac{dq}{dt} \quad \text{with} \quad q_{g,0} = u_{g,0} + iv_{g,0} \tag{23.56}$$

This rough approximation will be partly amended a little later. Successive differentiation of (23.56) with respect to time yields

$$\frac{dq}{dt} = \frac{dq_{g,0}}{dt} - \frac{1}{if_0}\frac{d^2q}{dt^2}, \qquad \frac{d^2q}{dt^2} = \frac{d^2q_{g,0}}{dt^2} - \frac{1}{if_0}\frac{d^3q}{dt^3}, \qquad \ldots \tag{23.57}$$

On substituting these expressions into (23.56) we obtain the series

$$\begin{aligned} q &= q_{g,0} - \frac{1}{if_0}\frac{dq_{g,0}}{dt} + \frac{1}{(if_0)^2}\frac{d^2q_{g,0}}{dt^2} - \frac{1}{(if_0)^3}\frac{d^3q_{g,0}}{dt^3} + \cdots \\ &= \sum_{n=0}^{\infty}(-if_0)^{-n}\frac{d^nq_{g,0}}{dt^n} \end{aligned} \tag{23.58}$$

For the convergence properties of this series, see the original paper of Philipps (1939). If the series is discontinued after the second term we obtain the scalar form of the wind equation:

$$u = u_{g,0} - \frac{1}{f_0}\frac{dv_{g,0}}{dt}, \qquad v = v_{g,0} + \frac{1}{f_0}\frac{du_{g,0}}{dt} \tag{23.59}$$

For brevity we write the vector form

$$\mathbf{v}_h = \mathbf{v}_{g,0} + \frac{1}{f_0}\mathbf{k} \times \frac{d\mathbf{v}_{g,0}}{dt} \tag{23.60}$$

This equation expresses the actual wind by means of the approximated geostrophic wind

$$\mathbf{v}_{g,0} = \frac{1}{f_0}\mathbf{k} \times \nabla_h\phi = \frac{f}{f_0}\mathbf{v}_g \tag{23.61}$$

and its time derivative. Writing the Coriolis parameter f in the form $f = f_0 + \Delta f$ with $\Delta f \ll f_0$, the factor f_0/f can be approximated as

$$\frac{f_0}{f} = 1 - \frac{\Delta f}{f} \approx 1 - \frac{\Delta f}{f_0} = 2 - \frac{f}{f_0} \tag{23.62}$$

From (23.61) we then obtain

$$\mathbf{v}_g = \left(2 - \frac{f}{f_0}\right)\mathbf{v}_{g,0} \tag{23.63}$$

Philipps made two basic assumptions.

(i) In order to partly restore the loss of the latitudinal variability of the Coriolis parameter, he replaced the first term on the right-hand side of (23.60) by the original geostrophic wind $\mathbf{v}_g$.

(ii) The individual derivative d/dt is approximated by $d_{g,0}/dt$ as given by

$$\frac{d_{g,0}}{dt} = \frac{\partial}{\partial t} + \mathbf{v}_{g,0} \cdot \nabla_h \tag{23.64}$$

Introduction of these two assumptions into equation (23.60) gives the approximation

$$\boxed{\mathbf{v}_h = \left(2 - \frac{f}{f_0}\right)\mathbf{v}_{g,0} + \frac{1}{f_0}\mathbf{k} \times \frac{d_{g,0}\mathbf{v}_{g,0}}{dt}} \tag{23.65}$$

which is known as the *Philipps wind.*

Now we are ready to derive the ageostrophic component of the horizontal wind in the p system. For simplicity we omit the suffix p. First we evaluate the second term on the right-hand side of (23.65) as

$$\begin{aligned}
\frac{1}{f_0}\mathbf{k} \times \frac{d_{g,0}\mathbf{v}_{g,0}}{dt} &= \frac{1}{f_0}\mathbf{k} \times \frac{\partial \mathbf{v}_{g,0}}{\partial t} + \frac{1}{f_0}\mathbf{k} \times (\mathbf{v}_{g,0} \cdot \nabla_h \mathbf{v}_{g,0}) \\
&= -\frac{1}{f_0^2}\nabla_h \frac{\partial \phi}{\partial t} + \frac{1}{2f_0}\mathbf{k} \times \nabla_h \mathbf{v}_{g,0}^2 - \frac{1}{f_0}\mathbf{k} \times [\mathbf{v}_{g,0} \times (\nabla_h \times \mathbf{v}_{g,0})] \\
&= -\frac{1}{f_0^2}\nabla_h \frac{\partial \phi}{\partial t} + \frac{1}{2f_0^3}\mathbf{k} \times \nabla_h(\nabla_h \phi)^2 - \frac{1}{f_0^2}\mathbf{v}_{g,0}\,\nabla_h^2 \phi \\
\text{since} \quad \mathbf{v}_{g,0}^2 &= \frac{1}{f_0^2}(\mathbf{k} \times \nabla_h \phi) \cdot (\mathbf{k} \times \nabla_h \phi) = \frac{1}{f_0^2}(\nabla_h \phi)^2 \\
\text{and} \quad \mathbf{k} \times [\mathbf{v}_{g,0} \times (\nabla_h \times \mathbf{v}_{g,0})] &= \frac{1}{f_0}\mathbf{v}_{g,0}\,\nabla_h^2 \phi
\end{aligned} \tag{23.66}$$

In this expression use of (M1.48) and of the Lamb development (M3.75) has been made. By utilizing (23.66) we obtain a suitable expression for the ageostrophic wind component:

$$\boxed{\begin{aligned}
\mathbf{v}_{ag} &= \mathbf{v}_h - \mathbf{v}_{g,0} \\
&= \left(1 - \frac{f}{f_0}\right)\mathbf{v}_{g,0} - \frac{1}{f_0^2}\nabla_h \frac{\partial \phi}{\partial t} + \frac{1}{2f_0^3}\mathbf{k} \times \nabla_h(\nabla_h \phi)^2 - \frac{1}{f_0^2}\mathbf{v}_{g,0}\,\nabla_h^2 \phi \\
&= \mathbf{v}_{ag}(\mathrm{I}) + \mathbf{v}_{ag}(\mathrm{II}) + \mathbf{v}_{ag}(\mathrm{III}) + \mathbf{v}_{ag}(\mathrm{IV})
\end{aligned}} \tag{23.67}$$

This geostrophic deviation is largely responsible for the occurrence of large-scale changes in weather. The four terms of equation (23.67) will now be discussed individually.

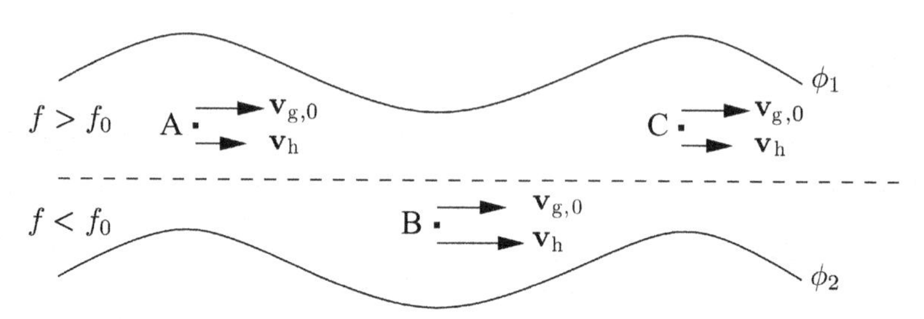

Fig. 23.6 A schematic interpretation of the latitude effect, term $\mathbf{v}_{ag}(\mathrm{I})$ of equation (23.67).

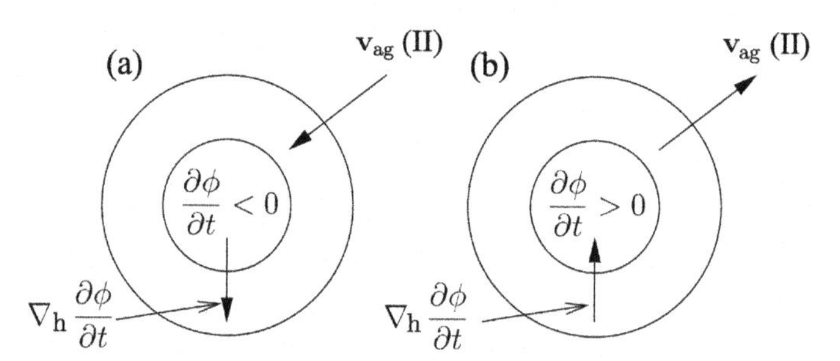

Fig. 23.7 A schematic interpretation of the pressure-tendency effect, term $\mathbf{v}_{ag}(\mathrm{II})$ of equation (23.67).

23.9.1 The latitude effect

Figure 23.6 shows idealized contour lines of ϕ and the corresponding geostrophic wind $\mathbf{v}_{g,0}$ at points A, B, and C. The Coriolis parameter f changes with latitude in such a way that $f < f_0$ in the south, that is below the dashed line, while in the north $f > f_0$. From (23.67) we see that the first term on the right-hand side yields $|\mathbf{v}_h| > |\mathbf{v}_{g,0}|$ at point B while $|\mathbf{v}_h| < |\mathbf{v}_{g,0}|$ at points A and C. Owing to its latitudinal dependency, this term is called the *latitude effect*.

23.9.2 The pressure-tendency effect

The effect of the second term on the right-hand side of (23.67) is easily illustrated. For simplicity we assume that circular isallobaric tendencies pertain. In a region of falling pressure the gradient of $\partial\phi/\partial t$ is directed away from the center of the system so that the ageostrophic component resulting from the second term has the opposite direction; see Figure 23.7(a). This implies that, due to mass transport, the system is filling or decreasing in strength. In a region of rising pressure the opposite effect will be observed; see Figure 23.7(b). With the help of (M4.51) it is easy to show that height tendencies of pressure surfaces may be converted to pressure tendencies $\partial p/\partial t$ at fixed levels by means of $(\partial p/\partial t)_z = \rho(\partial\phi/\partial t)_p$. Thus

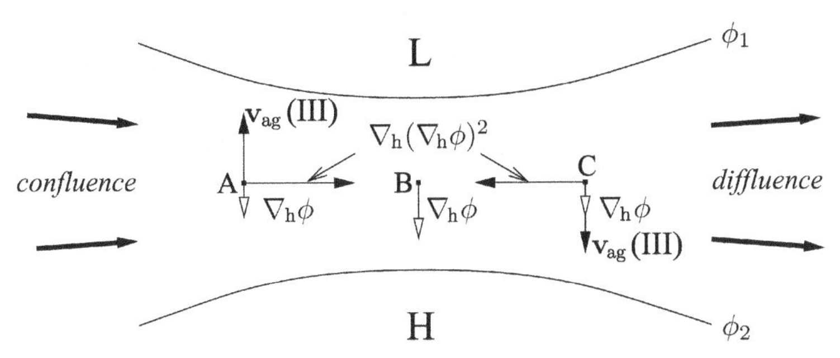

Fig. 23.8 A schematic interpretation of the confluence effect, term $\mathbf{v}_{ag}$(III) of equation (23.67).

isolines of $\partial\phi/\partial t$ may be relabelled as isallobars, which are isolines of $\partial p/\partial t$ or vice versa.

The term $-(1/f_0^2)\,\nabla_h(\partial\phi/\partial t)$ is known as the *isallobaric* or the *Brunt–Douglas wind* after the two scientists who first formulated this expression in the year 1928.

23.9.3 The confluence and diffluence effect

This kinematic effect is best demonstrated for regions of confluence and diffluence, which must not be confused with regions of convergence and divergence. Confluence and diffluence refer to converging and diverging contour lines or isobars. Let us consider an air parcel moving from the west to the east; see Figure 23.8. We assume that initially the wind is in geostrophic balance. As the air approaches a region with a stronger geopotential gradient, the geopotential gradient force $-\nabla_h\phi$ will exceed the Coriolis force so that the ageostrophic component is directed toward the low pressure; see point A of Figure 23.8. Let us assume that, at the point B, the geostrophic balance has been restored so that $\mathbf{v}_{ag}(\text{III}) = 0$. Because the air is approaching a region with a weaker geopotential gradient, the Coriolis force exceeds the geopotential gradient force, so the ageostrophic wind is directed toward the higher pressure; see point C of Figure 23.8. In summary, in a region of confluence the ageostrophic wind is directed toward the low pressure, whereas in a region of diffluence the ageostrophic wind is directed toward the high pressure. This effect is known as the *confluence and diffluence effect.*

23.9.4 The curvature effect

The fourth term on the right-hand side describes the so-called *curvature effect* and is schematically illustrated in Figure 23.9. The Laplacian of ϕ is negative in

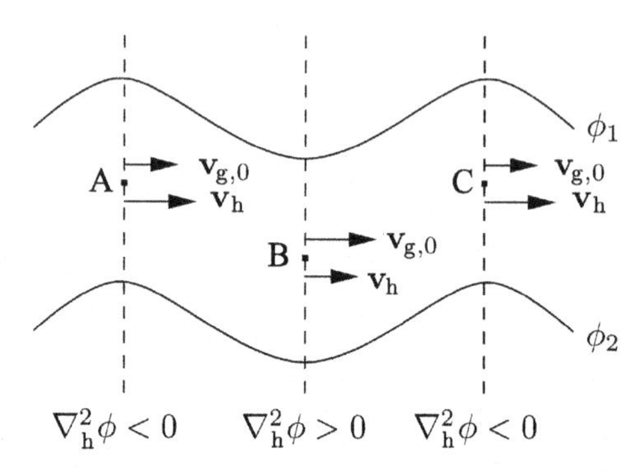

Fig. 23.9 A schematic interpretation of the curvature effect, term $\mathbf{v}_{ag}(\mathrm{IV})$ of equation (23.67).

a high-pressure region but positive in a low-pressure region. Since in (23.67) the term appears with a negative sign, we find that the horizontal wind $|\mathbf{v}_h| > |\mathbf{v}_{g,0}|$ at points A and C while $|\mathbf{v}_h| < |\mathbf{v}_{g,0}|$ at point B.

23.10 Applications of the Philipps wind

We multiply equation (23.65) scalarly by the horizontal gradient of the geopotential and obtain

$$\begin{aligned}\mathbf{v}_h \cdot \nabla_h \phi &= \left(2 - \frac{f}{f_0}\right)\left(\frac{1}{f_0}\mathbf{k} \times \nabla_h \phi\right) \cdot \nabla_h \phi + \frac{1}{f_0}\mathbf{k} \times \frac{d_{g,0}\mathbf{v}_{g,0}}{dt} \cdot \nabla_h \phi \\ &= -\left(\mathbf{k} \times \frac{d_{g,0}\mathbf{v}_{g,0}}{dt}\right) \cdot (\mathbf{k} \times \mathbf{v}_{g,0}) = -\frac{d_{g,0}\mathbf{v}_{g,0}}{dt} \cdot \mathbf{v}_{g,0} = -\frac{d_{g,0}}{dt}\left(\frac{\mathbf{v}_{g,0}^2}{2}\right)\end{aligned} \tag{23.68}$$

This is the selective geostrophic approximation of the kinetic energy of the horizontal motion. It should be observed that the horizontal advection of the geopotential involves the actual wind. Had we replaced the actual wind in the advection term by the geostrophic wind, this term would have vanished. Thus, the change in kinetic energy is given by the horizontal advection of the geopotential with the ageostrophic wind component.

In Section 23.7 we have discussed the geostrophic approximation of the vorticity equation, regarding which we had to use a number of arguments to justify the procedure. With the help of the Philipps wind this approximation follows very easily. Utilizing the Lamb development (M3.75), we expand (23.65) with the help of (23.66) and obtain

$$\begin{aligned}\mathbf{v}_{\rm h} &= 2\mathbf{v}_{\rm g,0} - \frac{f}{f_0}\mathbf{v}_{\rm g,0} + \frac{1}{f_0}\mathbf{k} \times \frac{\partial \mathbf{v}_{\rm g,0}}{\partial t} + \frac{1}{f_0}\mathbf{k} \times \left[\nabla_{\rm h}\left(\frac{\mathbf{v}_{\rm g,0}^2}{2}\right) + (\nabla_{\rm h} \times \mathbf{v}_{\rm g,0}) \times \mathbf{v}_{\rm g,0}\right] \\ &= 2\mathbf{v}_{\rm g,0} - \frac{f}{f_0}\mathbf{v}_{\rm g,0} + \frac{1}{f_0}\mathbf{k} \times \frac{\partial \mathbf{v}_{\rm g,0}}{\partial t} + \frac{1}{f_0}\mathbf{k} \times \nabla_{\rm h}\left(\frac{\mathbf{v}_{\rm g,0}^2}{2}\right) - \frac{1}{f_0}\zeta_{\rm g,0}\mathbf{v}_{\rm g,0}\end{aligned} \tag{23.69}$$

with $\zeta_{\rm g,0} = \mathbf{k} \cdot \nabla_{\rm h} \times \mathbf{v}_{\rm g,0}$. Now we take the horizontal divergence of (23.69) and recall that the divergence of $\mathbf{v}_{\rm g,0}$ is zero. The resulting equation

$$\boxed{\nabla_{\rm h} \cdot \mathbf{v}_{\rm h} = -\frac{1}{f_0}\mathbf{v}_{\rm g,0} \cdot \nabla_{\rm h} f - \frac{1}{f_0}\mathbf{k} \cdot \nabla_{\rm h} \times \frac{\partial \mathbf{v}_{\rm g,0}}{\partial t} - \frac{1}{f_0}\mathbf{v}_{\rm g,0} \cdot \nabla_{\rm h}\zeta_{\rm g,0}} \tag{23.70}$$

is already the desired result which is particularly suitable for physical interpretation. The usual form (23.46) of the geostrophic approximation can be obtained by rewriting (23.70) as

$$-f_0\,\nabla_{\rm h} \cdot \mathbf{v}_{\rm h} = \frac{d_{\rm g,0}\,\eta_{\rm g,0}}{dt} \tag{23.71}$$

Utilizing (23.17) and (23.18) gives the final form

$$\nabla_{\rm h}^2 \frac{\partial \phi}{\partial t} + J\left(\phi, \frac{1}{f_0}\nabla_{\rm h}^2\phi + f\right) = -f_0^2\,\nabla_{\rm h} \cdot \mathbf{v}_{\rm h} \tag{23.72}$$

which is identical with (23.46). This partial differential equation describing the tendency of the geopotential field apparently involves the three field quantities (u, v, ϕ). On replacing the divergence term by means of the continuity equation we recognize that this equation involves only the two variables ω and ϕ. The vertical velocity is found from the ω equation.

Now we will present a physical interpretation of the geostrophic approximation of the vorticity equation (23.70).

23.10.1 Westward displacement of a pressure system

Let us consider the idealized contour pattern of Figure 23.10. The gradient of the Coriolis parameter, of course, is pointing to the north. In region A we have $\cos(\mathbf{v}_{\rm g,0}, \nabla_{\rm h} f) < 0$ while in region B scalar product is positive. Owing to the negative sign, the first term on the right-hand side of (23.70) produces divergence in region A and convergence in region B. This results in a westward displacement of the wave, which is known as the β *effect* since $|\nabla_{\rm h} f| = \beta$.

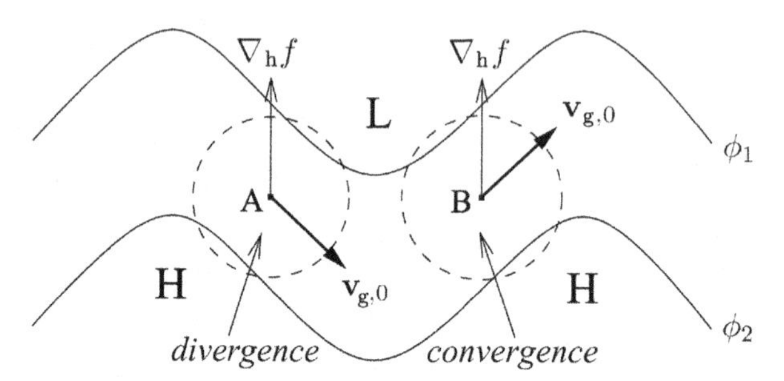

Fig. 23.10 A schematic interpretation of the term $-(1/f_0)\mathbf{v}_{g,0} \cdot \nabla_h f$ of equation (23.70).

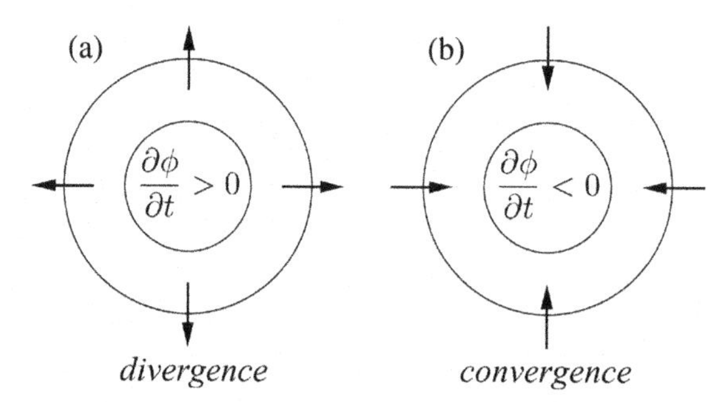

Fig. 23.11 A schematic interpretation of the term $-(1/f_0)\mathbf{k} \cdot \nabla_h \times \partial \mathbf{v}_{g,0}/\partial t = -(1/f_0^2)\nabla_h^2\, \partial\phi/\partial t$ of equation (23.70).

23.10.2 The pressure-tendency effect

Let us assume that we have an idealized circular pattern of the geopotential tendency whose magnitude is largest at the center. The sign of the term determines whether divergence or convergence occurs. For a positive tendency the Laplacian of this quantity is negative. The resulting divergence opposes the effect of the pressure tendency, as displayed in Figure 23.11(a). In case of a negative tendency the Laplacian of the pressure tendency is positive, so the opposite situation occurs.

23.10.3 The curvature effect

Again we consider sinusoidal contour lines, see Figure 23.12. In region A we observe $\cos(\mathbf{v}_{g,0}, \nabla_h \zeta_{g,0}) > 0$ resulting in convergence while in region B the scalar product is negative, thus producing divergence. This term, often called the curvature effect, results in an eastward displacement of the Rossby wave. Hence, the curvature effect opposes the β effect. Usually the shorter Rossby waves move from west to east, so the curvature effect dominates over the β effect. On the other hand, very long Rossby waves move from east to west, so in this case the β effect dominates

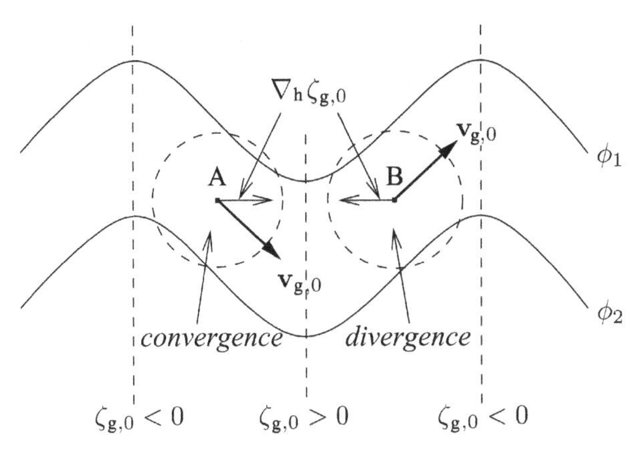

Fig. 23.12 A schematic interpretation of the term $-(1/f_0)\mathbf{v}_{g,0} \cdot \nabla_h \zeta_{g,0}$ of equation (23.70).

over the curvature effect. There exists a critical wavelength, see Chapter 16, at which standing Rossby waves form, resulting in blocking high- and low-pressure systems.

23.11 Problems

23.1: Show that

$$\sigma_0 \begin{cases} >0 & \text{stable atmosphere,} & \partial\theta/\partial z > 0 \\ =0 & \text{neutral atmosphere,} & \partial\theta/\partial z = 0 \\ <0 & \text{unstable atmosphere,} & \partial\theta/\partial z < 0 \end{cases}$$

Start your derivation with the definition of σ_0 given in (23.9).

23.2:
(a) Consider the heat equation $c_p\, dT/dt - (1/\rho)\, dp/dt = B/\rho$, where B represents the non-adiabatic terms. With the help of the hydrostatic approximation and the continuity equation written in the (x, y, σ) system ($\sigma = p/p_s$), show that the heat equation can be written in the form

$$\frac{\partial}{\partial t}\left(\frac{\partial \phi}{\partial \sigma}\right) + \mathbf{v}_h \cdot \nabla_{h,\sigma}\left(\frac{\partial \phi}{\partial \sigma}\right) + \dot{\sigma}\left(\frac{\partial^2 \phi}{\partial \sigma^2} + \frac{\partial \phi}{\partial \sigma}\frac{1-k}{\sigma}\right)$$
$$+ k\frac{\partial \phi}{\partial \sigma}\nabla_{h,\sigma} \cdot \mathbf{v}_h + k\frac{\partial \phi}{\partial \sigma}\frac{\partial \dot{\sigma}}{\partial \sigma} = \frac{k}{\sigma p_s}\frac{\partial \phi}{\partial \sigma} B$$

with $k = R_0/c_p$.
(b) Show that

$$\frac{\partial^2 \phi}{\partial \sigma^2} + \frac{\partial \phi}{\partial \sigma}\frac{1-k}{\sigma} = \frac{1}{g}\left(\frac{\partial \phi}{\partial \sigma}\right)^2 \frac{\partial \ln \theta}{\partial z}$$

23.3: The *Richardson equation* is a diagnostic equation for the vertical velocity w if the fields of the horizontal wind and the pressure are known. This equation can be derived with the help of the continuity equation and the equation of an adiabat which is given by $p = \text{constant} \times \rho^{\kappa}$, where $\kappa = c_p/c_v$.
(a) Show that the Richardson equation in the z system can be written in the form

$$\frac{\partial}{\partial z}\left(p\,\frac{\partial w}{\partial z}\right) = -\frac{\partial}{\partial z}(p\,\nabla_{\rm h}\cdot\mathbf{v}_{\rm h}) + \frac{1}{\kappa}\left(\frac{\partial p}{\partial z}\,\nabla_{\rm h}\cdot\mathbf{v}_{\rm h} - \frac{\partial\mathbf{v}_{\rm h}}{\partial z}\cdot\nabla_{\rm h}p\right)$$

if the hydrostatic equation is assumed to be valid.
(b) Compare the Richardson equation with the ω equation. Hint: First show that

$$\frac{dp}{dt} = \frac{1}{g}\frac{\partial p}{\partial z}\left(\nabla_{\rm h}\cdot\mathbf{v}_{\rm h} + \frac{\partial w}{\partial z}\right)$$

23.4:
(a) Find the solution to the differential equation (23.55) by assuming that $f = f_0$. The initial conditions are $q = q_0$ and $q_{\rm g} = q_{\rm g}(0)$.
(b) State the conditions under which the wind is geostrophic at all times. Give the form of the corresponding geopotential field.

24

A two-level prognostic model, baroclinic instability

24.1 Introduction

In this chapter we are going to discuss the *two-level quasi-geostrophic prediction model*. This model divides the atmosphere into four layers as shown in Figure 24.1. The vorticity equation is applied to levels $l = 1$ and $l = 3$ while the heat equation is applied to level $l = 2$. By eliminating the vertical velocity ω it becomes possible to determine the tendency of the geopotential $\partial\phi/\partial t$. Initially only the geopotential $\phi(x, y, p, t_0 = 0)$ for the entire vertical pressure range $0 \leq p \leq p_0$ must be available. The discussion will be facilitated by resolving the dependent variables in the vertical direction only. The remaining differentials will be left in their original forms, which may be approximated by finite differences whenever desired.

In the second part of this chapter we are going to discuss the concept of baroclinic instability. In a rotating atmosphere this type of instability, which was first investigated by Charney (1947) and Eady (1949), arises from the vertical wind shear if the static stability is not too large. The stability properties of the Charney model are difficult to analyze. The two-level model, however, makes it possible to obtain the stability criteria in a rather simple way, with results consistent with Charney's model. Details, for example, are given by Haltiner and Williams (1980).

24.2 The mathematical development of the two-level model

The basic system consists of the vorticity equation (23.44) and the first law of thermodynamics (23.19). These equations are restated in slightly changed forms as

$$\begin{aligned}\left(\frac{\partial}{\partial t} + \mathbf{v}_{\mathrm{g}}\cdot\nabla_{\mathrm{h}}\right)(\zeta_{\mathrm{g}} + f) &= f_0\frac{\partial\omega}{\partial p} + K_{\mathrm{h}}\nabla_{\mathrm{h}}^2\zeta_{\mathrm{g}}\\ \left(\frac{\partial}{\partial t} + \mathbf{v}_{\mathrm{g}}\cdot\nabla_{\mathrm{h}}\right)\frac{\partial\psi}{\partial p} &= -\frac{\overline{\sigma}_0\omega}{f_0} - \frac{R_0}{c_{p,0}f_0 p}\frac{d\!\!\!^{-}q}{dt}\end{aligned} \tag{24.1}$$

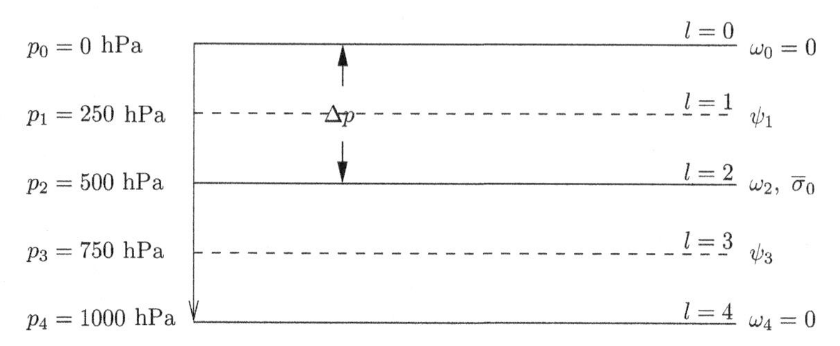

Fig. 24.1 The vertical grid structure of the two-level model.

In (24.1) we have also added the term $K_h \nabla_h^2 \zeta_g$ to simulate large-scale horizontal diffusion, where K_h is a large-scale diffusion constant. Furthermore, heat sources have been included on the right-hand side of (24.1). The vertical derivative of the arbitrary field function B will be approximated by

$$\left(\frac{\partial B}{\partial p}\right)_l = \frac{B_{l+1} - B_{l-1}}{\Delta p} \tag{24.2}$$

see Figure 24.1, so the vertical derivative of ω at levels $l = 1, 3$ is given by

$$\left(\frac{\partial \omega}{\partial p}\right)_1 = \frac{\omega_2 - \omega_0}{\Delta p} = \frac{\omega_2}{\Delta p}, \qquad \left(\frac{\partial \omega}{\partial p}\right)_3 = \frac{\omega_4 - \omega_2}{\Delta p} = -\frac{\omega_2}{\Delta p} \tag{24.3}$$

Thus, the basic system can also be written as

$$\begin{aligned}
&\text{(a)} \quad \left(\frac{\partial}{\partial t} + \mathbf{v}_{g,1} \cdot \nabla_h\right)(\zeta_{g,1} + f) = \frac{f_0 \omega_2}{\Delta p} + K_h \nabla_h^2 \zeta_{g,1} \\
&\text{(b)} \quad \left(\frac{\partial}{\partial t} + \mathbf{v}_{g,3} \cdot \nabla_h\right)(\zeta_{g,3} + f) = -\frac{f_0 \omega_2}{\Delta p} + K_h \nabla_h^2 \zeta_{g,3} \\
&\text{(c)} \quad \left(\frac{\partial}{\partial t} + \mathbf{v}_{g,2} \cdot \nabla_h\right)(\psi_3 - \psi_1) = -\frac{\overline{\sigma}_0 \omega_2 \, \Delta p}{f_0} - \frac{R_0 \, \Delta p}{c_{p,0} f_0 p_2}\left(\frac{dq}{dt}\right)_2
\end{aligned} \tag{24.4}$$

For the simple theory which is presented here this type of parameterization will be sufficient. The heating term dq/dt simulates all essential heat sources, including radiative effects.

In order to simplify the mathematical analysis of the problem we assume that we have a linear distribution of the geostrophic wind within the range $p_1 < p < p_3$. Hence, we obtain

$$\begin{aligned}
&\mathbf{v}_{g,2} = \mathbf{v}_{g,1} + \frac{\partial \mathbf{v}_g}{\partial p}\frac{\Delta p}{2}, \qquad \frac{\partial \mathbf{v}_g}{\partial p} = \frac{\mathbf{v}_{g,3} - \mathbf{v}_{g,1}}{\Delta p} = \frac{1}{\Delta p}\mathbf{k} \times \nabla_h(\psi_3 - \psi_1) \\
&\Longrightarrow \mathbf{v}_{g,2} = \mathbf{v}_{g,1} + \tfrac{1}{2}\mathbf{k} \times \nabla_h(\psi_3 - \psi_1) = \mathbf{v}_{g,3} - \tfrac{1}{2}\mathbf{k} \times \nabla_h(\psi_3 - \psi_1)
\end{aligned} \tag{24.5}$$

Utilizing this expression, the advection term of the first law of thermodynamics is given by

$$\mathbf{v}_{g,2}\cdot\nabla_h(\psi_3 - \psi_1) = \mathbf{v}_{g,1}\cdot\nabla_h(\psi_3 - \psi_1) = \mathbf{v}_{g,3}\cdot\nabla_h(\psi_3 - \psi_1) \tag{24.6}$$

since the scalar triple product must vanish whenever two identical vectors appear. This expression shows that the thermal wind is orthogonal to the gradient of the relative topography:

$$(\mathbf{v}_{g,3} - \mathbf{v}_{g,1})\cdot\nabla_h(\psi_3 - \psi_1) = 0 \tag{24.7}$$

Our next task is to eliminate the generalized vertical velocity ω_2 in the predicitve system (24.4). To accomplish this we multiply the heat equation by the term

$$\lambda^2 = f_0^2/[\overline{\sigma}_0(\Delta p)^2] \tag{24.8}$$

which is positive since $\overline{\sigma}_0$ is positive for large-scale motion. Furthermore, we use the fact that, due to (24.6) in the advection term of the heat equation, $\mathbf{v}_{g,2}$ may be replaced by $\mathbf{v}_{g,1}$ or $\mathbf{v}_{g,3}$. Therefore, (24.4c) can also be written as

$$\begin{aligned}\left(\frac{\partial}{\partial t} + \mathbf{v}_{g,1}\cdot\nabla_h\right)[\lambda^2(\psi_3 - \psi_1)] &= -\frac{f_0\omega_2}{\Delta p} - \frac{R_0\lambda^2}{c_{p,0}f_0}\left(\frac{đq}{dt}\right)_2\\ \left(\frac{\partial}{\partial t} + \mathbf{v}_{g,3}\cdot\nabla_h\right)[\lambda^2(\psi_3 - \psi_1)] &= -\frac{f_0\omega_2}{\Delta p} - \frac{R_0\lambda^2}{c_{p,0}f_0}\left(\frac{đq}{dt}\right)_2\end{aligned} \tag{24.9}$$

We observe that we have the same term on the right-hand sides of (24.4a) (24.4b) and (24.9). Therefore, ω_2 can be eliminated by adding and subtracting the heat equation (24.9) from (24.4a) and (24.4b), respectively. The result is

$$\begin{aligned}\left(\frac{\partial}{\partial t} + \mathbf{v}_{g,1}\cdot\nabla_h\right)[\zeta_{g,1} + f - \lambda^2(\psi_1 - \psi_3)] &= K_h\,\nabla_h^2\zeta_{g,1} - \frac{R_0\lambda^2}{c_{p,0}f_0}\left(\frac{đq}{dt}\right)_2\\ \left(\frac{\partial}{\partial t} + \mathbf{v}_{g,3}\cdot\nabla_h\right)[\zeta_{g,3} + f + \lambda^2(\psi_1 - \psi_3)] &= K_h\,\nabla_h^2\zeta_{g,3} + \frac{R_0\lambda^2}{c_{p,0}f_0}\left(\frac{đq}{dt}\right)_2\end{aligned} \tag{24.10}$$

In order to have a concise notation we introduce the following definitions:

$$\begin{aligned}q_1 &= \zeta_{g,1} - \lambda^2(\psi_1 - \psi_3) = \nabla_h^2\psi_1 - \lambda^2(\psi_1 - \psi_3)\\ q_3 &= \zeta_{g,3} + \lambda^2(\psi_1 - \psi_3) = \nabla_h^2\psi_3 + \lambda^2(\psi_1 - \psi_3)\end{aligned} \tag{24.11}$$

into (24.10). On momentarily setting $K_{\rm h}$ and $d\!\!\!{}^{-}q/dt$ equal to zero, the basic prediction equations are given by

$$\boxed{\left(\frac{\partial}{\partial t}+\mathbf{v}_{\rm g,1}\cdot\nabla_{\rm h}\right)(q_1+f)=0,\qquad \left(\frac{\partial}{\partial t}+\mathbf{v}_{\rm g,3}\cdot\nabla_{\rm h}\right)(q_3+f)=0} \tag{24.12}$$

This form is particularly suitable for demonstrating the principle of the numerical integration procedure.

24.2.1 The principle of the numerical integration procedure

Step 1
Prescribe at the initial time t_0 the two independent variables $\psi_1(x, y, t_0 = 0)$ and $\psi_3(x, y, t_0 = 0)$. Calculate $\mathbf{v}_{\rm g,1} = \mathbf{k} \times \nabla_{\rm h}\psi_1$, $\mathbf{v}_{\rm g,3} = \mathbf{k} \times \nabla_{\rm h}\psi_3$, and q_1, q_3.
Step 2
Compute the tendencies $\partial q_1/\partial t$, $\partial q_3/\partial t$ and find q_1, q_3 at $t_1 = t_0 + \Delta t$ by solving (24.12) according to

$$\begin{aligned} q_1(t_1) &= q_1(t_0) - \Delta t\,[\mathbf{v}_{\rm g,1}\cdot\nabla_{\rm h}(q_1+f)]_{t_0}\\ q_3(t_1) &= q_3(t_0) - \Delta t\,[\mathbf{v}_{\rm g,3}\cdot\nabla_{\rm h}(q_3+f)]_{t_0} \end{aligned} \tag{24.13}$$

Step 3
From q_1, q_3 at time t_1 calculate ψ_1, ψ_3 in order to iterate. We proceed by defining the quantities

$$\psi_+ = \frac{\psi_1+\psi_3}{2},\qquad \psi_- = \frac{\psi_1-\psi_3}{2},\qquad q_+ = \frac{q_1+q_3}{2},\qquad q_- = \frac{q_1-q_3}{2} \tag{24.14}$$

Addition and subtraction of the two equations (24.11) yields

$$q_+ = \tfrac{1}{2}\,\nabla_{\rm h}^2(\psi_1+\psi_3),\qquad q_- = \tfrac{1}{2}\,\nabla_{\rm h}^2(\psi_1-\psi_3) - \lambda^2(\psi_1-\psi_3) \tag{24.15}$$

When they are stated in the form

$$\nabla_{\rm h}^2\psi_+ = q_+,\qquad \nabla_{\rm h}^2\psi_- - 2\lambda^2\psi_- = q_- \tag{24.16}$$

we recognize that these two equations are decoupled. They represent two boundary-value problems of the Poisson and Helmholtz type, which can be solved by well-known numerical methods if suitable boundary conditions have been provided.
Step 4
From ψ_+ and ψ_- we find $\psi_1 = \psi_+ + \psi_-$ and $\psi_3 = \psi_+ - \psi_-$. In order to iterate we again start with step 1, where now t_0 is replaced by t_1.

We would like to remark that the complete decoupling stated in (24.16) is possible only for the two-level model, where by a three-dimensional problem could be reduced to the solution of two-dimensional partial differential equations. Each of these contains only one dependent variable. In multilevel models coupled systems must be solved simultaneously.

The model expressed by the prediction equations (24.12) is a typical short-range-forecast model. In order to extend the time interval of the forecast, we must also include heat sources and large-scale diffusion.

24.3 The Phillips quasi-geostrophic two-level circulation model

We employ the basic predictive equations (24.9) but now we include the large-scale diffusion and heating terms by taking $K_{\mathrm{h}} \neq 0$ and $(\text{đ}q/dt)_2 \neq 0$. We consider channel flow with solid walls at the southerly and northerly boundaries of the model as shown in Figure 24.2. For f we take the β approximation $f = f_0 + \beta y$; the large-scale exchange coefficient K_{h} is assumed to be a constant. The heating term $(\text{đ}q/dt)_2$ represents the sum of radiative heating $(\text{đ}q/dt)_{2,\mathrm{rad}}$ and of heat sources resulting from turbulent heat fluxes $(\text{đ}q/dt)_{2,\mathrm{tur}}$. To represent the radiative effects we assume that we have a heat source at $y = -d/2$ and a sink at $y = d/2$ with a linear variation between the walls:

$$\left(\frac{\text{đ}q}{dt}\right)_{2,\mathrm{rad}} = -\frac{2y}{d} Q_{\mathrm{rad}}, \qquad Q_{\mathrm{rad}} > 0 \tag{24.17}$$

The turbulent heat flux will be approximated with the help of (11.85):

$$\mathbf{J}_{\mathrm{t}}^{\theta} = -\overline{\rho}\widehat{c}_p K_{\mathrm{h}} \nabla_{\mathrm{h}} \widetilde{\theta} \tag{24.18}$$

Since the heat equation applies to the level p_2, we may replace the potential temperature $\widetilde{\theta}$ by the actual temperature T_2 at this level and absorb the constant

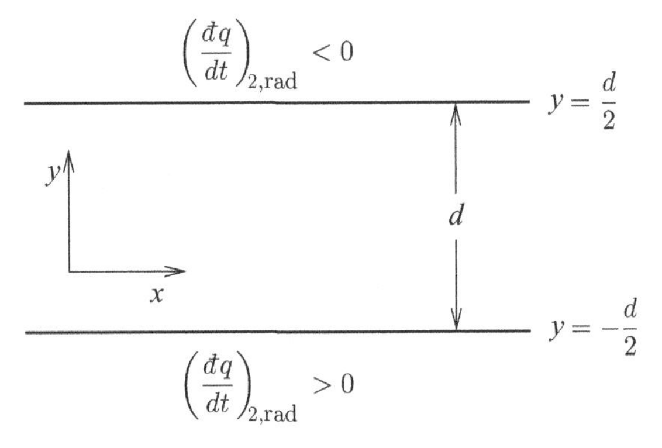

Fig. 24.2 The geometry of the channel flow.

$(p_0/p_2)^{R_0/c_{p,0}}$ into the large-scale turbulent exchange coefficient $K_{\rm h}$. However, it is still open to debate whether cyclonic and anticyclonic activities may be viewed as atmospheric macroturbulence so that large- and small-scale turbulent fluxes may be treated analogously. The turbulent heat flux is approximated as

$$\begin{aligned}
&\left(\frac{\not{d}q}{dt}\right)_{2,\rm tur} = -\frac{1}{\overline{\rho}}\,\nabla_{\rm h}\cdot\mathbf{J}_{\rm t}^{\theta} = \widehat{c}_p K_{\rm h}\,\nabla_{\rm h}^2 T_2 \\
&\text{with}\quad T_2 = -\frac{p_2}{R_0}\left(\frac{\partial\phi}{\partial p}\right)_2 = -\frac{p_2 f_0(\psi_3-\psi_1)}{R_0\,\Delta p} = \frac{2 f_0\psi_-}{R_0}
\end{aligned} \tag{24.19}$$

Substitution of (24.17) and (24.19) together with the definitions (24.11) into (24.10) yields the final form of the prediction equations:

$$\begin{aligned}
&\left(\frac{\partial}{\partial t} + \mathbf{v}_{\rm g,1}\cdot\nabla_{\rm h}\right)(q_1 + \beta y) = \frac{2\lambda^2 R_0 y}{c_{p,0} f_0 d}\,Q_{\rm rad} + K_{\rm h}\,\nabla_{\rm h}^2 q_1 \\
&\left(\frac{\partial}{\partial t} + \mathbf{v}_{\rm g,3}\cdot\nabla_{\rm h}\right)(q_3 + \beta y) = -\frac{2\lambda^2 R_0 y}{c_{p,0} f_0 d}\,Q_{\rm rad} + K_{\rm h}\,\nabla_{\rm h}^2 q_3
\end{aligned} \tag{24.20}$$

These equations may be integrated analogously to the procedure described above by including the inhomogeneous terms which were left out of equations (24.12) for simplicity.

Phillips (1956) used the system (24.20) to carry out the first successful numerical experiment to simulate the general circulation. He added another term to the second equation of (24.20). For details see the original paper.

24.4 Baroclinic instability

As stated in the introduction to this chapter, the solutions of baroclinic models may become physically unstable when the baroclinicity, i.e. the vertical wind shear, is sufficiently large and the static stability is sufficiently small. This instability behavior is known as baroclinic instability and is common to all baroclinic models. Baroclinic instability is responsible for the development of atmospheric cyclones and anticyclones.

The analysis proceeds by linearizing the fundamental equations (24.12). The stationary basic state and the disturbances are designated with an overbar and primes, respectively. Application of the Bjerkness linearization rule immediately results in the following system:

$$\begin{aligned}
&\left(\frac{\partial}{\partial t} + \overline{\mathbf{v}}_{\rm g,1}\cdot\nabla_{\rm h}\right)q_1' + \mathbf{v}_{\rm g,1}'\cdot\nabla_{\rm h}(\overline{q}_1 + f) = 0 \\
&\left(\frac{\partial}{\partial t} + \overline{\mathbf{v}}_{\rm g,3}\cdot\nabla_{\rm h}\right)q_3' + \mathbf{v}_{\rm g,3}'\cdot\nabla_{\rm h}(\overline{q}_3 + f) = 0
\end{aligned} \tag{24.21}$$

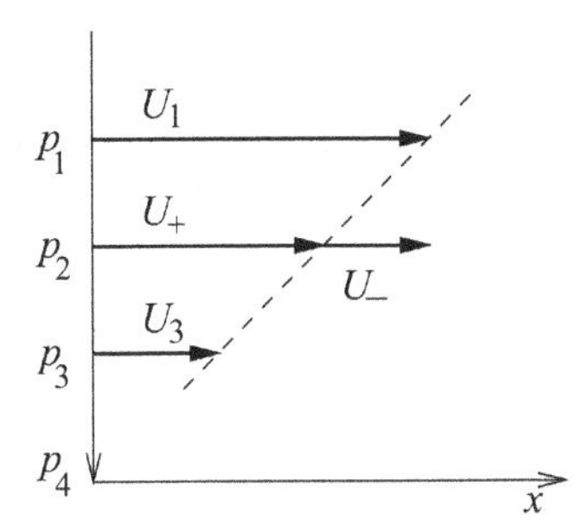

Fig. 24.3 The vertical profile of the horizontal wind.

Next we must describe the basic state:

$$
\begin{aligned}
&f = f_0 + \beta y, && \beta = \frac{\partial f}{\partial y} = \text{constant} \\
&\overline{u}_g = U = -\frac{\partial \overline{\psi}}{\partial y} = \text{constant}, && \overline{v}_g = 0 \\
&\overline{\psi} = \overline{\psi}(y), && \nabla_h^2 \overline{\psi} = 0 \\
&\overline{q}_1 = -\lambda^2(\overline{\psi}_1 - \overline{\psi}_3), && \overline{q}_3 = \lambda^2(\overline{\psi}_1 - \overline{\psi}_3) \\
&\frac{\partial \overline{q}_1}{\partial y} = \lambda^2(U_1 - U_3), && \frac{\partial \overline{q}_3}{\partial y} = -\lambda^2(U_1 - U_3)
\end{aligned}
\tag{24.22}
$$

The mean values U_+, ψ'_+, q'_+ and the differences U_-, ψ'_-, q'_- are defined as

$$
\begin{aligned}
U_+ &= \frac{U_1 + U_3}{2}, & \psi'_+ &= \frac{\psi'_1 + \psi'_3}{2}, & q'_+ &= \frac{q'_1 + q'_3}{2} \\
U_- &= \frac{U_1 - U_3}{2}, & \psi'_- &= \frac{\psi'_1 - \psi'_3}{2}, & q'_- &= \frac{q'_1 - q'_3}{2}
\end{aligned}
\tag{24.23}
$$

Figure 24.3 shows the vertical distribution of the horizontal wind and the corresponding values of U_+ and U_-. To facilitate the analysis we introduce the Q operators by putting

$$
\begin{aligned}
&Q_1 = \frac{\partial}{\partial t} + U_1 \frac{\partial}{\partial x}, \qquad Q_3 = \frac{\partial}{\partial t} + U_3 \frac{\partial}{\partial x}, \qquad Q_+ = \frac{\partial}{\partial t} + U_+ \frac{\partial}{\partial x} \implies \\
&\qquad Q_1 = \frac{\partial}{\partial t} + U_1 \frac{\partial}{\partial x} = \frac{\partial}{\partial t} + U_+ \frac{\partial}{\partial x} + U_- \frac{\partial}{\partial x} = Q_+ + U_- \frac{\partial}{\partial x} \\
&\qquad Q_3 = \frac{\partial}{\partial t} + U_3 \frac{\partial}{\partial x} = \frac{\partial}{\partial t} + U_+ \frac{\partial}{\partial x} - U_- \frac{\partial}{\partial x} = Q_+ - U_- \frac{\partial}{\partial x}
\end{aligned}
\tag{24.24}
$$

By employing these definitions, the linearized fundamental equations (24.21) can be written as

$$
\begin{aligned}
&\text{(a)} \qquad Q_1 q_1' + \frac{\partial \psi_1'}{\partial x}(2\lambda^2 U_- + \beta) = Q_+ q_1' + U_- \frac{\partial q_1'}{\partial x} + \frac{\partial \psi_1'}{\partial x}(2\lambda^2 U_- + \beta) = 0 \\
&\text{(b)} \quad Q_3 q_3' + \frac{\partial \psi_3'}{\partial x}(-2\lambda^2 U_- + \beta) = Q_+ q_3' - U_- \frac{\partial q_3'}{\partial x} + \frac{\partial \psi_3'}{\partial x}(-2\lambda^2 U_- + \beta) = 0
\end{aligned}
\tag{24.25}
$$

In order to derive a suitable form for the solution of the predictive equations we add and subtract (24.25a) and (24.25b) and find

$$
\begin{aligned}
Q_+ q_+' + U_- \frac{\partial q_-'}{\partial x} + 2\lambda^2 U_- \frac{\partial \psi_-'}{\partial x} + \beta \frac{\partial \psi_+'}{\partial x} = 0 \\
Q_+ q_-' + U_- \frac{\partial q_+'}{\partial x} + 2\lambda^2 U_- \frac{\partial \psi_+'}{\partial x} + \beta \frac{\partial \psi_-'}{\partial x} = 0
\end{aligned}
\tag{24.26}
$$

Next we eliminate q_+' and q_-' by introducing the above definitions. Observing (24.16), we obtain the expressions

$$
\begin{aligned}
q_+' &= \frac{q_1' + q_3'}{2} = \frac{1}{2} \nabla_{\rm h}^2 (\psi_1' + \psi_3') = \nabla_{\rm h}^2 \psi_+' \\
q_-' &= \frac{q_1' - q_3'}{2} = \frac{1}{2} \nabla_{\rm h}^2 (\psi_1' - \psi_3') - \lambda^2 (\psi_1' - \psi_3') = \nabla_{\rm h}^2 \psi_-' - 2\lambda^2 \psi_-'
\end{aligned}
\tag{24.27}
$$

to be substituted into equation (24.26). This gives the final forms of the prognostic equations:

$$
\boxed{
\begin{aligned}
\left(Q_+ \nabla_{\rm h}^2 + \beta \frac{\partial}{\partial x} \right) \psi_+' + U_- \frac{\partial}{\partial x} (\nabla_{\rm h}^2 \psi_-') = 0 \\
\left(Q_+ (\nabla_{\rm h}^2 - 2\lambda^2) + \beta \frac{\partial}{\partial x} \right) \psi_-' + U_- \frac{\partial}{\partial x} (\nabla_{\rm h}^2 + 2\lambda^2) \psi_+' = 0
\end{aligned}
}
\tag{24.28}
$$

which are linear in ψ_+' and ψ_-'. In order to solve this system we use the trial solutions

$$
\begin{pmatrix} \psi_+' \\ \psi_-' \end{pmatrix} = \begin{pmatrix} A_+ \\ A_- \end{pmatrix} \exp[ik_x(x - ct)]
\tag{24.29}
$$

where we assume that the perturbations depend on x and t only. The quantities A_+ and A_- are constant amplitudes, $k_x = 2\pi / L_x$ is the wavenumber in the x-direction corresponding to the wavelength L_x, and c is the phase velocity of the waves. From (24.29) we easily obtain the operators

$$
\begin{aligned}
&\frac{\partial}{\partial t} = -ik_x c, \qquad \frac{\partial}{\partial x} = ik_x, \qquad \nabla_{\rm h}^2 = -k_x^2 \implies \\
&Q_+ = \frac{\partial}{\partial t} + U_+ \frac{\partial}{\partial x} = -ik_x (c - U_+)
\end{aligned}
\tag{24.30}
$$

Using (24.30) in (24.28) yields the homogeneous system

$$\begin{aligned} &\text{(a)} \qquad \left[(c - U_+)k_x^2 + \beta\right]A_+ - U_- k_x^2 A_- = 0 \\ &\text{(b)} \quad \left(2\lambda^2 - k_x^2\right)U_- A_+ + \left[(c - U_+)\left(2\lambda^2 + k_x^2\right) + \beta\right]A_- = 0 \end{aligned} \tag{24.31}$$

where we have canceled out the exponential term representing the form of the wave. To find nontrivial solutions of this homogeneous system, i.e. $A_+ \neq 0$ and $A_- \neq 0$, the determinant of the coefficient matrix must vanish:

$$\begin{vmatrix} (c - U_+)k_x^2 + \beta & -U_- k_x^2 \\ \left(2\lambda^2 - k_x^2\right)U_- & (c - U_+)\left(2\lambda^2 + k_x^2\right) + \beta \end{vmatrix} = 0 \tag{24.32}$$

The expansion of this determinant results in the frequency equation

$$(c - U_+)^2 + 2(c - U_+)\frac{\beta\left(\lambda^2 + k_x^2\right)}{k_x^2\left(2\lambda^2 + k_x^2\right)} + \frac{\beta^2 + U_-^2 k_x^2\left(2\lambda^2 - k_x^2\right)}{k_x^2\left(2\lambda^2 + k_x^2\right)} = 0 \tag{24.33}$$

for the determination of the phase velocity c. The solution of this quadratic equation is

$$\boxed{\begin{aligned} c_{1,2} &= U_+ - \frac{\beta\left(\lambda^2 + k_x^2\right)}{k_x^2\left(2\lambda^2 + k_x^2\right)} \pm \sqrt{\delta} \\ \text{with} \quad \delta &= \frac{\beta^2\lambda^4}{k_x^4\left(2\lambda^2 + k_x^2\right)^2} - U_-^2 \frac{2\lambda^2 - k_x^2}{2\lambda^2 + k_x^2} \end{aligned}} \tag{24.34}$$

Inspection of (24.34) shows that we have either two real roots ($\delta > 0$) or two complex conjugated roots ($\delta < 0$), resulting in baroclinic stability or instability, respectively.

24.4.1 The instability condition

For a fixed wavenumber k_x = constant the two roots $c_{1,2}$ can now be used to construct the solution to (24.28) as

$$\begin{pmatrix} \psi'_+ \\ \psi'_- \end{pmatrix} = \begin{pmatrix} A_{+,1} \\ A_{-,1} \end{pmatrix} \exp[ik_x(x - c_1 t)] + \begin{pmatrix} A_{+,2} \\ A_{-,2} \end{pmatrix} \exp[ik_x(x - c_2 t)] \tag{24.35}$$

We note that the amplitudes A_+ and A_- are not independent. The relation between A_+ and A_- can be found from (24.31). We select (24.31a) and find the ratio

$$\frac{A_{-,j}}{A_{+,j}} = \frac{k_x^2(c_j - U_+) + \beta}{k_x^2 U_-}, \qquad j = 1, 2 \tag{24.36}$$

We could have also chosen (24.31b) to form the ratio, but the ratio is more complicated. From matrix theory it is known that every eigenvalue has an infinite number of eigenvectors. If **X** is an eigenvector then any scalar multiple of **X** is also an eigenvector with the same eigenvalue. Therefore, we may write for the components of the two eigenvectors in (24.35) the product

$$A_{+,j} = B_j \psi'_{+,j}, \qquad A_{-,j} = B_j \psi'_{-,j}, \qquad j = 1, 2 \tag{24.37}$$

where B_1 and B_2 are two arbitrary integration constants that may be determined from general initial conditions. Thus, we may write the solution in the form

$$\boxed{\begin{pmatrix} \psi'_+ \\ \psi'_- \end{pmatrix} = B_1 \begin{pmatrix} \psi'_{+,1} \\ \psi'_{-,1} \end{pmatrix} \exp[ik_x(x - c_1 t)] + B_2 \begin{pmatrix} \psi'_{+,2} \\ \psi'_{-,2} \end{pmatrix} \exp[ik_x(x - c_2 t)]} \tag{24.38}$$

By assigning the value 1 to one of the components, from (24.36) we obtain for the second component

$$\psi'_{+,j} = 1, \qquad \psi'_{-,j} = \frac{k_x^2(c_j - U_+) + \beta}{k_x^2 U_-}, \qquad j = 1, 2 \tag{24.39}$$

A particular example of the determination of an eigenvector was given in Section 4.2.2.

Suppose that the two roots $c_{1,2}$ are real. In this case the solution (24.38) is bounded and the two partial waves move along the x-direction without change of amplitude, i.e. without change of shape. If the roots are complex conjugate, one of the two partial waves approaches infinity with increasing time so that the solution is physically unstable.

Inspection of equation (24.34) shows that the sign of the discriminant determines whether the solution is physically stable or unstable, or even neutral. If $\delta > 0$ we have stability since $c_{1,2}$ are real quantities. If $\delta < 0$ we have instability since $c_{1,2}$ are complex conjugate; $\delta = 0$ denotes the neutral case.

24.4.2 Stable and unstable regions

Of special significance is the transition from the stable to the unstable region, which is characterized by the condition $\delta = 0$. By treating β as a constant, we may view the equation

$$\delta(U_-, \lambda^2, k_x) = \delta(U_-, \overline{\sigma}_0, L_x) = 0 \tag{24.40}$$

as a surface in space with the three coordinates U_-, $\overline{\sigma}_0 = f_0^2/[(\Delta p)^2\lambda^2]$, and $L_x = 2\pi/k_x$ separating the stable from the unstable region. This surface is schematically

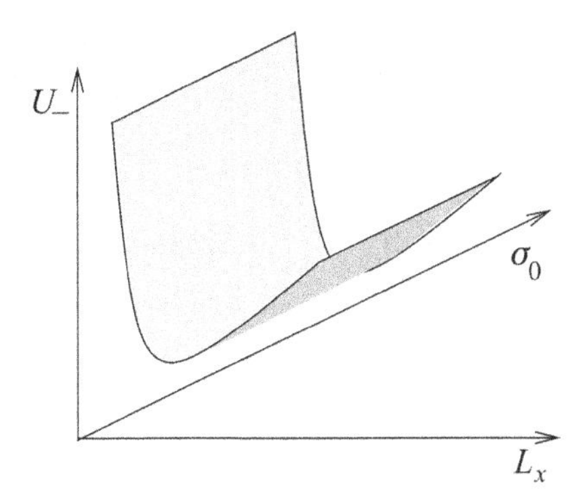

Fig. 24.4 The neutral surface separating the stable from the unstable region.

shown in Figure 24.4. Thus, a particular choice of the coordinates $(U_-, \overline{\sigma}_0, L_x)$ determines whether the corresponding solution (24.38) is stable or unstable. The separating surface is known as the *neutral surface*.

A convenient mathematical form of the neutral surface will be derived now. For $\delta = 0$ from (24.34) we find by obvious steps

$$k_x^4(4\lambda^4 - k_x^4) = \frac{\beta^2\lambda^4}{U_-^2} \implies \left(\frac{k_x^4}{2\lambda^4} - 1\right)^2 = 1 - \frac{\beta^2}{4U_-^2\lambda^4} \tag{24.41}$$

The latter equation motivates the introduction of the following (x, y)-coordinates:

$$x = \frac{k_x^4}{2\lambda^4} = \frac{k_x^4(\Delta p)^4}{2f_0^4}\overline{\sigma}_0^2, \qquad y = \frac{2U_-\lambda^2}{\beta} = \frac{2f_0^2}{(\Delta p)^2\beta}\frac{U_-}{\overline{\sigma}_0} \tag{24.42}$$

Therefore, the coordinate x gives the quantity k_x^4 in units of $2\lambda^4$ while y expresses the quantity U_- in units of $\beta/(2\lambda^2)$. From (24.41) and (24.42) we obtain the equation of the neutral surface:

$$(x-1)^2 = 1 - \frac{1}{y^2} \implies y = \pm\sqrt{\frac{1}{1-(x-1)^2}} \tag{24.43}$$

whose graph is shown in Figure 24.5. Reflecting this curve on the x-axis gives the representation of the neutral surface for negative values of y. In the following we will discuss positive y values only. Since according to (24.42) $x \geq 0$, we recognize that there is no need to consider negative x values.

From (24.43) we recognize that no real values of y exist for $x > 2$ and that the ordinate y approaches infinity for $x = 0$ and $x = 2$. Furthermore, for $x = 1$ we find $y = 1$, which is the minimum value of the curve, as may be verified by differentiation.

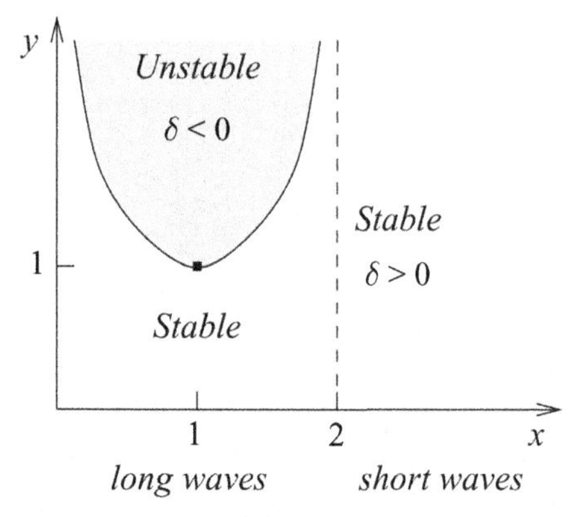

Fig. 24.5 A schematic representation of the stable ($\delta > 0$) and unstable ($\delta < 0$) regions.

Now we need to ask on which side of the neutral surface the unstable region is located. We consider the straight line $x = 1$ and investigate the behavior of δ along this line for increasing y. For $x = 1$ we find $k_x^4 = 2\lambda^4$. Next we substitute this identity into the instability condition (24.34) and then multiply the resulting expression by the positive quantity

$$M^2 = \frac{2(2\lambda^2 + k_x^2)^2}{\beta^2} > 0 \tag{24.44}$$

yielding

$$M^2\delta = 1 - y^2 \tag{24.45}$$

For $y < 1$ the discriminant $\delta > 0$; for $y > 1$ the discriminant $\delta < 0$. Thus, the stable and unstable regions are located as indicated in Figure 24.5. The boundary point ($x = 1$, $y = 1$) corresponds to a point on the neutral curve.

24.4.3 Unconditional stability regions, the dominating wavelength

Inspection of Figure 24.5 shows that the solution is stable in the region $y < 1$ for all x values, i.e. independently of the wavelength L_x. This situation relates the vertical shear of the horizontal wind field U_- to the atmospheric stability $\overline{\sigma}_0$ as

$$U_- < \frac{(\Delta p)^2\beta}{2f_0^2}\overline{\sigma}_0 \tag{24.46}$$

The region $x > 2$ is stable for all values of y, i.e. stability is guaranteed to be independent of U_-, which is a measure of the baroclinicity of the atmosphere. From (24.42) we obtain the corresponding stability condition

$$L_x < \frac{\pi\sqrt{2}\,\Delta p}{f_0}\sqrt{\overline{\sigma}_0} \tag{24.47}$$

showing that all waves with wavelengths shorter than the given value are stable, independently of the baroclinicity U_-.

From (24.42) we see that, for fixed values of β and $\overline{\sigma}_0$, instability will be generated if the baroclinicity U_- increases to large enough values. Instability will also be generated if $\overline{\sigma}_0$ becomes small enough for fixed values of β and U_-. The wavelength corresponding to $x = 1$ is of particular interest since, for increasing y, the unstable region is reached faster than at any other point on the x-axis. This condition yields

$$L_x = \frac{2^{3/4}\pi\,\Delta p}{f_0}\sqrt{\overline{\sigma}_0} \tag{24.48}$$

as follows from (24.42). The wavelength defined by this equation is known as the *dominating wavelength* and is proportional to the static stability $\overline{\sigma}_0$. It follows that, with increasing static stability, waves of increasing wavelength will dominate.

24.4.4 Neutral surfaces in the (U_-, L_x)-plane

The representation of the neutral curve in the (U_-, L_x)-plane is particularly easy to visualize. Now $\overline{\sigma}_0$ and β serve as parameters with which to label curves belonging to the same family. Suppose that we set $\beta =$ constant. Using the definition (24.42), the equation of the neutral curve can now be written as

$$U_- = \pm\frac{\beta}{2\lambda^2}\sqrt{\frac{1}{1-\left[\left(\dfrac{k_x^4}{2\lambda^4}\right)-1\right]^2}} \tag{24.49}$$

In the limiting case that $\overline{\sigma}_0 = 0$ or $\lambda \to \infty$, we find after a few easy steps that U_- is the parabola

$$U_- = \frac{\beta L_x^2}{8\pi^2} \tag{24.50}$$

Thus, for $\overline{\sigma}_0 = 0$ we find the parabola depicted in Figure 24.6. The area above the curve represents the unstable region. For values $\overline{\sigma}_0 > 0$ we obtain from (24.49) the family of curves shown in Figure 24.6. Hence, with increasing static stability the area of the unstable region decreases in size.

Reference to equation (23.20b) shows that, in the filtered model, for $\overline{\sigma}_0 = 0$ changes in temperature are caused by horizontal advection only. Models in which the static stability is set equal to zero are called *advection models*. In these models the short waves are always unstable, as follows from Figure 24.6. Using the terminology of optics, very short waves of the solar spectrum are located in the ultra-violet spectral region. Therefore, the unconditional instability for short waves in the advection models is called the *ultra-violet catastrophy*.

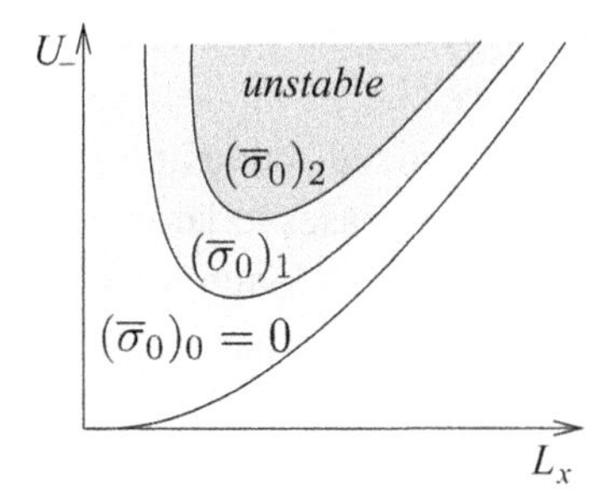

Fig. 24.6 A schematic representation of the neutral curves for β = constant and varying static stability $\overline{\sigma}_0$ with $(\overline{\sigma}_0)_2 > (\overline{\sigma}_0)_1$.

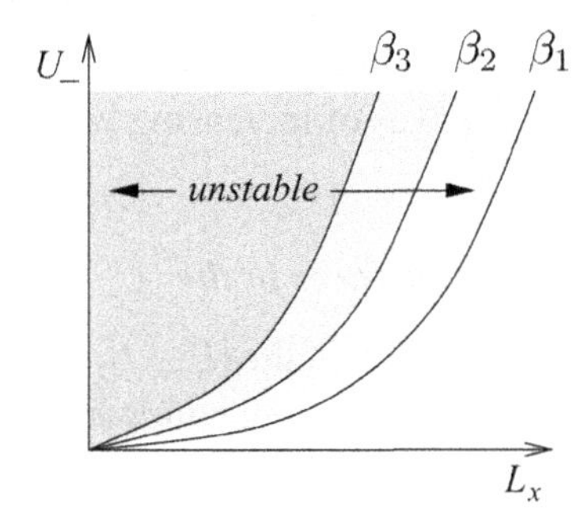

Fig. 24.7 A schematic representation of the effect of β on the region of instability for the limiting case $\overline{\sigma}_0 = 0$ with $\beta_3 > \beta_2 > \beta_1$.

Suppose that we vary the Rossby parameter β. From (24.50) we then obtain a family of parabolas as shown in Figure 24.7. The unstable region for each β is on the left-hand side of the corresponding parabola. This is best recognized by considering the stability condition (24.34) whereby increasing values of β prevent δ from becoming negative.

Let us now consider the special case that $\beta = 0$. From (24.34) we find for $\delta = 0$

$$2\lambda^2 = k_x^2 \implies L_x = \frac{\sqrt{2}\pi\ \Delta p}{f_0}\sqrt{\overline{\sigma}_0} \tag{24.51}$$

which represents a family of vertical lines. It stands to reason that, with increasing static stability, the stable region must increase in size, as shown in Figure 24.8.

From the previous figures we may conclude that short and long waves are stable in the quasi-geostrophic theory provided that $\beta \neq 0$ and $\overline{\sigma}_0 \neq 0$. Moreover, we observe that increasing values of β and $\overline{\sigma}_0$ have a stabilizing effect. By combining the results of this section we find Figure 24.9.

We wish to point out that the results derived from the two-level model are not sufficiently representative to permit us to draw general conclusions about the

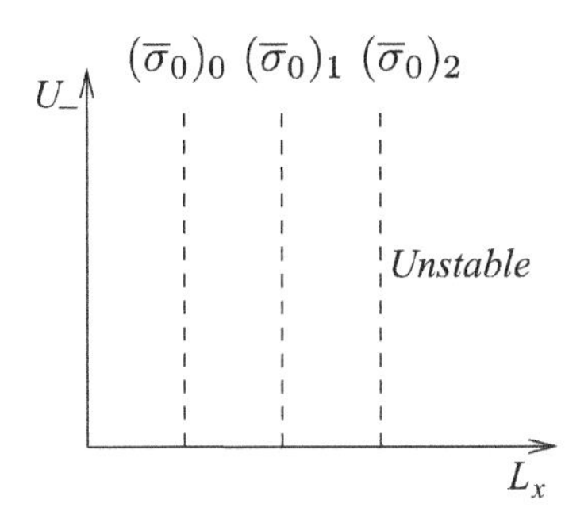

Fig. 24.8 A schematic display of neutral curves for $\beta = 0$ and variable static stability $\overline{\sigma}_0$ with $(\overline{\sigma}_0)_2 > (\overline{\sigma}_0)_1 > (\overline{\sigma}_0)_0$.

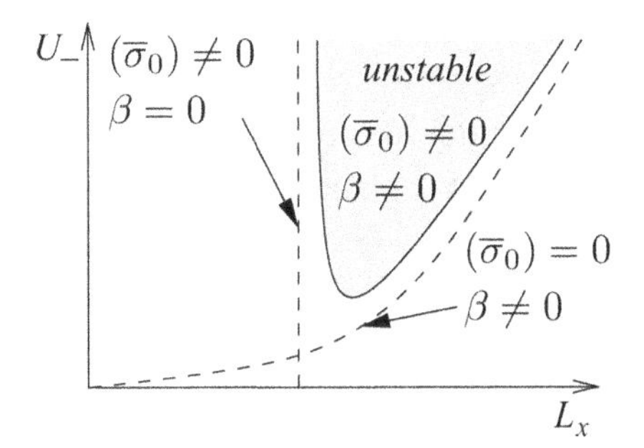

Fig. 24.9 The region of instability in the (L_x, U_-)-plane for the general case $\beta \neq 0$ and $\overline{\sigma}_0 \neq 0$.

baroclinic instability of the real atmosphere. Nevertheless, the two-level model provides at least qualitatively the major characteristics of baroclinic instability and yields results consistent with the more general model proposed by Charney (1947).

24.5 Problems

24.1: Show in detail the steps between equation (24.25) and (24.28).

24.2: Use calculus to verify the minimum of Figure 24.5.

25

An excursion concerning numerical procedures

In order to solve the atmospheric equations it is necessary to apply numerical methods. It is not our intention to present a detailed discussion on numerical methods since this would require a separate textbook of many chapters. An early book on numerical methods applicable to atmospheric dynamics was written by Thompson (1961). It is quite suitable as a first introduction to numerical weather analysis. A more modern and extensive account of numerical methods is that by Haltiner and Williams (1980). Both books may be consulted regarding the following discussion. There also exist many papers and reports on the subject, which are too numerous to be quoted here. In this chapter we wish to point out some problems that may arise when one is using finite-difference methods. In fact, many numerical problems become apparent even in treating the one-dimensional advection equation.

25.1 Numerical stability of the one-dimensional advection equation

25.1.1 Introduction

The typical appearance of numerical instability in analytic form can be easily recognized from the discretized form of the one-dimensional advection equation:

$$\frac{\partial \psi}{\partial t} + U \frac{\partial \psi}{\partial x} = 0 \quad \text{with} \quad U = \text{constant} \tag{25.1}$$

This is a linear first-order partial differential equation whose solution is given by $f(x - Ut)$, where f is an arbitrary function. A special solution is the harmonic wave

$$\psi = A \exp[ik(x - Ut)] \tag{25.2}$$

where $k = 2\pi/L$ is the wavenumber and L the wavelength. Equation (25.1) shall now be discretized in various ways in order to compare the solutions of the

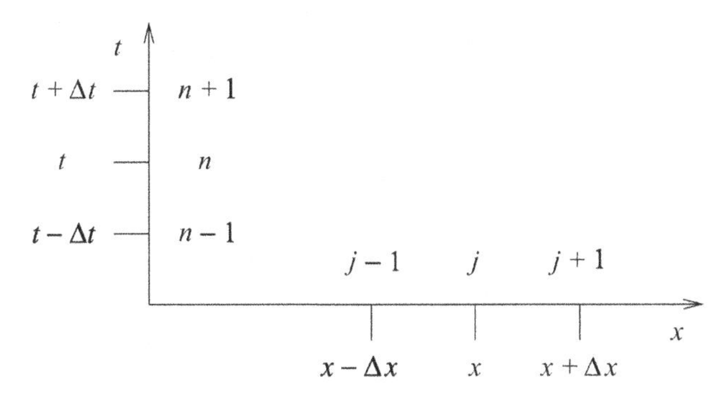

Fig. 25.1 The grid structure in space and time for one spatial dimension. The spatial distance x is counted in units of $j\,\Delta x$ while time t is advanced in units of $n\,\Delta t$.

difference equations with the solution (25.2). This procedure will permit us to study the stability behavior in terms of time for various numerical schemes.

The finite-difference equations can be constructed in various ways. We will obtain these by expanding the function $f(x - Ut)$ in a Taylor series of only one spatial dimension as shown in Figure 25.1. Let us first obtain finite-difference expressions with which to approximate the partial derivative $\partial\psi/\partial x$.

The Taylor-series expansions about the point j in forward and backward directions are given by

$$\begin{aligned}
&\text{(a)}\quad \psi_{j+1}^n = \psi_j^n + \left(\frac{\partial\psi}{\partial x}\right)_j \Delta x + \frac{1}{2}\left(\frac{\partial^2\psi}{\partial x^2}\right)_j (\Delta x)^2 + \frac{1}{6}\left(\frac{\partial^3\psi}{\partial x^3}\right)_j (\Delta x)^3 + \cdots \\
&\text{(b)}\quad \psi_{j-1}^n = \psi_j^n - \left(\frac{\partial\psi}{\partial x}\right)_j \Delta x + \frac{1}{2}\left(\frac{\partial^2\psi}{\partial x^2}\right)_j (\Delta x)^2 - \frac{1}{6}\left(\frac{\partial^3\psi}{\partial x^3}\right)_j (\Delta x)^3 + \cdots
\end{aligned} \tag{25.3}$$

From (25.3a) we find, for fixed time t, indicated by the superscript n, the finite-difference equation in the forward direction:

$$\left(\frac{\partial\psi}{\partial x}\right)_j = \frac{\psi_{j+1}^n - \psi_j^n}{\Delta x} - \frac{1}{2}\left(\frac{\partial^2\psi}{\partial x^2}\right)_j \Delta x + \cdots \tag{25.4a}$$

The terms following the finite-difference terms are dominated by a second-order derivative, which is multiplied by Δx. By ignoring this term and all other terms involving higher derivatives we have introduced a first-order truncation error $O(\Delta x)$ since Δx appears as a first power. From (25.3b) we could have obtained a similar approximation in the backward direction, which involves the point $j - 1$ instead of $j + 1$. Since we do not use the backward approximation, we will not write it down. The first term on the right-hand side of (25.4a) is known as the *forward-in-space difference approximation* to $\partial\psi/\partial x$. On subtracting (25.3b) from (25.3a) the second-order derivatives will vanish. The remaining terms are dominated by a term involving the third-order derivative at point j multiplied by Δx^3. On dividing by Δx we obtain the so-called *central finite-difference approximation* to the first-order derivative:

$$\left(\frac{\partial \psi}{\partial x}\right)_j = \frac{\psi_{j+1}^n - \psi_{j-1}^n}{2\,\Delta x} - \frac{1}{6}\left(\frac{\partial^3 \psi}{\partial x^3}\right)_j (\Delta x)^2 + \cdots \tag{25.4b}$$

which is the first term on the right-hand side of (25.4b). By ignoring all terms involving higher-order derivatives we obtain a better approximation to the derivative $\partial\psi/\partial x$ since now the truncation error is of second order, $O\left((\Delta x)^2\right)$.

In order to obtain approximations to the time derivative we proceed in a similar manner. Then we obtain expressions analogous to (25.3), where now j is fixed and n is varied. We will now apply the various difference approximations.

25.1.2 The numerical phase speed

The advection equation (25.1) in discretized form using central differences in time and space now reads

$$\psi_j^{n+1} - \psi_j^{n-1} + U\frac{\Delta t}{\Delta x}\left(\psi_{j+1}^n - \psi_{j-1}^n\right) = 0 \tag{25.5}$$

The solution to this finite-difference equation is found by substituting a trial grid-point function of the type (25.2) of the analytic solution into (25.5):

$$\begin{aligned}
&\text{(a)} \quad \psi_j^n = A\exp[ik(j\,\Delta x - cn\,\Delta t)] \\
&\text{(b)} \quad \psi_j^{n+1} = A\exp\{ik[j\,\Delta x - c(n+1)]\,\Delta t\} = \psi_j^n \exp(-ikc\,\Delta t) \\
&\text{(c)} \quad \psi_j^{n-1} = \psi_j^n \exp(ikc\,\Delta t) \\
&\text{(d)} \quad \psi_{j+1}^n = \psi_j^n \exp(ik\,\Delta x) \\
&\text{(e)} \quad \psi_{j-1}^n = \psi_j^n \exp(-ik\,\Delta x)
\end{aligned} \tag{25.6}$$

Here U has been replaced by the phase speed $c = \omega/k$, where ω is the circular frequency. The trial solution is identical with (25.2) only if $c = U$. Substitution of (25.6) into (25.5), after canceling out the common factor ψ_j^n, gives

$$\exp(ikc\,\Delta t) - \exp(-ikc\,\Delta t) = \frac{U\,\Delta t}{\Delta x}[\exp(ik\,\Delta x) - \exp(-ik\,\Delta x)] \tag{25.7}$$

Dividing both sides of this equation by $2i$ ($i = \sqrt{-1}$) and using the Euler expansion gives the frequency equation

$$\sin(kc\,\Delta t) = \frac{U\,\Delta t}{\Delta x}\sin(k\,\Delta x) \tag{25.8}$$

The *numerical phase speed* follows immediately and is given by

$$c = \frac{1}{k\,\Delta t}\arcsin\left(\frac{U\,\Delta t}{\Delta x}\sin(k\,\Delta x)\right) \tag{25.9}$$

Using

$$\lim_{\Delta x \to 0}\frac{\sin(k\,\Delta x)}{\Delta x} = k, \qquad \lim_{\epsilon \to 0}\frac{\arcsin\epsilon}{\epsilon} = 1 \tag{25.10}$$

we find for $(\Delta x, \Delta t) \longrightarrow 0$ the consistent result

$$\lim_{\Delta x, \Delta t \to 0} c = U \tag{25.11}$$

25.1.3 Numerical instability

We now reconsider equation (25.9) and assume that

$$U\frac{\Delta t}{\Delta x} > 1 \tag{25.12}$$

At the same time we require that $\sin(k\,\Delta x)$ in the argument of (25.9) reaches the maximum possible value 1. Because $k\,\Delta x = \pi/2$ this corresponds to the wavelength $L = 4\,\Delta x$. In this case the argument of arcsin( ) is greater than 1 and (25.9) can be satisfied only by complex values of the numerical phase speed,

$$c = c_r \pm i c_i \tag{25.13}$$

In order to evaluate the real part c_r and the imaginary part $c_i \geq 0$ of the phase speed, assuming the validity of (25.12), we go back to the frequency equation (25.8). This equation now reads

$$\sin(kc\,\Delta t) = \sin(\alpha_r \pm i\alpha_i) = 1 + a, \qquad a > 0 \tag{25.14}$$

The complex number $\alpha_r \pm i\alpha_i$ arises because of (25.13). The quantity a on the right-hand side must be positive. Application of the trigonometric addition theorems gives

$$\begin{aligned} &\text{(a)} \quad \sin\alpha_r\cos(i\alpha_i) \pm \sin(i\alpha_i)\cos\alpha_r = 1 + a \\ &\text{(b)} \quad \sin\alpha_r\cosh\alpha_i \pm i\sinh\alpha_i\cos\alpha_r = 1 + a \end{aligned} \tag{25.15}$$

where we have introduced the hyperbolic functions in (25.15b). Comparison of real and imaginary parts results in

$$\begin{aligned} &\text{(a)} \quad \text{Real part:} \quad \sin\alpha_r\cosh\alpha_i = 1 + a \\ &\text{(b)} \quad \text{Imaginary part:} \quad \cos\alpha_r\sinh\alpha_i = 0 \end{aligned} \tag{25.16}$$

Since we require that $\alpha_i = kc_i \, \Delta t \neq 0$ it follows from (25.16b) that $\cos \alpha_r = 0$, so

$$\alpha_r = kc_r \, \Delta t = \pm \frac{\pi}{2}, \pm \frac{3}{2}\pi, \ldots \tag{25.17}$$

For further discussion we select the representative value

$$c_r = \frac{\pi}{2k \, \Delta t} \tag{25.18}$$

On the other hand, with $\alpha_r = \pi/2$, from (25.16a) it follows that

$$\cosh \alpha_i = \cosh(kc_i \, \Delta t) = 1 + a \implies c_i = \frac{\cosh^{-1}(1 + a)}{k \, \Delta t} > 0 \tag{25.19}$$

From (25.6a), (25.18), and (25.19), assuming the validity of (25.12), we obtain the general solution to (25.5):

$$\begin{aligned} \psi_j^n &= A_1 \exp[ik(j \, \Delta x - c_r n \, \Delta t - ic_i n \, \Delta t)] \\ &\quad + A_2 \exp[ik(j \, \Delta x - c_r n \, \Delta t + ic_i n \, \Delta t)] \end{aligned} \tag{25.20}$$

Owing to (25.18) this expression may be rewritten in such a way that

$$\psi_j^n = A_1 \exp(ikj \, \Delta x) \, (-i)^n \exp(kc_i n \, \Delta t) + A_2 \exp(ikj \, \Delta x) \, (-i)^n \exp(-kc_i n \, \Delta t) \tag{25.21}$$

since in this case $\exp(-ikc_r n \, \Delta t) = \exp(-in\pi/2) = (-i)^n$. Because $c_i > 0$ the first term with alternating sign approaches infinity for large n, thus indicating numerical instability. The second term with alternating sign in (25.21) approaches zero. We conclude that the numerical solution in the present case using central difference quotients in time and space is stable only if the condition (25.12) is changed to

$$U \frac{\Delta t}{\Delta x} \leq 1 \quad \text{or} \quad \Delta t \leq \frac{\Delta x}{U} \tag{25.22}$$

In this case the phase velocity is real and the solution remains stable. Equation (25.22) is the well-known *Courant–Friedrichs–Lewy (CFL) stability criterion* for the case of the linear difference equation which we have considered here.

Equation (25.22) can be usefully interpreted. Numerical stability occurs whenever the time step Δt is smaller than the time required for the wave moving with phase speed U to cover the grid distance Δx.

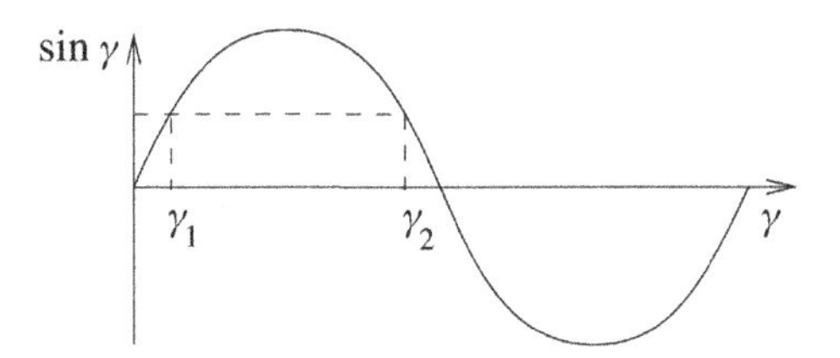

Fig. 25.2 Ambiguity of phase velocities.

25.1.4 The so-called weak instability

Even if the CFL criterion (25.22) is obeyed we still do not obtain an absolutely stable numerical behavior. The phase velocity c according to (25.9) now assumes real values, as desired. The arcsine function, however, admits two different real phase velocities $c_{1,2}$, in contrast to the analytic solution (25.2) of (25.1), which admits the phase velocity $c = U$ only. Owing to $\sin\gamma_1 = \sin\gamma_2$ with $\gamma_2 = \pi - \gamma_1$, see Figure 25.2, the frequency equation (25.8) is satisfied by

$$\begin{aligned} \gamma_1 &= c_1 k\,\Delta t = \arcsin\left(U\frac{\Delta t}{\Delta x}\sin(k\,\Delta x)\right) \\ \gamma_2 &= c_2 k\,\Delta t = \pi - c_1 k\,\Delta t \end{aligned} \tag{25.23}$$

With decreasing $\Delta x, \Delta t \longrightarrow 0$ these two phase velocities become

$$\lim_{\Delta t,\Delta x\to 0} c_1 = U, \qquad \lim_{\Delta t,\Delta x\to 0} c_2 = \infty \tag{25.24}$$

One part of the solution with $c_1 = U$ is physically real; the other part with $c_2 = \infty$ is a wave due to the numerical procedure and, therefore, is counted as a *weak instability*. The artificial numerical wave is caused by the difference equation (25.5), which requires for the forecast knowledge of ψ at two initial times $\psi^n_{j-1}, \psi^n_{j+1}, \psi^{n-1}_j$ whereas for the analytic solution of (25.1) only one initial condition is needed. The requirement of a second initial condition is responsible for the existence of the numerical wave with phase velocity c_2.

In order to discuss the behavior of the numerical wave we substitute c_1 and c_2 according to (25.23) into the complete numerical solution. According to the principle of superposition we find

$$\psi^n_j = A_1\exp[ik(j\,\Delta x - c_1 n\,\Delta t)] + A_2(-1)^n\exp[ik(j\,\Delta x + c_1 n\,\Delta t)] \tag{25.25}$$

since $\exp(-in\pi) = (-1)^n$. The second term on the right-hand side is the numerical wave whose amplitude changes sign with each time step. Moreover, by recalling that the analytic solution is given by $f(x - Ut)$, it is easily seen that the numerical wave moves in a direction opposite to the physical wave.

The amplitudes A_1 and A_2 will be determined from the initial conditions with the goal of eliminating the numerical wave. From (25.25) we find

$$\begin{aligned} \psi_j^{n=0} &= A_1 \exp(ikj\,\Delta x) + A_2 \exp(ikj\,\Delta x) \\ \psi_j^{n=1} &= A_1 \exp[ik(j\,\Delta x - c_1\,\Delta t)] - A_2 \exp[ik(j\,\Delta x + c_1\,\Delta t)] \end{aligned} \tag{25.26}$$

The first predicted value of ψ then refers to $n = 2$. In order to eliminate A_2 we multiply the first equation by $\exp(-ikc_1\,\Delta t)$ and then subtract the second equation. From this it follows that

$$A_2 = \frac{\psi_j^{n=0} \exp(-ikc_1\,\Delta t) - \psi_j^{n=1}}{\exp(ikj\,\Delta x)\,[\exp(ikc_1\,\Delta t) + \exp(-ikc_1\,\Delta t)]} \tag{25.27}$$

For A_2 to vanish we must set

$$\psi_j^{n=1} = \psi_j^{n=0} \exp(-ikc_1\,\Delta t) \tag{25.28}$$

In this particular simple case we were able to relate the initial conditions. In practical numerical weather prediction the differential equations to be solved numerically are much more complicated, so the phase speed c is unknown. In a later section we will show how to eliminate the numerical wave in a different manner.

25.2 Application of forward-in-time and central-in-space difference quotients

In contrast to the previous sections we will now replace the local time derivative by means of a forward-in-time difference quotient. Instead of (25.5) we now obtain

$$\psi_j^{n+1} - \psi_j^n + \frac{U\,\Delta t}{2\,\Delta x}\left(\psi_{j+1}^n - \psi_{j-1}^n\right) = 0 \tag{25.29}$$

In comparison with (25.5) the second initial value at time step $n - 1$ is not needed, resulting in only one phase velocity c. Now only one initial condition for $t = 0$ or $n = 0$ is required. Thus weak instability does not occur. We will now investigate whether the remaining wave is numerically stable. We substitute the trial solution (25.6a) into (25.29) and obtain after canceling out of ψ_j^n the expression

$$\exp(-ikc\,\Delta t) - 1 + \frac{U\,\Delta t}{2\,\Delta x}[\exp(ik\,\Delta x) - \exp(-ik\,\Delta x)] = 0 \tag{25.30}$$

Using $c = c_\mathrm{r} + ic_\mathrm{i}$ and the relation $\sin x = [\exp(ix) - \exp(-ix)]/(2i)$, we find

$$\exp(kc_\mathrm{i}\,\Delta t)\exp(-ikc_\mathrm{r}\,\Delta t) - 1 + \frac{iU\,\Delta t}{\Delta x}\sin(k\,\Delta x) = 0 \tag{25.31}$$

Application of Euler's formula permits us to separate the real and imaginary parts, i.e.

$$\begin{aligned} &\text{(a)} \qquad \text{Real part:} \quad \exp(kc_\mathrm{i}\,\Delta t)\cos(kc_\mathrm{r}\,\Delta t) = 1 \\ &\text{(b)} \qquad \text{Imaginary part:} \quad \exp(kc_\mathrm{i}\,\Delta t)\sin(kc_\mathrm{r}\,\Delta t) = \frac{U\,\Delta t}{\Delta x}\sin(k\,\Delta x) \end{aligned} \tag{25.32}$$

Squaring (25.32a) and (25.32b) and adding the results gives

$$\exp(2kc_\mathrm{i}\,\Delta t) = 1 + \left(\frac{U\,\Delta t}{\Delta x}\right)^2 \sin^2(k\,\Delta x) \tag{25.33}$$

Solving for c_i results in

$$c_\mathrm{i} = \frac{1}{2k\,\Delta t}\ln\left[1 + \left(\frac{U\,\Delta t}{\Delta x}\right)^2 \sin^2(k\,\Delta x)\right] > 0 \tag{25.34}$$

Dividing the imaginary part in (25.32) by the real part gives the real part of the phase speed:

$$c_\mathrm{r} = \frac{1}{k\,\Delta t}\arctan\left(\frac{U\,\Delta t}{\Delta x}\sin(k\,\Delta x)\right) \approx U \tag{25.35}$$

for sufficiently small Δt and Δx, see equation (25.11). By substituting the complex phase velocity (25.34) and (25.35) into the trial solution (25.6a) we find

$$\psi_j^n = A\exp[ik(j\,\Delta x - c_\mathrm{r} n\,\Delta t)]\exp(c_\mathrm{i} n\,\Delta t) \tag{25.36}$$

Inspection shows that the numerical solution to the difference equation (25.29) approaches infinity with increasing time since $c_\mathrm{i} > 0$. This means that the numerical solution is absolutely unstable.

In conclusion we will investigate the behavior of c_i as the time step approaches zero. From (25.34) it follows that

$$c_\mathrm{i} = \frac{1}{2k\,\Delta t}\ln[1 + B(\Delta t)^2] \implies \lim_{\Delta t \to 0} c_\mathrm{i} = 0 \tag{25.37}$$

with $B > 0$. This means that, for a very small first time step, the solution (25.36) to the difference equation (25.29) is sufficiently accurate.

25.3 A practical method for the elimination of the weak instability

The combination of the results of the previous sections leads to a practical computational method, which will be outlined now. For the first-order partial differential equation (25.1) a unique solution is guaranteed whenever the variable ψ is specified at $t = 0$. The corresponding difference equation (25.5), however, requires an arbitrary specification of the variable ψ not only at time $t = t^0 = 0$ but also at $t = t^1 = \Delta t$. The initial data for the two times t^0 and t^1 are not harmonized in any way by the difference equation. This freedom in the choice of the initial data resulted in the weak instability.

The application of the forward-in-time difference scheme (25.29), in agreement with the differential equation, requires specification of the variable only for the time $t = 0$. The weak instability does not occur in this case, but the numerical scheme is unstable. Thus we apply equation (25.29) only for the very first time step.

This suggests that we should use a combination of these two procedures, as will be explained now. The numerical calculations are started by applying the forward-in-time difference method for a fraction, say $\Delta t' = \frac{1}{8}\,\Delta t$, of the regular time step Δt. In this way we calculate with a high degree of stability, without the presence of the numerical wave, the variable ψ at the time $t = \Delta t/8$. In the second step the normal solution scheme is applied, using the centered difference quotients with $\Delta t' = \frac{1}{8}\,\Delta t$. In the third step one doubles the time step, i.e. $\Delta t' = \frac{1}{4}\,\Delta t$, thereby always starting from $n = 0$. In the next step we use $\Delta t' = \frac{1}{2}\,\Delta t$ until, in the fifth step with $\Delta t' = \Delta t$, the value of ψ at time step $n = 2$ is obtained. The procedure, which is also known as the *leap-frog method,* is shown schematically in Figure 25.3. By means of this successive initialization the phenomenon of the weak instability is suppressed very efficiently. Much more could be said about this and other calculation procedures, but we have given sufficient evidence that great care must be taken in applying finite-difference schemes. Later we will discuss an entirely different instability, which is associated with the numerical treatment of the nonlinear advection equation.

25.4 The implicit method

It is possible to give a numerical scheme for the solution of (25.1) that is absolutely stable. In this case we proceed as follows. Time and spatial derivatives are discretized by

$$\begin{aligned}
\frac{\partial \psi}{\partial t} &= \frac{\psi_j^{n+1} - \psi_j^n}{\Delta t} \\
\frac{\partial \psi}{\partial x} &= \frac{1}{2}\left(\frac{\psi_{j+1}^{n+1} - \psi_{j-1}^{n+1}}{2\,\Delta x} + \frac{\psi_{j+1}^{n} - \psi_{j-1}^{n}}{2\,\Delta x}\right)
\end{aligned} \tag{25.38}$$

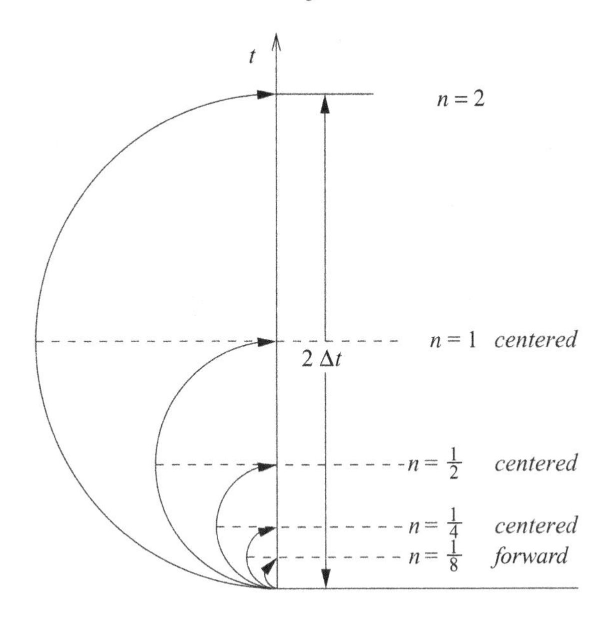

Fig. 25.3 The leap-frog method.

The partial derivative with respect to time is discretized by a forward-in-time difference quotient while the partial spatial derivative is expressed by a mean value in time. Therefore, the discretized form of equation (25.1) is given by

$$\psi_j^{n+1} - \psi_j^n + \frac{U\,\Delta t}{4\,\Delta x}\left(\psi_{j+1}^{n+1} - \psi_{j-1}^{n+1} + \psi_{j+1}^n - \psi_{j-1}^n\right) = 0 \tag{25.39}$$

In contrast to the methods which we have discussed so far, the required grid function at $n+1$ now appears at three different places and cannot be determined explicitly as before. Therefore, this difference method is called the *implicit scheme*. In order to find the solution of the advection problem, (25.39) must be written down for all grid points j, including the boundary points $j = 0$ and $j = J$. This leads to a band matrix that can be solved by known methods for the required values ψ_j^{n+1} with $j = 1, 2, \ldots, J-1$. We will not describe the numerical procedure, but the numerical effort by far exceeds the computational labor of the explicit schemes. This is the price to be paid for the stability of the numerical method.

In order to prove the stability of the scheme, we introduce the trial solution (25.6a) into (25.39) and obtain

$$\exp(-ikc\,\Delta t) - 1 + \frac{U\,\Delta t}{4\,\Delta x}[\exp(ik\,\Delta x) - \exp(-ik\,\Delta x)][\exp(-ikc\,\Delta t) + 1] = 0 \tag{25.40}$$

Dividing this equation by the expression within the second set of brackets gives

$$\frac{\exp(-ikc\,\Delta t)-1}{\exp(-ikc\,\Delta t)+1}+\frac{U\,\Delta t}{4\,\Delta x}[\exp(ik\,\Delta x)-\exp(-ik\,\Delta x)]=0 \tag{25.41}$$

On multiplying the numerator and denominator of the first term by $\exp(ikc\,\Delta t/2)$ and using well-known trigonometric relations we obtain without difficulty the frequency equation

$$-i\tan\left(\frac{kc\,\Delta t}{2}\right)+i\frac{U\,\Delta t}{2\,\Delta x}\sin(k\,\Delta x)=0 \tag{25.42}$$

The required expression for the phase velocity follows immediately:

$$c=\frac{2}{k\,\Delta t}\arctan\left(\frac{U\,\Delta t}{2\,\Delta x}\sin(k\,\Delta x)\right) \tag{25.43a}$$

The argument of the arc tangent may become arbitrarily large since the tangent may assume any value between minus and plus infinity. This means that the phase velocity is real for all values of Δt and Δx. For this reason the solution of the difference equation (25.39) remains absolutely stable. Since $\arctan x\approx x$ for $|x|\ll 1$ we find

$$\lim_{\Delta x,\Delta t\to 0} c=U \tag{25.43b}$$

so that in the limiting case we obtain the required phase velocity.

The implicit method is used in many practical applications, since the CFL criterion does not have to be obeyed in the implicit method. However, the smaller the values of Δt and Δx the closer the agreement with the analytic solution, in general.

The implicit treatment of the meteorological equations is very time-consuming. In order to reduce the numerical effort Robert (1969) introduced the so-called *semi-implicit method*. This is a procedure that treats implicitly only those terms which are mainly responsible for the propagation of the high-speed waves requiring very small time steps in the explicit treatment. The remaining terms are treated explicitly. This leads to a considerable increase of Δt, which is now limited only by the slower waves of meteorological significance.

25.5 The aliasing error and nonlinear instability

In order to demonstrate the existence of a different type of instability, we consider the one-dimensional nonlinear advection equation

$$\frac{\partial u}{\partial t} + u\,\frac{\partial u}{\partial x} = 0 \tag{25.44}$$

which is always a part of the equations of motion. The analytic solution, as is easily verified, is given by

$$u = f(x - ut) \tag{25.45}$$

where f is an arbitrary function. We now consider the nonlinear term which results from the multiplication of u and its spatial derivative. When the calculation is performed in finite differences, we obtain an error due to the inability of the grid to resolve wavelengths shorter than $2\,\Delta x$, or wavenumbers larger than $k_{\max} = \pi/\Delta x$. Mesinger and Arakawa (1976) give an illuminating example by considering the function

$$u = \sin(kx) \tag{25.46}$$

where $k < k_{\max}$. Substituting (25.46) into the nonlinear term gives

$$u\,\frac{\partial u}{\partial x} = k\sin(kx)\cos(kx) = \frac{1}{2}k\sin(2kx) \tag{25.47}$$

The wavenumber appearing in the sine wave of (25.47) is twice as large as the original wavenumber in (25.46). Suppose that the wavenumber in (25.46) is in the interval $\frac{1}{2}k_{\max} < k \le k_{\max}$. It follows that the nonlinear term produces a wave that cannot be resolved by the grid, which leads to an improper evaluation of the finite-difference calculation. To gain further insight, consider a wave whose wavenumber exceeds $k_{\max}$. This, for example, is the case if the wavelength is $4\,\Delta x/3$, as shown by the full line in Figure 25.4.

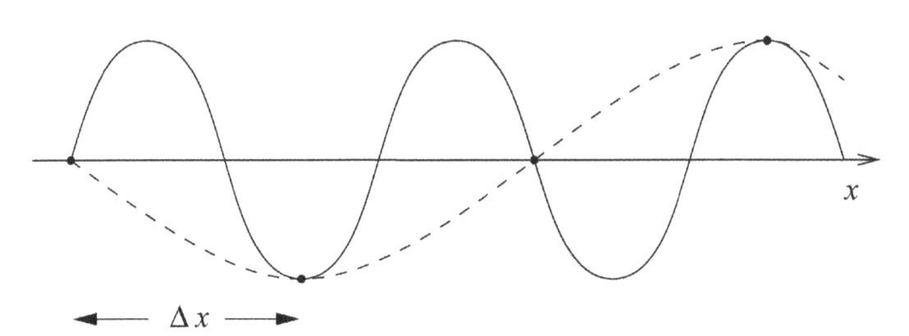

Fig. 25.4 Misrepresentation of a wave of length $4\,\Delta x/3$ (full curve) as a wave of length $4\,\Delta x$ due to the use of the finite-difference grid.

All we can know about a variable is its values at the grid points. Therefore, we cannot distinguish this wave from the dashed-line wave of wavelength $4\,\Delta x$. This misrepresentation is known as the *aliasing error*.

In the more general case the variable u will be represented by an infinite series of harmonic components of the type (25.46):

$$u = \sum_n u_n \quad \text{with} \quad u_n = \sin(k_n x) \tag{25.48}$$

so that many nonlinear terms of the type $\sin(k_1 x)\sin(k_2 x)$ will appear due to the nonlinear term. This, of course, will falsify the energy spectrum of the process to be studied. Since aliasing is due to nonlinear effects one speaks of *nonlinear instability*. Apparently, this effect was first encountered by Phillips (1956) in his famous numerical experiment modeling the general circulation of the atmosphere.

Arakawa (1966) and Arakawa and Lamb (1977) constructed finite-difference equations that suppress the effect of nonlinear instability by restricting the interaction between the resolved and unresolved scales. We refer to Kasahara (1977), who used the simple nonlinear advection equation (25.44) to demonstrate the basic property of the Arakawa scheme. First of all, we consider the integrated linear momentum M_u and the kinetic energy K_u per unit mass

$$M_u = \int_0^L u\,dx, \qquad K_u = \frac{1}{2}\int_0^L u^2\,dx \tag{25.49}$$

The dependent variable u is defined on a cyclic continuous domain from $x = 0$ to $x = L$ so that $u(0) = u(L)$. In the integrated form the advection equations for the linear momentum and the kinetic energy can be written as

$$\begin{aligned}
&\text{(a)} \quad \int_0^L \frac{\partial u}{\partial t}\,dx = -\int_0^L u\frac{\partial u}{\partial x}\,dx = -\int_0^L \frac{\partial}{\partial x}\left(\frac{u^2}{2}\right)dx = 0\\
&\text{(b)} \quad \int_0^L \frac{\partial}{\partial t}\left(\frac{u^2}{2}\right)dx = -\int_0^L u\frac{\partial}{\partial x}\left(\frac{u^2}{2}\right)dx = -\int_0^L \frac{\partial}{\partial x}\left(\frac{u^3}{3}\right)dx = 0
\end{aligned} \tag{25.50}$$

since $u(0) = u(L)$. Equation (25.50a) implies the well-known space-centered difference scheme

$$\frac{\partial u_j}{\partial t} = -\frac{(u_{j+1})^2 - (u_{j-1})^2}{4\,\Delta x} \tag{25.51}$$

Using (25.51), we may approximate (25.50a), to obtain

$$\frac{\partial}{\partial t}\sum_{j=1}^{J} u_j = -\frac{1}{4\,\Delta x}\sum_{j=1}^{J}[(u_{j+1})^2 - (u_{j-1})^2] = 0 \tag{25.52}$$

where $J = L/\Delta x$ must be a natural number. Hence it is seen that the finite-difference scheme (25.51) conserves the momentum. It will be left as an excercise to show that the kinetic energy is not conserved by using (25.51) as indicated by

$$\frac{\partial}{\partial t}\sum_{j=1}^{J}\frac{u_j^2}{2} \neq 0 \tag{25.53}$$

The conservation of the momentum by itself does not ensure computational stability.

Both momentum and energy may be conserved by using an alternate scheme. By writing (25.44) in the form

$$\frac{\partial u}{\partial t} = -u\,\frac{\partial u}{\partial x} = -\frac{1}{3}\left(u\,\frac{\partial u}{\partial x} + \frac{\partial u^2}{\partial x}\right) \tag{25.54}$$

we find the finite-difference approximation

$$\frac{\partial u_j}{\partial t} = -\frac{1}{3}\left(u_j\frac{u_{j+1} - u_{j-1}}{2\,\Delta x} + \frac{(u_{j+1})^2 - (u_{j-1})^2}{2\,\Delta x}\right) \tag{25.55}$$

It will be left as an exercise to show that the scheme (25.55) conserves the momentum and the kinetic energy. A difference scheme may be stable if it conserves both the momentum and the kinetic energy. Since no change in the potential energy of the system occurs, the total energy is conserved. This prevents the spurious growth of energy which may occur on using (25.51) and various other advection schemes. Additional details are given by Washington and Parkinson (1986), where further references may be found, particularly with respect to the use of spatial-difference schemes.

It is well known that most numerical modeling of all scales of atmospheric flow is based on the approximation of horizontal derivatives by finite differences. The basic concepts of finite-difference approximations are relatively simple to understand. Indeed, much sophistication has gone into the construction of many finite-difference schemes, which are needed in order to handle various flow problems. It is fair to say that numerical modeling of atmospheric flow since the pioneering work by Charney *et al.* (1950) has been a story of great success. Most of the numerical work represents the meteorological variables in space and time on a finite-difference grid. However, there are other methods for describing atmospheric fields. It has been shown by various modelers of the atmospheric flow that hemispheric and global modeling can be advantageously carried out by means of a spectral model that makes use of the orthogonality properties of the spherical functions. Various versions of spectral models have been devised. In the next chapter we will briefly discuss the philosophy of the spectral model and show how the model equations may be obtained.

25.6 Problems

25.1: Show that equation (25.39) can be expressed with the help of a band matrix multiplying the vector $\left(\psi_1^{n+1}, \psi_2^{n+1}, \ldots\right)^{\mathrm{T}}$, where T denotes the transpose. The boundary points are assumed to be independent of time; all ψ_j^n are considered to be known.

25.2: Show all steps between (25.41) and (25.43a).

25.3: Verify equation (25.53) by using equation (25.51).

25.4: Show that the momentum and the kinetic energy are conserved by using (25.55).

26

Modeling of atmospheric flow by spectral techniques

26.1 Introduction

The representation of atmospheric flow fields by means of spherical functions has a long history. Haurwitz (1940) represented the movement of Rossby waves by means of spherical functions. The development of the spectral method for the numerical integration of the equations of atmospheric motion goes back to Silberman (1954), who integrated the barotropic vorticity equation in spherical geometry. The spectral method attracted the attention of others and studies were performed, for example, by Lorenz (1960), Platzman (1960), Kubota *et al.* (1961), Baer and Platzman (1961), and Elsaesser (1966). Lorenz demonstrated that, for nondivergent barotropic flow, the truncated spectral equations have some important properties. Just like the exact differential equations, they preserve the mean squared vorticity, called *enstrophy*, and the mean kinetic energy. Platzman pointed out that this very desirable property automatically eliminated nonlinear instability, which at that time was a substantial difficulty in grid-point models. The early work made use of the so-called *interaction coefficients* to handle nonlinearity. This cumbersome procedure was replaced by the efficient transform technique for solving the spectral equations, which was devised independently by Orszag (1970) and by Eliasen *et al.* (1970). In compressed form the essential information on spectral modeling is given by Haltiner and Williams (1980). Much valuable information about spectral techniques – which is usually not readily available – can be extracted from the "gray" literature. We refer to an excellent report by Eliasen *et al.* (1970). Finally, we refer the reader to an excellent article on "Global modelling of atmospheric flow by spectral methods", by Bourke *et al.* (1977). This article states the merits of the spectral relative to finite-difference models. We will briefly repeat these.

Foremost among the advantages of the spectral method relative to finite-difference methods are

(i) the intrinsic accuracy of evaluation of horizontal advection,
(ii) the elimination of aliasing arising from quadratic nonlinearity,
(iii) ease of modeling flow over the entire globe, and
(iv) the ease of incorporation of semi-implicit integration over time.

These characteristics of the spectral method result in highly accurate and stable numerics and efficient and simple computer coding. The spectral transform technique described in this chapter follows the description in Technical Report No. 6 "The ECHAM 3 Atmospheric General Circulation Model", DKRZ, Hamburg, 1993, where the operational forecast model is presented. The model equations are based on various suitable approximations to simplify the required parameterizations of fluxes and the numerical procedures. These approximations include simplified forms of the continuity equation and the heat equation. We do not discuss these approximations but use the equations in the forms given in the previous chapters. Since the model equations ignore the diffusion flux of the dry air, it becomes possible to reduce the system of continuity equations of the partial masses to a single prediction equation for the moisture variable $q = m^{\mathrm{H_2O}}$, which here refers to the sum of the specific humidity for the vapor (m^1), the specific liquid (m^2), and the specific ice content (m^3). All physical effects are thought to be described by the symbol Q_q. The moisture equation in the model is structured in such a way that it is compatible with the heat equation.

26.2 The basic equations

In this section we will present the basic equations used for the spectral representation. These are the horizontal equations of motion, the continuity equation and the hydrostatic equation, the prognostic equations for temperature, and the moisture variable q. These equations have been derived in previous chapters but need to be rewritten to a certain extent. First of all we repeat the equation for the individual derivative in the spherical system, see equation (19.15),

$$\frac{d}{dt} = \frac{\partial}{\partial t} + \frac{u}{a\cos\varphi}\frac{\partial}{\partial\lambda} + \frac{v}{a}\frac{\partial}{\partial\varphi} + \dot{\xi}\frac{\partial}{\partial\xi} \tag{26.1}$$

where we have used the unspecified general vertical coordinate ξ and r has been replaced by the constant earth radius a. Robert (1966) has pointed out that it is appropriate for the spectral representation of the flow field on the globe to replace the original velocity components by the transformation

$$u = \frac{U}{\cos\varphi}, \qquad v = \frac{V}{\cos\varphi} \tag{26.2}$$

For notational convenience it is costumary to introduce the symbol $\mu = \cos\theta = \sin\varphi$, where $\theta = \pi/2 - \varphi$ is the co-latitude. From this definition follows a differential relation given by

$$\frac{\partial}{\partial \varphi} = \cos\varphi \frac{\partial}{\partial \mu} \tag{26.3}$$

Using the transformation (26.2), we find for the individual time derivative

$$\frac{d}{dt} = \frac{\partial}{\partial t} + \frac{U}{a(1-\mu^2)}\frac{\partial}{\partial \lambda} + \frac{V}{a}\frac{\partial}{\partial \mu} + \dot{\xi}\frac{\partial}{\partial \xi} \tag{26.4}$$

We have discussed in an earlier chapter the fact that, due to ever-existing computational limitations and insufficient data to specify the initial conditions, the atmosphere cannot be resolved to the extent necessary to include all scales of motion. Therefore, it becomes necessary to average the pertinent atmospheric equations. We know from our earlier work that the averaging process produces a number of correlations but otherwise retains the form of the unaveraged equations. These correlations represent the unresolved and, therefore, unknown subgrid fluxes which need to be parameterized. Various research groups use different parameterizations for the same correlations. Furthermore, these parameterizations are subject to revision as more and better observational data become available. For these reasons we will not discuss the parameterizations in the present context. Instead, we assume that the averaging process has been carried out already and we simply assign symbols to the correlations. Moreover, to keep the notation simple we also leave out the various averaging symbols.

Using (26.4) the equation for the U-component of the horizontal motion can then be written as

$$\begin{aligned}&\frac{\partial U}{\partial t} + \frac{U}{a(1-\mu^2)}\frac{\partial U}{\partial \lambda} + \frac{V}{a}\frac{\partial U}{\partial \mu} + \dot{\xi}\frac{\partial U}{\partial \xi} - fV \\ &\quad = -\frac{1}{a\rho}\frac{\partial p}{\partial \lambda} - \frac{1}{a}\frac{\partial \phi}{\partial \lambda} + P_U + K_U\end{aligned} \tag{26.5a}$$

The subgrid fluxes are simply denoted by P_U and K_U representing the vertical change of the momentum flux and the tendency of U due to horizontal diffusion. The description of the subgrid fluxes given here is rather vague but this has no effect on the discussion of the spectral method. Similarly, we obtain for the V-component

$$\begin{aligned}&\frac{\partial V}{\partial t} + \frac{U}{a(1-\mu^2)}\frac{\partial V}{\partial \lambda} + \frac{1}{a}\frac{\partial}{\partial \mu}\left(\frac{V^2}{2}\right) + \dot{\xi}\frac{\partial V}{\partial \xi} + \frac{U^2+V^2}{a}\frac{\mu}{1-\mu^2} + fU \\ &\quad = -\frac{1-\mu^2}{a\rho}\frac{\partial p}{\partial \mu} - \frac{1-\mu^2}{a}\frac{\partial \phi}{\partial \mu} + P_V + K_V\end{aligned} \tag{26.5b}$$

By omitting the P and K terms and reverting to the original variables u and v we return to the original form of the horizontal equations of motion in the spherical coordinate system.

Instead of transforming the U- and V-components directly, the motion will be described in terms of the vorticity and divergence equations. This approach was successfully practiced by Bourke (1972). To this end, first of all, we introduce the vorticity into equation (26.5) by means of

$$\zeta = \mathbf{e}_r \cdot \nabla \times \mathbf{v} = \frac{1}{a\cos\varphi}\left(\frac{\partial v}{\partial \lambda} - \frac{\partial}{\partial \varphi}(u\cos\varphi)\right) = \frac{1}{a}\left(\frac{1}{1-\mu^2}\frac{\partial V}{\partial \lambda} - \frac{\partial U}{\partial \mu}\right) \tag{26.6}$$

After a few easy mathematical steps we obtain

$$\begin{aligned} &\frac{\partial U}{\partial t} - (f+\zeta)V + \dot{\xi}\frac{\partial U}{\partial \xi} + \frac{R_0 T_{\rm v}}{a}\frac{\partial \ln p}{\partial \lambda} + \frac{1}{a}\frac{\partial}{\partial \lambda}(\phi + E) = P_U + K_U \\ &\qquad \text{with} \quad E = \frac{1}{1-\mu^2}\frac{U^2+V^2}{2} \end{aligned} \tag{26.7a}$$

$$\frac{\partial V}{\partial t} + (f+\zeta)U + \dot{\xi}\frac{\partial V}{\partial \xi} + (1-\mu^2)\frac{R_0 T_{\rm v}}{a}\frac{\partial \ln p}{\partial \mu} + \frac{1-\mu^2}{a}\frac{\partial}{\partial \mu}(\phi + E) = P_V + K_V \tag{26.7b}$$

$T_{\rm v}$ is the virtual temperature.

To formulate the prognostic equation of temperature we refer to equation (23.4). We replace the individual derivative d/dt by means of (26.4) and use $T_{\rm v}$ instead of T in the second ω term and obtain

$$\frac{\partial T}{\partial t} + \frac{U}{a(1-\mu^2)}\frac{\partial T}{\partial \lambda} + \frac{V}{a}\frac{\partial T}{\partial \mu} + \dot{\xi}\frac{\partial T}{\partial \xi} - \frac{R_0 T_{\rm v}}{c_p p}\omega = Q_{\rm h} \tag{26.8a}$$

or

$$\begin{aligned} &\frac{\partial T}{\partial t} - F_T = Q_{\rm h} \quad \text{with} \\ &F_T = -\frac{U}{a(1-\mu^2)}\frac{\partial T}{\partial \lambda} - \frac{V}{a}\frac{\partial T}{\partial \mu} - \dot{\xi}\frac{\partial T}{\partial \xi} + \frac{R_0 T_{\rm v}}{c_p p}\omega \end{aligned} \tag{26.8b}$$

The radiative flux divergence and the release of latent heat are thought to be included in $Q_{\rm h}$. For the reason given in the introduction to this chapter we may write for the moisture equation

$$\frac{\partial q}{\partial t} + \frac{U}{a(1-\mu^2)}\frac{\partial q}{\partial \lambda} + \frac{V}{a}\frac{\partial q}{\partial \mu} + \dot{\xi}\frac{\partial q}{\partial \xi} = Q_q \tag{26.9a}$$

or

$$\frac{\partial q}{\partial t} - F_q = Q_q \quad \text{with} \quad F_q = -\frac{U}{a(1-\mu^2)}\frac{\partial q}{\partial \lambda} - \frac{V}{a}\frac{\partial q}{\partial \mu} - \dot{\xi}\frac{\partial q}{\partial \xi} \tag{26.9b}$$

The continuity equation involving the generalized vertical coordinate ξ with the scaling factor $m_0 = 1$ follows from (22.16) and is given by

$$\frac{\partial}{\partial \xi}\left(\frac{\partial p}{\partial t}\right) + \nabla_{\mathrm{h}} \cdot \left(\mathbf{v}_{\mathrm{h}} \frac{\partial p}{\partial \xi}\right) + \frac{\partial}{\partial \xi}\left(\dot{\xi} \frac{\partial p}{\partial \xi}\right) = 0 \tag{26.10}$$

Now, ∇_{h} refers to spherical coordinates. Obviously, the hydrostatic equation in terms of ξ is given by

$$\frac{\partial \phi}{\partial \xi} = -\frac{R_0 T_{\mathrm{v}}}{p} \frac{\partial p}{\partial \xi} \tag{26.11}$$

Next, we wish to obtain a suitable expression for the generalized vertical velocity $\omega = dp/dt$. This is accomplished by rewriting (26.10),

$$\begin{aligned} &\frac{\partial}{\partial \xi}\left(\frac{\partial p}{\partial t} + \mathbf{v}_{\mathrm{h}} \cdot \nabla_{\mathrm{h}} p + \dot{\xi} \frac{\partial p}{\partial \xi} - \mathbf{v}_{\mathrm{h}} \cdot \nabla_{\mathrm{h}} p\right) + \nabla_{\mathrm{h}} \cdot \left(\mathbf{v}_{\mathrm{h}} \frac{\partial p}{\partial \xi}\right) = 0 \\ \text{or} \quad &\frac{\partial}{\partial \xi}(\omega - \mathbf{v}_{\mathrm{h}} \cdot \nabla_{\mathrm{h}} p) + \nabla_{\mathrm{h}} \cdot \left(\mathbf{v}_{\mathrm{h}} \frac{\partial p}{\partial \xi}\right) = 0 \end{aligned} \tag{26.12}$$

Integrating this expression between the limits shown yields

$$\int_0^{\xi} \frac{\partial \omega}{\partial \xi'}\, d\xi' = \int_0^{\xi} \frac{\partial}{\partial \xi'}(\mathbf{v}_{\mathrm{h}} \cdot \nabla_{\mathrm{h}} p)\, d\xi' - \int_0^{\xi} \nabla_{\mathrm{h}} \cdot \left(\mathbf{v}_{\mathrm{h}} \frac{\partial p}{\partial \xi'}\right) d\xi' \tag{26.13}$$

So far we have not specified the direction of ξ. The reason for this is that in the expression $\dot{\xi}\,\partial/\partial \xi$ it does not make any difference whether ξ increases from the top of the atmosphere or from the ground. If we start the integration from the ground then (26.13) necessarily involves the evaluation of ω at the earth's surface. Instead, we let ξ increase in the downward direction from the top of the atmosphere to get a simpler expression since $\omega(\xi = 0) = 0$, or

$$\omega(\xi) = (\mathbf{v}_{\mathrm{h}} \cdot \nabla_{\mathrm{h}} p)_{\xi} - \int_0^{\xi} \nabla_{\mathrm{h}} \cdot \left(\mathbf{v}_{\mathrm{h}} \frac{\partial p}{\partial \xi'}\right) d\xi' \tag{26.14}$$

Since the surface pressure is not fixed in general but varies with time, we need to derive a suitable expression for the surface-pressure tendency. By integrating (26.10) we obtain

$$\frac{\partial p_{\mathrm{s}}}{\partial t} = -\int_0^1 \nabla_{\mathrm{h}} \cdot \left(\mathbf{v}_{\mathrm{h}} \frac{\partial p}{\partial \xi}\right) d\xi \quad \text{or} \quad \frac{\partial \ln p_{\mathrm{s}}}{\partial t} = -\frac{1}{p_{\mathrm{s}}} \int_0^1 \nabla_{\mathrm{h}} \cdot \left(\mathbf{v}_{\mathrm{h}} \frac{\partial p}{\partial \xi}\right) d\xi \tag{26.15}$$

The predictive equations require knowledge of $\dot{\xi}$. By integrating the continuity equation (26.10) from the top of the atmosphere where $p = 0$ to the arbitrary level ξ we find

$$\dot{\xi} \frac{\partial p}{\partial \xi} = -\left(\frac{\partial p}{\partial t}\right)_{\xi} - \int_0^{\xi} \nabla_{\mathrm{h}} \cdot \left(\mathbf{v}_{\mathrm{h}} \frac{\partial p}{\partial \xi'}\right) d\xi' \tag{26.16}$$

If $\xi = 1$ we again obtain equation (26.15).

We are now ready to derive the vorticity and the divergence equations in terms of the transformed velocity variables U and V. To this end we apply the operators $a^{-1}\,\partial/\partial\mu$ to (26.7a) and $[a(1-\mu^2)]^{-1}\,\partial/\partial\lambda$ to (26.7b) and subtract one of the results from the other to obtain

$$\begin{aligned}
&\frac{1}{a}\frac{\partial}{\partial t}\left(\frac{1}{1-\mu^2}\frac{\partial V}{\partial\lambda}-\frac{\partial U}{\partial\mu}\right)\\
&\quad+\frac{1}{a(1-\mu^2)}\frac{\partial}{\partial\lambda}\left[(f+\zeta)U+\dot{\xi}\frac{\partial V}{\partial\xi}+(1-\mu^2)\frac{R_0T_{\mathrm{v}}}{a}\frac{\partial\ln p}{\partial\mu}\right]\\
&\quad+\frac{1}{a}\frac{\partial}{\partial\mu}\left[(f+\zeta)V-\dot{\xi}\frac{\partial U}{\partial\xi}-\frac{R_0T_{\mathrm{v}}}{a}\frac{\partial\ln p}{\partial\lambda}\right]=\frac{1}{a(1-\mu^2)}\\
&\quad\times\left(\frac{\partial P_V}{\partial\lambda}+\frac{\partial K_V}{\partial\lambda}\right)-\frac{1}{a}\left(\frac{\partial P_U}{\partial\mu}+\frac{\partial K_U}{\partial\mu}\right)
\end{aligned}\tag{26.17}$$

The expressions in the square brackets which are differentiated with respect to λ and μ will be abreviated by $-F_V$ and F_U. Reference to equation (26.6), which is the definition of the vorticity, shows that (26.17) can be written as

$$\begin{aligned}
\frac{\partial\zeta}{\partial t}&=\frac{1}{a(1-\mu^2)}\frac{\partial}{\partial\lambda}(F_V+P_V)-\frac{1}{a}\frac{\partial}{\partial\mu}(F_U+P_U)+K_\zeta\\
\text{with}\quad K_\zeta&=\frac{1}{a}\left(\frac{1}{1-\mu^2}\frac{\partial K_V}{\partial\lambda}-\frac{\partial K_U}{\partial\mu}\right)\\
F_U&=(f+\zeta)V-\dot{\xi}\frac{\partial U}{\partial\xi}-\frac{R_0T_{\mathrm{v}}}{a}\frac{\partial\ln p}{\partial\lambda}\\
F_V&=-(f+\zeta)U-\dot{\xi}\frac{\partial V}{\partial\xi}-(1-\mu^2)\frac{R_0T_{\mathrm{v}}}{a}\frac{\partial\ln p}{\partial\mu}
\end{aligned}\tag{26.18}$$

Now we wish to introduce the horizontal divergence by means of

$$D=\frac{1}{a(1-\mu^2)}\frac{\partial U}{\partial\lambda}+\frac{1}{a}\frac{\partial V}{\partial\mu}\tag{26.19}$$

The derivation of this equation will be left as an exercise. By applying the operators $[a(1-\mu^2)]^{-1}\,\partial/\partial\lambda$ to (26.7a) and $a^{-1}\,\partial/\partial\mu$ to (26.7b) and then adding the results we obtain

$$\begin{aligned}
&\frac{1}{a}\frac{\partial}{\partial t}\left(\frac{1}{1-\mu^2}\frac{\partial U}{\partial\lambda}+\frac{\partial V}{\partial\mu}\right)\\
&\quad-\frac{1}{a(1-\mu^2)}\frac{\partial}{\partial\lambda}\left((f+\zeta)V-\dot{\xi}\frac{\partial U}{\partial\xi}-R_0T_{\mathrm{v}}\frac{\partial\ln p}{\partial\lambda}+\frac{1}{a}\frac{\partial}{\partial\lambda}(\phi+E)\right)\\
&\quad+\frac{1}{a}\frac{\partial}{\partial\mu}\left((f+\zeta)U+\dot{\xi}\frac{\partial V}{\partial\xi}+(1-\mu^2)\frac{R_0T_{\mathrm{v}}}{a}\frac{\partial\ln p}{\partial\mu}+\frac{1-\mu^2}{a}\frac{\partial}{\partial\mu}(\phi+E)\right)\\
&=\frac{1}{a(1-\mu^2)}\frac{\partial}{\partial\lambda}(P_U+K_U)+\frac{1}{a}\frac{\partial}{\partial\mu}(P_V+K_V)
\end{aligned}\tag{26.20}$$

On introducing the divergence according to (26.19) we find

$$\begin{aligned}\frac{\partial D}{\partial t} &= \frac{1}{a(1-\mu^2)}\frac{\partial}{\partial\lambda}(F_U+P_U)+\frac{1}{a}\frac{\partial}{\partial\mu}(F_V+P_V)+K_D-\nabla_{\rm h}^2 G \\ \text{with}\quad K_D &= \frac{1}{a(1-\mu^2)}\frac{\partial K_U}{\partial\lambda}+\frac{1}{a}\frac{\partial K_V}{\partial\mu}, \qquad G=\phi+E \\ \nabla_{\rm h}^2 &= \frac{1}{a^2(1-\mu^2)}\frac{\partial^2}{\partial\lambda^2}+\frac{1}{a^2}\frac{\partial}{\partial\mu}\left((1-\mu^2)\frac{\partial}{\partial\mu}\right)\end{aligned} \tag{26.21}$$

We now recall that, according to Helmholtz's theorem (Section 6.1); the velocity vector $\mathbf{v}$ can be decomposed into the sum of the rotational and the divergent parts, $\mathbf{v} = \mathbf{v}_{\rm ROT} + \mathbf{v}_{\rm DIV}$. In the two-dimensional situation we may write, see equation (6.12a),

$$\begin{aligned}u\mathbf{e}_\lambda + v\mathbf{e}_\varphi &= \frac{U}{\cos\varphi}\mathbf{e}_\lambda + \frac{V}{\cos\varphi}\mathbf{e}_\varphi = \mathbf{e}_r\times\nabla_{\rm h}\psi + \nabla_{\rm h}\chi \\ &= \mathbf{e}_\varphi\frac{1}{a\cos\varphi}\frac{\partial\psi}{\partial\lambda} - \mathbf{e}_\lambda\frac{\cos\varphi}{a}\frac{\partial\psi}{\partial\mu} + \mathbf{e}_\lambda\frac{1}{a\cos\varphi}\frac{\partial\chi}{\partial\lambda} + \mathbf{e}_\varphi\frac{\cos\varphi}{a}\frac{\partial\chi}{\partial\mu}\end{aligned} \tag{26.22}$$

where ψ is the stream function and χ the velocity potential. Successive scalar multiplication by the unit vectors $\mathbf{e}_\lambda$ and $\mathbf{e}_\varphi$ gives

$$\begin{aligned}&\text{(a)}\quad U = -\frac{1-\mu^2}{a}\frac{\partial\psi}{\partial\mu}+\frac{1}{a}\frac{\partial\chi}{\partial\lambda} \\ &\text{(b)}\quad V = \frac{1}{a}\frac{\partial\psi}{\partial\lambda}+\frac{1-\mu^2}{a}\frac{\partial\chi}{\partial\mu}\end{aligned} \tag{26.23}$$

Using the horizontal Laplacian as shown in equation (26.21) leads to the well-known expressions for the vorticity ζ and the divergence D as in (6.13):

$$\zeta = \nabla_{\rm h}^2\psi, \qquad D = \nabla_{\rm h}^2\chi \tag{26.24}$$

The Laplacian in spherical ccordinates is defined in (26.21).

26.3 Horizontal discretization

The prognostic equations are formulated in terms of the vorticity ζ (26.18), the divergence D (26.21), the temperature T (26.8), the moisture q (26.9), and the surface pressure $\ln p_{\rm s}$ (26.15). These quantities as well as the surface geopotential ϕ will be represented in terms of the spherical functions. Before proceeding, we will briefly review those properties of the spherical functions which are needed in our work. More detailed information can be found in any standard textbook on this subject. We refer to Lense (1950).

26.3.1 Surface spherical harmonics

A function $f(x, y, z)$ is said to be homogeneous of degree n if the following relation holds:

$$f(\lambda x, \lambda y, \lambda z) = \lambda^n f(x, y, z) \tag{26.25}$$

It is required that n is a real number, not necessarily an integer. Partial differentiation of this identity with repect to λ, afterwards setting $\lambda = 1$, gives *Euler's relation*

$$x \frac{\partial f}{\partial x} + y \frac{\partial f}{\partial y} + z \frac{\partial f}{\partial z} = nf \tag{26.26}$$

A *spherical function* W_n of degree n is defined to be a homogeneous potential function of degree n. Solutions of Laplace's equation, in general, are called *potential* or *harmonic functions*. Therefore, the function W_n satisfies the Euler relation

$$x \frac{\partial W_n}{\partial x} + y \frac{\partial W_n}{\partial y} + z \frac{\partial W_n}{\partial z} = nW_n \tag{26.27}$$

and the Laplace equation

$$\nabla^2 W_n = 0 \tag{26.28}$$

Introduction of the spherical coordinates

$$x = r \sin\theta \cos\lambda, \qquad y = r \sin\theta \sin\lambda, \qquad z = r \cos\theta \tag{26.29}$$

into the expression $r\, \partial W_n(x, y, z)/\partial r$ gives

$$\begin{aligned} r \frac{\partial W_n}{\partial r} &= r\left(\frac{\partial W_n}{\partial x}\frac{\partial x}{\partial r} + \frac{\partial W_n}{\partial y}\frac{\partial y}{\partial r} + \frac{\partial W_n}{\partial z}\frac{\partial z}{\partial r}\right) \\ &= x \frac{\partial W_n}{\partial x} + y \frac{\partial W_n}{\partial y} + z \frac{\partial W_n}{\partial z} = nW_n \end{aligned} \tag{26.30}$$

Laplace's equation in spherical coordinates is given by

$$\frac{\partial}{\partial r}\left(r^2 \frac{\partial W_n}{\partial r}\right) + \frac{1}{\sin\theta}\frac{\partial}{\partial\theta}\left(\sin\theta \frac{\partial W_n}{\partial\theta}\right) + \frac{1}{\sin^2\theta}\frac{\partial^2 W_n}{\partial\lambda^2} = 0 \tag{26.31}$$

Since W_n is homogeneous of degree n, we may write

$$\begin{aligned} W_n(x, y, z) &= W_n(r \sin\theta \cos\lambda, r \sin\theta \sin\lambda, r \cos\theta) \\ &= r^n W_n(\sin\theta \cos\lambda, \sin\theta \sin\lambda, \cos\theta) \\ &= r^n Y_n(\lambda, \theta) \end{aligned} \tag{26.32}$$

The part depending only on the angular coordinates (λ, θ) is known as the *spherical function* Y_n.

In order to find a solution for Y_n, we first use (26.30) to obtain

$$r^2 \frac{\partial W_n}{\partial r} = nr W_n \tag{26.33}$$

and

$$\frac{\partial}{\partial r}\left(r^2 \frac{\partial W_n}{\partial r}\right) = n(n+1) W_n \tag{26.34}$$

so that (26.31) can be written as

$$\frac{1}{\sin\theta} \frac{\partial}{\partial\theta}\left(\sin\theta \frac{\partial Y_n}{\partial\theta}\right) + \frac{1}{\sin^2\theta} \frac{\partial^2 Y_n}{\partial\lambda^2} + n(n+1) Y_n = 0 \tag{26.35}$$

If the function $Y_n(\lambda, \theta)$ depends on θ only, we will call this function $P_n(\theta)$, which is known as the *zonal spherical function*. Thus the partial differential equation (26.35) reduces to

$$\frac{1}{\sin\theta} \frac{d}{d\theta}\left(\sin\theta \frac{\partial P_n(\theta)}{\partial\theta}\right) + n(n+1) P_n(\theta) = 0 \tag{26.36}$$

Since θ is the only independent variable, we have obtained an ordinary differential equation. On introducing the definition $\mu = \cos\theta$, equation (26.36) assumes the form

$$\boxed{(1-\mu^2) \frac{d^2 P_n(\mu)}{d\mu^2} - 2\mu \frac{d P_n(\mu)}{d\mu} + n(n+1) P_n(\mu) = 0} \tag{26.37}$$

which is known as *Legendre's differential equation*. The particular solution

$$\boxed{P_n(\mu) = \frac{1}{2^n n!} \frac{d^n}{d\mu^n}(\mu^2 - 1)^n, \qquad -1 \le \mu \le 1} \tag{26.38}$$

is called the *Legendre polynomial* of degree n. It is well known that these polynomials form a system of orthogonal functions on the interval $-1 \le \mu \le 1$. Tables of $P_n(\mu)$ can be found in many textbooks and mathematical handbooks.

By differentiating (26.37) m times with respect to μ, we easily find

$$(1-\mu^2) \frac{d^2 P_n^{(m)}(\mu)}{d\mu^2} - 2(m+1)\mu \frac{d P_n^{(m)}(\mu)}{d\mu} + [n(n+1) - m(m+1)] P_n^{(m)}(\mu) = 0 \tag{26.39}$$

where $m = 0, 1, \ldots$ and $P_n^{(m)}(\mu)$ represents the mth derivative of $P_n(\mu)$. We leave the verification of (26.39) to the exercises. We now introduce the definition

$$\boxed{P_n^m(\mu) = (-1)^m (1-\mu^2)^{m/2} P_n^{(m)}(\mu) = (-1)^m \frac{(1-\mu^2)^{m/2}}{2^n n!} \frac{d^{n+m}}{d\mu^{n+m}}(\mu^2-1)^n} \tag{26.40}$$

into (26.39) and obtain

$$\boxed{(1-\mu^2)\frac{d P_n^m(\mu)}{d\mu^2} - 2\mu \frac{d P_n^m(\mu)}{d\mu} + \left(n(n+1) - \frac{m^2}{1-\mu^2}\right) P_n^m(\mu) = 0} \tag{26.41}$$

which is known as the *associated Legendre equation*. The $P_n^m(\mu)$ are the *associated Legendre polynomials* satisfying (26.41). Details regarding going from (26.39) to (26.41) will be left to the problems. The associated Legendre polynomials also form a system of orthogonal functions on the interval $-1 \leq \mu \leq 1$. Some details will be given shortly.

We will now obtain a suitable expression for the spherical surface function Y_n, also known as *Laplace's spherical function*, by separating the variables θ and λ. By substituting

$$Y_n(\lambda, \theta) = \Theta(\theta)\Phi(\lambda) \tag{26.42}$$

into (26.35), we find

$$\frac{\sin\theta}{\Theta} \frac{d}{d\theta}\left(\sin\theta \frac{d\Theta}{d\theta}\right) + n(n+1)\sin^2\theta = -\frac{1}{\Phi} \frac{d^2\Phi}{d\lambda^2} \tag{26.43}$$

whose left-hand side depends on θ only, so we immediately obtain

$$\frac{d}{d\lambda}\left(\frac{1}{\Phi} \frac{d^2\Phi}{d\lambda^2}\right) = 0 \tag{26.44}$$

The solution to equation (26.41) is given by

$$\frac{1}{\Phi} \frac{d^2\Phi}{d\lambda^2} = -m^2 \tag{26.45}$$

where m is a constant. Moreover, the solution to (26.45) is

$$\Phi(\lambda) = A_m \cos(m\lambda) + B_m \sin(m\lambda) \tag{26.46}$$

By introducing $\mu = \cos\theta$ into (26.43) we obtain

$$(1-\mu^2)\frac{d^2\Theta}{d\mu^2} - 2\mu \frac{d\Theta}{d\mu} + \left(n(n+1) - \frac{m^2}{1-\mu^2}\right)\Theta = 0 \tag{26.47}$$

If m is an integer and $0 \leq m \leq n$, the comparison of (26.47) with (26.41) shows that Θ is the associated Legendre polynomial $P_n^m(\mu)$, i.e.

$$\Theta(\mu) = P_n^m(\mu) \tag{26.48}$$

Since the differential equation (26.35) is linear and homogeneous, we obtain from (26.42) the desired solution

$$\boxed{Y_n(\lambda, \theta) = \sum_{m=0}^{n} [A_m \cos(m\lambda) + B_m \sin(m\lambda)] P_n^m(\cos\theta)} \tag{26.49}$$

We will now give the orthogonality relation for the assciated Legrende polynomials:

$$\int_{-1}^{+1} P_n^m(\mu) P_l^m(\mu)\, d\mu = \frac{2}{(2l+1)} \frac{(l+m)!}{(l-m)!} \delta_{n,l} \tag{26.50}$$

As usual, the symbol $\delta_{n,l}$ is the Kronecker delta. The proof of this relation is not particularly difficult and can be found in any textbook on the subject. As will be seen, the normalized form of the associated Legendre polynomials $\widetilde{P}_n^m(\mu)$ is given by

$$\widetilde{P}_n^m(\mu) = \sqrt{\frac{2n+1}{2} \frac{(n-m)!}{(n+m)!}}\, P_n^m(\mu) \tag{26.51}$$

from which it follows that

$$\int_{-1}^{+1} \widetilde{P}_n^m(\mu) \widetilde{P}_l^m(\mu)\, d\mu = \delta_{n,l} \tag{26.52}$$

Various useful identities involving the Legendre polynomials are listed next:

$$\begin{aligned}
P_l^m(\mu) &= 0 \quad \text{if} \quad m > l \\
P_l^m(-\mu) &= (-1)^{l+m} P_l^m(\mu) \\
P_l^{-m}(\mu) &= (-1)^m \frac{(l-m)!}{(l+m)!} P_l^m(\mu) \\
\widetilde{P}_l^{-m}(\mu) &= (-1)^m \widetilde{P}_l^m(\mu) \\
P_l^{m=0}(\mu) &= P_l(\mu)
\end{aligned} \tag{26.53}$$

The linearly independent parts of (26.49) are written down separately together with the zeros of the function:

$$\begin{aligned}
&\text{(a)} \quad Y_n^m(\lambda, \theta)_c = \cos(m\lambda)\, P_n^m(\cos\theta) \quad \text{with zeros at} \quad \lambda = \frac{\pi}{2m}, \frac{3\pi}{2m}, \ldots \\
&\text{(b)} \quad Y_n^m(\lambda, \theta)_s = \sin(m\lambda)\, P_n^m(\cos\theta) \quad \text{with zeros at} \quad \lambda = 0, \frac{\pi}{m}, \frac{2\pi}{m}, \ldots
\end{aligned} \tag{26.54}$$

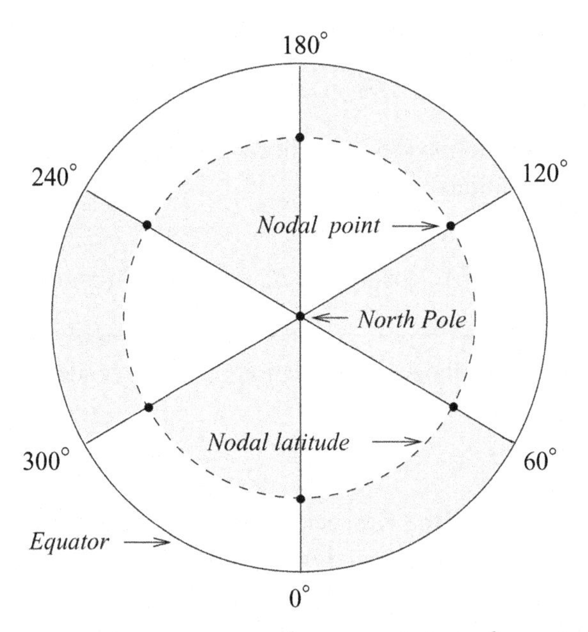

Fig. 26.1 Nodes of the surface harmonics $Y_5^3(\lambda, \theta) = \sin(3\lambda)\, P_5^3(\cos\,\theta)$. The function has negative values in the shaded regions.

They are also known as the surface *spherical harmonics of the first kind*, tesseral for $m < n$ and sectoral for $m = n$. Their usefulness follows from the fact that everywhere they are single-valued and continuous functions on the surface of the sphere. The number of waves m around the latitude circle, defined either by (26.54a) or by (26.54b), is known as the *east–west planetary wavenumber*. Moreover, the function $P_n^m(\mu)$ has $n - m$ zeros that are symmetric with respect to the equator at $\theta = \pi/2$. The number $n - m$ is the so-called *meridional wavenumber*, which is defined as the number of nodal zeros between the poles but excluding these. A typical example for the tesseral harmonics is given in Figure 26.1 for $n = 5$ and $m = 3$. The so-called nodal latitude is 19.5°. The nodal latitudes are also known as the *Gaussian latitudes*. Values for other combinations of m and n are given by Haurwitz (1940).

The alternating pattern of low and high values of variables may be thought of as negative and positive deviations from the mean values. A specific and interesting example involving the geopotential derived from the shallow-water equations is given by Washington and Parkinson (1986).

Instead of using equation (26.49) in terms of $\cos(m\varphi)$ and $\sin(m\varphi)$ it is more convenient to use the normalized form

$$Y_n^m(\lambda, \theta) = \sqrt{\frac{2n+1}{4\pi}\frac{(n-m)!}{(n+m)!}}P_n^m(\cos\theta)\exp(im\lambda) = \frac{1}{\sqrt{2\pi}}\widetilde{P}_n^m(\cos\theta)\exp(im\lambda) \tag{26.55}$$

Observing the orthogonality relation

$$\int_0^{2\pi} \exp(im\lambda)\exp(-im'\lambda)\,d\lambda = 2\pi\,\delta_{m,m'} \tag{26.56}$$

and the normalized form (26.51) of the associated Legendre polynomials, we may easily verify the orthogonality relation

$$\int_0^{2\pi} d\lambda \int_{-1}^{1} Y_n^m(\lambda, \mu)\left[Y_{n'}^{m'}(\lambda, \mu)\right]^* d\mu = \delta_{m,m'}\delta_{n,n'} \tag{26.57}$$

The superscript * denotes the complex conjugate of the function Y_n^m.

Equation (26.49) may be written in the more convenient form

$$Y_n(\lambda, \theta) = \sum_{m=-n}^{n} A^m Y_n^m(\lambda, \theta) \tag{26.58}$$

Equations (26.49) and (26.58) are completely equivalent, each requiring $2n+1$ expansion coefficients.

26.3.2 Spectral representation

According to Sommerfeld (1964), equation (26.58) represents the most general surface spherical harmonics. If we wish to represent a function f on the sphere we must sum over all n so that

$$f(\lambda, \mu) = \sum_{n=0}^{\infty} Y_n = \sum_{n=0}^{\infty}\sum_{m=-n}^{n} A_n^m Y_n^m(\lambda, \mu) \tag{26.59}$$

This expression, also known as *Laplace's series*, is a uniformly convergent double series of spherical harmonics. Now we wish to apply (26.59) to the prognostic variables $X = \zeta, D, T, q$, and $\ln p_s$, where the variables ζ and D are used instead of U and V. Obviously, the meteorological variables vary on the surface of the earth not only with the longitude λ and the latitude φ but also with the generalized height ξ and with time. In order to use the series representation in the numerical integrations it is necessary to truncate the series and retain only a finite number of terms. We restrict the series expansion to $m \leq M$ and $n - |m| \leq J$, where M and

J are positive integers. Thus, we write

$$X(\lambda, \mu, \xi, t) = \sum_{m=-M}^{M} \sum_{n=|m|}^{|m|+J} X_n^m(\xi, t) Y_n^m(\lambda, \mu) \tag{26.60}$$

In order to obtain expressions for the expansion coefficients X_n^m we make use of the orthogonality relations (26.57). We obtain immediately the complex-valued expression

$$X_n^m(\xi, t) = \int_{-1}^{+1} \int_0^{2\pi} X(\lambda, \mu, \xi, t) \big[Y_n^m(\lambda, \mu)\big]^* \, d\lambda \, d\mu \tag{26.61}$$

Since X is real we recognize with the help of (26.53) that

$$X_n^{-m} = (-1)^m \big(X_n^m\big)^* \tag{26.62}$$

In the complex representation of Fourier series, the Fourier coefficients of X are defined by

$$X^m(\mu, \xi, t) = \frac{1}{2\pi} \int_0^{2\pi} X(\lambda, \mu, \xi, t) \exp(-im\lambda) \, d\lambda \tag{26.63}$$

Substitution of (26.60) into (26.63) and using the orthogonality condition (26.56) yields

$$X^m(\mu, \xi, t) = \frac{1}{\sqrt{2\pi}} \sum_{n=|m|}^{|m|+J} X_n^m(\xi, t) \widetilde{P}_n^m(\mu) \tag{26.64}$$

Therefore, (26.60) can be written as

$$X(\lambda, \mu, \xi, t) = \sum_{m=-M}^{M} X^m(\mu, \xi, t) \exp(im\lambda) \tag{26.65}$$

Equations (26.60) and (26.43) can now be used to obtain the horizontal derivatives of X entirely analytically:

$$\begin{aligned}
\frac{\partial X(\lambda, \mu, \xi, t)}{\partial \lambda} &= \sum_{m=-M}^{M} \sum_{n=|m|}^{|m|+J} X_n^m(\xi, t) \frac{im}{\sqrt{2\pi}} \widetilde{P}_n^m(\mu) \exp(im\lambda) \\
\frac{\partial X(\lambda, \mu, \xi, t)}{\partial \mu} &= \sum_{m=-M}^{M} \sum_{n=|m|}^{|m|+J} X_n^m(\xi, t) \frac{1}{\sqrt{2\pi}} \frac{d\widetilde{P}_n^m(\mu)}{d\mu} \exp(im\lambda)
\end{aligned} \tag{26.66}$$

The derivative of the associated Legendre polynomials involved in (26.66) can be expressed by the following recurrence relation:

$$(1-\mu^2)\frac{d}{d\mu}\widetilde{P}_n^m(\mu) = -n\epsilon_{n+1}^m \widetilde{P}_{n+1}^m(\mu) + (n+1)\epsilon_n^m \widetilde{P}_{n-1}^m(\mu)$$
$$\text{with} \quad \epsilon_n^m = \sqrt{\frac{n^2-m^2}{4n^2-1}} \tag{26.67}$$

We now multiply equation (26.35) by the factor $1/a^2$, where a is the mean radius of the earth. The first two terms are equivalent to the horizontal Laplacian operator in spherical coordinates. Thus we obtain

$$\nabla_{\mathrm{h}}^2 Y_n(\lambda,\mu) = -\frac{n(n+1)}{a^2} Y_n(\lambda,\mu) \tag{26.68}$$

This important relation will be used shortly.

Next we need to formulate the spectral coefficients for the vorticity and the divergence. The mathematical process is not difficult but is somewhat lengthy. Therefore, the derivation will be done in several steps. The first step is the formulation of the Fourier coefficients for U and V as stated by (26.23). Application of equation (26.63) to U gives immediately

$$U^m(\mu,\xi,t) = \frac{1}{2\pi a}\int_0^{2\pi}\left(-(1-\mu^2)\frac{\partial\psi}{\partial\mu} + \frac{\partial\chi}{\partial\lambda}\right)\exp(-im\lambda)\,d\lambda \tag{26.69}$$

Note that the stream function ψ and the velocity potential χ appear in differentiated form. On replacing X in (26.66) by ψ and then by χ we find

$$U^m(\mu,\xi,t) = \frac{1}{a(2\pi)^{3/2}}\sum_{m'=-M}^{M}\sum_{n=|m'|}^{|m'|+J}\left[-(1-\mu^2)\psi_n^{m'}(\xi,t)\frac{d}{d\mu}\widetilde{P}_n^{m'}(\mu)\right.$$
$$\left. + im'\chi_n^{m'}(\xi,t)\widetilde{P}_n^{m'}(\mu)\right]\int_0^{2\pi}\exp[i(m'-m)\lambda]\,d\lambda \tag{26.70}$$

For compactness of notation we will introduce

$$H_n^m(\mu) = -(1-\mu^2)\frac{d}{d\mu}\widetilde{P}_n^m(\mu) \tag{26.71}$$

into (26.70). Thus we find

$$U^m(\mu,\xi,t) = \frac{1}{\sqrt{2\pi}}\sum_{n=|m|}^{|m|+J}\frac{1}{a}\left[\psi_n^m(\xi,t)H_n^m(\mu) + im\chi_n^m(\xi,t)\widetilde{P}_n^m(\mu)\right] \tag{26.72}$$

Our next goal is to involve the vorticity and the divergence (see (26.24)) instead of the stream function and the velocity potential. On expanding the vorticity according to (26.60) we find without difficulty

$$\begin{aligned}\zeta(\lambda, \mu, \xi, t) &= \sum_{m=-M}^{M} \sum_{n=|m|}^{|m|+J} \zeta_n^m(\xi, t) Y_n^m(\lambda, \mu) \\ &= \sum_{m=-M}^{M} \sum_{n=|m|}^{|m|+J} \psi_n^m(\xi, t)\, \nabla_{\mathrm{h}}^2\left[Y_n^m(\lambda, \mu)\right] \\ &= \sum_{m=-M}^{M} \sum_{n=|m|}^{|m|+J} \psi_n^m(\xi, t)\left(-\frac{n(n+1)}{a^2}\right) Y_n^m(\lambda, \mu)\end{aligned} \tag{26.73}$$

where we have used (26.68) to eliminate the Laplacian operator. Inspection of (26.73) shows that

$$\zeta_n^m(\xi, t) = -\psi_n^m(\xi, t)\frac{n(n+1)}{a^2} \tag{26.74}$$

Since ζ and D have the same mathematical structure, we obtain immediately

$$D_n^m(\xi, t) = -\chi_n^m(\xi, t)\frac{n(n+1)}{a^2} \tag{26.75}$$

Substituting (26.74) and (26.75) into (26.73) expresses the Fourier coeffient U^m in terms of the spectral coefficients ζ_n^m and D_n^m,

$$U^m(\mu, \xi, t) = -\frac{1}{\sqrt{2\pi}} \sum_{n=|m|}^{|m|+J} \frac{a}{n(n+1)}\left[\zeta_n^m(\xi, t) H_n^m(\mu) + im D_n^m(\xi, t)\widetilde{P}_n^m(\mu)\right] \tag{26.76}$$

After introducing the definitions

$$\begin{aligned}U_\zeta^m(\mu, \xi, t) &= -\frac{1}{\sqrt{2\pi}} \sum_{n=|m|}^{|m|+J} \frac{a}{n(n+1)} \zeta_n^m(\xi, t) H_n^m(\mu) \\ U_D^m(\mu, \xi, t) &= -\frac{1}{\sqrt{2\pi}} \sum_{n=|m|}^{|m|+J} \frac{iam}{n(n+1)} D_n^m(\xi, t)\widetilde{P}_n^m(\mu)\end{aligned} \tag{26.77}$$

we may write

$$U^m(\mu, \xi, t) = U_\zeta^m(\mu, \xi, t) + U_D^m(\mu, \xi, t) \tag{26.78}$$

Inspection of equations (26.23a) and (26.23b) reveals a certain symmetry in the structure of the equations involving the stream function ψ and the velocity potential χ. This allows us to write down the spectral coefficients for the velocity components V_ζ^m and V_D^m without any additional work:

$$
\begin{aligned}
V_\zeta^m(\mu,\xi,t) &= -\frac{1}{\sqrt{2\pi}}\sum_{n=|m|}^{|m|+J}\frac{iam}{n(n+1)}\zeta_n^m(\xi,t)\widetilde{P}_n^m(\mu) \\
V_D^m(\mu,\xi,t) &= \frac{1}{\sqrt{2\pi}}\sum_{n=|m|}^{|m|+J}\frac{a}{n(n+1)}D_n^m(\xi,t)H_n^m(\mu)
\end{aligned}
\tag{26.79}
$$

In analogy to (26.78) we find

$$
V^m(\mu,\xi,t) = V_\zeta^m(\mu,\xi,t) + V_D^m(\mu,\xi,t) \tag{26.80}
$$

The function $H_n^m(\mu)$ is defined by (26.71) and can be evaluated by means of the recurrence relation (26.67).

Much work has gone into the formulation of a suitable truncation of the series expression and various truncation schemes have been proposed. One rather simple scheme is the so-called *triangular scheme* in which $J = M$. At present, this scheme is being used operationally with J exceeding 100.

26.3.3 The spectral tendencies

The general form of the spectral model is based on the work of Bourke (1974) and Hoskins and Simmons (1975). The formulation of the spectral tendencies for the prognostic variables $X = \zeta$, D, T, q, and $\ln p_s$ given by (26.18), (26.21), (26.8b), (26.9b), and (26.15) is accomplished with the help of equation (26.61). We will demonstrate the procedure for the vorticity equation. The remaining prognostic equations are handled likewise. We multiply both sides of equation (26.18) by $[Y_n^m(\lambda,\mu)]^*$ and then integrate over the sphere. For the left-hand side of the equation we find

$$
\int_{-1}^{+1}\int_0^{2\pi}\frac{\partial}{\partial t}\zeta(\lambda,\mu,\xi,t)\left[Y_n^m(\lambda,\mu)\right]^*\,d\lambda\,d\mu = \frac{\partial}{\partial t}\zeta_n^m(\xi,t) \tag{26.81}
$$

so that the spectral tendency equation for the vorticity is given by

$$
\begin{aligned}
\frac{\partial}{\partial t}\zeta_n^m(\xi,t) &= \frac{1}{a}\int_{-1}^{+1}\int_0^{2\pi}\left(\frac{1}{1-\mu^2}\frac{\partial}{\partial\lambda}(F_V+P_V)\right. \\
&\qquad \left. -\frac{\partial}{\partial\mu}(F_U+P_U)\right)\left[Y_n^m(\lambda,\mu)\right]^*\,d\lambda\,d\mu + (K_\zeta)_n^m \\
&\text{with}\quad (K_\zeta)_n^m = \int_{-1}^{+1}\int_0^{2\pi}K_\zeta\left[Y_n^m(\lambda,\mu)\right]^*\,d\lambda\,d\mu
\end{aligned}
\tag{26.82}
$$

Likewise, we find

$$\frac{\partial}{\partial t}D_n^m(\xi,t)=\frac{1}{a}\int_{-1}^{+1}\int_0^{2\pi}\left(\frac{1}{1-\mu^2}\frac{\partial}{\partial\lambda}(F_U+P_U)+\frac{\partial}{\partial\mu}(F_V+P_V)\right.$$
$$\left.-\,a\,\nabla_{\rm h}^2G\right)\left[Y_n^m(\lambda,\mu)\right]^*\,d\lambda\,d\mu+(K_D)_n^m$$
$$\text{with}\quad (K_D)_n^m=\int_{-1}^{+1}\int_0^{2\pi}K_D\left[Y_n^m(\lambda,\mu)\right]^*\,d\lambda\,d\mu \tag{26.83}$$

$$\frac{\partial}{\partial t}T_n^m(\xi,t)=\int_{-1}^{+1}\int_0^{2\pi}(F_T+Q_{\rm h})\left[Y_n^m(\lambda,\mu)\right]^*\,d\lambda\,d\mu \tag{26.84}$$

$$\frac{\partial}{\partial t}q_n^m(\xi,t)=\int_{-1}^{+1}\int_0^{2\pi}(F_q+Q_q)\left[Y_n^m(\lambda,\mu)\right]^*\,d\lambda\,d\mu \tag{26.85}$$

$$\frac{\partial}{\partial t}(\ln p_{\rm s})_n^m=\int_{-1}^{+1}\int_0^{2\pi}F_p\left[Y_n^m(\lambda,\mu)\right]^*\,d\lambda\,d\mu$$
$$\text{with}\quad F_p=-\frac{1}{p_{\rm s}}\int_0^1\nabla_{\rm h}\cdot\left(\mathbf{v}_{\rm h}\frac{\partial p}{\partial\xi}\right)d\xi \tag{26.86}$$

Equations (26.82) and (26.83) contain partial derivatives, which can be eliminated by partial integration to bring them into the form of the remaining prognostic equations. Observing that $P_n^m(\pm 1)=0\,(m>0)$ and using the cyclic boundary condition $(0,2\pi)$ which applies when one is integrating along a fixed latitude circle, we easily obtain

$$\frac{\partial}{\partial t}\zeta_n^m(\xi,t)=\frac{1}{a}\int_{-1}^{+1}\int_0^{2\pi}\left(\frac{im}{1-\mu^2}(F_V+P_V)\left[Y_n^m(\lambda,\mu)\right]^*\right.$$
$$\left.+\,(F_U+P_U)\frac{d}{d\mu}\left[Y_n^m(\lambda,\mu)\right]^*\right)d\lambda\,d\mu+(K_\zeta)_n^m \tag{26.87}$$

and

$$\frac{\partial}{\partial t}D_n^m(\xi,t)=\frac{1}{a}\int_{-1}^{+1}\int_0^{2\pi}\left(\frac{im}{1-\mu^2}(F_U+P_U)\left[Y_n^m(\lambda,\mu)\right]^*\right.$$
$$-\,(F_V+P_V)\frac{d}{d\mu}\left[Y_n^m(\lambda,\mu)\right]^*$$
$$\left.+\,\frac{n(n+1)}{a}G\left[Y_n^m(\lambda,\mu)\right]^*\right)d\lambda\,d\mu+(K_D)_n^m \tag{26.88}$$

Every forecast requires a set of consistent initial data. In the ideal case these data should be free from any imbalances between the wind and the mass field. Sophisticated procedures for providing such initial data sets have been worked out by various research groups. However, these initialization techniques are model-specific.

Here we will simply assume that such data sets exist for the prognostic variables $X = \zeta, D, T, q$, and $\ln p_s$ on a suitable grid of points over the sphere. With the help of equation (26.61) the quantities $X_n^m\big|_{t=0} = \zeta_n^m, D_n^m, T_n^m, q_n^m$, and $\ln p_s$ can be computed. These quantities are required for the first time step in order to evaluate the tendency equations (26.84)–(26.88). Moreover, the $X_n^m\big|_{t=0}$ are sufficient for calculating $U_\zeta^m, U_D^m, V_\zeta^m$, and V_D^m, so U^m and V^m can be found from (26.78) and (26.80). With the help of (26.65) the velocity components U and V which are needed for the evaluation of F_U, F_V in (26.18) and F_T, F_q in (26.8) and (26.9) can then be found. The horizontal derivatives $\partial T/\partial\lambda$, $\partial T/\partial\mu$, $\partial q/\partial\lambda$, $\partial q/\partial\mu$, $\partial \ln p_s/\partial\lambda$, and $\partial \ln p_s/\partial\mu$ can be found analytically by summing (26.66) over m. This information should also be sufficient for computing F_p in (26.86) and G in (26.21) as well as all parameterized quantities. Since all field quantities are evaluated at the same grid points, the forecast can be started. Whenever desired, the conversion to the variables $X = \zeta, D, T, q$, and $\ln p_s$ can be achieved with the help of (26.60).

To carry out the forecast, the model must contain a suitable vertical finite-difference scheme to evaluate the terms involving the generalized vertical coordinate ξ. There exist various suitable vertical schemes. In the earlier phases of modeling the σ system ($\xi = \sigma = p/p_s$) which satisfies the condition that $\sigma = 0$ at the top of the atmosphere and $\sigma = 1$ at the surface was used. We will omit any discussion of the specific numerical procedures such as the evaluation of $(K_\xi)_n^m$ and $(K_D)_n^m$ since these are subject to continual revision. As stated before, the spectral model is well suited for the application of the semi-implicit method. Following Robert *et al.* (1972), the model uses such a scheme for the equations of divergence, temperature, and the surface pressure. In addition, the model uses a semi-implicit method for the zonal advection terms in various equations.

Many details, too numerous to be described here, are required for proper handling of the model. For further information the original literature should be consulted.

26.4 Problems

26.1: Omitting the underlined terms in equation (19.15) and replacing the radius r by the constant radius a, show that the horizontal components of the equation of motion can be written as stated in (26.5a) and (26.5b). Ignore frictional effects.

26.2: Use equation (1.77) and the formulas stated in problem (1.8) to show that the vorticity can be written in the form (26.6). Set $r = a$ whenever r appears in undifferentiated form.

26.3: Prove the validity of equation (26.19).

26.4: Show that the definition of ∇_h^2 stated in (26.21) is correct.

26.5: Find $P_3(\mu)$ and show that this function satisfies equation (26.37).

26.6: Differentiate Legendre's equation (26.37) m times to prove the validity of (26.39).

26.7: Show that equation (26.39) can be transformed to give equation (26.41) by using the transformation (26.40).

26.8: Use the recurrence relations

$$(1-\mu^2)\frac{dP_n^m}{d\mu} = -n\mu P_n^m + (n+m)P_{n-1}^m$$

and

$$\mu P_n^m = \frac{n+m}{2n+1}P_{n-1}^m + \frac{n-m+1}{2n+1}P_{n+1}^m$$

to prove the validity of equation (26.67).

27

Predictability

The *convection equations* in the form presented by Lorenz (1963) have been investigated very thoroughly because they exhibit chaotic behavior for a wide range of values of model parameters. Lorenz discovered that his simple-looking three-dimensional deterministic system yielded solutions that oscillate irregularly, never exactly repeating but always remaining in a bounded region of phase space. This aperiodic long-term behavior exhibits a very sensitive dependence on initial conditions. Since the *Lorenz system* consists of nonlinear differential equations, he found the trajectories describing the convection by numerical integration. By plotting the trajectories in three dimensions, he discovered that they settled onto a very complicated set, which is now called a *strange attractor*. The expression *deterministic system* implies that the system has no random or noisy inputs or parameters. The irregular behavior results from the nonlinear terms.

27.1 Derivation and discussion of the Lorenz equations

Lorenz obtained his mathematical system by drastically simplifying the convection equations of Saltzman (1962) which describe convection rolls in the atmosphere. We will now derive and discuss the Lorenz equations in some detail, beginning with the basic model equations for two-dimensional flow in the (x, z)-plane. We apply the *Boussinesq approximation* by assuming that no divergence takes place and that the rotation of the earth can be neglected. In the (x, z)-plane the velocity vector $\mathbf{v}_{\mathrm{v}}$, the gradient operator ∇_{v}, the divergence D_{v}, and the vorticity ζ_{v} are defined by

$$\mathbf{v}_{\mathrm{v}} = u\mathbf{i} + w\mathbf{k}, \qquad \nabla_{\mathrm{v}} = \mathbf{i}\frac{\partial}{\partial x} + \mathbf{k}\frac{\partial}{\partial z}, \qquad D_{\mathrm{v}} = \frac{\partial u}{\partial x} + \frac{\partial w}{\partial z}, \qquad \zeta_{\mathrm{v}} = \frac{\partial w}{\partial x} - \frac{\partial u}{\partial z} \tag{27.1}$$

where the index v denotes the vertical direction. The equation of motion in this plane is given by

$$\begin{aligned}
&\text{(a)} \quad \frac{\partial u}{\partial t} + \mathbf{v}_{\mathrm{v}} \cdot \nabla_{\mathrm{v}} u + \alpha \frac{\partial p}{\partial x} - \nu \nabla_{\mathrm{v}}^2 u = 0 \\
&\text{(b)} \quad \frac{\partial w}{\partial t} + \mathbf{v}_{\mathrm{v}} \cdot \nabla_{\mathrm{v}} w + \alpha \frac{\partial p}{\partial z} - \nu \nabla_{\mathrm{v}}^2 w = 0
\end{aligned} \tag{27.2}$$

where ν is the kinematic viscosity and α is the specific volume. By differentiating (27.2a) and (27.2b) with respect to z and x, and subtracting one of the resulting equations from the other, we obtain the vorticity equation

$$\frac{\partial \zeta_{\mathrm{v}}}{\partial t} + u \frac{\partial \zeta_{\mathrm{v}}}{\partial x} + w \frac{\partial \zeta_{\mathrm{v}}}{\partial z} + \frac{\partial \alpha}{\partial x} \frac{\partial p}{\partial z} - \frac{\partial \alpha}{\partial z} \frac{\partial p}{\partial x} - \nu \nabla_{\mathrm{v}}^2 \zeta_{\mathrm{v}} = 0 \tag{27.3}$$

The rotational axis of the vorticity is orthogonal to the (x, z)-plane, thus pointing in the y-direction. Since the divergence is assumed to be zero, the velocity field and the vorticity can be expressed in terms of the stream function $\psi(x, z, t)$:

$$u = -\frac{\partial \psi}{\partial z}, \qquad w = \frac{\partial \psi}{\partial x}, \qquad \zeta_{\mathrm{v}} = \frac{\partial^2 \psi}{\partial x^2} + \frac{\partial^2 \psi}{\partial z^2} = \nabla_{\mathrm{v}}^2 \psi \tag{27.4}$$

Thus the vorticity equation can also be written as

$$\frac{\partial}{\partial t}(\nabla_{\mathrm{v}}^2 \psi) + J(\psi, \nabla_{\mathrm{v}}^2 \psi) = J(\alpha, -p) + \nu \nabla_{\mathrm{v}}^2 (\nabla_{\mathrm{v}}^2 \psi) \tag{27.5}$$

The Jacobian operator $J(\alpha, -p)$ is the *solenoidal term* representing the number of unit solenoids in the (x, z)-plane. The solenoids cause an acceleration of the circulation driving the convection.

With Saltzman (1962), we make the following assumptions about the temperature and the specific volume fields. The temperature distribution is expressed by

$$T(x, z, t) = T_m(t) + T_1(x, z, t) \tag{27.6}$$

where $T_{\mathrm{m}}(t)$ is a mean value over the entire convection region and $T_1(x, z, t)$ is the deviation from the mean value. The *anelastic assumption* is assumed to apply, so the relative temperature deviations are large in comparison with the relative pressure variations. Therefore, the Taylor expansion of the specific volume α about an equilibrium point (T_0, p_0) can be approximated by

$$\begin{aligned}
\alpha(T, p) &= \alpha(T_0, p_0) + \left.\frac{\partial \alpha}{\partial T}\right|_{T_0, p_0} (T - T_0) + \left.\frac{\partial \alpha}{\partial p}\right|_{T_0, p_0} (p - p_0) + \cdots \\
&\approx \alpha_0 + \frac{\alpha_0}{T_0}(T - T_0) \quad \text{with} \quad \alpha_0 p_0 = R_0 T_0
\end{aligned} \tag{27.7}$$

On replacing T_0 in this expression by T_m and α_0 by α_m we obtain the equation of state for the specific volume:

$$\alpha(T) = \alpha_\mathrm{m}(1 + \epsilon T_1) \tag{27.8}$$

where α_m is the mean value of the specific volume of the entire region of convection and $\epsilon = 1/T_\mathrm{m}$ is the cubic expansion coefficient of the air.

Next we need a suitable expression for the pressure $p(x, z, t)$, which may be expressed as the sum of the hydrostatic pressure $p_\mathrm{h}(x, z, t)$ and the deviation $p'(x, z, t)$ from this value. Let $\overline{p}_\mathrm{h}(z, t)$ represent the horizontal average of the hydrostatic pressure at height z and time t and $\overline{p}_\mathrm{h}(z, 0)$ be the initial value at time $t = t_0$. Furthermore, we introduce a new pressure variable P according to

$$\begin{aligned} P(x, z, t) &= [p(x, z, t) - \overline{p}_\mathrm{h}(z, 0)]\alpha_\mathrm{m} \\ &= [p(x, z, t) - \overline{p}_\mathrm{h}(z_\mathrm{t}, 0)]\alpha_\mathrm{m} - g(z_\mathrm{t} - z) \\ \text{with} \quad p(x, z, t) &= p_\mathrm{h}(x, z, t) + p'(x, z, t), \\ \overline{p}_\mathrm{h}(z, 0) &= \overline{p}_\mathrm{h}(z_\mathrm{t}, 0) + \frac{g}{\alpha_\mathrm{m}}(z_\mathrm{t} - z) \end{aligned} \tag{27.9}$$

where z_t denotes the top of the convective layer. In reality P has dimensions of energy per unit mass. Thus we may express the Jacobian $J(\alpha, -p)$ by

$$J(\alpha, -p) = J\left(\alpha_\mathrm{m}(1 + \epsilon T_1), -\frac{P}{\alpha_\mathrm{m}} - p_\mathrm{h}(z_t, 0) - \frac{g}{\alpha_\mathrm{m}}(z_\mathrm{t} - z)\right) \tag{27.10a}$$

Expansion of the Jacobian yields

$$J(\alpha, -p) = \epsilon J(T_1, -P) + \epsilon g \frac{\partial T_1}{\partial x} \approx \epsilon g \frac{\partial T_1}{\partial x} \tag{27.10b}$$

The first term on the right-hand side of this expression is of second order in comparison with the second term and has, therefore, been neglected. By utilizing (27.10b) in (27.5) we obtain Saltzman's approximate form of the vorticity equation:

$$\boxed{\frac{\partial}{\partial t}(\nabla_\mathrm{v}^2 \psi) + J(\psi, \nabla_\mathrm{v}^2 \psi) = \epsilon g \frac{\partial T_1}{\partial x} + \nu \nabla_\mathrm{v}^2(\nabla_\mathrm{v}^2 \psi)} \tag{27.11}$$

This is an equation in two unknowns involving the stream function ψ and the temperature deviation T_1. Therefore, a second prognostic equation is needed in order to determine T_1. In order to avoid the appearance of new dependent variables Saltzman chooses the very simplified form of the heat equation

$$\frac{\partial T_1}{\partial t} + \mathbf{v}_\mathrm{v} \cdot \nabla_\mathrm{v} T_1 = K \nabla_\mathrm{v}^2 T_1 \tag{27.12}$$

where K is the exchange coefficient for heat. Introducing the stream function ψ according to (27.4), the heat equation can be written in the concise form

$$\boxed{\frac{\partial T_1}{\partial t} + J(\psi, T_1) = K\,\nabla_{\rm v}^2 T_1} \tag{27.13}$$

Thus equations (27.11) and (27.13) describe the model convection in the (x, z)-plane. It should be realized that, due to the appearance of clouds and phase transitions, the actual convective processes are much more complicated.

The temperature deviation T_1 may be decomposed and expressed as the sum of an average value $\overline{T}_1(z, t)$ along the x-axis plus the deviation $T_1'(x, z, t)$:

$$T_1(x, z, t) = \overline{T}_1(z, t) + T_1'(x, z, t) \tag{27.14}$$

Furthermore, $\overline{T}_1(z, t)$ can be decomposed into a part representing a linear variation between the upper and lower boundary and a departure T'' from the linear variation:

$$\overline{T}_1(z, t) = \left(\overline{T}_1(0, t) - \Delta T_0(t)\,\frac{z}{z_{\rm t}}\right) + \overline{T}_1''(z, t) \tag{27.15}$$

with $\Delta T_0(t) = \overline{T}_1(0, t) - \overline{T}_1(z_{\rm t}, t)$. Now equation (27.14) can be written in the following form:

$$T_1(x, z, t) = \left(\overline{T}_1(0, t) - \Delta T_0(t)\,\frac{z}{z_{\rm t}}\right) + \vartheta(x, z, t) \tag{27.16}$$

with $\vartheta(x, z, t) = \overline{T}_1''(z, t) + T_1'(x, z, t)$. Saltzmann assumes that the temperatures at the lower and upper model boundaries are kept constant by external heating. This implies

$$\overline{T}_1(0, t) = \overline{T}_1(0), \quad \overline{T}_1(z_{\rm t}, t) = \overline{T}_1(z_{\rm t}) \implies \Delta T_0 = \overline{T}_1(0) - \overline{T}_1(z_{\rm t}) = \text{constant} \tag{27.17}$$

so that

$$\frac{\partial T_1}{\partial t} = \frac{\partial \vartheta}{\partial t}, \qquad \frac{\partial T_1}{\partial x} = \frac{\partial \vartheta}{\partial x}, \qquad \frac{\partial T_1}{\partial z} = -\frac{\Delta T_0}{z_{\rm t}} + \frac{\partial \vartheta}{\partial z}, \qquad \nabla_{\rm v}^2 T_1 = \nabla_{\rm v}^2 \vartheta \tag{27.18}$$

follow from (27.16) and (27.17). From (27.11) and (27.13) we finally find a pair of partial differential equations describing the convection

$$\boxed{\begin{aligned} \frac{\partial}{\partial t}\left(\nabla_{\rm v}^2 \psi\right) + J\left(\psi, \nabla_{\rm v}^2 \psi\right) &= g\epsilon\,\frac{\partial \vartheta}{\partial x} + \nu\,\nabla_{\rm v}^2(\nabla_{\rm v}^2 \psi) \\ \frac{\partial \vartheta}{\partial t} + J(\psi, \vartheta) &= \frac{\Delta T_0}{z_{\rm t}}\,\frac{\partial \psi}{\partial x} + K\,\nabla_{\rm v}^2 \vartheta \end{aligned}} \tag{27.19}$$

The first terms on the right-hand sides of (27.19) represent the *driving forces* of the convection while the second terms cause damping due to viscosity and heat conduction.

We wish to introduce the dimensionless *Rayleigh number*, which is defined by

$$\boxed{Ra = \frac{g\epsilon z_t^3 \, \Delta T_0}{\nu K}} \tag{27.20}$$

Ra is a measure of convective instability. The numerator of Ra involves the buoyancy force per unit mass $g\epsilon\Delta T_0$ while the denominator denotes damping expressed by the kinematic viscosity ν and the heat-conduction coefficient K. If z_t and l stand for the height and the width of the convection rolls, the aspect ratio $a = z_t/l$ defines the *critical Rayleigh number*

$$Ra_c = \frac{\pi^4(1+a^2)^3}{a^2} \tag{27.21}$$

The critical Rayleigh number depends on the geometric parameters of the model. Whenever the ratio $r = Ra/Ra_c > 1$, convection is expected to occur.

Saltzman (1962) assumes that the stream function and the temperature departure in equation (27.19) can be expressed as sums of double Fourier components. Lorenz (1963), motivated by Saltzman's work, assumed the validity of the simple trial solutions

$$\begin{aligned} \psi &= \frac{\sqrt{2}(1+a^2)K}{a} X(t) \sin\left(\frac{\pi a x}{z_t}\right) \sin\left(\frac{\pi z}{z_t}\right) \\ \vartheta &= \frac{\Delta T_0}{\pi r}\left[Y(t)\sqrt{2}\cos\left(\frac{\pi a x}{z_t}\right)\sin\left(\frac{\pi z}{z_t}\right) - Z(t)\sin\left(\frac{2\pi z}{z_t}\right)\right] \end{aligned} \tag{27.23}$$

containing only three development coefficients $X(t)$, $Y(t)$, $Z(t)$, which are slowly varying amplitudes in time. According to the trial solution, the X mode represents the flow pattern, the Y mode the temperature cells, and the Z mode the temperature stratification. When equations (27.23) are substituted into (27.19), ignoring trigonometric terms that do not occur in the trial solution, we obtain the famous *Lorenz system* consisting of three ordinary differential equations in dimensionless time t^*:

$$\boxed{\begin{aligned} \frac{dX}{dt^*} &= \dot{X} = -\sigma X + \sigma Y = f_1(X, Y, Z) \\ \frac{dY}{dt^*} &= \dot{Y} = rX - Y - XZ = f_2(X, Y, Z) \\ \frac{dZ}{dt^*} &= \dot{Z} = XY - bZ = f_3(X, Y, Z) \end{aligned}}$$

$$\text{with} \quad t^* = \frac{\pi^2(a^2+1)K}{z_t^2}t, \qquad \sigma = \frac{\nu}{K}, \qquad b = \frac{4}{a^2+1} > 0, \qquad r = \frac{Ra}{Ra_c} \tag{27.24}$$

The term σ is the *Prandtl number*. The consequence of the simplification is that the advection term in the first equation of (27.19) is suppressed.

The Lorenz system is a three-dimensional autonomous dynamic system, which may formally be expressed by

$$\dot{\mathbf{X}} = \mathbf{f}(\mathbf{X}) \quad \text{with} \quad \mathbf{X} = (X, Y, Z) \tag{27.25}$$

If the function $\mathbf{f}$ depends not only on the vector $\mathbf{X}$ but also explicitly on time t, we speak of a nonautonomous dynamic system. This type of system is more difficult to handle.

Owing to the simplifications introduced by Lorenz, this approach does not realistically model convection rolls. What Lorenz had in mind was to show that the low-order nonlinear dynamic system (27.24) with only three degrees of freedom reacts very sensitively to small variations in the initial conditions. This simple fact has many important consequences for weather prediction.

The discussion of simple properties of the Lorenz system requires that the reader is somewhat familiar with a few basic concepts of nonlinear dynamics. Some readers may be quite familiar with this subject; others may wish to consult one of the many textbooks which are available at all levels of sophistication. For the reader entirely unfamiliar with the subject, we gave a brief and incomplete introduction in Chapter M7. There we extract most of the required information from Strogatz's excellent book on *Nonlinear Dynamics and Chaos* (1994). His very informative book is much more than a first introduction. It provides pleasant reading at a level of mathematical sophistication that does not exceed the usual background of a senior student in meteorology. The next four sections again follow his book.

Inspection of (27.24) shows that the only nonlinearities are the quadratic terms XY and XZ. Lorenz found that, over a wide range of parameters, the numerical solutions oscillate irregularly in time, but they never exactly repeat. This led him to title his now famous paper "Deterministic Non-Periodic Flow". Moreover, he found that the solutions to (27.24) always remain in a bounded region of phase space. When he plotted the trajectories in three dimensions, there emerged a strange figure that resembled a pair of butterfly wings. This figure is now called a strange attractor. In the following we list important simple properties of the Lorenz equations.

27.1.1 Nonlinearity

The Lorenz system consists of the two nonlinearities XY and XZ.

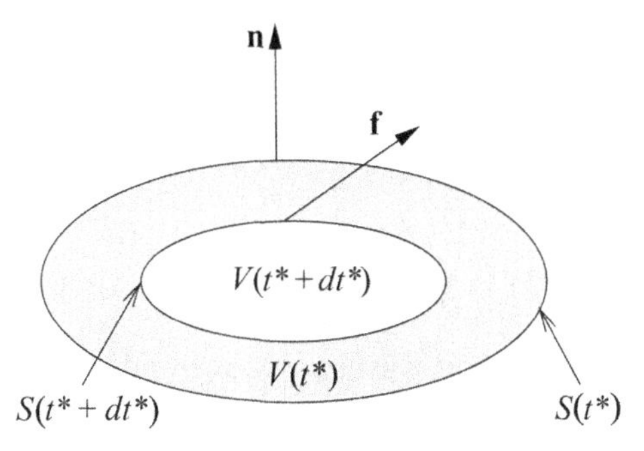

Fig. 27.1 Contraction in volume of a dissipative system.

27.1.2 Symmetry

On replacing (X, Y) in (27.24) by $(-X, -Y)$, the equations do not change. Thus, if $[X(t^*), Y(t^*), Z(t^*)]$ is a solution to (27.24), so is $[-X(t^*), -Y(t^*), Z(t^*)]$.

27.1.3 The Lorenz system is dissipative

In general, a dynamic system is either *conservative* or *dissipative*. An important property of phase space is the preservation of phase volume of conservative or constant-energy systems. For dissipative systems, the volume in phase space contracts under the flow. In this case the divergence of the phase velocity is negative. In contrast, if the divergence is zero, the dynamic system is conservative.

To illuminate the idea, we consider an arbitrary closed surface $S(t^*)$ of the volume $V(t^*)$ as shown in Figure 27.1. We may think of points on S as initial conditions for trajectories that evolve during the time interval dt^*, thus resulting in a change in volume. What is the volume at time $t^* + dt^*$? As usual, the vector $\mathbf{n}$ denotes the outward normal on S. We do not have to restrict ourselves to the Lorenz system, but we consider the arbitrary three-dimensional system $\dot{\mathbf{X}} = \mathbf{f}(\mathbf{X})$. The scalar product $\mathbf{f} \cdot \mathbf{n}$ is the outward normal component of the velocity vector $\dot{\mathbf{X}}$. During the time interval dt^* a patch of surface area $d\mathbf{S} = \mathbf{n}\,dS$ sweeps out a volume element $dV = (\mathbf{f} \cdot \mathbf{n}\,dt^*)\,dS = V(t^* + dt^*) - V(t^*)$. Integrating over all patches yields

$$\dot{V} = \frac{V(t^* + dt^*) - V(t^*)}{dt^*} = \int_S \mathbf{f} \cdot \mathbf{n}\,dS = \int_V \nabla \cdot \mathbf{f}\,dV \tag{27.26}$$

where we have used Gauss' divergence theorem (M6.31). For the Lorenz system the divergence $\nabla \cdot \mathbf{f}$ turns out to be

$$\nabla \cdot \mathbf{f} = \frac{\partial f_1}{\partial X} + \frac{\partial f_2}{\partial Y} + \frac{\partial f_3}{\partial Z} = -(\sigma + 1 + b) < 0 \tag{27.27}$$

On substituting this expression into (27.26) we obtain for the change of the volume with time

$$\dot{V}(t^*) = -(\sigma + 1 + b)V(t^*) \tag{27.28a}$$

from which it follows that

$$V(t^*) = V(0)\exp[-(\sigma + 1 + b)t^*] \tag{27.28b}$$

Thus the volume in phase space shrinks exponentially fast. This implies that a large blob of initial conditions eventually shrinks to a limiting set of zero volume. Stating it differently, all trajectories starting in the blob end up somewhere in the limiting set. We recognize that volume contraction imposes stringent constraints on possible candidate solutions.

27.1.4 Fixed points

The Lorenz system has two types of *fixed points.*

(1) The origin is a fixed point for all parameter values of r, σ, b:

$$(X^*, Y^*, Z^*) = (0, 0, 0) \tag{27.29a}$$

This state at the origin of the phase space does not describe any convection, but rather describes the equalization of temperature resulting from conduction of heat. The system remains at rest.

(2) For $r > 1$ there is a pair of symmetric fixed points. Setting $\dot{X} = 0$, $\dot{Y} = 0$, $\dot{Z} = 0$ in (27.24) gives

$$X^* = \pm\sqrt{b(r-1)}, \qquad Y^* = X^*, \qquad Z^* = r - 1 \tag{27.29b}$$

Lorenz called these symmetric points C^+ and C^-. As $r \to 1$, the symmetric fixed points collide with the origin in a *pitchfork bifurcation.* Of meteorological interest is that the symmetric fixed points represent left- or right-turning convection rolls.

27.1.5 Linear stability at the origin

The three-dimensional Jacobian matrix of the system is given by

$$A = \begin{pmatrix} \dfrac{\partial f_1}{\partial X} & \dfrac{\partial f_1}{\partial Y} & \dfrac{\partial f_1}{\partial Z} \\ \dfrac{\partial f_2}{\partial X} & \dfrac{\partial f_2}{\partial Y} & \dfrac{\partial f_2}{\partial Z} \\ \dfrac{\partial f_3}{\partial X} & \dfrac{\partial f_3}{\partial Y} & \dfrac{\partial f_3}{\partial Z} \end{pmatrix} = \begin{pmatrix} -\sigma & \sigma & 0 \\ r - Z & -1 & -X \\ Y & X & -b \end{pmatrix} \tag{27.30}$$

By evaluating this matrix at the origin we obtain for the linearized system

$$\begin{pmatrix} \dot{X} \\ \dot{Y} \\ \dot{Z} \end{pmatrix} = \begin{pmatrix} -\sigma & \sigma & 0 \\ r & -1 & -0 \\ 0 & 0 & -b \end{pmatrix} \begin{pmatrix} X \\ Y \\ Z \end{pmatrix} \tag{27.31}$$

From this equation we recognize that the Z-component is uncoupled from the remaining system, so

$$\begin{pmatrix} \dot{X} \\ \dot{Y} \end{pmatrix} = \begin{pmatrix} -\sigma & \sigma \\ r & -1 \end{pmatrix} \begin{pmatrix} X \\ Y \end{pmatrix}, \qquad \dot{Z} = -bZ \tag{27.32}$$

Hence $Z(t^*)$ is proportional to $\exp(-bt^*)$, i.e. $Z(t^*)$ is decreasing exponentially fast to zero as $t^* \to \infty$. The trace and the determinant of the square matrix in (27.32) are given by

$$\tau = \lambda_1 + \lambda_2 = -(\sigma + 1), \qquad \Delta = \lambda_1 \lambda_2 = \sigma(1 - r) \tag{27.33}$$

According to Figure M7.13, for $r > 1$ the origin is a saddle point since the determinant $\Delta < 0$. It is easy to show that, in this case, one of the eigenvalues of the two-dimensional system is negative, while the other one is positive. Counting the Z-component, we have two incoming and one outgoing directions. If $r < 1$ all eigenvalues are negative. This means that all directions are incoming, so the origin is a sink. In this case $\tau^2 - 4\Delta = (\sigma - 1)^2 + 4\sigma r > 0$. Thus Figure M7.13 reveals that, for $r < 1$, the origin is a stable node.

We will now verify that, for $r < 1$, every trajectory approaches the origin as the time t^* approaches infinity. This implies that the origin is a *globally stable point.* Hence there can be no limit cycle or chaos for $r < 1$. We verify this statement with the help of the *Liapunov function*

$$V(X, Y, Z) = \frac{1}{\sigma} X^2 + Y^2 + Z^2 \tag{27.34}$$

Surfaces $V =$ constant are concentric ellipsoids about the origin. First we wish to show that, for $(X, Y, Z) \neq (0, 0, 0)$ and $r < 1$, the time derivative of the volume is negative. Carrying out the differentiation dV/dt^* using (27.24), we find

$$\begin{aligned} \frac{1}{2}\dot{V} &= \frac{1}{\sigma} X\dot{X} + Y\dot{Y} + Z\dot{Z} \\ &= (XY - X^2) + (rXY - Y^2 - XYZ) + (XYZ - bZ^2) \\ &= (r+1)XY - X^2 - Y^2 - bZ^2 \end{aligned} \tag{27.35}$$

Completing squares results in

$$\frac{1}{2}\dot{V} = -\left(X - \frac{r+1}{2}Y\right)^2 - \left[1 - \left(\frac{r+1}{2}\right)^2\right]Y^2 - bZ^2 \tag{27.36}$$

Inspection shows that, for $r < 1$, this expression cannot be positive. The question of whether $\dot{V}$ could be zero arises. This is possible only if each term on the right-hand side vanishes separately. This happens if $(X, Y, Z) = (0, 0, 0)$. The total conclusion is that $\dot{V} = 0$ only if $(X, Y, Z) = (0, 0, 0)$; otherwise $\dot{V} < 0$ and the origin is globally stable.

27.1.6 Stability of C^+ and C^-

By evaluating the left-hand side of (27.30) at $(X^*, Y^*, Z^*) = (C^{\pm}, C^{\pm}, r-1)$ we obtain

$$\begin{aligned}&\begin{pmatrix} -\sigma & \sigma & 0 \\ r - Z & -1 & -X \\ Y & X & -b \end{pmatrix}_{C^{\pm}, C^{\pm}, r-1} \\ &= \begin{pmatrix} -\sigma & \sigma & 0 \\ 1 & -1 & \mp\sqrt{b(r-1)} \\ \pm\sqrt{b(r-1)} & \pm\sqrt{b(r-1)} & -b \end{pmatrix}\end{aligned} \tag{27.37}$$

The three eigenvalues of this linear system are found by solving the equation for the determinant $|A - \lambda E| = 0$, resulting in the characteristic equation

$$\lambda^3 + (\sigma + b + 1)\lambda^2 + b(\sigma + r)\lambda + 2b\sigma(r-1) = 0 \tag{27.38}$$

Inspection of the fixed points $C^{\pm} = \pm\sqrt{b(r-1)}$ shows that the condition $r \geq 1$ must hold, otherwise C^+ and C^- cannot exist.

First let us consider the special case that $r = 1$, that is $Ra = Ra_c$, marking the onset of convection. It is not difficult to recognize from (27.38) that the eigenvalues are given by $\lambda_1 = 0$, $\lambda_2 = -b$, $\lambda_3 = -(\sigma + 1)$. In this case one speaks of the *margin of stability* since no eigenvalue is larger than zero.

In order to study the characteristic equation in the more general case, we write equation (27.38) in the form

$$\lambda^3 + a_2\lambda^2 + a_1\lambda + a_0 = 0 \tag{27.39}$$

According to Vieta's theorem, see for example Abramowitz and Segun (1968), there exists the relation

$$\lambda_1 \lambda_2 \lambda_3 = -a_0 \tag{27.40}$$

By substituting $\lambda_2 = -b$ and $\lambda_3 = -(\sigma + 1)$ into this equation, we obtain an approximate value for λ_1, which may be used if $r \approx 1$:

$$\lambda_1 \approx -\frac{2\sigma(r-1)}{1+\sigma} \quad \text{for} \quad r \approx 1 \tag{27.41}$$

Thus, for $r > 1$ and in the limit $r \rightarrow 1$, stability is lost because the approximate negative eigenvalue λ_1 approaches 0.

Now suppose that $r > 1$ so that all coefficients in the characteristic equation (27.38) are positive. The theory of cubic equations provides several intricate relationships for the roots of this equation involving the coefficients a_0, a_1, and a_2. With some patience one can show that there exists one real root and two complex-conjugate roots. We have shown already by an approximate method that the real root $\lambda_1 < 0$ so that $\lambda_{2,3} = \lambda_r \pm i\lambda_i$. Convection rolls remain stable as long as $\lambda_r < 0$ and begin to be unstable if $\lambda_r = 0$ so that $\lambda_{2,3}$ are purely imaginary. In this case r assumes a critical value r_H where the subscript H has been used to show that C^+ and C^- lose stability at the *Hopf bifurcation* at $r = r_H$.

We will now find an expression for $r = r_H$. According to a theorem attributed to Vieta, the sum of the roots is given by

$$\lambda_1 + \lambda_2 + \lambda_3 = -a_2 = -(\sigma + b + 1) \tag{27.42}$$

Since the complex conjugates are purely imaginary, the sum of the complex conjugates $\lambda_2 + \lambda_3$ is zero, so we obtain

$$\lambda_1 = -(\sigma + b + 1) \tag{27.43}$$

Substituting (27.43) into (27.38) results in

$$\begin{aligned} &-(\sigma + b + 1)^3 + (\sigma + b + 1)(\sigma + b + 1)^2 \\ &- b(\sigma + r)(\sigma + b + 1) + 2b\sigma(r-1) = 0 \end{aligned} \tag{27.44}$$

After some elementary algebra we find the desired expression for r_H:

$$r = r_H = \frac{\sigma(\sigma + b + 3)}{\sigma - b - 1} \tag{27.45}$$

To repeat, the quantity r_H is that value of r for which the complex-conjugate roots are purely imaginary. Instability can arise only for such σ, b that $r_H > 1$. Thus the fixed points C^+ and C^- remain stable if and only if either of the following condition holds:

$$\text{(i)}\ \sigma < b+1 \quad \text{and} \quad r > 1, \qquad \text{(ii)}\ \sigma > b+1 \quad \text{and} \quad 1 < r < r_{\mathrm{H}}.$$

Both conditions can be readily understood. The parameter r must be larger than 1 so that the fixed points C^+ and C^- can exist. (i) For $\sigma < b+1$ there is no positive value of r to satify (27.45). (ii) $\sigma > b+1$ causes equation (27.45) to be positive but the value of r is still less than the critical value r_{H}.

In meteorological terms, stationary convection rolls remain stable as long as either condition is satified. For large enough Rayleigh numbers the convection becomes unstable. If the critical value r_{H} is exceeded then the real parts λ_{r} of the eigenvalues $\lambda_{2,3}$ will be positive and irregular convective motion takes place, and the so-called *deterministic chaos* may occur.

Following Lorenz, we study the particular case $\sigma = 10$. Selecting $a^2 = \frac{1}{2}$, we find from (27.24) $b = \frac{8}{3}$, so $r_{\mathrm{H}} = 24.74$. Hence, on choosing $r = 28$, which is just past the Hopf bifurcation, we expect something strange to happen. Steady-state convection may be calculated from (27.29) to give $(X^*, Y^*, Z^*) = (6\sqrt{2}, 6\sqrt{2}, 27)$ and $(-6\sqrt{2}, -6\sqrt{2}, 27)$ while the state of no convection corresponds to $(X^*, Y^*, Z^*) = (0, 0, 0)$.

As we know, the complete set (27.24) cannot be integrated by analytic methods. Nonstationary solutions can be found only numerically. Lorenz began integrating from the initial condition (0, 1, 0) which is close to the saddle point at the origin. The result is depicted in Figure 27.2. The trajectory starts near the origin and immediately swings to the right and then dives into the center of the spiral on the left. From there the trajectory spirals outward very slowly and then, all of a sudden, shoots back over to the right-hand side. There it spirals around, and so on indefinitely. The number of circuits made on either side varies unpredictably from one cycle to the next one. The spiral leaves of the attractor simulate rising and descending air of the convective motion.

Figure 27.2 shows what is now called a *strange* or *chaotic attractor*. The strange attractor consists of an infinite number of closely spaced sheets having zero volume but an infinite surface area. Numerical experiments have shown that the fractional dimension of the Lorenz attractor is about 2.06 if the Hausdorff definition is applied. The fractional dimension of the strange attractor implies a fractional structure of many length scales. If one magnifies a small part of the strange attractor, new substructures will emerge.

Figure 27.3 shows as an example the evolution of Y versus t^*. After reaching an early peak, irregular oscillations persist as time increases. Since the motion never repeats exactly, we speak of *aperiodic motion.*

Let us now consider the importance of Lorenz's discoveries for weather prediction. The character of chaotic dynamics can be recognized very easily by imagining that the system is started twice but from slightly different initial conditions. We

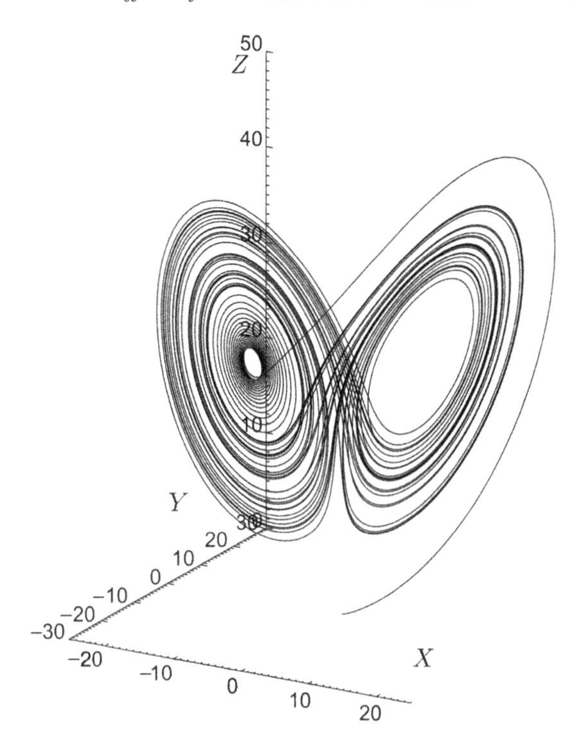

Fig. 27.2 Aperiodic motion of the Lorenz system, showing trajectories in the (X, Y, Z)-space.

may think of this small initial difference as simulating measurement errors, which can never be avoided entirely. For nonchaotic systems this uncertainty in the initial conditions leads to a prediction error that grows linearly with time. For chaotic systems this error grows exponentially with time, so that, after a short time, the state of the system is essentially unknown. This means that long-term prediction is impossible for Lorenz-type systems.

27.2 The effect of uncertainties in the initial conditions

We wish to discuss the situation more quantitatively. Let the uncertainties in the initial conditions be represented by two points on two neighboring trajectories. Suppose that the transient has decayed so that the trajectory is already on the attractor. Let $X(t)$ represent a point on the attractor at time t. We consider a nearby point $X(t) + \delta(t)$ on a neighboring trajectory, where $\delta(t)$ is a tiny separation vector of initial length $|\delta_0|$; see Figure 27.4. Numerical studies of the Lorenz attractor show that the separation vector changes approximately according to

$$\delta(t) = |\delta_0| \exp(\lambda_L t), \qquad \lambda_L \approx 0.9 \tag{27.46}$$

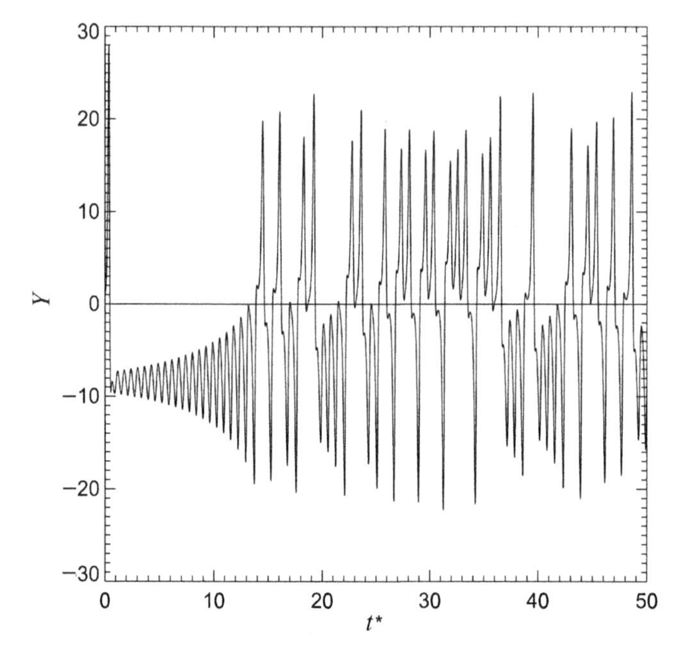

Fig. 27.3 Aperiodic motion of the Lorenz system, showing the Y-component versus dimensionless time t^*.

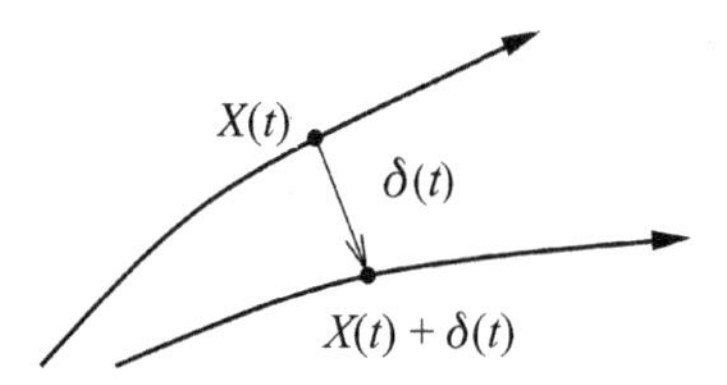

Fig. 27.4 Two points in phase space on neighboring trajectories.

where λ_L is the Liapunov exponent. Hence neighboring trajectories separate exponentially fast. Actually there are n different Liapunov exponents for an n-dimensional system but $\lambda_L \approx 0.9$ turns out to be the largest value. Whenever a system has a positive Liapunov exponent, there is a time horizon beyond which prediction breaks down. This is shown schematically in Figure 27.5. Beyond $t = t_{hor}$ the predictability breaks down. The two initial conditions represented by points on the plot are so closely spaced that they cannot be distinguished.

Let us assume that the initial conditions of some physical experiment have been determined very accurately. Nevertheless, a small measurement error will always occur so that the measured initial condition differs from the true initial condition by $|\delta_0|$. After time t of the prediction period the discrepancy has increased according

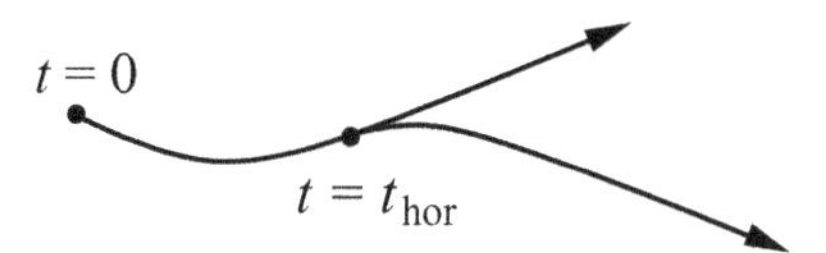

Fig. 27.5 A schematic plot of the time horizon where the two trajectories diverge rapidly.

to (27.46). We must permit a certain acceptable tolerance in the accuracy of the prediction. Let a be a measure of this tolerance, i.e. if a prediction is within a of the true state, then the prediction is tolerable. Thus the prediction becomes intolerable if $|\delta(t)| \geq a$. According to (27.46) the order of magnitude $\mathcal{O}$ of the time interval t_{hor} may be expressed as

$$t_{hor} \approx \mathcal{O}\left[\frac{1}{\lambda_L}\ln\left(\frac{a}{|\delta_0|}\right)\right] \tag{27.47}$$

Let us consider the following idealized example, which has little to do with actual weather prediction. In a first attempt, we are trying to predict a chaotic system within a tolerance of $a = 10^{-3}$, assuming that the initial state is uncertain to within $|\delta_0| = 10^{-7}$. For what time period can we predict the state of the system and still remain within the given tolerance? In this case the time horizon is given by

$$t_{hor}(1) \approx \frac{1}{\lambda_L}\ln\left(\frac{10^{-3}}{10^{-7}}\right) = \frac{4\ln 10}{\lambda_L} \tag{27.48}$$

In a second attempt we succeed in measuring the initial state a million times better so that $|\delta_0| = 10^{-13}$. How much longer can we predict within the tolerance limits? The result is disappointing, as follows from

$$t_{hor}(2) \approx \frac{1}{\lambda_L}\ln\left(\frac{10^{-3}}{10^{-13}}\right) = \frac{10\ln 10}{\lambda_L} = 2.5t_{hor}(1) \tag{27.49}$$

The conclusion is that a million-fold improvement in the initial uncertainty permits us to predict for only 2.5 times longer. Lorenz suggested that this logarithmic dependency on $|\delta_0|$ is what makes long-term weather prediction so difficult.

27.3 Limitations of deterministic predictability of the atmosphere

In the previous section we gave a rather artificial example of the Liapunov time horizon of a chaotic system that is difficult to apply to weather prediction. It is at present not possible to state this time horizon in a general form since the time span of predictability depends on the scale of the fields to be predicted and also on the synoptic situation.

Table 27.1. *Spatial and time scales of atmospheric motion*

Motion	Spatial scale	Time scale
Turbulence	0.1–100 m	1–1000 s
Convection	0.1–10 km	1–60 min
Mesoscale motion	1–1000 km	30 min to several days
Large-scale motion	1000–10 000 km	1–10 days

After and even before Lorenz (1963) published his celebrated paper, a number of publications attempting to assess the predictability of the atmosphere appeared in the literature. To a large extent, the prediction of the future state of the atmosphere for an extended time period is very difficult due to the width of the spectrum of atmospheric processes. To give an impression of the scales of motion, we present in Table 27.1 a very rough classification of the spatial and time scales of atmospheric motion.

More refined classifications are avaliable, see e.g. Orlanski (1975). For large-scale motion the characteristic scales are determined by the wavelength and the period of the planetary waves. On smaller scales the dimensions and life times of turbulent vortices determine the characteristic scales. Nonlinear interactions of physical processes between different scales take place continually and vary strongly in time and space.

Weather prediction on a mathematical–physical basis proceeds as follows. In order to study weather phenomena of a certain scale, we must construct a mathematical model by employing the methods of scale analysis. The model is made deterministic by parameterizing the interactions with neighboring scales in terms of available model variables. Since a general parameterization theory is not yet available, it is necessary to employ empirical knowledge to determine the parameter functions. This procedure introduces nondeterministic elements into the prediction model. In order to develop a global circulation model, for example, it would be necessary to parameterize the interactions with the complete spectrum of the subsynoptic processes. So far this problem has not been solved satisfactorily.

In order to carry out short- and medium-range numerical weather forecasts, it is necessary to initialize the model with consistent input data in order to integrate the predictive system of differential equations forward in time. The field of the initial data may be obtained with the help of an objective analyis. Procedures for how to carry out this analysis are given by Haltiner and Williams (1980) and in more recent references. Various sophisticated methods have been devised in order to make the observed data physically and computationally as consistent with the numerical model as possible. Methods have been proposed for producing an objective analysis by statistically extracting the maximum amount of information from observations, climatological data, spatial correlations between meteorological variables, and other available data. This type of procedure, known as the *optimum interpolation*, requires knowledge of the statistical structure of the fields of meteorological variables. Nevertheless, upon completion of the objective analysis, mass and motion fields are still not precisely balanced.

Naturally, questions about the sensitivity of the weather-prediction model arise due to uncertainties in the parameterized interaction with neighboring scales, the uncertainties in the initial conditions, and the particular properties of the numerical scheme. In other words, for what prediction time interval is the deterministic character of the model equations prevalent before the nondeterministic elements begin to dominate the prediction? We can well imagine that there must be limitations to deterministic predictability, which vary from one mathematical model to the next. In the remaining part of this chapter we will investigate the uncertainties in the initial conditions.

Among various attempts to investigate the predictability of the atmosphere, Lorenz (1969) modeled the atmosphere with the help of the divergence-free barotropic vorticity equation in which the horizontal velocity is expressed in terms of the stream function. Assuming "exact" initial conditions and ignoring possible errors due to the numerical procedure, the solution of this deterministic equation should result in an "exact" deterministic prediction over an arbitrarily long period of time. Exact initial conditions, however, do not exist, so the predicted fields are expected to be at variance with nature after a certain time span. Moreover, it should be realized that the model equations are too simple to approximate the actual atmospheric behavior over an extended time period.

To simulate the effect of uncertainties in the initial conditions, two predictions may be carried out. The first prediction uses initial conditions that are defined to be exact. The prediction on the basis of the model equations is deterministic and "exact," thus correctly representing the nature of the model. The second prediction uses somewhat different initial conditions, which simulate imperfect measurements and other errors. After a longer prediction time the results differ so much that, even on the largest scales, they are not even similar. In fact, Lorenz showed that, on the basis of hardly discernible differences in the initial conditions of the subsynoptic scales, but identical initial conditions in the synoptic scale, after three weeks of prediction time the two forecasts were so different that they could not be compared in a reasonable way. We may conclude from this numerical investigation that the atmosphere has forgotten the initial conditions altogether after a time span of no more than four weeks. The short memory of the atmosphere is caused by the nonlinear interactions taking place, which are accompanied by a propagation of errors relative to the initial conditions. This memory property has been well known for many years from the numerical experiments simulating the general circulation. Even on starting the numerical integration with the simplest and most unrealistic initial conditions, that is a resting atmosphere and isothermal vertical temperature stratification, due to an influx of energy and to friction a reasonably realistic atmosphere has evolved after a time span of only one month. Again, this implies that the atmosphere has completely forgotten the unrealistic initial conditions.

Proponents of the so-called *stochastic dynamic method* of weather prediction and others have criticized the conclusions resulting from Lorenz's numerical investigation. They argued that the Lorenz model, ignoring even the important β effect, is too simple to give a realistic estimate of the limits of atmospheric predictability. The prediction range can be extended by applying the stochastic dynamic method to the Lorenz scheme and to more complete prediction models. The stochastic dynamic approach apparently was initiated by Obuchov (1967). Just like any other dynamic prediction, the method is based on the deterministic model equations generally used to describe the atmosphere. The novel part is that this method introduces the concept of uncertainty in the initial conditions by means of statistical characteristics.

Even if the initial conditions cannot be precisely known, from experience we have developed some ideas about the most probable forms of the fields of variables describing the initial state of the atmosphere. The typical operational forecast requires one field of initial data for each variable with unknown errors. Instead of prescribing a single set of initial fields of the variables, one prescribes a set of initial fields of variables by means of a probability distribution of initial fields. Instead of a single forecast resulting from a single set of initial fields of the variables, a great number of forecasts can be produced. How one might obtain an approximation of the initial joint probability distribution will not be discussed here. For our purposes we simply assume that it can be constructed in some way.

Any probability distribution may be characterized by the usual statistical measures: expectation, variance, and higher moments. Thus we may describe the joint probability distribution of the initial fields of variables by these statistical measures. Moreover, we may also calculate the same statistical measures from the collection of the predicted fields. For various times and spatial points the statistical measures of the predicted fields may be related to the corresponding measures of the initial fields. The prediction of the statistical measures of the variables is the core of the stochastic dynamic prediction. This forecasting procedure makes it possible to extend the predictability horizon. By employing the barotropic vorticity equation in the form used by Lorenz, it is possible to increase the duration of predictability significantly.

In the following we will give a brief mathematical description of the stochastic dynamic method. Lorenz (1963) had already recognized that all meteorological prediction equations, after transformation into the spectral domain, can be written in the form

$$\dot{X}_k = \sum_m \sum_n A_{kmn} X_m X_n - \sum_m B_{km} X_m + C_k \tag{27.50}$$

where the X_k are general time-dependent variables describing the state of the atmosphere. The time t is the only independent variable; A, B, C are constants describing the nonlinear interaction, external forces, and dissipative mechanisms. The symbols m and n are dummy indices. Fortak (1973) verified the form (27.50) by employing the following five dependent variables: the three components of the wind vector (u, v, w), the Exner function $\Pi = c_p(p/p_0)^{R_0/c_{p,0}}$, and the potential temperature θ. By developing the five scalar variables in the form

$$\begin{pmatrix} u \\ v \\ w \end{pmatrix} = \sum_{m=1}^{N} F_m \begin{pmatrix} u_m \\ v_m \\ w_m \end{pmatrix}, \qquad \Pi = \sum_{m=1}^{N} F_m \Pi_m, \qquad \theta = \sum_{m=1}^{N} F_m \theta_m \tag{27.51}$$

where $F_k = F_k(x, y, z)$ is a suitable normalized three-dimensional orthogonal function, we obtain a set of five ordinary differential equations $(\dot{u}_k, \dot{v}_k, \dot{w}_k, \dot{\Pi}_k, \dot{\theta}_k)$ of relatively simple structure.

In order to proceed with our discussion, we do not need to know the mathematical form of the F_k and need not repeat Fortak's analysis. The functions F_k may be obtained in generalization to the two-dimensional case, see Abramowitz and Segun (1968). Suppose, for example, that the spatial fields are resolved to a wavenumber $N = 20$. In this case we would have to evaluate deterministically $5N = 100$ nonlinear coupled time-dependent ordinary differential equations for the coefficient functions X_k assuming the existence of exact initial conditions of the variables. Imagining that this $5N$-dimensional space is spanned by the totality of the variables X_k then the state of the system in phase space at a certain time is represented by a point. As we have previously discussed, any changes in the state of this system are represented by the motion of this point or by a trajectory in phase space.

We know that the initial conditions cannot be specified exactly. This is equivalent to saying that we do not know the exact location of the point in phase space specifying the initial state $t = t_0$. Suppose that the probability density $P(X_1, X_2, \ldots, X_{5N}, t_0)$ for the position of the point is available with $P > 0$ for all

X_k and all t. By definition, the normalization condition

$$\begin{aligned}\int_{-\infty}^{\infty}\cdots\int P(X_1, X_2, \ldots, X_{5N}, t)\, dX_1\, dX_2 \cdots dX_{5N}\\ = \int_{-\infty}^{\infty}\cdots\int P(X_1, X_2, \ldots, X_{5N}, t)\, d\tau = 1\end{aligned} \tag{27.52}$$

must be valid for all t. The evolution with time of the probability density follows from the differentiation with respect to time of the normalization condition. By employing Lagrange's method we obtain

$$\frac{d}{dt}\left(\int_{-\infty}^{\infty}\cdots\int P\, d\tau\right) = \int_{-\infty}^{\infty}\cdots\int \frac{dP}{dt}\, d\tau + \int_{-\infty}^{\infty}\cdots\int P\left(\frac{1}{d\tau}\frac{d}{dt}(d\tau)\right) d\tau = 0 \tag{27.53}$$

The expression in parentheses on the right-hand side represents the relative change in volume or the divergence:

$$\frac{1}{d\tau}\frac{d}{dt}(d\tau) = \sum_{m=1}^{5N}\frac{\partial \dot{X}_m}{\partial X_m} \tag{27.54}$$

which is a generalization of the three-dimensional case; see (M6.38b). Next we replace dP/dt in (27.53) by the generalized Euler expansion

$$\frac{dP}{dt} = \frac{\partial P}{\partial t} + \sum_{m=1}^{5N}\dot{X}_m\frac{\partial P}{\partial X_m} \tag{27.55}$$

Substitution of this expression together with (27.54) into (27.53) yields

$$\frac{d}{dt}\left(\int_{-\infty}^{\infty}\cdots\int P\, d\tau\right) = \int_{-\infty}^{\infty}\cdots\int\left(\frac{\partial P}{\partial t} + \sum_{m=1}^{5N}\frac{\partial}{\partial X_m}(P\dot{X}_m)\right) d\tau = 0 \tag{27.56}$$

In differential form this equation may be written as

$$\frac{\partial P}{\partial t} + \sum_{m=1}^{5N}\frac{\partial}{\partial X_m}(P\dot{X}_m) = 0 \tag{27.57}$$

which is known as the *Liouville equation* or the *continuity equation for the probability density*.

Equation (27.57) can be evaluated, at least in principle, if we replace $\dot{X}_k$ by (27.50). This results in

$$\frac{\partial P}{\partial t} + \sum_{m=1}^{5N}\frac{\partial}{\partial X_m}\left(P\sum_{r=1}^{5N}\sum_{s=1}^{5N}A_{mrs}X_rX_s - \sum_{r=1}^{5N}B_{mr}X_r + C_m\right) = 0 \tag{27.58}$$

which describes the evolution with time of the probability density for a specific meteorological problem as formulated by a model. The latter equation was originally and independently derived by Epstein (1969) and by Tatarsky (1969). In principle, this equation can be solved numerically if $P(t = t_0)$ is known and if the boundary conditions for P (vanishing at infinity) can be satisfied.

At present the numerical effort required in order to solve (27.58) is prohibitively expensive and practically impossible. Tatarsky and Epstein pointed out that complete knowledge of P includes much superfluous information. So it is usually sufficient to know the simplest moments describing essential characteristics of the distribution instead of knowing the complete distribution. These are the first and the second moment. The first moment is the expectation or mean value of the variable while the second moment is related to the covariance. Knowledge of these two statistical measures is sufficient for approximating the actual probability density by the Gaussian distribution.

27.4 Basic equations of the approximate stochastic dynamic method

We will now derive the prognostic equations for the expectation value and for the covariance tensor. These equations were given independently by Tatarsky and Epstein. It turns out that we will have to deal with the same type of closure problem as that which is known to us from the theory of turbulence.

The first moment or the expectation value of X, also known as the mean value μ_k, is defined by

$$E(X_k) = \mu_k = \int_{-\infty}^{\infty} \cdots \int X_k P(X_1, X_2, \ldots, X_{5N}, t)\, d\tau \tag{27.59}$$

The second moment ρ_{kl} is given by

$$\rho_{kl}(t) = \int_{-\infty}^{\infty} \cdots \int X_k X_l P(X_1, X_2, \ldots, X_{5N}, t)\, d\tau \tag{27.60}$$

This quantity is related to the covariance σ_{kl} as will be shown next. In analogy to the theory of turbulence, we split the variable X_k into the mean value μ_k plus the deviation X'_k

$$X_k = \mu_k + X'_k \quad \text{with} \quad E(X'_k) = 0 \tag{27.61}$$

By definition, the expectation value of the deviation X'_k is zero. Equation (27.60) involves the product $X_k X_l$. By introducing (27.61) into this product we find

$$X_k X_l = \mu_k \mu_l + X'_k \mu_l + X'_l \mu_k + X'_k X'_l \tag{27.62}$$

On applying the expectation operator to (27.62) we immediately find

$$E(X_k X_l) = \mu_k \mu_l + \sigma_{kl} \quad \text{with} \quad \sigma_{kl} = E(X'_k X'_l) \tag{27.63}$$

Since $E(X_k X_l) = \rho_{kl}$ we have found the relation between the second moment and the covariance. The expectation value of a function is defined analogously to (27.59) simply by replacing X_k by the function $f(X_1, X_2, \ldots, X_{5N})$.

Of primary importance to stochastic dynamic forecasting is the time derivative of the expectation value. With the help of the Liouville equation (27.57) it is not very difficult to prove the validity of

$$\frac{d}{dt}[E(f)] = E\left(\frac{df}{dt}\right) \tag{27.64}$$

showing that the mathematical expectation operator and the time derivative may be interchanged. We will leave the proof to the exercises.

To find the prognostic equation of the mean value μ_k we first apply the expectation operator to equation (27.50) and obtain

$$E\left(\frac{dX_k}{dt}\right) = \sum_{m=1}^{5N} \sum_{n=1}^{5N} A_{kmn} E(X_m X_n) - \sum_{m=1}^{5N} B_{km} E(X_m) + C_k E(1) \tag{27.65}$$

With the help of (27.63) and (27.64) it is not difficult to obtain the desired equation

$$\dot{\mu}_k = \sum_{m=1}^{5N} \sum_{n=1}^{5N} A_{kmn} \mu_m \mu_n - \sum_{m=1}^{5N} B_{km} \mu_m + C_k + \sum_{m=1}^{5N} \sum_{n=1}^{5N} A_{kmn} \sigma_{mn} \tag{27.66}$$

Comparison of (27.66) with (27.50) shows the similarity between these two equations. The first three terms are identical in form. In place of X_k in (27.50) we now have the mean value μ_k. The decisive difficulty with equation (27.66) arises from the last term, which requires knowledge of the covariance σ_{kl}. Thus we need to derive a prognostic equation for the covariance. If there were no uncertainties, the covariance tensor σ_{kl} would be zero and the final term would vanish so that there would be no need to obtain a prognostic equation for σ_{kl}. In practical situations this idealized case of zero uncertainties does not occur. The analogy with averaging the equation of motion is apparent since the averaging procedure produces the Reynolds stress tensor as an additional term.

Now we direct our attention to the evaluation of the covariance term in (27.66). There is no formal difficulty in finding the prognostic equation for σ_{kl}. All we need to do is to differentiate (27.63) with respect to time, yielding

$$\dot{\sigma}_{kl} = E(\dot{X}_k X_l) + E(X_k \dot{X}_l) - \dot{\mu}_k \mu_l - \mu_k \dot{\mu}_l \tag{27.67}$$

The terms involving $\dot{X}_k$ and $\dot{\mu}_k$ can be eliminated with the help of (27.50) and (27.66). This gives a fairly involved expression, which we are not going to state explicitly. The interested reader may find this equation in explicit form in Epstein (1969), Fortak (1973), or elsewhere.

Had we written out this equation, we would have found that new unknown quantities appear in the form of the next higher moments τ_{ijk}. Formally, we could easily obtain a differential equation for τ_{ijk} containing as new unknowns still higher moments. We could go on indefinitely deriving prognostic equations for any order of moments, but these would always contain the unknown next-higher generation of moments. The same type of closure problem is already known from the theory of turbulence. Since the deterministic prediction equations are nonlinear, it is impossible to derive a closed finite set of prognostic equations for the moments. The numerical evaluation of the third and even higher moments requires prohibitively large amounts of computer time, so Epstein (1969) was compelled to close the system of predictive equations consisting of the μ_k and the σ_{kl} by ignoring the occurrence of the third-order moments τ_{ijk} in the σ_{kl} equations. Fleming (1971a) discussed the closure problem very thoroughly. For details we refer to his paper.

Moreover, Fleming (1971a) also applied the stochastic dynamic method to investigate the effect of uncertainties in the initial conditions on the energetics of the atmosphere. When one is investigating certain phenomena, the method makes it possible also to study in a systematic fashion the uncertainties resulting from the parameterization of neighboring scales. By introducing the quadratic ensemble mean $\sum \mu_k^2$ one may estimate the most probable energy distribution, which is called the *certain energy*. With the help of the variance $\sum \sigma_{kk}$ or uncertainty, the so-called *uncertain energy* can be determined. An atmospheric scale corresponding to a certain wavenumber k becomes unpredictable whenever the uncertain energy is as large as or even larger than the certain energy. For details we must refer to the original literature.

We may easily recognize the advantages of the stochastic dynamic method over the deterministic forecast. Stochastic dynamic forecasts produce a significantly smaller mean square error than do deterministic forecasts. As discussed above, the range of useful forecasts can be extended by applying the stochastic dynamic method. The main disadvantage of this method is that forecasts require a much higher level of computational effort than does the deterministic procedure.

Many research papers on the stochastic dynamic method have appeared in the literature since Fortak (1973). The interested reader will have no difficulty in finding the proper references. It is beyond the scope of this book to discuss newer developments aiming to increase the accuracy of the forecast and the period of useful predictability. Suffice it to say that, at present, even with the best available weather-prediction models, the predictability on the synoptic scale hardly exceeds two weeks. A brief account of the predictability problem is also given by Pichler (1997).

27.5 Problems

27.1: Show that equation (27.64) is valid.

Answers to problems

Chapter M1

M1.1: Yes.

M1.2: (a) linearly dependent, (c) linearly independent.

M1.3:

$$|\mathbf{A}_2| = 0.683\,|\mathbf{A}|, \qquad \alpha = 7.25^\circ$$

M1.4:

$$\text{(a)}\quad (A^1, A^2, A^3) = (3, -2, 3), \qquad \text{(b)}\quad (A_1, A_2, A_3) = (7, 8, 8)$$

M1.8:

$$[\widetilde{\mathbf{A}}, \widetilde{\mathbf{B}}, \widetilde{\mathbf{C}}] = \frac{1}{[\mathbf{A}, \mathbf{B}, \mathbf{C}]}$$

Chapter M2

M2.1:

$$\Phi'' = -\sqrt{g}\big[A^1(\mathbf{q}^2\mathbf{q}^3 - \mathbf{q}^3\mathbf{q}^2) + A^2(\mathbf{q}^3\mathbf{q}^1 - \mathbf{q}^1\mathbf{q}^3) + A^3(\mathbf{q}^1\mathbf{q}^2 - \mathbf{q}^2\mathbf{q}^1)\big]$$

M2.4: The eigenvalues are

$$(\lambda_1, \lambda_2, \lambda_3) = (2, 2 + \sqrt{2}, 2 - \sqrt{2})$$

The corresponding eigenvectors are

$$\mathbf{A}_1 = C_1(\mathbf{i}_1 - \mathbf{i}_3), \quad \mathbf{A}_2 = C_2(\mathbf{i}_1 - \sqrt{2}\mathbf{i}_2 + \mathbf{i}_3), \quad \mathbf{A}_3 = C_3(\mathbf{i}_1 + \sqrt{2}\mathbf{i}_2 + \mathbf{i}_3)$$

where the C_i are constants.

Chapter M3

M3.2:

$$\nabla\theta = \frac{\theta}{T}\nabla T - \frac{\theta}{\Pi}\nabla\Pi$$

M3.3:

(a) $2\boldsymbol{\Omega}\cdot\mathbf{r}$, (b) $2\mathbf{r}\times\boldsymbol{\Omega}$, (c) $6\boldsymbol{\Omega}$, (d) $2\mathbf{v}\cdot\mathbf{r}\boldsymbol{\Omega}$

M3.4:

(a) $\nabla^2\chi$, (b) $\nabla^2\boldsymbol{\Psi}$, (c) $-\nabla\times\nabla^2\boldsymbol{\Psi}+\nabla(\nabla^2\chi)$

(d) $\nabla(\nabla^2\chi)$, (e) $-\frac{1}{2}\nabla\times(\nabla^2\boldsymbol{\Psi})+\frac{2}{3}\nabla(\nabla^2\chi)$

M3.6:

(b) $$(J_{ij}) = \frac{\eta}{2}\begin{pmatrix} 0 & 0 & \dfrac{\partial u}{\partial z} \\ 0 & 0 & \dfrac{\partial v}{\partial z} \\ \dfrac{\partial u}{\partial z} & \dfrac{\partial v}{\partial z} & 0 \end{pmatrix}$$

(c) $\lambda_1 = 0, \quad \lambda_{2,3} = \pm\eta A/2$

(d) $$\boldsymbol{\Psi}^1 = \frac{1}{A}\left(\frac{\partial v}{\partial z}\mathbf{i} - \frac{\partial u}{\partial z}\mathbf{j}\right), \quad \boldsymbol{\Psi}^2 = \frac{1}{\sqrt{2}A}\left(\frac{\partial u}{\partial z}\mathbf{i} + \frac{\partial v}{\partial z}\mathbf{j} + A\mathbf{k}\right)$$

$$\boldsymbol{\Psi}^3 = \frac{1}{\sqrt{2}A}\left(\frac{\partial u}{\partial z}\mathbf{i} + \frac{\partial v}{\partial z}\mathbf{j} - A\mathbf{k}\right) \quad \text{with} \quad A = \sqrt{\left(\frac{\partial u}{\partial z}\right)^2 + \left(\frac{\partial v}{\partial z}\right)^2}$$

(e) $$z = \frac{1}{x}\frac{\mathbf{r}\cdot\mathbb{J}\cdot\mathbf{r}}{\eta\partial u/\partial z}$$

Chapter M4

M4.1:

(a) $g_{11} = r^2\cos^2\varphi, \quad g_{22} = r^2, \quad g_{33} = 1,$

$g_{ij} = 0 \quad \text{if} \quad i \neq j, \quad \sqrt{\underset{q}{g}} = r^2\cos\varphi$

(b) $g^{11} = \dfrac{1}{r^2\cos^2\varphi}, \quad g^{22} = \dfrac{1}{r^2}, \quad g^{33} = 1, \quad g^{ij} = 0 \quad \text{if} \quad i \neq j$

M4.2:

$$\underset{q}{W}^1 = \Omega, \quad \underset{q}{W}^2 = 0, \quad \underset{q}{W}^3 = 0$$

M4.3:

$$\sqrt{\underset{\xi}{g}} = \sqrt{\underset{q}{g}}\left(\frac{\partial q^3}{\partial\xi}\right)_{q^1,q^2}$$

Chapter M6

M6.4:

$$\text{(a)}\quad \mathbf{k}\int_0^L \big[A_y(x, y=0) + A_y(x, y=M)\big]\,dx - \mathbf{k} \times \int_0^M \big[A_x(x=0, y) + A_x(x=L, y)\big]\,dy$$

$$\text{(b)}\quad \mathbf{k}\left(\frac{L^2}{2} - 2 + \cos M - \sin M\right)$$

M6.5:

$$\text{(a)}\quad C[r(2) - r(1)], \qquad \text{(b)}\quad 4C\pi a^2$$

M6.6:

$$\text{(a)}\quad 2\pi a^2\Omega, \qquad \text{(b)}\quad 2\boldsymbol{\Omega}\cdot\int_S d\mathbf{S}$$

Chapter 1

1.3:

$$\text{(b)}\quad \nabla\cdot\mathbb{J}(\mathbf{v}) = \mu\frac{\partial^2 u}{\partial z^2}\mathbf{i} \implies \nabla\cdot\mathbb{J}(\mathbf{v}_1) = -\frac{\mu C_1}{z^2}\mathbf{i}, \quad \nabla\cdot\mathbb{J}(\mathbf{v}_2) = 0$$

1.4:

$$\text{(b)}\quad \frac{\partial p}{\partial x} = \mu\left(\frac{\partial^2 u}{\partial y^2} + \frac{\partial^2 u}{\partial z^2}\right), \qquad \text{(c)}\quad u = \left|\frac{\partial p}{\partial x}\right|\frac{\left|R^2 - R_0^2\right|}{4\mu},$$

$$\text{(d)}\quad Q = \left|\frac{\partial p}{\partial x}\right|\frac{\pi\rho R_0^4}{8\mu}$$

Chapter 2

2.3:

$$M_\lambda = \frac{uw}{a} - \frac{uv}{a}\tan\varphi, \qquad M_\varphi = \frac{vw}{a} + \frac{u^2}{a}\tan\varphi, \qquad M_r = -\frac{u^2 + v^2}{a}$$

Chapter 3

3.2: The flow is nonstationary since the time t appears explicitly.

(a) $y = x$, (b) $x = y^{1+\ln y}$.

3.3:

$$\text{(a)}\quad y - y_0 = \frac{C(x - x_0)}{(A + Bt_0)}, \qquad \text{(b)}\quad x - x_0 = \frac{A}{C}(y - y_0) + \frac{B}{2C^2}(y - y_0)^2$$

Chapter 5

5.1:

$$\nabla \cdot \mathbb{J} = \mu[\nabla^2 \mathbf{v}_A + \tfrac{1}{3}\nabla(\nabla \cdot \mathbf{v}_A)]$$

5.3:

(a) Parabolic velocity profile with

$$u(z) = \frac{g\rho \sin\alpha\,(2H - z)z}{2\mu}$$

α is the angle of inclination.

(b) Isobars are parallel to the x-axis with $p(z) = p_0 - g\rho z \cos\alpha$.

(c) The amount of fluid is

$$q = \frac{g\rho^2 H^3 \sin\alpha}{3\mu}$$

(d) $\displaystyle\int_0^H \rho\epsilon\, dz = \frac{\rho(g\rho \sin\alpha)^2 H^3}{3\mu}$

5.4:

(a) $u(z) = u(z = H)z/H$.

(b) $u(z) = \left|-\dfrac{1}{\rho}\dfrac{\partial p}{\partial x}\right| \dfrac{\rho(H - z)z}{2\mu}$, parabolic flow, $u\left(\dfrac{H}{2}\right) = u_{\max}$

5.5:

$$\mathbf{v}_A = C\mathbf{r}, \quad C = \text{constant} \Longrightarrow \nabla\mathbf{v}_A = C\,\nabla\mathbf{r} = C\mathbb{E}, \quad \nabla \cdot \mathbf{v}_A = 3C$$

$$\mathbb{D}_{ai} = \frac{\nabla\mathbf{v}_A + \mathbf{v}_A\overset{\curvearrowleft}{\nabla}}{2} - \frac{\nabla \cdot \mathbf{v}_A}{3}\mathbb{E} = C\mathbb{E} - C\mathbb{E} = 0$$

$$\mathbb{J} = \left(\frac{2\mu}{3} - \lambda\right)\nabla \cdot \mathbf{v}_A\mathbb{E}, \qquad \mathbf{p}_2(\mathbf{v}_A) = \mathbf{n} \cdot \mathbb{J} = 3C\mathbf{n}\left(\frac{2\mu}{3} - \lambda\right)$$

$$\mathbf{r} \cdot \mathbb{J} \cdot \mathbf{r} = 3C\mathbf{r}^2\left(\frac{2\mu}{3} - \lambda\right) = 2F = \pm 1$$

depending on the sign of C. The tensor ellipsoid is a spherical surface.

Chapter 6

6.1:

(a) $$(\nabla\mathbf{v})_I = \nabla^2\phi, \quad (\nabla\mathbf{v})_\times = \nabla^2\mathbf{\Psi},$$

$$(\nabla\mathbf{v})'' = \frac{1}{2}\left(-\nabla(\nabla \times \mathbf{\Psi}) + \nabla \times \mathbf{\Psi}\overset{\curvearrowleft}{\nabla}\right)$$

(b) $$\frac{d}{dt}(\nabla^2\mathbf{\Psi}) = -(\nabla^2\mathbf{\Psi})\,\nabla^2\phi - (\nabla^2\mathbf{\Psi}) \cdot \nabla(\nabla \times \mathbf{\Psi}) + (\nabla^2\mathbf{\Psi}) \cdot \nabla(\nabla\phi) - \nabla\left(\frac{1}{\rho}\right) \times \nabla p$$

Chapter 7

7.1:

(a)
$$v_r = \frac{\partial \chi}{\partial r}, \qquad v_t = \frac{1}{r}\frac{\partial \chi}{\partial \theta}$$

(b)
$$\frac{\partial^2 \chi}{\partial r^2} + \frac{1}{r}\frac{\partial \chi}{\partial r} + \frac{1}{r^2}\frac{\partial^2 \chi}{\partial \theta^2} = 0$$

7.2: $\chi_B - \chi_A$, where A and B are the endpoints of L.

7.3:

(a) $\chi = \chi_{A_1} + 2n\pi$, where n is an arbitrary integer.

(b) $\chi_{B_2} = \chi_{A_2} = 0$. Reason: After one complete circulation the angle θ returns to its original value.

7.4:

(a) $|\mathbf{v}_h| = \sqrt{a^2 + b^2}, \ u = a, \ v = b$

(b) $\psi = -ay + bx = \text{constant}$

Chapter 9

9.1:

$$\omega_s = \rho_s \left[\left(\frac{\partial \phi}{\partial t} \right)_p + \mathbf{v}_h \cdot \nabla_{h,p}\phi \right]_s - \rho_s \mathbf{v}_h \cdot \nabla_{h,p}\phi_s$$

In the case that at all times the earth's surface is identical with the pressure surface p_0, we may set $\nabla_{h,p}\phi_s = \nabla_h \phi_s$.

9.2:

$$\tan\alpha = \frac{f}{g} \frac{T_v^{(2)} v_g^{(1)} - T_v^{(1)} v_g^{(2)}}{T_v^{(2)} - T_v^{(1)}}$$

Chapter 12

12.1: In both cases you obtain $6[V(1)V(5) + V(2)V(4) + V(3)V(3) + V(4)V(2) + V(5)V(1)]$.

Chapter 13

13.2:

$$\Pi = \left(\frac{\partial \widetilde{\theta}}{\partial z} \right) \left(\frac{H}{\overline{\rho} c_{p,0}} \right)^{-2/3} \left(\frac{g}{\overline{T}} \right)^{1/3} z^{4/3}$$

13.3:

(a) $\epsilon = \kappa_K \overline{\epsilon}_M^{2/3} k^{-5/3}$, $\Pi = \dfrac{\epsilon(k) k^{5/3}}{\overline{\epsilon}_M^{2/3}}$

(b) $\Pi_1 = \dfrac{\epsilon(k)}{\nu^{5/4}\overline{\epsilon}^{1/4}}$, $\Pi_2 = k\left(\dfrac{\nu^3}{\epsilon_M}\right)^{1/4}$

13.4:

(b) $S_l(0) = 1$

13.5:

(13.84): $S_l = 0.143$, (13.88): $S_L = 0.578$

13.7:

$$\widehat{u} = u_g\left[1 - \sqrt{2}\sin\alpha_0 \cos(Az + \pi/4 - \alpha_0)\exp(-Az)\right]$$

$$\widehat{v} = u_g\sqrt{2}\sin\alpha_0 \sin(Az + \pi/4 - \alpha_0)\exp(-Az)$$

$$z_g = \left(\frac{3}{4\pi} + \alpha_0\right)\sqrt{\frac{2k}{f}}$$

Chapter 14

14.1: Reflection of $U = U_0 \sin(kx - \omega t)$ on the x-axis.

14.2:

$$U = 2A \sin\left(\frac{2\pi x}{L}\right)\cos(\omega t)$$

14.4: No. In order to obtain (14.37d) you must assume that $\alpha_0 \neq \alpha_0(z)$.

14.6:

$$H \to \infty, \qquad \tanh(k_x H) \to 1$$

14.7: The influence of the earth's rotation is small.

14.8: 2200 m.

Chapter 15

15.5:

(a) From the plot it follows that the curvature changes sign. $\Longrightarrow$ Barotropic instability.

(b) $-2 \le y_0^2 \beta / u_0 \le \frac{2}{3}$.

Chapter 16

16.1:

$$c = u_0 - \frac{\beta}{a^2}, \qquad a^2 = k_x^2(1+\mu^2), \qquad \overline{u} = u_0 - Aa\cos(ay+\epsilon)$$

Chapter 17

17.2:

$$\xi = \xi_0 + \frac{f_0}{2v_0}y^2$$

Chapter 18

18.3:

$$\text{covariant:} \quad \frac{d}{dt} = \frac{\partial}{\partial t} + \frac{v_1}{r^2\cos^2\varphi}\frac{\partial}{\partial\lambda} + \frac{v_2}{r^2}\frac{\partial}{\partial\varphi} + v_3\frac{\partial}{\partial r}$$

$$\text{contravariant:} \quad \frac{d}{dt} = \frac{\partial}{\partial t} + \dot{\lambda}\frac{\partial}{\partial\lambda} + \dot{\varphi}\frac{\partial}{\partial\varphi} + \dot{r}\frac{\partial}{\partial r}$$

$$\text{physical:} \quad \frac{d}{dt} = \frac{\partial}{\partial t} + \frac{u}{r\cos\varphi}\frac{\partial}{\partial\lambda} + \frac{v}{r}\frac{\partial}{\partial\varphi} + w\frac{\partial}{\partial r}$$

18.4:

$$\text{(a)} \qquad \frac{1}{\rho}\frac{d\rho}{dt} + \frac{1}{r\cos\varphi}\left(\frac{\partial u}{\partial\lambda} + \frac{\partial}{\partial\varphi}(v\cos\varphi)\right) + \frac{\partial w}{\partial r} + \frac{2w}{r} = 0$$

$$\text{(b)} \qquad \frac{1}{\rho}\frac{d\rho}{dt} + \frac{1}{r\cos\varphi}\left(\frac{\partial}{\partial\lambda}(r\cos\varphi\dot{\lambda}) + \frac{\partial}{\partial\varphi}(r\dot{\varphi}\cos\varphi)\right) + \frac{\partial\dot{r}}{\partial r} + \frac{2\dot{r}}{r} = 0$$

18.5:

(a)
$$\underset{\Omega}{W_1} = \Omega r^2\cos^2\varphi, \qquad \underset{\Omega}{W^1} = \Omega$$

$$\Gamma^1_{12} = -\tan\varphi, \qquad \Gamma^1_{13} = \frac{1}{r}, \qquad \Gamma^2_{11} = \cos\varphi\sin\varphi$$

$$\Gamma^2_{23} = \frac{1}{r}, \qquad \Gamma^3_{11} = -r\cos^2\varphi, \qquad \Gamma^3_{22} = -r$$

(b)
$$\underset{\Omega}{\omega_{12}} = \Omega r^2\cos\varphi\sin\varphi, \qquad \underset{\Omega}{\omega_{13}} = -\Omega r\cos^2\varphi,$$

$$\underset{\Omega}{\overset{*}{\omega}_{12}} = \Omega\sin\varphi, \qquad \underset{\Omega}{\overset{*}{\omega}_{13}} = -\Omega\cos\varphi$$

with $\Gamma^k_{ij} = \Gamma^k_{ji}$ and $\underset{\Omega}{\omega_{ij}} = -\underset{\Omega}{\omega_{ji}}$. The remaining terms of $\underset{\Omega}{W_i}$, $\underset{\Omega}{W^i}$, $\underset{\Omega}{\omega_{ij}}$, and Γ^k_{ij} vanish.

Chapter 19

19.4:

$$g_{11} = \epsilon^2 \cosh^2 r' \cos^2 \varphi', \qquad g_{22} = g_{33} = \epsilon^2 (\cosh^2 r' - \cos^2 \varphi'), \qquad g_{ij} = 0, \quad i \neq j$$

Chapter 20

20.3:

(a) For constant values of x, y, λ, and φ the ratios $ds_{\mathrm{E},x}/d_{\mathrm{sphere},\lambda}$ and $ds_{\mathrm{E},y}/d_{\mathrm{sphere},\varphi}$ give the same result. Thus, the projection is conformal. $m = 1/\cos\varphi$.

(b)

$$\begin{aligned}
(d\mathbf{r})^2 &= \frac{1}{m^2}(dx^2 + dy^2) + dz^2 \\
K_{\mathrm{A}} &= \frac{1}{2m^2}(\dot{x}^2 + \dot{y}^2) + \frac{\dot{z}^2}{2} + \frac{\Omega r}{m^2}\dot{x} + \frac{\Omega r^2}{m^2} \\
g_{11} &= 1/m^2, \quad g_{22} = 1/m^2, \quad g_{33} = 1, \quad g_{ij} = 0, \quad i \neq j \\
g^{11} &= m^2, \quad g^{22} = m^2, \quad g^{33} = 1, \quad \sqrt{g} = 1/m^2 \\
v_1 &= \dot{x}/m^2, \quad v_2 = \dot{y}/m^2, \quad v_3 = \dot{z} \\
\overset{*}{v}{}^1 &= \dot{x}/m, \quad \overset{*}{v}{}^2 = \dot{y}/m, \quad \overset{*}{v}{}^3 = \dot{z}
\end{aligned}$$

(c) $\sqrt{\underset{p}{g}} = -\frac{1}{m^2 g \rho}$

Chapter 21

21.1:

$$\frac{dv_1}{dt} - \frac{v_n^2}{g_{nn}}\frac{\partial}{\partial x}(h g_{nn}) - \frac{f v_2}{g_{22}} = -\frac{1}{\rho}\frac{\partial p}{\partial x} + g\eta\,\frac{\partial h}{\partial x} + \frac{1}{\rho}\mathbf{q}_1 \cdot \nabla \cdot \mathbb{J}$$

Chapter 22

22.3:

(a) The same as (22.58) but replace $\partial M/\partial x$ and $\partial M/\partial y$ as follows:

$$\frac{\partial M}{\partial x} \longrightarrow \frac{\partial N}{\partial x} + \dot{\rho}\,\frac{\partial v_1}{\partial \rho}, \qquad \frac{\partial M}{\partial y} \longrightarrow \frac{\partial N}{\partial y} + \dot{\rho}\,\frac{\partial v_2}{\partial \rho}$$

The hydrostatic equation is

$$\frac{\partial N}{\partial \rho} = -\frac{R_0 T}{\rho}$$

The individual derivative is

$$\frac{d}{dt} = \frac{\partial}{\partial t} + m_0^2\left(v_1\,\frac{\partial}{\partial x} + v_2\,\frac{\partial}{\partial y}\right) + \dot{\rho}\,\frac{\partial}{\partial \rho}$$

(b) Replace θ in (22.58) by ρ and add the term $(\partial/\partial\rho)(\dot{\rho}\partial p/\partial\rho)$.

(c) Lower boundary condition: Replace θ_s in (22.59) by ρ_s, but $\rho_s \neq 0$. Upper boundary condition: $\dot{\rho} = 0$.

(d) Replace θ in (22.61) by ρ and add the right-hand side term $\dot{\rho}\,\partial p/\partial\rho$. The integration limits are $(0, \rho)$. Also multiply $\partial p/\partial t$ by -1.

22.4:

$$\frac{\partial z_H}{\partial t} = -\nabla_h \cdot [\mathbf{v}_h(z_H - z_s)]$$

Chapter 23

23.4:

(a) $$q(t) = [q_0 - q_g(0)]\exp(-if_0t) + q_g(t) - \int_0^t \frac{dq_g(t')}{dt'} \exp[-if_0(t-t')]\,dt'$$

(b) $q_0 = q_g(0)$ and $dq_g(t)/dt = 0$ or $(d/dt)\,\partial\phi/\partial x = 0$, $(d/dt)\,\partial\phi/\partial y = 0$. Hence $\phi = ax + by + F(p,t)$, where $F(p,t)$ is an arbitrary function.

Chapter 26

26.5: $P_3(\mu) = 5\mu^3/2 - 3\mu/2$.

List of frequently used symbols

Symbol	Meaning	Unit
a^i:	Lagrangian enumeration coordinates	
c:	displacement speed	$(\mathrm{m\ s^{-1}})$
c_p:	specific heat at constant pressure	$(\mathrm{m^2\ s^{-2}\ K^{-1}})$
c_v:	specific heat at constant volume	$(\mathrm{m^2\ s^{-2}\ K^{-1}})$
D:	divergence	$(\mathrm{s^{-1}})$
$\mathbb{D}$:	deformation part of the velocity dyadic	$(\mathrm{s^{-1}})$
d_{g}:	geometric differential	
D_3/Dt:	budget operator	$(\mathrm{s^{-1}})$
e:	specific internal energy	$(\mathrm{m^2\ s^{-2}})$
$\mathbf{e}_i$, $\mathbf{e}^i$:	covariant and contravariant unit vectors	
$\mathbb{E}$:	unit dyadic	
Eu:	Euler number	
f:	vertical Coriolis parameter	$(\mathrm{s^{-1}})$
Fr:	Froude number	
$\mathbf{F}_{\mathrm{R}}$:	radiative flux	$(\mathrm{kg\ s^{-3}})$
g:	acceleration due to gravity	$(\mathrm{m\ s^{-2}})$
g_{ij}, g^{ij}:	covariant and contravariant metric fundamental quantities	
$\sqrt{g}$:	functional determinant	
h:	specific enthalpy	$(\mathrm{m^2\ s^{-2}})$
I^k:	phase-transition rate of substance k	$(\mathrm{kg\ m^{-3}\ s^{-1}})$
$\mathbb{J}$:	viscous-stress tensor	$(\mathrm{kg\ m^{-1}\ s^{-2}})$
$\mathbf{J}^h$:	enthalpy flux	$(\mathrm{kg\ s^{-3}})$
$\mathbf{J}^h_{\mathrm{s}}$:	sensible enthalpy flux	$(\mathrm{kg\ s^{-3}})$
$\mathbf{J}^k$:	diffusion flux	$(\mathrm{kg\ m^{-2}\ s^{-1}})$
$\mathbf{J}^k_{\mathrm{t}}$:	turbulent diffusion flux of substance k	$(\mathrm{kg\ s^{-3}})$
$\mathbf{J}^\theta_{\mathrm{t}}$:	turbulent heat flux	$(\mathrm{kg\ s^{-3}})$
K:	exchange coefficient	$(\mathrm{m^2\ s^{-1}})$
K:	specific kinetic energy for relative motion	$(\mathrm{m^2\ s^{-2}})$
K_{A}:	specific kinetic energy for absolute motion	$(\mathrm{m^2\ s^{-2}})$

k: wavenumber (m^{-1})
L: Lagrangian function ($m^2\ s^{-2}$)
l: horizontal Coriolis parameter (s^{-1})
l: mixing length (m)
L_*: Monin–Obukhov length (m)
M: mass (kg)
M^k: partial mass of substance k (kg)
m^k: concentration of substance k
p: pressure ($kg\ m^{-1}\ s^{-2}$)
P_E: Ertel's potential vorticity ($kg^{-1}\ m^4\ s^{-3}\ K^{-1}$)
P_R: Rossby's potential vorticity ($kg^{-1}\ m\ s\ K$)
Q: entropy production ($kg\ m^{-1}\ s^{-3}\ K^{-1}$)
q: specific humidity
$\mathbf{q}_i$, $\mathbf{q}^i$: covariant and contravariant basis vectors
q_i, q^i: covariant and contravariant position coordinates of a vector
R: distance from the earth's axis (m)
R_0: gas constant for dry air ($m^2\ s^{-2}\ K^{-1}$)
Re: Reynolds number
Ri: Richardson number
Ri_f: flux Richardson number
Ro: Rossby number
$\mathbf{r}$: position vector
S: surface (m^2)
s: specific entropy ($m^2\ s^{-2}\ K^{-1}$)
St: Strouhal number
T: temperature (K)
t: time (s)
$\mathbf{T}$: stress vector ($kg\ m^{-1}\ s^{-2}$)
$\mathbb{T}$: general stress tensor ($kg\ m^{-1}\ s^{-2}$)
u: velocity component in x-direction ($m\ s^{-1}$)
u_*: frictional velocity ($m\ s^{-1}$)
V: volume (m^3)
v: velocity component in y-direction ($m\ s^{-1}$)
$\mathbf{v}$: relative velocity ($m\ s^{-1}$)
$\mathbf{v}_A$: absolute velocity ($m\ s^{-1}$)
$\mathbf{v}_D$: deformation velocity ($m\ s^{-1}$)
$\mathbf{v}_g$: velocity of the geostrophic wind ($m\ s^{-1}$)
$\mathbf{v}_h$: velocity of the horizontal wind ($m\ s^{-1}$)
$\mathbf{v}_P$: velocity of the point ($m\ s^{-1}$)
$\mathbf{v}_T$: translatory velocity ($m\ s^{-1}$)

$\mathbf{v}_\Omega$:	rotational velocity	(m s^{-1})
w:	velocity component in z-direction	(m s^{-1})
W_i, W^i:	covariant and contravariant measure numbers of $\mathbf{v}_P$	
α:	specific volume	(m^3 kg^{-1})
β:	Rossby parameter	(m^{-1} s^{-1})
Γ_{ijk},Γ_{ij}^k:	Christoffel symbols of the first and the second kind	
γ:	lapse rate	(K m^{-1})
ϵ:	energy dissipation	(kg m^{-1} s^{-3})
ζ:	vorticity	(s^{-1})
η:	absolute vorticity	(s^{-1})
λ:	degrees of longitude	
μ_k:	specific chemical potential of substance k	(m^2 s^{-2})
ρ:	mass density	(kg m^{-3})
ρ^k:	partial density of substance k	(kg m^{-3})
$\Vert$:	surface Hamilton operator	
σ_0:	static stability	(kg^{-2} m^4 s^2)
τ:	stress vector	(kg m^{-1} s^{-2})
ϕ:	geopotential	(m^2 s^{-2})
ϕ_a:	gravitational potential	(m^2 s^{-2})
ϕ_z:	centrifugal potential	(m^2 s^{-2})
$\mathbf{\Phi}$:	general dyadic	
$\mathbf{\Phi}'$:	symmetric dyadic	
$\mathbf{\Phi}''$:	antisymmetric dyadic	
$\mathbf{\Phi}_a$:	adjoint dyadic	
$\mathbf{\Phi}_\times$:	vector of a dyadic	
$\mathbf{\Phi}_I$:	first scalar of a dyadic	
$\mathbf{\Phi}_{II}$:	second scalar of a dyadic	
$\mathbf{\Phi}_{III}$:	third scalar of a dyadic	
$\mathbf{\Phi}^{-1}$:	inverse dyadic	
φ:	degrees of latitude	
χ:	velocity potential	(m^2 s^{-1})
ψ:	stream function	(m^2 s^{-1})
$\{\psi\}$:	jump of the field function ψ	
$\mathbf{\Omega}$:	angular velocity of the earth	(s^{-1})
$\mathfrak{R}$:	rotational part of the velocity dyadic	(s^{-1})
ω:	angular speed	(s^{-1})
ω:	generalized vertical speed	(kg m^{-1} s^{-3})

List of constants

β:	Rossby parameter (55° latitude)	$(1.313 \times 10^{-11}\ \mathrm{m^{-1}\ s^{-1}})$
c_{L}:	speed of sound at 273 K	$(331\ \mathrm{m\ s^{-1}})$
γ:	gravitational constant	$(6.672 \times 10^{-11}\ \mathrm{N\,m^2\,kg^{-2}})$
$C_{p,0}$:	specific heat at constant pressure, dry air	$(1005\ \mathrm{J\ kg^{-1}\ K^{-1}})$
$C_{p,1}$:	specific heat at constant pressure, water vapor	$(1847\ \mathrm{J\ kg^{-1}\ K^{-1}})$
$C_{v,0}$:	specific heat at constant volume, dry air	$(718\ \mathrm{J\ kg^{-1}\ K^{-1}})$
$C_{v,1}$:	specific heat at constant volume, water vapor	$(1386\ \mathrm{J\ kg^{-1}\ K^{-1}})$
c_2:	specific heat of liquid water	$(4190\ \mathrm{J\ kg^{-1}\ K^{-1}})$
c_3:	specific heat of ice	$(2090\ \mathrm{J\ kg^{-1}\ K^{-1}})$
f:	vertical Coriolis parameter (55° latitude)	$(1.195 \times 10^{-4}\ \mathrm{s^{-1}})$
l:	horizontal Coriolis parameter (55° latitude)	$(0.836 \times 10^{-4}\ \mathrm{s^{-1}})$
l_{21}:	latent heat of vaporization	$(2.5 \times 10^{6}\ \mathrm{J\ kg^{-1}})$
l_{31}:	latent heat of sublimation	$(2.834 \times 10^{6}\ \mathrm{J\ kg^{-1}})$
l_{32}:	latent heat of melting	$(0.334 \times 10^{6}\ \mathrm{J\ kg^{-1}})$
M_{E}:	mass of the earth	$(5.973 \times 10^{24}\ \mathrm{kg})$
μ:	Lamé coefficient (293 K)	$(1.815 \times 10^{-5}\ \mathrm{kg\ m^{-1}\ s^{-1}})$
R^*:	universal gas constant	$(8.314\,32\ \mathrm{J\ mol^{-1}\ K^{-1}})$
R_0:	gas constant of dry air	$(287.05\ \mathrm{J\ kg^{-1}\ K^{-1}})$
R_1:	gas constant of water vapor	$(461.51\ \mathrm{J\ kg^{-1}\ K^{-1}})$

References and bibliography

Abdullah, A. T., 1955: The atmospheric solitary wave. *Bull. Am. Meteorol. Soc.*, **36**, 511–518.

Abramowitz, M., and I. A. Segun, 1968: *Handbook of Mathematical Functions.* Dover Publications, Inc., New York, 1046 pp.

Arakawa, A., 1966: Computational design for long-term numerical integrations of the equations of atmospheric motion. *J. Comput. Phys.*, **1**, 119–143.

Arakawa, A., and V. R. Lamb, 1977: Computational design of the basic dynamical processes of the UCLA general circulation model. *Methods in Computational Physics, Volume 17.* Academic Press, New York, pp. 174–265.

Arfken, G., 1970: *Mathematical Methods for Physicists.* Academic Press, New York, 815 pp.

Baer, F., and G. W. Platzman, 1961: A procedure for numerical integration of the spectral vorticity equation. *J. Meteorol.*, **18**, 393–401.

Blackadar, A. K., 1962: The vertical distribution of wind and turbulent exchange in a neutral atmosphere, *J. Geophys. Res.*, **67**, 3095–3102.

Bolin, B., 1950: On the influence of the earth's orography on the general character of the westerlies. *Tellus*, **2**, 184–195.

Bourke, W., 1972: An efficient, one level, primitive-equation spectral model. *Mon. Weather Rev.*, **100**, 683–689.

Bourke, W., 1974: A multi-level spectral model. I. Formulation and hemispheric integrations. *Mon. Weather Rev.*, **102**, 687–701.

Bourke, W., B. McAvaney, K. Puri, and R. Thurling, 1977: Global modelling of atmospheric flow by spectral methods. *Methods in Computational Physics, Volume 17.* Academic Press, New York, pp. 267–324.

Bretherton, F. P. 1966: The propagation of groups of internal gravity waves in shear flow. *Q. J. R. Meteorol. Soc.*, **92**, 466–480.

Burger, A., 1958: Scale considerations of planetary motions of the atmosphere. *Tellus*, **10**, 195–205.

Byers, H. R., 1959: *General Meteorology.* McGraw-Hill Book Company, New York, 461 pp.

Charney, J. G., 1947: The dynamics of long waves in a baroclinic westerly current. *J. Meteorol.*, **4**, 135–162.

Charney, J. G., 1948: *On the scale of dynamic motions.* Geofys. Publik, 17 pp.

Charney, J. G., 1963: A note on large scale motions in the tropics. *J. Atmos. Sci.*, **20**, 607–609.

Charney, J. G., and A. Eliassen, 1949: A numerical method for predicting the perturbations on the middle latitude westerlies. *Tellus*, **1**, 38–54.
Charney, J. G., Fjörtoft, and J. von Neumann, 1950: Numerical integration of the barotropic vorticity equation. *Tellus*, **2**, 237–254.
Christie, D. R., K. J. Muirhead, and A. L. Hales, 1978: On solitary waves in the atmosphere. *J. Atmos. Sci.*, **35**, 805–825.
Currie, I. G., 1974: *Fundamental Mechanics of Fluids.* McGraw-Hill Book Company, New York, 441 pp.
Deardorff, J. W., 1978: Efficient prediction of ground surface temperature and moisture, with inclusion of a layer of vegetation. *J. Geophys. Res.*, **83**, 1889–1903.
Drazin, P. G., 1992: *Nonlinear Systems.* Cambridge University Press, Cambridge, 316 pp.
Dutton, J. A., 1976: *The Ceaseless Wind. An Introduction to the Theory of Atmospheric Motion.* Mc.Graw Hill Book Company, New York. 579 pp.
Eady, E. T., 1949: Long waves and cyclonic waves. *Tellus*, **1**, 33–52.
Edelmann, W., 1973: Barotrope Modelle mit Lärmfilterung. Deutscher Wetterdienst. *Promet*, 11–15.
Edelmann, W., and H. Reiser, 1973: Das Anfangswertproblem bei ungefilterten barotropen Modellen, Deutscher Wetterdienst, *Promet*, 15–16.
Egger, J., and C. Schär, 1994: Comments on 'A solenoid-mass-continuity invariance criterion related to the potential vorticity conservation' by F. Herbert and H. Pichler. *Meteorol. Atmos. Phys.*, **59**, 257–258.
Eliasen, E., B. Machenhauer, and E. Rasmussen, 1970: On the numerical method for integration of the hydrodynamical equations with the spectral representation of the horizontal fields. *Report No. 2, Institute for Theoretical Meteorology.* Copenhagen University, Copenhagen, 35 pp.
Elsaesser, H. W., 1966: Evaluation of spectral versus grid method of hemispheric numerical weather prediction. *J. Appl. Meteorol.*, **5**, 246–262.
Ellison T. H., 1957: Turbulent transport of heat and momentum from an infinite rough plate. *J. Fluid Mech.*, **2**, 456–466.
Epstein, E. S., 1969: Stochastic dynamic prediction. *Tellus*, **21**, 739–757
Ertel, H., 1941: Über neue atmosphärische Bewegungsgleichungen und eine Differentialgleichung des Luftdruckfeldes. *Meteorol. Z.*, **58**, 77–78.
Ertel, H., 1942: Ein neuer hydrodynamischer Wirbelsatz. *Meteorol. Z.*, **59**, 277–281.
Ertel, H., 1943: Die Westgebiete der Troposphäre als Instabilitätszonen. *Meteorol. Z.*, **59**, 397–400.
Ertel, H., and C. G. Rossby, 1949: Ein neuer Erhaltungssatz der Hydrodynamik. *Sitz. Ber. Deutsch. Akad. Wiss. Berlin, Kl. Math. allgem. Naturwiss., Nr. 1.*
Ertel, H., 1954: Ein neues Wirbeltheorem der Hydrodynamik. *Sitz. Ber. Deutsch. Akad. Wiss. Berlin, Kl. Math. allgem. Naturwiss., Nr. 5.*
Ertel, H., 1960: Relación entre la derivada individual y una cierta divergencia espacial en hidrodinamica. *Gerlands Beitr. Geophys.*, **69**, 357–361.
Etling, D., 1996: *Theoretische Meteorologie*, Friedrich Viehweg & Sohn Verlagsgesellschaft mbH, Braunschweig and Wiesbaden, 318 pp.
Fleming, R. J., 1971a: On stochastic dynamic prediction: I. The energetics of uncertainty and the question of closure. *Mon. Weather Rev.*, **99**, 851–872.
Fleming, R. J., 1971b: On stochastic dynamic prediction: II. Predictability and utility. *Mon. Weather Rev.*, **99**, 927–938.
Fließbach, T., 1998: *Allgemeine Relativitätstheorie.* Spektrum Akademischer Verlag, Heidelberg, 335 pp.

Fortak, H., 1956: Zur Frage allgemeiner hydrodynamischer Wirbelsätze. *Gerlands Beitr. Geophys.*, **65**, 283–294.
Fortak, H., 1973: Prinzipielle Grenzen der deterministischen Vorhersagbarkeit atmosphärischer Prozesse. *Ann. Meteorol.*, **6**, 111–120.
Fowles, G., 1968: *Introduction to Modern Optics.* Holt, Rinehart and Winston, Inc., New York, 304 pp.
Gal-Chen, T., and R. C. J. Somerville, 1975: On the use of coordinate transformation for the solution of the Navier–Stokes equations. *J. Comput. Phys.*, **17**, 209–228.
Godske, C. L., T. Bergeron, J. Bjerkness, and R. C. Bundgaard, 1957: *Dynamic Meteorology and Weather Forecasting.* American Meteorological Society, Boston and Carnegie Institute of Washington, Washington.
Gossard, E. E., and W. H. Hooke, 1975: *Waves in the Atmosphere.* American Elsevier Publishing Company, New York, 456 pp.
Haltiner, G. J., and F. L. Martin, 1957: *Dynamical and Physical Meteorology.* McGraw-Hill Book Company, New York, 470 pp.
Haltiner, G. J., and R. T. Williams, 1980: *Numerical Weather Prediction and Dynamical Meteorology.* John Wiley and Sons, New York, 477 pp.
Haugen, D. A., 1973: *Workshop on Micrometeorology.* American Meteorological Society, New York, 392 pp.
Haurwitz, B., 1931: Wogenwolken und Luftwogen. *Meteorol. Z.*, **46**, 483–484.
Haurwitz, B., 1940: The motion of atmospheric disturbances on the spherical earth. *J. Marine Res.*, **3**, 254–267.
Herbert, F., and H. Pichler, 1994: A solenoid-mass continuity invariance criterion related to the potential vorticity equation. *Meteorol. Atmos. Phys.*, **53**, 123–130.
Hess, S. L., 1959: *Introduction to Theoretical Meteorology.* Henry Holt and Company, New York, 362 pp.
Hinkelmann, K. H., 1959: *Ein numerisches Experiment mit den primitiven Gleichungen.* Rossby Memorial Volume. Rockefeller Institute Press, New York, and Oxford University Press, Oxford.
Hinze, J. O., 1959: *Turbulence.* McGraw-Hill Book Company, New York, 586 pp.
Hollmann, G., 1965: Ein vollständiges System hydrodynamischer Wirbelsätze. *Arch. Meteorol. Geophys. Biokl.*, **14**, 1–13.
Holton, J. R., 1972: *An Introduction to Dynamic Meteorology.* Academic Press, New York, 319 pp.
Hopf, E., 1942: Abzweigung einer periodischen Lösung von einer stationären Lösung eines Differentialgleichungssystems. *Ber. Math. Phys. Kl. Sächs. Akad. Wiss., Leipzig*, **94**, 1.
Hoskins, B. J., and A. J. Simmons, 1975: A multi-layer spectral model and the semi-implicit method. *Q. J. R. Meteorol. Soc.*, **101**, 637–655.
Jordan, W., and P. Smith, 1987: *Nonlinear Ordinary Differential Equations.* Clarendon Press, Oxford, 381 pp.
Kasahara, A., 1977: Computational aspects of numerical models for weather prediction and climate simulation. *Methods Comput. Phys.*, **17**, 2–66.
Keener, J. P., 1988: *Principles of Applied Mathematics.* Addison-Wesley Publishing Company, Redwood City, California, 560 pp.
Kleinschmidt, E., 1941a: Stabilitätstheorie des geostrophischen Windfeldes. *Ann. Hydrogr.*, **69**, 305–325.
Kleinschmidt, E., 1941b: Zur Theorie der labilen Anordnung. *Meteorol. Z.*, **58**, 57–163.
Korn, G., and T. Korn, 1968: *Mathematical Handbook for Scientists and Engineers.* McGraw-Hill Book Company, New York, 1130 pp.

Kubota, S., M. Hirose, Y. Kikuchi, and Y. Kurihara, 1961: Barotropic forecasting with the use of surface spherical harmonic representation. *Pap. Meteorol. Geophys.*, **12**, 199–215.

Kuo, J. L., 1951: Dynamical aspects of the general circulation and the stability of zonal flow. *Tellus*, **3**, 268–284.

Lagally M., and W. Franz, 1956: *Vorlesungen über Vektorrechnung*. Akademische Verlagsgesellschaft, Geest & Portig K.-G., Leipzig, 462 pp.

Langhaar, H. L., 1967: *Dimensionless Analysis and Theory of Models*. John Wiley & Sons, Inc., New York, 166 pp.

Lass, H., 1950: *Vector and Tensor Analysis*. McGraw-Hill Book Company, New York, 347 pp.

LeMéhaute, B., 1976: *An Introduction to Hydrodynamics & Water Waves*. Springer-Verlag, New York, 322 pp.

Lense J., 1950: *Kugelfunktionen*. Akademische Verlagsgesellschaft, Geest & Portig, K.-G., Leipzig, 294 pp.

Lohr E., 1939: *Vektor und Dyadenrechnung für Physiker und Techniker*. Walter De Gruyter & Co., Berlin, 411 pp.

Lorenz, E. N., 1955: Available potential energy and the maintenance of the general circulation. *Tellus*, **7**, 157–167.

Lorenz, E .N., 1960: Energy and numerical weather prediction. *Tellus*, **12**, 364–373.

Lorenz, E. N., 1963: Deterministic nonperiodic flow. *J. Atmos. Sci.*, **20**, 130–141.

Lorenz, E. N., 1969: The predictability of a flow which possesses many scales of motion. *Tellus*, **21**, 289–302.

Lowell S. C., 1951: Boundary conditions at the troposphere, *Tellus*, **3**, 78–81.

Magnus W., and F. Oberhettinger, 1948: *Formeln und Sätze für die speziellen Funktionen der mathematischen Physik*. Springer-Verlag. Berlin, 230 pp.

Mandelbrot B., 1991: *Die fraktale Geometrie der Natur*. Birkhäuser Verlag, Basel, 491 pp.

Mason, P. J., 1999: Large-eddy simulation: A critical review of the technique. *Q. J. R. Meteorol. Soc.*, **120**, 1–26.

Matsuno, T., 1966: Quasi-geostrophic motions in the equatorial area. *J. Meteorol. Soc. Japan*, **44**, 25–42.

Mellor, G. L., and T. Yamada, 1974: A hierarchy of turbulence closure models for the planetary boundary layer. *J. Atmos. Sci.*, **31**, 1971–1806.

Mesinger, F., and A. Arakawa, 1976: *Numerical methods used in atmospheric models*. WMO/ICSU, Joint Organizing Committee, GARP Publication Series No. 17, World Meteorological Organization, Geneva, 64 pp.

Monin, A. S., and A. M. Obukhov, 1954: Basic laws of turbulent mixing in the atmosphere near the ground. *Tr. Akad. Nauk, SSR Geofiz. Inst.*, **24**, 1963–1987.

Moritz H., and B. Hofmann-Wellenhof, 1993: *Geometry. Relativity, Geodesy*. Herbert Wichmann Verlag GmbH, Karlsruhe, 367 pp.

Neamtan, S. M., 1946: The motion of harmonic waves in the atmosphere. *J. Meteorol.*, **2**, 53–56.

Obukhov, A. M., 1967: *Weather and Turbulence*. IAMAP Presidential Address, Fourteenth General Assembly of the IUGG, Lucerne.

Obukhov, A. M., 1971: Turbulence in an atmosphere with a non-uniform temperature. *Boundary-Layer Meteorol.*, **38**, 7–29.

Orlanski, I., 1975: A rational subdivision of scales for atmospheric processes. *Bull. Am. Meteorol. Soc.*, **56**, 527–530.

Orszag, S. A., 1970: Transform method for calculation of vector-coupled sums. Application to the spectral form of the vorticity equation. *J. Atmos. Sci.*, **27**, 890–895.

O'Brien, J. J., 1970: A note on the vertical structure of the eddy exchange coefficient in the planetary boundary layer. *J. Atmos. Sci.*, **27**, 1213–1215.

Panchev S., 1985: *Dynamic Meteorology.* D. Reidel Publishing Company, Dordrecht, 360 pp.

Panhans, W.-G., 1976: *Analytische und numerische Bestimmung von atmosphärischen Zustandsparametern und Flüssen im Bereich und nahe der Erdoberfläche.* Dissertation, Johannes Gutenberg-Universität Mainz, 264 pp.

Pedlosky, J., 1979: *Geophysical Fluid Dynamics.* Springer-Verlag, New York, 625 pp.

Petterssen, S., 1956: *Weather Analysis and Forecasting, Volume 1.* McGraw-Hill Book Compnany, Inc., New York, 428 pp.

Philipps, H., 1939: Die Abweichung vom geostrophischen Wind. *Meteorol. Z.*, **56**, 460–475.

Phillips, N. A., 1956: The general circulation of the atmosphere: A numerical experiment. *Q. J. R. Meteorol. Soc.*, **82**, 123–164.

Phillips, N. A., 1957: A coordinate system having some special advantages for numerical forecasting. *J. Meteorol.*, **14**, 184–185.

Phillips, N. A., 1963: Geostrophic motion, *Rev. Geophys.*, **1**, 123–176.

Pichler, H., 1997: *Dynamik der Atmosphäre.* Spektrum Akademischer Verlag, Heidelberg, 572 pp.

Pielke, R. A., 1984: *Mesoscale Meteorological Modeling.* Academic Press, Orlando, Florida, 612 pp.

Platzman, G. W., 1960: The spectral form of the vorticity equation. *J. Meteorol.*, **17**, 635–644.

Prandtl, L., 1925: Bericht über die Untersuchungen zur ausgebildeten Turbulenz. *Z. angew. Math. Mech.*, **5**, 136–139.

Promet, 1973. Edited by the Deutscher Wetterdienst.

Queney, P., 1948: The problem of air flow over mountains. A summary of theoretical studies. *Bull. Am. Meteorol. Soc.*, **29**, 16–26.

Queney, P., G. A. Corby, N. Gerbier, H. Koschmieder, and J, Zierep, 1960: The air flow over mountains. *World Meteorol. Organization, General Techn. Note, 34*, 135 pp.

Rayleigh, Lord, 1880: On the stability or instability of certain fluid motions. *Scientific Papers, Volume 3.* Cambridge University Press, Cambridge, 594–596.

Research Manual 2, ECMF Research Department.

Reiser H., 1973: Barotrope Vorhersagemodelle ohne Lärmfilterung. Deutscher Wetterdienst, *Promet*, 6–10.

Richardson, L. F., 1922: *Weather Prediction by Numerical Process.* Cambridge University Press, Cambridge; reprinted Dover, New York, 1965, 236 pp.

Richtmeyer R. D., and K. W. Morton, 1967: *Difference Methods for Initial-Value Problems.* Interscience Publishers, a division of John Wiley & Sons, New York, 405 pp.

Robert, A. J., 1966: The integration of low order spectral form of the primitive equations. *J. Meteorol. Soc. Japan*, **2**, 4237–4245.

Robert, A. J., 1969: The integration of a spectral model of the atmosphere by implicit method. *Proc. WMO/IUGG Symposium on Numerical Weather Prediction*, Meteorological Society of Japan, Tokyo, pp. VII-19–VII-24.

Robert, A. J., H. Henderson, and C. Turnbull, 1972: An implicit time integration scheme for baroclinic models of the atmosphere. *Mon. Weather Rev.*, **100**, 329–335.

Rodi, W., 1993: On the simulation of turbulent flow past bluff bodies. *J. Wind Engineering and Industrial Aerodynamics*, **46**, 3–19.

Rossby, C. G., 1939: Relation between variations of the intensity of the zonal circulation of the atmosphere and the displacement of semi-permanent centers of action. *J. Marine Res.*, **2**, 38–55.

Rossby, C. G., 1940: Planetary flow patterns in the atmosphere. *Q. J. R. Meteorol. Soc.*, **66**, 68–87.

Rotta, J. C., 1972: *Turbulente Strömungen*. B. G. Teubner, Stuttgart, 267 pp.

Saltzmann, B., 1962: Finite amplitude free convection as an initial value problem. *J. Atmos. Sci.*, **19**, 329–341.

Saucier, W. J., 1955: *Principles of Meteorological Analysis*. The University of Chicago Press, Chicago, 438 pp.

Schlichting, H., 1968: *Boundary Layer Theory*, McGraw-Hill Book Company, New York, 748 pp.

Schuster, H. 1984: *Deterministic Chaos*. Physik Verlag, Weinheim, 220 pp.

Seidel R., 1988: *From Equilibrium to Chaos*, Elsevier, New York, 367 pp.

Sellers, P. J., Y. Mintz, V. C. Sud, and A. Dalcher, 1986: A simple model (Si-B) for use within general circulation models. *J. Atmos. Sci.*, **43**, 505–531.

Siebert, J., U. Sievers, and W. Zdunkowski, 1992: A one-dimensional simulation of the interaction between land surface processes and the atmosphere. *Boundary-Layer Meteorol.*, **59**, 1–34.

Sievers, U., 1982: Thermodynamics of the turbulent atmosphere and parameterization of fluxes. *Beitr. Phys. Atmos.*, **55**, 189–200.

Sievers, U., 1984: The turbulent atmosphere and the inclusive system of model equations. *Beitr. Phys. Atmos.*, **57**, 324–345.

Sievers, U., 1995: Verallgemeinerung der Stromfunktionsmethode auf drei Dimensionen. *Meteorol. Z.*, **4**, 3–15.

Silberman, I. S., 1954: Planetary waves in the atmosphere. *J. Meteorol.*, **11**, 27–34.

Smith, R. B., 1979: The influence of mountains on the atmosphere. *Adv. Geophys.*, **21**, 87–230.

Sokolnikoff, I., R. Redheffer, 1966: *Mathematics of Physics and Modern Engineering*. McGraw-Hill Book Company, New York, 752 pp.

Sommerville, R. C. J., and T. Gal-Chen, 1974: Numerical solution of the Navier–Stokes equations with topography, *J. Comput. Phys.*, **17**, 276–309.

Sommerfeld, A., 1964: *Partial Differential Equations in Physics. Lectures on Theoretical Physics, Volume 6*. Academic Press, New York, 335 pp.

Staege, S., 1979: *Representation of the Equation of Motion in Coordinate Systems*. Diploma Thesis, Institute of Meteorology, University of Mainz, in German, 97 pp.

Steeb, W., and A. Kunick, 1989: *Chaos in dynamischen Systemen*. Wissenschaftsverlag Mannheim, Vienna, 239 pp.

Strogatz, S., 1994: *Nonlinear Dynamics and Chaos*. Addison-Wesley Publishing Companay, Reading, Massachusetts, 498 pp.

Stull, R., 1989: *An Introduction to Boundary Layer Meteorology*. Kluwer Academic Publishers, Dordrecht, 666 pp.

Takeuchi, K., and O. Yokoyama, 1963: Scale of turbulence and the wind profile in the surface boundary layer. *Meteorol. Soc. Japan, Tokyo, Journal, Ser. 2*, **41**, 108–117.

Tatarsky, V. L., 1969: The use of dynamic equations in the probability prediction of the pressure field. *Izv. Acad. Sci. USSR. Atmosph. Oceanic Phys.*, **5**, 162–164.

Taylor, G. I., 1915: Eddy motion in the atmosphere. *Phil. Trans. R. Soc., London, A*, **215**, pp. 1–26.

Thompson, P. D., 1961: *Numerical Weather Prediction*. MacMillan, New York, 170 pp.

Tiedtke, M., 1973: Energieänderungen im barotropen Modell. Deutscher Wetterdienst, *Promet*, 17–20.

Van Mieghem, J. H., 1951: *Hydrodynamic stability. AMS, Compendium of Meteorology*, 434–453.

Van Mieghem, J. H., 1973: *Atmospheric Energetics*. Clarendon Press, Oxford.

Voelker, D., and G. Doetsch, 1950: *Die zweidimensionale Laplace-Transformation*. Birkhäuser Verlag, Basel, 257 pp.

Washington, W. M., and C. L. Parkinson, 1986: *An Introduction to Three-Dimensional Climate Modelling*. University Science Books. Mill Valley, California, 422 pp.

Wiin-Nielsen, A. C., 1975: *European Centre for Medium Range Forecasts, Seminar Part 1. Reading 1–2 Sept.*, pp. 139–202.

Wylie, C. R., 1966: *Advanced Engineering Mathematics*. McGraw-Hill Book Company, New York, 813 pp.

Zdunkowski, W., and T. Kandelbinder, 1997: An analytic solution to nocturnal cooling. *Beilr. Phys. Atmos.*, **70**, 337–348.

Index